ROBOTIC SYSTEMS – APPLICATIONS, CONTROL AND PROGRAMMING

Edited by **Ashish Dutta**

Robotic Systems – Applications, Control and Programming
Edited by Ashish Dutta

Published by InTech
Janeza Trdine 9, 51000 Rijeka, Croatia

Publishing Process Manager Anja Filipovic

Technical Editor Teodora Smiljanic

Cover Designer InTech Design Team

Additional hard copies can be obtained from orders@intechopen.com

Robotic Systems – Applications, Control and Programming,
Edited by Ashish Dutta
p. cm.
ISBN 978-953-307-941-7

For EU product safety concerns:
IN TECH d.o.o., Prolaz Marije Krucifikse Kozulić 3, 51000 Rijeka, Croatia,
info@intechopen.com or visit our website at intechopen.com.

Meet the editor

Dr. Ashish Dutta obtained his PhD in Systems Engineering from Akita University, Japan. From 1994 to 2000 he was with the Bhabha Atomic Research Center (India) where he worked on telemanipulator design and control for nuclear applications. Since 2002 he has been working with the department of mechanical engineering in the Indian Institute of Technology Kanpur, in addition to working as an assistant professor in Nagoya University, Japan from 2006 to 2007 in the department of Mechanical Science and Engineering. His research interests are in the areas of humanoid robotics, micro sensors and actuators, intelligent control systems and rehabilitation engineering.

Contents

Preface

Over the last few decades the focus of robotics research has moved beyond the traditional area of industrial applications to newer areas, including healthcare and domestic applications. These newer applications have given a strong impetus to the development of advanced sensors, control strategies and algorithms. The first section of this book contains advanced applications of robotics in surgery, rehabilitation, modular robotics among others. This is followed by sections on control and sensors, while the fourth section is devoted to robot algorithms.

I would like to thank all the authors for entrusting us with their work and specially the editorial members of InTech publishing.

Dr. Ashish Dutta
Department of Mechanical Engineering
Indian Institute of Technology, Kanpur
India

Part 1

Applications

1

Modular Robotic Approach in Surgical Applications – Wireless Robotic Modules and a Reconfigurable Master Device for Endoluminal Surgery –

Kanako Harada, Ekawahyu Susilo, Takao Watanabe, Kazuya Kawamura, Masakatsu G. Fujie, Arianna Menciassi and Paolo Dario
[1]*The University of Tokyo*
[2]*Scuola Superiore Sant'Anna*
[3]*Italian Institute of Technology*
[4]*Waseda University*
[1,4]*Japan*
[2,3]*Italy*

1. Introduction

The trend in surgical robots is moving from traditional master-slave robots to miniaturized devices for screening and simple surgical operations (Cuschieri, A. 2005). For example, capsule endoscopy (Moglia, A. 2007) has been conducted worldwide over the last five years with successful outcomes. To enhance the dexterity of commercial endoscopic capsules, capsule locomotion has been investigated using legged capsules (Quirini, M. 2008) and capsules driven by external magnetic fields (Sendoh, M. 2003; Ciuti, G. 2010; Carpi, F. 2009). Endoscopic capsules with miniaturized arms have also been studied to determine their potential for use in biopsy (Park, S.-K. 2008). Furthermore, new surgical procedures known as natural orifice transluminal endoscopic surgery (NOTES) and Single Port Access surgery are accelerating the development of innovative endoscopic devices (Giday, S. 2006; Bardaro, S.J. 2006). These advanced surgical devices show potential for the future development of minimally invasive and endoluminal surgery. However, the implementable functions in such devices are generally limited owing to space constraints. Moreover, advanced capsules or endoscopes with miniaturized arms have rather poor dexterity because the diameter of such arms must be small (i.e. a few millimeters), which results in a small force being generated at the tip.
A modular surgical robotic system known as the ARES (Assembling Reconfigurable Endoluminal Surgical system) system has been proposed based on the aforementioned motivations (Harada, K. 2009; Harada, K. 2010; Menciassi, A. 2010). The ARES system is designed for screening and interventions in the gastrointestinal (GI) tracts to overcome the intrinsic limitations of single-capsules or endoscopic devices. In the proposed system,

miniaturized robotic modules are ingested and assembled in the stomach cavity. The assembled robot can then change its configuration according to the target location and task. Modular surgical robots are interesting owing to their potential for application as self-reconfigurable modular robots and innovative surgical robots. Many self-reconfigurable modular robots have been investigated worldwide (Yim, M. 2007; Murata, S. 2007) with the goal of developing systems that are robust and adaptive to the working environment. Most of these robots have been designed for autonomous exploration or surveillance tasks in unstructured environments; therefore, there are no strict constraints regarding the number of modules, modular size or working space. Because the ARES has specific applications and is used in the GI tract environment, it raises many issues that have not been discussed in depth in the modular robotic field. Modular miniaturization down to the ingestible size is one of the most challenging goals. In addition, a new interface must be developed so that surgeons can intuitively maneuver the modular surgical robot.

The purpose of this paper is to clarify the advantages of the modular approach in surgical applications, as well as to present proof of concept of the modular robotic surgical system.

The current paper is organized as follows: Section 2 describes the design of the ARES system. Section 3 details the design and prototyping of robotic modules, including the experimental results. Section 4 describes a reconfigurable master device designed for the robotic modules, and its preliminary evaluation is reported.

2. Design of the modular surgical system

2.1 Clinical indications and proposed procedures

The clinical target of the ARES system is the entire GI tract, i.e., the esophagus, stomach, small intestine, and colon. Among GI tract pathologies that can benefit from modular robotic features, biopsy for detection of early cancer in the upper side of the stomach (the fundus and the cardia) was selected as the surgical task to be focused on as a first step. Stomach cancer is the second leading cause of cancer-related deaths worldwide (World Health Organization 2006), and stomach cancer occurring in the upper side of the stomach has the worst outcome in terms of the 5-year survival ratio (Pesic, M. 2004). Thus, early diagnosis of cancer utilizing an advanced endoluminal device may lead to better prognosis. The stomach has a large volume (about 1400 ml) when distended, which provides working space to assemble the ingested robotic modules and change the topology of the assembled robot inside (i.e. reconfiguration). Each robotic module should be small enough to be swallowed and pass through the whole GI tract. Because the size of the commercial endoscopic capsules (11 mm in diameter and 26 mm in length (Moglia, A. 2007)) has already been shown to be acceptable for the majority of patients as an ingestible device, each module needs to be miniaturized to this size before being applied to clinical cases.

The surgical procedures proposed for the ARES system (Harada, K. 2010) are shown in Fig. 1. Prior to the surgical procedure, the patient drinks a liquid to distend the stomach to a volume of about 1400 ml. Next, the patient ingests 10-15 robotic modules that complete the assembly process before the liquid naturally drains away from the stomach in 10-20 minutes. The number of the modules swallowed depends on the target tasks and is determined in advance based on the pre-diagnosis. Magnetic self-assembly in the liquid using permanent magnets was selected for this study since its feasibility has already been demonstrated (Nagy, Z. 2007). Soon after the assembly, the robot configures its topology according to preoperative planning by repeated docking and undocking of the modules (the undocking mechanism and electrical contacts between modules are necessary for reconfiguration, but they have not been implemented in the presented design). The robotic modules are controlled via wireless bidirectional communication with a master device operated by the surgeon, while the progress in procedure is observed using intraoperative imaging devices such as fluoroscopy and cameras mounted on the modules. After the surgical tasks are completed, the robot reconfigures itself to a snake-like shape to pass through the pyloric sphincter and travel to examine the small intestine and the colon, or it completely disassembles itself into individual modules so that it can be brought out without external aid. One of the modules can bring a biopsy tissue sample out of the body for detailed examination after the procedure is complete.

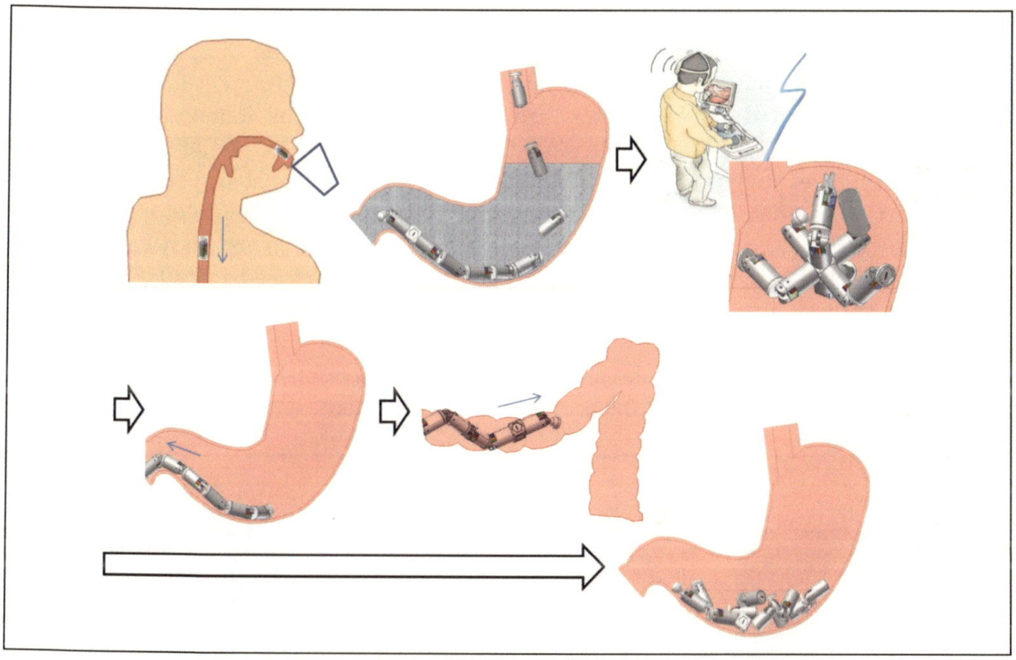

Fig. 1. Proposed procedures for the ARES system

2.2 Advantages of the modular approach in surgical applications

The modular approach has great potential to provide many advantages to surgical applications. These advantages are summarized below using the ARES system as shown in Fig.2. The numbering of the items in Fig.2 is correlated with the following numbering.

i. The topology of the modular surgical robot can be customized for each patient according to the location of the disease and the size of the body cavity in which the modular robot is deployed. A set of functional modules such as cameras, needles and forceps can be selected for each patient based on the necessary diagnosis and surgical operation.
ii. The modular approach facilitates delivery of more components inside a body cavity that has small entrance/exit hole(s). As there are many cavities in the human body, the modular approach would benefit treatment in such difficult-to-reach places. Because

several functional modules can be used simultaneously, the modular robot may perform rather complicated tasks that a single endoscopic capsule or an endoscopic device is not capable of conducting. For example, if more than two camera modules are employed, the surgeon can conduct tasks while observing the site from different directions.

iii. Surgical tools of relatively large diameter can be brought into the body cavity. Conventionally, small surgical forceps that can pass through an endoscopic channel of a few millimeters have been used for endoluminal surgery. Conversely, surgical devices that have the same diameter as an endoscope can be used in the modular surgical system. Consequently, the force generated at the tip of the devices would be rather large, and the performance of the functional devices would be high.

iv. The surgical system is more adaptive to the given environment and robust to failures. Accordingly, it is not necessary for the surgical robot to equip all modules that might be necessary in the body because the surgeons can decide whether to add modules with different functionalities, even during the surgical operation. After use, the modules can be detached and discarded if they are not necessary in the following procedures. Similarly, a module can be easily replaced with a new one in case of malfunction.

As these advantages suggest, a modular surgical robot would be capable of achieving rather complicated tasks that have not been performed using existing endoluminal surgical devices. These advantages are valid for modular robots that work in any body cavity with a small entrance and exit. Moreover, this approach may be introduced to NOTES or Single Port Access surgery, in which surgical devices must reach the abdominal cavity through a small incision.

In Section 3, several robotic modules are proposed, and the performance of these modules is reported to show the feasibility of the proposed surgical system.

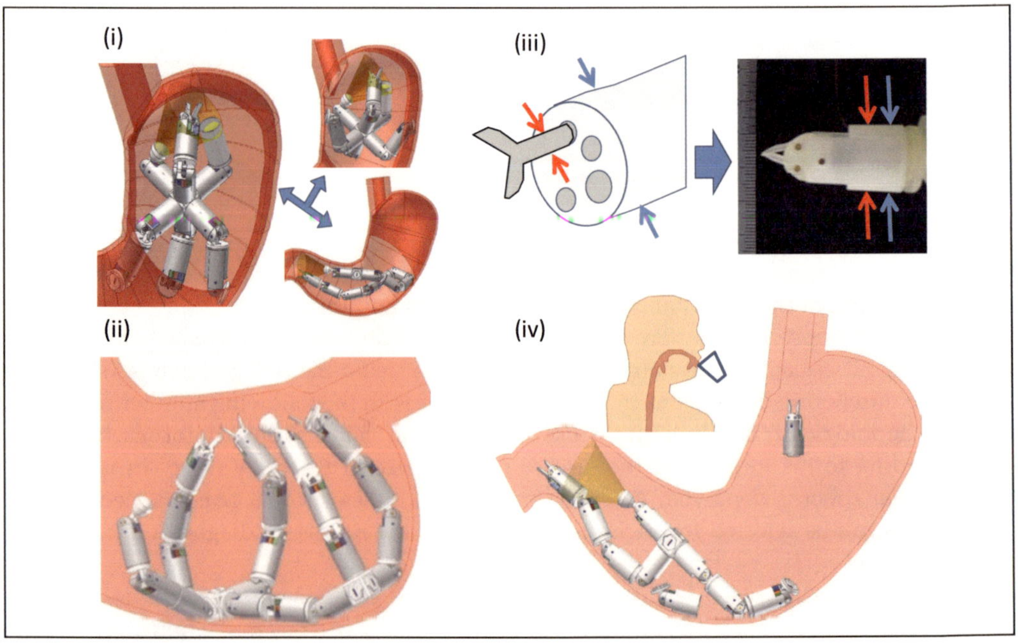

Fig. 2. Advantages of the modular approach in surgical applications

3. Robotic modules

3.1 Design and prototyping of the robotic modules

Figure 3 shows the design and prototypes of the Structural Module and the Biopsy Module (Harada, K. 2009, Harada, K. 2010). The Structural Module has two degrees of freedom (±90° of bending and 360° of rotation). The Structural Module contains a Li-Po battery (20 mAh, LP2-FR, Plantraco Ltd., Canada), two brushless DC geared motors that are 4 mm in diameter and 17.4 mm in length (SBL04-0829PG337, Namiki Precision Jewel Co. Ltd., Japan) and a custom-made motor control board capable of wireless control (Susilo, E. 2009). The stall torque of the selected geared motor is 10.6 mNm and the speed is 112 rpm when controlled by the developed controller. The bending mechanism is composed of a worm and a spur gear (9:1 gear reduction), whereas the rotation mechanism is composed of two spur gears (no gear reduction). All gears (DIDEL SA, Switzerland) were made of nylon, and they were machined to be implemented in the small space of the capsule. Two permanent magnets (Q-05-1.5-01-N, Webcraft GMbH, Switzerland) were attached at each end of the module to help with self-alignment and modular docking. The module is 15.4 mm in diameter and 36.5 mm in length; it requires further miniaturization before clinical application. The casing of the prototype was made of acrylic plastic and fabricated by 3D rapid prototyping (Invison XT 3-D Modeler, 3D systems, Inc., USA). The total weight is 5.6 g.
Assuming that the module would weigh 10 g with the metal chassis and gears, the maximum torque required for lifting two connected modules is 5.4 mNm for both the bending DOF and rotation DOF. Assuming that the gear transmission efficiency for the bending mechanism is 30%, the stall torque for the bending DOF is 28.6 mNm. On the other hand, the stall torque for the rotation DOF is 8.5 mNm when the transmission efficiency for the rotation mechanism is 80%. The torque was designed to have sufficient force for surgical operation, but the transmission efficiency of the miniaturized plastic gears was much smaller than the theoretical value as explained in the next subsection.

- Controller

The aforementioned brushless DC motor came with a dedicated motor driving board (SSD04, Namiki Precision Jewel Co., Ltd., 19.6 mm × 34.4 mm × 3 mm). This board only allows driving of one motor; hence, two boards are required for a robotic module with 2 DOFs. Because there was not sufficient space for the boards in the robotic module, a custom made high density control board was designed and developed in-house. This control board consisted of one CC2430 microcontroller (Texas Instrument, USA) as the main wireless controller and three sets of A3901 dual bridge motor drivers (Allegro MicroSystem, Inc., USA). The fabricated board is 9.6 mm in diameter, 2.5 mm in thickness and 0.37 g in weight, which is compatible with swallowing. The A3901 motor driver chip was originally intended for a brushed DC motor, but a software commutation algorithm was implemented to control a brushless DC motor as well. An IEEE 802.15.4 wireless personal area network (WPAN) was introduced as an embedded feature (radio peripheral) of the microcontroller. The implemented algorithm enables control of the selected brushless DC motor in Back Electro-Motive Force (BEMF) feedback mode or slow speed stepping mode. When the stepping mode is selected, the motor can be driven with a resolution of 0.178°.
For the modular approach, each control board shall be equipped with a wired locating system for intra-modular communication in addition to the wireless communication. Aside from wireless networking, the wired locating system, which is not implemented in the presented design, would be useful for identification of the sequence of the docked modules

in real time. The wired locating system is composed of three lines, one for serial multidrop communication, one for a peripheral locator and one as a ground reference. When the modules are firmly connected, the intra-modular communication can be switched from wireless to wired to save power while maintaining the predefined network addresses. When one module is detached intentionally or by mistake, it will switch back to wireless mode.

- Battery

The battery capacity carried by each module may differ from one to another (e.g. from 10 mAh to 50 mAh) depending on the available space inside the module. For the current design, a 20 mAh Li-Po battery was selected. Continuous driving of the selected motor on its maximum speed using a 20 mAh Li-Po battery was found to last up to 17 minutes. A module does not withdraw power continuously because the actuation mechanisms can maintain their position when there is no current to the motor owing to its high gear reduction (337:1). A module consumes power during actuation, but its power use is very low in stand-by mode.

- Biopsy Module

The Biopsy Module is a Functional Module that can be used to conduct diagnosis. The grasping mechanism has a worm and two spur gears, which allows wide opening of the grasping parts. The grasping parts can be hidden in the casing at the maximum opening to prevent tissue damage during ingestion. The motor and other components used for the Biopsy Module are the same as for the Structural Module. The brushless DC geared motors (SBL04-0829PG337, Namiki Precision Jewel Co. Ltd., Japan), the control board, a worm gear and two spur gears (9:1 gear reduction) were implemented in the Biopsy Module. A permanent magnet (Q-05-1.5-01-N, Webcraft GMbH, Switzerland) was placed at one side to be connected to another Structural Module.

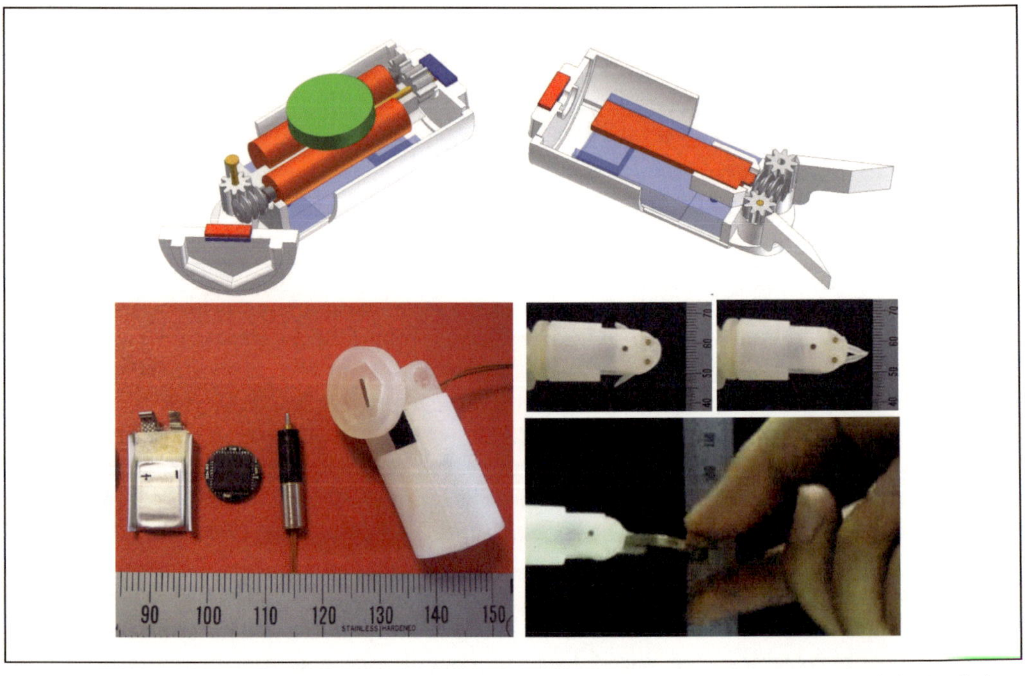

Fig. 3. Design and prototypes of the structural module (left) and the biopsy module (right)

The Biopsy Module can generate a force of 7.1 N at its tip, and can also open the grasping parts to a width of 19 mm with an opening angle of 90 degrees. These values are much larger than those of conventional endoscopic forceps, which are 2-4 mm in diameter. As a demonstration, Figure 3 shows the Biopsy Module holding a coin weighing 7.5 g.
In conventional endoscopy, forceps are inserted through endoscopic channels that are parallel to the direction of the endoscopic view, which often results in the forceps hiding the target. Conversely, the Biopsy Module can be positioned at any angle relative to the endoscopic view owing to the modular approach, thereby allowing adequate approach to the target.

3.2 Performance of the Structural Module

The mechanical performance of the bending and rotation DOFs of the Structural Module was measured in preliminary tests (Menciassi, A. 2010), and the results are summarized in Fig.4. The bending angle was varied by up to ± 90° in steps of 10° three times in succession. The measured range of the bending angle was -86.0° to +76.3°, and the maximum error was 15.8°. The rotation angle was increased from 0° to 180° in steps of 45° three times in succession, and the measured range of the rotational angle was between 0° and 166.7° with a maximum error of 13.3°. The difference between the driven angle and the measured angle was due to backlash of the gears and the lack of precision and stiffness of the casing made by 3D rapid prototyping. Regardless of the errors and the hysteresis, the repeatability was sufficient for the intended application for both DOFs. These results indicate that the precision of each motion can be improved by changing the materials of the gears and the casing. Since the motor can be controlled with a resolution of 0.178°, very precise surgical tasks can be achieved using different manufacturing processes.

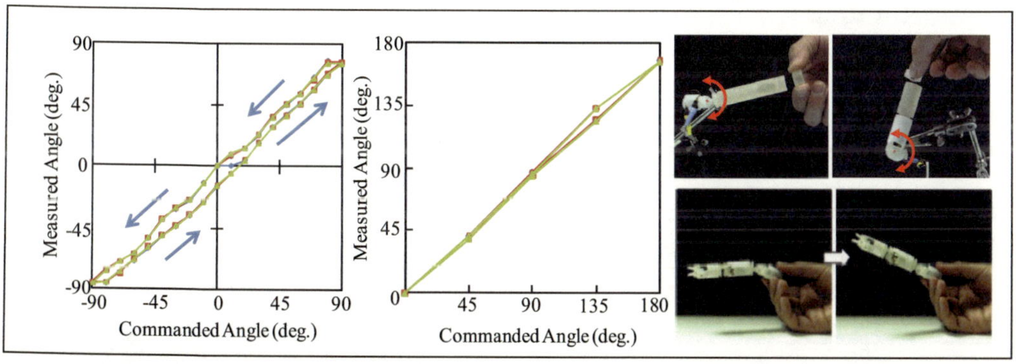

Fig. 4. Bending angle measurement (left), rotation angle measurement (middle), and torque measurement (right) (Menciassi, A. 2010)

In addition to the angle measurements, both bending and rotation torque were measured. The torque was measured by connecting cylindrical parts with permanent magnets at both ends until the bending/rotational motion stopped. The length and weight of each cylinder was designed in advance, and several types of cylinders were prepared. The measured bending torque was 6.5 mNm and the rotation torque was 2.2 mNm. The figure also shows one module lifting up two modules attached to its bending mechanism as a demonstration. The performance in terms of precision and generated torque, which are very important for reconfiguration and surgical tasks, was sufficient; however, the precision was limited owing to the aforementioned fabrication problems. The thin walls of the casing made of acrylic plastic were easily deformed, which caused friction between the parts. The casing made of metal or PEEK and tailor-made metal gears with high precision will improve the mechanism rigidity and performance, thus producing the optimal stability.

3.3 Possible designs of robotic modules

Figure 5 shows various designs of robotic modules that can be implemented in the modular surgical robot. The modules can be categorized into three types: structural modules, functional modules, and other modules. Structural modules are used to configure a robotic topology. Functional modules are used for diagnosis or intervention, while other modules can be added to enhance the performance and robustness of the robotic system. Obviously, an assembled robot made of different types of modules (i.e. a robot with high heterogeneity) may provide high dexterity, but the self-assembly in the stomach and control of the modules would become more difficult. To optimize the level of heterogeneity, self-assembly of the robotic modules must be developed so that the reconfiguration of the robotic topology following the self-assembly can be planned in advance. Employing pre-assembled modular arms or tethered modules can be another option to facilitate assembly in a body cavity; however, this would require further anesthesia, and it would hinder the promotion of massive screening.

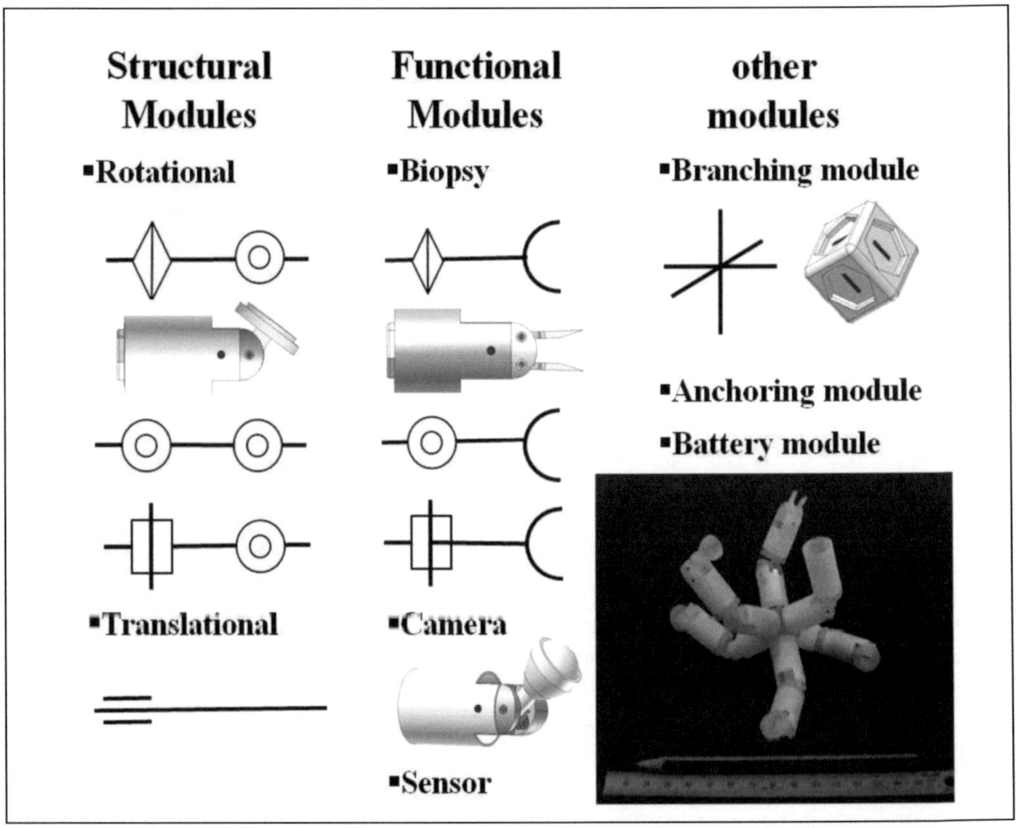

Fig. 5. Various designs of the robotic modules

4. Reconfigurable master device

4.1 Design and prototyping of the reconfigurable master device

One main advantage of using a modular approach in surgical applications is the adaptivity to the given environment as mentioned in Section 2.2. Wherever the robotic platform is deployed in the GI tract, the robotic topology can be changed based on preoperative plans or the in-situ situation to fit in any particular environment. This dynamic changing and reshaping of the robotic topology should be reflected on the user interface. Since it is possible for a robotic topology to have redundant DOFs, the master device for the modular surgical system needs to be able to handle the redundancy that is inherent to modular robots. Based on these considerations, we propose a reconfigurable master device that resembles the robotic platform (Fig.6). When the assembled robot changes its topology, the master device follows the same configuration. The robotic module shown in Fig. 6 has a diameter of 15.4 mm, while a module of the reconfigurable master device has a diameter of 30 mm. The master modules can be easily assembled or disassembled using set screws, and it takes only a few seconds to connect one module to another.

Each robotic module is equipped with two motors as described in the previous section; thus, each master module is equipped with two potentiometers (TW1103KA, Tyco Electronics) that are used as angular position sensors. Calculating the angular position of each joint of the reconfigurable master device is quite straightforward. A common reference voltage is sent from a data acquisition card to all potentiometers, after which the angular position can be calculated from the feedback readings. Owing to the identical configuration, the angle of each joint of the robotic modules can be easily determined, even if the topology has redundancy.

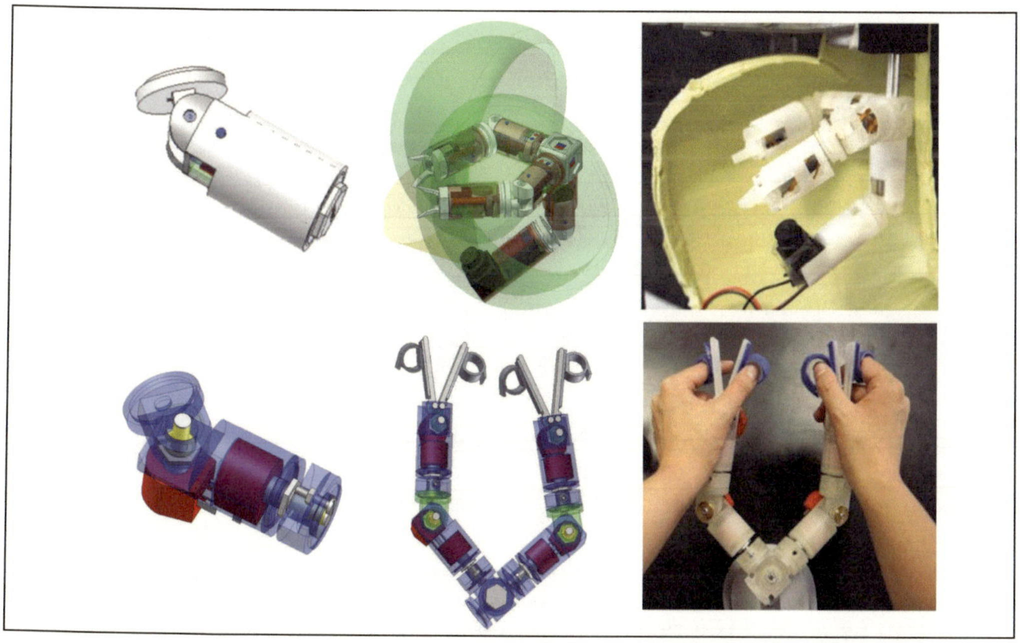

Fig. 6. Robotic modules (top line) and the reconfigurable master device (bottom line): one module (left), assembled modules (middle) and prototypes (right)

The advantages of the proposed master device include intuitive manipulation. For example, the rotational movement of a structural module used to twist the arm is limited to ± 180°, and the master module also has this limitation. This helps surgeons intuitively understand the range of the motion and the reachable working space of the modules. Using a conventional master manipulator or an external console, it is possible that the slave manipulator cannot move owing to its mechanical constraints, while the master manipulator can still move. However, using the proposed master device, the surgeon can intuitively understand the mechanical constraints by manipulating the master device during practice/training. Furthermore, the position of the master arm can indicate where the robotic modules are, even if they are outside of the camera module's view. These characteristics increase the safety of the operation. This feature is important because the entire robotic system is placed inside the body. In other surgical robotic systems, the position or shape of the robotic arms is not important as they are placed outside of the body and can be seen during operation. Unlike other master devices, it is also possible for two or more surgeons to move the reconfigurable master device together at the same time using multi arms with redundant DOFs.

4.2 Evaluation

A simulation-based evaluation setup was selected to simplify the preliminary evaluation of the feasibility of the reconfigurable master device. The authors previously developed the Slave Simulator to evaluate workloads for a master-slave surgical robot (Kawamura, K. 2006). The Slave Simulator can show the motion of the slave robot in CG (Computer Graphics), while the master input device is controlled by an operator. Because the simulator can virtually change the parameters of the slave robot or its control, it is easy to evaluate the parameters as well as the operability of the master device. This Slave Simulator was appropriately modified for the ARES system. The modified Slave Simulator presents the CG models of the robotic modules to the operator. The dimension and DOFs of each module in CG were determined based on the design of the robotic modules. The angle of each joint is given by the signal from the potentiometers of the reconfigurable master device, and the slave modules in CG move in real time to reproduce the configuration of the master device. This Slave Simulator is capable of altering joint positions and the number of joints of the slave arms in CG so that the workspace of the reconfigurable master device can be reproduced in a virtual environment for several types of topologies. The simulator is composed of a 3D viewer that uses OpenGL and a physical calculation function. This function was implemented to detect a collision between the CG modules and an object placed in the workspace.
To simplify the experiments to evaluate the feasibility of the proposed master device and usefulness of the developed simulator, only one arm of the reconfigurable master device was used. Three topologies that consist of one Biopsy Module and one or two Structural Module(s) were selected as illustrated in Fig.7. Topology I consists of a Structural Module and a Biopsy Module, and the base is fixed so that the arm appears with an angle of 45 degrees. One Structural Module is added to Topology I to configure Topology II, and Topology III is identical to Topology II, but placed at 0 degrees. Both Topology II and Topology III have redundant DOFs. The projection of the workspace of each arm and the shared workspace are depicted in Fig.8. A target object on which the arm works in the experiments must be placed in this shared area, which makes it easy to compare topologies.

A bar was selected as the target object instead of a sphere because the height of the collision point is different for each topology when the object appears in the same position in the 2D plane.
The experiment was designed so that a bar appears at random in the shared workspace. The bar represents a target area at which the Biopsy Module needs to collect tissue samples, and this experiment is a simple example to select one topology among three choices given that the arm can reach the target. We assumed that this choice may vary depending on the user, and this experiment was designed to determine if the reconfigurability of the master device, i.e. customization of the robot, provides advantages and improves performance.

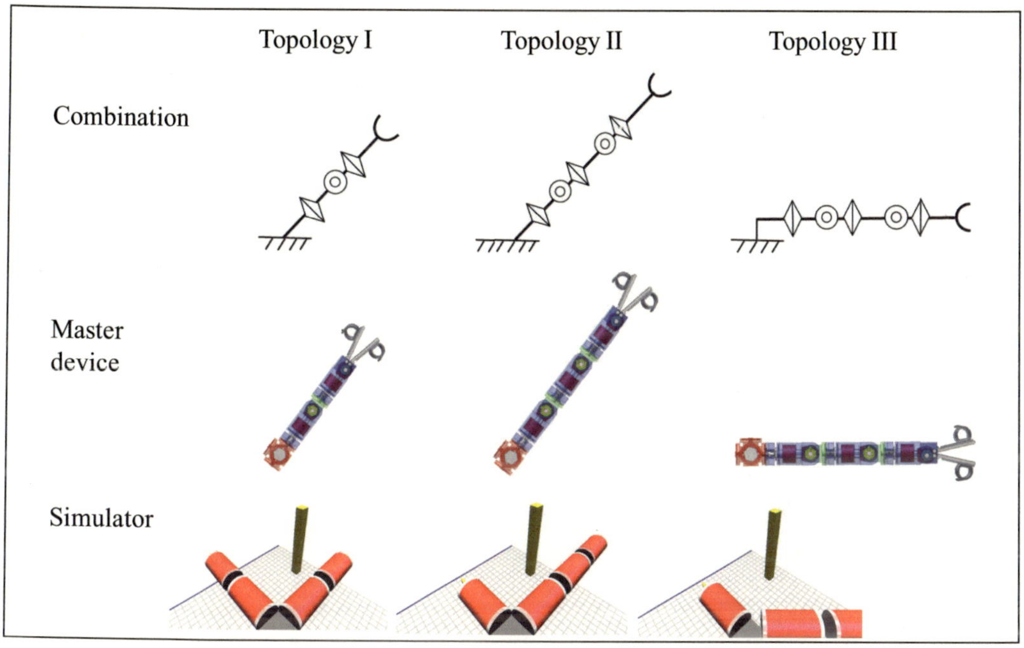

Fig. 7. Three topologies used in the experiments

During the experiment, the operator of the reconfigurable master device could hear a beeping sound when the distal end of the arm (i.e. the grasping part of the biopsy module) touched the bar. The task designed for the experiments was to move the arm of the reconfigurable master device as quickly as possible, touch the bar in CG, and then maintain its position for three seconds. The plane in which the bar stands is shown in grids (Fig.9), and the operator obtains 3D perception by observing these grids. The plane with the grids is the same for all topologies. The angle of the view was set so that the view is similar to that from the camera module in Fig.6.
Five subjects (a-e) participated in the experiments, none of whom were surgeons. Each subject was asked to freely move the master device to learn how to operate it; however, this practice was allowed for one minute before starting the experiments. Each subject started from Topology I, then tested Topology II and finally Topology III. The time needed to touch the bar and maintain it for three seconds was measured. This procedure was repeated ten times for each topology with a randomized position of the bar. During the procedure, the bar appeared at random; however, it always appeared in the shared workspace to ensure

that the arm could reach it. After finishing the experiment, the subjects were asked to fill in a questionnaire (described below) for each topology. The subjects were also asked which topology they preferred.

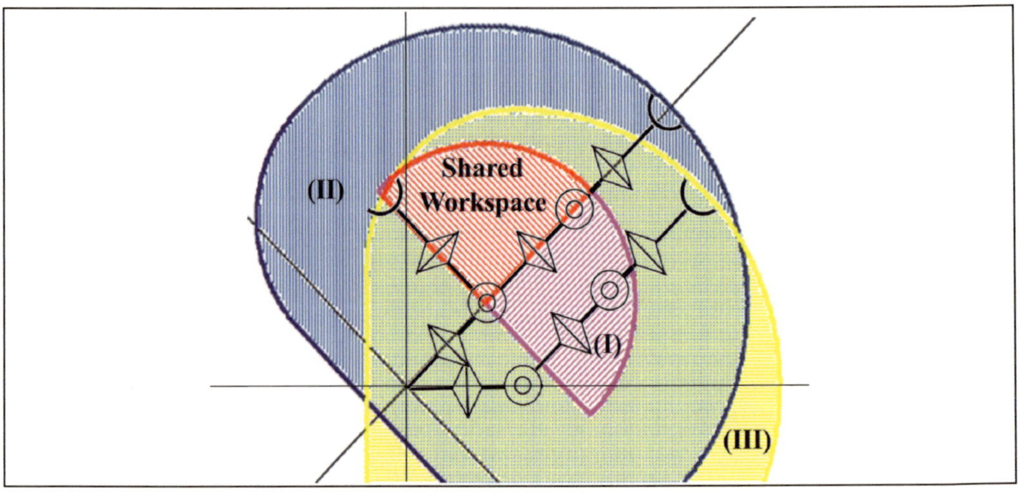

Fig. 8. Workspace of each topology and the shared workspace

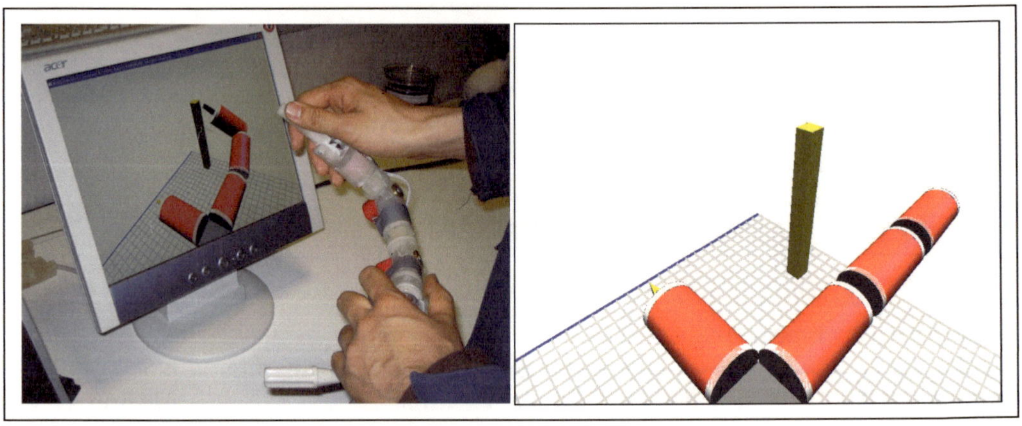

Fig. 9. Simulator and the master device during one test

A NASA TLX questionnaire (NASA TLX (website)) was used to objectively and quantitatively evaluate the workload that the subjects felt during the experiments. This method has versatile uses, and we selected this method also because it was used to evaluate the workload in a tele-surgery environment (Kawamura, K. 2006). This method evaluates Metal Demand, Physical Demand, Temporal Demand, Performance, Effort and Frustration, and gives a score that represents the overall workload that the subject felt during the task.

4.3 Results

The time spent conducting the given task, the workload score evaluated using the NASA TLX questionnaire and the preference of the topology determined by the subjects are summarized in Table 1. For each item, a smaller value indicates a more favorable evaluation by the subject.

Considering the results of the time and workload score, Topology II was most difficult. The difference between Topology I and III was interesting. Two of the subjects ((b) and (c)) preferred Topology I, which did not have a redundant DOF. Conversely, three of the subjects ((a), (d) and (e)) preferred Topology III because they could select the path to reach the target owing to the redundant DOF. The average scores of the NASA TLX parameters shown in Fig.10 suggest that the Physical Demand workload was high for Topology I, while the Effort workload was high for Topology III.

The two subjects who preferred Topology I rather than Topology III claimed that it was not easy to determine where the bar was located when Topology III was used owing to the lack of 3D perception. In addition, they reported that the bar seemed to be placed far from the base. However, the bar appeared randomly, but in the same area; therefore, the bar that appeared in the experiment that employed Topology III was not placed farther from the base when compared to the experiments that used Topology I or Topology II. Accordingly, these two subjects may have had difficulty obtaining 3D perception from the gridded plane. In Topology III, the arm was partially out of view in the initial position; thus, the operator needed to obtain 3D perception by seeing the grids. It is often said that most surgeons can obtain 3D perception even if they use a 2D camera, and our preliminary experimental results imply that this ability might differ by individual. Some people appear to obtain 3D perception primarily by seeing the relative positions between the target and the tool they move. Redundant DOFs may also be preferred by operators with better 3D perception capability.

Although the experiments were preliminary, there must be other factors that accounted for the preference of the user. Indeed, it is likely that the preferable topology varies depending on the user, and the developed simulator would be useful to evaluate these variations. The proposed reconfigurable master device will enable individual surgeons to customize the robot and interface as they prefer.

	Topology	Subject					Average
		a	b	c	d	e	
Time (s)	I	5.7	4.1	4.0	5.8	5.0	4.9
	II	7.6	6.2	4.8	5.5	6.7	6.1
	III	4.9	4.3	5.6	4.4	4.7	4.8
Work Load: NASA-TLX Score	I	30.7	11.3	28.3	32.0	73.3	35.1
	II	47.6	26.7	28.0	53.0	68.3	44.7
	III	37.0	5.0	24.3	14.3	61.3	28.4
Preference	I	3	1	1	3	3	2.2
	II	2	3	2	2	2	2.2
	III	1	2	3	1	1	1.6

Table 1. Experimental results

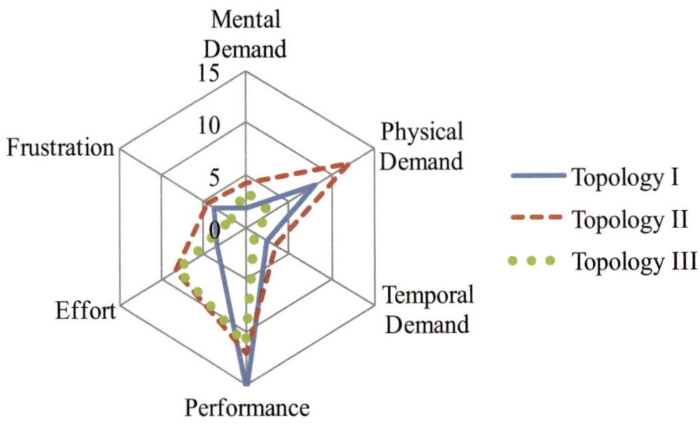

Fig. 10. NASA TLX parameters for three topologies

5. Conclusion

A modular robot was proposed for endoluminal surgery. The design, prototyping and evaluation of the modules were reported. Although there are some issues related to the fabrication problems, the results of the performance tests show the feasibility of the modular surgical system. A reconfigurable master device has also been proposed, and its feasibility was evaluated by simulation-based experiments. The preliminary results showed that the preferred topology may vary depending on the user. Moreover, the reconfigurable master device would enable each surgeon to customize the surgical system according to his/her own preferences. Development of the robotic modules and the reconfigurable master device provided proof of concept of the modular robotic system for endoluminal surgery, suggesting that the modular approach has great potential for surgical applications.

6. Acknowledgments

This study was supported in part by the European Commission in the framework of the ARES Project (Assembling Reconfigurable Endoluminal Surgical System, NEST-2003-1-ADVENTURE/15653), by the European Union Institute in Japan at Waseda University (EUIJ Waseda, http://www.euij-waseda.jp/eng/) within the framework of its Research Scholarship Programme, and by the Global COE Program "Global Robot Academia" from the Ministry of Education, Culture, Sports, Science and Technology of Japan. The authors are grateful to Mr. Nicodemo Funaro for manufacturing the prototypes and Ms. Sara Condino for her invaluable technical support.

7. References

Bardaro, S. J. & Swanström, L. (2006). Development of advanced endoscopes for Natural Orifice Transluminal Endoscopic Surgery (NOTES). In: *Minim. Invasive Ther. Allied Technol.*, 15(6), pp. 378–383.

Carpi, F. & Pappone, C. (2009). Magnetic maneuvering of endoscopic capsules by means of a robotic navigation system. In: *IEEE Trans Biomed Eng*, 56(5), pp. 1482-90.

Ciuti, G.; Valdastri, P., Menciassi, A. & Dario, P. (2010). Robotic magnetic steering and locomotion of capsule endoscope for diagnostic and surgical endoluminal procedures. In: *Robotica*, 28, pp. 199-207.

Cuschieri, A (2005). Laparoscopic surgery: current status, issues and future developments. In: *Surgeon*, 3(3), pp. 125–138.

Giday, S.; Kantsevoy, S. & Kalloo, A. (2006). Principle and history of natural orifice translumenal endoscopic surgery (notes). In: *Minim. Invasive Ther. Allied Technol.*, 15(6), pp. 373–377.

Harada, K.; Susilo, E., Menciassi, A. & Dario, P. (2009). Wireless reconfigurable modules for robotic endoluminal surgery. In: *Proc. IEEE Int. Conf. on Robotics and Automation*, pp. 2699-2704.

Harada, K.; Oetomo, D., Susilo, E., Menciassi, A., Daney, D., Merlet, J.-P. & Dario, P. (2010). A Reconfigurable Modular Robotic Endoluminal Surgical System: Vision and preliminary results, In: *Robotica*, 28, pp. 171-183.

Kawamura, K.; Kobayashi, Y & Fujie, M. G. (2006). Development of Real-Time Simulation for Workload Quantization in Robotic Tele-surgery. In: *Proc. IEEE Int. Conf. on Robotics and Biomimetics*, pp.1420-25.

Menciassi, A.; Valdastri, P., Harada, K. & Dario, P. (2010). Single and Multiple Robotic Capsules for Endoluminal Diagnosis and Surgery. In: *Surgical Robotics - System Applications and Visions*, Rosen, J., Hannaford, B. & Satava, M., pp. 313-354., Springer-Verlag, 978-1441911254.

Moglia, A.; Menciassi, A. Schurr, M. & Dario, P. (2007). Wireless capsule endoscopy: From diagnostic devices to multipurpose robotic systems. In: *Biomed Microdevices*, 9, pp. 235–243.

Murata, S. & Kurokawa, H. (2007). Self-reconfigurable robots. In: *IEEE Rob. Autom Mag.*, 14(1), pp. 71–78.

Nagy, Z.; Abbott, J. J. & Nelson, B. J. (2007). The magnetic self-aligning hermaphroditic connector: A scalable approach for modular microrobotics. In: *Proc. IEEE/ASME Int. Conf. Advanced Intelligent Mechatronics*. pp. 1–6.

NASA TLX. available from http://humansystems.arc.nasa.gov/groups/TLX/

Park, S. K.; Koo, K. I., Bang, S. M., Park J. Y., Song, S. Y., & Cho, D. G. (2008). A novel microactuator for microbiopsy in capsular endoscopes. In: *J. Micromech. Microeng.*, 18 (2), 025032.

Pesic, M.; Karanikolic, A., Djordjevic, N., Katic, V., Rancic, Z., Radojkovic, M., Ignjatovic N. & Pesic, I. (2004). The importance of primary gastric cancer location in 5-year survival rate. In: *Arch Oncol*, 12, pp. 51–53.

Quirini, M.; Menciassi, A., Scapellato, S., Stefanini, C., and Dario, P. (2008). Design and fabrication of a motor legged capsule for the active exploration of the gastrointestinal tract. In: *Proc. IEEE/ASME Trans. Mechatronics*, 13, pp. 169–179.

Sendoh, M.; Ishiyama, K. & Arai, K. (2003). Fabrication of magnetic actuator for use in a capsule endoscope. In: *IEEE Trans. Magnetics*, 39(5), pp. 3232–34.

Susilo, E.; Valdastri, P., Menciassi, A. & Dario, P. (2009). A miniaturized wireless control platform for robotic capsular endoscopy using advanced pseudokernel approach. In: *Sensors and Actuators A*, 156(1), pp.49-58.

World Health Organisation, "Fact sheet n.297," Available from: http://www.who.int/mediacen-ter/factsheets/ fs297, 2006.

Yim, M.; Shen, W., Salemi, B., Rus, D., Moll, M., Lipson, H., Klavins, E. & Chirikjian, G. (2007). Modular self-reconfigurable robot systems [Grand Challenges of Robotics]. In: *IEEE Rob. Autom Mag.*, 14(1), pp. 865–872.

2

Target Point Manipulation Inside a Deformable Object

Jadav Das and Nilanjan Sarkar
Vanderbilt University, Nashville, TN
USA

1. Introduction

Target point manipulation inside a deformable object by a robotic system is necessary in many medical and industrial applications such as breast biopsy, drug injection, suturing, precise machining of deformable objects etc. However, this is a challenging problem because of the difficulty of imposing the motion of the internal target point by a finite number of actuation points located at the boundary of the deformable object. In addition, there exist several other important manipulative operations that deal with deformable objects such as whole body manipulation [1], shape changing [2], biomanipulation [3] and tumor manipulation [4] that have practical applications. The main focus of this chapter is the target point manipulation inside a deformable object. For instance, a positioning operation called linking in the manufacturing of seamless garments [5] requires manipulation of internal points of deformable objects. Mating of a flexible part in electric industry also results in the positioning of mated points on the object. In many cases these points cannot be manipulated directly since the points of interest in a mating part is inaccessible because of contact with a mated part. Additionally, in medical field, many diagnostic and therapeutic procedures require accurate needle targeting. In case of needle breast biopsy [4] and prostate cancer brachytherapy [6], needles are used to access a designated area to remove a small amount of tissue or to implant radio-active seed at the targeted area. The deformation causes the target to move away from its original location. To clarify the situation we present a schematic of needle insertion for breast biopsy procedure as shown in Figure 1. When tip of the needle reaches the interface between two different types of tissue, its further insertion will push the tissue, instead of piercing it, causing unwanted deformations. These deformations move the target away from its original location as shown in Figure 1(b). In this case, we cannot manipulate the targeted area directly because it is internal to the organ. It must be manipulated by controlling some other points where forces can be applied as shown in Figure 1(c). Therefore, in some cases one would need to move the positioned points to the desired locations of these deformable objects (e.g., mating two deformable parts for sewing seamlessly) while in other cases one may need to preserve the original target location (e.g., guiding the tumor to fall into the path of needle insertion). In either of these situations, the ability of a robotic system to control the target of the deformable object becomes important, which is the focus of this chapter.

To control the position of the internal target point inside a deformable object requires appropriate contact locations on the surface of the object. Therefore, we address the issue of determining the optimal contact locations for manipulating a deformable object such that the internal target point can be positioned to the desired location by three robotic fingers using minimum applied forces. A position-based PI controller is developed to control the motion of the robotic fingers such that the robotic fingers apply minimum force on the surface of the object to position the internal target point to the desired location. However, the controller for target position control is non-collocated since the internal target point is not directly actuated by the robotic fingers. It is known in the literature that non-collocated control of a deformable object is not passive, which may lead to instability [7]. In order to protect the object and the robotic fingers from physical damage and also in order to diminish the deterioration of performance caused by unwanted oscillation, it is indispensable to build stable interaction between the robotic fingers and the object. Here we consider that the plant (i.e., the deformable object) is passive and does not generate any energy. So, in order to have stable interaction, it is essential that the controller for the robotic fingers must be stable. Thus, we present a new passivity-based non-collocated controller for the robotic fingers to ensure safe and accurate position control of the internal target point. The passivity theory states that a system is passive if the energy flowing in exceeds the energy flowing out. Creating a passive interface adds the required damping force to make the output energy lower than the input energy. To this end we develop a passivity observer (PO) and a passivity controller (PC) based on [8] for individual robotic finger where PO monitors the net energy flow out of the system and PC will supply the necessary damping force to make PO positive. Our approach extends the concept of PO and PC in [8] to multi-point contacts with the deformable object.

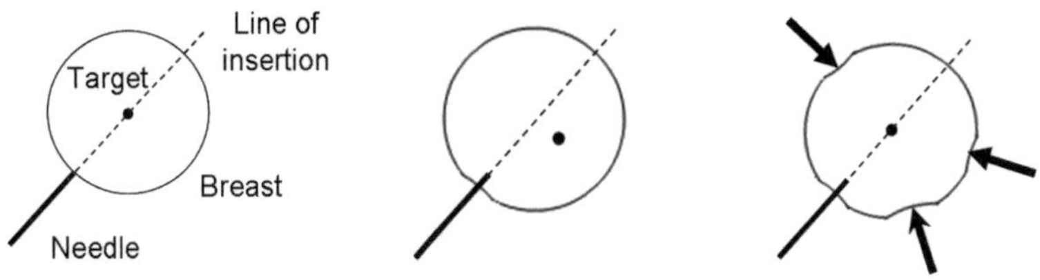

(a) (b) (c)

Fig. 1. Schematics of needle breast biopsy procedure: (a) needle insertion, (b) target movement, and (c) target manipulation

The remainder of this chapter is organized as follows: we discuss various issues and prior research in Section 2. The problem description is stated in Section 3. Section 4 outlines the mathematical modelling of the deformable object. A framework for optimal contact locations is presented in Section 5. The control methods are discussed in Section 6. The effectiveness of the derived control law is demonstrated by simulation in Section 7. Finally, the contributions of this work and the future directions are discussed in Section 8.

2. Issues and prior research

A considerable amount of work on multiple robotic systems has been performed during the last few decades [9-11, 12-15]. Mostly, the position and/or force control of multiple manipulators handling a rigid object were studied in [9-11]. However, there were some works on handling deformable object by multiple manipulators presented in [12-15]. Saha and Isto [12] presented a motion planner for manipulating deformable linear objects using two cooperating robotic arms to tie self-knots and knots around simple static objects. Zhang et al. [13] presented a microrobotic system that is capable of picking up and releasing operation of microobjects. Sun et al. [14] presented a cooperation task of controlling the reference motion and the deformation when handling a deformable object by two manipulators. In [15], Tavasoli et al. presented two-time scale control design for trajectory tracking of two cooperating planar rigid robots moving a flexible beam. However, to the best of our knowledge the works on manipulating an internal target point inside a deformable object are rare [4, 5]. Mallapragada et al. [4] developed an external robotic system to position the tumor in image-guided breast biopsy procedures. In their work, three linear actuators manipulate the tissue phantom externally to position an embedded target inline with the needle during insertion. In [5] Hirai et al. developed a robust control law for manipulation of 2D deformable parts using tactile and vision feedback to control the motion of the deformable object with respect to the position of selected reference points. These works are very important to ours present application, but they did not address the optimal locations of the contact points for effecting the desired motion.

A wide variety of modeling approaches have been presented in the literature dealing with computer simulation of deformable objects [16]. These are mainly derived from physically-based models to produce physically valid behaviors. Mass-spring models are one of the most common forms of deformable objects. A general mass-spring model consists of a set of point masses connected to its neighbors by massless springs. Mass-spring models have been used extensively in facial animation [17], cloth motion [18] and surgical simulation [19]. Howard and Bekey [20] developed a generalized method to model an elastic object with the connections of springs and dampers. Finite element models have been used in the computer simulation to model facial tissue and predict surgical outcomes [21, 22]. However, the works on controlling an internal point in a deformable object are not attempted.

In order to manipulate the target point to the desired location, we must know the appropriate contact locations for effecting the desired motion. There can be an infinite number of possible ways of choosing the contact location based on the object shapes and task to be performed. Appropriate selection of the contact points is an important issue for performing certain tasks. The determination of optimal contact points for rigid object was extensively studied by many researchers with various stability criteria. Salisbury [23] and Kerr [24] discussed that a stable grasp was achieved if and only if the grasp matrix is full row rank. Abel et al. [25] modelled the contact interaction by point contact with Coulomb friction and they stated that optimal grasp has minimum dependency on frictional forces. Cutkosky [26] discussed that the size and shape of the object has less effect on the choice of grasp than by the tasks to be performed after examining a variety of human grasps. Ferrari et al. [27] defined grasp quality to minimize either the maximum value or sum of the finger forces as optimality criteria. Garg and Dutta [28] shown that the internal forces required for grasping deformable objects vary with size of object and finger contact angle. In [29], Watanabe and Yoshikawa investigated optimal contact points on an arbitrary shaped object in 3D using the concept of required external force set. Ding et al. proposed an algorithm for computing form closure grasp on a 3D polyhedral object using local search strategy in [30]. In [31, 32], various concepts and methodologies of robot grasping of rigid objects were reviewed. Cornella et al. [33] presented a mathematical approach to obtain optimal solution of contact points using the dual theorem of nonlinear programming. Saut et al. [34]

presented a method for solving the grasping force optimization problem of multi-fingered dexterous hand by minimizing a cost function. All these works are based on grasp of rigid objects.
There are also a few works based on deformable object grasping. Like Gopalakrishnan and Goldberg [35] proposed a framework for grasping deformable parts in assembly lines based on form closure properties for grasping rigid parts. Foresti and Pellegrino [36] described an automatic way of handling deformable objects using vision technique. The vision system worked along with a hierarchical self-organizing neural network to select proper grasping points in 2D. Wakamatsu et al. [37] analyzed grasping of deformable objects and introduced bounded force closure. However, position control of an internal target point in a deformable object by multi-fingered gripper has not been attempted. In our work, we address the issue of determining the optimal contact locations for manipulating a deformable object such that the internal target point can be positioned to the desired location by three robotic fingers using minimum applied forces.
The idea of passivity can be used to guarantee the stable interaction without exact knowledge of model information. Anderson and Spong [38] published the first solid result by passivation of the system using scattering theory. A passivity based impedance control strategy for robotic grasping and manipulation was presented by Stramigioli et al. [39]. Recently, Hannaford and Ryu [40] proposed a time-domain passivity control based on the energy consumption principle. The proposed algorithm did not require any knowledge about the dynamics of the system. They presented a PO and a PC to ensure stability under a wide variety of operating conditions. The PO can measure energy flow in and out of one or more subsystems in real-time by confining their analysis to system with very fast sampling rate. Meanwhile the PC, which is an adaptive dissipation element, absorbs exactly net energy output measured by the PO at each time sample. In [41], a model independent passivity-based approach to guarantee stability of a flexible manipulator with a non-collocated sensor-actuator pair is presented. This technique uses an active damping element to dissipate energy when the system becomes active. In our work we use the similar concept of PO and PC to ensure stable interaction between the robotic fingers and the deformable object. Our work also extends the concept of PO and PC for multi-point contact with the deformable object.

3. Problem description

Consider a case in which multiple robotic fingers are manipulating a deformable object in a 2D plane to move an internal target point to a desired location. Before we discuss the design of the control law, we present a result from [42] to determine the number of actuation points required to position the target at an arbitrary location in a 2D plane. The following definitions are given according to the convention in [42].
Manipulation points: are defined as the points that can be manipulated directly by robotic fingers. In our case, the manipulation points are the points where the external robotic fingers apply forces on the deformable object.
Positioned points: are defined as the points that should be positioned indirectly by controlling manipulation points appropriately. In our case, the target is the position point.
The control law to be designed is non-collocated since the internal target point is not directly actuated by the robotic fingers. The following result is useful in determining the number of actuation points required to accurately position the target at the desired location.

Result [42]: The number of manipulated points must be greater than or equal to that of the positioned points in order to realize any arbitrary displacement.
In our present case, we assume that the number of positioned points is one, since we are trying to control the position of the target. Hence, ideally the number of contact points would also be one. But practically, we assume that there are two constraints: (1) we do not want to apply shear force on the deformable object to avoid the damage to the surface, and (2) we can only apply control force directed into the deformable object. We cannot pull the surface since the robotic fingers are not attached to the surface. Thus we need to control the position of the target by applying only unidirectional compressive force.
In this context, there exists a theorem on the force direction closure in mechanics that helps us determining the equivalent number of compressive forces that can replace one unconstrained force in a 2D plane.
Theorem [43]: A set of wrenches $\mathbf{w}$ can generate force in any direction if and only if there exists a three-tuple of wrenches $\{\mathbf{w}_1, \mathbf{w}_2, \mathbf{w}_3\}$ whose respective force directions $\mathbf{f}_1$, $\mathbf{f}_2$, $\mathbf{f}_3$ satisfy:

i. Two of the three directions $\mathbf{f}_1$, $\mathbf{f}_2$, $\mathbf{f}_3$ are independent
ii. A strictly positive combination of the three directions is zero.

$$\alpha\mathbf{f}_1 + \beta\mathbf{f}_2 + \gamma\mathbf{f}_3 = 0 \tag{1}$$

where α, β, and γ are constants. The ramification of this theorem for our problem is that we need three control forces distributed around the object such that the end points of their direction vectors draw a non-zero triangle that includes their common origin point. With such an arrangement we can realize any arbitrary displacement of the target point. Thus the problem can be stated as:
Problem statement: Given the number of actuation points, the initial target and its desired locations, find appropriate contact locations and control action such that the target point is positioned to its desired location by controlling the boundary points of the object with minimum force.

4. Deformable object modelling

Consider a schematic in Figure 2 where three robotic fingers are positioning an internal target (point A) in a deformable object to the desired location (point B). We assume that all the end-effectors of the robotic fingers are in contact with the deformable object such that they can apply only push on the object as needed.
The coordinate systems are defined as follows: Σ_w is the task coordinate system, Σ_o is the object coordinate system, fixed on the object and Σ_i is the *i*-th robotic finger coordinate system, fixed on the *i*-th end-effectors located at the grasping point. In order to formulate the optimal contact locations, we model the deformable object using mass-spring-damper systems. The point masses are located at the nodal points and a Voigt element [20] is inserted between them. Figure 3 shows a single layer of the deformable object. Each element is labeled as E_j for $j = 1,2,\ldots,NE$ where NE is total number of elements in a single layer. Position vector of the *i*-th mesh point is defined as $\mathbf{p}_i = [x_i \quad y_i]^T$, $i = 1,2,3,\ldots,N$ where, N is total number of point masses. k and c are the spring stiffness and the damping coefficient, respectively. Assume that no moment exerts on each mesh point. Then, the resultant force exerted on the mesh point, $\mathbf{p}_i$, can be calculated as

$$\mathbf{w}_i = -\frac{\partial U}{\partial \mathbf{p}_i} \tag{2}$$

where, U denotes the total potential energy of the object

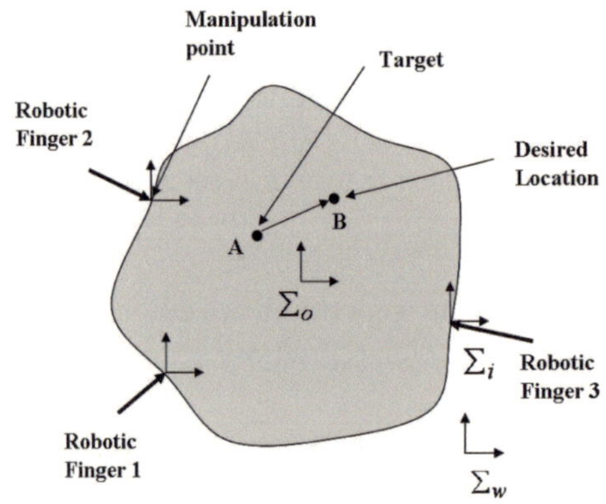

Fig. 2. Schematic of the robotic fingers manipulating a deformable object

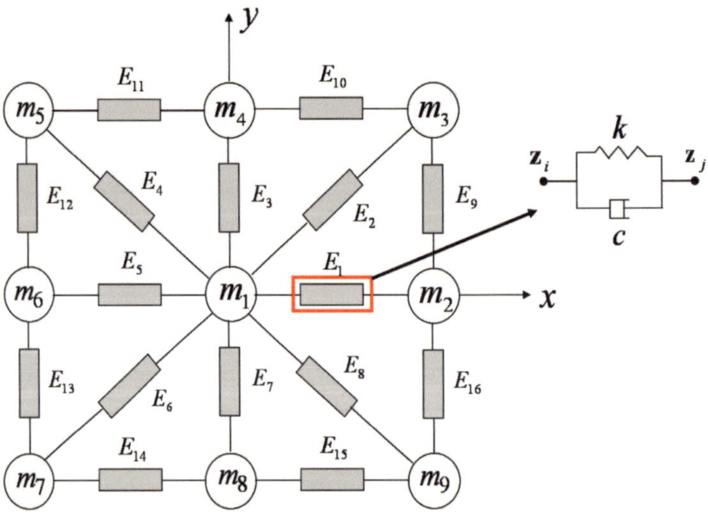

Fig. 3. Model of a deformable object with interconnected mass-spring-damper

5. Framework for optimal contact locations

We develop an optimization technique that satisfies the force closure condition for three fingers planar grasp. The resultant wrench for the contacts of three robotic fingers is given by

$$\mathbf{w} = \sum_{i=1}^{3} f_i \mathbf{n}_i(\mathbf{r}_i)\,,\ (\forall \mathbf{w} \in \Re^2)\ (\exists f_i \geq 0,\ 1 \leq i \leq 3) \tag{3}$$

where, $\mathbf{n}_i(\mathbf{r}_i)$ is the unit inner normal of i-th contact and f_i denotes the i-th finger's force. We assume that the contact forces should exist in the friction cone to manipulate objects without slip of the fingertip. Now we need to find three distinct points, $\mathbf{r}_1(\theta_1)$, $\mathbf{r}_2(\theta_2)$, and $\mathbf{r}_3(\theta_3)$, on the boundary of the object such that Equation (3) is satisfied. Here, θ_1, θ_2, and θ_3 are the three contact point locations measured anti-clockwise with respect to the x axis as shown in Figure 4. In addition, we assume that the normal forces have to be non-negative to avoid separation and slippage at the contact points, i.e.,

$$f_i \geq 0, i = 1,2,3 \tag{4}$$

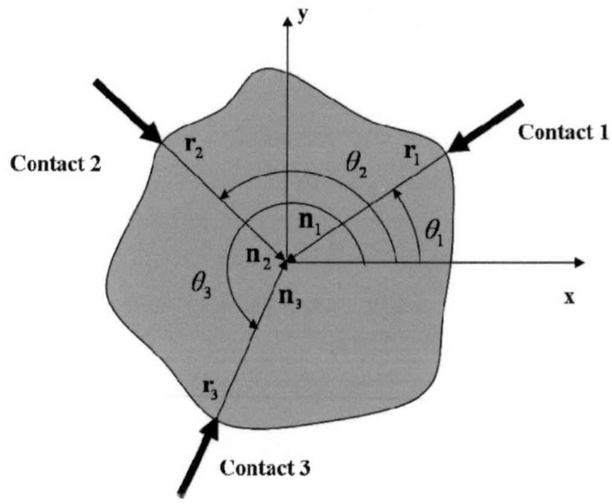

Fig. 4. Three fingers grasp of a planar object

A physically realizable grasping configuration can be achieved if the surface normals at three contact points positively span the plane so that they do not all lie in the same half-plane [44]. Therefore, a realizable grasp can be achieved if the pair-wise angles satisfy the following constraints

$$\theta_{\min} \leq |\theta_i - \theta_j| \leq \theta_{\max}\ , \theta_{\text{low}} \leq \theta_i \leq \theta_{\text{high}}\ , \ i,j = 1,2,3\ , \ i \neq j \tag{5}$$

A unique solution to realizable grasping may not always exist. Therefore, we develop an optimization technique that minimizes the total force applied on to the object to obtain a particular solution. The optimal locations of the contact points would be the solution of the following optimization problem.

min $\mathbf{f}^T\mathbf{f}$

sunject to

$$\begin{aligned} \mathbf{w} &= \sum_{i=1}^{3} f_i \mathbf{n}_i(\mathbf{r}_i) \\ \theta_{\min} \leq \left|\theta_i - \theta_j\right| &\leq \theta_{\max}\ , \ i,j = 1,2,3\ , \ i \neq j \\ f_i &\geq 0\ , \ i = 1,2,3 \\ 0 \leq \theta_i &\leq 360^\circ\ , \ i = 1,2,3 \end{aligned} \tag{6}$$

Once we get the optimal contact locations, all three robotic fingers can be located at their respective positions to effect the desired motion at those contact points.

6. Design of the controller

In this section, a control law for the robotic fingers is developed to guide a target from any point A to an arbitrary point B within the deformable object as shown in Figure 2.

6.1 Target position control

At any given time-step, point A is the actual location of the target and point B is the desired location of the target. $\mathbf{n}_1$, $\mathbf{n}_2$ and $\mathbf{n}_3$ are unit vectors which determine the direction of force application of the actuation points with respect to the global reference frame Σ_w. Let assume, $\mathbf{p}_d$ is the position vector of point B and $\mathbf{p}$ is the position vector of point A. Referring to Figure 2, the position vector of point A is given by

$$\mathbf{p} = [x \quad y]^T \tag{7}$$

where, x and y are the position coordinates of point A in the global reference frame Σ_w. The desired target position is represented by point B whose position vector is given by

$$\mathbf{p}_d = [x_d \quad y_d]^T \tag{8}$$

where, x_d and y_d are the desired target position coordinates. The target position error, **e,** is given by

$$\mathbf{e} = \mathbf{p}_d - \mathbf{p} \tag{9}$$

Once the optimal contact locations are determined from Equation (6), the planner generates the desired reference locations for these contact points by projecting the error vector between the desired and the actual target locations in the direction of the applied forces, which is given by

$$\mathbf{e}_i^* = \mathbf{e} \cdot \mathbf{n}_i \tag{10}$$

where,

$$\mathbf{n}_i = [n_{xi} \quad n_{yi}]^T \tag{11}$$

All robotic fingers are controlled by their individual controllers using the following proportional-integral (PI) control law

$$f_i = K_{Pi}\mathbf{e}_i^* + K_{Ii}\int \mathbf{e}_i^* dt, \quad i-1,2,3 \tag{12}$$

where, K_{Pi}, and K_{Ii} are the proportional and integral gains. Note that in the control law (12), mechanical properties of the deformable object are not required. Forces applied by the fingers on the surface of the deformable object are calculated by projecting the error vector in the direction of the applied forces. But the Equation (12) does not guarantee that the system will be stable. Thus a passivity-based control approach based on energy monitoring is developed to guarantee the stability of the system.

6.2 Passivity-based control

A passivity-based control approach based on energy monitoring is developed for deformable object manipulation to guarantee passivity (and consequently stability) of the system. The main reason to use passivity-based control is to ensure stability without the need of having an accurate model of the deformable object. It is not possible to develop a precise dynamic model of a deformable object due to complex nonlinear internal dynamics as well as variation in geometry and mechanical properties. Thus passivity based control is an ideal candidate to ensure stability since it is a model independent technique. The basic idea is to use a PO to monitor the energy generated by the controller and to dissipate the excess energy using a PC when the controller becomes active [41], without the need for modeling the dynamics of the plant (deformable object).

Passivity Observer (PO)

We develop a network model with PO and PC similar to [41] as shown in Figure 5. The PO monitors the net energy flow of the individual finger's controller. When the energy becomes negative, PC dissipates excess energy from the individual controller. Similar to [41] energy is defined as the integral of the inner product between conjugate input and output, which may or may not correspond to physical energy. Definition of passivity [41] states that the energy applied to a passive network must be positive for all time. Figure 5 shows a network representation of the energetic behavior of this control system. The block diagram in Figure 5 is partitioned into three elements: the trajectory generator, the controller and the plant. Each controller corresponds to one finger. Since three robotic fingers are used for planar manipulation, three individual controller transfer energy to the plant.

The connection between the controller and the plant is a physical interface at which conjugate variables (f_i , v_i ; where f_i is the force applied by *i*-th finger and v_i is the velocity of *i*-th finger) define physical energy flow between controller and plant. The forces and velocities are given by

$$\mathbf{f} = [f_1 \quad f_2 \quad f_3]^T \tag{13}$$

$$\mathbf{v} = [v_1 \quad v_2 \quad v_3]^T \tag{14}$$

The desired target velocity is obtained by differentiating (8) with respect to time and is given by

$$\dot{\mathbf{p}}_d = [\dot{x}_d \quad \dot{y}_d]^T \tag{15}$$

where, $\dot{x}_d$ and $\dot{y}_d$ are the desired target velocities, respectively. The desired target velocity along the direction of actuation of the *i*-th robotic finger is given by

$$v_{di} = \dot{\mathbf{p}}_d \cdot \mathbf{n}_i \tag{16}$$

The trajectory generator essentially computes the desired target velocity along the direction of actuation of the robotic fingers. If the direction of actuation of the robotic fingers, $\mathbf{n}_i$, and desired target velocity, $\dot{\mathbf{p}}_d$, are known with respect to a global reference frame then the trajectory generator computes the desired target velocity along the direction of actuation of the fingers using Equation (16).

The connections between the trajectory generator and the controller, which traditionally consist of a one-way command information flow, are modified by the addition of a virtual feedback of the conjugate variable [41]. For the system shown in Figure 5, output of the trajectory generator is the desired target velocity, v_{di}, along direction of *i*-th finger and output of the controller is calculated from Equation (12).

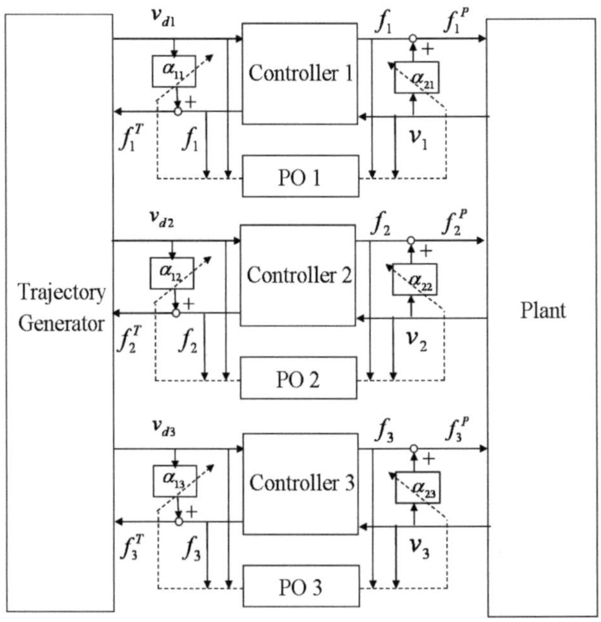

Fig. 5. Network representation of the control system. α_{1i} and α_{2i} are the adjustable damping elements at each port, *i*=1,2,3

For both connections, virtual feedback is the force applied by the robotic fingers. Integral of the inner product between trajectory generator output (v_{di}) and its conjugate variable (f_i) defines "virtual input energy." The virtual input energy is generated to give a command to the controller, which transmits the input energy to the plant through the controller in the form of "real output energy." Real output energy is the physical energy that enters to the plant (deformable object) at the point where the robotic finger is in contact with the object. Therefore, the plant is a three-port system since three fingers manipulate the object. The conjugate pair that represents the power flow is f_i, v_i (the force and the velocity of *i*-th finger, respectively) The reason for defining virtual input energy is to transfer the source of energy from the controllers to the trajectory generator. Thus the controllers can be represented as two-ports which characterize energy exchange between the trajectory generator and the plant. Note that the conjugate variables that define power flow are discrete time values and so the analysis is confined to systems having a sampling rate substantially faster than the system dynamics.

For regulating the target position during manipulation, $v_{di} = 0$. Hence the trajectory generator is passive since it does not generate energy. However, for target tracking, $v_{di} \neq 0$

and $f_i \neq 0$. Therefore the trajectory generator is not passive because it has a velocity source as a power source. It is shown that even if the system has an active term, stability is guaranteed as long as the active term is not dependent on the system states [45]. Therefore, passivity of the plant and controllers is sufficient to ensure system stability.

Here, we consider that the plant is passive. Now we design a PO for sufficiently small time-step ΔT as:

$$E_i(t_k) = \Delta T \sum_{j=0}^{k} (f_i(t_j) v_{di}(t_j) - f_i(t_j) v_i(t_j)) \tag{17}$$

where, ΔT is the sampling period and $t_j = j \times \Delta T$. In normal passive operation, $E_i(t_j)$ should always be positive. In case when $E_i(t_j) < 0$, the PO indicates that the *i*-th controller is generating energy and going to be active. The sufficient condition to make the whole system passive can be written as

$$\Delta T \sum_{j=0}^{k} f_i(t_j) v_{di}(t_j) \geq \Delta T \sum_{j=0}^{k} f_i(t_j) v_i(t_j), \ \forall t_k \geq 0, i = 1,2,3 \tag{18}$$

where k means the *k*-th step sampling time.

The active and passive port can be recognized by monitoring the conjugate signal pair of each port in real time. A port is active if $fv < 0$ that means energy is flowing out of the network system and it is passive if $fv \geq 0$, that means energy is flowing into the network system. The input and output energy can be computed as [46]

$$E_{1i}^T(k) = \begin{cases} E_{1i}^T(k-1) + f_i(k) v_{di}(k) & \text{if } f_i(k) v_{di}(k) > 0 \\ E_{1i}^T(k-1) & \text{if } f_i(k) v_{di}(k) \leq 0 \end{cases} \tag{19}$$

$$E_{2i}^T(k) = \begin{cases} E_{2i}^T(k-1) - f_i(k) v_{di}(k) & \text{if } f_i(k) v_{di}(k) < 0 \\ E_{2i}^T(k-1) & \text{if } f_i(k) v_{di}(k) \geq 0 \end{cases} \tag{20}$$

$$E_{1i}^P(k) = \begin{cases} E_{1i}^P(k-1) - f_i(k) v_i(k) & \text{if } f_i(k) v_i(k) < 0 \\ E_{1i}^P(k-1) & \text{if } f_i(k) v_i(k) \geq 0 \end{cases} \tag{21}$$

$$E_{2i}^P(k) = \begin{cases} E_{2i}^P(k-1) + f_i(k) v_i(k) & \text{if } f_i(k) v_i(k) > 0 \\ E_{2i}^P(k-1) & \text{if } f_i(k) v_i(k) \leq 0 \end{cases} \tag{22}$$

where, $E_{1i}^T(k)$ and $E_{2i}^T(k)$ are the energy flowing in and out at the trajectory side of the controller port, respectively, whereas $E_{1i}^P(k)$ and $E_{2i}^P(k)$ are the energy flowing in and out at the plant side of the controller port, respectively. So the time domain passivity condition is given by

$$E_{1i}^T(k) + E_{1i}^P(k) \geq E_{2i}^T(k) + E_{2i}^P(k), \ \forall k \geq 0 \tag{23}$$

Net energy output of an individual controller is given by

$$E_i(k) = E_{1i}^T(k) - E_{2i}^P(k) + E_{1i}^P(k) - E_{2i}^T(k) \\ + \alpha_{1i}(k-1)v_{di}(k-1)^2 + \alpha_{2i}(k-1)v_i(k-1)^2 \tag{24}$$

where, the last two terms are the energy dissipated at the previous time step. $\alpha_{1i}(k-1)$ and $\alpha_{2i}(k-1)$ are the damping coefficient calculated based on PO discussed below.

Passivity Controller (PC)

In order to dissipate excess energy of the controlled system, a damping force should be applied to its moving parts depending on the causality of the port. As it is well known, such a force is a function of the system's velocities giving the physical damping action on the system. Mathematically, the damping force is given by

$$f_d = \alpha v \tag{25}$$

where α is the adjustable damping factor and v is the velocity. From this simple observation, it seems necessary to measure and use the velocities of the robotic fingers in the control algorithm in order to enhance the performance by means of controlling the damping forces acting on the systems. On the other hand, velocities measurements are not always available and in these cases position measurements can be used to estimate velocities and therefore to inject damping.

When the observed energy becomes negative, the damping coefficient is computed using the following relation (which obeys the constitutive Equation (25)). Therefore, the algorithm used for a 2-port network with impedance causality (i.e., velocity input, force output) at each port is given by the following steps:

1. Two series PCs are designed for several cases as given below:

   Case 1: If $E_i(k) \geq 0$, i.e., if the output energy is less than the input energy, there is no need to activate any PCs.

   Case 2: If $E_i(k) < 0$, i.e., if the output energy is more than the input energy, i.e., $E_{2i}^P(k) > E_{1i}^T(k)$, then we need to activate only the plant side PC.

$$\alpha_{1i}(k) = 0 \\ \alpha_{2i}(k) = -E_i(k) / v_i(k)^2 \tag{26}$$

   Case 3: Similarly, if $E_i(k) < 0$, $E_{2i}^T(k) > E_{1i}^P(k)$, then we need to activate only the trajectory side PC.

$$\alpha_{1i}(k) = -E_i(k) / v_{di}(k)^2 \\ \alpha_{2i}(k) = 0 \tag{27}$$

2. The contributions of PCs are converted into power variables as

$$f_i^t(k) = \alpha_{1i}(k) v_{di}(k) \\ f_i^p(k) = \alpha_{2i}(k) v_i(k) \tag{28}$$

3. Modified outputs are

$$f_i^T(k) = f_i(k) + f_i^t(k) \\ f_i^P(k) = f_i(k) + f_i^p(k) \tag{29}$$

where, $f_i^t(k)$ and $f_i^p(k)$ are the PCs' outputs at trajectory and plant sides of the controller ports, respectively. $f_i^T(k)$ and $f_i^P(k)$ are the modified outputs at trajectory and plant sides of the controller ports, respectively.

7. Simulation and discussion

We perform extensive simulations of positioning an internal target point to a desired location in a deformable object by external robotic fingers to demonstrate the feasibility of the concept. We discretize the deformable object with elements of mass-spring-damper. We choose m=0.006 kg for each point mass, k=10 N/m for spring constant and c=5 Ns/m for damping coefficient. With this parameter set up, we present four different simulation tasks.

Task 1:

In Task 1, we present the optimal contact locations of various objects using three robotic fingers such that an internal target point is positioned to the desired location with minimum force. The optimal contact locations are computed using Equation (6) as shown in Figures 6 to 8. In these figures, the base of the arrow represents the initial target location and the arrow head denotes the desired location of the target point. The contact locations are depicted by the bold red dots on the periphery of the deformable object. Note that in determining the optimal contact locations, we introduced minimum angle constraints between any two robotic fingers to achieve a physically realizable grasping configuration.

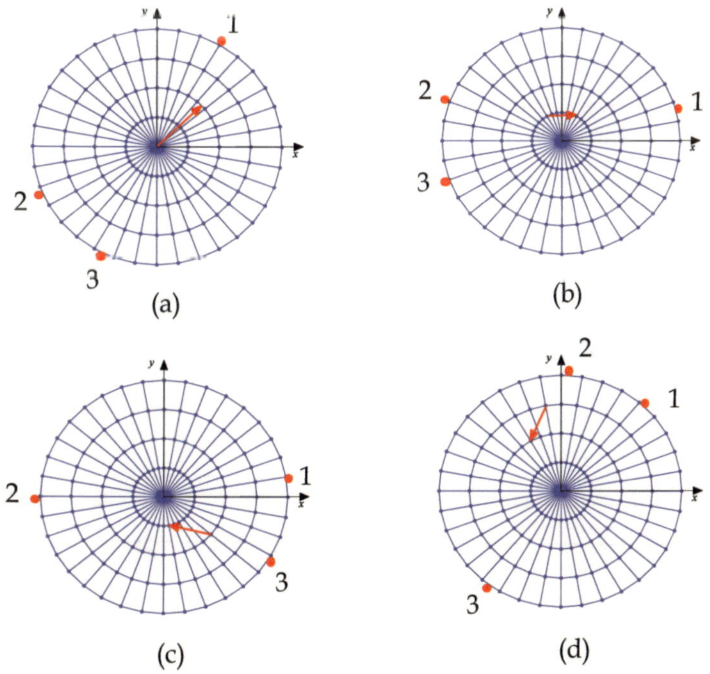

Fig. 6. Optimal contact locations (θ_1, θ_2, θ_3): (a) 59.98°, 204.9°, 244.9°, (b) 14.96°, 159.9°, 199.9°, (c) 7.54°, 182.54°, 327.54°, and (d) 48.59°, 88.59°, 234.39°

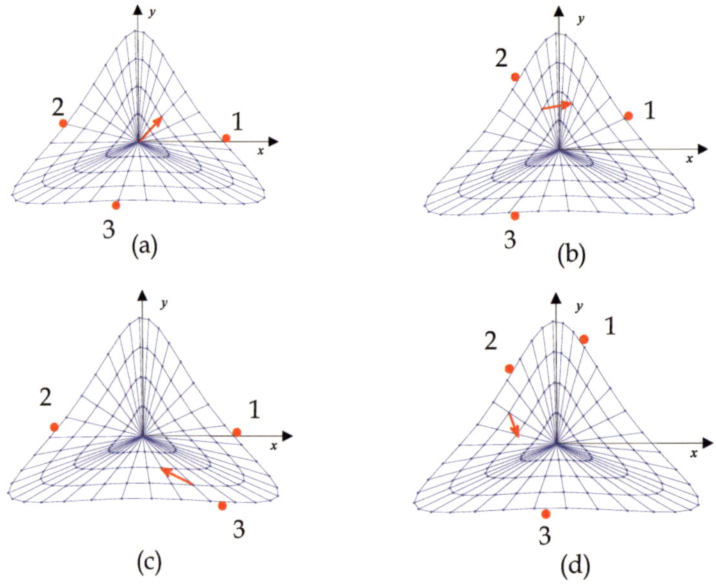

Fig. 7. Optimal contact locations (θ_1, θ_2, θ_3): (a) 0°, 170°, 253.8°, (b) 29.07°, 116.93°, 233.86°, (c) 0°, 175°, 320°, and (d) 76.93°, 116.93°, 261.94°

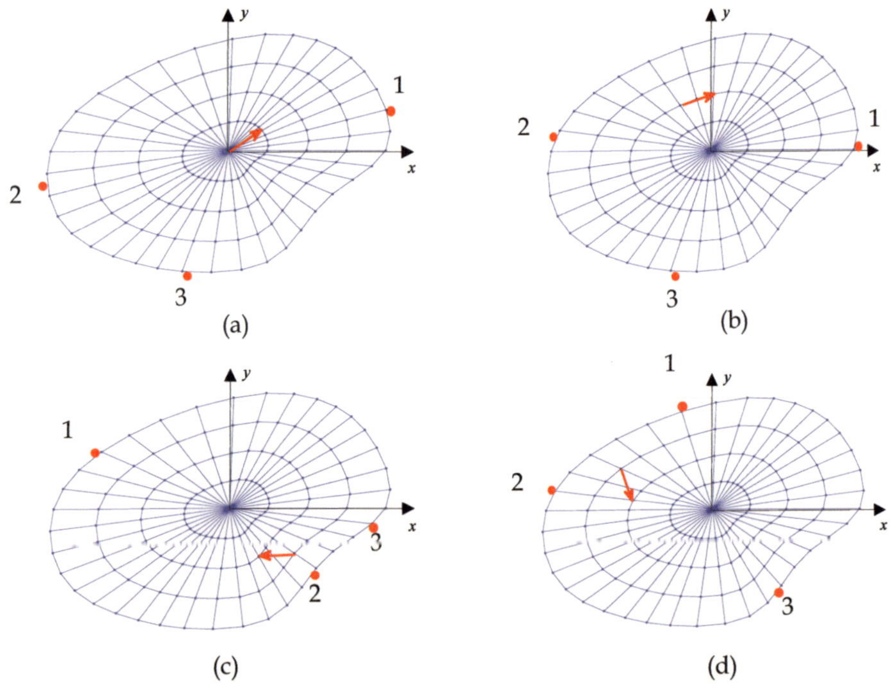

Fig. 8. Optimal contact locations (θ_1, θ_2, θ_3): (a) 25.18°, 199.48°, 262.22°, (b) 0°, 175°, 262.62°, (c) 141.05°, 303.66°, 343.66° and (d) 96.37°, 169.35°, 288.29°

Task 2:

In Task 2, we present a target positioning operation when the robotic fingers are not located at their optimal contact locations. For instance, we choose that the robotic fingers are located at 0, 120, and 240 degrees with respect to the x-axis as shown in Figure 9. We assume that the initial position of the target is at the center of the section of the deformable object, i.e., (0, 0) mm. The goal is to position the target at the desired location (5, 5) mm with a smooth

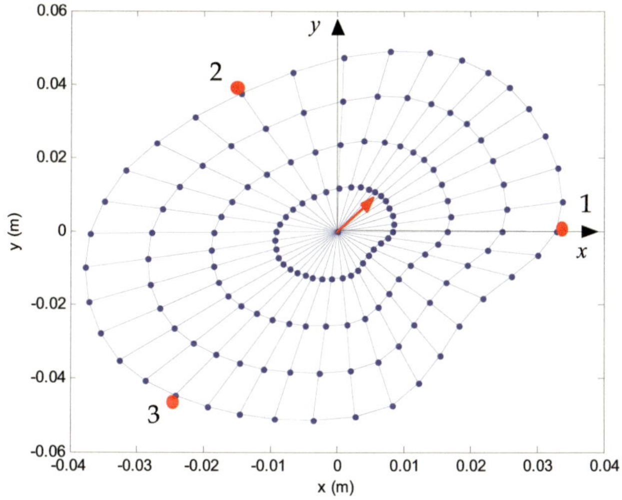

Fig. 9. Deformable object with contact points located at 0, 120 and 240 degrees with respect to *x*-axis

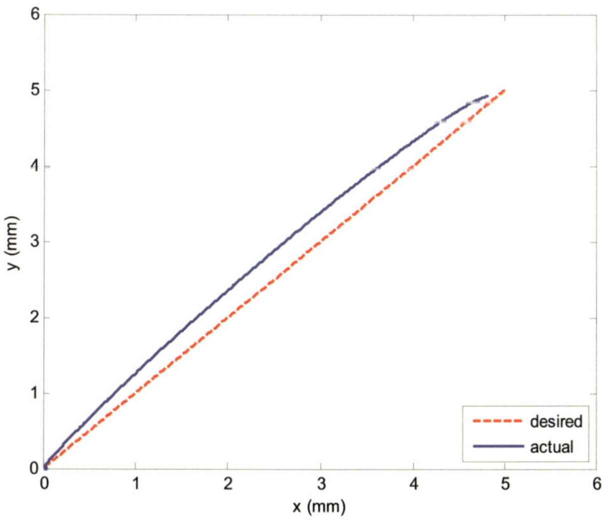

Fig. 10. The desired (red dashed) and the actual (blue solid) straight lines when robotic fingers are located at 0, 120, and 240 degrees with respect to *x*-axis

straight line trajectory. In this simulation, we choose $K_{Pi}=1000$ and $K_{Ii}=1000$, i=1,2,3. Figure 10 shows the actual and desired position trajectories of the target point. It is noticed that there is some error present in the tracking of the desired trajectory. Robotic fingers forces generated by the PI controller are presented in Figure 11 and the POs are falling to negative as shown in Figure 12. Negative values of POs signify that the interaction between the robotic fingers and the deformable object is not stable.

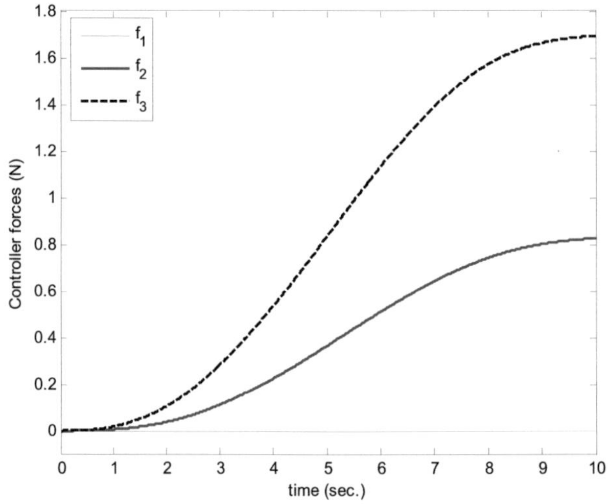

Fig. 11. Controller forces when robotic fingers are located at 0, 120, and 240 degrees with respect to x-axis

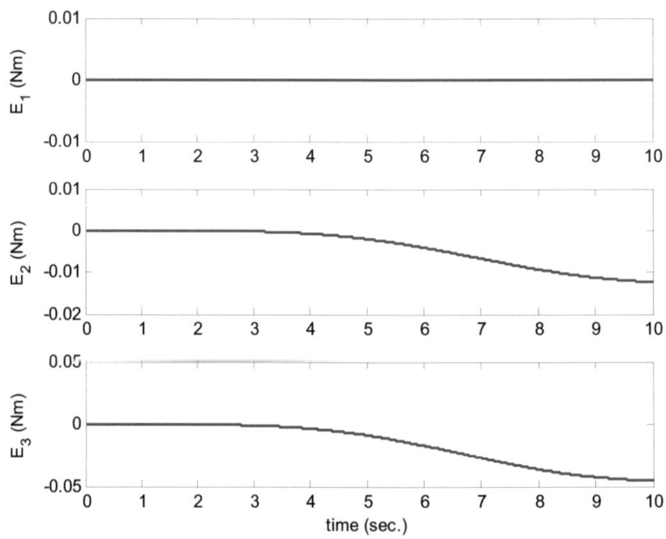

Fig. 12. POs when robotic fingers are located at 0, 120, and 240 degrees with respect to x-axis

Task 3:

In Task 3, we consider the same task as discussed above under Task 2 but the robotic fingers are positioned at their optimal contact locations (Figure 8(a)) and the target is following the desired straight line trajectory. In this case, PCs are not turned on while performing the task. A simple position based PI controller generates the control command based on the error between the desired and the actual location of the target. Figure 13 shows that the target

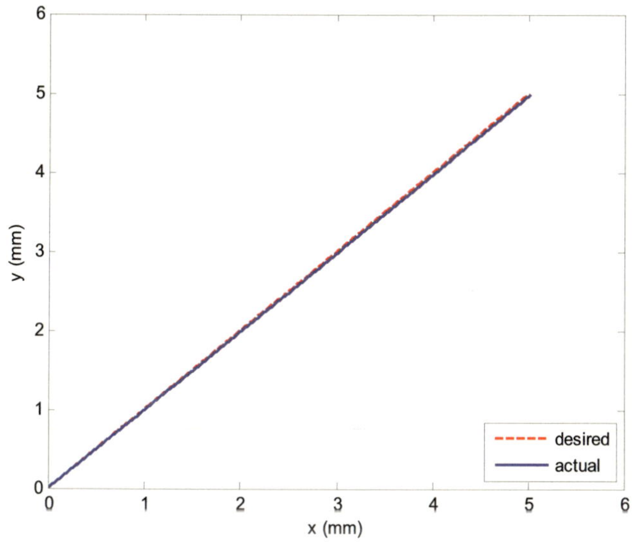

Fig. 13. The desired (red dashed) and the actual (blue solid) straight lines when PCs are not turned on

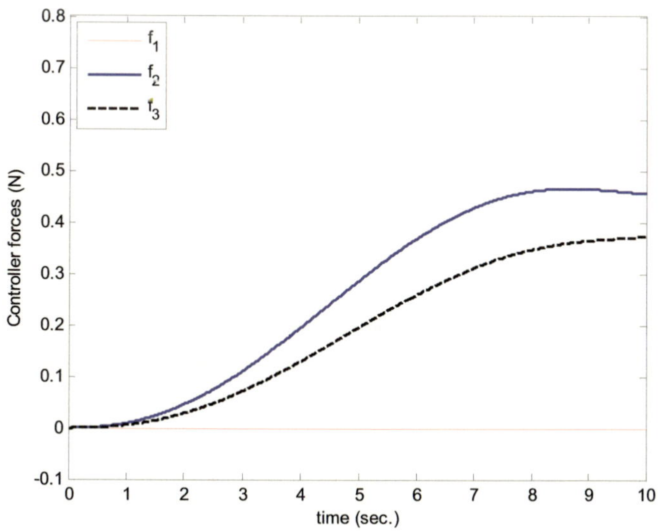

Fig. 14. Controller forces when PCs are not turned on

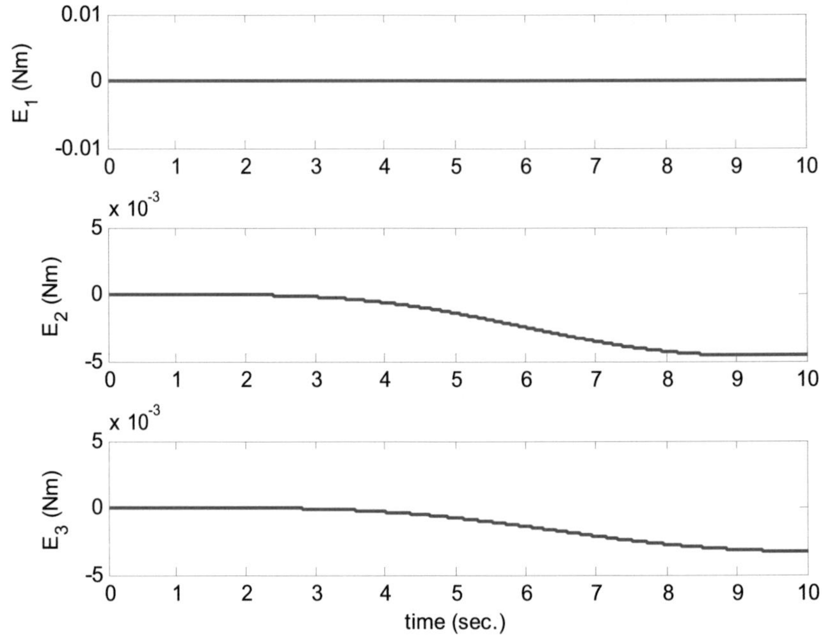

(a)

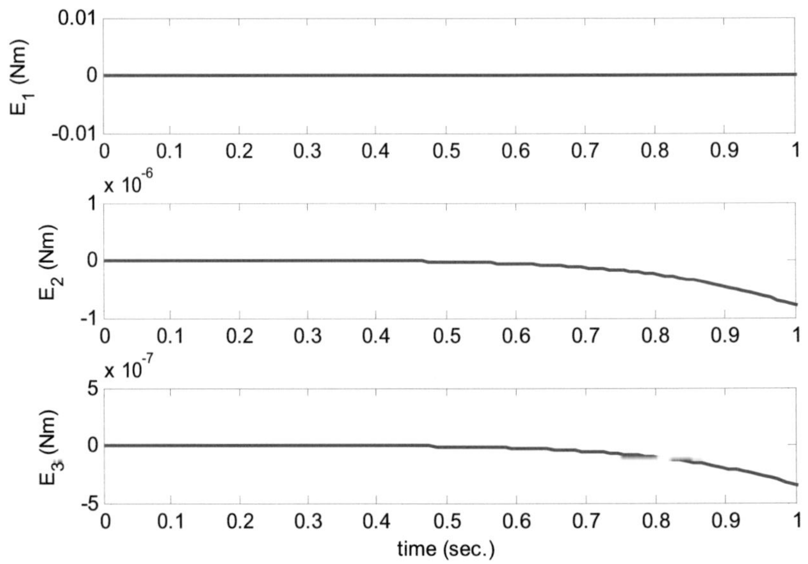

(b)

Fig. 15. (a) POs for three robotic fingers when PCs are not turned on, (b) a magnified version of (a) for first few seconds tracked the desired position trajectory. Robotic fingers forces generated by the PI controller are presented in Figure 14. Force values in Figure 14 are quite less than those in Figure 11 because of the optimal contact location of the robotic fingers. However, the POs for robotic fingers 2 and 3 are become negative as shown in Figure 15. Negative values of the POs signify that the output energy of the 2-port network is greater than the input energy. Since the plant is considered to be passive, the only source of generating extra energy is the controller that makes the whole system unstable. So we must engage passivity controller to modify the controller output by dissipating the extra amount of energy.

Task 4:

In Task 4, the PCs are turned on and the robotic fingers are commanded to effect the desired motion of the target. The PCs are activated when the POs cross zero from a positive value. The required damping forces are generated to dissipate only the excess amount of energy generated by the controller. In this case, the target tracks the desired straight line trajectory well with the POs remaining positive. Figure 16 represents the actual and the desired trajectories of the target when PCs are turned on. For this case, the PCs on the plant side are only activated whereas the PCs on the trajectory side remain idle. Figure 17 shows the PCs forces generated at the plant side when the POs cross zero. The POs become positive during interaction between the robotic fingers and the object as shown in Figure 18. Hence, the stability of the overall system is guaranteed. The PCs on the trajectory side are shown in Figure 19, which are all zeros. The modified controller outputs to move the target point are shown in Figure 20.

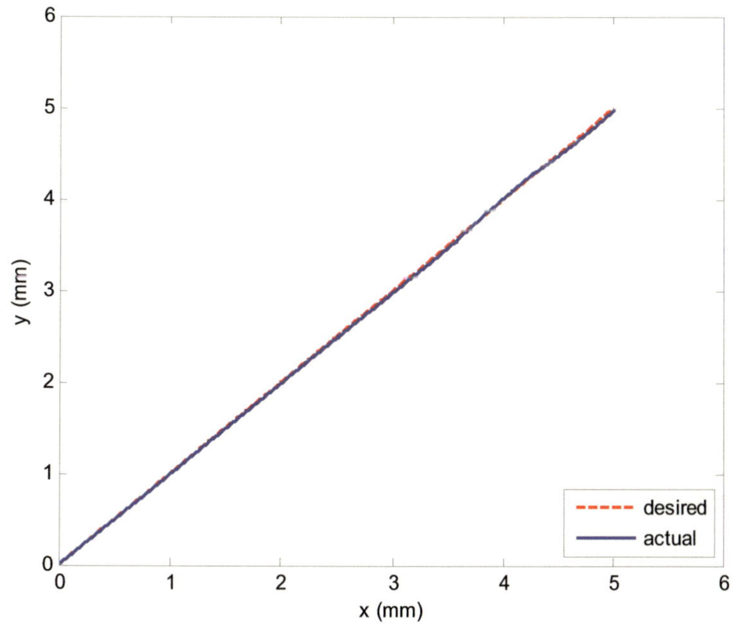

Fig. 16. The desired (red dashed) and the actual (blue solid) straight lines when PCs are turned on

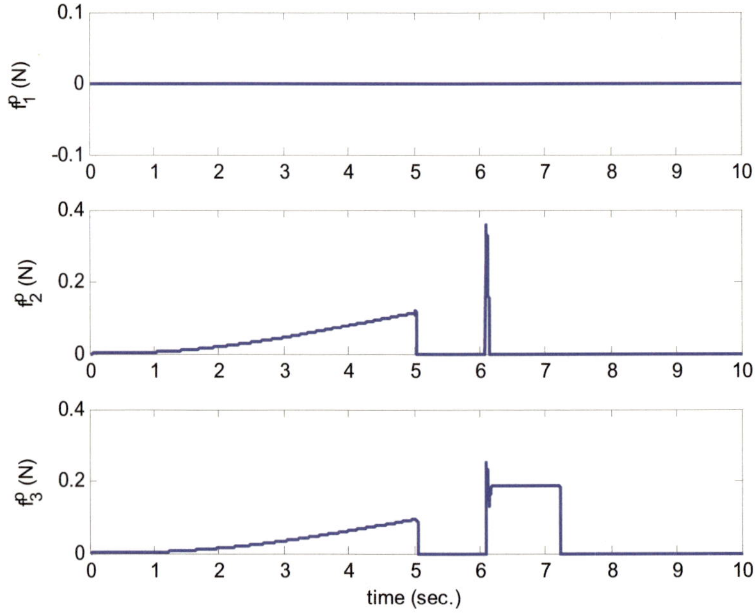

Fig. 17. Required forces supplied by PCs at the plant side when PCs are turned on

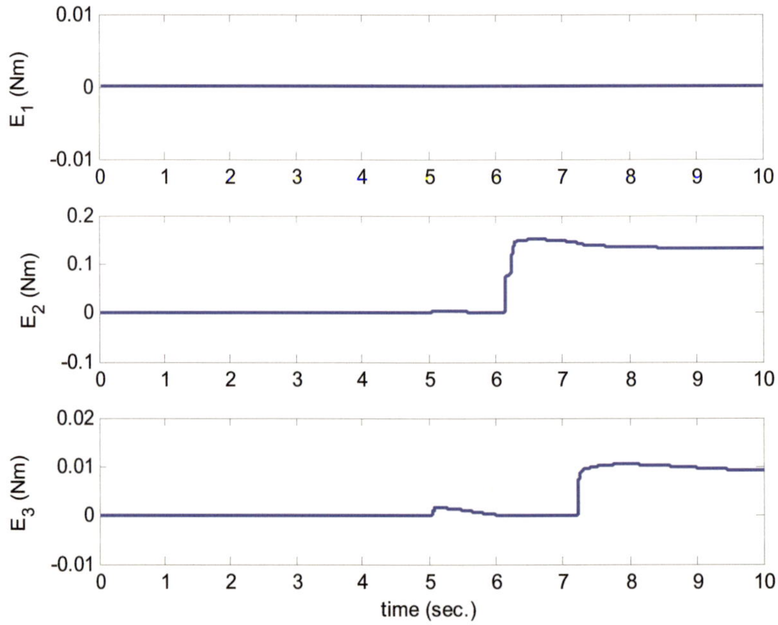

Fig. 18. POs for three robotic fingers when PCs are turned on

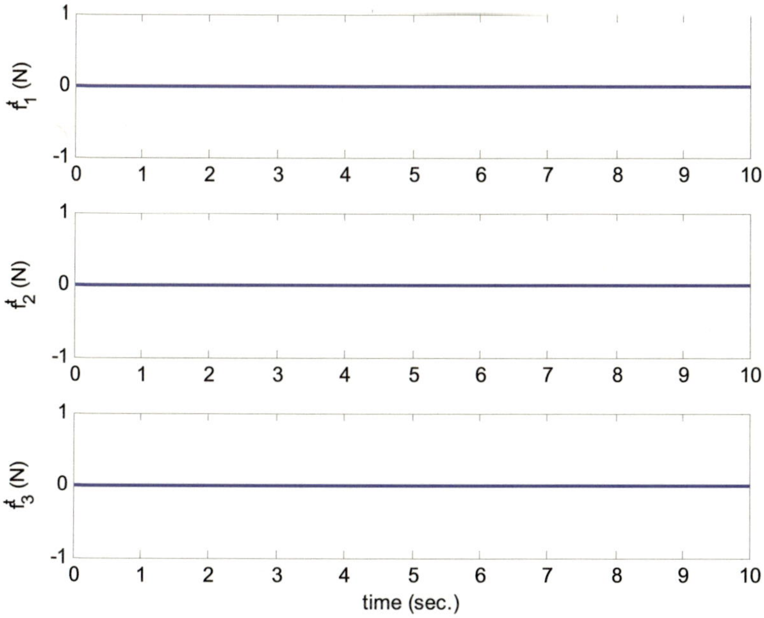

Fig. 19. PCs forces at the trajectory side when PCs are turned on

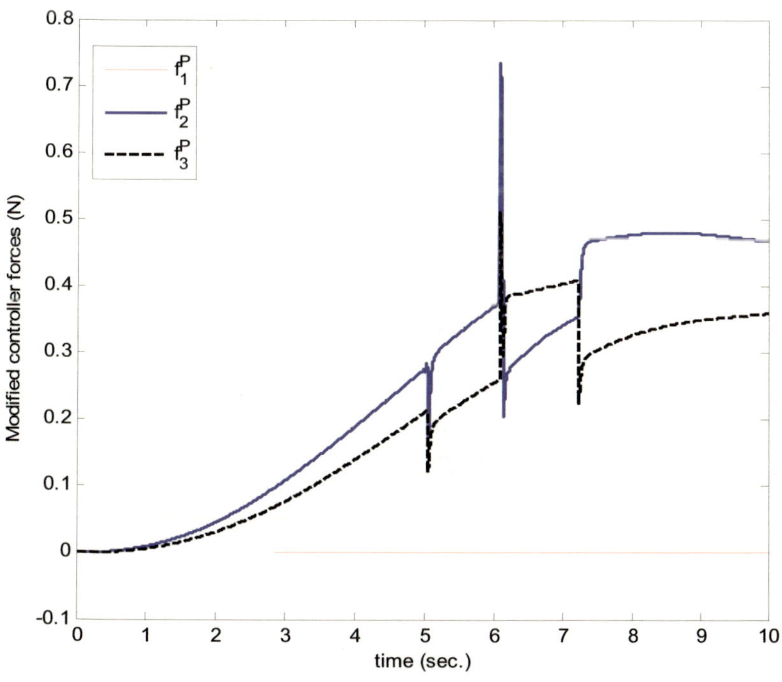

Fig. 20. Modified controller forces when PCs are turned on

8. Conclusion and future work

In this chapter, an optimal contact formulation and a control action are presented in which a deformable object is manipulated externally by three robotic fingers such that an internal target point is positioned to the desired location. First, we formulated an optimization technique to determine the contact locations around the periphery of the object so that the target can be manipulated with minimum force applied on the object. The optimization technique considers a model of the deformable object. However, it is difficult to build an exact model of the deformable object in general due to nonlinear elasticity, friction, parameter variations, and other uncertainties. Therefore, we considered a coarse model of the deformable object to determine the optimal contact locations which is more realizable. A time-domain passivity control scheme with adjustable dissipative elements has been developed to guarantee the stability of the whole system. Extensive simulation results validate the optimal contact formulation and stable interaction between the robotic fingers and the object.

9. References

[1] D. Sun and Y. H. Liu, Modeling and impedance control of a two-manipulator system handling a flexible beam, ASME Journal of Dynamic Systems, Measurement, and Control, vol. 119, pp. 736-742, 1997.

[2] P. Dang, F. L. Lewis, K. Subbarao and H. Stephanou, Shape control of flexible structure using potential field method, 17th IEEE International Conference on Control Applications, Texas, USA, pp. 540-546, 2008.

[3] X. Liu, K. Kim, Y. Zhang and Y. Sun, Nanonewton force sensing and control in microrobotic cell manipulation, The International Journal of Robotics Research, vol. 28, issue 8, pp. 1065-1076, 2009.

[4] V. G. Mallapragada, N. Sarkar and T. K. Podder, Robot-assisted real-time tumor manipulation for breast biopsy, IEEE Transactions on Robotics, vol. 25, issue 2, pp. 316-324, 2009.

[5] S. Hirai and T. Wada, Indirect simultaneous positioning of deformable objects with multi pinching fingers based on uncertain model, Robotica, Millennium Issue on Grasping and Manipulation, vol. 18, pp. 3-11, 2000.

[6] S. Nath, Z. Chen, N. Yue, S. Trumpore and R. Peschel, Dosimetric effects of needle divergence in prostate seed implant using 125I and 103Pd radioactive seeds, Medical Physics, vol. 27, pp. 1058-1066, 2000.

[7] A. Albu-Schaffer, C. Ott and G. Hirzinger, Constructive energy shaping based impedance control for a class of underactuated Euler-Lagrange systems, IEEE International Conference on Robotics and Automation, pp. 1387-1393, 2005.

[8] B. Hannaford and R. Jee-Hwan, Time-domain passivity control of haptic interfaces, IEEE Transactions on Robotics and Automation, vol. 18, pp. 1-10, 2002.

[9] S. Ali A. Moosavian and R. Rastegari, Multiple-arm space free-flying robots for manipulating objects with force tracking restrictions, Robotics and Autonomous System, vol. 54, issue 10, pp. 779-788, 2006.

[10] Z. Li, S. S. Ge and Z. Wang, Robust adaptive control of coordinated multiple mobile manipulators, Mechatronics, vol. 18, issues 5-6, pp. 239-250, 2008.

[11] M. Namvar and F. Aghili, Adaptive force-motion control of coordinated robots interacting with geometrically unknown environments, IEEE Transactions on Robotics, vol. 21, issue 4, pp. 678-694, 2005.
[12] M. Saha and P. Isto, Manipulation planning for deformable linear objects, IEEE Transactions on Robotics, vol. 23, issue 6, pp. 1141-1150, 2007.
[13] Y. Zhang, B. K. Chen, X. Liu and Y. Sun, Autonomous robotic pick-and-place of microobjects, IEEE Transactions on Robotics, vol. 26, issue 1, pp. 200-207, 2010.
[14] D. Sun and Y. H. Liu, Modeling and impedance control of a two-manipulator system handling a flexible beam, ASME Journal of Dynamic Systems, Measurement, and Control, vol. 119, no. 4, pp. 736-742 ,1997.
[15] A. Tavasoli, M. Eghtesad and H. Jafarian, Two-time scale control and observer design for trajectory tracking of two cooperating robot manipulators moving a flexible beam, Robotics and Autonomous Systems, vol. 57, issue 2, pp. 212-221, 2009.
[16] S. F. F. Gibson and B. Mirtich, A survey of deformable modeling in computer graphics, MERL Technical Report, TR97-19, 1997.
[17] Y. Zhang, E. C. Prakash and E. Sung, A new physical model with multilayer architecture for facial expression animation using dynamics adaptive mesh, IEEE Transactions on Visualization and Computer Graphics, vol. 10, issue 3, pp. 339-352, 2004.
[18] M. Meyer, G. Debunne, M. Desbrun and A. H. Barr, Interactive animation of cloth-like objects in virtual reality, The Journal of Visualization and Computer Animation, vol. 12, no. 1, pp. 1-12(12), 2001.
[19] A. Joukhadar and C. Laugier, Dynamic simulation: model, basic algorithms, and optimization, In J.-P. Laumond and M. Overmars, editors, Algorithms for Robotic Motion and Manipulation, pp. 419-434, A. K. Peters Ltd., 1997.
[20] A. M. Howard, and G. A. Bekey, Recursive learning for deformable object manipulation, Proc. of International Conference on Advanced Robotics, pp. 939-944, 1997.
[21] R. M. Koch, M. H. Gross, F. R. Carls, D. F. Von Buren, G. Fankhauser, Y. I. H. Parish, Simulating facial surgery using finite element models, In ACM SIGGRAPH 96 Conf. Proc., pp. 421-428, 1996.
[22] S. D. Pieper, D. R. Laub, Jr., and J. M. Rosen, A finite element facial model for simulating plastic surgery, Plastic and Reconstructive Surgery, 96(5), pp. 1100-1105, Oct. 1995.
[23] J. K. Salisbury, Kinematic and force analysis of articulated hands, PhD Thesis, Department of Mechanical Engineering, Standford University, Standford, CA, 1982.
[24] J. Kerr, Analysis of multifingered hand, PhD Thesis, Department of Mechanical Engineering, Standford University, Standford, CA, 1984.
[25] J. M. Abel, W. Holzmann and J. M. McCarthy, On grasping planar objects with two articulated fingers, IEEE Journal of Robotics and Automation, vol. 1, pp. 211-214, 1985.
[26] M. R. Cutkosky, Grasping and fine manipulation for automated manufacturing, PhD Thesis, Department of Mechanical Engineering, Carnegie Mellon University, Pittsburgh, PA, 1985.
[27] C. Ferrari and J. Canny, Planning optimal grasp, IEEE International Conference on Robotics and Automation, pp. 2290-2295, 1992.
[28] S. Garg and A. Dutta, Grasping and manipulation of deformable objects based on internal force requirements, International Journal of Advanced Robotic Systems, Vol. 3, No. 2, pp. 107-114, 2006.

[29] T. Watanabe and T. Yoshikawa, Grasping optimization using a required external force set, IEEE Transactions on Automation Science and Engineering, Vol. 4, pp. 52-66, 2007.
[30] D. Ding, Y. H. Liu and S. Wang, Compuattion of 3D form-closure grasps, IEEE Transactions on Robotics and Autonamtion, Vol. 17, pp. 515-522, 2001.
[31] A. Bicchi and V. Kumar, Robotic grasping and contact: a review, Proc. IEEE International Conference on Robotics and Automation, pp. 348-353, 2000.
[32] M. T. Mason, Mechanics of robotic manipulation, The MIT Press.
[33] J. Cornella, R. Suarez, R. Carloni and C. Melchiorri, Dual programming based approach for optimal grasping force distribution, Journal of Mechatronics, Vol. 18, pp. 348-356, 2008.
[34] P. Saut, C. Remond, V. Perdereau and M. Drouin, Online computation of grasping force in multi-fingered hands, Proc. IEEE/RSJ International Conference on Intelligent Robots and Systems, pp. 1223-1228, 2005.
[35] K. Gopalakrishanan and K. Goldberg, D-space and deform closure grasps of deformable parts, International Journal of Robotics Research, vol. 24, pp. 889-910, 2005.
[36] G. L. Foresti and F. A. Pellegrino, Automatic visual recognition of deformable objects for grasping and manipulation, IEEE Transaction on Systems, Man, and Cybernetics: Applications and Reviews, vol. 34, pp. 325-333, 2004.
[37] H. Wakamatsu, S. Hirai and K. Iwata, Static analysis of deformable object grasping based on bounded force closure, IEEE International Conference on Robotics and Automation, vol.4, pp. 3324-3329, 1996.
[38] R. J. Anderson and M. W. Spong, Asymptotic stability for force reflecting teleoperators with time delays, in, 1989. IEEE International Conference on Robotics and Automation, vol.3, pp. 1618-1625, 1989.
[39] S. Stramigioli, C. Melchiorri and S. Andreotti, A passivity-based control scheme for robotic grasping and manipulation, The 38th Conference on Decision and Control, Phoenix, Arizona, USA, 1999.
[40] B. Hannaford and J.-H. Ryu, Time domain passivity control of haptic interfaces, IEEE International Conference on Robotics and Automation, vol.2, pp. 1863-1869, 2001.
[41] J.-H. Ryu, D.-S. Kwon and B. Hannaford, Control of a flexible manipulator with noncollocated feedback: time-domain passivity approach, IEEE Transactions on Robotics, vol. 20, pp. 776-780, 2004.
[42] T. Wada, S. Hirai, S. Kawamura and N. Kamiji, Robust manipulation of deformable objects by a simple PID feedback, IEEE International Conference on Robotics and Automation, vol.1, pp. 85-90, 2001.
[43] V. D. Nguyen, Constructing force-closure grasps, IEEE International Conference on Robotics and Automation, pp. 1368-1373, 1986.
[44] J. Ponce and B. Faverjon, On computing three-finger force-closure grasps of polygonal objects, Fifth International Conference on Advanced Robotics, vol.2, pp. 1018-1023, 1991.
[45] J. E. Colgate and N. Hogan, Robust control of dynamically interacting systems, International Journal of Control, vol. 48, pp. 65 - 88, 1988.
[46] J. J.-H. Ryu and C. Preusche, Stable bilateral control of teleoperators under time-varying communication delay: time domain passivity approach, IEEE International Conference on Robotics and Automation, pp. 3508-3513, 2007.

3

Novel Assistive Robot for Self-Feeding

Won-Kyung Song and Jongbae Kim
Korea National Rehabilitation Research Institute, Korea National Rehabilitation Center
Korea

1. Introduction

Assistive robots, with which users can interact directly, have attracted worldwide attention. They can assist people with disabilities and older persons in the activities of daily living. Assistive robots could be employed for improving quality of life as they can be adjusted according to demographic changes. There are several crucial issues to be considered with regard to these robots, such as customizing them according to the specific culture of the users as well as ensuring cost-effectiveness (Mann, 2005).

In Korea, the official number of registered people with disabilities due to illnesses, injuries, and the natural aging process has already exceeded two million (Employment Development Institute, 2009). More than one-third of these disabled people are the elderly. Moreover, due to longer life spans and a decline in birthrate, the elderly make up over 10% of the population in Korea. As a result, effective caregiving with restricted resources is an urgent problem.

In order to achieve efficient caregiving for people with disabilities and elderly persons, caregivers should physically interact with the people. For example, caregivers have to assist people in performing the routine activities of their daily lives, such as eating, changing clothes, changing their posture, moving from one location to another, and bathing. Among these activities, eating meals is one of the most essential daily activities. In this regard, caregivers must interact with people frequently to assist with food selection, feeding interval, etc. Existing robotic technologies can be utilized to take over the functions of the caregivers. Thus, assistive robots represent one of the solutions by which disabled or elderly people can receive support for performing the activities of daily life.

The design of assistive robots to help with self-feeding depends strongly on the specific culture of the user. Korean food consists chiefly of boiled rice, soup, and side dishes such as Kimchi. The procedure of having a meal is as follows: the user eats the boiled rice first and then the side dishes. These steps are performed repetitively. In comparison with foreign boiled rice, Korean boiled rice sticks together very well after cooking. Handling this sticky boiled rice can be problematic. In addition, Korean soup includes meat, noodles, and various vegetables, thus the existing feeding robots find it difficult to handle Korean foods.

Various assistive robots have been developed since the late 1980s, as shown in Fig. 1. Handy1 (Topping & Smith, 1999) is an assistive robot for daily activities such as eating, drinking, washing, shaving, teeth cleaning, and applying make-up. Handy1 consists of a five-DOF (degree-of-freedom) robot, a gripper, and a tray unit. The major function of Handy1 is to help with eating. Handy1 allows a user to select food from any part of the tray. A cup is attached to enable users to drink water with their meal. The walled columns of a food dish serve an important purpose: the food can be scooped on to the dish without any resultant mixing of food items.

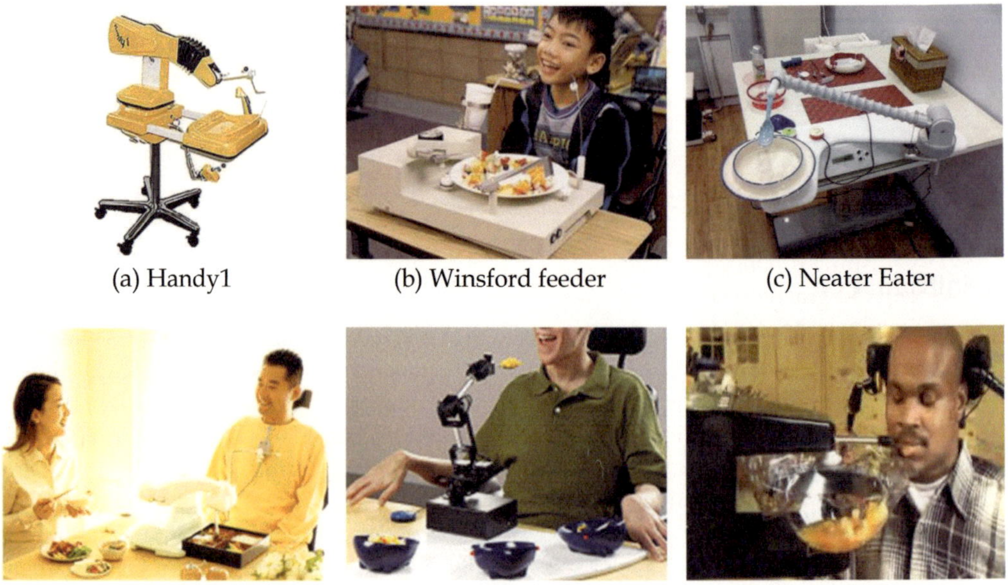

(a) Handy1 (b) Winsford feeder (c) Neater Eater

(d) My Spoon (e) Meal Buddy (f) Mealtime Partner Dining System

Fig. 1. Feeding systems

The Winsford feeder (Sammons Preston; Hermann et al., 1999) is a mechanical self-feeding system. It uses a mechanical pusher to fill a spoon and a pivoting arm to raise the spoon to the user's mouth that is at a preset position. The plate is rotated to place more food in front of the pusher. The user can choose from two input devices: a chin switch and a rocker switch.

Neater Eater (Neater Solutions) has two versions: a manual-operation-type and an automatic-operation-type system. Neater Eater consists of a two-DOF arm and one dish. Two types of food can be present on the one dish. The manual type can be used to suppress the tremors of a user's upper limbs while he or she eats.

My Spoon (Soyama et al., 2003) is suitable in the case of Japanese food. It consists of a five-DOF manipulator, a gripper, and a meal tray. The meal tray has four rectangular cells. My Spoon combines several pre-programmed motions: automatic operation, semiautomatic operation, and manual operation. The semiautomatic operation allows a user to select food. The manual operation can change the position in which the food is held. The input device can be selected from among the following: the chin joystick, reinforcement joystick, and switch. The end-effector of the robotic arm has one spoon and one fork, which move together to realize the grasping motion. During the grasping process, the gap between the spoon and the fork changes and thus the end-effector grasps the food. Then the robot moves to a predefined position in front of the user's mouth, and the fork moves backward to enable the user to eat the food off the spoon.

Meal Buddy (Sammons Preston) has a three-DOF robotic arm and three bowls that can be mounted on a board using magnets. After the system scoops the food, the robotic arm scrapes the surplus food off the spoon with the rod on the bowls.

The Mealtime Partner Dining System (Mealtime Partners) is positioned in front of a user's mouth. Three bowls can rotate in front of the mouth. The spoon picks up the food and then moves a short distance toward the preset location of the mouth. This system reduces the chances of the spoon slipping on wet food because the underside of the spoon is wiped off after scooping. Because of the way the system is positioned, the user does not need to lean toward the feeder. In some systems, a beverage straw is located beside the spoon (Pourmohammadali, 2007). Other systems are designed for multiple users (Guglielmelli, 2009).

Most feeding systems scoop the food with a spoon. Those systems are not suitable for use in the case of boiled rice, which is a staple food in Korea. In addition, some systems have a single dish, and thus different types of food might be mixed during scooping. My Spoon uses the grasping function to pick up food, but this system has difficulty serving Korean rice due to its fixed grasping strength and the grasping openness of the gripper. As a result, My Spoon's gripper sometimes gets a lot of rice attached to its surface. The previously mentioned self-feeding robotic systems also have difficulty scooping this staple Korean food.

Feeding robots enable users to enjoy food independently during mealtimes. After preparing food, users can choose when they want to eat the desired food. We developed an assistive robot for self-feeding by taking into consideration the feedback of user candidates and clinical experts. We evaluated the self-feeding robot by performing a series of user tests. The overall process, i.e., formulating a concept, design, and evaluation involves feedback from users and clinical experts. The development process is performed on the basis of the philosophy of participatory action design (Ding et al., 2007).

In this paper, we introduce a newly designed self-feeding robot that will be suitable in the case of Korean food, including sticky rice, and we report the results of tests that were performed with several disabled people. In Section 2, we will present an overview of the new self-feeding robot for Korean food. Basic operational procedures of the self-feeding robot will be presented in Section 3. Section 4 contains the results and discussions of tests performed with users with spinal cord injuries. Finally, we will present the conclusion in Section 5.

2. Self-feeding robot system

In this section, we will present an overview of the users, requirements, and system configurations of the self-feeding robot system.

2.1 Users

The primary users of self-feeding robots are people with physical disabilities who have difficulty moving their upper limbs. Such people include those suffering from high-level spinal cord injuries, cerebral palsy, and muscular diseases. For example, people with cervical level-4 spinal cord injuries have difficulty moving their upper limbs and retain full movement only above their necks. Some people afflicted with cerebral palsy cannot move their arms and hands, and they often have difficulty moving their necks. When the spoon of a self-feeding robot approaches such a user's mouth, that person has a hard time putting the food in his or her mouth. People with muscular diseases have weak muscle movements. Even though they can move their hands, they have limited motor functions in their elbows and shoulder joints. We can also include senior citizens who have difficulties with the motor functions of their upper limbs, e.g., the fragile elderly, among the abovementioned disabled people. It is clear that the number of overall target users of self-feeding robots will be growing in the near future.

2.2 Requirements of a self-feeding robot

We surveyed a group of people with disabilities as well as clinical experts to learn about the requirements of a feeding robot (Song et al., 2010a, 2010b). The focus group consisted of a person with a spinal cord injury and a person with cerebral palsy. The clinical experts included occupational therapists and medical doctors of physical medicine and rehabilitation.

The major findings of the survey are as follows. Firstly, a user should be able to control the feeding interval for the desired food. In the case of caregiving, one of the common problems is the difficulty in controlling the feeding interval. People with spinal cord injury are able to talk quickly and can therefore manage a short feeding interval. However, people with cerebral palsy have difficulty representing their intentions quickly when the feeding interval is too short.

Fig. 2. Korean food on a standard food container. (From the lower left-hand side going counterclockwise: rice, soup, and side dishes.)

Secondly, the specialists and the user candidates believe that the feeding systems are designed more for western-style food. Those systems are not suitable for Korean food, which includes boiled rice, soup, and side dishes. A user eats one of the side dishes and then the boiled rice in turn. These steps are performed repetitively during mealtime. In comparison with foreign boiled rice, Korean boiled rice sticks together very well after cooking. One of the problems of self-feeding systems is handling this sticky boiled rice. In addition, Korean soup includes meat, noodles, and various vegetables. Therefore, existing feeding robots have difficulty handling Korean foods (Fig. 2).

Thirdly, a feeding robot should be suitable for use in private homes and facilities. From an economic point of view, a feeding robot is effective in facilities that have many persons with upper-limb disability. Such facilities do not have enough caregivers to help with feeding due to the heavily time-consuming nature of this task. Thus, a robot reduces the burden of caregiving for feeding. A feeding robot can also be used in an ordinary home to improve the quality of life of the users and their families. Members of a family can face each other and freely enjoy talking. The other members of the family can go out for a few hours because they are freed from the burden of having to help with feeding.

The location of bowls or a tray is another important factor. According to Korean culture, the location of bowls or a tray is strongly related to the dignity of the person. A simple feeding system can be made with the bowls located in front of a user's mouth. However, some senior user candidates hate having the bowls right in front of their mouth; they prefer to eat the food like ordinary people. Thus, we focus mainly on using a tabletop tray.

Other comments of user candidates are as follows: plain machines that can serve simple dishes with water are required. When a caregiver goes out for a while, a user needs to be able to eat cereal with milk with the help of a self-feeding robot. The water supply tool should be located next to the user's body. The meal tray must have a cover to protect the food from dust contamination. The cost should be reasonable, e.g., the price should be between US$1800 and $2700. Obviously, a low-cost system is preferred. The feeding robot should be able to deal with noodles. The feeding robot should be able to accommodate the posture of the user. Finally, the feeding robot should be lightweight.

We concentrate on rice handling. We do not take into account the handling of soup in this development. We will handle the requirements of feeding Korean soup in a future version's design. Instead of handling Korean soup via a self-feeding robot, a user drinks soup stored in a cup. Generally, we assume that a user drinks soup or water through a straw.

Technically, we considered four types of feeding robots in order to ensure that it can grip and release boiled rice effectively, as shown in Fig. 3.

In the first concept, a number of bowls are located in front of a user's mouth, and the food is presented by the spoon with a short traveling distance. For example, if there are three bowls, one bowl has rice and two bowls have side dishes. However, two side dishes are not enough to constitute a satisfying meal. In general, Korean people eat three or four side dishes with boiled rice at a time. Therefore, we need four or five bowls.

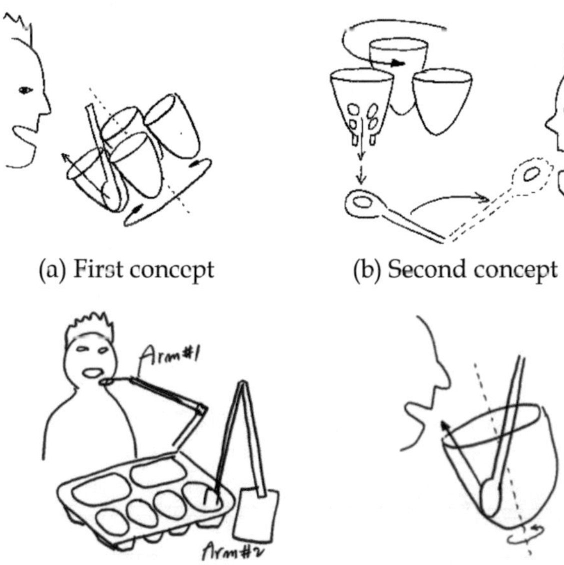

(a) First concept (b) Second concept

(c) Third concept (d) Fourth concept

Fig. 3. Design concepts of a feeding robot. (Third concept is chosen.)

In the second concept, the bowls are located in the upper front of a user's mouth, and then the food drops from the bottom of a bowl. The food is placed in the spoon by a dropping motion, and then the spoon approaches the user's mouth. This method requires the mechanism of food dropping on the spoon. This method could be suitable for a bite-sized rice cake.

In the third concept, the system with a food tray is located on a table. The robotic arm picks up food and then moves it to a user's mouth. These tasks are divided into two steps: one is picking up the food and the other is moving the food to the user's mouth. Two arms can be used to perform the above two tasks, respectively. One of the user candidates pointed out the easy installation of the feeding robots, especially a dual-arm manipulator. This is significant because some caregivers might be elderly people who are not familiar with brand-new machines.

Finally, one bowl is located in front of the user's mouth. The mixed food with rice is loaded in that bowl. Some users do not like the mixed food, even though they do prefer a simple feeding system.

We decided on the third concept, which is located on a table, based on the opinions of specialists and user candidates.

2.3 Design of the feeding robot

We have developed a simple robotic system that has a dual-arm manipulator that can handle Korean food such as boiled rice in an ordinary food container, as shown in Fig. 4. We divide a self-feeding task into two subtasks: picking up/releasing food and transferring food to a user's mouth. The first robotic arm (a spoon-arm, Arm #1) uses a spoon to transfer the food from a container on a table to a user's mouth. The second robotic arm (a grab-arm, Arm #2) picks food up from a container and then puts it on the spoon of a spoon-arm.

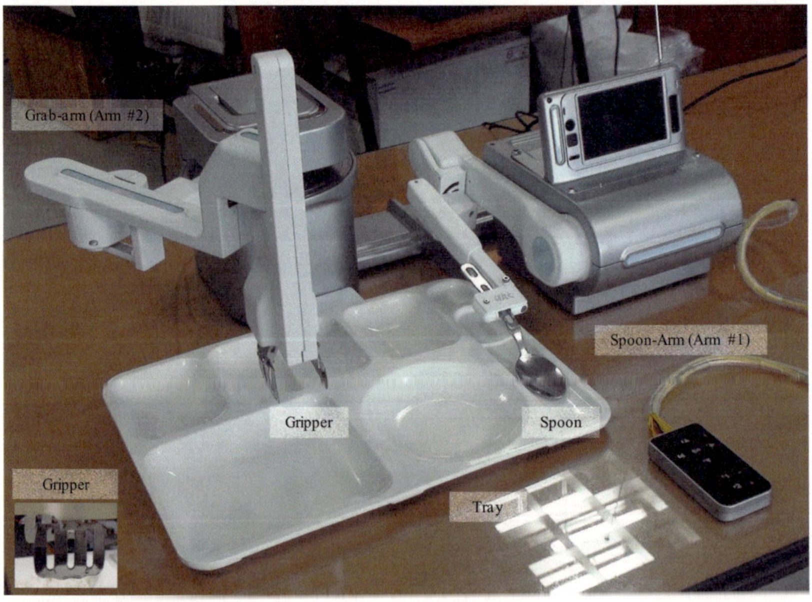

Fig. 4. Assistive robot for self-feeding. Spoon-arm (Arm #1) uses a spoon to transfer the food from a container to a user's mouth. Grab-arm (Arm #2) picks up the food from a container and then loads it onto the spoon of Arm #1

The two arms have different functions. The design of the end-effectors of the two arms could be chosen effectively. To pick up or release the food stably, a grab-arm can use an odd-shaped gripper, as shown in the bottom left-hand side of Fig. 4, because that gripper does not need to approach a user's mouth. However, the end-effector of a spoon-arm has an

intact round-shaped spoon to serve food to a user's mouth safely. If an end-effector has an unusual shape, then it might pose a danger to the user as it approaches his or her face.
The two proposed arms with their different end-effectors mimic Korean eating behavior. Specifically, Koreans use a spoon and steel chopsticks during mealtime, as shown in Fig. 5 (a) and (b). Some people use a spoon and chopsticks simultaneously [Fig. 5 (c)]. In the designed system, the gripper of a grab-arm and the spoon of a spoon-arm take on the roles of chopsticks and a spoon, respectively. Many Korean caregivers pick up food with chopsticks and then put it on a spoon in order to serve food to users such as children and patients. In that sense, the proposed two-armed system stems from Korean eating tools.

(a) (b) (c)

Fig. 5. (a) A spoon for scooping food. (b) Chopsticks for picking up food. (c) A person using a spoon and chopsticks

A spoon-arm has two degrees of freedom (DOF) in order to transfer food on the spoon without changing the orientation of the spoon. A grab-arm includes a three-DOF SCARA joint for the planar motion, a one DOF prismatic joint for the up and down motion, and a gripper. The overall number of DOFs of a dual-arm without a gripper is six, as shown in Fig. 6. The feeding robot can use an ordinary cafeteria tray.

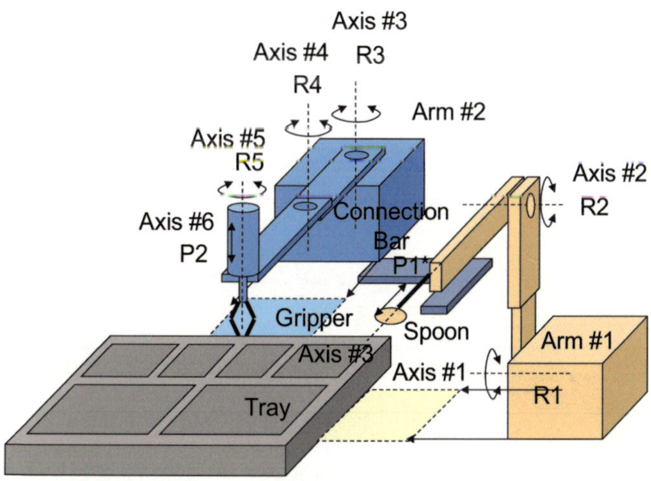

Fig. 6. The joint configuration of a novel feeding robot for Korean foods. P1 (Prismatic Joint #1) is optionally applied. R = Revolute. P = Prismatic

The feeding robot uses a microcontroller unit to control a spoon-arm and a grab-arm, as shown in Fig. 7. We add a small-sized PC with a touch screen to enable the user to enjoy entertainment and to test various kinds of user interfaces. During mealtime, a user wants to enjoy multimedia such as movies or music. In addition, the small-sized PC has a Windows operating system, and we can effectively add assistive devices for human computer interaction, i.e., switches, a joystick, and biosignal interface devices.

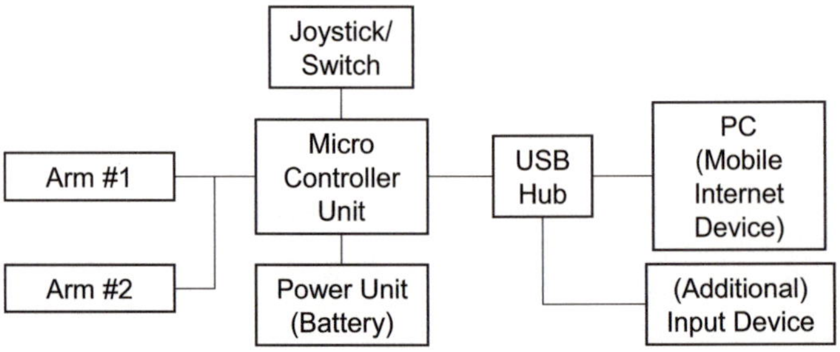

Fig. 7. Block diagram of feeding robot

The microcontroller unit allows a user or a caregiver to select the following: operation modes (automatic/semiautomatic/manual mode), the shape and size of a container, the location of the mouth, the robot's speed, the time to stay in front of the mouth, and so on. Depending on the types of food, a user also selects the divided grasping region in each container and the grab strength of the gripper of the grab-arm. Our system will be capable of selecting the above parameters. A user can save the parameters of various kinds of food. We expect that in a community, different members will be able to exchange their robots effectively by exchanging their individual parameters via the Internet.
The grasping regions of boiled rice in a bowl could be defined in 3D space because the bowl should be over 50 mm in height. The grasping volume of dishes could be defined as shown in Fig. 8. Our team is making the prototype of the proposed feeding robot. Fig. 9 shows the designed appearance of the proposed self-feeding robot.

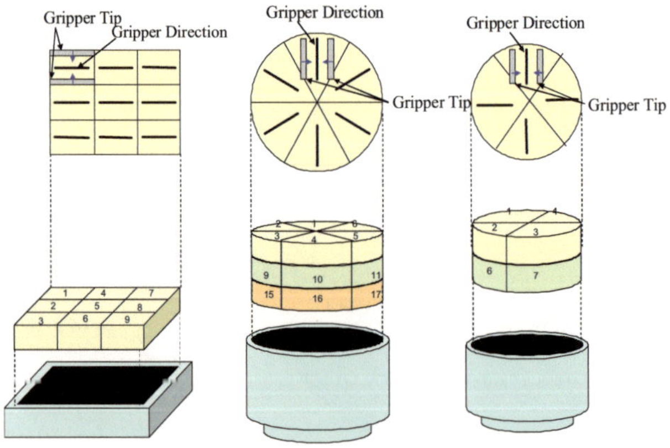

Fig. 8. The definition of grasping volume in containers

In order to use a conventional food container, we decided that the length of links of the grab-arm should cover the whole area of a food container. A grab-arm is located behind a container or on the left-hand side of a container. A grab-arm can chiefly be located behind a food container on a standard table, as shown in Fig. 10. The lap board of a bed does not provide enough space behind a food container; therefore, the grab-arm should be located on the left hand side of a food container.

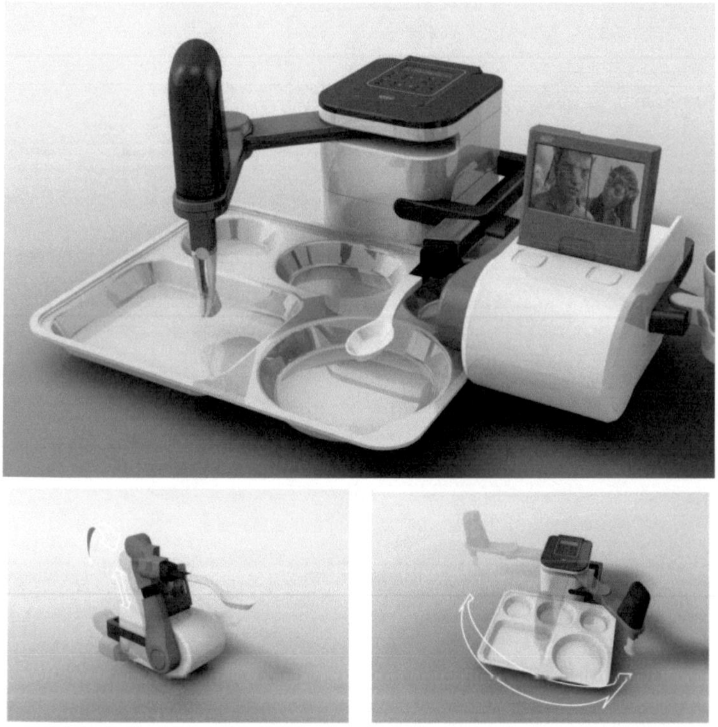

Fig. 9. The design of a novel feeding robot for Korean food. A spoon-arm (lower left-hand-side figure) for transferring food and a grab-arm (lower right-hand-side figure) for picking up and releasing food

The spoon-arm has two additional variables, namely the motorized prismatic motion toward a user's mouth, and the manual change of the link length between the first axis and the second axis of the grab-arm. Fig. 11 shows the overall workspace of the spoon of a grab-arm. In accordance with the position of a user's mouth, the predefined location in front of the mouth is adjusted when the system is installed.

The height of the spoon on the spoon-arm is 250–381 mm with respect to the surface of a table. The height of the spoon-arm depends on the table height. We assume that the height of a table is 730–750 mm. The spoon could be located at a height of 980–1131 mm with respect to the ground. According to the statistics of the Korean disabled, the height of a user's mouth could be 1018 mm (female) and 1087 mm (male), as shown in Table 1. Thus, the height of the spoon corresponds with the height of the user's mouth.

Item	Sitting height on a wheelchair	The distance from a crown to a mouth	Mouse height on a wheelchair
Male	1261	174	1087
Female	1187	169	1018

Table 1. Statistics of wheelchair users (unit: mm)

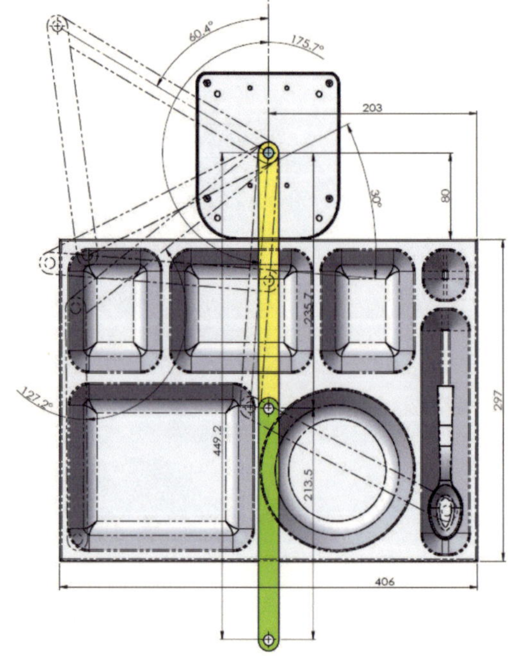

Fig. 10. The configuration of a grab-arm

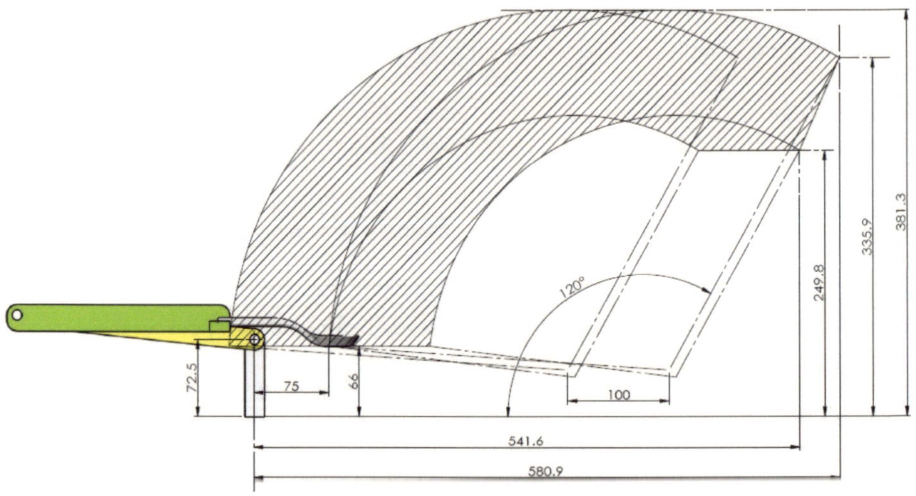

Fig. 11. The work space of a spoon-arm

3. Basic operation of the self-feeding robot

We built two arm configurations of the developed self-feeding robot: a dual-arm configuration and a single-arm configuration. A dual-arm configuration follows an original design concept using both a spoon-arm and a grab-arm. A single-arm configuration uses only the spoon-arm, and a caregiver takes the role of the grab-arm. We explain the two arm configurations of the feeding robot in the following sections.

3.1 Dual-arm configuration

A dual-arm robotic system is applied in accordance with an original design concept. If a caregiver prepares food, users can eat the food on the basis of their intentions. A grab-arm picks up the desired food on a food container, and the arm releases the food on the spoon of a spoon-arm. The spoon-arm moves the spoon to the user's mouth. Then the user can eat the food on the spoon.

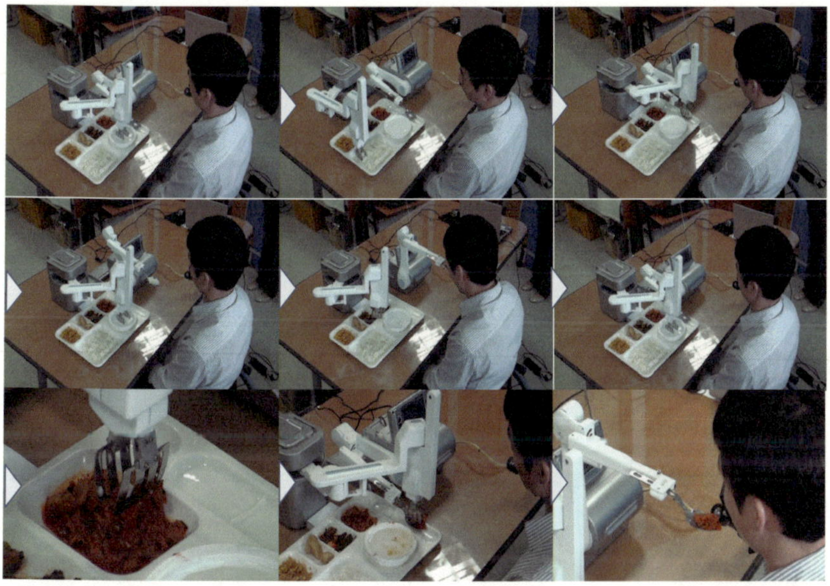

Fig. 12. The sequential motions of the self-feeding robot in dual-arm configuration. From the top left-hand side, the robot puts a gripper into a bowl of water and then grasps rice

The self-feeding robots have three operation modes: an automatic mode, a semiautomatic mode, and a manual mode. The automatic mode has a fixed serving sequence of dishes; users only push a start button when they want to eat the next food on a spoon. In a semiautomatic mode, a user can choose the dishes on the basis of their intention. In this mode, a user can have the dishes that they want to eat. In a manual mode, the user can choose food and control the posture of the robot. In all three modes, the user can select the feeding timing when they want eat.

In experiments on handling boiled rice, we observed that releasing rice is as important as picking up rice. The stickiness of the boiled rice can change depending on its temperature. Slightly cool rice is difficult to release from the gripper. In order to solve this problem, the feeding robot automatically puts the gripper of the grab-arm in water before grasping the food. The water is located in a bowl next to the rice. When this is done, the gripper can release the rice on the spoon because the stickiness of the rice has decreased. Fig. 12 shows the whole operation of the self-feeding robot.

The amount of rice picked up is adjusted on the basis of actual experiments on rice grasping. A gripper's mechanism is the simple opening/closing of gripper fingers via a linear actuator. The weight of the rice corresponding to one grasping motion increases depending on the open/close width (Fig. 13) of the fingers of the gripper when grasping begins, as shown in Fig. 13. The default open/close width of the gripper is 32 mm in order to grasp an average of 10 g rice. The close width of the gripper makes the grasping force to food. Thus, we can grasp various foods by adjusting the open/close width of the gripper.

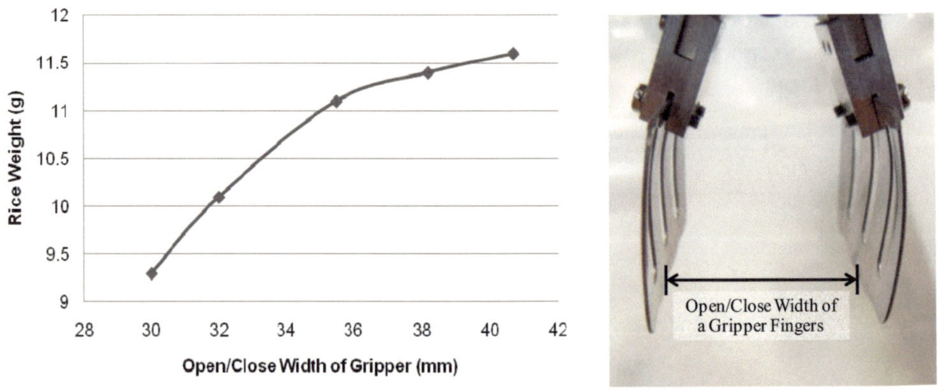

Fig. 13. Amount of rice in a single grasp

3.2 Single-arm configuration

A spoon-arm could be used independently without a grab-arm, as shown in Fig. 14. The caregiver manually picks up the food on a tray and puts the food on the spoon of a spoon-

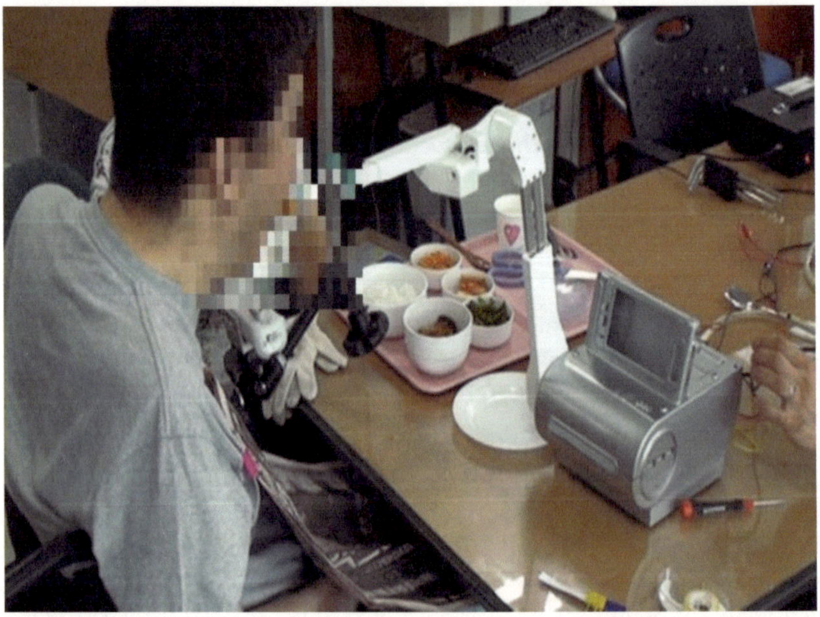

Fig. 14. The motions of the self-feeding robot in single-arm configuration. The caregiver picks food up on the spoon of the self-feeding robot, and then the user presses the button to receive the food

arm. The next step is similar to that of the dual-arm robotic arm. A caregiver only provides a user with food when the spoon is empty in the home position. From the perspective of the caregiver, he or she can reduce the amount of time needed to check or wait while the user is chewing. From the perspective of the user, the food can be eaten when he or she wants to eat. Although a user may have difficulty choosing food in an automatic manner, he or she can chew the food sufficiently without considering the next spoon serving from a caregiver.
From an economical point of view, the single-arm configuration has a lower cost in comparison with the dual-arm configuration. Applying a spoon-arm has advantages in facilities such as hospitals or nursing homes. One caregiver supports more than one consumer. That means one caregiver can serve food on the spoons of multiple users' spoon-arms in turn. This configuration is especially useful if the labor cost of a caregiver is not high, as in developing countries.

4. User test

At first, we designed the self-feeding robot on the basis of users' opinions in order to develop a practical system. After constructing the prototype of the self-feeding robot, we performed a user test using seven people with disabilities via the developed self-feeding robot. Participants used the self-feeding robot to eat Korean food, and we collected feedback.

4.1 Users' overall feedback on the self-feeding robot

In the users' opinions, a self-feeding robot could be useful when a user stays at home alone. The self feeding robot can be used in two situations: one is solitary eating and the other is eating together. The most important role is in supporting self-feeding without a caregiver when people with disabilities stay in their homes alone and caregivers prepare the food in advance.
Some users prefer using a large spoon. For example, a spoon could be a Chinese-style spoon. If a spoon is large, then it can be used to feed wet food. However, a female user could prefer to use a small-sized spoon. The size of the spoon should be customized according to the user preferences. We will consider several spoons with various sizes as well as various depths.
Users sometimes request quick motion of the robotic arm. A user wants to be able to adjust the motion speed of the self-feeding robot. The adjusting speed should be customized to the user.
The spoon should tilt to the user's mouth in order to unload the food on a spoon easily when the spoon arm is positioned in the user's mouth. Technically, the spoon should tilt to a user with cerebral palsy because such a user can move his/her head to a limited extent. If the self-feeding robot does not have a tilting function, then the user will struggle to eat the food on the spoon. Specifically, if the robot has a range detection sensor around the spoon, then the robot could move more intelligently in front of the user's mouth. That means a spoon automatically tilts in front of a user's mouth. If a user's mouth moves, the preprogrammed motion is not suitable. If a spoon tilts in the wrong position, food could drop down on the floor. Some people with cerebral palsy or muscular disease have trouble moving their neck, and thus the tilting function of a spoon should be an option.
Users and experts want to eat food comfortably using a spoon. For instance, the proposed system moves the spoon in front of the user's face. At that time, the spoon moves to the user's mouth along the sagittal plane, as shown in Figs. 15 (a) and (b). Some users complain

about a threatening feeling when the spoon moves on the sagittal plane. If the spoon approaches a user's mouth from the side, then a user can feel safer. Those motions are similar to most people's motions when they eat food. In addition, a user wants to touch the side surface of a spoon. As a remedy, we will consider how the spoon approaches, as shown in Figs. 15 (c), (d), and (e). However, the sideways approach may require a large installation area for the self-feeding robot.

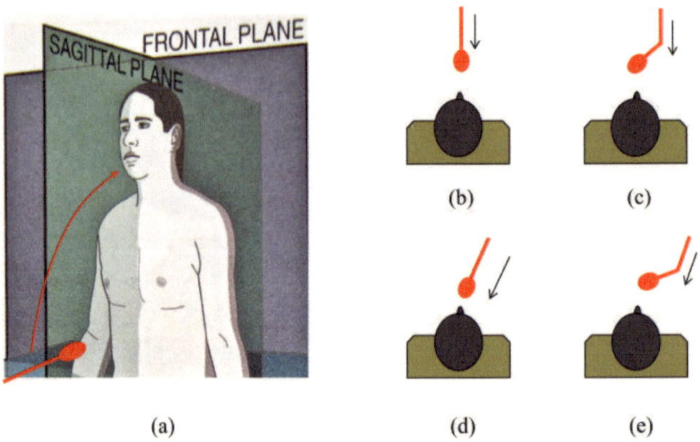

Fig. 15. Spoon approaching modes. (a) The sagittal plane on which a spoon moves. (b), (c), (d), and (e) represent the top view of a spoon when a robot approaches a user's mouth. The red object means a spoon. (c), (d) and (e) are more comfortable then (b)

A spoon should have a safety function because the spoon frequently comes in contact with a user's mouth. A spoon can be fixed at the tip of an arm with a spring as a component to guarantee safety. The spoon can be connected to magnets. When a large force acts on the spoon, the magnet's connection with the spoon could be detached for the user's safety. A user who has spasticity needs the compliance of a spoon.

Users request a smoother motion of a spoon when it contains food. In addition, the vibration of a spoon should be reduced at that time.

Users want a small-sized self-feeding robot for easy installation on a desk. In addition, a user wants to be able to adjust the spoon level of the self-feeding robot. For example, a user who uses a power wheelchair has various height levels from a table. When users first use the self-feeding robot, the level of the spoon on the robotic arm should be adjustable.

As a basic function, food rarely drops on a table when a user fails to eat the round-shaped food. When a user eats food that is too stiff, grasping failure occurs. Therefore, we consider the motion, i.e., the speed, of a spoon as well as the shape of a spoon.

Some users want to eat rice with a side dish simultaneously. In general, the disabled person who lives alone and receives a caregiving service usually eats rice and a side dish on one spoon at the same time. Some people eat rice mixed with side dishes. The self-feeding robot should mimic that task. The self-feeding robot optionally puts two kinds of food, i.e., rice and a side dish, on a large spoon simultaneously.

The spoon should be returned to the home position after a predefined time interval. The robot has two ways to return to the home position of a spoon: one is by making an input signal, and the other is determining a time interval.

Sometimes a robot should regrasp food in order to avoid grabbing too much of a side dish. When a robot tries to grab some curry rice, the gripper shape should be changed. The

gripper should be cleaned frequently. The amount of grabbed food could be adjustable. A stop or pause function is required.
Other comments are as follows: the robots should be less noisy, have a small-sized input device, serve water, and enable users to eat Korean soup. Users who do not have experience eating food by themselves have a hard time using a self-feeding robot for the first time. Such a person needs to be trained on how to use it. Some users want to eat noodles, such as spaghetti. Most users wish to use a low-cost feeding robot.
The filtering of an involuntary input signal of a user should be considered. For instance, in order to reduce malfunctions, a user's input could be ignored immediately after a previous input. One user who can push a button by himself prefers to use buttons rather than joysticks. One user prefers using a small-sized joystick. Two joysticks are better than one joystick with buttons. The length of a joystick should be adjustable. Buttons should be large sized. The operation without a PC is required. A user who has limited head motion likes to use the long joystick because that user cannot make a fine motion. The unification of an input device of a wheelchair and a self-feeding robot should be considered. Wireless input devices are preferred.

4.2 Discussion of a single-arm configuration

In the single-arm configuration, the caregiver picks food up instead of the grab-arm and then loads food on a spoon. The user makes an input signal to move the spoon to the user's mouth. After the user eats food on a spoon, it returns to the home position upon receiving a command or some time interval, as with a dual-arm configuration.
The single-arm configuration is useful to a caregiver as well as a user. From the perspective of a user, the feeding timing could be changed freely on the basis of a user's intention. The user can chew food sufficiently. When a user watches television, she leisurely eats food. Some users complain about a single-arm configuration. That means a caregiver must stay with a user even though the single arm is used. The single-arm configuration is useful for a caregiver when a user has a meal with his family because the caregiver does not need to move food to the user's mouth. Thus, a caregiver likes to use the single-arm configuration.
The users and caregivers are satisfied even though picking up food should be performed manually. Food frequently drops down on the floor when a caregiver serves food to the user's mouth. However, a user can reduce the instances of dropping food in the case of a single-arm configuration because the spoon is fixed on the spoon-arm and thus the user can estimate the spoon posture correctly.

4.3 Input devices

A system should be easy to control for people with disabilities and the elderly. Many users have only head motion, and thus the importance of input devices was mentioned.
In this study, we analyzed the self-feeding robot on the basis of two input devices, namely buttons and joysticks, as shown in Fig. 16. Most input devices are developed for hand manipulation. If a user has hand function, then the easiest way is to use the system is with his or her hands. That means the simplest input device is a set of buttons. However, when a user only uses neck motion to make a control input, he or she has difficulty handling input devices with dexterity. Table 2 shows the properties of the input devices.

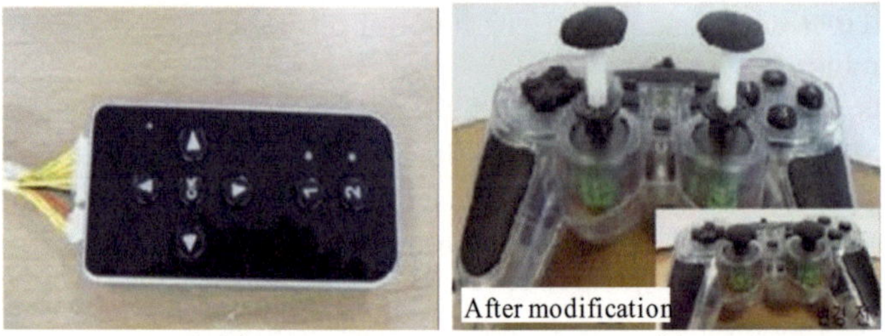

(a) (b)

Fig. 16. Input devices of a self-feeding robot. (a) Buttons, (b) Joysticks

Input Device	Buttons	Dual Shock Type Joypad
# of Used Buttons	7	N/A
# of Used Joysticks	N/A	2
Length of Joystick (mm)	N/A	53
Gap between Joysticks (mm)	N/A	44
Angles of Joystick (degrees)	N/A	±25
Connection	Wire	USB

Table 2. Input devices for the self-feeding robot

4.3.1 Buttons

The self-feeding robot has a basic input device consisting of a set of buttons. We usually use six buttons that correspond with start, return, and four directions. The buttons were originally developed to check the basic operations of the self-feeding robot. However, quadriplegics who can only move their neck and head would have difficulty pushing the button with their chin. Because the buttons are out of the field of view, the user has a hard time knowing where the buttons are and whether or not they are pushed. Additionally, pushing these buttons requires excessive force and can result in muscle fatigue in a user's neck. Thus, a user who uses his or her neck would have difficulty pushing the buttons. Users prefer to use joysticks rather than buttons. On the basis of users' opinions, we tested input devices that have joysticks.

4.3.2 Joysticks

Two joysticks are employed. Originally, it was determined that a user wants to use two joysticks rather than one joystick and buttons. The length of the joystick is modified from 10 to 53 mm according to users' opinions. Because of the gap between the two joysticks, a user can easily manipulate one of the joysticks without any resulting malfunction of the other joystick.

The length of the joystick depends on user preference. Some users prefer a long joystick while others like a short one. Most users prefer the wide gap between the two joysticks because a short gap can result in the malfunction of the unused joystick. The moving angles of a joystick are ±25°. Users express satisfaction with the flexibility of the joystick and its silicon cover, which can be connected to the user's skin.

4.4 Satisfaction evaluation of the self-feeding robot

We carried out the user tests with user candidates, including people with spinal cord injuries. After they actually ate food using the developed self-feeding robot, we collected their feedback to determine their satisfaction score in accordance with input devices. The users ate food using the self-feeding robot with each of the input devices. The results of the users' feedback pertained to the self-feeding robot as well as the input devices. The users rated their satisfaction with the input device activity on a scale of 1 to 10, as with the Canadian Occupational Performance Measure (Pendleton, 2001). A score of 10 indicates the highest level of satisfaction. Most users were satisfied with the self-feeding system that had a dual joystick, as shown in Table 3. This indicates that the use of joysticks is more comfortable than that of buttons. In addition, the self-feeding system operates well. Based on the analysis of users' feedback, the key factors affecting the handling of joysticks with regard to a user's neck motion are as follows: the distance between the joysticks, the moving angle of the joysticks, and the length of the joysticks. We will perform a comparison study between commercialized feeding systems and the developed system.

Input Device	Buttons	Joysticks
SCI #1	6	8
SCI #2	8	8
SCI #3	1	8
SCI #4	1	8
SCI #5	1	6
SCI #6	1	7
SCI #7	1	8
Average Score	2.7	7.6

Table 3. Satisfaction score of input devices when users eat food via a self-feeding robot (max = 10)

5. Concluding remarks

We have developed a novel assistive robot for self-feeding that is capable of handling Korean food, including sticky rice. This paper presents the overall operation of the self-feeding robot. The proposed robot has three distinguishing points: handling sticky rice, using an ordinary meal tray, and a modular design that can be divided into two arms. Users are people with physical disabilities who have limited arm function. During the development of the robot, we considered the feedback provided by several users and experts. In addition, the user candidates tested the actual the self-feeding robot. It was determined that the input device has the most important role. Many users prefer a dual joystick for self-feeding. Most of the users who participated in the experiments gave us positive feedback. Some users were impressed that they were able to eat their desired food when they wanted to eat it. In future work, we will add several functions to the robot, including improving the reliability of basic operations and adding a safety feature. We will also simplify the system components and perform user evaluations.

6. Acknowledgment

This research was supported by a grant (code #10-A-01, #11-A-04) from the Korea National Rehabilitation Research Institute, Korea National Rehabilitation Center, Korea. The authors acknowledge the input of consumer candidates, including Mr. Hongki Kim and Mr. Kwangsup Lee. In addition, the authors gratefully acknowledge the work of Mr. Won-Jin Song and the comments of clinical specialists, including Dr. Bum-Suk Lee, Dr. Sung-Il Hwang, and Ms. Mi-Ok Son at the Korea National Rehabilitation Center.

7. References

Ding, D.; Cooper, R. & Pearlman, J. (2007). *Incorporating Participatory Action Design into Research and Education*, International Conference on Engineering Education (ICEE) 2007, Coimbra, Portugal.

Employment Development Institute (2009). *2009 Disability Statistics (in Korean)*, Survey & Statistics Team , Korea, ISBN 978-89-5813-737-5.

Guglielmelli, E.; Lauro, G.; Chiarugi, F.; Giachetti, G.; Perrella, Y.; Pisetta, A. & Scoglio, A. (2009). *Self-feeding Apparatus*, US Patent2009/0104004.

Hermann, R.; Phalangas, A.; Mahoney, R. & Alexander, M. (1999). *Powered Feeding Devices: An Evaluation of Three Models*, Arch Phys Med Rehabil, Vol. 80, pp. 1237–1242, 1999.

Mann, W. (2005). *Smart Technology for Aging, Disability, and Independence: The State of the Science*, Wiley, Hoboken, New Jersey, ISBN 978-0-471-69694-0.

Mealtime Partners, http://www.mealtimepartners.com.

Neater Solutions, http://www.neater.co.uk.

Sammons Preston, http://www.sammonspreston.com.

Song, W.-K.; Kim, J.; An, K.-O.; Lee, I.-H.; Song, W.-J.; Lee, B.-S.; Hwang, S.-I.; Son, M.-O. & Lee, E.-C. (2010a). *Design of Novel Feeding Robot for Korean Food*, ICOST2010, LNCS 6159, Yeunsook Lee *et al.*, Eds. Springer, pp. 152–159.

Song, W.-K.; Kim, J.; An, K.-O.; Lee, I.-H.; Song, W.-J. & Lee, B.-S. (2010b). *New Dual-Arm Assistive Robot for Self-Feeding*, Second International Symposium on Quality of Life Technology, Las Vegas, USA.

Soyama, R.; Ishii, S. & Fukase, A. (2003). *The Development of Meal-assistance Robot MySpoon*, International Conference on Rehabilitation Robotics, pp. 88–91, Daejeon, Korea.

Topping, M. & Smith, J. (1999) *The Development of HANDY 1, a Robotic System to Assist the Severely Disabled*, International Conference on Rehabilitation Robotics, 1999.

Pendleton, H. & Schultz-Krohn, W. (2001). *Pedretti's Occupational Therapy*, Elsevier.

Pourmohammadali, H.; Kofman, J. & Khajepour, A. (2007). *Design of a Multiple-user Intelligent Feeding Robot for Elderly and Disabled People*, 30th Canadian Medical and Biological Engineering Conference (CMBEC30), Toronto, Ontario, Canada.

4

Robot Handling Fabrics Towards Sewing Using Computational Intelligence Methods

Zacharia Paraskevi
University of Patras, Department of Mechanical Engineering & Aeronautics
Greece

1. Introduction

In cloth making industry, the need for high flexibility in automation is really imperative due to the extensive style and material variation of the products and the fast fashion changes. The automated handling of fabric materials is one of the most challenging problems in the field of robotics, since it contains a vast potential for high productivity and cost reduction. The main difficulty to get full use of this potential is the fact that it is rather impossible to predict the behaviour of fabrics towards handling due to their low bending rigidity. Moreover, it is difficult to be modelled due to their unpredictable, non-linear and complex mechanical behaviour.

Sewing is the most important joining technology in garments and textiles manufacturing. The fact that sewing fabrics is a "sensitive" operation due to the erratic behaviour of fabrics poses barriers in the development of automated sewing systems. A solution to manipulation tasks that are afflicted with uncertainty, subjectivity, ambiguity and vagueness is an intelligent control approach. The development of intelligent control systems with vision feedback enables robots to perform skilful tasks in more realistic environments and make the research efforts for flexibility and automation really promising. Thus, industrial robots supported by machine vision and sensors can contribute towards advanced automation in apparel manufacturing.

On the other hand, sewing fabrics with the exclusive use of robots, without human intervention, is a complicated task. The stitching process is special in the sense that the error cannot be corrected after a part of the cloth has been stitched, implying that the stitching process is not reversible. This limitation implies that clothes that do not conform to the specifications are defective and should be withdrawn from the production. This imposes a great demand on precision during the sewing process. In practice, there is a predefined tolerance considered to be acceptable, as errors resulting from robot accuracy and camera resolution cannot be eliminated.

The main focus of this work lies on the design of an innovative intelligent robot controller based on visual servoing that aims to enable robots to handle flexible fabric sheets towards sewing. A great challenge is the design of a robot controller showing robustness against deformations that are likely to appear on the fabric towards robot handling. Special emphasis is given on robot handling fabrics comprised of curved edges with arbitrary curvature. This task is even harder, because up-todate industrial robots present limitations in their movement, owing to the fact that they can only be programmed to make straight or circular motions.

2. Literature review on automated sewing systems

To the best of my knowledge, the published research work on robot handling fabrics with curved edges towards sewing is limited to few papers. A first approach to automating fabric manipulation was introduced by (Torgerson & Paul, 1988) including robotic motion control of various fabric shapes. The determination of robot motion paths is based on visual information defining the location of the fabric edges in world coordinates. However, no visual feedback was employed during the robot motion, making the method less accurate. The vision system detects the workpiece edge, extracts the boundary points, determines the parameters defining the geometry of the interior points, and computes the coordinates of the robot path points along subsequent linear and circular path segments. The error deviations between the desired seam line and the actual path line ranged between 3 and 5 mm. However, the flexibility of the method is limited owing to the fact that the path determination algorithms require a geometrical relationship between the robot end-effector and the workpiece to extract the edge parameters that are necessary for determining the robot motion.

The FIGARO system (Gershon and Porat, 1988, 1986) performed handling and assembling operations using two superimposed servo control systems. The first system maintains a small tension moving the robot forward with the cloth and the second one rotates the cloth about the sewing needle to produce a seam parallel to the edges. The system performed best with shirting and worsted woven fabrics, which have a reasonable resistance in buckling. Robotic manipulation strategies have been investigated (Gershon, 1993) for handling limp materials proposing a parallel process decomposition of the robot sewing task (Gershon, 1990). A robot arm manipulates the fabric, modeled as a mass-spring-damper system, to modify its orientation and control the fabric tension during the sewing process, which was decomposed into four concurrent processes within a superposition parallel architecture.

After FIGARO system, an automated sewing system including two robots handling the fabric on the table was developed (Kudo et al., 2000). The nominal trajectories for both robots are defined through the definition of the points and are necessary for the programming of the robots. Visual information was used to improve tracking accuracy. Sewing along a curved line is achieved by translating the fabric in the sewing direction using the internal command for straight-line motion and simultaneously rotating about the needle according to the visual feedback information. The rotation angle is computed making use of the tangent of the cloth panel based on the method in reference (Gershon & Porat, 1988). The trajectory error was within the range of ±0.5mm.

A position-based visual servoing system for edge trimming of fabric embroideries by laser was proposed by (Amin-Nejad et al.). A tracking controller, decomposed into a feedforward controller in the tangential direction of the seam and a feedback controller in the normal direction, was implemented. The aim of the controller is to move the cutting beam along the seam with constant velocity and with a constant offset from it as a cutting tolerance. In the experimental results, three types of seam pattern, straight line, sinusoidal, and circular, were chosen for edge trimming. The accuracy achieved with this method was within ±0.5mm.

3. Sewing fabrics with straight edges

The present work is part of a project for the development of a robotic workcell for sewing fabrics. The project includes handling tasks (ply separation, translation, placement, folding, feeding and orientation), tension control and quality control of fabrics and stitches

(Koustoumpardis & Aspragathos, 2003; Koustoumpardis et al., 2006; Moulianitis et al., 1999; Zoumponos & Aspragathos, 2008). After the fabric has been placed at a random location on the working table, a number of sub-tasks should be performed before the sewing process starts. These preliminary sub-tasks are: the recognition of the fabric's shape, the extraction of the 'seam line', the detection of the edges targeted for sewing, planning of the stitching process and the location of the end-effector on the fabric (Koustoumparis et al., 2007). After the preprocess planning, the robot sewing process is considered and divided into three sub-tasks: the manipulation of the fabric towards the needle, the sewing of the seam segment and the rotation around the needle.

Concerning visual servoing, the developed vision system is a combination of image-based and position-based control system. The image-based analysis is used for the identification of the fabric's shape. After the image acquisition of the fabric, the features (vertices of the edges), the needle-point and the sewing line's orientation are derived from the image analysis. Besides, the position of the needle is also known in the robot coordinate system. The end-effector's position is unknown in the image coordinate system; however, the robot system gives feedback to the control system of the current end-effector's position in the robot base coordinate system. Moreover, the relation between the robot- and the image-coordinate system is known from the calibration of the camera. This approach makes the system more flexible and limits the calibration errors.

3.1 The sewing process

Initially, the fabric lies free of wrinkles at a random location on the work table. The camera captures the image of the fabric without the gripper on it in order to obtain the shape of the fabric and use it as the reference shape towards handling. The stitching process is performed on the *seam line* situated parallel to the specified fabric edges. The distance between the outer edge and the seam line, called *seam allowance*, depends on the cloth part and is defined by the apparel manufacturer. In practice, it usually ranges between 1/4 inch and 5/8 inch. After the seam edges of the fabric (dashed-line in Fig. 1) have been determined, the sewing process is ready to start. The sewing line is determined by the feeding mechanism. The sewing process can be decomposed in three separate tasks: transfer, stitching and rotation.

The movement towards the needle. After the robot touches the fabric at a proper location, the features, namely the vertices of the fabric, are extracted from the image taken from the camera. After the starting seam edge is identified, the vision system captures the distance (r) between the starting seam vertex and the sewing needle and the orientation angle (θ) formed by the starting seam segment and the sewing line (Fig. 1 (a)). The linear (u) and angular velocity (ω) of the robot end-effector are derived through the designed fuzzy decision system, described in Section 3.2. Given the time step dt and the angle φ, i.e. the orientation of r in the image coordinate system, the new position and orientation of the end-effector is computed by:

$$
\begin{aligned}
x' &= x + u \times \cos(\varphi) \times dt \\
y' &= y + u \times \sin(\varphi) \times dt \\
\theta' &= \theta - \omega \times dt
\end{aligned}
\tag{1}
$$

Therefore, the robot end-effector moves along the direction of r (defined by φ) and simultaneously rotates around the end-effector's z-axis, which is vertical to the table, until the angle becomes θ' (Zacharia et al., 2005). The fabric is transferred to a new position with new orientation as a result of the movement and rotation of the end-effector, which is stuck on the fabric to avoid slipping between the gripper and the fabric. This motion stops when the seam segment reaches the needle with the desired orientation within an acceptable tolerance.

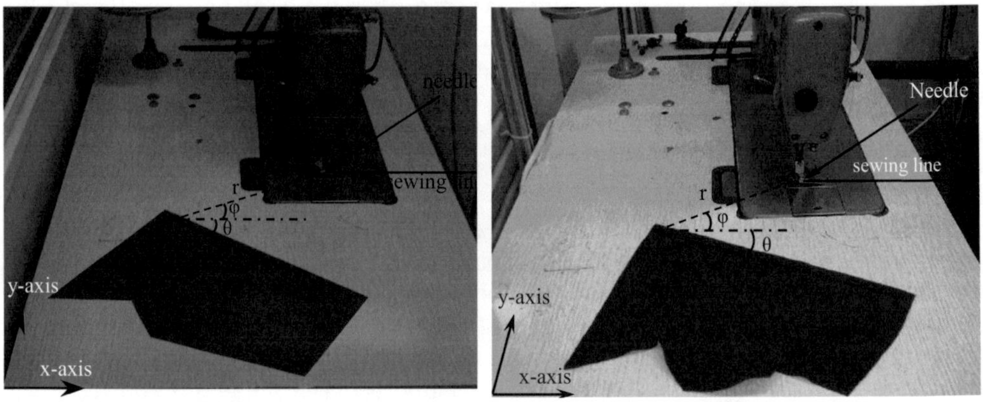

(a) (b)

Fig. 1. Scene of the fabric lying on the work table (a) without deformations (b) with deformations

The stitching process. During sewing, the fabric is guided along the sewing line with a constant velocity, which should be the same with the velocity of the sewing machine, so that good seam quality is ensured. When the end-effector's velocity is higher than the sewing velocity, puckering will appear, whereas in the case where it is lower, low seam quality will be produced due to the tension increase. At each time step, the current image of the fabric is captured in order that the orientation error is determined. The orientation error is fed back to be corrected by rotating the fabric around the needle, while simultaneously the robot moves the fabric towards the direction of the sewing line. To circumvent the problem of uncertainty due to the distortions of the fabric's shape, fuzzy logic control is necessary. The inputs are the orientation error and its rate and the output is the rotation angle around the needle.

The rotation around the needle. When the seam segment coincides with the sewing line, the robot rotates the fabric around the needle until the next seam segment is aligned to the sewing line. It is worth noting that the end-effector is enforced to make a circular motion around the needle, since it has penetrated into the fabric. The orientation error (e_θ) of the next seam segment in relation to the sewing line and its time rate are the inputs to the fuzzy system that controls the rotation of the fabric around the needle, whereas the output is the angular velocity (ω) of the end-effector's motion around an axis perpendicular to the table at the needle-point.

3.2 The fuzzy robot control system

Since modeling the mechanical behavior of fabrics for real-time applications is rather difficult due to their low bending rigidity, an approach based on a fuzzy logic controller (FLC) is developed (Zacharia et al., 2009) considering the robustness and the fast response requirements. The block diagram for the control system is shown in Fig. 2, where the symbols in parentheses stand for the fuzzy system that controls the orientation of the end-effector. The controller inputs are the position error (e_r) and the orientation error (e_θ), which are computed on the image, and their time rates (e'_r) and (e'_θ). The linear (u) and angular velocity (ω) of the end-effector around z-axis are the outputs.

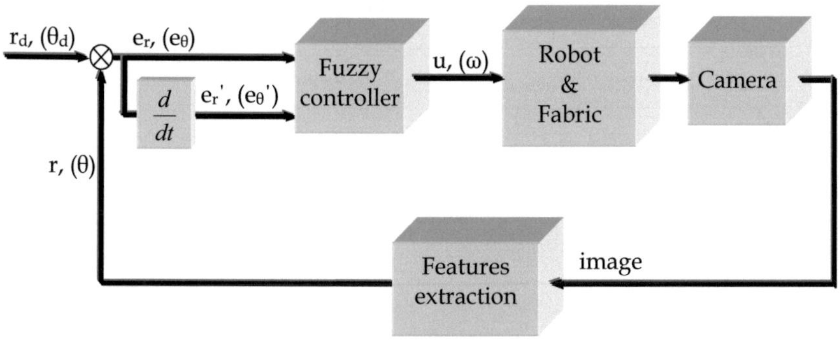

Fig. 2. The block diagram of the fuzzy logic control

The membership functions for the two inputs (Fig. 3 (a) and (b)) and the output of the system that controls the translation of the fabric towards the needle (Fig. 3 (c)) are experimentally tuned with various fabrics. Expert knowledge, often afflicted with uncertainty, is summarized in the proposition: *The larger the distance/angle and the smaller its change rate is, the faster the fabric should move/rotate*. The fuzzy associative memory (FAM) of the system that controls the translation of the fabric (Table 1) is derived after studying the behavior of the human workers towards sewing. For the rule evaluation, the "min" operator is used and for the aggregation mechanism the "max" operation is used. The centroid defuzzification method is used to determine the linear and angular velocity. Owing to space limitation, only the study for the system that controls the translation is presented.

position error rate	*position error*		
	SMALL	**MEDIUM**	**LARGE**
SMALL	Very Small	Medium	Very Large
MEDIUM	Very Small	Medium	Very Large
LARGE	Very Small	Small	Large

Table 1. Fuzzy Associative Memory (FAM) of the sewing system

An analytical model of the robot and the fabric is not necessary, since no geometrical characteristics are taken into account and there is no need for special mathematical computations for different fabrics. The proposed technique is feature-based, since only the features, r and u, are necessary for handling the fabric and can cope with the uncertainties that arise due to deformations. The proposed controller can handle possible deformations without changes in its structure achieving the desired accuracy on condition that the seam segment targeted for sewing is undistorted. Fig. 1 (b) shows the fabric of Fig. 1 (a), which has been deformed except for the seam segment targeted for sewing. It is clear from Fig. 1 (b) that the presence of deformations does not affect the control system, despite the fact that the shape of the fabric has significantly changed.

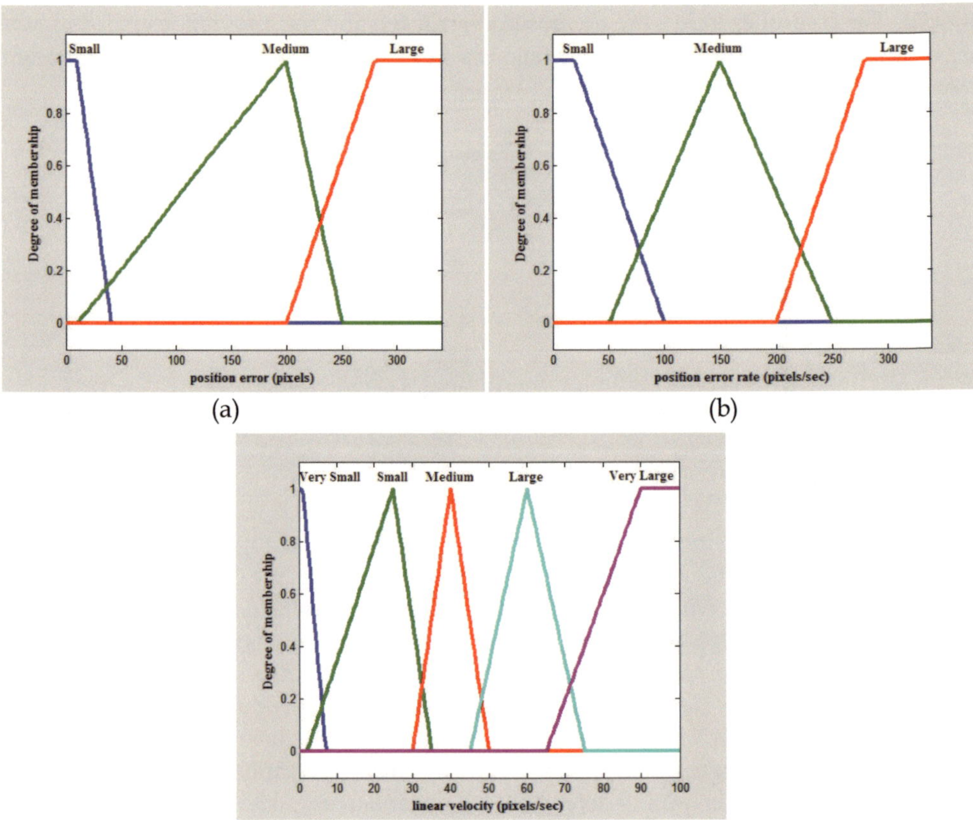

(a) (b) (c)

Fig. 3. The membership functions of (a) the position error (b) the position error rate (c) the linear velocity after manual tuning

4. Sewing fabrics with curved edges through a Genetic-based approach

Real fabric pieces used for cloth making consist of edges with arbitrary curvature. This fact gave an impetus to tackle the problem of robot-handling fabrics with curved edges. To robot-sew fabrics of different shapes and materials is a rather difficult task that requires system flexibility and good quality of the final product. To compensate for ensuing high machine-hour rates, a high throughput is vital if product costs are to be kept low. To enhance system flexibility and its efficiency to handle various fabrics, fuzzy logic and visual servoing control has been employed.

Curved edges pose additional difficulties in robot handling compared to straight edges, owing to the fact that up-to-date robots present limitations in their movements. The current industrial robots can only be programmed to perform straight or circular motions using internal commands. Sewing a fabric with arbitrary curvatures is a complicated task and can only be performed by approximating the curve with movements along straight-line segments (Zacharia et al., 2006). The robot makes a straight-line motion along the sewing line and a rotation around the needle-point simultaneously. This motion results in a smooth stitch that resembles the stitch produced by human operators.

4.1 Related work

In the field of pattern recognition, considerable research work has been done on the polygonal approximation of digital contours. Two main families of technique concerning the polygonal approximation of digital curves have been published: those that try to obtain an error-bounded polygonal approximation and those that search for a set of dominant points as vertices of the approximating polygon.

In the first family of methods, the main idea is to approximate a given curve with a minimal number of line segments such that the approximation error between the original curve and the corresponding line segments is less than a pre-specified tolerance. Several methods have been used for polygonal approximation, but most of them have the disadvantage that the results are dependent on the selection of the initial points and the arbitrary initial solution. To circumvent these drawbacks, nature-inspired algorithms have been proposed for polygonal approximation.

In the genetic algorithm presented in reference (Zhang & Guo, 2001), each chromosome is a binary string of fixed length. The chromosome represents a subset of the curve points, Where the i-th bit denoting the ith original curve point is '1' if the corresponding curve point is taken as a vertex of the approximation polygon and '0' otherwise. The algorithm is improved based on Pareto optimal solution, and the results show that it is efficient and achieves less processing time compared to dynamic programming algorithms.

In reference (Yin, 2003), an ant colony search (ACS) algorithm for optimal polygonal approximation is developed. To apply the ACS, the underlying problem should be represented in terms of a graph. Apparently, for the polygonal approximation problem, each point on the curve should be represented as a node of the graph. A number of artificial ants are distributed on the graph and communicate with one another through the pheromone trails that are a form of the long-term memory guiding the future exploration of the graph. The authors compared the performance of the proposed method to those of genetic-based and tabu-search-based methods, and concluded that the numerical results are very encouraging.

A polygonal approximation approach of digital curves based on the particle swarm optimization (PSO) algorithm was presented by (Yin, 2004). In PSO - which belongs to the class of natural algorithms - each particle is presented as a binary vector corresponding to a candidate solution to the polygonal approximation problem. A swarm of particles are initiated and fly through the solution space for targeting the optimal solution. The experimental results manifest that their devised PSO algorithm is comparable to the genetic algorithm (GA), and it also outperforms other PSO versions in the literature for polygonal approximation.

In the second family of methods, the approximation does not depend on the starting point and is insensitive to changes of orientation and scale. A representative technique is the dominant point approach (Teh & Chin, 1989). This method focused on the determination of the region of support of every point of the curve, which is the region used in the calculation of each point's curvature. Thus, the method overcomes the problem of various scales in the curve's features. However, the Teh–Chin approach is not robust in the presence of noise. For the noise compensation, Wu proposed a modification of the method based on a dynamic determination of the region of support (Wu, 2002).

4.2 Sewing using polygonal approximation

The apparel manufacturer's requirements focus on time minimization for the accomplishment of the sewing process and satisfactory quality of seam, which is mainly based on stitch accuracy and smoothness. To fulfil these requirements, an algorithm based on the polygonal approximation of the curved edges of the fabric is developed for robot handling fabrics towards the sewing process.

During the stitching process the robot's end-effector is commanded to make two motions simultaneously: a straight motion with constant speed (equal to the sewing machine's speed to ensure good seam quality) in the direction of the sewing line, and a rotation around the needle-point. Therefore, the final stitch is not a straight line, but a line that is yielded as a result of the combined end-effector's motion. After all the straight-line segments have been stitched, the deviation error is computed. Then, a new image is taken and the process is iterated. The stitching process stops when the whole curve has been sewn, i.e. when the last point reaches the needle.

The problem is formulated as follows: The curve section of the fabric is defined by N data points acquired by the vision system in the form of pixel coordinates in clockwise order, and can be described by a sequence of these points: $G=\{p_1, p_2, \ldots, p_N\}$. Given the N data-points, find the optimal polygon that approximates the curve satisfying two criteria: the criterion for sewing time minimization and the criterion for acceptable accuracy. In this case, the time minimization is translated to the requirement for minimal length of the approximating polygon. Initially, a local search technique is developed to extract the dominant points and then a global search technique is applied taking the dominant points as inputs (Zacharia et al., 2008). The idea behind this is to filter out the initial curve points in an attempt to speed up the convergence of the algorithm. The proposed approach combines two algorithms to benefit from the advantages of each one.

The contribution of the work addressed in this Section is twofold. Firstly, a strategy for handling fabrics with curved edges towards the sewing process using a visual servoing controller is developed based on polygonal approximation. Secondly, the algorithms are further extended based on polygonal approximation that exhibited superiority in many other applications, and a new algorithm is proposed based on the dominant point detection approach and the genetic algorithm.

4.3 The proposed genetic-based approach

Genetic algorithms (GAs) (Michalewitz, 1996) have attracted attention in solving combinatorial optimization problems of high complexity because of their intrinsic parallelism and their effectiveness. The contribution of the proposed GA is a variable-length chromosome encoding scheme to reduce the computational time and memory requirement and the use of micro-GAs (Krishnakumar, 1989) as an approach for shortening the computational time. The traditional fixed length chromosomes could also be used to solve this problem, but it would lead to computational time increase, as the length of the chromosome would be too large to incorporate all the points comprising the curve. On the other hand, micro-GAs are ideally suited to optimization problems, where the emphasis is on quickly identifying a near-optimal region. In contrast to the traditional GA, the population size is small, yielding a much quicker convergence. Moreover, the micro-GA uses elitism and convergence checking with re-initialization to obtain the near-optimal solution. As a consequence, the proposed micro-GA with variable-length chromosomes can be used in a real-time application.

The evaluation mechanism: Given a maximum acceptable error ε_0, find the polygonal approximation with the minimal number of vertices (corresponding to the minimum total length of the polygon), such that the polygon is distant from the curve by no more than ε_0. It is worth noting that the algorithm searches in a set of N_s points that are the dominant points

defined by the dominant point detection approach. Consequently, the input for the micro-GA is the output of the dominant point detection approach.
Let $G^{**}=\{P_1, P_2,\ldots, P_m\}$ be a subset of $G^{*}=\{P_1, P_2,\ldots, P_{Ns}\}$, $G^{**}\subseteq G^{*}$, where $m\,(m<N_s)$ is the number of the vertices $P_i=(x_i, y_i)$ of the polygon that approximate the curve section captured in the current image. The function expressing the total length of a polygon with m vertices can be written as:

$$L_{total} = \sum_{i=1}^{m-1} L_i \tag{2}$$

and L_i is the length of the i_{th} edge of the polygon given by:

$$L_i = \left\| P_{i+1} - P_i \right\|^2 \tag{3}$$

The constraint that should be satisfied is expressed as the maximum deviation between the curve section and the approximating polygon section. For two points $P_i=(x_i, y_i)$ and $P_{i+1}=(x_{i+1}, y_{i+1})$ defining an edge of the polygon and a data point $p_j=(x_j, y_j)$ between P_i and P_{i+1}, the perpendicular distance of the arc point p_j to the chord P_iP_{i+1} is computed by

$$\varepsilon_j = \frac{\left|(P_{i+1} - P_i)\times(P_i - p_j)\right|}{\left|P_{i+1} - P_i\right|} \tag{4}$$

Consequently, the maximum deviation ε is found by:

$$\varepsilon = \max\left(\varepsilon_1, \varepsilon_2, \ldots, \varepsilon_j, \ldots, \varepsilon_q\right) \tag{5}$$

where q is the number of the data points between P_i and P_{i+1}.
The fitness function is given by:

$$\text{fitness} = \begin{cases} \dfrac{1}{L_{total}}, & \text{if } \varepsilon \le \varepsilon_0 \\ 0, & \text{otherwise} \end{cases} \tag{6}$$

where L_{total} is given by Eq.(2).
The representation mechanism: In this algorithm, variable-length chromosomes are defined to represent a polygon with m vertices approximating the curve. The maximum number of polygon vertices, defined by the user, is equal to ℓ, which is the maximum possible length of the chromosome. The encoding mechanism maps each approximating polygon to a string composed of integers is in the following form:

$$\underbrace{1\ 10\ 24\ 49\ 67\ 92\ 104\ldots i\ldots N_s}_{\text{m numbers}}$$

where $1 \le i \le N_s$, N_s is the number of the integer-coordinate data-points of curve section captured by the camera and m, where $m \le \ell$, is the length of the chromosome representing the number of vertices of the approximated polygon. The integers i represent the rows of the matrix consisting of all the points $P_i=(x_i, y_i)$ in ascending order. Consequently, the genes of

the chromosome represent the curve data-points used to construct the polygon, i.e. the vertices of the polygon. It should also be mentioned that the first (1) and the last (m) row of the matrix are fixed for each chromosome.
Crossover: Crossover is a recombination operator and follows the reproduction. The individuals are randomly selected according to a predefined probability (crossover rate). The one-point crossover is modified, so that the produced offspring are feasible chromosomes. The cut-point lies between the first and the last gene of the parent chromosome with the minimum length. Next, the numbers at each string are reordered in order to appear in ascending order, so that the polygonal section is created by joining the successive points.
Population size: The population size depends on the nature and the complexity of the current problem. In this work, the proposed algorithm was tested for various small population sizes, so that both quick convergence and near-optimum solution are achieved. Finally, the selected population size is equal to 20.

5. Sewing fabrics with curved edges through ANFIS

One of the difficulties when applying the fuzzy approach of Section 4 is to obtain the fuzzy rules and the membership functions. In this Section, a neuro-fuzzy approach is introduced with the purpose of extracting fuzzy rules and the membership functions automatically. The proposed neuro-fuzzy approach combines the main components of soft computing (fuzzy logic, neural networks and genetic algorithms) that have shown great ability in solving complex control problems to benefit from the advantages of each one.

5.1 ANFIS framework

Fuzzy systems have the capability of translating the expert knowledge into linguistic rules inside a robust mathematical framework in order to draw conclusions and generate responses. However, a fuzzy model consisting of large number of if-then rules to map inputs to outputs is not desired due to the phenomenon of overfitting. Thus, the Takagi–Sugeno–Kang type fuzzy models (Sugeno & Kang, 1988; Takagi & Sugeno, 1985), known as TSK models, are widely used for control and modeling because of their high accuracy and relatively small models (Delgado et al., 2001; Männle, 2001).
When the system complexity is high, fuzzy modeling from input/output data is a useful way of creating FIS models. Because it is a more compact and computationally efficient representation than a Mamdani system, the Sugeno system using cluster information lends itself to the use of adaptive techniques for constructing fuzzy models with a minimum number of rules. Thus, data clustering algorithms were elaborated in order to construct FIS models from data, since they partition a collection of data into clusters based on similarity between data.
An adaptive neuro-fuzzy inference system, ANFIS, has been proposed in (Jang, 1993) to effectively deal with multivariable complex systems. ANFIS uses techniques like least squares or back propagation algorithms to determine the membership functions for a given input/output data set. These adaptive techniques can be used to customize the membership functions so that the fuzzy system best models the data. The effectiveness of using ANFIS for control and modeling has been pointed in (Jang et al., 1997; Rizzi et al., 1999).
This Section proposes an innovative approach for robot handling pieces of fabrics with curved edges towards sewing, which is based on learning. The fact that adaptive neuro-

fuzzy approaches have been successfully used for other applications in robotics (Marichal et al., 2001; Rusu et al., 2003; Hui et al., 2006) gave rise to the idea of using it for robot sewing fabrics of curved edges. The main purpose is to overcome difficulties arising from the fact that the fabric pieces have curved edges with arbitrary curvatures; thus, a learning technique is used. The proposed ANFIS based on genetic-oriented clustering combines the concepts of fuzzy logic and neural networks to form a hybrid intelligent system that enhances the ability to automatically learn and adapt. The system learns from the information obtained from the fabrics used for training and is able to respond to new fabrics, which had not been included in the training process.

5.2 Sewing tracking approach

The main scope of this Section is to provide a control system capable of dealing with different curvatures and thus, flexible to changes in fabrics' shape. To enhance system's flexibility and its efficiency to handle fabric edges with various curvatures, Artificial Intelligent techniques and visual servoing control are employed. Since there is no standardization for curved edges, a method is necessary for estimating the curve. However, searching for the equation that approximates the curve is a time-consuming procedure. In addition, such an approach would require curve estimation for each one curved edge for different fabrics.

The proposed system does not need any geometrical computations for estimating the curves, which would lead to computational burden and would, therefore, be prohibitive for using it in an on-line process. The developed control scheme is able to learn from the existed knowledge and respond to new curved edges.

Now, assume a piece of fabric with a curved edge of arbitrary curvature, as the one depicted in Fig. 4, which is guided for sewing. The camera captures an image that covers a small area in the vicinity of the sewing needle. The curved section captured in the image is approximated by an edge (AB) that joins the two endings of the curved section. Then, the distance d and the angle φ are extracted from the image. The feature d is the minimum distance between the needle-point and the straight edge that approximates the curved section. If the seam allowance is α, then the distance D, where D=d-α, is the minimum distance between the seam line and the needle. The angle φ is defined by the straight edge that approximates the curve section and the sewing line (Zacharia, 2010).

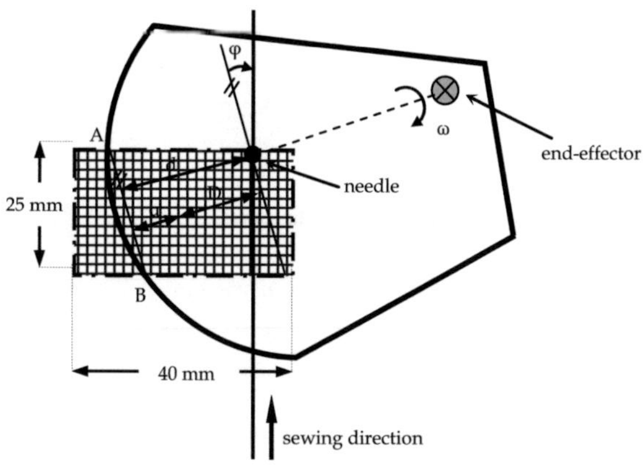

Fig. 4. Features extracted from the camera image

The position error is defined as e_D=D-D_0, where the desired distance is D_0=0 and the orientation error is defined as e_φ=φ-φ_0, where the desired angle is φ_0=0. To compensate for the seam errors (e_D and e_φ), the robot should be assigned to rotate around the needle with an angular velocity (ω) around z-axis. The angular velocity (ω) of the robot's end-effector is derived through a Sugeno-type fuzzy decision system, which takes as inputs the position error e_D and the orientation error e_φ.

The end-effector guides the fabric towards the sewing direction and the camera monitors the fabric to compensate for the seam error. The fuzzy controller, which is tuned through ANFIS, inputs the position and orientation error (e_D and e_φ) of the new fabric piece and outputs the predicted angular velocity (ω) of the robot's end-effector. Next, the robot is commanded to make a circular motion around the sewing needle by an angle θ and a translational motion in the sewing direction. The angle θ is derived by the relationship $\theta \approx \omega \cdot \Delta t$, where Δt is the sampling period.

5.3 Training using the modified ANFIS

The basic idea behind using neuro-adaptive learning techniques is that it is very simple and allows implementation of multi-input–single-output first order Sugeno-type FIS. This technique provides a method for the fuzzy modeling procedure to learn information about a data set, in order to compute the membership function parameters that best allow the associated fuzzy inference system to track the given input/output data. This learning approach takes place prior to the operation of the control system. The ANFIS methodology can be decomposed into four steps:

Data acquisition (Step 1): A number of different fabric pieces of various arbitrary curvatures are selected for experimental tests. During the sewing process, the position and orientation error in the image, as well as the end-effector's angular velocity are measured. As a result, a number of data sets, each one consisting of 3 attributes (e_D, e_φ and ω) obtained from the robot sewing process. These data sets are divided into training and checking data sets. It is worth noting that the position error (e_D), which is derived from the image, is computed in pixels and not in millimetres.

Genetic-oriented Clustering (Step 2): Clustering is used to generate a Sugeno-type fuzzy inference system that best models the data behavior using a minimum number of rules. The rules partition themselves according to the fuzzy qualities associated with each of the data clusters. The initial Sugeno-type FIS models are built by means of a genetic-oriented clustering approach using the training data set. In this approach, a Genetic Algorithm using variable-length chromosomes is applied to find the optimum number of cluster centers as well as the partitioning of the data. A comprehensive description of this algorithm can be found in (Zacharia & Aspragathos, 2008). This technique has the advantage over other clustering algorithms that the selection of the cluster centers is not limited to the data points and that it is able to escape from local optima, which is an inherent ability of Genetic Algorithms (GAs). Another advantage is the ability of automatically evolving the number of clusters due to the use of variable-length chromosomes in GA's structure. After applying the genetic-oriented clustering approach, the training data is partitioned into groups, called clusters, and as a result, simpler optimized FIS models with the minimum number of rules are obtained.

ANFIS architecture (Step 3): The purpose of this step is to optimize the initial FIS created by using the genetic-oriented clustering technique. A network-type structure similar to that of a neural network, which maps inputs through input membership functions and associated parameters, and then through output membership functions and associated parameters to outputs can be used to interpret the input-output map. Considering a first-order Sugeno fuzzy model with two inputs x and y and one output f, a typical rule set with two fuzzy if-then rules can be expressed as

- Rule 1: *If (x is A1) and (y is B1) then (f1 = p1x + q1y + r1)*
- Rule 2: *If (x is A2) and (y is B2) then (f2 = p2x + q2y + r2)*

where x, y are inputs, A_i, B_i are membership functions and p_i, q_i, r_i are consequent parameters and i is the node number. The entire system architecture consists of five layers, namely, the fuzzy layer (Layer 1), product layer (Layer 2), normalized layer (Layer 3), consequence layer (Layer 4) and total output layer (Layer 5).

Model Validation (Step 4): The validation process is used to evaluate the model generalization capability. One problem with model validation is selecting a data set that is both representative of the data the trained model is intended to emulate, yet sufficiently distinct from the training data set so as not to render the validation process trivial. Especially in noisy measurements, it is possible that the training data set does not include all the representative features needed for the model. The idea behind using a checking data set for model validation is that after a certain point in the training, the model begins overfitting the training data set. Overfitting can be detected when the checking error starts increasing while the training error is still decreasing. In practice, the experimental data is divided into training data and checking data. The training data is used in both genetic-oriented clustering process and ANFIS training. The checking data is only used in ANFIS training to prevent the model from overfitting.

6. Experimental results

The experiments were carried out using a robotic manipulator with 6 rotational degrees of freedom (RV4A) and controlled by a PC. The robot is programmed in Melfa-Basic language in Cosirop environment, while the analysis of the visual information is performed with Matlab 7.1. The vision system consists of a Pulnix analog video camera at 768×576 pixels resolution RGB and a analog camera of the same resolution, which are fixed above the working table in a vertical position (Fig. 5). The vision is transferred to the second camera when the position error becomes less than 10 pixels (≈ 12.3 mm) and the orientation error is less than 5°. Using the second camera, the accepted position and orientation error are set equal to 10 pixels (≈ 1.389 mm) and 1°, respectively. A simple gripper has been designed, so that the arm of the robot is distant from the fabric, as shown in Fig. 5. The fabric is stuck on the gripper to avoid slipping; however, the placement of the gripper onto the fabric is out of the reach of this work.

In this work, buckling modes during robot handling are supportable on the condition that it does not appear in the segment to be sewed. However, it still remains a problem that should be avoided. Thus, the position where the gripper touches the fabric is estimated in terms of reducing the possibilities for deformation appearance, taking into account the properties of the fabric (Koustoumpardis & Aspragathos, 2003). It is worth noting that the intent of this work deals with buckling in context of achieving a successful seam tracking and not the correction strategy against folding or wrinkling problems. The experimental tests also showed that deformations are likely to appear close to the gripper position on the fabric, and not on the edge.

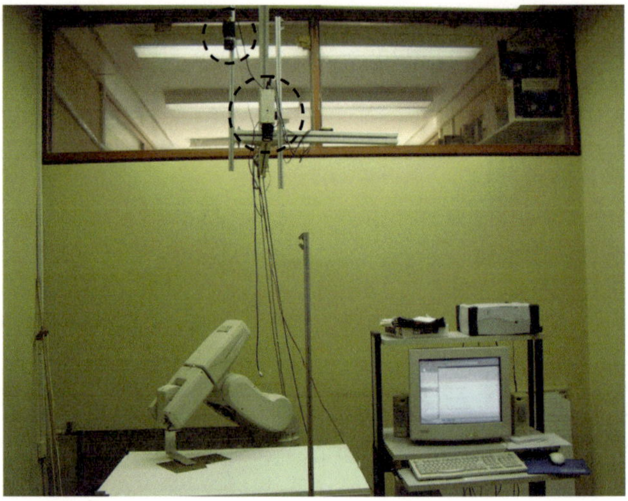

Fig. 5. The experimental stage

Since there is no standardization for deformations, it is difficult to be quantified. However, the greater deformation observed in buckling is about 30 mm, which is the height between the highest and the lowest point of the fabric. In some cases, fabrics were 'partially folded'. In other cases, the wrinkles induced the fabric to fold due to its own weight forming a loop at one side of the fabric. In some experiments, the fabric presented simultaneously both wrinkles, and folds.

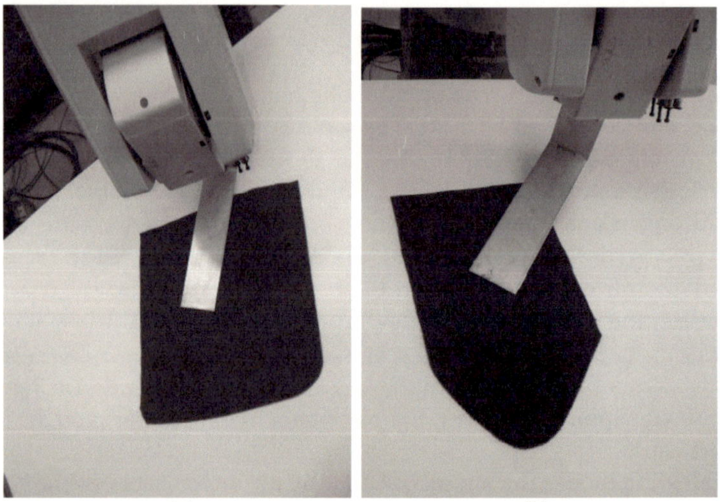

(a) (b)

Fig. 6. Fabric A with curved edge of small curvature (a) Fabric B with curved edge of large curvature

In the experiments, a considerable number of fabric pieces of different materials, shape and colour have been used. Both simulation and experimental tests are conducted to verify the efficiency of the proposed approach. A fabric piece is indicatively presented to show the results for both simulation and experimental tests. The shape of this fabric consists of two straight-line segments and an arbitrary curve, its colour is red and it has medium bending rigidity (Fig. 6 (a)).

Initially, simulation tests were carried out in order to approve the effectiveness of the proposed polygonal approximation based on a micro-Genetic Algorithm (micro-GA). A number of points along a section of the curved edge are captured by the camera that focuses on the area of the needle-point. The array that contains the curve data-points has dimensions 185×2 and is the input to the micro-GA. The maximum length of the chromosomes is set to 6 and the maximum acceptable deviation is arbitrarily set to 8 pixels (≈1 mm).

Fig. 7 (a) shows the optimum polygon section resulted from the micro-GA, which approximates the given data-points of the curved edge. The curve section is approximated by a polygon section consisted of three sides and each of them deviates from the corresponding arc section 6.87, 5.55 and 7.07 pixels. Increasing the maximum acceptable deviation from 8 pixels to 12 pixels, the arc section is approximated by a polygon section with three sides, as shown in Fig. 7 (b), where the maximum deviation errors between the curve and each polygon side are 4.48, 5.88 and 10.37 pixels, respectively. Decreasing the maximum acceptable deviation to 4 pixels, the arc is approximated by a polygon section with six sides, as shown in Fig. 7 (c), where the maximum deviation errors are 3.37, 2.52, 2.32, 0.88, 3.15 and 3.08 pixels. Fig. 7 (d) zooms in the first two sides of the polygon shown in Fig. 7 (c).

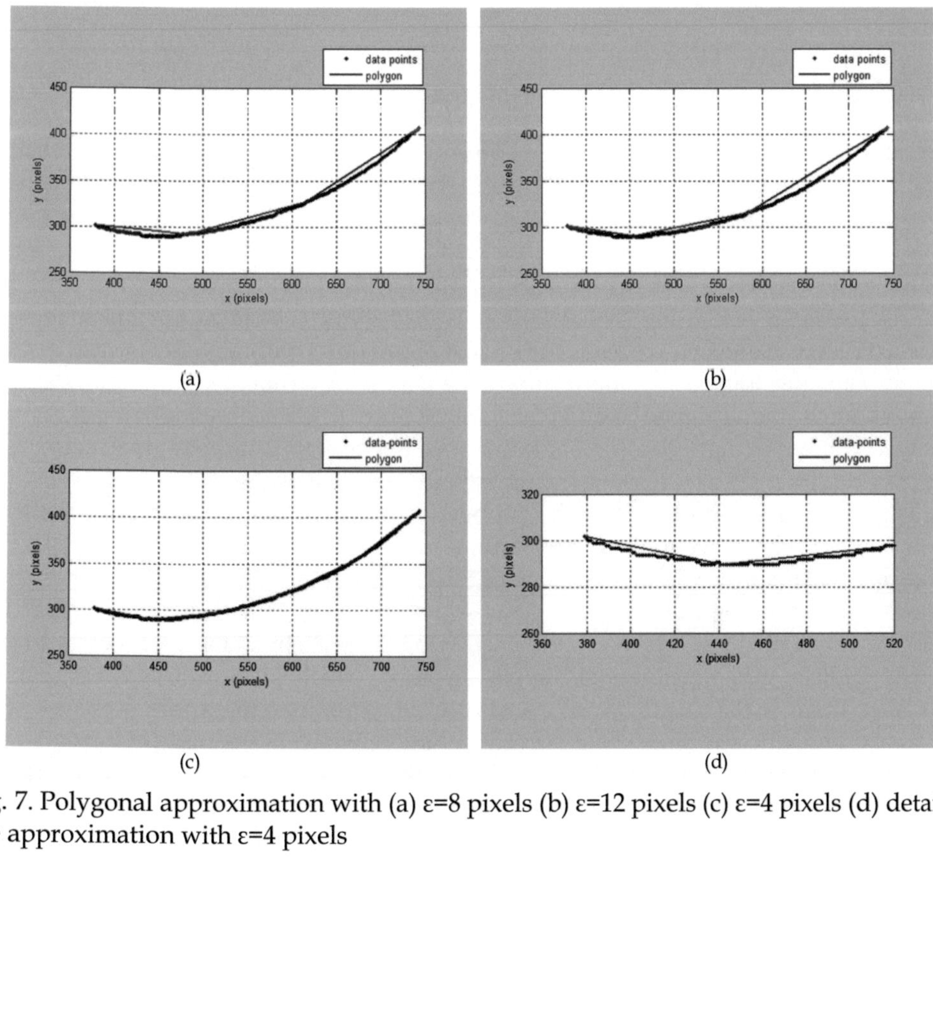

Fig. 7. Polygonal approximation with (a) ε=8 pixels (b) ε=12 pixels (c) ε=4 pixels (d) detail of the approximation with ε=4 pixels

These simulation tests demonstrate that the polygon approximation of the curve serves as a trade-off between rapidity and smoothness affected by the tolerance for imprecision. The results show that the approximation leads to a satisfactory seam approximation, while simultaneously the time for the entire process is minimized.

In practice, the sewing process is repeated many times and the deviation errors are collected and processed. The maximum acceptable deviation error for all tested cases is set to 8 pixels (≈1 mm). Fig. 8 (a) shows the deviation errors for Fabric A, a fabric piece with small curvature (Fig. 6 (a)) and Fig. 8 (b) shows the deviation errors for Fabric B, a fabric piece with small curvature (Fig. 6 (b)).

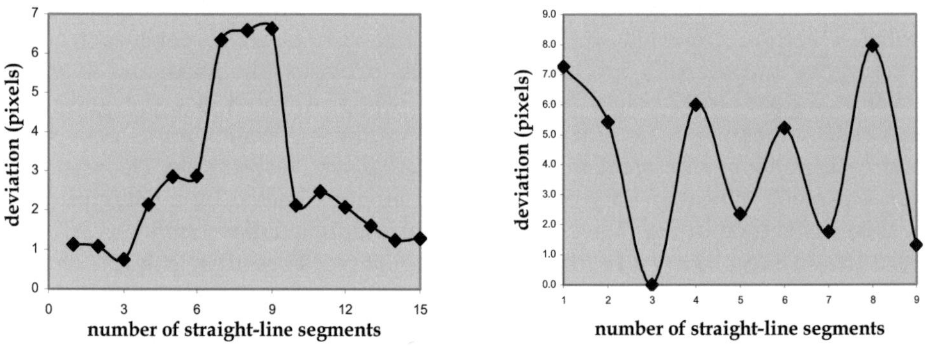

(a) (b)

Fig. 8. Deviation between real and desired path using the micro-GA for (a) fabric A (b) fabric B

Fabric A: The proposed micro-GA is experimentally tested using the dominant points as input. The maximum deviation (in pixels) between the needle-point and polygon approximation is computed from the image captured. The sewing process for the curved edge is accomplished after 15 steps and the results are shown in Fig. 8 (a). The maximum deviation is 6.63 pixels (≈0.83 mm) and is lower than the maximum acceptable limit of 8 pixels. The average value for the deviation is 2.75 pixels (≈0.34 mm), which is really satisfactory. The steps 6-10 correspond to the part of the curve with high curvature.

Fabric B: Using the dominant points as input, the micro-GA terminates with 9 steps and the average value for the deviation is 4.14 pixels (≈0.52 mm). The maximum deviation for each method is depicted in Fig. 8 (b). The deviation error for this fabric piece is greater compared to the previous one, which is reasonable, since the curvature is higher.

More fabric pieces are used to test the proposed approach, but detailed results are not presented due to the space limit. Table 2 shows indicatively the total steps, as well as the average and maximum deviations for some of the tested fabrics. The task is accomplished in less steps (corresponding to lower total time) when the micro Genetic based approach that uses the dominant points as input is applied, satisfying the predefined the boundary of 1 mm.

In apparel industry, the maximum allowable deviation, which is empirically evaluated, lies in the range between 0.5-5 mm. For all tested cases, the proposed algorithm is proved to be quite robust and efficient, since the achieved deviation, which is the maximum predefined deviation set by the user, is less than 1 mm. Bearing in mind that the robot accuracy is within ± 0.03 mm, the deviations resulting from the vision- and the robot position errors are very satisfactory.

Fabric type/color	Steps	Maximum deviation (mm)	Average deviation (mm)
fleece/red	15	0.83	0.34
cotton/green	9	1.00	0.52
jeans/blue	10	0.99	0.65
baize/green	10	0.79	0.34
tweed	10	0.96	0.58

Table 2. Experimental results for the tested fabrics

(a) (b)

Fig. 9. (a) Robot-camera-sewing machine system (b) zoom in the camera

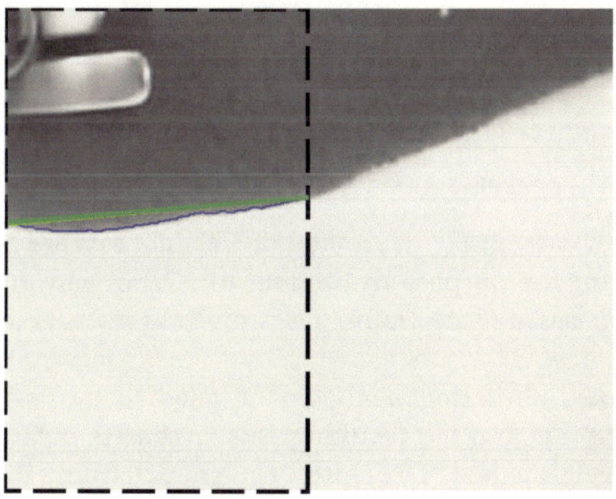

Fig. 10. Identification of a curved section

To test the ANFIS system, the robot-camera-sewing system is restructured to achieve even better accuracy (Fig. 9). Now, the vision system consists of a web camera at 480×640 pixels resolution RGB, which is mounted on the sewing machine, so that the area in the vicinity of the sewing needle is in the field of view of the web camera. The area captured by the web camera is about 40×55 mm, but the seam errors (e_D and e_φ) are computed in a smaller area

40×25 mm for a better approximation. Fig. 10 presents indicatively a curved section of a fabric, which is identified and approximated by a straight edge.

The proposed seam control strategy is tested using real fabric pieces of different materials, shapes and colours. Fig. 11 shows four fabric pieces with curved edges that were used for training and Fig. 12 shows three fabric pieces that were used for testing. Several experimental tests are conducted to verify the efficiency and approve the effectiveness of the proposed adaptive neuro-fuzzy approach for robot sewing fabrics of curved edges.

(a) (b) (c) (d)

Fig. 11. (a) jacket's sleeve (b) shirt's sleeve (c) trouser's leg (d) short sleeve

(a) (b) (c)

Fig. 12. (a) skirt's piece (b) pocket (c) hood

Using the fabric pieces shown in Fig. 11, a total of 350 data sets are obtained. A total of 300 data sets are selected for the purpose of training in ANFIS and the rest 50 data sets are selected for testing purposes after the training is completed in order to verify the accuracy of the predicted values.

Next, the genetic-oriented clustering method is applied to the training data set. Fig. 13 shows the training data sets and the resulting cluster centers obtained after applying the genetic-oriented clustering method. The cluster centers determine the number of the fuzzy sets and the parameters (mean values) μ of the Gaussian membership functions of the antecedent part, as well as the number of fuzzy rules of the Sugeno-type FIS. The number of the resulted clusters for r_a=0.4 is seven. As a result, each input variable is characterized by seven fuzzy sets with the linguistic values {Extremely Small (ES), Very Small (VS), Small (S), Medium (M), Large (L), Very Large (VL), Extremely Large (EL)}. The consequent parameters of each rule of the Sugeno-type FIS are determined by using the linear least-squares algorithm. The membership functions for the two inputs resulting from these cluster centers are shown in Fig. 14 (a). The rule base obtained through the genetic-oriented clustering approach consists of 7 rules, shown in Table 3.

1. If (e_D *is Small*) and (e_φ *is Extremely Small*) then (ω *is Extremely Small*) (1)
2. If (e_D *is Very Small*) and (e_φ *is Small*) then (ω *is Very Small*) (1)
3. If (e_D *is Extremely Small*) and (e_φ *is Extremely Large*) then (ω *is Small*) (1)
4. If (e_D *is Large*) and (e_φ *is Very Small*) then (ω *is Medium*) (1)
5. If (e_D *is Medium*) and (e_φ *is Large*) then (ω *is Large*) (1)
6. If (e_D *is Very Large*) and (e_φ *is Medium*) then (ω *is Very Large*) (1)
7. If (e_D *is Extremely Large*) and (e_φ *is Very Large*) then (ω *is Extremely Large*) (1)

Table 3. Fuzzy rule base

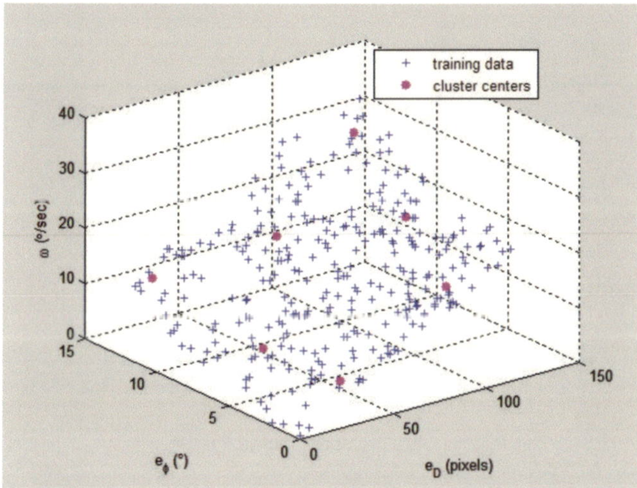

Fig. 13. Cluster centers resulting from the genetic-oriented clustering method

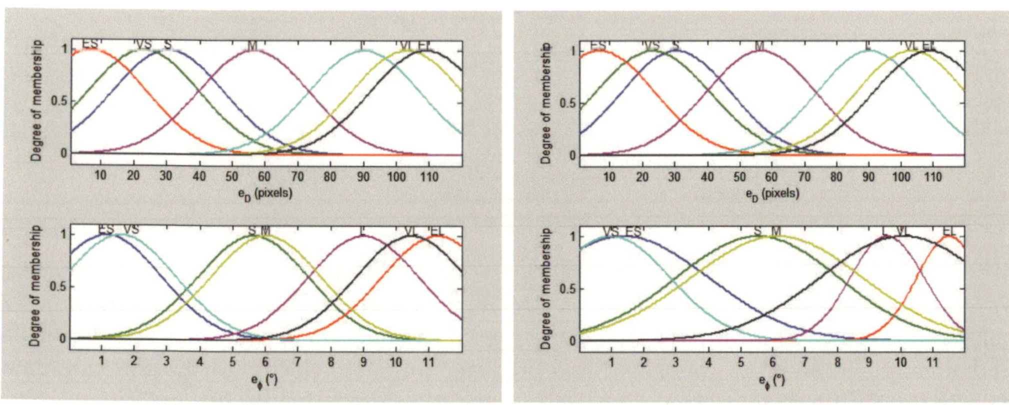

(a) (b)

Fig. 14. The membership functions for inputs using (a) genetic-based clustering (b) ANFIS

The next step is the training process that aims at tuning the fuzzy inference system. Fig. 14 (b) shows the final Gaussian membership functions derived after training the system. In contrast to the first input, there is a considerable change in the final membership functions concerning the second input, since the supports of all fuzzy sets are broadened. The root mean squared errors of the output over 100 training epochs, which are obtained by using 300 training datasets and 50 checking data sets, are plotted in Fig. 15. It is obvious that both training and checking error gradually decrease versus the number of epochs.

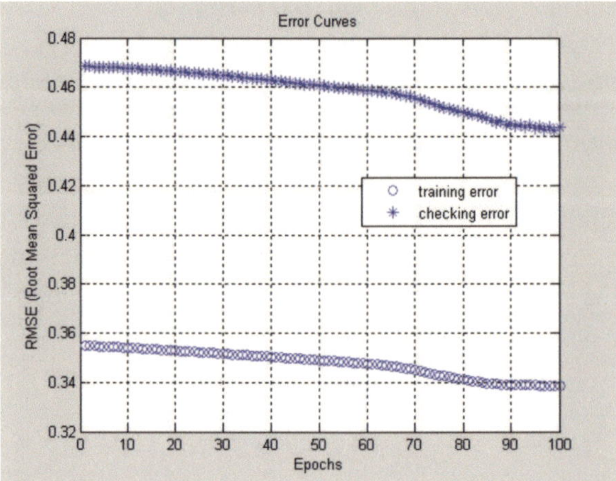

Fig. 15. ANFIS training process

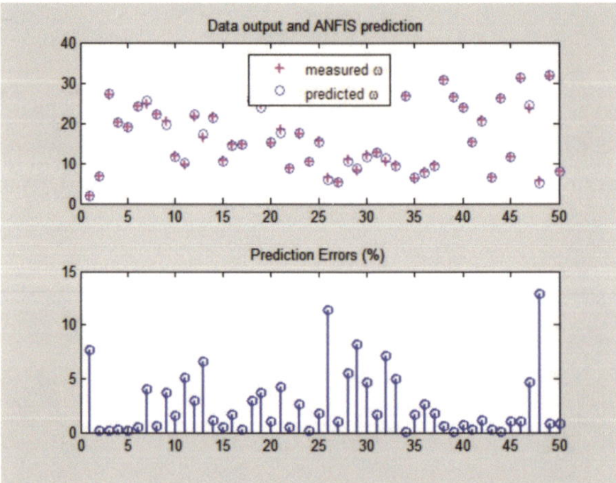

Fig. 16. Comparisons between measured & predicted angular velocity

After 50 checking data sets are entered in ANFIS, an output value can be obtained from the calculation results. This output value is the predicted value for the angular velocity of the end-effector. Fig. 16 shows measured and predicted angular velocity for the checking data, as well as the prediction error expressed as the absolute value of the percentage error between the measured and predicted angular velocity. These two diagrams demonstrate that the predicted values are close to the experimentally measured values, as many of the data points fall very close to the predicted points, indicating good correlation. The average error of the prediction of the angular velocity is around 2.57%, which means that the accuracy is as high as 97.43%.

An interesting and important conclusion established from the results is that the proposed approach is capable of well estimating data points outside the training space using the advantages of fuzzy logic, neural networks and genetic algorithms. This conclusion reflects the model's ability to predict the output based on the input data used for training. In practice, this implies that the method is effective in robot handling fabrics with curved edges, which have not been used in the training process.

The strategy for robot-sewing fabrics with curved edges described above has the advantage of performing well regardless of the fabric's deformations. This advantage is of major importance, as fabrics are limp materials that have a tendency to distort and change their shape when handled on a work table. The fact that this approach is based on the information taken from the image that captures the fabric in the neighbourhood of the needle amplifies the system's robustness against deformations. Therefore, deformations that may appear on the fabric during robot handling do not affect the effectiveness of the stitching process.

7. Conclusion

The main focus of this work lies on the design of an innovative visual servoing manipulator controller based on Fuzzy Logic that aims to enable robots to handle flexible fabric sheets lying on a work table. Visual servoing and fuzzy logic are used in robot motion control increases the intelligence of robots and the flexibility of the system. The designed visual servoing control system can deal with a variety of fabrics that are likely to deform and can cope with possible deformations due to buckling (wrinkling, folding) towards handling without degrading its performance, on condition that the seam segment to be sewed is undistorted. The desired accuracy is achieved in approaching the needle even in cases where deformations appeared.

This work focuses on the difficult case, where fabrics consist of edges with arbitrary curvatures. The need for approximating the curved edges through straight lines arises from the fact that current industrial robots can only be programmed to make straight or circular motions. From the standpoint of apparel industry manufacturing, it is important to assure that the fabric is sewn in the minimum time, while simultaneously satisfactory accuracy and smoothness are achieved.

In our approach, the curve is approximated through small straight segments applying a micro-GA approach that uses the dominant point detection method as input. The proposed approach aims at the minimization of the execution time satisfying a predefined maximum tolerance. Based on the results, some consideration is made concerning the trade-offs between running times and the quality of the final approximation.

To alleviate the computational burden of geometrical computations, an innovative method for robot handling fabrics with curved edges towards sewing has been proposed, which is based on a novel genetic-oriented clustering method and an adaptive neuro-fuzzy inference system. This work presents the design and tune of an ANFIS network with the minimum number of fuzzy rules for modelling the complex process of robot sewing fabrics. The proposed adaptive neuro-fuzzy inference system benefits from the advantages of fuzzy logic, neural networks and genetic algorithms. This feature makes this approach a powerful

tool to deal with uncertainty embedded in the curved edges of real cloth parts and to cope with new fabric pieces that have not been used for the training process.
The experimental results show that the proposed control scheme is effective and efficient in guiding the fabric towards the sewing needle, sewing it and rotating it around the needle. After extensive experimentation, it has been proved to be rather simple, flexible and robust due to its capability to respond to any position and orientation error for a variety of fabrics that are likely to present deformations. The experimental data were obtained using the robot-camera-sewing machine system and real parts of cloths. The proposed method presented good results when applied to fabrics with curved edges of unknown curvatures, which manifest the validity of proposed approach. This method is applicable to any piece of fabric with edges of arbitrary curvature, since it has been proved to be efficient in estimating the appropriate angular velocity of fabrics that were not included in the training process. Although no direct comparison with human stitching is possible, as the equipment deployed attained differ, the achieved accuracy is really promising for future use in industrial applications.

8. Acknowledgment

The author would like to acknowledge Robotics Group of Mechanical Engineering & Aeronautics Dep., University of Patras (http://robotics.mech.upatras.gr/www/) for help and cooperation.

9. References

Amin-Nejad, S.; Smith, J.-S. & Lucas, J. (2003). A visual servoing system for edge trimming of fabric embroideries by laser, *Mechatronics*, Vol.13, No.6, pp.533–551.

Delgado, M.R.; Von Zuben, F. & Gomide, F. (2001). Hierarchical genetic fuzzy systems, *Information Sciences*, Vol.136, No.1, pp.29–52.

Gershon, D. & Porat, I. (1986). Robotic sewing using multi-sensory feedback, *Proceedings of the 16th International Symposium of Industrial Robots*, Brussels, pp. 823-34.

Gershon, D. & Porat, I. (1988). Visual servo control of a robotic sewing system, *Proceedings of the IEEE International Conference on Robotics and automation*, pp.1830–1835, Philadelphia, PA, USA.

Gershon, D. (1990). Parallel process decomposition of a dynamic manipulation task: robotic sewing, *IEEE Transactions on Robotics and Automation*, Vol.6 No.3, pp. 357-366.

Gershon, D. (1993), Strategies for robotic handling of flexible sheet material, *Mechatronics*, Vol.3 No.5, pp. 611-23.

Hui, N.B.; Mahendar, V. & Pratihar, D.K. (2006). Time-optimal, collision-free navigation of a carlike mobile robot using neuro-fuzzy approaches, *Fuzzy Sets and Systems*, Vol.157, No.16, pp.2171–2204.

Jang, J.-S.R. (1993). ANFIS: Adaptive network-based fuzzy inference system, *IEEE Transactions on Systems, Man and Cybernetics*, Vol.23, No.3, pp.665–685.

Jang, J.-S.R.; Sun, C.-T. & Mizutani, E. (1997). Neuro-Fuzzy and Soft Computing: A Computation Approach to Learning and Machine Intelligence, Prentice Hall, Upper Saddle River.

Koustoumpardis P.; Zacharia P. & Aspragathos N. (2007). Intelligent robotic handling of fabrics for sewing, Industrial Robotics: Programming, Simulation and Applications, pIV pro literature Verlag Robert Mayer-Scholz, Chapter 28, pp.559-582, February.

Koustoumpardis, P. & Aspragathos, N. (2003). Fuzzy logic decision mechanism combined with a neuro-controller for fabric tension in robotized sewing process, *Journal of Intelligent and Robotic Systems*, Vol.36 No.1, pp.65-88.

Koustoumpardis, P.; Zoumponos, G.; Zacharia, P.; Mariolis, I.; Chatzis, I.; Evangelidis, G. & Zampetakis, A. (2006). Handling of non-rigid materials (XROMA), application in robotic sewing, *37th International Symposium on novelties in Textiles*, pp.528-533, Ljubljana, Slovenia, June 15–17, 2006.

Krishnakumar, K. (1989). Micro-genetic algorithms for stationary and non-stationary function optimization. *SPIE Intelligent Control and Adaptive Systems*, Vol.1196, pp.289–296.

Kudo, M.; Nasu, Y.; Mitobe, K. & Borovac, B. (2000). Multi-arm robot control system for manipulation of flexible materials in sewing operation, *Mechatronics*, Vol.10, No.8, pp.371–402.

Männle, M. (2001). FTSM—Fast Takagi-Sugeno Fuzzy Modeling, Institute for Computer Design and Fault Tolerance, University of Karlsruhe, Karlsruhe.

Marichal, G.N.; Acosta, L.; Moreno, L.; Méndez, J.A.; Rodrigo, J.J. & Sigut, M. (2001). Obstacle avoidance for a mobile robot: a neuro-fuzzy approach, *Fuzzy Sets and Systems*, Vol.124, No.2, pp.171–179.

Michalewitz, Z. (1996). Genetic Algorithms + Data Structures = Evolution Programs, 3rd edition, Springer-Verlag.

Moulianitis, V.; Dentsoras, A. & Aspragathos, N. (1999). A knowledge-based system for the conceptual design of grippers for handling robots, *Artificial Intelligence for Engineering Design, Analysis and Manufacturing*, Vol.13, No.1, pp. 3-25.

Rizzi, A.; Mascioli F.M.F. & Martinelli G. (1999). Automatic training of ANFIS networks, IEEE International Fuzzy Systems Conference, Seoul, Korea, Vol.3, pp.1655-1660, ISBN: 0-7803-5406-0.

Rusu, P.; Petriu, E.M.; Whalen, T.E.; Cornell, A. & Spoelder H.J.W. (2003). Behaviour-based neuro fuzzy controller for mobile robot navigation, *IEEE Transactions on Instrumentation and Measurement*, Vol.52, No.4, pp.1335–1340.

Sugeno, M. & Kang, G.T. (1988). Structure identification of fuzzy model, *Fuzzy Sets and Systems*, Vol.28, No.1, pp.15–33.

Takagi, T. & Sugeno, M. (1985). Fuzzy identification of systems and its applications to modeling and control, *IEEE Transactions on Systems, Man and Cybernetics*, Vol.15, No.1, pp.116–132.

Teh, C. H. & Chin, R. T. (1989). On the detection of dominant points in digital curves, *IEEE Transactions on Pattern Analysis and Machine Intelligence*, Vol.11, No.8, pp.859–872.

Torgerson, E. & Paul, F.-W. (1988). Vision-guided robotic fabric manipulation for apparel manufacturing, *IEEE Control Systemss Magazine*, Vol.8, No.1, pp.14–20.

Wu, W.-Y. (2002). A dynamic method for dominant point detection, *Graphical Models*, Vol.64, No.5, pp.304–315.

Yin, P.-Y. (2003). Ant colony search algorithms for optimal polygonal approximation of plane curves, *Pattern Recognition*, Vol.36, No.8, 1783–1797.

Yin, P.-Y. (2004). A discrete particle swarm algorithm for optimal polygonal approximation, *Journal of Visual Communication and Image Representation*, Vol.15, No.2, pp.241–260.

Zacharia P.Th. (2010). An adaptive neuro-fuzzy inference system for robot handling fabrics with curved edges towards sewing, *Journal of Intelligent and Robotic Systems*, Vol.58, No.3, pp.193-209.

Zacharia, P.; Mariolis, I., Aspragathos, N. & Dermatas, E. (2005). Visual servoing of a robotic manipulator based on fuzzy logic control for handling fabric lying on a table, *I*PROMS Virtual Conference on Innovative Production Machines and Systems*, , pp. 411–416, July 4–15 ,2005.

Zacharia, P.; Mariolis, I., Aspragathos, N. & Dermatas, E. (2006). Visual servoing controller for robot handling fabrics of curved edges, *I*PROMS Virtual Conference on Innovative Production Machines and Systems*, pp. 301–306, July3–14, 2006.

Zacharia, P.Th. & Aspragathos, N.A. (2008), Genetically oriented clustering using variable length chromosomes, *I*PROMS Virtual International Conference on Intelligent Production Machines and Systems*, pp.204-209, July 1-14, 2008.

Zacharia, P.Th.; Mariolis, I.G.; Aspragathos, N.A. & Dermatas E.S. (2008). Robot handling fabrics with curved edges based on visual servoing and polygonal approximation, *Special Issue of Proceedings of the Institution of Mechanical Engineers, Part B, Journal of Engineering Manufacture*, Vol.222, No.10, pp. 263-1274.

Zacharia, P.Th.; Mariolis, I.G.; Aspragathos, N.A. & Dermatas E.S. (2009). A robotic system based on fuzzy visual servoing for handling flexible sheets lying on a table, *Industrial Robot*, Vol. 36, No.5, pp.489-496.

Zhang, H. & Guo, J. (2001). Optimal polygonal approximation of digital planar curves using meta heuristics, *Pattern Recognition*, Vol.34, No.7, pp.1429–1436.

Zoumponos, G. & Aspragathos, N. (2008). Fuzzy logic path planning for the robotic placement of fabrics on a work table, *Robotics & Computer-integrated Manufacturing*, Vol.24, No.2, pp.174-186.

5

Robotic Systems for Radiation Therapy

Ivan Buzurovic[1], Tarun K. Podder[2] and Yan Yu[1]
[1]*Thomas Jefferson University, Philadelphia, Pennsylvania,*
[2]*East Carolina University, Greenville, North Carolina*
USA

1. Introduction

Medical robotics is an exciting and relatively new field. Robotics plays an important role in medical engineering. Medical robots were initially used in the 1980s, in the field of urology. Robotic arms were developed and used for prostate resection. They can also be highly specialized and assist in diagnosing and treating patients. While there is still much more work to be done, using robots can enhance medical treatments in terms of both the quality and accessibility of care. Using robots can help reduce human error and bring highly specialized information to remote areas without requiring physicians' direct intervention.

In radiation therapy, high-energy radiation from x-rays, gamma rays, neutrons, and other sources has been used to kill cancer cells and shrink tumors. Radiation may come from a machine outside the body (external-beam radiation therapy), or it may come from radioactive materials placed in the body near cancer cells (internal radiation therapy, implant radiation, or brachytherapy).

The usage of robotic systems to improve the cancer treatment outcome is a new field. This field overlaps with electronics, computer science, artificial intelligence, mechatronics, nanotechnology, and bioengineering. For this purpose, robots can be used in medical facilities to perform different tasks such as delivering radiation sources, real-time tumor tracking during radiation delivery or external beam delivery.

The only product in the market for robotic radiotherapy is CyberKnife Robotic Radiosurgery System. The robotic system has provision for so-called real-time tracking during beam delivery. The device itself is a 6MV linear accelerator mounted on a six degree-of-freedom (DOF) Keller und Knappich Augsburg (KUKA) industrial robot. This system has real-time image-guided control. Consequently, there is a significantly long time delay (about 200 ms) between the acquisition of tumor coordinates and repositioning to the linear accelerator. The CyberKnife-KUKA robot with linear accelerator end-effector is suited for radiation therapy to any body sites. Its field size is restricted to the limited geometry of 12 discrete circular fields ranging from 5mm to 60mm in diameter. Therefore, the workspace is confined and the radiation therapy community has not fully embraced the idea of using an industrial articulated robotic manipulator yet.

The details about CyberKnife robotic system are not included in this chapter. Consequently, the basic idea is to present the novel research results in the field of robotic radiation therapy and its applications.

2. Brachytherapy robotics

Brachytherapy is a method of treatment in which sealed radioactive sources are used to deliver radiation at a short distance by interstitial, intracavitary, or surface application. With this mode of therapy, a high radiation dose can be delivered locally to the tumor with rapid dose falloff in the surrounding normal tissue.

Recently, several research groups have reported the investigation and development of the robotic systems for the prostate seeds implants, (Stoianovici et al., 2003), (Wei et al., 2004), (Fichtinger et al., 2006), (Yu et al., 2007), (Meltsner, 2007), (Meltsner et al., 2007), (Stoianovici et al., 2007), (Podder et al., 2007), (Salcudean et al., 2008), (Lin et al., 2008), (Moerland et al., 2008). The potential advantages of the robotic seed implants include improving the accuracy of the needle placement and seed delivery, as well as improving the consistency of the seed implant. In medical, especially prostate brachytherapy, applications of robots, precise end-effector position, steady state and positioning accuracy is required. In such applications, even small positioning errors at the manipulator end-effector can have dangerous and costly consequences, (Mavrodis et al., 1997). To achieve the enhancements of the robotic seed delivery, the robots need to be calibrated. Properly calibrated robotic systems have higher absolute positioning accuracy and the deposited seeds positions correspond better to the ones calculated in the planning systems. The brachytherapy robots are usually ultrasound (US) or magnetic resonance imaging (MRI) guided. Generally, to improve needle placement and seed deposition in brachytherapy procedure several methods have been presented in the literature, such as parameter optimization, different needle rotation techniques, robotic insertion, force modeling, and needle steering techniques. Robot assisted therapeutic delivery systems are attractive for several reasons. The potential advantages are increased accuracy, reduced human variability, reduction of clinician's fatigue and reduction of operation time.

There can be two methods for robotic needle insertion and seed deposition: single-channel approach and multi-channel approach. In single-channel approach one needle can be inserted at a time and typically 2-5 seeds along the needle track are deposited in the prostate according to dosimetry plan. On the other hand, multi-channel system is capable of placing several needles at the time. Thereby, seed delivery can be faster. Since prostate is not rigidly mounted, it can move and rotate as well as unpredictably deform. When several needles are inserted concurrently, prostate will be uniformly pushed back symmetrically to more stable position and deformation of the tissue can be better estimated for precise delivery.

In the following sections two brachytherapy robotic systems has been presented. Both robotic systems: single-channel and multi-channel systems have been designed, developed and manufactured in our research laboratories.

2.1 Single-channel brachytherapy robotic system

We have designed and developed a robotic system, named **E**ndo-**U**ro **C**omputer **L**attice for **I**ntratumoral **D**elivery, **I**mplantation, and **A**blation with **N**anosensing (EUCLIDIAN). The system consists of a surgical module, a positioning module and a electronic housing, as in Fig. 1.a. Kinematic Denavit-Hartenberg (DH) parameters of the system are presented in Fig.1.b.

The platform connects the surgical module to the cart. The platform has provision for both translational and rotational motion. The vertical lift of the surgery module is motorized for ease of operation against gravitational effect. The supporting platform connects the surgical module to the cart. The surgical module consists of two robotic manipulators, i.e., two open kinematic chains called needling mechanism and an ultrasound probe driver with five and two DOF, respectively.

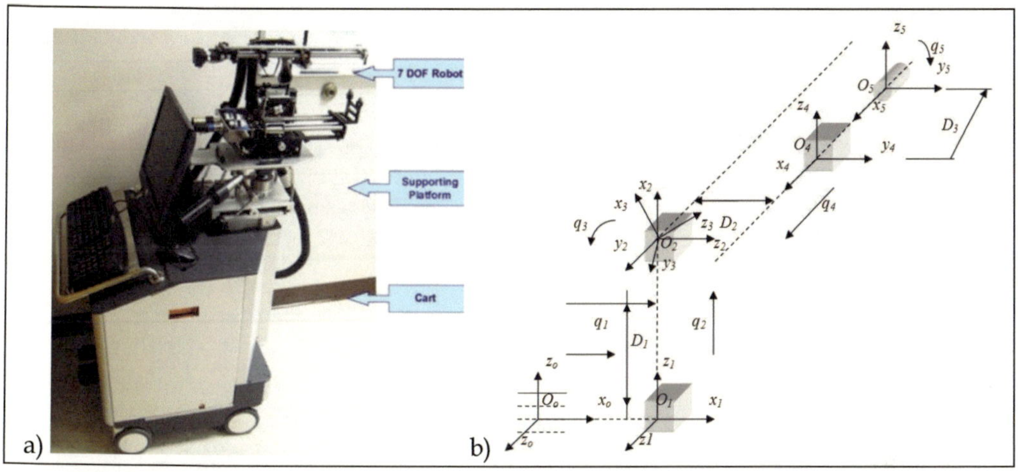

Fig. 1. a) EUCLIDIAN – image-guided brachytherapy robotic system; b) Kinematic DH schema with reference frames

2.1.1 System description

The three main subsystems of the surgical module, Fig. 2, are: 1) Two DOF transrectal ultrasound (TRUS) probe driver, three DOF robotic gantry, and two DOF needle inserter (*denote by needling mechanism* in Fig. 21.) with seed holder and seed pusher.

A brief description of these subsystems is provided in the following sections. The TRUS probe, which contains two transducers, transverse and sagittal planes can be translated and rotated separately by two motors fitted with high-resolution optical encoders. This enables imaging the prostate in transverse as well as in sagittal planes with variable slice thicknesses or intervals, as thin as 0.1 mm.

Working ranges of motion of the TRUS probe are 0–185 mm and -91° to +91° in translation and rotation, respectively. The TRUS probe can also be operated manually using the knobs; during this mode, the motors are disengaged automatically by electric clutches. There is a provision for a template holder at the end of the TRUS probe driver, enabling manual takeover if required. A pair of prostate stabilization needles can be placed with angulation in both sagittal and coronal planes to prevent the obstruction of robotic needle insertion. This approach has also been shown to produce significant improvement in prostate immobilization.

This subsystem, which features two translational motions, x and y direction, and one rotational motion (*pitch*, i.e., the rotation upward or downward about the x-axis), connects the needle driving module to the positioning platform. The motions are achieved by motors and optical encoders fitted with the motors, Figs. 1 and 2.

The range of motion is 62 mm - 67 mm in the x-y direction. The rotational range for angulating the needle is -5° to +5° to avoid pubic arch interference and to reach the target region close to the TRUS probe. The TRUS probe motion and the rest of the surgery module (gantry and needle driver) is decoupled, as they are two separate open kinematic chains, which allow independent motions of the TRUS probe and the needle.

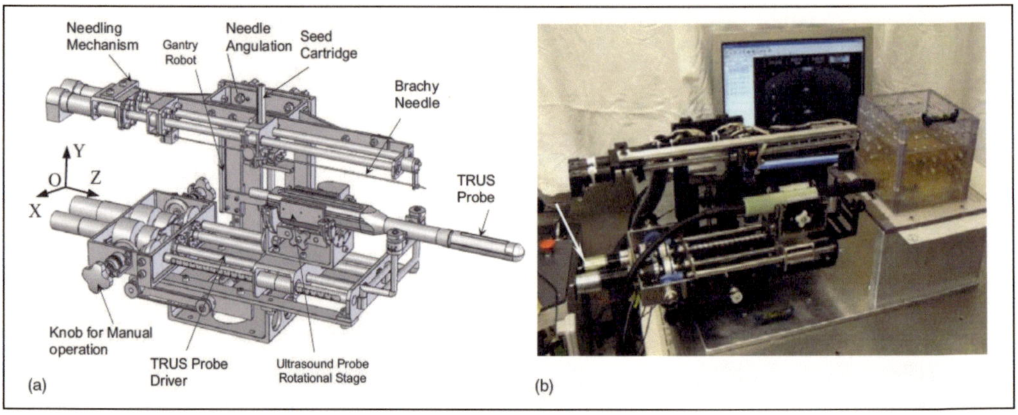

Fig. 2. Surgical module of EUCLIDIAN

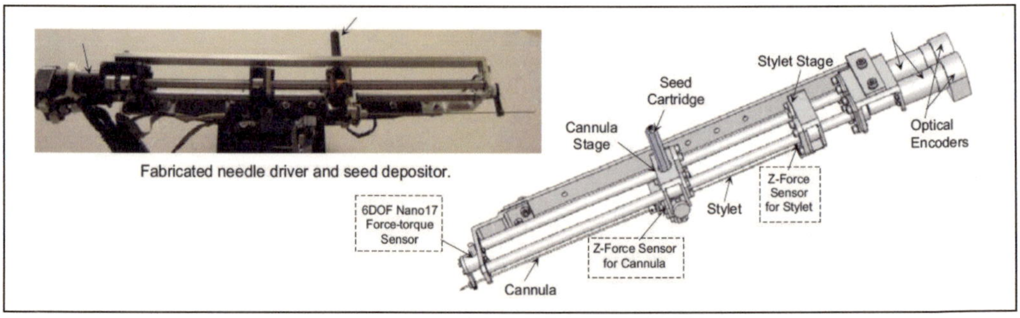

Fig. 3. Needle driver and seed depositor of EUCLIDIAN robot

The needle driver is the most complex subsystem of the EUCLIDIAN, Fig. 3, as the degree of complexity is increased with the introduction of three force-torque sensors, stylet sensor, cannula sensor, and whole needle sensor for cannula rotation and autonomous seed delivery system. The main components of this subsystem are: 1) stylet driver, 2) cannula driver, 3) seed-cartridge holder, and 4) force-torque sensors. The cannula and the stylet are driven separately by two motors. The travel ranges of both the cannula and the stylet are 0–312 mm. During the surgical procedure, the motions of the cannula as well as of the stylet are autonomous in accordance with the treatment plan; however, the clinician must approve the movements at critical points. Our custom-designed seed cartridge can hold 35 radioactive seeds. Accommodation of additional seeds posed some challenges for smooth functioning of the seed cartridge due to spring nonlinearity. However, after careful investigation and adjustments, the seed cartridge is working satisfactorily.

The seed pusher, a flat ended stylet, is deployed to expel the seed out of the cartridge and to deposit it at the planned location. Every motion during the sequence of seed delivery is fully automatic; however, the clinician is able to interrupt and/or manipulate the movements at any time using a hand pendant. By monitoring the force data from the sensor installed at the proximal end of the stylet, Fig. 4, the seed removal from the cartridge to the cannula can be verified. A ten-button hand pendant provides the clinician with the authority and freedom to assert control of the surgery module at any desired time. That is, the clinician can command each of the motors for manipulating the motion of various components of the surgical module, such as needle insertion, needle rotation, seed deposition, x-y movement of the gantry, and system abort.

The three DOF cart (two translations and a rotation about the vertical axis) provides gross movement of the robotic system to align with the patient, while the six DOF platform enables finer movement for desired positioning and orientation of the robot in three-dimensional (3D) space.

The computer, system electronics, and cable junctions are housed in the electronics housing, Fig. 1. The EUCLIDIAN's surgery module is fully autonomous; all the motors are fitted with high-resolution optical encoders and precision gear boxes. The robot is controlled by an industrial computer, which is proven to be robust and reliable for working in harsh industrial environments and military applications. It has a special metallic casing for minimizing electromagnetic interferences.

Two Galil control cards, Model DMC- 1842; Galil Motion Control, Inc., Rocklin, CA, are used: One card to control the TRUS probe driver and gantry motions and the other card to control the needle driver and the seed pusher. A robust and stable proportional, integral and derivative (PID) controller has been developed for controlling the motorized surgical module. We have tuned the PID gains in such a manner so that the system's stability is maintained when the needle changes its states from merely position control in the air to both position and force control mode in the patient's body. The needle can be driven at a maximum velocity of approximately 100 mm/s; however, a lower velocity, 60 mm/s, setting is used as the default. A frame grabber, FlashBus, Integrated Technologies, Indianapolis, IN, is used for TRUS image capturing. Three force-torque sensors , Nano17, ATI Industrial Automation, Apex, NC and M13, Honeywell, Morristown, NJ, are used for needle insertion force monitoring and robot control feedback. Each of the motors is fitted with an optical encoder, MicroMo Electronics, Inc., Faulhaber Group, Clearwater, FL, which can provide final motion resolutions, considering gear ratios and screw leads, of 0.0007 mm for gantry x-y translations, 0.004 mm for stylet and cannula motions, and 0.005 mm and 0.06° for TRUS probe translation and rotation, respectively.

2.1.2 Advanced control techniques

In the previous studies, it was concluded that the proper selection of the translational and rotational velocities may reduce tissue deformation and target movements by reducing insertion force, (Buzurovic et al., 2008.a). Therefore, it can be concluded that the insertion force has a dominant influence on the needle insertion, seed deposition precision, and dosimetry distribution in brachytherapy procedure. In our initial work, (Buzurovic et al., 2008.b) we have investigated tracking problem for the same robotic system using neural network approach. The force prediction model was introduced as well. In this article we have introduced novel methods to control the brachytherapy robot with the lowest tissue deformation and highest seed precision delivery as the ultimate goals. In order to achieve the desired dynamic behavior of the robotic system, the control strategies that we implemented in the robotic system have a force prediction module. For the first one we have used an artificial neural network (ANN) and neural network predictive control (NNPC), (Buzurovic et al., 2010.a).

In the following part the use of a feedforward model predictive control (MPC) was described. The purpose of this approach is to control the reactive force which is responsible for tissue displacement. Also, as the second control task was to predict and compensate the impact of the measured and unmeasured disturbances rather than waiting until the effect

appears at the output of the system on the other side. When the reactive force is minimized, tissue displacement decreases. The force prediction control was also used to minimize the effect of system time-delay. Because of the fact that this procedure required an accurate robot system model, it was necessary to obtain a more precise model. That is a reason for adopting the robotic system model as a singular system of differential equations.
Fig. 4 represents the contact surface where reactive force appears during surgery. A mathematical equation of contact surface is calculated using EUCLIDIAN robotic system software.

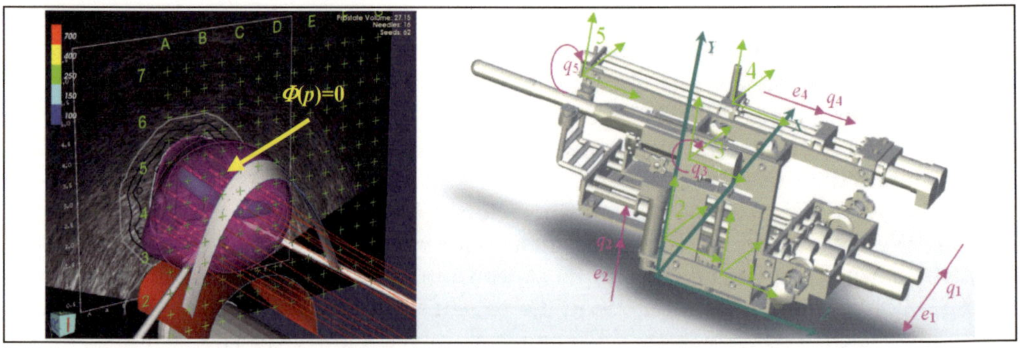

Fig. 4. a) 3D representation of the contact surface (prostate gland) during insertion. b) 5DOF surgical module. Absolute and local coordinate systems together with generalized coordinates q_i

Assume that the contact surface is a scalar function Φ: $R^m \rightarrow R^1$, Fig. 4a.

$$\Phi(p) = 0 \tag{1}$$

The contact force vector is:

$$f = D^T(p)\lambda \tag{2}$$

λ is a scalar multiplier for the constraint function and $D(p)$ is the gradient of the constraint function, described as in (3),

$$D(p) = \frac{\partial \Phi(p)}{\partial p} \tag{3}$$

In the previous equation $p \in \Re^3$ is a position vector from the fixed reference frame to the constrained surface. Additional constraint is $p = \Phi(q)$ at the contact point. The matrix function $J(q)$ is defined as the Jacobian matrix function.

$$J(q) = \frac{\partial \Phi(q)}{\partial q} \tag{4}$$

The influence of the contact force to the system can be described as

$$M(q)\ddot{q} + G(q,\dot{q}) = \tau + J^T(q) f \tag{5}$$

where $M(q)$ denotes the inertia matrix function and G is the vector function which describes coriolis, centrifugal and gravitational effects. q is a vector of the generalized coordinates and τ is the torque vector. Influence of the external contact forces is described by factor f. Combining equations (1)-(5) the mathematical model of the system is

$$M(q)\ddot{q} + G(q,\dot{q}) = \tau + J^T(q)D^T(M(q))\lambda(q,\dot{q},\tau) \tag{6}$$

After an appropriate transformation, equation (6) is transformed to its matrix form

$$\begin{bmatrix} I & 0 & 0 \\ 0 & M(q) & 0 \\ 0 & 0 & 0 \end{bmatrix} \begin{pmatrix} \dot{q} \\ \ddot{q} \\ \dot{\lambda} \end{pmatrix} = \begin{bmatrix} 0 & I & 0 \\ 0 & 0 & J^T D^T \\ 0 & 0 & 0 \end{bmatrix} \begin{pmatrix} q \\ \dot{q} \\ \lambda \end{pmatrix} + \begin{bmatrix} 0 \\ -G(q,\dot{q}) - M(q) \\ \Phi(p) \end{bmatrix} + \begin{bmatrix} 0 \\ I \\ 0 \end{bmatrix} \tau \tag{7}$$

System (7) is linearized at the surrounding point where the needle is inserted into the patient. The linearized mathematical model of the system is obtained, as follows

$$E\dot{x}(t) = Ax(t) + Bu(t) + d \tag{8}$$

with singular matrix $E \in \Re^{n \times n}$, and vector d which represents the unmeasured disturbances such as needle deflection, tissue deformation and displacement, $A \in \Re^{n \times n}$, $B \in \Re^{m \times r}$. x and u represent the system state-space vector and control vector, respectively, $x \in \Re^{n}$, $u \in \Re^{r}$.

The system defined by (8) is known as a singular system, descriptor system, semi-state system, system of differential-algebraic equations or generalized state space system. They arise naturally in many physical applications, such as electrical networks, aircraft and robotic systems, neutral delay and large-scale systems, economics, optimization problem and constrained mechanics. The matrices of linearized system (8) are defined as follows

$$A = \begin{bmatrix} 0 & I & 0 \\ \frac{\partial}{\partial q}(G - J^T D^T \lambda)|_0 & 0 & J^T D^T|_0 \\ DJ|_0 & 0 & 0 \end{bmatrix}, \quad E = \begin{bmatrix} I & 0 & 0 \\ 0 & M(q_0) & 0 \\ 0 & 0 & 0 \end{bmatrix},$$
$$B = \begin{bmatrix} 0 \\ I \\ 0 \end{bmatrix}, \quad u = \delta\tau, \quad d = \begin{bmatrix} 0 \\ \Delta\tau \\ 0 \end{bmatrix} \tag{9}$$

Now it is possible to find subspaces S and F, as in (Buzurovic et al., 2010.a), having in mind that for state space X the system equation (8) can be represented as the direct sum of S and F, $X = S \oplus F$. As a result of the matrix transformation **M** applied to system (8), the slow and fast subsystems can be described by a mathematical formulation

$$\begin{aligned} \dot{x}_s &= L_s x_s + B_s u + d_s \\ L_f \dot{x}_f &= x_f + B_f u + d_f \end{aligned} \tag{10}$$

with L_s = $\mathbf{M}A|S$, L_f = $\mathbf{M}A|F$, $B_s = P\mathbf{M}B$, $B_f = Q\mathbf{M}B$, $d_s = P\mathbf{M}d$ and $d_f = Q\mathbf{M}d$ for some linear transformations Q and P represented by matrices Q and P, respectively. For that case, the initial conditions are chosen to satisfy $x_{0s}=P\ x_0$ and $x_{0f}=Q\ x_0$. The solution of system (10) is $x=x_s+ x_f$

$$x_s = e^{L_s t}x_{0s} + \int_0^t e^{L_s(t-\tau)}B_s u(\tau)d\tau + \int_0^t e^{L_s(t-\tau)}d(\tau)d\tau, \quad x_f = -\sum_{i=0}^{v-1} L_f^i B_f u^i \tag{11}$$

Now it is possible to apply the MPC control, Fig. 5, and to investigate dynamical behavior of the system. Disturbance d is given to MPC and its effect is predicted and compensated before the effect appears at the system output.

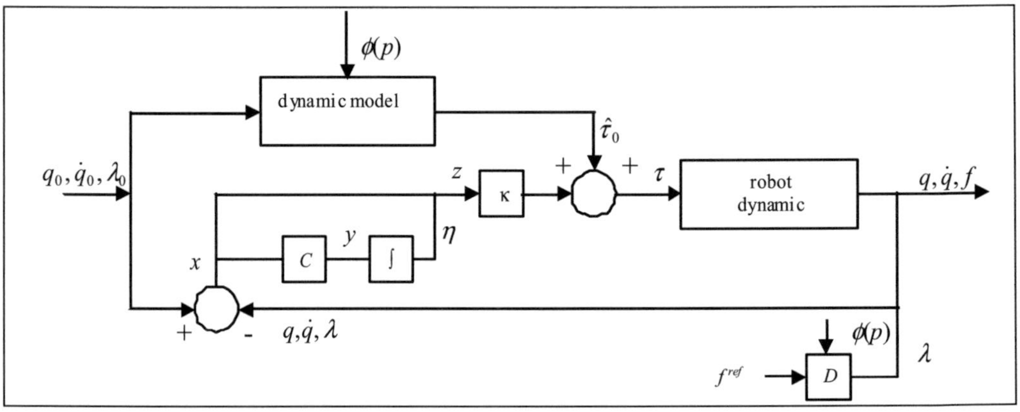

Fig. 5. Insertion force control scheme for the slow subsystem

In (Mills, 1988) is shown that the impulsive behavior of the system can be avoided using appropriate initial conditions, defined by $x_{0s}=P\ x_0$ and $x_{0f}=Q\ x_0$. By using the described approach the fast subsystem will not induce impulsive behavior. Moreover, it can be concluded as stated previously and from equation (11) that there is little need to find fast feedback to eliminate the impulsive behavior. The necessary task was to find an appropriate feedback for the slow subsystem. The control scheme for the slow subsystem is represented in Fig. 5, as suggested in (Cobb, 1983).

Furthermore, when the MPC controller is applied, the main objective is to hold the insertion force at the predefined acceptable reference value, by adjusting the control signal from actuators in order to minimize the prostate displacement, i.e. the reactive force which acts upon the tissue. Using this approach it is possible to decrease the insertion force during insertion trajectory. The needle displacement is an unmeasured disturbance and the controller provides feedback compensation for such disturbances. For the insertion force the controller provides feedforward compensation. Various noise effects can corrupt the measurements. The noise could vary randomly with a zero mean, or could exhibit a non-zero, drifting bias. The MPC uses a filtering method for removing estimated noise component. At the beginning of each sampling instant, the controller estimates the current system state. Accurate knowledge of the state improves prediction accuracy which improves controller performances.

During the insertion passive force control becomes active and keeps passive insertion force close to the predefined minimal value. When the MPC control approach was implemented it is possible to decrease the insertion force, as it is shown in Fig. 6. Also, peaks during the insertion are minimized. These conclusions give us a reason to believe that tissue deformation can be decreased better than using the traditional PID control.

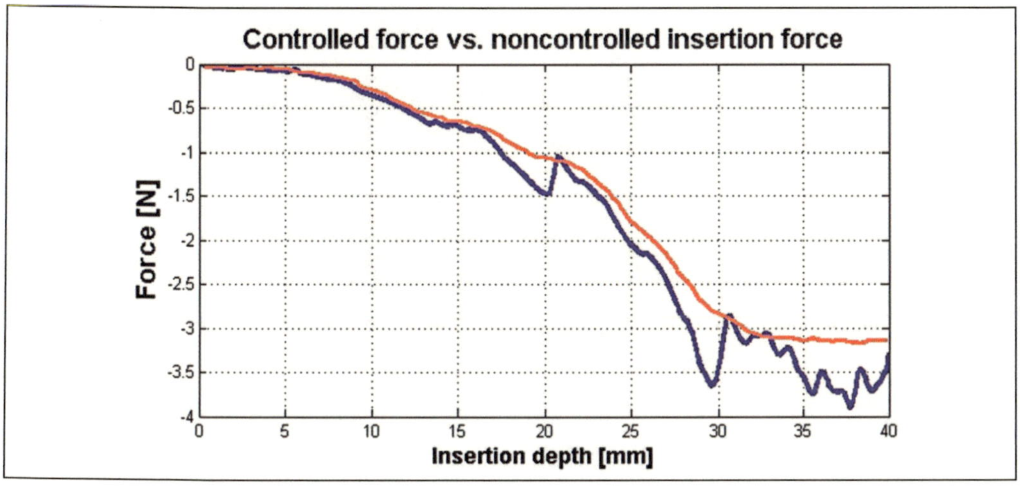

Fig. 6. Controlled force with MPC (red) and insertion force with PID controller (blue)

2.2 Multi-channel brachytherapy robotic system

Multi-channel system is capable of placing several needles or even all needles at a time and thereby, it can be faster in delivering the seeds required for the treatment. A multi-channel delivery system can effectively avoid the problem of gradual prostate swelling (i.e., edema) and deformation, which occurs while depositing the seeds by single needle. Since the prostate is not rigidly mounted, the prostate can move and rotate as well as deform quite unpredictably, at every time a needle is inserted. But, when several needles are inserted concurrently, the prostate will be uniformly pushed back symmetrically to a more stable position and the deformation can better be estimated for precise delivery of seeds. Thus, the multi-channel system can overcome some of the drawbacks that may be encountered by the single-channel robotic systems. In this part, we present our **M**ultichannel **I**mage-guided **R**obotic **A**ssistant for **B**rachytherapy (MIRAB), which is designed and fabricated for prostate seed implantation. Currently, MIRAB can simultaneously rotate and insert 16 needles. The MIRAB is capable of inserting more needles concurrently, if needle rotation is excluded.

2.2.1 System description

The MIRAB system shown in Fig. 7 consisted of five modules: (1) Rotary Needle Adapter, (2) Surgical x-y Carrier, 3) Mounting and Driving Mechanism, (4) Seed Applicator, and (5) Transrectal Ultrasound (TRUS) Driver, (Podder et al., 2010).

Rotary Needle Adapter can hold 16 needles with rotational capability. However, additional needles can be installed without provision for rotation. Two direct current (DC) motors rotate the needle using a spur- gear train. It is known that provision of needle rotation reduces the insertion force as well as organ (or target) deflection and deformation. For guiding the extended needles (brachytherapy needles are about 200mm in length and 1.27mm or 1.47mm in diameter) a regular brachytherapy template is installed at the distal end.

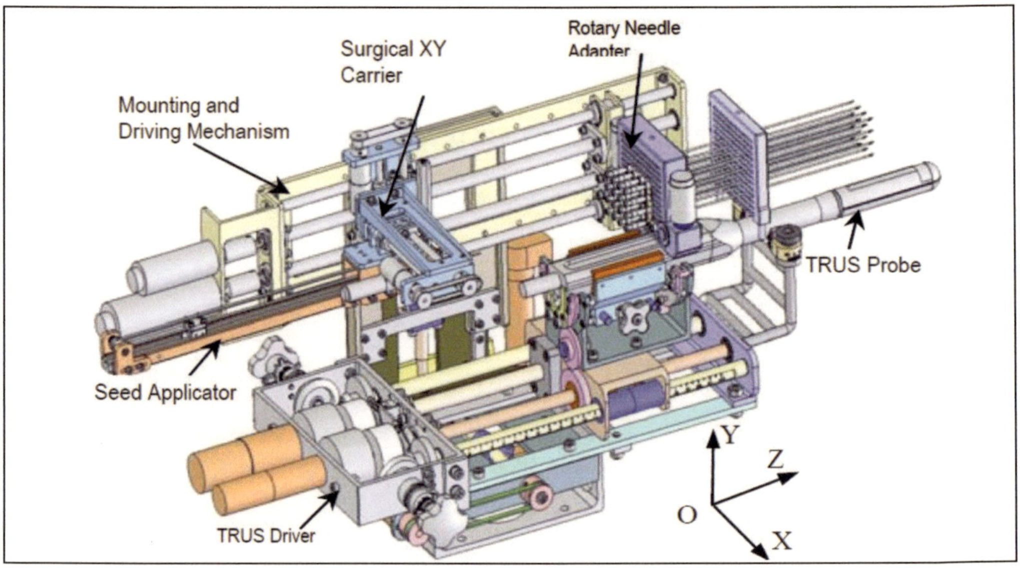

Fig. 7. MIRAB robotic system

The needles can be simultaneously inserted using one of the DC motors on the Mounting and Driving Module. Surgical x-y Carrier is 3-DOF module carries the Seed Applicator which delivers seeds and withdraws needle. Two DC servo motors on the x-y carrier provide motions to the Seed Applicator in x- and y-direction, while another DC servo motor on the Mounting and Driving module provide motion in the z-direction, i.e. the depth. Mounting and Driving Mechanism is driven my two DC servo motors to impart translational motion (along z-direction) to the Needle Adapter and Surgical x-y Carrier. Seed Applicator is a module which is attached to the Surgical x-y Carrier. A DC servo motor is employed to expel the seed from the seed cartridge. Transrectal Ultrasound (TRUS) Driver is a module was originally developed for EUCLIDIAN system, Fig. 2). However, the MIRAB is designed in a way so that it can be installed on the EUCLIDIAN by taking the needling module away from the EUCLIDIAN away. This interchangeability is of MIRAB and EUCLIDIAN will be very convenient to switch from a single-channel system to a multichannel system. TRUS can image the prostate and the relevant anatomies in transverse as well as sagittal planes.

All the DC servo motors are fitted with high-resolution (up to about 0.0007mm) optical encoder, MicroMo Electronics, Inc., Faulhaber Group, Clearwater, FL. An industrial computer, Plans, San Diego, CA, is used for controlling the whole system. Two Galil control cards, Model DMC-1842; Galil Motion Control, Inc., Rocklin, CA, are used. All the desired motions are achieved by deploying a Proportional, Derivative and Integral (PID) controller.

2.2.2 Experimental results

The purpose of the following experiments is to evaluate performances of multichannel robotic brachytherapy system designed for prostate seed implantation.

The developed multichannel robotic system is capable of inserting large number of needles concurrently and depositing seeds automatically.

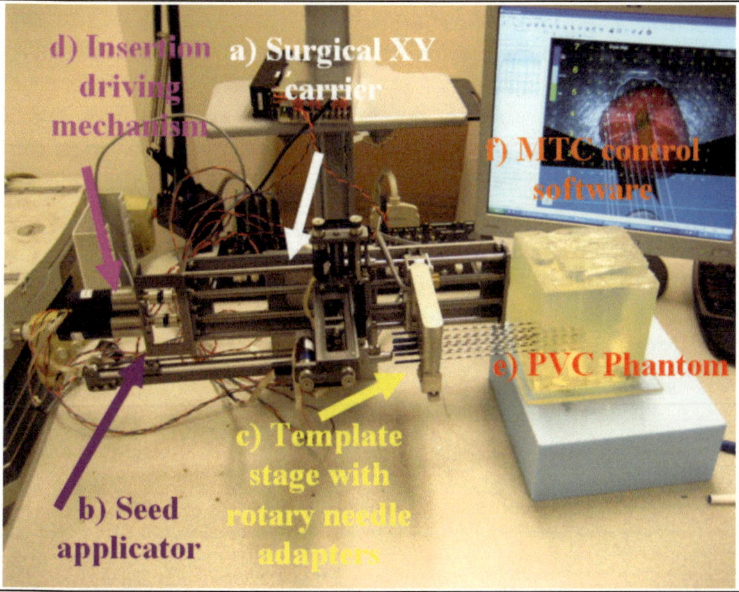

Fig. 8. MIRAB robotic system, experimental setup

The experimental procedure, Fig. 8, is described in the following part. For cannula, we used 18ga x 20cm Stainless steel BARD needle and for stylet Rev. A4, 304 Stainless steel needle Vita Needle Company. Phantom was prepared from polyvinylchloride (PVC) plastic and PVC+ hardener in the ratio 80% to 20%, respectively, MF Manufacturing, TX. Force measurements were performed using single-axis force sensor, Model 13, Honeywell Sensotech, Columbus, OH, installed on proximal end of the cannula. Deposited seeds were rounded stainless steel dummy seeds, BARD. Position of the tip of the needle and consequently depth of the deposition into the phantom was measured using optical encoders FAULHABER MicroMo HEDM5500J series 5500, attached to the template stage and seed applicator motors. The MIRAB was evaluated for insertion speed of 5mm/s, 10mm/s, 20mm/s, 40mm/s, 60mm/s and 80mm/s and stylet speed in the range of 20-60 mm/s. For each insertion speed we recorded force for 1, 2, 4, 8 and 16 needles installed together on the template stage.

Absolute seed placement error was 0.10mm (SD=0.11mm) in X and Y direction and 0.15mm (SD=0.12mm) in Z direction for plan with 16 needles and 64 seeds. Relative position error between seeds were 0.07mm (SD=0.05mm). It can be seen in Fig. 9.a that maximum insertion force for insertion speed of 40mm/s was in the case when 16 needles were inserted together. For that case, maximum force value was 43N. Fig. 9.b represents the insertion force for the insertion speed of 80mm/s. Maximum insertion force for 16 needles was 52N. It can be concluded that more needles were inserted in one insertion the force value was higher. But when 8 needles were inserted in the same time maximum insertion force did not change. For insertion speed of 40mm/s and 80mm/s the insertion force was around 35N. In Fig. 9.c and Fig. 9.d insertion force for whole range of the insertion speed were represented. In the former case 8 needles were inserted together while in the latter 16 needles were inserted. The insertion force for the latter case was about 7N higher due to bigger number of the inserted needles. However, it can be noticed that force does not significantly change in the range of insertion speed higher that 40mm/s (60mm/s or 80mm/s). The conclusion based on this fact is that insertion speed can be divided into two regions with different insertion parameters.

It can be concluded that the average maximum insertion force was less then 45N for moderate speed range (insertion speed 40mm/s and stylet speed 30mm/s, 16 needles in the stage) and 52N for high speed range (insertion speed greater then 40mm/s and stylet speed 50mm/s, 16 needles in the stage). Insertion time per seed was 5-8 seconds. Plan delivery time for high speed range was 8min and 12min for moderate speed range.

The summary of conclusion is as follows. It was observed that more needles were inserted together force value was higher. However, when 8 and 16 needles were inserted in the same time maximum insertion force did not change. Furthermore, force did not change significantly in the range of insertion speed higher than 40mm/s. Consequently, insertion speed range can be divided into two regions with different insertion parameters. Preliminary results reveal that MIRAB is efficacious for delivering seed accurately. MIRAB can potentially be used for image-guided robotic brachytherapy.

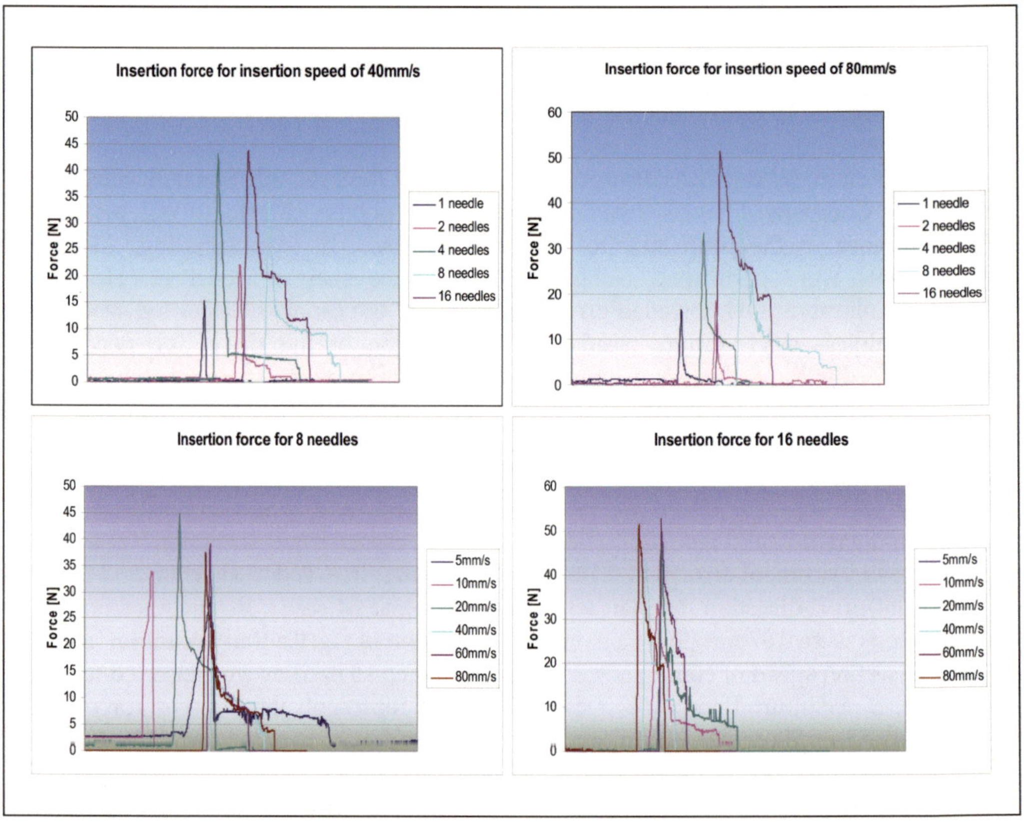

Fig. 9. a) Insertion force for different number of needles installed on the template stage, for insertion speed of 40mm/s. b) Insertion force for different number of needles installed on the template stage, for insertion speed of 80mm/s. c) Insertion force for different insertion speed;8 needles inserted in the same time. d) Insertion force for different insertion speed; 16 needles inserted in the same time

3. Robotic systems for real-time tumor tracking

Respiratory and cardiac motions induce displacement and deformation of the tumor-volumes in various internal organs. To accommodate this undesired movement and other errors, physicians incorporate a large margin around the tumor to delineate the Planning Target Volume (PTV), so that the Clinical Target Volume (CTV) receives the prescribed radiation dose under any scenario. Consequently, a large volume of healthy tissue is irradiated and sometimes it is difficult to spare critical organs adjacent to the tumor, (Buzurovic et al., 2010.b,c).

In this section we described a novel approach to the 4D Active Tracking and Dynamic Delivery (ATDD) incorporating tumor motion prediction technique. The proposed algorithm can predict the tumor position and the robotic systems are able to continuously track the tumor during radiation dose delivery. Therefore a precise dose is given to a moving target while the dose to nearby critical organs is reduced to improve patient treatment outcome. The efficacy of the proposed method has been investigated by extensive computer simulation, (Buzurovic et al., 2010.d, 2011.a).

Recently, several research groups are investigating various aspects of tumor tracking and developing tools to deliver precise dose to moving target-volumes, (Ozhasoglu & Murphy, 2001), (Keall et al., 2006.a), (Benchetrit, 2000), (Vedam et al., 2004), (Sharp et al., 2004), (Schweikard et al., 2000), (Kamino et al., 2006), (D'Souza & McAvoy, 2006), (Keall et al., 2006.b), (D'Souza et al., 2005), (Chung et al., 2006), (Podder et al., 2007, 2008), (Buzurovic et al., 2010.d, 2011.a). Generally, commonly practiced methods for compensating target/tumor motion can be structured as: (*i*) breath-hold techniques, (*ii*) gating techniques, and (*iii*) ATDD. The ATDD is the most effective technique, but it is the most challenging one. The ATDD can be accomplished in three different ways: (1) using the multi-leaf collimator (MLC), (2) using the treatment couch, and (3) using the MLC and the couch simultaneously. However, each of them has its own unique limitations.

For instance, MLC gating technique using internal fiducials requires kilovoltage x-ray which delivers unwanted radiation dose to the patient, and additionally gating suffers from severely low duty-cycle (only 30%-50%) and intensity modulated radiation therapy (IMRT) efficiency (only 20%-50%); all these lead to a 4- to 15-fold increase in delivery time over conventional treatment, (Keall et al., 2006.a). Therefore, it is of tremendous clinical interest, if the radiation beam can be delivered with higher duty-cycle (or almost continuously) while compensating for the target movement without exposing the patient to kilovoltage x-ray. Robotic systems can help in a great deal solving this problem.

In the following part we present developed dynamic equations of motion HexaPOD robotic couch. We applied the similar approach to one standard 3DOF robotic couch for radiation treatment (Buzurovic et al., 2010.d). In addition, we presented the control approach and prediction module which is capable of compensating time delay of the mechanical system.

For system dynamics, we used energy based Lagrangian formulation. The Lagrangian function of dynamic systems can be expressed as:

$$L = \text{Kinetic energy (K)} - \text{Potential energy (P)} \tag{12}$$

Thus, the general form of dynamic equations is

$$\frac{d}{dt}\left(\frac{\partial L}{\partial \dot{q}}\right)-\frac{\partial L}{\partial q}=\tau \tag{13}$$

where, $q\in R^n$ is the vector of generalized coordinates, and τ is the generalized force (or torque) applied to the system through the actuators.

The dynamical behavior of a 6DOF robotic couch, has been investigated. Proposed approach is explained below.

3.1 Parallel robotic platform

The HexaPOD is a special type of Stewart Platform, i.e., a parallel robotic manipulator. The parallel robotic mechanism consists of a rigid body top plate, or mobile plate, connected to a fixed base plate and is defined by at least three stationary points on the grounded base connected to six independent kinematic legs. Typically, the six legs are connected to both the base plate and the top plate by universal joints in parallel located at both ends of each leg. The legs are designed with an upper body and lower body that can be adjusted, allowing each leg to be varied in length as in Fig. 10.

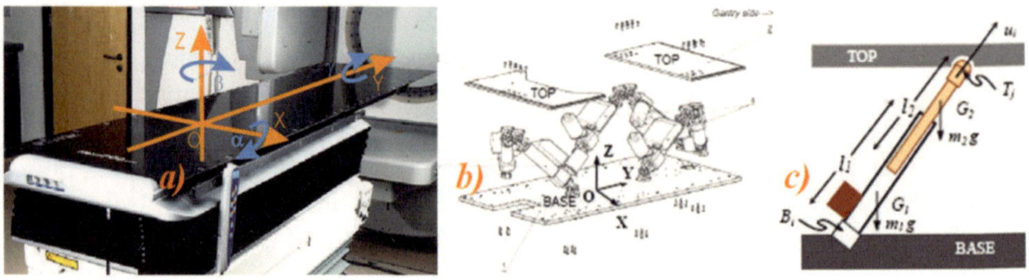

Fig. 10. HexaPOD robotic couch, a) External isometric view with moving coordinate system, b) External schematic view with fixed coordinate system, c) Schematic of the leg

In this study, we have used the following notations for modeling the HexaPOD, i.e. the Stewart platform. Referring figure 2 we have assigned an inertial frame (X,Y,Z) at the center O of the lower platform, i.e. the BASE, and assigned another moving coordinate system (x,y,z) at the center of the top platform, i.e. the TOP. The BASE frame is assigned and called as fixed frame and the TOP frame is moving and is called as moving frame.

To specify the configuration of the 6DOF Stewart Platform, six independent position-orientation variables are needed. Denote the position of the origin of the TOP frame with respect to the BASE frame $[p_x \;\; p_y \;\; p_z]^T$. The orientation is not defined by the standard Euler angles, but by rotating the TOP frame first about the fixed X-axis by α, then about fixed Y-axis by β and finally about Z-axis by γ. We denote $R_X(\alpha)$, $R_Y(\beta)$ and $R_Z(\gamma)$ as the three matrices that represent three basic rotations as follows:

$$R_X(\alpha),=\begin{bmatrix}1 & 0 & 0\\ 0 & C\alpha & -S\alpha\\ 0 & S\alpha & C\alpha\end{bmatrix}, R_Y(\beta),=\begin{bmatrix}C\beta & 0 & S\beta\\ 0 & 1 & 0\\ -S\beta & 0 & C\beta\end{bmatrix},\; R_Z(\gamma),=\begin{bmatrix}C\gamma & -S\gamma & 0\\ S\gamma & C\alpha & 0\\ 0 & 0 & 1\end{bmatrix} \tag{14}$$

where $C(.)$ =cos(.), and $S(.)$ = sin(.). This definition of the orientation not only provides us with a clear physical meaning but also avoids violating the one-to-one relationship between the system configuration and the values of X_{p-0}, which may cause the Jacobian matrix to lose its rank, even if the system is not in a singular configuration. Thus, the position and orientation of the upper platform is specified by the Cartesian coordinate system as $X_{p-o} = [p_x, p_Y, p_Z, \alpha, \beta, \gamma]^T$.

Now, the dynamic equations in Cartesian-space (or task-space) are expressed as:

$$M(X_{p-o})\ddot{X}_{p-o} + V_m(X_{p-o}, \dot{X}_{p-o})\dot{X}_{p-o} + G(X_{p-o}) = J^T(X_{p-o})F_{td} \tag{15}$$

where, $F_{td} = (f_1, f_2, f_3, f_4, f_5, f_6)$ is the vector of forces applied by the actuators of the legs, J is the full system Jacobian matrix. For convenience, we divide the HexaPOD robotic couch into two subsystems: the upper platform (TOP) and the six legs, Fig. 10. We compute the kinetic energy and potential energy of these subsystems and then derive the global dynamic equations.

$$K_{up} = K_{up(trans)} + K_{up(rot)} = \frac{1}{2}m_u(\dot{p}_X^2 + \dot{p}_Y^2 + \dot{p}_Z^2) + \frac{1}{2}\Omega_{up(mf)}^T I \Omega_{up(mf)} \tag{16}$$

is a expression for kinetic energy of the upper platform. Potential energy of the upper platform:

$$P_{up} = m_u g\, p_Z \tag{17}$$

Kinetic energy of the legs:

$$K_{L_i} = \frac{1}{2}(m_1 + m_2)[h_i V_{T_j}^T V_{T_j} - k_i V_{T_j}^T (u_i)(u_i)^T V_{T_j}] \tag{18}$$

where, L_i is the length of the leg at the current configuration, $(m_1+m_2)$is the mass of the leg, V_{Tj} is the velocity of the leg, and :

$$h_i = \left(\frac{\hat{l}}{L_i} + \frac{m_2}{m_1 + m_2}\right)^2, \ \hat{l} = \frac{\delta m_1 - \frac{1}{2}m_2 l_2}{m_1 + m_2}, k_i = h_i - \left(\frac{m_2}{m_1 + m_2}\right)^2 \tag{19}$$

Potential energy of the legs:

$$P_{Legs} = (m_1 + m_2)g\sum_{i=1}^{3}\left[\hat{l}\left[\frac{1}{L_{2i}} + \frac{1}{L_{2i-1}}\right] + \frac{2m_2}{m_1 + m_2}\right](p_Z + Z_{T_i}) \tag{20}$$

where, $Z_{T_j} = [0 \quad 0 \quad 1][R_Z(\gamma)^T R_X(\alpha)^T R_Y(\beta)^T GT_{j(mf)}]$; $GT_{j(mf)}$ is obtained from the geometry of the platform. Substituting the expression from equation (15)-(20) into equations (12), and (13), we can obtain the dynamic equations for the HexaPOD (leg plus top platform). Recalling equation (15), more precisely it can be written,

$$M(X_{p-o}) = M_{up} + M_{Legs}, V_m(X_{p-o}, \dot{X}_{p-o}) = V_{m_{up}} + V_{m_{Legs}} \ G(X_{p-o}) = G_{up} + G_{Legs} \tag{21}$$

The dynamic equations of motion for HexaPOD robotic couch have been discussed above. These equations are essential in developing our proposed dynamics-based coordinated control system. In the following part it can be seen that the speeds for tracking are slow. However, it was necessary to use dynamical model approach to fulfill strong requirements regarding the tracking accuracy. Since the tracking is impaired with a real-time radiation delivery, it was of the significant importance to have accurate model and to decrease possibilities for inaccurate radiation delivery during tracking.

3.2 Control and prediction

To track the tumor trajectory for optimal dose delivery to the CTV while sparing normal tissue, we propose PID control for HexaPOD parallel robotic platform. The basic goal of the controller was to specify the desired trajectory of the couch for all tumor positions. Block diagram of the decentralized coordinated dynamics-based closed-loop control strategy for HexaPOD robotic couch using prediction module PM, controller C and robotic couch T, is presented in Fig.11.

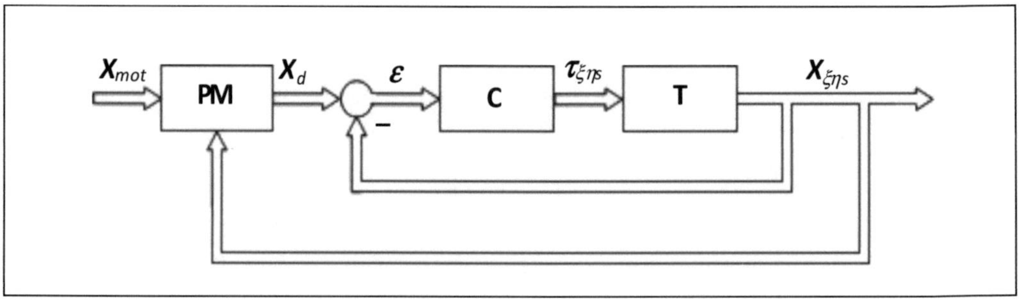

Fig. 11. Control schema for parallel robotic platform

By taking this approach it is possible to compare the dynamical behavior of the one of the most suitable system for 4D tracking. In the proposed approach, the trajectories of the tumor motion in the x, y and z directions were obtained from the 4D Computer Tomography (CT) data of real patients. These task-space trajectories were used to generate joint-space motion trajectories for the robotic systems.

Referring Fig. 11, we have denoted the following parameters: X_{mot} is raw data of the tumor motion. X_{mot} is the input into the prediction module (PM). Output of the PM is predicted value of the tumor motion X_d. Dynamic-based controller is denoted by C. Robotic couch is denoted by T. Input to the T module are desired forces values and output is the motion of the robotic table , denoted by $X_{\xi\eta s}$, which compensate tumor motion.

3.2.1 Dynamic-based controller

To implement the proposed algorithms, we have used a PID control scheme. Dropping the subscripts, we can rewrite equation (21) as

$$M(X)\ddot{X} + \xi(X,\dot{X}) = \Im \,, \tag{22}$$

where, $\xi(X,\dot{X}) = V(X,\dot{X})\dot{X} + G(X)$, and $\Im = J^T(X_{p-o})F_{td}$. Now, the computed torque can be written as follows:

$$\hat{M}[\ddot{X}_d + K_V(\dot{X}_d - \dot{X}) + K_P(X_d - X)] + \hat{\xi} = \Im\,, \tag{23}$$

where, $\hat{M}$ and $\hat{\xi}$ are the estimates of the system's model parameters, K_V is the derivative gain matrix and K_P is the proportional gain matrix. For HexaPOD $K_V = diag(k_{v1}\ k_{v2}\ k_{v3}\ k_{v4}\ k_{v5}\ k_{v6})$ and $K_P = diag(k_{p1}\ k_{p2}\ k_{p3}\ k_{p4}\ k_{p5}\ k_{p6})$. If the estimates of the model parameters are close approximation of the actual system's parameter, from equations (22) and (23) it can be written,

$$\ddot{X} = \ddot{X}_d + K_V(\dot{X}_d - \dot{X}) + K_P(X_d - X) \tag{24}$$

Now, denoting the position, velocity, and acceleration errors as $\varepsilon = X_d - X$, $\dot{\varepsilon} = \dot{X}_d - \dot{X}$, $\ddot{\varepsilon} = \ddot{X}_d - \ddot{X}$, we can rewrite equation (24) as

$$\ddot{\varepsilon} + K_V\dot{\varepsilon} + K_P\varepsilon = 0 \tag{25}$$

Equation (25) guarantees asymptotic reduction of the tumor tracking errors or at least keeps error in the acceptable margins. However, to reduce any steady-state errors, an integral control part was also incorporated into equation (26). Thus the final control equation becomes,

$$\ddot{\varepsilon} + K_V\,\dot{\varepsilon} + K_P\,\varepsilon + K_I\int_0^t \varepsilon\,dt = 0\,, \tag{26}$$

where, K_I is the integral gain. Thus, equation (26) ensures asymptotic decay of the transient errors as well as reduction of steady-state errors.

3.2.2 Prediction module

Prediction module PM is developed to predict tumor motion and to compensate errors due to delay in the system response. Prediction module algorithm uses X_{mot} to calculate predicted value X_d. The system output $X_{\xi\eta\varsigma}$ is used for accuracy checking and fine tuning purposes.

Normalized Least Mean Square (nLMS) algorithm is adapted for the dynamic system to predict the position in three dimensions. System delay t_d is a parameter in the developed algorithm.

The nLMS algorithm belongs to the family of Least Mean Square LMS algorithms. The LMS algorithm is an adaptive algorithm presented by Widrow and Hoff as in (Widrow & Walach, 1994) and (Chen, 1993). The LMS algorithm use iterative procedures to make successive corrections to the weight vector towards the negative direction of the gradient vector. As a result minimal mean square error will be minimized. The LMS algorithm can be summarized as follows:

Input vector is $X(n) = [x(n), x(n\text{-}1),\ldots, x(n\text{-}t\text{+}1)]^T$, where n is the current algorithm iteration and t is the tap number. Desired vector is $D(n) = [d(n), d(n\text{-}1),\ldots, d(n\text{-}t\text{+}1)]^T$. Consequently, predicted vector is $Y(n) = [y(n), y(n\text{-}1),\ldots, y(n\text{-}t\text{+}1)]^T$. Error vector is $E(n) = [e(n), e(n\text{-}1),\ldots, e(n\text{-}t\text{+}1)]^T$. Filter vector $W(n)$ is used to calculate predicted value $y(n)$. It is of the form $W(n) = [w(n), w(n\text{-}1),\ldots, w(n\text{-}t\text{+}1)]^T$. Predicted value is

$$y(n) = W(n-1)^H X(n) \tag{27}$$

where $W(n\text{-}1)^H$ is the hermitian transpose of $W(n\text{-}1)$. Thus, algorithm error is

$$e_1(n) = d(n) - y(n). \tag{28}$$

Now, filter vector becomes $W(n) = W(n\text{-}1)+\mu X(n)e(n)$, where μ is the learning rate. In order to ensure the stability of the algorithm it is necessary to adjust LMS to its normalized form nLMS. For the nLMS algorithm the filter vector is:

$$W(n) = W(n-1) + \frac{\mu X(n)e(n)}{X(n)^H X(n)} \tag{29}$$

Furthermore, to ensure accuracy of the prediction process, algorithm checks difference between predicted value $y(t)$ and system output $X_{\xi\eta s}$. We defined

$$e_2(n) = d(n) - X_{\xi\eta s}(n). \tag{30}$$

Finally, prediction module calculates next position of the tumor, based on algorithm which incorporates nLMS error $e_1(n)$ and physical error $e_2(n)$. Therefore, the resultant error is sum of the prediction and tracking error.

3.3 Simulation

The computer simulation results for HexaPOD robotic couch have been presented in the Fig. 12 through Fig. 13 below. Fig. 12 shows position of the each HexaPOD leg during tracking task. Combined motion of the robotic legs results in tracking the desired trajectory. This means that the robotic couch will start moving the legs according to the desired trajectory, obtained from 4D-CT. Combined motion of the HexaPOD legs will result in patient motion which has opposite direction of the tumor. Consequently, tumor will appear steady and the beam will irradiate smaller PTV which does not include all possible tumor position during the respiratory cycle.

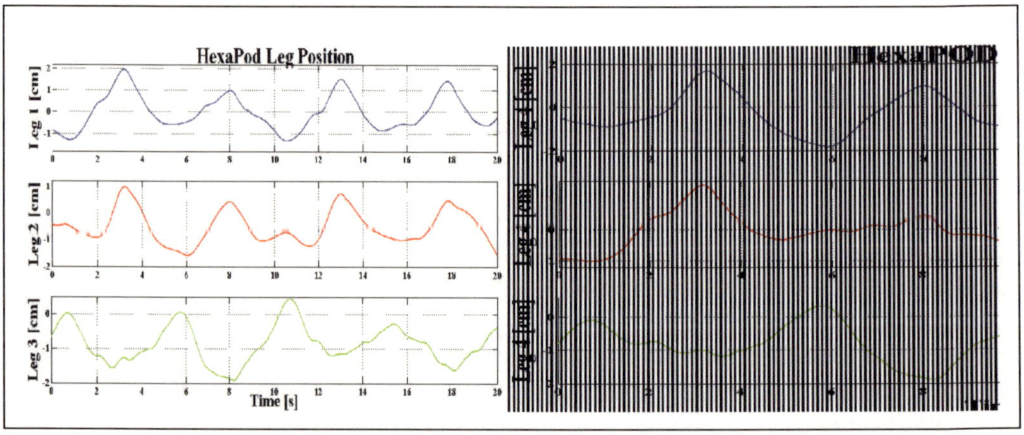

Fig. 12. HexaPOD legs positions during tracking

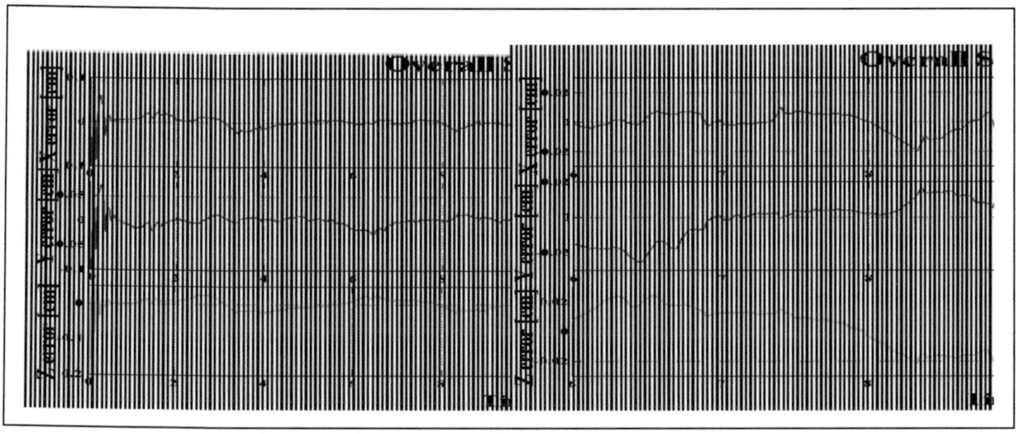

Fig. 13. Overall system errors; a) ε_X, ε_Y,ε_Z in X, Y and Z directions for HexaPOD couch.
b) The error amplitudes for steady states

An average system error after the transient time was ε_{XH}=ε_{YH} =ε_{ZH}= ±0.2mm and 0.4mm for irregular motion pattern, Fig.13. The tracking error cannot be zero, but it can be kept within the tolerable limit.

From the simulation results it appeared that proposed methods could yield superior prediction and tracking of the tumor motion induced by respiratory motion.

Analyzing the simulation results it can be concluded that the robotic system show the same maximum tracking error which was 1mm. Based on dosimetric studies, (Buzurovic et al., 2010.b,c) it was noticed that implementation of real-time tracking techniques can minimize irradiation to healthy tissues and improve sparing of critical organs. It was shown in the studies that the dosimetric effect on PTV and CTV coverage, caused by the prediction error in tumor tracking procedure, is insignificant. Consequently, tumor tracking error for the described proposed method will not compromise patient treatment outcome. Implementation of the proposed technique can potentially improve real-time tracking of the tumor-volume to deliver highly conformal precise radiation dose at almost 100% duty cycle while minimizing irradiation to health tissues and sparing critical organs. This, in turn, will potentially improve the quality of patient treatment by lowering the toxicity level and increasing survival.

In this study, we have deployed a closed-loop PID control. Adaptive control can be a good choice because of the variability in the payload on the system, i.e., the weight of the patient. Results of the adaptive control applied to the proposed system can be found in (Buzurovic et al., 2011.b).

4. Conclusion

In this chapter the use of the robotic systems in radiation therapy has been presented. It was shown that robotic systems can greatly influence to the current method of radiation therapy in both delivering radioactive material inside the cancerous tissue and in compensation of moving tumors during the external beam radiation therapy. The presented systems are novel and the clinical applications of such systems will be near future in modern cancer treatments. It was shown many times that modern robotics can improve many aspects of human work. The presented robotics systems are the example of the previous statement.

5. References

Benchetrit, G. (2000). Breathing pattern in humans: diversity and individuality, *Respir. Physiol.*, 22(2-3), pp. 123–129

Buzurovic, I.; Podder, T.; Yan, K.; et al. 2008. Parameter optimization for brachytherapy robotic needle insertion and seed deposition, *Medical Physics*, Vol. 35, No. 6, p. 2865

Buzurovic, I.; Podder, T.K.; and Yu, Y. (November 2008). Force prediction and tracking for image-guided robotic system using neural network approach, *Proceedings of IEEE Biomedical Circuits and Systems Conference* (*BIOCAS '08*), Baltimore, MD, USA, pp. 41–44

Buzurovic, I.; Podder, T.K.; and Yu, Y. (2010). Prediction Control for Brachytherapy Robotic System, *Journal of Robotics*, Vol. 2010, Article ID 581840, doi:10.1155/2010/581840

Buzurovic, I.; Werner-Wasik, M.; Biswas. T.; Galvin J.; Dicker, A.P.; Yu, Y.; Podder, T. (2010). Dosimetric Advantages of Active Tracking and Dynamic Delivery. *Med. Phys.*, 37, p. 3191

Buzurovic, I.; Huang, K.; Werner-Wasik, M.; Biswas. T.; Galvin J.; Dicker, A.P.; Yu, Y.; Podder, T. (2010). Dosimetric Evaluation of Tumor Tracking in 4D Radiotherapy, *Int. J. Radiat. Oncol. Biol. Phys.*, 78(3) Suppl.1, p. S689

Buzurovic, I.; Huang, K.; Yu, Y.; Podder, T. (2011). Robotic Approach to 4D Real-time Tumor Tracking for Radiotherapy, *Phys. Med. Biol.*, 56(5), pp.1299-1320

Buzurovic, I.; Huang, K.; Yu, Y.; Podder, T. (May-June, 2010). Tumor Motion Prediction and Tracking in Adaptive Radiotherapy, *Proc. of 2010 IEEE Int. Conf. on Bioinformatics and Bioengineering* (*BIBE*), pp. 273-278

Buzurovic, I., Yu, Y.; Podder, T.K. (September, 2011). Active Tracking and Dynamic Dose Delivery for Robotic Couch in Radiation Therapy, *Proc. of 2011 IEEE Int. Conf. on Engineering in Medicine and Biology* (*EMBC*), Boston, MA, USA

Chen, W.K. (1993). Linear Networks and Systems: Algorithms and Computer-Aided Implementations, *Belmont Wadsworth*, pp. 123–135

Chung, H.; et al. (2006). Mechanical Accuracy of a Robotic Couch, *Medical Physics*, 33(6), p.2041

Cobb, D. (May, 1983). Descriptor Variable Systems and State Regulation", *IEEE Transaction on Automatic Control*, Vol. AC-28., No.5

D'Souza, W.; McAvoy, T.J.; Analysis of the treatment couch and control system dynamics for respiration-induced motion compensation. Medical Physics, 2006, 33(12):4701-4701.

D'Souza, W.; Naqvi, S.A.; Yu, C.X.; (2005). Real-time intra-fraction-motion tracking using the treatment couch: a feasibility study, *Phys. Med. and Biol.*, 50, pp. 4021-4033

Fichtinger, G.; Burdettem, E.C.; Tanacsm, A.; et al. (2006). Robotically assisted prostate Brachytherapy with transrectal ultrasound guidance - Phantom experiments, *Brachytherapy* 5, pp. 14-26

Kamino Y.; Takayama, K.; Kokubo, M.; et al. (2006). Development of a four-dimensional image-guided radiotherapy system with a gimbaled x-ray head, *Int. J. Radiation Oncology Biol. Phys.*, 66(1), pp. 271–278

Keall, P.J.; Mageras, G.S.; Balter, J.M.; et al. (2006). The management of respiratory motion in radiation oncology report of AAPM Task Group 76, *Medical Physics*, 33(10), pp. 3874-3900

Keall, P.J.; Cattell, H.; Pokhrel, D.; et al. (2006). Geometric accuracy of a real-time target tracking system with dynamic multileaf collimator tracking system, *Int. J. Radiation Oncology, Biol. Phys.*, 65(5), pp. 1579-1584

Lin, A.; Trejos, A.L.; Patel R.V.; and Malthaner, R.A. (2008). Robot-Assisted Minimally Invasive Brachytherapy for Lung Cancer, *Teletherapy* (book chapter), ISBN: 978-3-540-72998-3, Springer, Berlin, Germany, pp. 33-52

Mavrodis, C.; Dubowsky, S.; et al. (April 19-23, 1997). A Systematic Error Analysis of Robotic Manipulators: Application to a High Performance Medical Robot, *IEEE Int. Conf. Robotics and Automation (ICRA)*, Albuquerque, NM, pp. 2975-2981

Meltsner, M.A. (2007). Design and optimization of a brachytherapy robot, *PhD Thesis*, Department of Medical Physics, University of Wisconsin-Madison

Meltsner, M.A.; Ferrier N.J.; and Thomadsen, B.R. (2007). Observations on rotating needle insertions using a brachytherapy robot, *Phys. Med. Biol.* 52, pp. 6027-6037

Mills, J.K. (1988). Constrained Manipulator Dynamic Models and Hybrid Control, *Proc. of the IEEE International Symposium on Intelligent Control*, Vol. I, Arlington, VA, USA, pp. 424-429

Moerland, M; Van den Bosch, M.; Lagerburg, V.; et al. (2008). An MRI scanner compatible implant robot for prostate brachytherapy, *Brachytherapy*, 7, p.100

Ozhasoglu, C.; Murphy, M.J. (2002). Issues in respiratory motion compensation during external-beam radiotherapy, *Int. J Radiat Oncol Biol Phys.*, 52(5), pp. 1389–1399

Podder, T.K.; Ng, W.S.; and Yu, Y. (August 23-26, 2007). Multi-channel robotic system for prostate brachytherapy, *Int. Conf. of the IEEE Engineering in Medicine and Biology (EMBS)*, Lyon, France, pp. 1233-1236

Podder, T.K.; Buzurovic, I.; and Yu, Y. (May-June 2010). Multichannel Robot for Image guided Brachytherapy, *Proc. of 10th IEEE International Conference on Bioinformatics& Bioengineering (BIBE)*, Philadelphia, PA, USA, pp.209-213

Podder, T.K.; Buzurovic, I.; Hu, Y.; Galvin, J.M.; Yu, Y. (2007). Partial transmission high-speed continuous tracking multi-leaf collimator for 4D adaptive radiation therapy, *IEEE Int. Conf. on Bioninformatics and Bioengineering (BIBE)*, Boston, MA, USA, pp. 1108-1112

Podder, T.K.; Buzurovic, I.; Hu, Y.; Galvin, J.M.; Yu, Y. (2008). Dynamics-based decentralized control of robotic couch and multi leaf collimators for tracking tumor motion, *IEEE Int. Conf. on Robotics and Automation (ICRA)*, Pasadena, CA, USA, pp. 2496-2502.

Salcudean, S.E.; Prananta, T.D.; et al. (May 19-23, 2008). A robotic needle guide for prostate brachytherapy, *IEEE Int. Conf. Robotics and Automation (ICRA)*, Anaheim, CA, pp. 2975-2981

Schweikard, A.; Glosser, G.; Bodduluri, M.; et al. (2000). Robotic motion compensation for respiratory movement during radiosurgery, *Comput. Aided Surg.*, 5(4), pp. 263–277

Sharp, G.C.; Jiang, S.B.; Shimizu, S.; Shirato, H. (2004). Prediction of respiratory tumour motion for real-time image-guided radiotherapy, *J.Phys. Med Biol.*, 49(3), pp. 425–440

Stoianovici, K.; Cleary, A.; Patriciu, M.; et al. (2003). AcuBot: A robot for radiological percutaneous Interventions, *IEEE Trans. on Robotics and Automation*, Vol. 19, pp. 927 930

Stoianovici, D.; Song, S.; Petrisor, D.; et al. (2007). MRI Stealth robot for prostate interventions, *Minimally Invasive Therapy* 16, pp. 241-248

Vedam, S.S.; Keall, P.J.; Docef, A.; et al. (2004). Predicting respiratory motion for four-dimensional radiotherapy, *J. Medical Physics*, 31(8), pp. 2274–2283

Wei, Z; Wan, G; Gardi, L; Fenster, A.; et al. (2004). Robot-assisted 3D-TRUS guided prostate brachytherapy: system integration and validation, *Med. Phys.* 31, pp. 539-548

Widrow, B.; Walach, E. (1994). Adaptive Inverse Control, *Prentice-Hall,* NJ, USA

Yu Y.; Podder, T.K.; Zhang, Y.D.; Ng, W.S.; et al. (2007). Robotic system for prostate brachytherapy, *J. Comp. Aid. Surg.* 12, pp. 366-370

6

Robotic Urological Surgery: State of the Art and Future Perspectives

Rachid Yakoubi, Shahab Hillyer and Georges-Pascal Haber
Center for Laparoscopic and Robotic Surgery, Glickman Urological and Kidney Institute, Cleveland Clinic, Cleveland, Ohio, USA

1. Introduction

Minimally invasive surgery has gained popularity over the last decade by offering shorter convalescences, improved peri-opertive outcomes as well as enhanced cosmesis.

Community urologists have typically performed standard laparoscopic nephrectomies secondary to its short learning curve, low complication rate, and limited requirement for sophisticated laparoscopic skills. However, advanced minimally invasive operations such as partial nephrectomies, pyeloplasties and prostatectomies necessitate advanced laparoscopic adeptness. The emergence of robotics, with 3D vision and articulated instruments, has allowed wider applications of minimally invasive techniques for more complex urological procedures.

Though robotics has overcome some shortcomings of laparoscopic surgery, there remains a limitation with its assertion as a standard amongst the urological community. The lack of tactile feedback, displacement of the surgeon from the bedside, fixed-port system, longer operating room times and cost remain barriers to widespread acceptance of robotics. In addition, the deficiencies within the robotic platform have propagated an evolution in the field with micro and nano-robotics. This chapter highlights the history of robotic surgery along with current and future applications.

2. The Da Vinci surgical system and urology

The da Vinci robot (Intuitive Surgical, Sunnyvale, CA, USA) (Figure 1) remains the only commercially available robotic surgical system since the fusion of computer motion and intuitive surgical system (Table 1). It is a master–slave system in which the surgeon operates the robot from a remote console. The articulating laparoscopic instruments in the da Vinci robot offer six degrees of freedom simulating the human wrist movement during open surgery. This facilitates intra-corporeal suturing especially in reconstructive surgery (Yohannes et al., 2002).

Robotic-assisted laparoscopic radical prostatectomy was first reported in 2000 using the da Vinci robot system. Since then, the robotic platform has been applied for a variety of minimally invasive procedures, such as partial nephrectomy, pyeloplasty, cyctectomy and adrenalectomy.

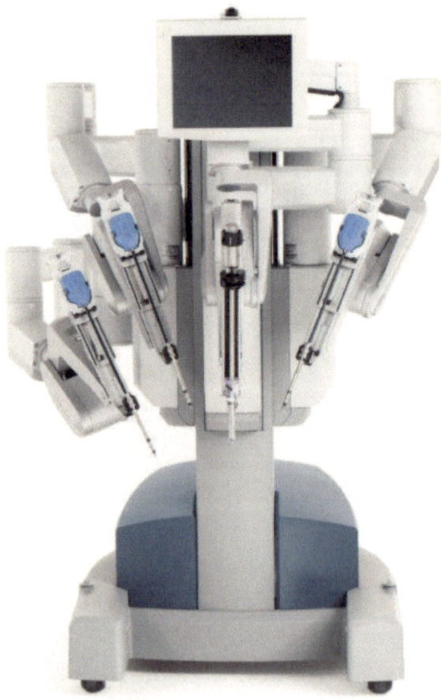

Fig. 1. Da Vinci robot

1985	First surgical robot utilization (Neurosurgery)
1989	First urologic robot (Probot)
1993	First commercially available robot approved by the FDA (AESOP)
1998	Zeus system commercially available
2000	First robotic radical prostatectomy
2001	FDA clearance for Da Vinci system
2003	The Zeus system and Intuitive Surgical fusion

Table 1. Robotic surgery timeline

2.1 Prostatectomy

Robotic-assisted laparoscopic radical prostatectomy (RARP) was first reported in 2000 (Abbou et al., 2000). Since then the number of patients undergoing RARP for prostate cancer has steadily increased. Early comparisons between radical retropubic prostatectomy (RRP) and RARP show encouraging results. Menon et al. in a prospective nonrandomized study, compared results of 30 consecutive patients undergoing (RRP) and 30 initial patients undergoing (RARP). Estimated blood loss (EBL), blood transfusions, pain score, hospital stay, and mean duration of postoperative catheterization were improved in the RARP group. However, the mean operating time increased for RARP (Menon et al. 2002).

In a recent review, Ficcara et al. found that the mean OR time for RARP ranged between 127 and 288 minutes, corroborating a longer operative time with robotics versus open. Nevertheless, transfusion rates and hospital stay were lower with RARP then open RRP. In addition there were comparable complication rates between the RARP and open RRP patients (Ficarra et al., 2009).

Positive surgical margins are a surrogate for oncological outcomes reported for RARP. A significant advantage in positive margin rates was demonstrated for RALP over RRP. This same difference amongst the robotic and open groups does not exist when comparing laparoscopic radical prostatectomy (LRP). The positive surgical margin rates ranged from 11% to 37% after RRP, from 11% to 30% after LRP, and from 9.6% to 26% after RALP (Ficarra et al., 2009).

Recently, survival outcomes have been reported after RARP. Barocas et al. compared 491 open RRP with 1,413 patients undergoing RARP, over a median follow-up of 10 months (Barocas et al., 2010). Robotic group had lower pathological stage (80.5% pT2 vs 69.6% pT2, p <0.01). The 3-year biochemical recurrence-free survival rate was similar between the 2 groups, even after adjusting for pathological stage, grade and margin status. More recently, comparisons of 522 consecutive RARP were matched to patients who underwent LRP and RRP evaluating oncological outcomes (Magheli et al., 2011). Positive surgical margin rates were higher in the robotic group (19%), compared to LRP (13%) and RRP (14%). This difference was not significant for pT2 disease. The mean follow-up was 2.5, 1.4 and 1.3 years for RRP, LRP and RARP respectively, with no statistically significant difference in biochemical-free survival between groups.

Erectile function and continence are major outcomes evaluated after prostatectomy. In a matched cohort, comparing 294 RARP for clinically localized prostate cancer with 588 RRP, showed no difference in continence between the 2 approaches at the 1-year follow-up (Krambeck et al. 2009). Furthermore, Roco et al showed 1year continence rates of 97% vs 88% after RARP and RRP, respectively (P = 0.014). The 1 year overall potency recovery rate was 61% vs 41%, after RARP and RRP, respectively (P = 0.003). Overall, RRP seems to be a faster procedure. However, EBL, hospitalization time, and functional outcomes were superior with RARP. Early oncological outcome seemed to be equivalent in the two groups (Rocco et al., 2009).

The drawback of RARP is the cost related to purchasing and maintaining the instruments of the robotic system. Bolenz et al. compared the cost of 262 RALP, 220 LRP, and 161 RRP performed at the same institution. The direct cost was higher for the robotic approach than laparoscopic or open. The median direct cost was US$ 6752, US$ 5687, and US$ 4437 for RALP, LRP, and RRP respectively. The most important difference was due to surgical supply and operating room cost (Bolenz et al., 2010). However, the surgical volume may reduce this difference. Scales et al. showed that the cost of RALP is volume dependent, and cost equivalence is achievable with RRP at a surgical volume of 10 cases weekly (Scales et al., 2005). Even the cost of robotic surgery is a difficult question to assess; the cost-effectiveness of robotics will probably continue to improve with time (Wilson & Torrey, 2011).

The available data demonstrates improvements in blood loss, hospital stay, and pain control with RALP. However the lack of long term cancer specific mortality limits robust oncological comparisons of RARP to RRP.

2.2 Radical nephrectomy

Robotic radical and simple nephrectomy is a feasible and safe procedure. A comparison of 46 laparoscopic nephrectomies, 20 hand-assisted laparoscopic nephrectomy, and 13 robotic nephrectomies showed no significant advantage of robotics over traditional laparoscopy or hand-assisted approaches. However, cost analysis illustrated far more cost with robotics among the three groups (Boger et al., 2010).

Platforms where robotics may be applied to radical nephrectomies are complex tumors with caval thrombosis. Abaza reported five patients with a renal tumor and inferior vena cava (IVC) thrombus who underwent robotic nephrectomy. The robotic system allowed for venotomy into the vena cava with suture of the defect and no complications. (Abaza, 2011)
Robotic-assisted nephrectomy was also applied for live kidney donor. A series of 35 first cases was reported with a mean warm ischemia time of 5.9 minutes. (Louis et al., 2009). However, the cost of the robotic approach may limit its applicability for radical nephrectomies except in complex renal cases with caval thromsis.

2.3 Partial nephrectomy

The da Vinci Surgical System provides advantages during robotic partial nephrectomy (RPN) such as 3-D vision, articulating instruments, scaling of movement, tremor filtration, fourth robotic arm assistance, and the TilePro™ software (Intuitive surgical, Sunnyvale, CA), a live intra-operative ultrasound platform. All these tools are helpful during partial nephrectomy and may overcome the technical challenges of laparoscopic partial nephrectomy (LPN).
Since the first report by Gettman et al. showing the feasibility of robotic assisted partial nephrectomy (RPN) (Gettman et al., 2004), a steadily increasing number of series have been reported.
Forty consecutive RPN and 62 LPN were retrospectively compared in a study by wang and colleagues showing no significant difference in EBL, collecting system repair (56% for each group), or positive margin rate (1 case in each group). Furthermore, the mean operative time, warm ischemia time (19 vs 25 minutes, for RPN and LPN respectively), and length of stay decreased significantly in the robotic group (Wang & Bhayani, 2009).
Haber et al. compared results of 150 consecutive patients who underwent RPN (n = 75) or LPN (n = 75) by a single surgeon. There was no significant difference between the 2 groups in the mean operative time (200 and 197 minutes) (P = .75)), warm ischemia time (18.2 minutes vs 20.3 minutes, P = .27), length of hospital stay (P = .84), change in renal function, or adverse events in the RPN and LPN groups respectively. The mean EBL was higher in the RPN group (323 vs 222 mL); surprisingly fewer patients required a blood transfusion in either group. The higher EBL with RPN may be explained by the surgeons learning curve for RPN compared with LPN. Overall these findings demonstrated comparable outcomes to LPN, regardless of the vast experience the primary laparoscopic surgeon possessed (Haber et al., 2010a).
The safety and effectiveness of RPN regarding functional and oncologic outcomes were evaluated in a multi-institutional review with 183 patients who underwent RPN (Benway et al., 2010). The means of peri-operative demographics and outcomes were analyzed illustrating a tumor size 2.87 cm, total operative time 210 min, warm ischemic time 23.9 min, and EBL of 131.5 ml. Sixty-nine percent of excised tumors were malignant, of which 2.7% had positive surgical margins. The incidence of major complications was 8.2%. At up to 26 months follow-up, there have been no recurrences and no significant change in renal function.
Additionally, indications for RPN have significantly expanded to include RPN for complex renal tumors (Rogers et al., 2008a; White et al., 2011). White et al. reviewed 67 patients who underwent RPN for a moderately or highly complex renal mass according to the R.E.N.A.L. nephrometry score (≥7). The median tumor size was 3.7 cm, median operative time was 180 minutes, median EBL was 200 mL, and the warm ischemia time was 19.0 minutes (range 15-

26). After a mean follow-up of 10 months, no recurrences had occurred indicating that RPN is a safe and feasible option for highly complex renal masses (White et al., 2011).
RPN seems to be as an effective and safe alternative to LPN. Surgical technique for RPN has improved and indications have been expanded to more challenging tumors. Currently available comparative studies are retrospective and with a limited follow-up. Future trials are expected to confirm encouraging findings from early reported series.

2.4 Adrenalectomy

Robotic-assisted adrenalectomy (RAA) was first reported in 1999 and 2001 using the Aesop (Hubens et al., 1999) and Da Vinci systems, respectively (Horgan & Vanuno, 2001). Since then, the few studies published were about RAA using Da Vinci system.
Brunaud et al. prospectively evaluated 100 consecutive patients who underwent RAA. The mean operative time was 95 minutes with a conversion rate 5%. Complication and mortality rates were 10% and 0%, respectively. The mean operative time decreased by 1 minute every 10 cases. Operative time improved more for junior surgeons than for senior surgeons after the first 50 cases. Surgeon's experience, first assistant level and tumor size were independent predictors of operative time. The robotic procedure was 2.3 times more costly than laparoscopic adrenalectomy (Brunaud et al., 2008a).
The same authors, compared prospectivly perioperative data of 50 patients who underwent RAA with 59 patients who underwent laparoscopic adrenalectomy (LA). RAA was associated with lower blood loss but longer operative times. However, the difference in operative time was not significant after the learning curve of 20 cases. Operative time increased, only in the LA group for obese patients (body mass index >30 kg/m2) and patients with large tumors (>55mm). Length of hospital stay, complication and conversion rates were equivalent in the groups (Brunaud et al., 2008b).
Recently, Giulianotti et al. examined 42 patients who underwent RAA by a single surgeon. Median hospital stay was 4 days with postoperative complication rate of 2.4% and mortality rate of 2.4% (Giulianotti et al., 2011).
Suggestions have been made that robot assistance may be beneficial for obese patients with large tumors (Brunaud et al., 2008a; Giulianotti et al., 2011), as well as for surgeons with limited laparoscopic experience. Regardless of its feasibility, Prospective studies must focus on potential improve in learning curve with robotic utilization along with a cost analysis evaluating RAA compared to current standards. (Brunaud et al., 2008a)

2.5 Pyeloplasty

The first clinical experience of robot-assisted pyeloplasty (RAP) was reported in 2002 (Gettman et al., 2002a). Since then, numerous studies have evaluated the efficiency of RAP. Gupta et al. prospectively evaluated results of 85 consecutive patients who had transperitoneal RAP, using four or five ports. Based on anatomic considerations, different types of pyeloplasty were completed. The mean operative time was 121 min, 47 min of which was for the anastomosis. Mean EBL was 45 mL, with hospital stay of 2.5 days. Three patients had stent migrations, and delayed drainage. Mean follow-up was 13.6 months with an overall success rate of 97%, based on imaging assessment (Gupta et al., 2010).
In a comparative non-randomized study, 98 RAP were compared with 74 LP with a mean operative time of 189.3 and 186.6 minutes for RAP and LP respectively. Complication rate was similar with 5.1% and 2.7% for RAP and LP respectively. The suturing time was shorter

for the RAP but without statistical significance (48.3 and 60 minutes (P =0.30) for RAP and LP respectively). RAP had a success rate of 93.4% versus 95% for LP based on renal scintigraphy (Bird et al., 2011).

Hemal and colleagues illustrated successful application of robotics to pyeoplasty surgery. A nonrandomized study, comparing results of 30 RAP with 30 LP, performed in a transperitoneal approach by a single surgeon. The mean total operating times were 98 minutes and 145 minutes, the mean EBL were 40 mL and 101 mL, and the mean hospital stay of the patients were 2 days and 3.5 days, for RAP and LP, respectively. At follow up, one patient in LP group had obstruction managed by balloon dilation (Hemal et al., 2010).

Insufficent evidence exists for the retroperitoneal approach to RAP. Kaouk et al. reported results of 10 patients who underwent retroperitoneal RAP. Four ports were placed for the robot and a successful operation was accomplished. Operative time was 175 mins with minimal complications. The advantage to the retroperitoneal approach was the direct access to the UPJ; however, retroperitoneal surgery has limited working space with unfamiliar anatomy to most urologist (Kaouk et al., 2008).

Cestari et al. compared 36 patients who underwent retroperitoneal RAP and 19 transperitoneal RAP for UPJO. Median operative time and hospital stay were similar. Complication rates were comparable. (Cestari et al, 2010).

Studies of RAP demonstrate feasibility, efficacy and safety. However, the cost of robotic surgery continues to limit the widespread application of this platform.

2.6 Cystectomy

Radical cystectomy with pelvic lymphadectomy (PLND) remains the gold-standard treatment for patients with muscle-invasive blader cancer. However, morbidity related to open radical cystectomy (ORC) presents a real challenge. Thus, robotic-assisted radical cystectomy (RRC) represents a potentiel alltenative to the open approach. Similar to other robotic procedures, the potential advantages of RRC over ORC are decrease in blood loss, pain, and hospital stay. On the other hand, the oncologic outcomes of the RRC remain largely unknown.

In a series of 100 consecutive patients who underwent RRC for clinically localized bladder cancer, the mean operative time was 4.6 hours and an EBL 271 ml. Mean hospital stay was 4.9 days with 36 patients experiencing postoperative complications; 8% major complications. Urinary diversion included, ileal conduits 61 % of the time (Pruthi et al., 2010).

Kauffman et al. presented results of 85 consecutive patients treated with RRC for bladder cancer. The median age was 73.5 years with high proportion of patients having comorbidities (46% of ASA class ≥3). Extended pelvic lymphadenectomy was performed in almost all patients (98%). Extravesical disease was found in 36.5% cases and positive surgical margins were present in 6% of patients. At a mean follow-up of 18 months, 20 (24%) patients had presented recurrence; three of them (4%) only had a local recurrence. The overall survival and disease-specific survival rates for the cohort at 2 years were 79% and 85%, respectively. Extravesical disease, positive lymph node, and lymphovascular invasion were associated with worse prognosis (Kauffman et al., 2011). Even with the encouraging results of this study, comparative analysis and long term outcomes are still needed.

Josephson et al., in his report of 58 RRC, found different results, with a overall survival rate of 54% and disease specific survival rates of 76% at 2-year (Josephson et al., 2010). However, stratification of survival outcomes by patholoical stage was not reported, making comparasions with other studies futile.

In an others series of 100 patients with a mean follow up of 21 months, 15 patients had disease recurrence and 6 died of bladder cancer (Pruthi et al., 2010). Comparisons between open and robotic cystectomy are lacking because of the heterogenity in disease burden and patient selection.

In a multi-institutional study, of the 527 patients who underwent RRC, 437 (82.9%) had a lymphadenectomy. Surgeons experience on the robot influenced whether a lymphadenectomy was performed (Hellenthal, 2010a). In the same data base, the Positive surgical margin was 6.5% (Hellenthal, 2010b). In a study of 100 consecutive patients with RRC, no positive margin was found (Pruthi et al., 2010). However in this series only 13% cases were pT3/T4 at the final pathology, which may explain this result.

RRC offers the potential for improving peri-operative outcomes with the advantages of minimally invasive surgery. It may offer shorter hospital stay, decrease blood loss and pain, and offer a smaller incision, with equal complication rates and oncological outcomes to the current gold standard, ORC. Future studies with longer follow up are essential to have robust data on the comparability of RRC to ORC

3. Robotic laparoendoscopic single-site surgery and natural orifice translumenal endoscopic surgery: Current status

Laparoendoscopic single-site surgery (LESS) (Figure. 2) and natural orifice translumenal endoscopic surgery (NOTES) exploit the use of a single incision or natural entry points into the body cavity. A trend towards scareless surgery with the advantages of laparoscopy has fueled the development of a novel surgical technique, LESS and NOTES.

The first laparoscopic transvaginal nephrectomy (NOTES) in a porcine model was described in 2002, introducing another achievable surgical approach (Gettman et al, 2002b). Moreover, the continued dynamism for workable new approaches sprung the inception of the first reported hybrid NOTES nephrectomy in a porcine model utilizing the da Vinci surgical system (Box et al., 2008).

Robotic technologies offer a potential improvement over the current flexible instruments and endoscopes. Other limitations include the indirect transmission of forces and space constraints for the operating surgeon (Table 2).

In 2008, 30 urologic robotic NOTES procedures on 10 porcine models were performed using the current da Vinci robotic system (Haber et al., 2008a). A 12 and an 8-mm port were placed through a single transumbilical incision to introduce the robotic camera along with a robotic arm using a single port (Uni-X (TM) P-navel Systems, Morganville, NJ). A flexible 12-mm cannula 20cm long (US Endoscopy, Mentor, Ohio, USA) served as a transvaginal port, through which the second robotic arm was docked. All interventions were conducted without complications or conversion. However, problems encountered included the inadequate length of transvaginal instruments, essentially prolonging surgical time. In order for widespread use of newer approaches, development of appropriate instruments should simultaneously occur.

The Da Vinci® S robotic system using a LESS approach has also been applied to prostate surgery. A transvesical robotic radical prostatectomy was performed in a cadaver model (Desai et al., 2008a) (Figure. 3). Articulated instruments and 3D vision facilitated the dissection of the prostate through the single-port. Moreover, architectural advances with the robotic system will allow accessibility to a wider range of surgeons interested in LESS and NOTES approaches. An example of robotic modifications helping the growth of LESS came with the advent of the novel robotic platform (VeSPA, Intuitive Surgical, California, USA) (Figures. 4 & 5). These curved cannulae and semi-rigid instruments have been designed to compensate for the limitations encountered with conventional LESS surgery. The VeSPA curved cannulae and semirigid instrument design allow the cannulae and instruments to be inserted in close proximity while allowing approximate triangulation intra-abdominally. The VESPA platform was successfully applied in a porcine model without conversions or addition of ports (Haber et al., 2010b). However, it is recognized that Robotic-LESS for the kidney is technically more challenging in humans compared to animal model. Therefore, further clinical research is required to confirm these early experimental results.

	Standard LESS	**Da Vinci robotic LESS**
Scope	Two-dimensional vision, high resolution, unstable vision	Three-dimensional vision, high definition, stable vision
Instruments	Straight and articulating/flexible instruments	Articulating instruments
Triangulation	Lacking (crossed hand surgery)	Lacking (chopstick surgery)
Instrument collision	Significant	Significant (due to bulky robotic arms)
Range of motion	Limited	Enhanced surgeon dexterity because of the Endowrist technology
Tissue dissection	Challenging	Limited at steep angles
Suturing	Extremely challenging	Accurate
Ergonomics	Reduced	Enhanced (surgeon sitting at console)
Main assistant's role	Camera manoeuvring	Managing collisions
Steepness of learning curve	Very high	Medium high

Table 2. From laparoendoscopic single-site surgery to robotic laparoendoscopic single-site surgery: technical advances with da Vinci system

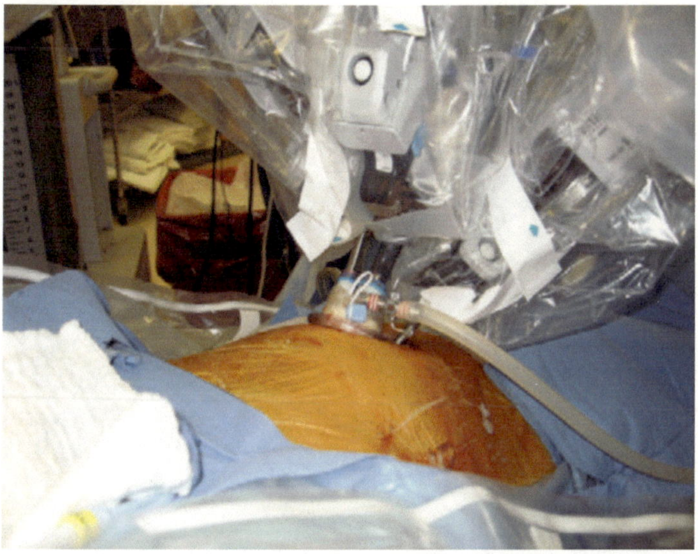

Fig. 2. The robot scope and two arms are inserted through a 2.5-cm incision in the umbilicus

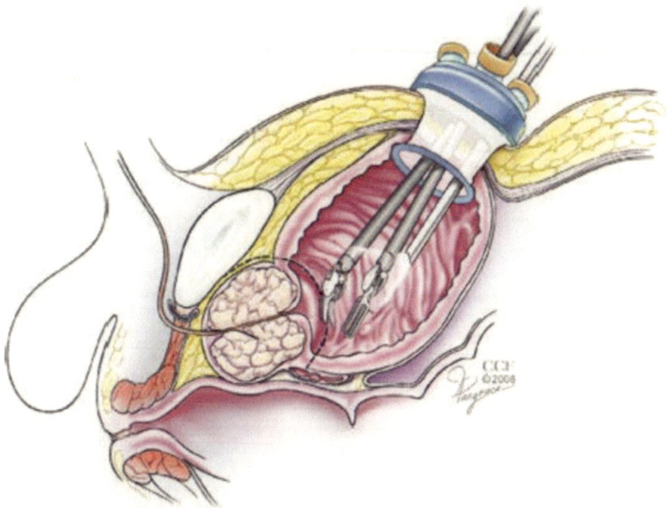

Fig. 3. Single-port and robot instruments through the bladder

The initial clinical experience with robotic single-port transumbilical surgery was reported in 2009 (Kaouk et al, 2009a). A multichannel single port (R-port; Advanced Surgical Concepts, Dublin, Ireland) was inserted through a 2-cm umbilical incision into the abdomen. Three procedures were performed, including radical prostatectomy, dismembered pyeloplasty, and right sided radical nephrectomy. All procedures were completed without intraoperative complications. The radical prostatectomy was completed in 5 h, with 45 min required for the anastomosis. The pathology reported negative margins. The pyeloplasty was completed in 4.5 h, and the radical nephrectomy was completed in 2.5 h. R-LESS procedures were accomplished without additional ports or significant differences in peri-operative outcomes compared to standard robotic approaches.

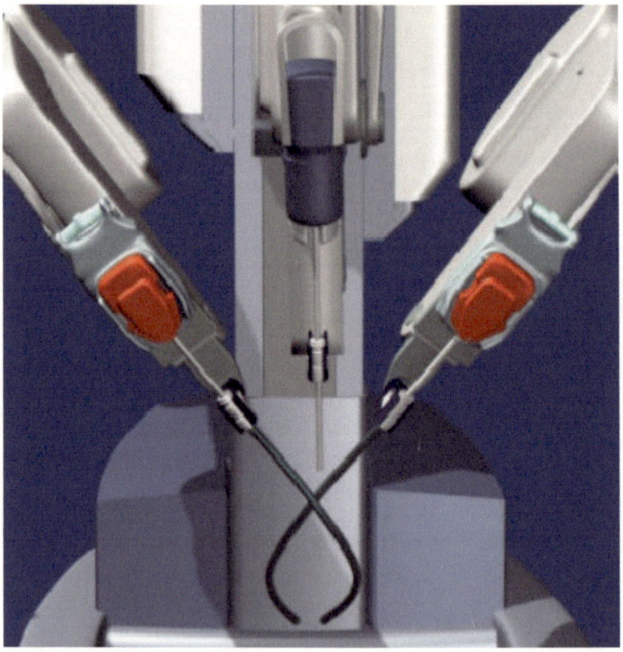

Fig. 4. Da Vinci® robotic with VeSPA instruments

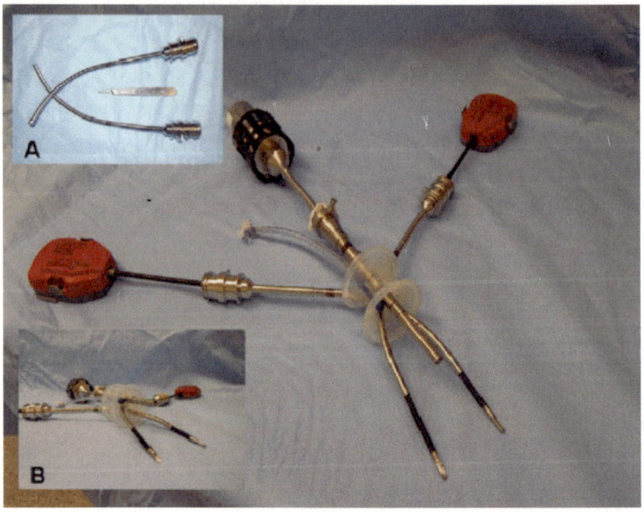

Fig. 5. VeSPA instruments and accessories. (A) Curved cannulae; (B) multichannel single-port, 8.5-mm robotic scope, flexible instruments passed through the cannulae

We recently reported our initial experience with single port robotic partial nephrectomy in two patients without conversions or complications (Kaouk et al. 2009b). A multichannel port (Triport; Advanced Surgical Concepts, Bray, Co Wicklow, Ireland) was utilized. Pediatric 5-mm robotic instruments were used. A 30° robotic lens placed in the upward configuration minimized clashing between the scope and instruments. A 2.8 cm left lower pole tumor and a 1.1 cm right lower pole tumor were excised without renal hilar clamping using the harmonic scalpel. EBL was 100 ml, operative time was 170 min, length of stay was 3.5 days, and visual analog pain scale at discharge was 1.0/10.

Stein et al. reported robotic LESS using a gel port (Applied Medical, Rancho Santa Margarita, California, USA) as the access platform (Stein et al., 2010). Four clinical procedures were performed, including two pyeloplasties, one radical nephrectomy, and one partial nephrectomy. The gel port system was used to allow for a larger working platform. The partial nephrectomy was completed in 180 min. The mass was excised without hilar clamping, using the harmonic scalpel and Hem-o-lok clips (Weck Closure Systems, Research Triangle Park, North Carolina, USA). A radical nephrectomy was performed for a 5-cm left-sided lower pole mass. The renal vein and artery were secured with an endoscopic stapler, and the remainder of the dissection was completed with hook cautery. The specimen was retrieved with an entrapment sac and removed via the umbilical incision extended to 4 cm. EBL was 250 ml, operative time was 200 min, and the hospital stay was 2 days.

The Da Vinci system has several advantages over conventional laparoscopy allowing an increasing number of urologist adopt this method. Preliminary results of these minimal invasive approaches are encouraging in LESS and NOTES techniques. Further refinement in instrumentation, improved triangulation and development of robotic systems specific to LESS and NOTES may define the new horizons in single site surgery.

4. Flexible robots

The development of robotic technologies has also flourished in endocopic surgeries. A flexible robotic catheter manipulator (Sensei, Hansen Medical, Mountain View, CA, USA) thought to allow enhanced ergonomics with improved efficiency over standard flexible ureteroscopy was developed by Sensei, Hansen medical. Initial experience in the porcine model showed promising outcomes.

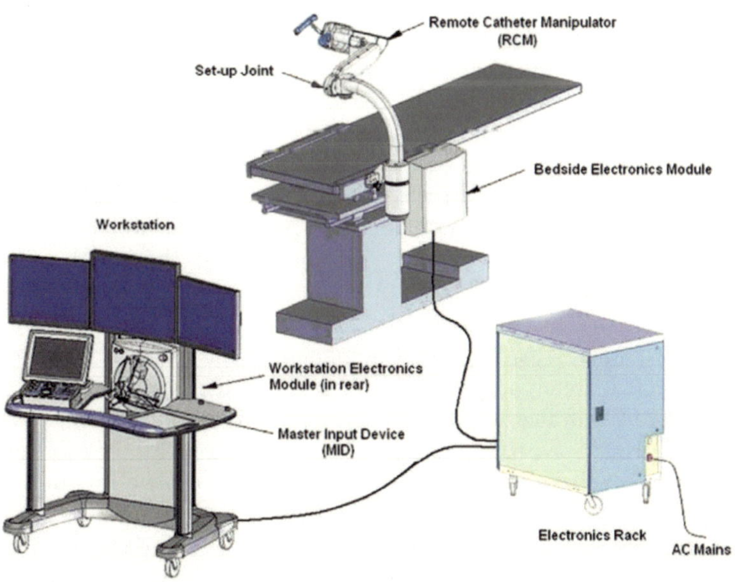

Fig. 6. Pictorial depiction of components of flexible robotic catheter control system. Surgeon console (workstation) showing three LCD screens, one touch screen, and MID

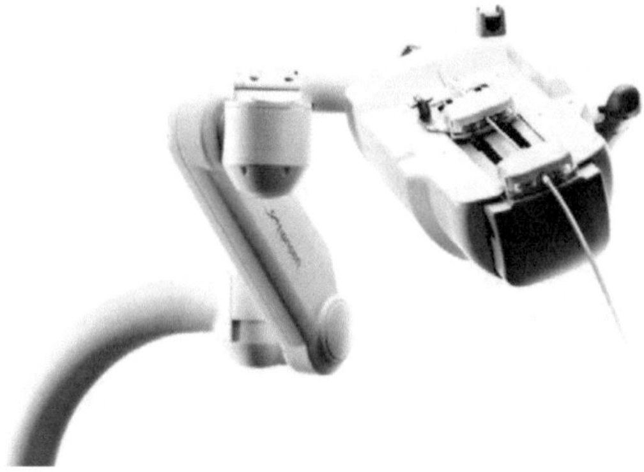

Fig. 7. Robotic arm and steerable catheter

The novel robotic catheter system comprises the following components (Figure. 6 & 7), (a) surgeon console, including the LCD display and master input device, (b) steerable catheter system, (c) remote catheter manipulator, and (d) electronic rack. The master input device (MID) is a three-dimensional joystick the surgeon uses to remotely maneuver the catheter tip. The real-time image of ureteroscopy, fluoroscopy and representation in space of the catheter position is projected simultaneous on the screens. Furthermore, images from the CT scan or an ultrasound probe images display.
The steerable catheter system contains an outer catheter sheath (14F/12F) and an inner catheter guide (12F/10F). The movement of the MID intuitively controls the tip of the catheter guide. The robotic arm, steerable catheter, is attached to the edge of the table. The catheter can be controlled either by fluoroscopy mode, direct vision from the ureteroscope or by a combination of both. In addition, a 3D reconstruction of the extremity the catheter is displayed on the screen, assisting in the identifying the location. Moreover, manoeuvrability is not diminished when the use of fiber laser 200 to 365 microns.
The technical feasibility of the robotic catheter system was evaluated in retrograde renoscopy in the porcine model. Authors concluded that the robotic catheter system could easily manoeuvre the ureteroscope into 83 (98%) of the 85 calices tested in a reproducible manner. The ease and reproducibility of intra-renal navigation was rated at a mean of 10 of 10 on the VAS (Desai et al., 2008b). Clinical investigation for flexible robotic system in ureteroscopy was performed in 18 patients who had renal calculi. Mean stone size was 11.9 mm. All procedures were done unilaterally and all patients had ureteral stents placed for 2 weeks. The robotic catheter system was introduced into the renal collecting system manually using fluoroscopic control along a guide wire. All intra-renal manoeuvres were performed completely by the surgeon from the robotic console. All procedures were technically successful, and all calculi were fragmented to the surgeon's satisfaction, without conversion to standard ureteroscopy. Mean operative time was 91 minutes, including a robot docking time of 7 minutes, and stone localization time of 9 minutes. The mean visual analog scale rating (from 1, worst, to 10, best) for ease of stone localization was 8.3, ease of maneuvering was 8.5, and ease of fragmentation was 9.2. Complete fragment clearance was achieved in 56% of patients at 2 months, and in 89% of patients at 3 months. One patient required secondary ureteroscopy for a residual stone (Desai et al, 2008c)
Improvements in flexible instrumentation have brought a revolution with retrograde intra-renal surgery. Advancements in flexible ureteroscopes, laser lithotripsy, and ureteroscope accessories are contributors to the metamorphosis with robotics in endo-urology.

Miniaturization has been the key advancement for the progression of robotics in endoscope. Further studies are needed to corroborate these early clinical results.

5. Robotic prostate biopsy

Incorporation of real-time radiographic imaging for biopsy or ablative treatment of urological cancer is currently in development. A robotic arm Vicky® (Endo-Control, Grenoble, France) and an ultra-sound probe (B-K Medical, Denmark), were utilized to perform tranrectal prostate biopsies. The software was modified to save up to 24 spatial coordinates and answer the constraints of transrectal ultrasound. Tests on phantom and human cadaver prostates demonstrated the feasibility of robotic transrectal ultrasound, with precision biopsies performed on a target tumor ranging from (0.1 to 0.9 mm) on the prostate and (0.2 to 0.9 mm) on the phantom (Figure. 8). Initial results are encouraging with clinical trials forthcoming (Haber et al., 2008b).

Moreover, a robotic positioning system with a biplane ultrasound probe on a mobile horizontal platform was used to perform prostate biopsies in a phantom model. . The integrated software acquires ultrasound images for three-dimensional modeling, coordinates target planning and directs the robotic positioning system. A repeatable accuracy of <1 mm was obtained (Ho et al., 2009). This robotic prostate biopsy system can theoretically biopsy targets from MRI images after merging the ultrasound images, allowing treatment of positive areas found on biopsies.

Presently, a fully automated robot system, the MrBot, has been developed for transperineal prostate access. (Patriciu et al., 2007). It is mounted alongside the patient in the MR imager and is operated from the control room under image feedback. The robot presents 6 degrees of freedom: 5 for positioning and orienting the injector and 1 for setting the depth of needle insertion. The first driver was developed for fully automated low-dose (seed) brachytherapy. Compared with classic templates for needle guides, the robot enables additional freedom of motion for better targeting. As such, multiple needle insertions can be performed through the same skin entry point. In addition, angulations reduce the pubic arch interference, allowing better prostate access for sampling. The MrBot robot was constructed to be ubiquitous with medical imaging equipment such as ultrasound and MRI. An accuracy of 0.72 ± 0.36 mm was described for seed placement. Furthermore, multiple clinical interventions are possible with the MrBot system, such as biopsy, therapy injections, and radiofrequency ablations. Preclinical testing in cadaver and animal models has shown favorable results (Mozer et al., 2009).

6. Future trends in the design and application of surgical robots

Advances in robotic systems accompanied by trends toward miniature devices, microrobots and eventually nanorobots, are realistic near future advancements.

Prototypes of the Microrobot cameras (15 mm /3 inches) were placed inside the abdomen of a canine model during laparoscopic prostatectomy and nephrectomy allowing for additional views (360-degree) of the surgical field The microrobot was mobile, controlled remotely to desired locations, further aiding the laparoscopic procedures (Joseph et al., 2008).

More recently, Lehman et al. showed the feasibility of LESS cholecystectomy in a porcine model, using a miniature robot platform (Lehman et al., 2011). The robot platform incorporated a dexterous in-vivo robot and a remote surgeon interface console. In addition, multiple robots could be inserted through a single incision rather than the traditional use of multiple port sites. Capabilities such as tissue retraction, single incision surgery, supplementary visualization, or lighting can be delivered by these micro-robots.

A well-known limitation of the current robotic platform is the lack of tactile feedback. However, the improved visualization and accuracy of robotic instrumentation may overcome this limitation. Furthermore, newer technologies with multiple compact robots, tracking tools, and tactile feedback apparatus may further expand the application of robotic surgery.

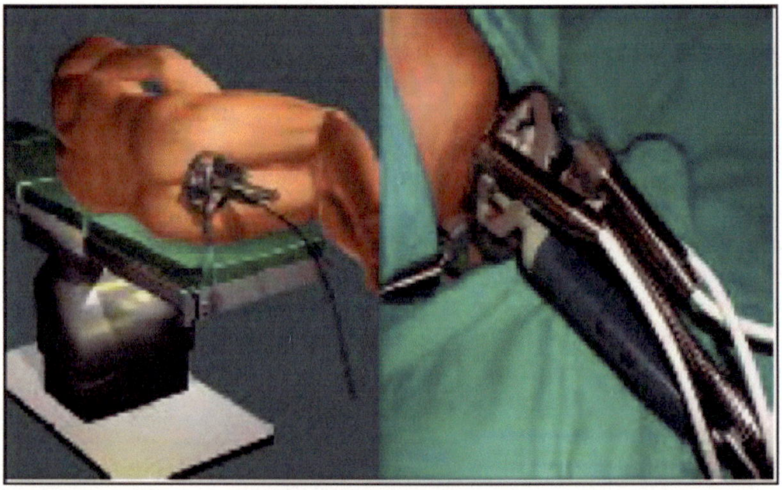

Fig. 8. Robot installation for tranrectal prostate biopsies

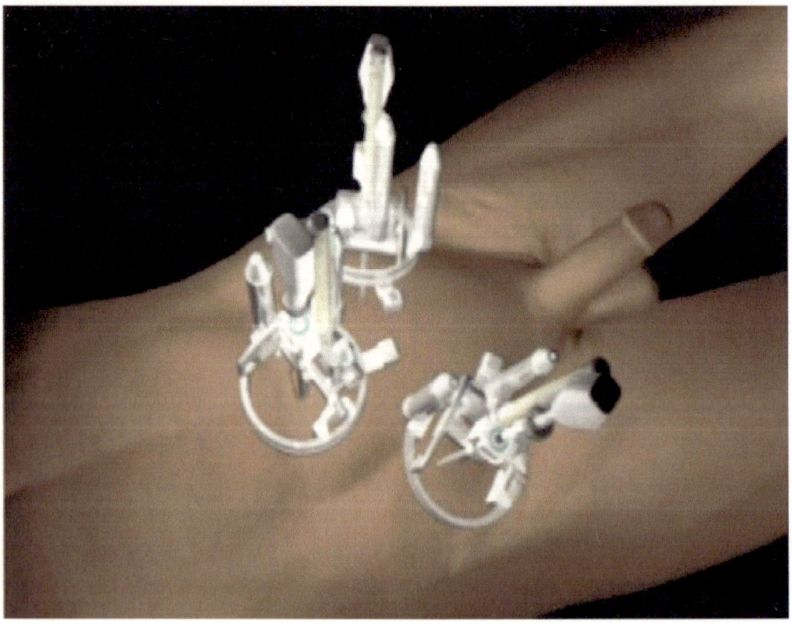

Fig. 9. The combination of multiple compact robots

7. Conclusion

Robotic applications in urology are effective and safe technologies. Surgical technique have evolved and indications have been expanded to more challenging scenarios. New technologies are constantly reshaping the urological sphere especially robotics. Further refinements in robotic systems along with a reduction in cost are key components for rapid assimilation amongst the urological community.

8. References

Abaza R. (2011) Initial series of robotic radical nephrectomy with vena caval tumor thrombectomy. *Eur Urol.* 2011 Apr;59(4):652-6.

Abbou CC, Hoznek A, Salomon L, Lobontiu A, Saint F, Cicco A, Antiphon P, & Chopin D. (2000) [Remote laparoscopic radical prostatectomy carried out with a robot.Report of a case]. *Prog Urol.* 2000 Sep;10(4):520-3.

Barocas DA, Salem S, Kordan Y, Herrell SD, Chang SS, Clark PE, Davis R, Baumgartner R, Phillips S, Cookson MS, & Smith JA Jr. (2010) Robotic assisted laparoscopic prostatectomy versus radical retropubic prostatectomy for clinically localized prostate cancer: comparison of short-term biochemical recurrence-free survival. *J Urol.* 2010 Mar;183(3):990-6.

Bird VG, Leveillee RJ, Eldefrawy A, Bracho J, & Aziz MS. (2011) Comparison of robot-assisted versus conventional laparoscopic transperitoneal pyeloplasty for patients with ureteropelvic junction obstruction: a single-center study. *Urology.* 2011 Mar;77(3):730-4.

Benway BM, Bhayani SB, Rogers CG, Porter JR, Buffi NM, Figenshau RS, & Mottrie A. (2010) Robot-assisted partial nephrectomy: an international experience. *Eur Urol.* 2010 May;57(5):815-20

Boger M, Lucas SM, Popp SC, Gardner TA, & Sundaram CP. (2010) Comparison of robot-assisted nephrectomy with laparoscopic and hand-assisted laparoscopic nephrectomy. *JSLS.* 2010 Jul-Sep;14(3):374-80.

Bolenz C, Gupta A, Hotze T, Ho R, Cadeddu JA, Roehrborn CG, Lotan Y. (2010) Cost comparison of robotic, laparoscopic, and open radical prostatectomy for prostate cancer. Eur Urol. 2010 Mar;57(3):453-8.

Box GN, Lee HJ, Santos RJ, Abraham JB, Louie MK, Gamboa AJ, Alipanah R, Deane LA, McDougall EM, & Clayman RV. (2008) Rapid communication: robot-assisted NOTES nephrectomy: initial report. *J Endourol.* 2008 Mar;22(3):503-6.

Brunaud L, Ayav A, Zarnegar R, Rouers A, Klein M, Boissel P, & Bresler L. (2008a) Prospective evaluation of 100 robotic-assisted unilateral adrenalectomies. *Surgery.* 2008a Dec;144(6):995-1001.

Brunaud L, Bresler L, Ayav A, Zarnegar R, Raphoz AL, Levan T, Weryha G, & Boissel P. (2008b) Robotic-assisted adrenalectomy: what advantages compared to lateral transperitoneal laparoscopic adrenalectomy? *Am J Surg.* 2008b Apr;195(4):433-8.

Cestari A, Buffi NM, Lista G, Sangalli M, Scapaticci E, Fabbri F, Lazzeri M, Rigatti P, & Guazzoni G. (2010) Retroperitoneal and transperitoneal robot-assisted pyeloplasty in adults: techniques and results. *Eur Urol.* 2010 Nov;58(5):711-8.

Desai MM, Aron M, Berger A, Canes D, Stein R, Haber GP, Kamoi K, Crouzet S, Sotelo R, Gill IS. (2008a) Transvesical robotic radical prostatectomy. *BJU Int.* 2008a Dec;102(11):1666-9.

Desai MM, Aron M, Gill IS, Haber GP, Ukimura O, Kaouk JH, Stahler G, Barbagli F, Carlson C, & Moll F. (2008b) Flexible robotic retrograde renoscopy: description of novel robotic device and preliminary laboratory experience. *Urology.* 2008b Jul;72(1):42-6.

Desai MM, Grover R, Aron M, Haber GP, Ganpule A, Kaouk I, Desai M, & Gill IS. (2008c) Remote robotic ureteroscopic laser lithotripsy for renal calculi: initial clinical experience with a novel flexible robotic system. *J Urol* 2008c; 179(4S):435 [Abstract # 1268].

Ficarra V, Novara G, Artibani W, Cestari A, Galfano A, Graefen M, Guazzoni G, Guillonneau B, Menon M, Montorsi F, Patel V, Rassweiler J, & Van Poppel H. (2009) Retropubic, laparoscopic, and robot-assisted radical prostatectomy: a systematic review and cumulative analysis of comparative studies. *Eur Urol.* 2009 May;55(5):1037-63.

Gettman MT, Neururer R, Bartsch G, & Peschel R. (2002a) Anderson–Hynes dismembered pyeloplasty performed using the da Vinci robotic system. *Urology* (2002a) 60(3), 509–513

Gettman MT, Lotan Y, Napper CA, & Cadeddu JA. (2002b) Transvaginal laparoscopic nephrectomy: development and feasibility in the porcine model. *Urology* 2002b; 59:446–450.

Gettman MT, Blute ML, Chow GK, Neururer R, Bartsch G, & Peschel R. (2004) Robotic-assisted laparoscopic partial nephrectomy: technique and initial clinical experience with DaVinci robotic system. *Urology.* 2004 Nov;64(5):914-8.

Giulianotti PC, Buchs NC, Addeo P, Bianco FM, Ayloo SM, Caravaglios G, & Coratti A. (2011) Robot-assisted adrenalectomy: a technical option for the surgeon? *Int J Med Robot.* 2011 Mar;7(1):27-32.

Gupta NP, Nayyar R, Hemal AK, Mukherjee S, Kumar R, & Dogra PN. (2010) Outcome analysis of robotic pyeloplasty: a large single-centre experience. *BJU Int.* 2010 Apr;105(7):980-3.

Haber GP, Crouzet S, Kamoi K, Berger A, Aron M, Goel R, Canes D, Desai M, Gill IS, & Kaouk J. (2008a) Robotic NOTES (Natural Orifice Translumenal Endoscopic Surgery) in reconstructive urology: initial laboratory experience. *Urology.* 2008a Jun;71(6):996-1000.

Haber GP, Kamoi K, Vidal C, Koenig P, Crouzet S, Berger A, Aron M, Kaouk J, Desai M, & Gill IS. (2008b) Development and Evaluation of a Transrectal Ultrasound Robot for Targeted Biopsies and Focal Therapy of Prostate Cancer. EUS 2008 Annual Meeting; 2008b. p. 85.

Haber GP, White WM, Crouzet S, White MA, Forest S, Autorino R, & Kaouk JH. (2010a) Robotic versus laparoscopic partial nephrectomy: single-surgeon matched cohort study of 150 patients. *Urology.* 2010a Sep;76(3):754-8.

Haber GP, White MA, Autorino R, Escobar PF, Kroh MD, Chalikonda S, Khanna R, Forest S, Yang B, Altunrende F, Stein RJ, & Kaouk J. (2010b) Novel robotic da Vinci instruments for laparoendoscopic single-site surgery. *Urology.* 2010b Dec;76(6):1279-82.

Hellenthal NJ, Hussain A, Andrews PE, Carpentier P, Castle E, Dasgupta P, Kaouk J, Khan S, Kibel A, Kim H, Manoharan M, Menon M, Mottrie A, Ornstein D, Palou J, Peabody J, Pruthi R, Richstone L, Schanne F, Stricker H, Thomas R, Wiklund P, Wilding G, & Guru KA. (2011) Lymphadenectomy at the time of robot-assisted radical cystectomy: results from the International Robotic Cystectomy Consortium. *BJU Int.* 2011 Feb;107(4):642-6.

Hellenthal NJ, Hussain A, Andrews PE, Carpentier P, Castle E, Dasgupta P, Kaouk J, Khan S, Kibel A, Kim H, Manoharan M, Menon M, Mottrie A, Ornstein D, Palou J, Peabody J, Pruthi R, Richstone L, Schanne F, Stricker H, Thomas R, Wiklund P, Wilding G, & Guru KA. (2010) Surgical margin status after robot assisted radical cystectomy: results from the International Robotic Cystectomy Consortium. *J Urol.* 2010 Jul;184(1):87-91.

Hemal AK, Mukherjee S, & Singh K. (2010) Laparoscopic pyeloplasty versus robotic pyeloplasty for ureteropelvic junction obstruction: a series of 60 cases performed by a single surgeon. *Can J Urol.* 2010 Feb;17(1):5012-6.

Ho HS, Mohan P, Lim ED, Li DL, Yuen JS, Ng WS, Lau WK, & Cheng CW. (2009) Robotic ultrasound-guided prostate intervention device: system description and results from phantom studies. *Int J Med Robot.* 2009 Mar;5(1):51-8

Horgan S, Vanuno D. (2001) Robots in laparoscopic surgery. *J Laparoendosc Adv Surg Tech* 2001;11:415-9. 15.

Hubens G, Ysebaert D, Vaneerdeweg W, Chapelle T, & Eyskens E. (1999) Laparoscopic adrenalectomy with the aid of the AESOP 2000. *Acta Chir Belg* 1999;99:125-9.

Joseph JV, Oleynikov D, Rentschler M, Dumpert J, & Patel HR. (2008) Microrobot assisted laparoscopic urological surgery in a canine model. *J Urol.* 2008 Nov;180(5):2202-5.

Josephson DY, Chen JA, Chan KG, Lau CS, Nelson RA, & Wilson TG. (2010) Robotic-assisted laparoscopic radical cystoprostatectomy and extracorporeal continent urinary diversion: highlight of surgical techniques and outcomes. *Int J Med Robot.* 2010 Sep;6(3):315-23.

Kaouk JH, Hafron J, Parekattil S, Moinzadeh A, Stein R, Gill IS, & Hegarty N. (2008) Is retroperitoneal approach feasible for robotic dismembered pyeloplasty: initial experience and long-term results. *J Endourol.* 2008 Sep;22(9):2153-9.

Kaouk JH, Goel RK, Haber GP, Crouzet S, Stein RJ. (2009a) Robotic single-port transumbilical surgery in humans: initial report. *BJU Int.* 2009a Feb;103(3):366-9.

Kaouk JH, Goel RK. (2009b) Single-port laparoscopic and robotic partial nephrectomy. *Eur Urol* 2009b; 55:1163–1169.

Kauffman EC, Ng CK, Lee MM, Otto BJ, Wang GJ, & Scherr DS. (2011) Early oncological outcomes for bladder urothelial carcinoma patients treated with robotic-assisted radical cystectomy. *BJU Int.* 2011 Feb;107(4):628-35.

Krambeck AE, DiMarco DS, Rangel LJ, Bergstralh EJ, Myers RP, Blute ML, & Gettman MT. (2009) Radical prostatectomy for prostatic adenocarcinoma: a matched comparison of open retropubic and robot-assisted techniques. *BJU Int.* 2009 Feb;103(4):448-53.

Lehman AC, Wood NA, Farritor S, Goede MR, & Oleynikov D. (2011) Dexterous miniature robot for advanced minimally invasive surgery. *Surg Endosc.* 2011 Jan;25(1):119-23.

Louis G, Hubert J, Ladriere M, Frimat L, & Kessler M. (2009) [Robotic-assisted laparoscopic donor nephrectomy for kidney transplantation. An evaluation of 35 procedures]. *Nephrol Ther.* 2009 Dec;5(7):623-30

Magheli A, Gonzalgo ML, Su LM, Guzzo TJ, Netto G, Humphreys EB, Han M, Partin AW, & Pavlovich CP. (2011) Impact of surgical technique (open vs laparoscopic vs robotic-assisted) on pathological and biochemical outcomes following radical prostatectomy: an analysis using propensity score matching. *BJU Int.* 2011 Jun;107(12):1956-62.

Menon M, Tewari A, Baize B, Guillonneau B, & Vallancien G. (2002) Prospective comparison of radical retropubic prostatectomy and robot-assisted anatomic prostatectomy: the Vattikuti Urology Institute experience. *Urology.* 2002 Nov;60(5):864-8.

Mozer PC, Partin AW, & Stoianovici D. (2009) Robotic image-guided needle interventions of the prostate. Rev Urol. 2009 Winter;11(1):7-15

Patriciu A, Petrisor D, Muntener M, Mazilu D, Schär M, & Stoianovici D. (2007) Automatic brachytherapy seed placement under MRI guidance. IEEE Trans Biomed Eng. 2007 Aug;54(8):1499-506

Pruthi RS, Nielsen ME, Nix J, Smith A, Schultz H, & Wallen EM. (2010) Robotic radical cystectomy for bladder cancer: surgical and pathological outcomes in 100 consecutive cases. *J Urol.* 2010 Feb;183(2):510-4.

Rocco B, Matei DV, Melegari S, Ospina JC, Mazzoleni F, Errico G, Mastropasqua M, Santoro L, Detti S, & de Cobelli O. (2009) Robotic vs open prostatectomy in a laparoscopically naive centre: a matched-pair analysis. *BJU Int.* 2009 Oct;104(7):991-5.

Rogers CG, Singh A, Blatt AM, Linehan WM, & Pinto PA. (2008a) Robotic partial nephrectomy for complex renal tumors: surgical technique. *Eur Urol.* 2008a Mar;53(3):514-21

Scales CD Jr, Jones PJ, Eisenstein EL, Preminger GM, & Albala DM. (2005) Local cost structures and the economics of robot assisted radical prostatectomy. *J Urol.* 2005 Dec;174(6):2323-9.

Stein RJ, White WM, Goel RK, Irwin BH, Haber GP, & Kaouk JH. (2010) Robotic laparoendoscopic single-site surgery using GelPort as the access platform. *Eur Urol.* 2010 Jan;57(1):132-6

Wang AJ, Bhayani SB. (2009) Robotic partial nephrectomy versus laparoscopic partial nephrectomy for renal cell carcinoma: single-surgeon analysis of >100 consecutive procedures. *Urology.* 2009 Feb;73(2):306-10.

White MA, Haber GP, Autorino R, Khanna R, Hernandez AV, Forest S, Yang B, Altunrende F, Stein RJ, & Kaouk JH. (2011) Outcomes of robotic partial nephrectomy for renal masses with nephrometry score of ≥7. *Urology.* 2011 Apr;77(4):809-13.

Wilson T, Torrey R. (2011) Open versus robotic-assisted radical prostatectomy: which is better? *Curr Opin Urol.* 2011 May;21(3):200-5.

Yohannes P, Rotariu P, Pinto P, Smith AD, & Lee BR. (2002) Comparison of robotic versus laparoscopic skills: is there a difference in the learning curve? *Urology.* 2002. Jul;60(1):39-45

7

Reconfigurable Automation of Carton Packaging with Robotic Technology

Wei Yao and Jian S. Dai
King's College London
United Kingdom

1. Introduction

In the highly competitive food market, a wide variety of cartons is used for packaging with attractive forms, eye-catching shapes and various structures from a logistical and a marketing point of view. These frequent changes require innovative packaging. Hence, for some complex styles like origami cartons and cartons for small batch products, most confectionery companies have to resort to using manual efforts, a more expensive process adding cost to the final products, particularly when an expensive seasonal labor supply is required. This manual operating line must go through an expensive induction and training program to teach employees how to erect a carton.

Current machines are only used for the same general style or are specifically designed for a fixed type of cartons and all new cartons require development and manufacture of new tooling for this machinery. New tooling is also required for each different pack size and format. The development and manufacture of this tooling can be very expensive and increases the manufacturer's lead time for introducing new products to market, therefore reducing the manufacturer's ability to match changes in consumer demands. Tooling changeovers, when changing production from one packaging format to another, also adds cost to the manufacturer. It is uneconomical to produce a dedicated machine for a small batch production. Hence, in this high seasonal and competitive market, manufacturer needs a dexterous and reconfigurable machine for their variety and complexity in attracting customers.

The machines are expected to reconfigure themselves to adapt to different types of cartons and to follow different instructions for various closing methods. Rapid expansion of robotics and automation technology in recent years has led to development of robots in packaging industry. Though pick and place robots were extensively used, complex tasks for erecting and closing origami-type cartons still resort to manual operations. This presents a challenge to robotics.

Some related areas with carton folding on the problem of creating 3-D sheet metal parts from sheet metal blanks by bending are explored by some researchers to provide a starting point for exploration in this field when discussing folding sequences based on their objective function, consisting of a high-level planner that determines the sequence of bends, and a number of low level planners for individual actions. deVin (de Vin *et al.,1994)* described a computer-aided process planning system to determine bending sequences for sheet-metal manufacturing. Shpitalni and Saddan (Shpitalni & Saddan., 1994) described a system to automatically generate bending sequences. The domain-specific costs for example the number of tool changes and part reorientations were used to guide the A* search algorithm. Radin *et al.*, (Radin *et al.*, 1997) presented a two-stage algorithm. This method generates a bending sequence using collision avoidance heuristics and then searches for lower cost solutions without violating time constrains. Inui *et al.* (Inui *et al., 1987)* developed a method to plan sheet metal parts and give a bending simulation. Gupta *et al.* (Gupta *et al., 1997*) described a fully automated process planning system with a state-space search approach. Wang and Bourne (Wang & Bourne, 1997) explored a way to shape symmetry to reduce planning complexity for sheet metal layout planning, bend planning, part stacking, and assembly, the geometric constraints in carton folding parallel those in assembly planning. Wang (Wang, 1997) developed methods to unfold 3-D products into 2-D patterns, and identified unfolding bend sequences that avoided collisions with tools.

In computational biology, protein folding is one of the most important outstanding problems that fold a one-dimensional amino acid chain into a three-dimensional protein structure. Song and Amato (Song & Amato, 2001a, 2001b) introduced a new method by modeling these foldable objects as tree-like multi-link objects to apply a motion planning technique to a folding problem. This motion planning approach is based on the successful *probabilistic roadmap* (PRM) method (Kavraki *et al.*, 2001a, 2001b) which has been used to study the related problem of ligands binding (Singh *et al.*, 1999, Bayazit *et al.*, 2001). Advantages of the PRM approach are that it efficiently covers a large portion of the planning space and it also provides an effective way to incorporate and study various initial conformations, which has been of interest in drug design. Molecular dynamics simulations are the other methods (Levitt *et al*, 1983, Duan. *et al.*, 1993) which tried to simulate the true dynamics of the folding process using the classic Newton's equations of motion. They provided a means to study the underlying folding mechanism, to investigate folding pathways, and provided intermediate folding states.

Rapid expansion of robotics and automation technology in recent years has led to development of robots in packaging industry. Automation of the carton folding process involves several subtasks including automatic folding sequence generation, motion planning of carton folding, and development of robotic manipulators for reconfigurability. Though pick and place robots have been extensively used, complex tasks for erecting and closing origami-type cartons still resort to manual operations. This presents a challenge to robotics.

Investigating origami naturally leads to the study of folding machinery. This evolves to folding a carton. Lu and Akella (Lu & Akella, 1999, 2000) in 1999 by exploring a fixture technique to describe carton folding based on a conventional cubical carton using a SCARA-type robot and on the similarity between carton motion sequences with robot operation sequences. This approach uses the motion planner to aid the design of minimal complexity hardware by a human designer. But the technology only focused on rectangular cartons not on origami-type cartons which include spherical close chains on vertexes. Balkcom and Mason (Balkcom & Mason, 2004, Balkcom et al., 2004) investigated closed chain folding of origami and developed a machine designed to allow a 4DOF Adept SCARA robot arm to make simple folds. However for complex cartons including multi-closed chains the flexibility and reconfigurability of this method are limited. This requires a quantitative description of a carton, its folding motion and operations. Dai (Dai & Rees Jones, 2001, 2002, 2005, Liu & Dai, 2002, 2003, Dubey &Dai, 2006) and Yao (Yao & Dai, 2008, 2010, 2011) at King's College London developed a multi-fingered robotic system for carton folding.

2. Equivalent mechanism of an origami carton

A carton is formed with crease lines and various geometric shapes of panels. Folding and manipulating a carton is a process of changing the position of panels and their relationship. Taking creases as revolute joints and cardboard panels as links, a carton can be modeled as a mechanism. The kinematic model of this equivalent mechanism provides motion of carton folding.

Carton folding starts from a flat cardboard consisting of a set of panels with creases connecting adjacent panels. All the required motions involve carton panels folding about a crease-line through an angle. There are many ways to fold the carton box into its final state, which are based on the location and orientation of the folding. For the simple cartons the manipulation can be assumed to be at one joint between adjacent boards and each joint moves to its goal configuration. The motion path and sequence can be easily calculated by the joint angles of the carton. Folding sequence can be checked by following steps, including checking the reduced graph, which is checking the equivalent mechanism structures of cartons, to see if they are symmetric or have any other geometric characteristics; identifying independent branches which result in independent operations; searching folding layers in turn; and manipulating from the layer which is closest to the base, from the symmetric panels and from the node which is present in the shortest branch.

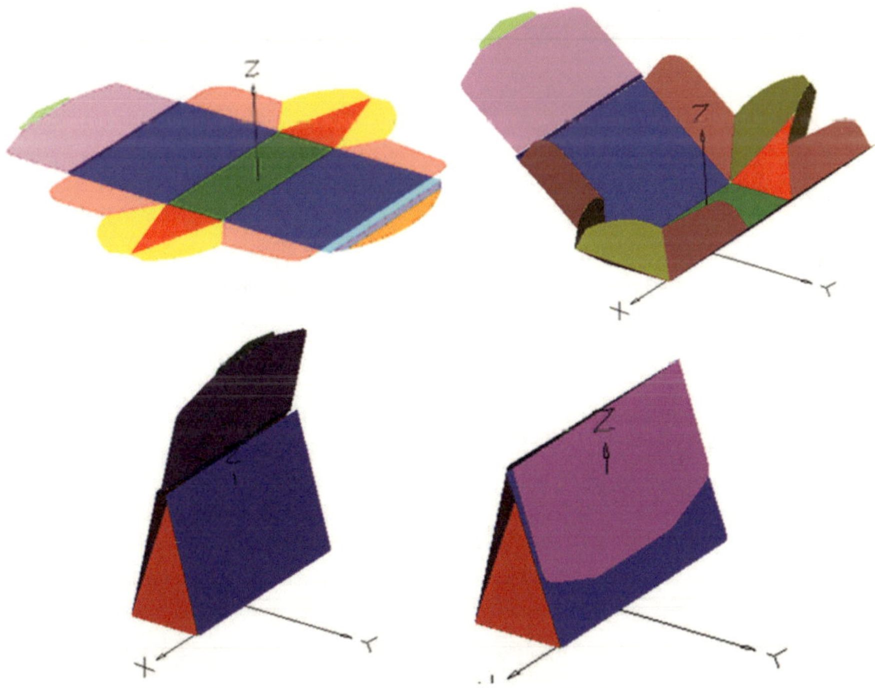

Fig. 1. A Carton Folding Sequence for an Origami Carton

During the process of the Origami carton folding the guiding joint can be identified. Four closed loops with four folding sub-mechanisms are found. For the sub-mechanism of a guiding joint, the mechanism can be described as a 5-Bar mechanism that the panels are described linkage and the creases are revolute joints. It is means that the reaction forces at the joint of the box paper do not produce unexpected deformation. Five revolute joints axes that intersect in the point that is the intersection of the creases define the spherical linkage and their movement is defined.

Figure 2 gives a half-erected origami carton which consists of four sections at vertexes located at four corners of the carton labeled as O_1, O_2, O_3, and O_4Folding manipulation of the carton is dominated by these sections. The rectangular base of the carton is fixed on the ground. Four creases along the rectangular base are given by four vectors $\boldsymbol{b_1}$, $\boldsymbol{b_2}$, $\boldsymbol{b_3}$ and $\boldsymbol{b_4}$ Four centrally movable creases are given as $\mathbf{s}$, $\mathbf{s}''$, $\mathbf{s}''$ and $\mathbf{s}'''$ each of which is active in a gusset corner. The carton is symmetric along line AA′.

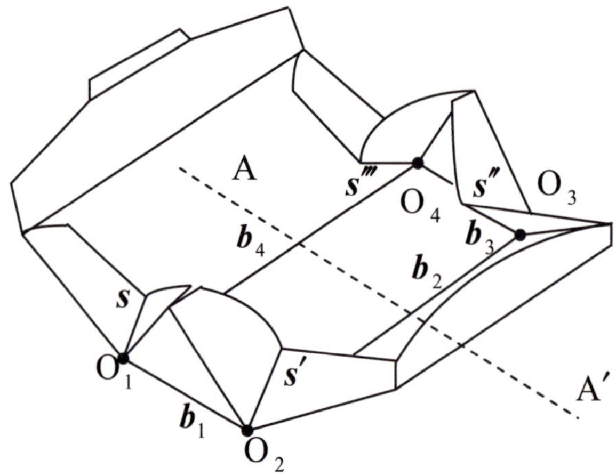

Fig. 2. A half-erected origami

Thus an equivalent mechanism is modeled in figure 3.4. The five bar mechanism's centres are located at four corners of the carton base O_1, O_2, O_3, and O_4.From the equivalent mechanism and folding process, the five-bar spherical mechanism can determine the configuration of the carton.

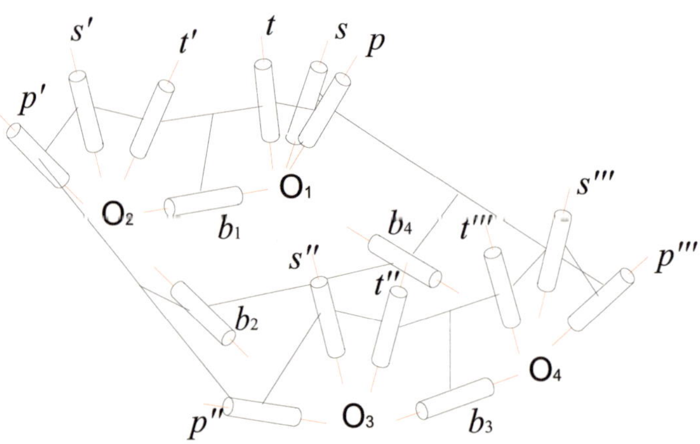

Fig. 3. The equivalent mechanism of the base of a double wall tray

One of the corners located at O_1 formed one spherical mechanism. When its folding joint *s* moves and controls the configuration of the mechanism and drives joints ***t***, and ***p***. When the two joints attached the mobility of the mechanism reduces from 2 to 1 and the mechanism gets its singularity. The joint ***s*** is active and called control joint. During the manually folding of this origami carton human fingers attaches the control joints and drives them into goal positions.

3.Geometry analysis of the mechanism

The geometric arrangement is defined in figure 4.

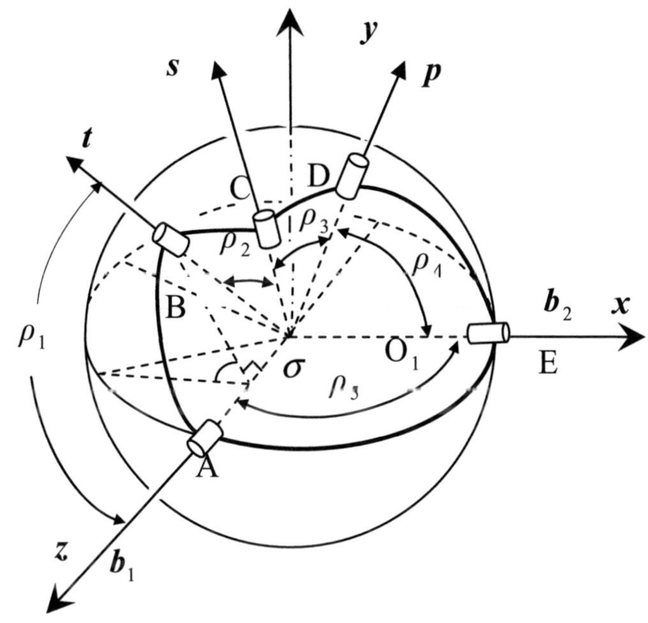

Fig. 4. Geometry of spherical five-bar mechanism

Among joint axes ***t, s, p*** and fixed axes $\boldsymbol{b_1}$ and $\boldsymbol{b_2}$ configuration control vector ***s*** provides the guiding joint axis that controls the carton section configuration. The spherical links between joint axes are labeled as AB, BC, CD, DE, and EA. Since angles ρ_i between revolute joints are constant throughout the movement of the linkage, the coordinates of the joint axes ***t, s,*** and ***p*** can be obtained by matrix operation from original joint axis $\boldsymbol{b_1}$ which is a unit vector. This relationship of the axes can be given in terms of geometric constraints as,

$$\rho_1 = \arccos(\boldsymbol{b} \cdot \boldsymbol{t})$$

$$\rho_2 = \arccos(\boldsymbol{t} \cdot \boldsymbol{s})$$

$$\rho_3 = \arccos(\boldsymbol{s} \cdot \boldsymbol{p})$$

$$\rho_4 = \arccos(\boldsymbol{p} \cdot \boldsymbol{d})$$

$$\rho_5 = \arccos(\boldsymbol{d} \cdot \boldsymbol{b}) \tag{1}$$

Further, the configuration control vector **s** can be determined as,

$$\mathrm{s} = \mathrm{R}(-{}^1\boldsymbol{y}, 90^0 + \phi)\mathrm{R}(-{}^1\boldsymbol{x}', \xi)\boldsymbol{b}_1 =$$

$$= \begin{bmatrix} \mathrm{c}(90^0 + \phi & 0 & -\mathrm{s}(90^0 + \phi) \\ 0 & 1 & 0 \\ \mathrm{s}(90^0 + \phi) & 0 & \mathrm{c}(90^0 + \phi) \end{bmatrix} \begin{bmatrix} 1 & 0 & 0 \\ 0 & \mathrm{c}\xi & \mathrm{s}\xi \\ 0 & -\mathrm{s}\xi & \mathrm{c}\xi \end{bmatrix} \begin{bmatrix} 0 \\ 0 \\ 1 \end{bmatrix} = \begin{bmatrix} -\mathrm{c}\phi\mathrm{c}\xi \\ \mathrm{s}\xi \\ -\mathrm{s}\phi\mathrm{c}\xi \end{bmatrix} \tag{2}$$

The side elevation vector $\boldsymbol{t}$ is given as,

$$\boldsymbol{t} = \mathrm{R}(-{}^1\boldsymbol{z}, \sigma)\mathrm{R}(-{}^1\boldsymbol{y}', \rho_1)\boldsymbol{b}_1 =$$

$$= \begin{bmatrix} \mathrm{c}\sigma & \mathrm{s}\sigma & 0 \\ -\mathrm{s}\sigma & \mathrm{c}\sigma & 0 \\ 0 & 0 & 1 \end{bmatrix} \begin{bmatrix} \mathrm{c}\rho_1 & 0 & -\mathrm{s}\rho_1 \\ 0 & 1 & 0 \\ \mathrm{s}\rho_1 & 0 & \mathrm{c}\rho_1 \end{bmatrix} \begin{bmatrix} 0 \\ 0 \\ 1 \end{bmatrix} = \begin{bmatrix} -\mathrm{c}\sigma\mathrm{s}\rho_1 \\ \mathrm{s}\sigma\mathrm{s}\rho_1 \\ \mathrm{c}\rho_1 \end{bmatrix} \tag{3}$$

Further the geometric constraint gives,

$$\boldsymbol{t} \cdot \boldsymbol{s} = \mathrm{c}\rho_2 \tag{4}$$

Substituting $\boldsymbol{t}$ of equation (2) and **s** of equation (3) into above yields

$$\mathrm{c}\rho_2 = c\phi\ c\xi\ c\sigma\ s\rho_1 + s\xi\ s\sigma\ s\rho_1 - s\phi\ c\rho_1\ c\xi \tag{5}$$

Collecting the same terms with $\mathrm{c}\sigma$ and $\mathrm{s}\sigma$, the above equation can be rewritten as,

$$A(\phi,\xi)\mathrm{c}\sigma + B(\phi,\xi)\mathrm{s}\sigma - C(\phi,\xi) = 0, \tag{6}$$

Where

$$\begin{aligned} A(\phi,\xi) &= s\rho_1\ c\xi\ c\phi, \\ B(\phi,\xi) &= s\xi\ s\rho_1, \\ C(\phi,\xi) &= -s\phi\ c\rho_1\ c\xi - \mathrm{c}\rho_2. \end{aligned} \tag{7}$$

The side elevation angle σ varying with the transmission angles ϕ and ξ can hence be given:

$$\sigma(\phi,\xi) = \tan^{-1}(\frac{A}{B}) \pm \cos^{-1}(\frac{C}{\sqrt{A^2 + B^2}}) \tag{8}$$

Further, the main elevation vector $\boldsymbol{p}$ can be obtain as,

$$\boldsymbol{p} = \mathrm{R}({}^1\boldsymbol{y}, \rho_5)\mathrm{R}({}^1\boldsymbol{z}', \varsigma)\mathrm{R}({}^1\boldsymbol{y}', \rho_4)\boldsymbol{b}_1 \tag{9}$$

Hence,

$$p = \begin{bmatrix} c\rho_5 & 0 & s\rho_5 \\ 0 & 1 & 0 \\ -s\rho_5 & 0 & c\rho_5 \end{bmatrix} \begin{bmatrix} c\varsigma & -s\varsigma & 0 \\ s\varsigma & c\varsigma & 0 \\ 0 & 0 & 1 \end{bmatrix} \begin{bmatrix} c\rho_4 & 0 & s\rho_4 \\ 0 & 1 & 0 \\ -s\rho_4 & 0 & c\rho_4 \end{bmatrix} \begin{bmatrix} 0 \\ 0 \\ 1 \end{bmatrix} \tag{10}$$

Since $\rho_4 = \rho_5 = \pi / 2$, the above becomes,

$$p = \begin{bmatrix} 0 \\ s\varsigma \\ -c\varsigma \end{bmatrix} \tag{11}$$

Substituting $\boldsymbol{p}$ of equation (4.15) and $\boldsymbol{s}$ of equation (4.6) into equation (4.3) gives,

$$c\rho_3 = c\varsigma\ s\phi\ c\xi + s\varsigma\ s\xi \tag{12}$$

Assembling the same terms of $c\varsigma$ and $s\varsigma$, the above can be rearranged as,

$$D(\sigma,\xi)c\varsigma + E(\sigma,\xi)s\varsigma + F = 0 \tag{13}$$

Where,

$$\begin{aligned} &D(\sigma,\xi) = s\phi\ c\xi \\ &E(\sigma,\xi) = s\xi \\ &F = -c\rho_3 \end{aligned} \tag{14}$$

The main elevation angle ς varying with the transmission angles ϕ and ξ can hence be obtained,

$$\varsigma(\phi,\xi) = \tan^{-1}(\frac{D}{E}) \pm \cos^{-1}(\frac{F}{\sqrt{D^2+E^2}}) \tag{15}$$

The control vector $\boldsymbol{s}$ also can be gotten from the transformation:

$$s = \boldsymbol{R}({}^1z',\sigma)\boldsymbol{R}({}^1y',\rho_1)\boldsymbol{R}({}^1z',\mu)\boldsymbol{R}({}^1y',\rho_2)b_1 \tag{16}$$

The vector can be expressed,

$$s = \begin{bmatrix} -c\rho_1 s\rho_2 c\mu c\sigma - c\rho_2 s\rho_1 c\sigma + s\sigma s\mu s\rho_2 \\ -s\sigma s\mu s\rho_2 + c\sigma s\mu s\rho_2 \\ -s\rho_2 s\rho_2 c\mu + c\rho_2 c\rho_1 \end{bmatrix} \tag{17}$$

Two columns are equal and can be easily got,

$$c\mu = \frac{c\rho_3 c\varsigma - c\rho_2 c\rho_1}{s\rho_2^2} \tag{18}$$

$$\mu(\sigma,\varsigma)=\arccos\frac{c\rho_3 c\varsigma - c\rho_2 c\rho_1}{s\rho_2^2} \tag{19}$$

Hence, the elevation angles are related to the transmission angles that control the configuration vector ***s***. Both sets of angles feature the carton configuration. The following Figure 5 shows the end-point trajectories of four configuration control vector.

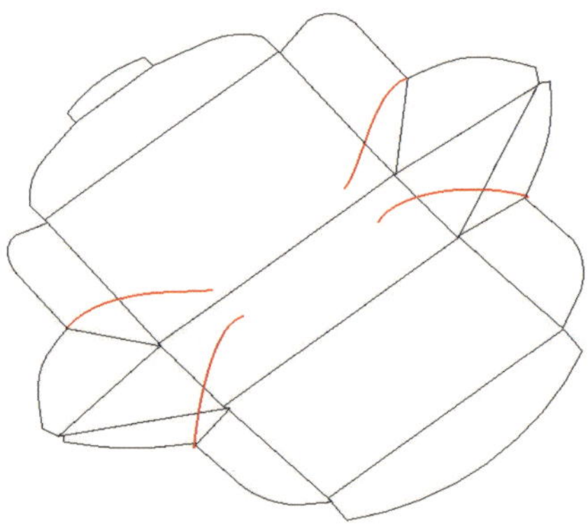

Fig. 5. End-point trajectories of four configuration control vectors

4. Velocity-control formulation of the single-vertex on the origami carton

The kinematic representation of the spherical mechanism is showed an above figure which is in the Cartesian space

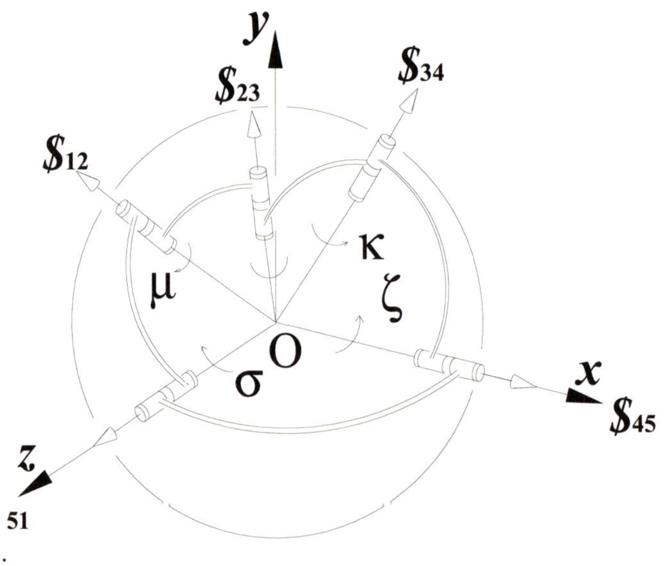

Fig. 6. Screw of the spherical five-bar mechanism

Since the system is spherical the fixed point o can be selected as represent point and the system only have 3 rotation parts in Plücker coordinates the infinitesimal screw are given by, $\$_{51} = (0,0,1;\vec{0})$, $\$_{12} = (-c\sigma s\rho_1, s\sigma s\rho_1, c\rho_1;\vec{0})$, $\$_{45} = (1,0,0;\vec{0})$, $\$_{34} = (0, s\varsigma, -c\varsigma;\vec{0})$, $\$_{23} = (\vec{s};\vec{0})$. $\$_{23} = (\vec{s};\vec{0})$.

Where, **s** is control vector that got from last section
Assume that a velocity of control vector *v* is,

$$v = \dot{\varsigma}\$_{45} + \dot{\kappa}\$_{34} = \dot{\sigma}\$_{15} + \dot{\mu}\$_{21} + \omega_s\$_{32} \tag{20}$$

Thus after canceling the dual parts of the infinitesimal screw the joint speeds can be computed as,

$$\begin{bmatrix} -\$_{34} & \$_{21} & \$_{32} \end{bmatrix} \begin{bmatrix} \dot{\kappa} \\ \dot{\mu} \\ \omega_s \end{bmatrix} = -\dot{\varsigma}\$_{45} + \dot{\sigma}\$_{15} \tag{21}$$

With the application of Cramer's rule,

$$\dot{\kappa} = \frac{\left| -\dot{\varsigma}\boldsymbol{b}_2 + \dot{\sigma}\boldsymbol{b}_1 \ \boldsymbol{t} \ \boldsymbol{p} \right|}{\left| \boldsymbol{t} \ \boldsymbol{p} \ \boldsymbol{s} \right|} \tag{22}$$

For the coefficients can be gotten,

$$G_1 = \frac{-s\kappa s\varsigma + s\rho_3 c\kappa c\varsigma + s\varsigma}{\begin{matrix} c\sigma c\varsigma s\rho_1 s\kappa s\rho_3 c\rho_3 - c\sigma c\varsigma s\rho_1 c\kappa s\varsigma \\ -s\kappa s\varsigma + s\rho_3 c\kappa c\varsigma + s\varsigma c\rho_1 - s\varsigma c\varsigma s\rho_1 \end{matrix}} \tag{23}$$

then,

$$G_2 = \frac{c\kappa c\varsigma s\varsigma - s\rho_3 s\kappa c\varsigma c\rho_3 + s\varsigma c\varsigma c\rho_3}{\begin{matrix} c\sigma c\varsigma s\rho_1 s\kappa s\rho_3 c\rho_3 - c\sigma c\varsigma s\rho_1 c\kappa s\varsigma \\ -s\kappa s\varsigma + s\rho_3 c\kappa c\varsigma + s\varsigma c\rho_1 - s\varsigma c\varsigma s\rho_1 \end{matrix}} \tag{24}$$

So, the velocity of the control joint **s** can be obtained,

$$\boldsymbol{v} = \dot{\varsigma}\boldsymbol{b}_2 + \dot{\varsigma}G_1 + \dot{\sigma}G_2 \tag{25}$$

Where,

$$\omega_s \begin{bmatrix} s\kappa s\varsigma - s\rho_3 c\kappa c\varsigma \\ c\kappa s\varsigma - s\rho_3 s\kappa c\rho_3 \\ -c\rho_3 c\varsigma \end{bmatrix} = \begin{bmatrix} -c\sigma s\rho_1 & 1 & 0 & 0 \\ s\sigma s\rho_1 & 0 & 0 & s\varsigma \\ c\rho_1 & 0 & 1 & -c\varsigma \end{bmatrix} \begin{bmatrix} \omega_t \\ \omega_{b_1} \\ \omega_{b_2} \\ \omega_p \end{bmatrix} \tag{26}$$

where, the matrix $\begin{bmatrix} -c\sigma s\rho_1 & 1 & 0 & 0 \\ s\sigma s\rho_1 & 0 & 0 & s\varsigma \\ c\rho_1 & 0 & 1 & -c\varsigma \end{bmatrix}$ is the Jacobean matrix for the control of the system.

5. Simulation of automatically folding origami carton using a multi-fingered robotic system

The robot has been designed based on the analysis of common motion and manipulation, which has four fingers in the following figure 7, two with three degrees of freedom noted finger 1 and finger 2, and two noted with finger 3 and finger4 with two degrees of freedom each. These fingers design are able to offer required folding trajectories by changing the control programs of the fingers' motors. Two horizontal jaws are arranged with the pushing direction parallel to the fingers' horizontal tracks. The joints of the fingers are actuated directly by motors and the whole system requires 14 controlled axes. This design is considered specifically based on providing high degree of reconfigurability with minimum number of motors to be controlled.

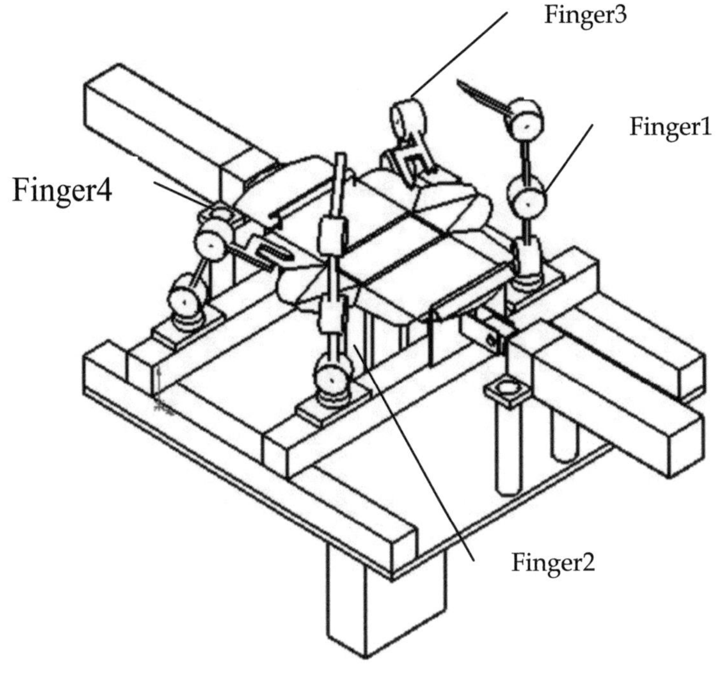

Fig. 7. A Multi-fingered robot for origami carton folding

To determine the control setting and speeds of the actuated joints for the multi-finger robotic system when fingertips are in special location and twists the *Jacobian* [J] need to be identified as.

$$[\$_s]=[J][\omega_i] \tag{27}$$

For corner1 which located in O_1 in configuration space the main parts of the screw is shown in equation 25.

Expanding to the four fingers robotic system, each finger related one Conner manipulation which their velocities is got from four five bar's control joints twist.

While ,

$$\begin{bmatrix} v_1 & v_2 & v_3 & v_4 \end{bmatrix} \tag{28}$$

Then the Jacobean matrix of the robotic system is determined as $\left[\$_{ij}\right]$, where i for finger's number and j for motors number.

$$\left[\omega_{ij}\right]=[J]^{-1}[v_i] \tag{29}$$

Then every motor's speeds and control can be set.
The whole folding process can be simulated by the multi-fingered robotic system in the following figure 8.
This methodology of multi-robotic manipulations using configuration space transformation in interactive configuration space is the theoretical base for the further development of multi-finger reconfigurable systems. This test rig has given experimental knowledge and information for further reconfigurable systems in packaging industry. In this robotic system, all motors of the robotic fingers and the two horizontal jaws are controlled by a computer. For different types of cartons, the system just need be changed the programs based on their folding trajectories without readjusting the hardwires of the system.

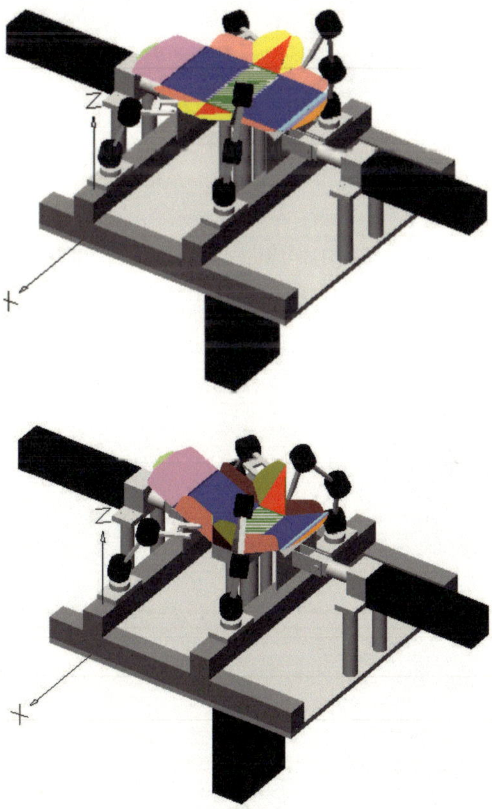

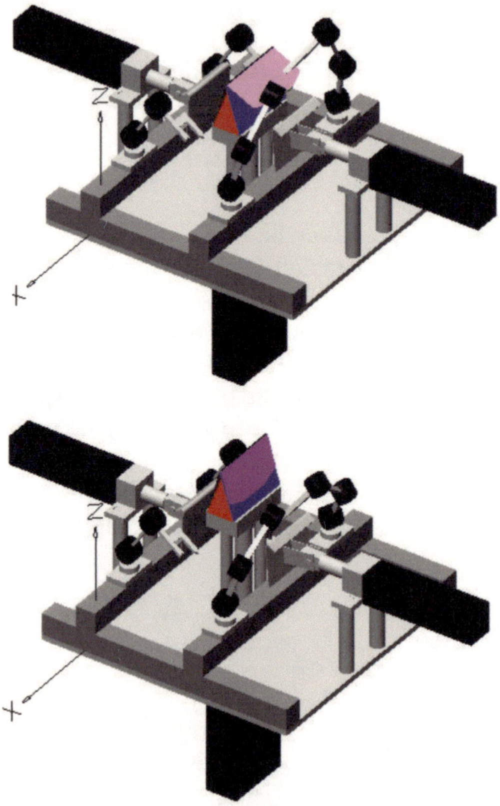

Fig. 8. Simulation of a multi-fingered manipulation on an origami carton

Fig. 9. A multi-fingered robotic system

6. Conclusion

This paper has presented new approaches to carton modelling and carton folding analysis. By investigating carton examples, with an analysis of the equivalent mechanism, the gusset vertexes of the cartons - being a common structure of cartons - were extracted and analyzed based on their equivalent spherical linkages and were identified as guiding linkages that determine folding.

A reconfigurable robotic system was designed based on the algorithms for multi-fingered manipulation and the principle of reconfigurability was demonstrated. The simulation results for the folding of an origami carton were given to prove the theory of the multi-fingered manipulation strategy and the concept of reconfigurable robotic systems. Test rigs were developed to demonstrate the principles of the reconfigurable packaging technology.

7. References

Balkcom, D.J., Demaine, E.D. and Demaine, M.L. Folding paper shopping bags, In *Proceedings of the 14th Annual Fall Workshop on Computational Geometry*, Cambridge, Mass., Nov. 19-20. 2004.

Balkcom, D.J.and Mason, M.T. Introducing Robotic Origami Folding, *Proc. 2004 IEEE International Conference on Robotics and Automation*, 2004.

Bayazit, O. B., Song, G. and Amato, N. M. Ligand binding with obprm and haptic user input: Enhancing automatic motion planning with virtual touch, In *Proc. IEEE Int. Conf. Robot. Autom. (ICRA)*, 2001. To appear. This work will be presented as a poster at RECOMB'01.

Dai, J. S. and Rees Jones, J. Interrelationship between screw systems and corresponding reciprocal systems and applications, *Mech.Mach.Theary*, 36(3), pp633-651, 2001.

Dai, J. S. and Rees Jones, J., Kinematics and Mobility Analysis of Carton Folds in Packing Manipulation Based on the Mechanism Equivalent, *Journal of Mechanical Engineering Science, Proc, IMechE, Part C*, 216(10): 959-970, 2002.

Dai, J. S. and Rees Jones, J., Matrix Representation of Topological Configuration Transformation of Metamorphic Mechanisms, *Transactions of the ASME: Journal of Mechanical Design*, 127(4): 837-840, 2005

de Vin, L. J., de Vries, J., Streppel, A. H., Klaassen, E. J. W. and Kals, H. J. J., The generation of bending sequences in a CAPP system for sheet metal components, *J. Mater. Processing Technol.*, vol. 41, pp. 331–339, 1994.

Duan, Y. and Kollman, P.A. Pathways to a protein folding intermediate observed in a 1-microsecond simulation in aqueous solution, *Science*, 282:740–744, 1998.

Dubey, V. N., and Dai, J.S., A Packaging Robot for Complex Cartons, *Industrial Robot: An International Journal*, ISSN: 0143-991X, 33(2): 82-87, 2006.

Gupta, S. K., Bourne, D. A., Kim, K. H. and Krishnan, S. S., Automated process planning for sheet metal bending operations, *J. Manufact. Syst.*, vol. 17, no. 5, pp. 338–360, 1998.

Inui, M., Kinosada, A., Suzuki, H., Kimura, F. and Sata, T., Automatic process planning for sheet metal parts with bending simulation, in *ASME Winter Annu. Meeting, Symp. on Intelligent and Integrated Manufacturing Analysis and Synthesis*, 1987, pp. 245–258.

Kavraki, L., Geometry and the discovery of new ligands, In *Proc. Int. Workshop on Algorithmic Foundations of Robotics (WAFR)*, pages 435–448, 1996.

Kavraki, L., Svestka, P., Latombe, J. C. and Overmars, M., Probabilistic roadmaps for path planning in high-dimensional configuration spaces, *IEEE Trans. Robot. Automat.*, 12(4):566–580, August 1996.

Levitt, M., Protein folding by restrained energy minimization and molecular dynamics, *J. Mol. Bio.*, 170:723–764, 1983.

Liu, H. and Dai, J.S., Carton Manipulation Analysis Using Configuration Trasformation, *Journal of Mechanical Science, Proc. IMechE.* 216(C5), pp. 543-555,2002.

Liu,H. and Dai, J.S., An Approach to Carton-Folding Trajectory Planning Using Dual Robotic Fingers, *Robotics and Autonomous Systems*, 42(1), pp. 47-63, 2003.

Lu, L and Akella, S., Folding Cartons with Fixtures: A Motion Planning Approach, *IEEE Trans on Robotics and Automation*, vol.16, no. 4, pp. 346-356, August 2000.

Lu, L. and Akella, S. Folding Cartons with Fixtures: A Motion Plannin Approach, *Proc of the IEEE International Conference on Robotics and Automation*, May 1999.

Radin, B., Shpitalni, M. and Hartman, I., Two-stage algorithm for determination of the bending sequence in sheet metal products, *ASME J. Mechan. Design*, vol. 119, pp. 259–266, 1997.

Radin, B., Shpitalni, M. and Hartman, I., Two-stage algorithm for determination of the bending sequence in sheet metal products, *ASME J. Mechan. Design*, vol. 119, pp. 259–266, 1997.

Shpitalni, M. and Saddan, D., Automatic determination of bending sequence in sheet metal products, *Ann. CIRP*, vol. 43, no. 1, pp. 23–26, 1994.

Singh, A.P., Latombe, J.C. and Brutlag, D.L. A motion planning approach to flexible ligand binding, In *7th Int. Conf. on Intelligent Systems for Molecular Biology (ISMB)*, pages 252–261, 1999.

Song, G. and Amato, N. M., A motion planning approach to folding: From paper craft to protein folding, In *Proc. IEEE Int. Conf. Robot. Autom. (ICRA)*, pages 948–953, 2001.

Song, G. and Amato, N. M., Using motion planning to study protein folding pathways, In *Proc. Int. Conf. Comput. Molecular Biology (RECOMB)*, pages 287–296, 2001.

Wang, C.-H. and Bourne, D. A., Using symmetry to reduce planning complexity: A case study in sheet-metal production, in *Proc. 1997 ASME Design Engineering Technical Conf.*, Sacramento, CA, Sept. 1997, DETC97DFM4327.

Wang, C.-H., Manufacturability-Driven Decomposition of Sheet Metal Products, Ph.D. dissertation, The Robotics Inst., Carnegie Mellon Univ., Pittsburgh, PA, Robotics Inst. Tech. Rep. CMU-RI-TR-97-35, Sept. 1997.

Yao,W. and Dai,J.S., Dexterous Manipulation of Origami Cartons with Robot Fingers in the Interactive Configuration Space, *Transactions of the ASME: Journal of Mechanical Design*, 130(2),pp.022303_1-8, 2008.

Yao,W., Cannella,F and Dai,J.S., Automatic Folding of Cartons Using a Reconfigurable Robotic System, *Robotics and Computer-Integrated Manufacturing*. 27(3), pp 604-613, 2011.

Yao,W., Dai,J.S., Medland,T. and Mullineux,G., A Reconfigurable Robotic Folding System for Confectionery Industry, *Industrial Robot: An International Journal*, 37(6), pp 542-551, 2010.

8

Autonomous Anthropomorphic Robotic System with Low-Cost Colour Sensors to Monitor Plant Growth in a Laboratory

Gourab Sen Gupta[1], Mark Seelye[1], John Seelye[2] and Donald Bailey[1]
[1]*School of Engineering and Advanced Technology (SEAT), Massey University, Palmerston North,*
[2]*The New Zealand Institute for Plant & Food Research Limited, Palmerston North*
New Zealand

1. Introduction

An autonomous anthropomorphic robotic arm has been designed and fabricated to automatically monitor plant tissue growth in a modified clonal micro-propagation system which is being developed for the New Zealand Institute for Plant & Food Research Limited. The custom-fabricated robotic arm uses a vertical linear ball shaft and high speed stepper motors to provide the movements of the various joints, with the arm able to swivel 180 degrees horizontally. Sensors located at the end of the arm monitor plant growth and the ambient growing environment. These include a compact colour zoom camera mounted on a pan and tilt mechanism to capture high resolution images, RGB (red, green and blue) colour sensors to monitor leaf colour as well as sensors to measure ambient atmospheric temperature, relative humidity and carbon dioxide. The robotic arm can reach anywhere over multiple trays (600mm x 600mm) of plantlets. Captured plant tissue images are processed using innovative algorithms to determine tissue, or whole plant, growth rates over specified time periods. Leaf colour sensors provide information on the health status of tissue by comparing the output with predetermined optimum values. Custom software has been developed to fully automate the operation of the robotic arm and capture data, allowing the arm to return to specified sites (i.e. individual plantlets) at set time intervals to identify subtle changes in growth rates and leaf colour. This will allow plant nutrient levels and the immediate environment to be routinely adjusted in response to this continuous sensing, resulting in optimised rapid growth of the plant with minimal human input.
These low cost colour sensors can be incorporated into a continuous automated system for monitoring leaf colour of growing plants. Subtle colour changes can be an early indication of stress from less than optimum nutrient concentrations. In this chapter we also detail the calibration technique for a RGB sensor and compare it with a high end spectrophotometer.

2. Robotic systems in agriculture

Robotic and automated systems are becoming increasingly common in all economic sectors. In the past decade there has been a push towards more automation in the horticulture

industry, and it is only now, as robots become more sophisticated and reliable, that we are beginning to see them used to undertake routine, often repetitive tasks, which are expensive to do using a highly paid labour force. With rapid strides in technological advancement, more and more applications have become possible. These include the development of a robotic system for weed control (Slaughter, et al., 2008), a system for automatic harvesting of numerous agri-science products such as cutting flowers grown in greenhouses (Kawollek & Rath, 2008) and automating cucumber harvesting in greenhouses (van Henten, et al., 2002). Advances in electronics have empowered engineers to build robots that are capable of operating in unstructured environments (Garcia, et al., 2009). Camera-in-hand robotic systems are becoming increasingly popular, wherein a camera is mounted on the robot, usually at the hand, to provide an image of the objects located in the robot's workspace (Kelly, et al., 2000). Increasingly, robots are being used to sort, grade, pack and even pick fruits. Fruits can be identified and classified on a continuously moving conveyer belt (Reyes & Chiang, 2009). An autonomous wheeled robot has been developed to pick kiwifruit from orchard vines (Scarfe, et al., 2009). Robotic techniques for production of seedlings have been developed, identifying a need to add a machine vision system to detect irregularities in seed trays and to provide supplementary sowing using a 5-arm robot (HonYong, et al., 1999).
Automation of micro propagation for the rapid multiplication of plants has been described for the micro propagation of a grass species that replaces the costly and tedious manual process (Otte, et al., 1996). A system has also been developed that combines plant recognition and chemical micro-dosing using autonomous robotics (Sogaard & Lund, 2007).
Colour as a means of assessing quality is also gaining popularity amongst researchers. These include evaluating bakery products using colour-based machine vision (Abdullah, et al., 2000), monitoring tea during fermentation (Borah & Bhuyan, 2005), grading specific fruits and vegetables (Kang & Sabarez, 2009; Miranda, et al., 2007; Omar & MatJafri, 2009) and in the health sector to determine blood glucose concentrations (Raja & Sankaranarayanan, 2006). Near infrared (NIR) sensors are also gaining popularity as non-destructive means of assessing fruit and plant material, including the measurements of plant nutrient status (Menesatti, et al., 2010) as well as testing of fruit quality (Hu, et al., 2005; Nicolaï, et al., 2007; Paz, et al., 2009).
Investigation into non-destructive methods to measure the health status of plants is a relatively new area of research. Subtle leaf colour changes can be used as a measure of plant health. Although limited work has been carried out in real time, a recent micro-propagation based system used potato tissue images captured via a digital camera, to identify the colour of selected pixels (Yadav, et al., 2010). Spectral reflectance, using a range of spectral bands, has been used as a non-destructive measure of leaf chlorophyll content in a range of species (Gitelson, et al., 2003). Alternative methods make use of spectroscopic systems using a fixed light source to record colour reflectance of multiple samples (Yam & Papadakis, 2004).
The introduction of machine vision as a means of investigating the environment allows for very complex systems to be developed. Over the years the conventional "robotic design types" have remained more or less the same; however modified designs are increasingly being employed for specific tasks. Designs of robotic arms have made huge progress in recent years, as motor controllers, sensors and computers have become more sophisticated. It is envisaged that as more sensors, such as NIR (near infra-red) and colour sensors, become readily available, these will be integrated in the robotic arm. One such integrated system, which is unique and different from off-the-shelf robots, is detailed in this chapter.

3. Overview of the robotic system

The reported robotic system has been developed to work in a specific environment using specific sensors - it is meant to monitor growth of plant tissues in a laboratory. The plantlets are growing in multiple trays (600mm x 600mm) which are arranged contiguously on a shelf and there are multiple shelves one above the other. The robot should thus be able to extend its reach vertically and monitor plants in each shelf and be capable of reaching each tray. The robotic system designed is based on a SCARA (Selective Compliant Assembly/Articulated Robot Arm) robot. However, SCARA robots are rigid in the Z-axis and pliable in the XY-axes only. This rigidity in the Z-axis is a serious limitation of a SCARA robot for the given scenario. In the proposed system the entire arm is able to move in the Z-axis, as opposed to just the end-effector.

Another point of difference between the proposed system and conventional off-the-shelf industrial robot is the mechanism to house the sensors. To monitor the growth of plants, colour camera and RGB sensors are used. To enable the robot to position itself at a desired distance from the plant surface, proximity sensors are also required. The sensors need to be housed in an enclosure at the end of the robotic arm. In order to capture images and take RGB sensor readings from any angle it should be possible to pan and tilt the sensor housing structure. Such a mechanism is not a standard part of a conventional off-the-shelf industrial robot. In general, the costs of industrial robotic systems are far greater than the proposed system, often a lot more bulky and it is hard to integrate additional components (i.e. Sensors).

Two prototypes of the camera-in-hand robotic system were designed and built. The initial prototype made use of servo motors, designed as a simple experiment to test the viability of the system and its control mechanism. The colour camera was incorporated in this prototype and its control was implemented. The captured images were stored in a database for subsequent retrieval and processing. The prototype also helped to experiment with the wireless remote control of the robotic arm and the remote setting of camera features such as zoom, gain and exposure. Having established the 'proof-of-concept', the second prototype was designed and built to industrial specifications. This version of the prototype made use of high-torque stepper motors and greatly improved the performance of the robotic arm. Additionally, this prototype incorporated low-cost RGB colour sensors for monitoring plant health together with a proximity sensor. Sensors to monitor the ambient atmosphere were also incorporated to measure temperature, humidity and CO_2 levels. Section 3.1 gives a concise overview of the first prototype and section 3.2 details the second prototype in greater depth.

3.1 Robotic arm using servo motors

The basic concept of this robotic system, from human input to the control of the arm and camera, is outlined in the functional block diagram shown in figure 1. A system engineering approach was employed to take the robotic arm from concept to reality, making use of standard components and integrating them together to make the final product. The robotic arm was designed using 5 servo motors and implemented a pan and tilt mechanism for the camera.

The operator uses a joystick to control the movement of the robotic arm. This joystick connects to the PC via a USB interface. Movements of the joystick, made by the operator, vary the slide bars on the Graphical User Interface (GUI) running on the PC and at the same time control the movement of the joints of the arm. Serial data is then sent via the USB to the wireless transmitter (Zigbee Pro) module which transmits the data to the wireless receiver module. The received data is then sent to the camera and the servo controller board. The user interface is shown in figure 2 and the completed robotic system, with plant trays, is shown in figure 3.

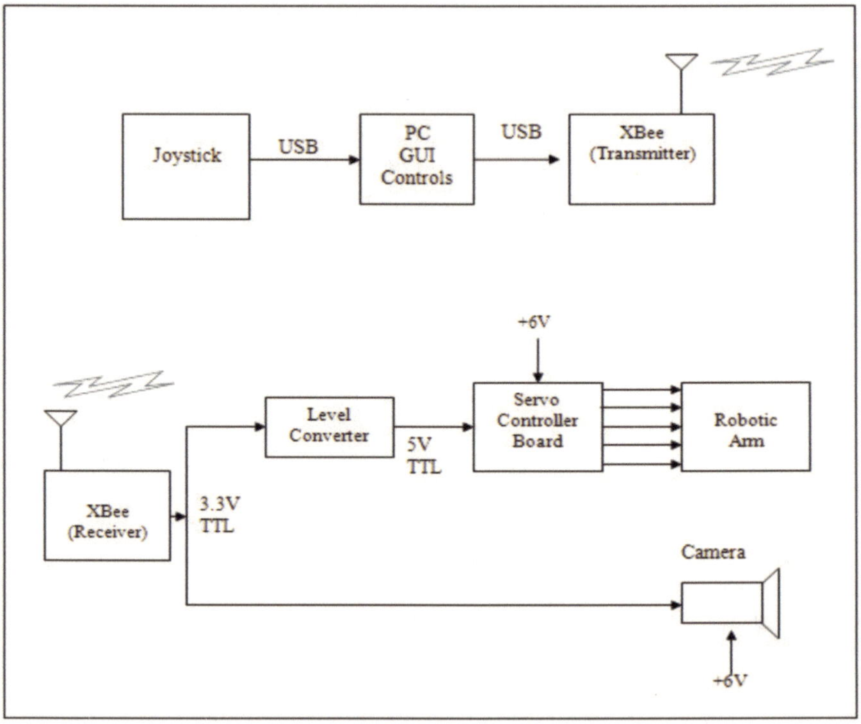

Fig. 1. Functional block diagram of the camera-in-hand robotic system

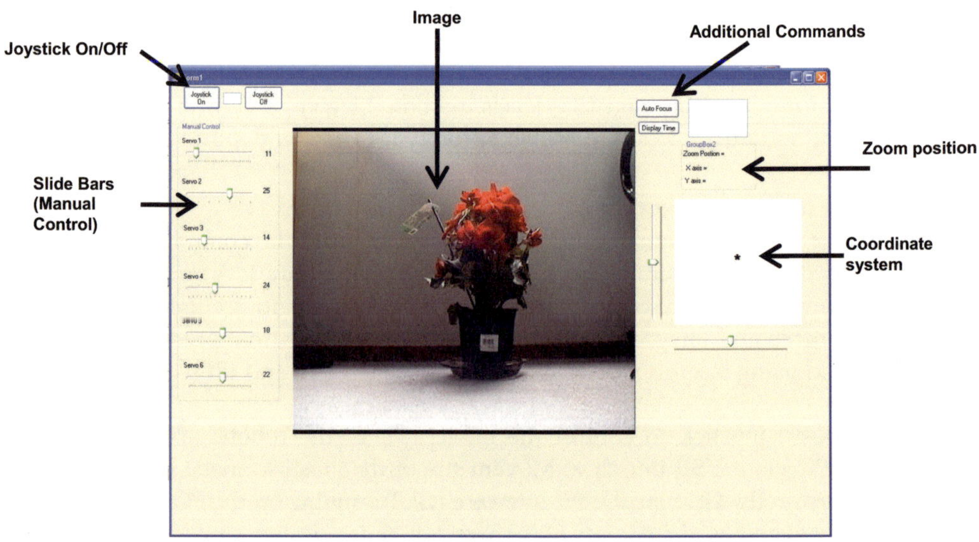

Fig. 2. The Graphical User Interface (GUI) of the robot control system

Fig. 3. The completed robotic arm with camera, joystick and plant trays

3.1.1 Control of servo motor based robotic arm

Using a modular approach the system was built and tested in parts. The sub-systems that had to be programmed, such as the servo-controller, joystick and camera, were tested separately. The testing setup and procedures are explained in this section. The servo controller board can take two types of serial mode signals - USB and 5V TTL UART. The controller board, together with the joystick, was tested using the connection diagram shown in figure 4.

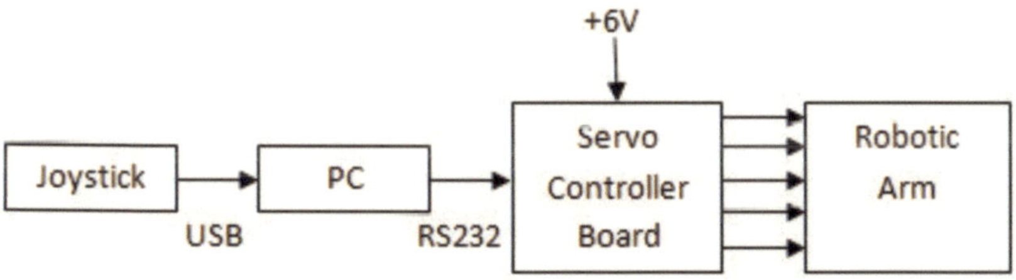

Fig. 4. Block diagram showing connection between PC, servo controller board and joystick

In the first instance a simple program was written in Visual Basic, allowing each servo motor to be controlled separately by clicking buttons. The motor parameters such as stepping rate and acceleration interval could be entered through the program's user interface. The user interface of the test program is shown in figure 5. This program sent the corresponding commands to the servo motor controller board. The next step was to control the servo motor by implementing a slide bar (figure 2). This allowed the operator to slide the bar, which incremented or decremented the position value, allowing simple movements based on the position byte. On successfully implementing controls for one servo motor, multiple servo motors could then be controlled in the same manner.

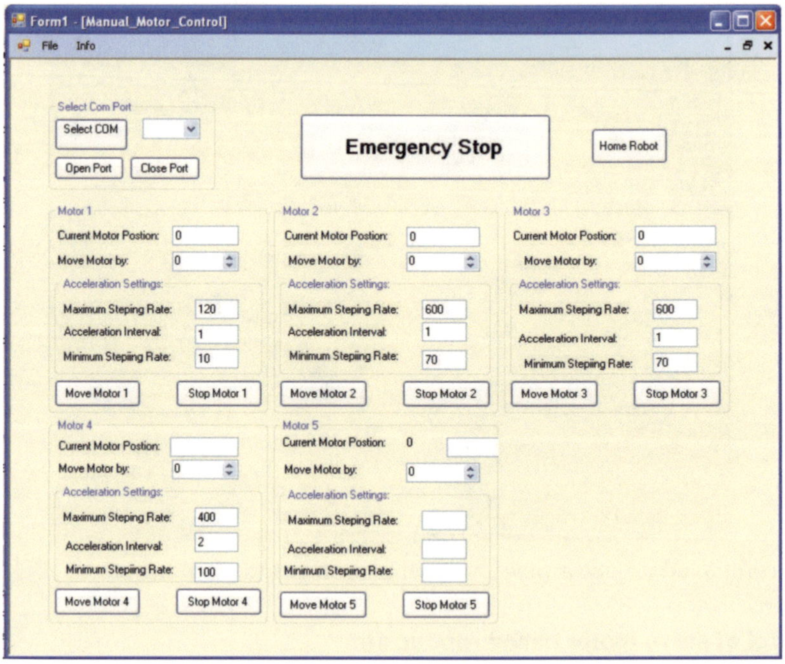

Fig. 5. User interface of the test program to control individual motors

3.2 Robotic arm using stepper motors

The second design of the robot was extensively influenced by the knowledge and insight gained from the servo motor based design of the robotic arm. The robotic arm was designed to move the end-effector over trays of plant material located on different levels of shelving units to capture images and the colour of the plant material and to use this information as a non-destructive measurement of plant health. To achieve this, a compact colour camera ("Sony Professional," 2010), proximity and colour sensor ("Parallax Home," 2010) are housed in the end-effector. Each tray measures approximately 600mm x 600mm, with each shelf located approximately 300mm above the previous shelf, with the top shelf approximately 1000mm high. The system is designed to move the arm into the trays, capture the required information and then move up to the next shelf and repeat the process on the next tray.

3.2.1 Mechanical design of the robotic arm

To allow the robotic arm to move vertically, a ball screw and shaft assembly is incorporated, converting rotational motion into vertical movement. The conceptual design is shown in figure 6. The arm contains a pan and tilt system at its distal end, which houses the camera, colour and proximity sensors. The operation of the arm is completely automated, continually gathering information from the sensors and capturing images for assessment and analysis.

The design is based on a modified SCARA robotic arm. Designed in the 3D CAD package, Solidworks, all components where machined in-house using a CNC machine. The arm itself has been through a number of design iterations to overcome unforeseen problems and to improve the efficiency of operation.

The robotic arm went through a number of design phases, with each design an improvement over the previous design iteration. In the initial concept it was intended to

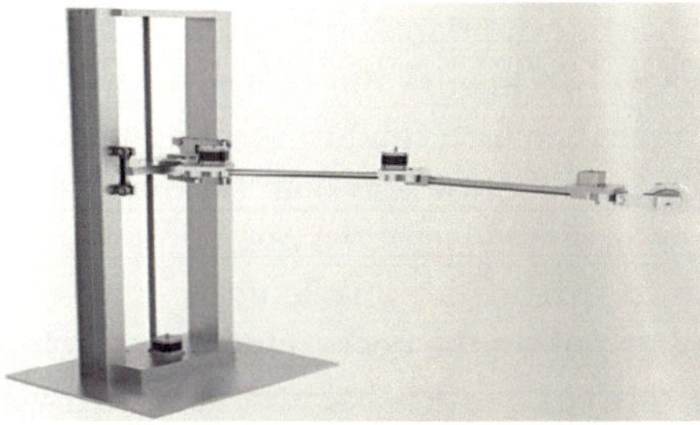

Fig. 6. SolidWorks rendered design of the robotic arm

Fig. 7. The system in various stages of development and integration

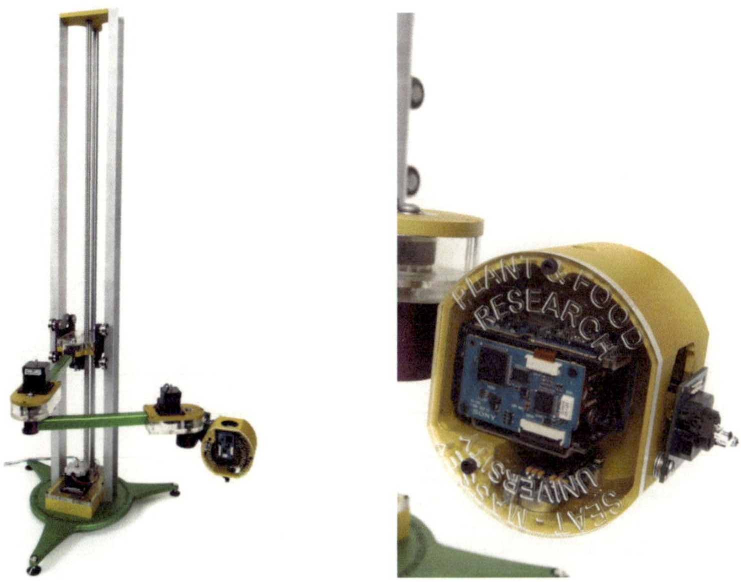

Fig. 8. (a) Completed robotic arm using stepper motors (b) Camera housing and RGB colour Sensors

make the links of the robotic arm extendable so that the robot can be flexible and adapted to operate in various workspaces. To ensure the motor torque ratings were not exceeded, a gearing system was investigated, which made use of spur gears to increase torque ratios. However, once constructed, a number of issues arose, including excessive weight (with the links extended) and slippage between gears. To overcome these issues a rethink of the arm design removed the ability to extend the link lengths, and a belt and pulley system was integrated to alleviate slippage within the gears. However, each design iteration maintained the overall original design concept. Figure 7 shows an initial version of the robotic arm in various stages of assembly and figure 8 shows the final design, with the various aluminium parts anodized. The completed robotic arm is shown in figure 8 (a). The close up view of the camera housing mechanism and the RGB colour sensors is shown in figure 8(b).

3.2.2 Actuation and control of the robotic arm joints using stepper motors

To allow the various joints to move, the arm uses bipolar, high torque stepper motors, which have varying amounts of torque, depending on the joint. The robotic arm uses five stepper motors that are controlled through a motor controller (KTA-190) and micro-step driver (M542) ("OceanControls," 2010). All the five motors have a step angle of 1.8 degrees and make use of a micro step driver that allows the user to select an even finer resolution (i.e. increasing the number of steps per revolution). Both a pulse signal and a direction signal are required for connecting a 4-wire stepper motor to the driver, with speed and torque depending on the winding inductance.

The KTA-190 motor controllers provide an interface between the computer and up to 4 stepper motors as well as the ability to control each motor independently or collectively. Utilizing a RS-232 9600, 8N1 ASCII serial communication protocol, up to 4 controller boards can be linked, giving control of up to 16 stepper motors (figure 9). A motor is controlled by a simple address, followed by the appropriate ASCII commands. The controller has as interface to allow limit switches to be used to prevent the motors from travelling out of range. With a total of 17 commands it is possible to tailor the operation and move the motors. Commands include: setting the position of the motors, returning the current positions of the motors, moving the motors by a relative or absolute amount and acceleration settings. A status command returns a 12-bit binary representation on the status of the controller board at any given time, providing information on the movement, direction and status of the limit switch respectively.

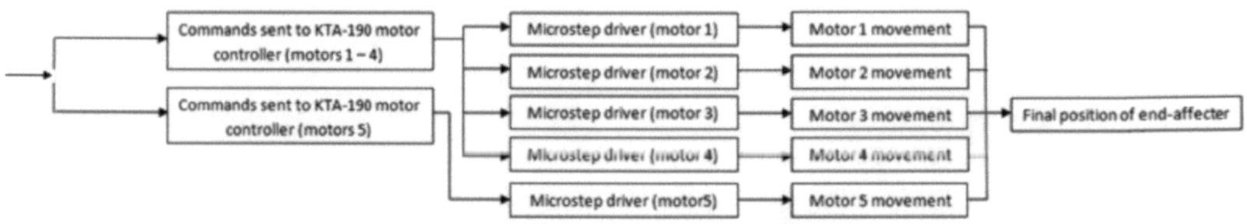

Fig. 9. Block diagram showing control of the 5 stepper motors

The angle each motor is required to move is calculated via an inverse kinematic algorithm. The user simply enters the desired tray that is required to be monitored, along with the number (and frequency) of readings within the tray. The software then calculates the required motor positions to enable the camera and sensors to capture the required information. A proximity sensor has been integrated into the system to ensure that the colour readings are taken at a fixed height of 20mm above the plant material.

4. Sensors

Sensors provide information on the surroundings, which vary depending on their intended applications. A number of sensors have been integrated into the system to provide information in a non-destructive method on plant growth, with the intention of using low cost sensors which are easily amendable into the system. The use of a camera provides information on how fast the plant is growing, by identifying the quantity of plant material from a 2D view. Colour sensors provide information on the colour of an object. When colour sensors are used to monitor plant leaves, subtle changes can be detected before the human eye can identify them. This allows for media and nutrient levels to be adjusted accordingly. A proximity sensor ensures colour readings are taken at a fixed height, while temperature, humidity and CO_2 sensors provide information on the ambient environment.
We detail the camera control and testing in section 4.1. Colour sensor selection, along with calibration techniques and results, are presented in detail in section 4.2.

4.1 Camera

A Sony colour camera (model: FCB-IX11AP) was used. It features a ¼" CCD (charge coupled device) image sensor using PAL (Phase Alternating Line) encoding system. The camera has a 40x zoom ratio (10x optical, 4x digital) that is controllable from a PC via Sony's VISCA (Video System Control Architecture) command protocol. Over 38,000 different command combinations are possible for controlling the cameras features. The camera's macro capability allows it to capture images as close as 10mm from the subject and it can operate in light conditions as low as 1.5 Lux. The electronic shutter speed is controllable from 1/10 to 1/10000 of a second allowing for clarity in photographs. In order for the camera's functions to be controlled, hexadecimal commands (as serial data) are sent to the camera. These serial commands require 8 data bits, 1 start bit, 1 (or 2) stop bit and no parity. They have a communication speed of 9.6, 19.2 or 38.4 kbps. The camera can be programmed and controlled using a TTL or RS232C signal level serial interface. To test the camera features, it was directly wired to the PC using the RS232C interface via a USB-to-RS232 converter as shown in figure 10. The video output signal from the camera was fed to a frame grabber/digitizer which is interfaced to the PC using USB. The image captured is displayed on the application's GUI (figure 12).

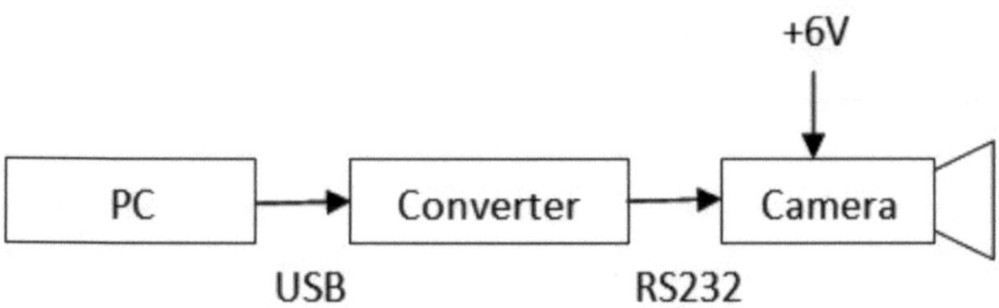

Fig. 10. Block diagram showing wired connection of PC to camera's RS232 inputs

To familiarize with the VISCA command structure and to test the various camera functions, especially the programming commands for controlling the zoom, a standard communication program (Terminal v1.9b) was used to send commands to the camera. To test the TTL interface, the system shown in figure 11 was employed. IC ST232 was used to convert the RS232 level signals to 5V TTL.

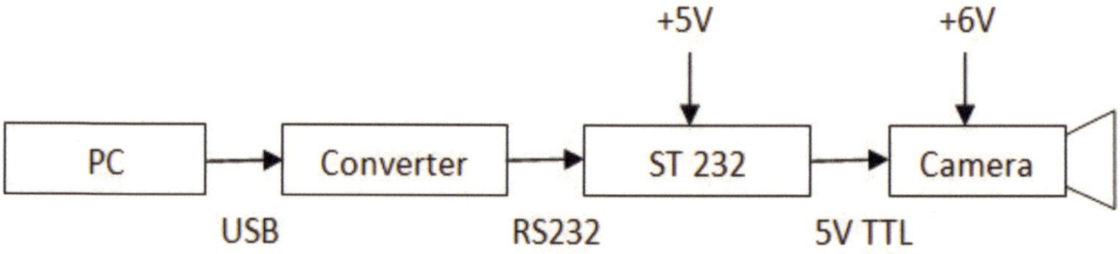

Fig. 11. Block diagram showing wired connection of PC to camera's TTL inputs

In the final design of the robotic system the camera was connected using the original RS-232 interface, with custom software created to control the features of the camera. Figure 12 shows the user interface which allows camera zoom, iris, gain, exposure settings and shutter features to be manually controlled. It also displays the image captured by the camera. Figure 13 shows the test program depicting the magnified image of the object under observation, with the camera set to 8x zoom.

Each of the controllable settings for the camera is controlled by sending arrays of hexadecimal commands to the camera, making use of Sony's VISCA protocol. Custom created software allows the user to control a number of settings to customize the camera to the user's desire.

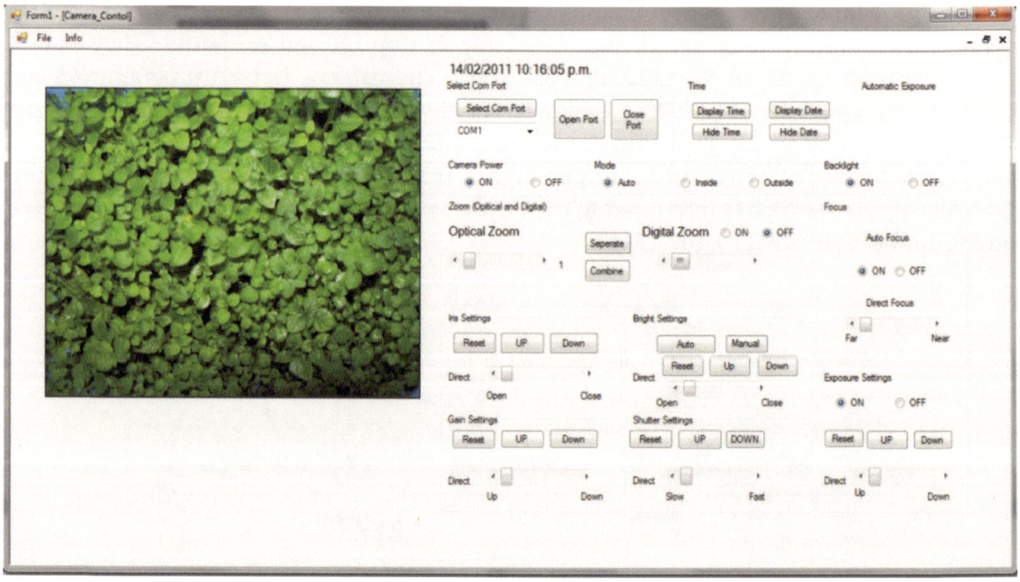

Fig. 12. User interface of the camera control program

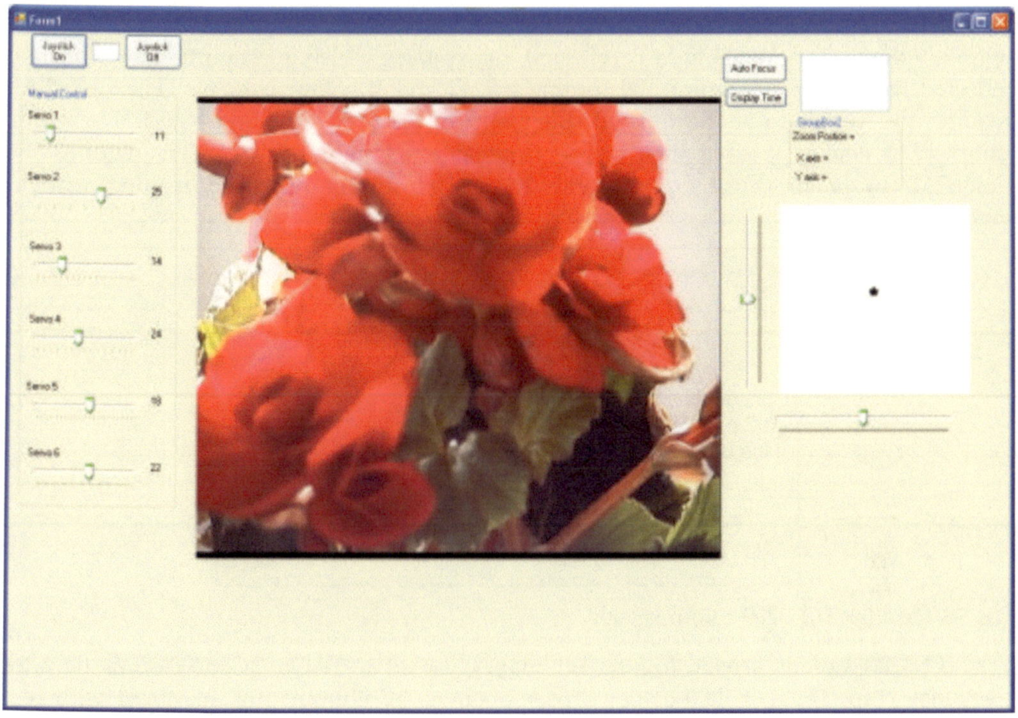

Fig. 13. Graphical User Interface (GUI) of the remote control application (camera zoom 8x)

4.2 RGB colour sensors

Currently there are a number of colour sensors on the market, with prices ranging from low cost light-to-frequency chips to sophisticated and very expensive spectrophotometers. Parallax (Parallax Inc, CA, USA) has two colour sensors that integrate seamlessly with their Basic Stamp microcontroller. Both the ColorPAL and TCS 3200 colour sensors are provided with some source code, making them amenable to integrating with our customised system.

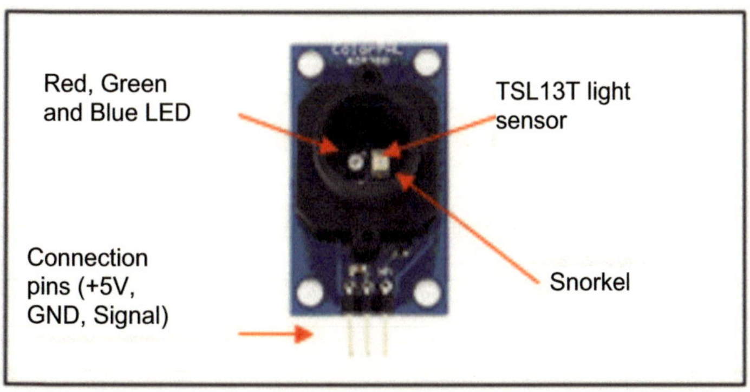

Fig. 14. Parallax ColorPAL colour sensor

The ColorPAL sensor (figure 14) illuminates a sample using in-built red, green and blue LED light sources (one colour at a time) and records the quantity of light reflected back from the object. The ColorPAL makes use of a TAOS (Texas Advanced Optoelectronic Solutions) light-to-voltage chip. When light is reflected, the voltage, which is proportional to the light reflected, is used to determine the sample's R, G and B colour contents. The ColorPAL requires the sample to be illuminated using each of the red, green and blue LEDs, with a 'snorkel' to shield possible interference from external light sources. This requires the ColorPAL to be in direct contact with the object for an optimum reading without interference.

Fig. 15. Parallax TCS3200 colour sensor

The TCS3200 Colour sensor (figure 15) makes use of a TAOS TCS3200 RGB light-to-frequency chip. The TCS3200 colour sensor operates by illuminating the object with two white LEDs, while an array of photo detectors (each with a red, green, blue and clear filter) interpret the colour being reflected by means of a square wave output whose frequency is proportional to the light reflected. The TSC3200 Colour sensor has a 5.6-mm lens, which is positioned to allow an area of 3.5 mm^2 to be viewed.
A USB4000 spectrometer (Ocean Optics Inc., FL, USA) was used to find the height at which the greatest intensity of light occurred when the RGB sensor was placed above a sample. As the two white LEDs are directed down at an angle, there is a point where the light intensity is the greatest. This position was 20 mm above the surface of the sample, as shown in figure 16.

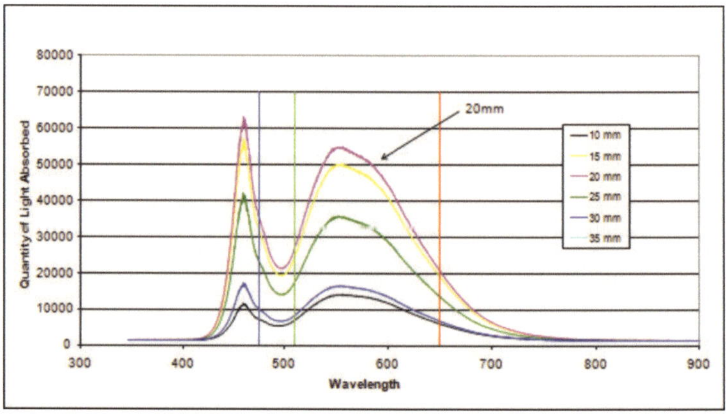

Fig. 16. Light absorbed from TCS3200 across the white LED light spectrum when the sensor is positioned at 6 different heights

Since the TCS3200 is mounted 20 mm above the sample, and therefore not in direct contact with the sample, it was more suited for our application than the full contact required by the ColorPAL sensor. A Konica Minolta CM-700D Spectrophotometer (Konica Minolta Sensing Americas, Inc., NJ, USA) was used to validate and calibrate the RGB sensors. For accurate measurements, the CM-700D was calibrated by taking both white and black readings by sampling a supplied white and black object respectively.
The CM-700D gives colour in the XYZ colour space, as well as L*a*b*, L*C*h, Hunter Lab, Yxy and Munsell. A linear transformation matrix was required to transform data from the XYZ colour space to the RGB colour space to enable comparisons and calibrations to the Parallax sensor. The linear transformation equation to be solved (Juckett, 2010) is:

$$\begin{pmatrix} X \\ Y \\ Z \end{pmatrix} = M \times \begin{pmatrix} R \\ G \\ B \end{pmatrix} \quad (1)$$

$$x = \frac{X}{X+Y+Z} \quad (2)$$

$$y = \frac{Y}{X+Y+Z} \quad (3)$$

$$z = \frac{Z}{X+Y+Z} \quad (4)$$

Equations (2 - 4) combined with the standard 1931 xy chromaticity diagram provided the foundation for the linear transformation (Eq. 1). This transformation converted the XYZ data initially to sRGB colour. The chromaticity values of x, y and z are shown in Table 1 (Lindbloom, 2010).

Colour	x	y	z
Red	0.64	0.33	0.212656
Green	0.30	0.60	0.715158
Blue	0.15	0.06	0.072186

Table 1. x, y, and z chromaticity values of red, green and blue converting xyz to sRGB

From the x, y and z chromaticity values (Table 1), the transformation matrix, M, is calculated (Eq. 5)

$$M \approx \begin{pmatrix} 0.721144 & 0.063298 & 0.166008 \\ 0.303556 & 0.643443 & 0.052999 \\ 0.000076 & 0.064689 & 1.024294 \end{pmatrix} \quad (5)$$

To calculate the R, G and B values the inverse is taken (Eq. 6 - 7).

$$\begin{pmatrix} R \\ G \\ B \end{pmatrix} = M^{-1} \times \begin{pmatrix} X \\ Y \\ Z \end{pmatrix} \quad (6)$$

$$M^{-1} \approx \begin{pmatrix} 1.436603 & -0.118534 & -0.226698 \\ -0.681279 & 1.618476 & 0.026671 \\ 0.042919 & -0.102206 & 0.974614 \end{pmatrix} \tag{7}$$

Colour	x	y	z
Red	0.7350	0.2650	0.176204
Green	0.2740	0.7170	0.812985
Blue	0.1670	0.0090	0.010811

Table 2. x, y, and z chromaticity values of red, green and blue converting xyz to CIE RGB

The x, y and z chromaticity values shown in Table 2, are again used to solve the transformation matrix, M (Eq. 8)
Testing showed that the TCS3200 produced light in the CIE RGB colour space, and although results would later show the colour sensor could be calibrated to the sRGB colour space, results from the calibration would be incorrect. Therefore a colour transformation to a CIE RGB colour space was more appropriate than the sRGB colour space; consequently a new linear transformation was required.

$$M \approx \begin{pmatrix} 0.4887180 & 0.3106803 & 0.2006017 \\ 0.1762044 & 0.8129847 & 0.0108109 \\ 0.0000000 & 0.0102048 & 0.9897952 \end{pmatrix} \tag{8}$$

Again calculating the R, G and B values the inverse is taken (Eq. 6).

$$M^{-1} \approx \begin{pmatrix} 2.3706743 & -0.9000405 & -0.4706338 \\ -0.5138850 & 1.4253036 & 0.0885814 \\ 0.0052982 & 0.0052982 & 1.0093968 \end{pmatrix} \tag{9}$$

4.2.1 Colour sensor calibration and results

In order to validate the TCS3200 colour sensor, it was necessary to calibrate and test it using the CM-700D Spectrophotometer. This involved taking 200 RGB readings with the TCS3200 using fifteen different coloured samples from the Royal Horticulture Society (RHS) colour charts and averaging them. The same samples were measured, each 20 times, with the CM-700D and again averaged. These tests were all completed in a constant temperature dark room. As the CM-700D uses the XYZ colour space, the linear transformation matrix was used to convert the XYZ values to a CIE RGB colour space (Eq. 9).
The TCS3200 was firstly calibrated through software by modifying the integration time, to allow the white object (used to calibrate the CMD-700) to have a RGB value as close as possible to 255,255,255 followed by scaling each of the RGB values, to ensure the white reading was that of the CMD-700.
In order to calculate a calibration factor the following equation was used:

$$R_N^{'} = R_N^{\gamma} \tag{10}$$

where: R'_N = CM-700D (desired RGB value)

R_N = TCS3200 RGB (Un-calibrated sensor data)

γ = Gamma (required calibration factor)

First the TCS3200 sensor data were scaled to ensure all values are offset, thus ensuring that the white reading is that of the CMD-700 for each of R, G and B reading (Eq. 11)

$$R_N = R \times \frac{R'_N}{R_{\max}}, G_N = G \times \frac{G'_N}{G_{\max}}, B_N = B \times \frac{B'_N}{B_{\max}} \tag{11}$$

where $R_{\max}, G_{\max}, B_{\max}$ represent the maximum R, G and B value of a white object from the TCS3200.

ID	TCS-3200						CMD-700			Output		
	Gain Adjusted			White Adjusted			RGB Equivalent (CIE)			Calibrated		
	R	G	B	R	G	B	R	G	B	R	G	B
123A	99	148	167	88	152	144	62	155	152	85	158	143
127C	38	79	75	34	81	65	17	89	69	31	89	64
129C	99	152	137	88	156	118	71	166	123	85	162	117
131C	25	41	35	22	42	30	10	43	27	20	49	29
133C	62	88	85	55	90	73	47	93	80	52	98	72
135C	42	51	35	37	52	30	36	78	39	35	60	29
137C	42	51	35	37	52	30	40	54	30	35	60	29
139C	68	82	58	61	84	50	63	88	48	57	92	49
141C	57	80	45	51	82	39	55	87	35	48	90	38
143C	71	88	48	63	90	41	72	91	32	60	98	41
145C	171	168	122	152	172	105	169	185	101	149	178	104
147C	84	86	62	75	88	53	84	91	51	71	96	53
149C	174	183	114	155	187	98	170	206	86	152	192	97
155D	255	249	258	227	255	222	227	255	222	226	255	222
202A	17	17	20	15	17	17	10	13	13	14	22	17

Table 3. Results obtained comparing the TCS3200 colour sensor (calibrated and un-calibrated) with the CM-700D spectrophotometer over a range of 15 RHS colours

Table 3 shows the data recorded from the colour sensors along with the equivalent results from the CMD-700 (using the CIE RGB transformation matrix) and the calibrated TCS3200 results. Table 4, shows the errors associated with the Table 3.

The calibration factors (γ) for each colour were calculated using normalized data. (Eq. 12)

$$\gamma_R = \frac{\log(R'_N / 255)}{\log(R_N / 255)}, \gamma_G = \frac{\log(G'_N / 255)}{\log(G_N / 255)}, \gamma_B = \frac{\log(B'_N / 255)}{\log(B_N / 255)} \tag{12}$$

For each colour sample measured, the calibration factor was calculated and averaged using a geometric mean (as opposed to the more general arithmetic mean function (Fleming &

Wallace, 1986), thus providing the γ factor for R, G and B individually. The (desired) calibrated values were then obtained using equation 13.

$$R'_{N(calibrated)} = (R_N / 255)^{\gamma} \times 255 \tag{13}$$

For a range of fifteen colours, measurements were taken using the TCS3200 RGB sensor and the CM-700D Spectrophotometer (Table 3). The gamma calibration factors calculated were:

$$(\text{Red})\lambda_R = 1.05,\ (\text{Green})\lambda_G = 0.92,\ (\text{Blue})\lambda_B = 1.00 \tag{14}$$

	TCS3200 (un-calibrated)			TCS3200 (calibrated)		
Colour	R	G	B	R	G	B
Error	9.691	6.806	5.107	10.161	6.162	4.966
Error %	3.8	2.669	2.003	3.985	2.416	1.947
S.D	7.423	7.298	3.485	6.631	4.757	3.699

Table 4. Average Error (0-255), percentage error and standard deviation for red, green and blue measurements of the TCS3200 colour sensor, calibrated and un-calibrated, compared with CM-700D spectrophotometer results across a range of colours

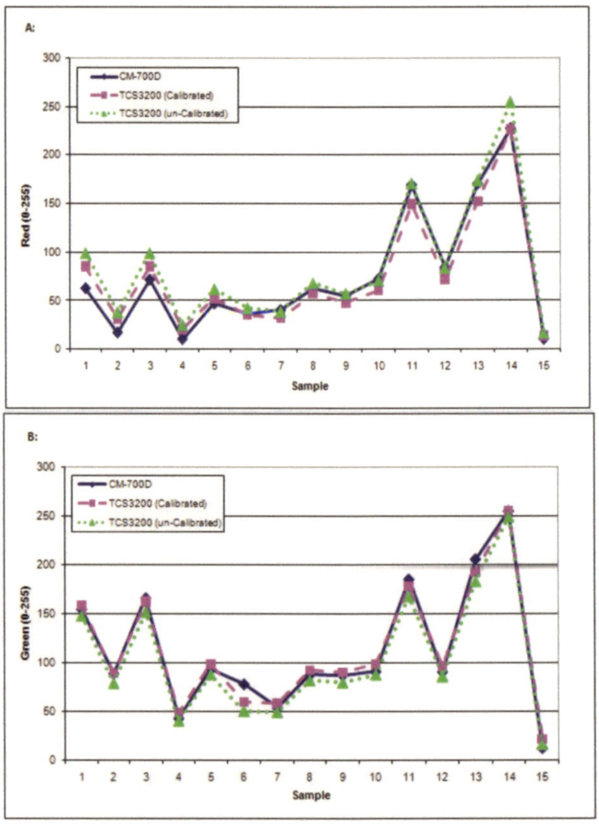

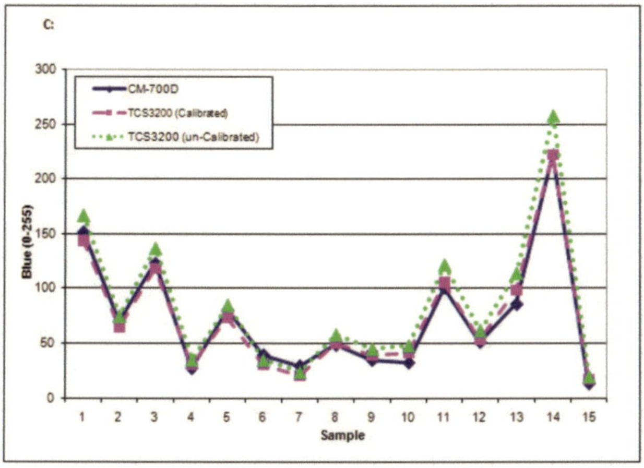

Fig. 17. TCS3200 sensor RGB readings, calibrated and un-calibrated, compared with the CM-700D readings of: Red (A); Green (B); Blue (C) using a CIE RGB colour transformation. (Colour samples are as given in Table 3)

An example of a green colour interpreted by the CM-700D and TCS3200 colour sensor before and after calibration is shown in figure 18.

TCS3200 (un-calibrated)	TCS3200 (calibrated)	CM-700D Spectrophotometer
RGB = 63,90,41	RGB = 60,98,41	RGB = 72,91,32

Fig. 18. Graphical representation of the colour differences between, calibrated and un-calibrated TCS3200 colour sensor

Results showed when calibrating the CM-700D XYZ values to CIE RGB instead of sRGB, the calibration results improved, as shown in Table 5, to have a much smaller error for R, G and B. A colour representation can be seen in Figure 19.

	CIE RGB			sRGB		
Colour	R	G	B	R	G	B
Error	10.289	6.117	5.683	14.777	7.055	9.564
Error %	4.035	2.399	2.229	5.795	2.767	3.751
S.D	6.562	4.739	3.357	12.314	7.54	5.772

Table 5. Comparisons between CIE RGB and sRGB transformation matrix, showing the CIE RGB results to be more accurate than the sRGB

TCS3200 (raw reading)	CM-700D Spectrophotometer (sRGB)	CM-700D Spectrophotometer (CIE RGB)
RGB 71,88,48	RGB = 149,166,81	RGB = 72,91,31

Fig. 19. An example of a green colour interpreted by the TCS3200 colour sensor with no calibration compared with the CM-700D with both a sRGB and CIE RGB

5. Conclusion

An anthropomorphic robotic arm has been designed, fabricated and fully tested to meet the requirements set out by The New Zealand Institute for Plant & Food Research Limited. The robotic system is able to reach and monitor plantlets growing in trays on a multi-level shelving unit. Custom software has been developed to fully automate the control of the robotic arm. The position of the robotic arm can be controlled with great precision using the microstepper controller to allow micro-motion of the stepper motors. The robot can be programmed to autonomously position itself and take readings at regular intervals.

Several sensors have been integrated in the robotic system, namely a high-end colour camera for capturing high resolution images of plantlets; proximity sensor to position the arm at a predetermined distance from the plant surface for taking measurements; temperature, humidity and CO_2 sensors to monitor the ambient atmospheric conditions and a low-cost RGB sensor to measure the colour of plant leaves.

Two different RGB sensors have been evaluated. Experimental results show that the Parallax TCS3200 RGB sensor is a useful low cost colour sensor, which when calibrated to an industry standard spectrophotometer, can provide accurate RGB readings. It is therefore a useful component for integrating into an automated system such as a robotic arm, with various other sensors, for monitoring plants growing in a modified plant micro-propagation system.

The system has the potential for not only monitoring plant material in a laboratory environment but other applications as well where non-destructive measurements of colour are required.

6. Acknowledgment

The hardware cost of this research has been funded by The New Zealand Institute for Plant & Food Research. The authors greatly appreciate the financial and technical workshop support given by the School of Engineering and Advanced Technology, Massey University and The New Zealand Institute for Plant & Food Research Limited.

7. References

Abdullah, M. Z., Aziz', S. A., & Mohamed, A. M. D. (2000). Quality Inspection Of Bakery Products Using A Color-Based Machine Vision System. Journal of Food Quality, 23 (1), 39-50.

Borah, S., & Bhuyan, M. (2005). A computer based system for matching colours during the monitoring of tea fermentation. International Journal of Food Science and Technology, 40(6), 675-682.

Lindbloom, B. (2010). Useful colour Equations. Retrieved 18/8/2010, from http://www.brucelindbloom.com/

Fleming, P. J., & Wallace, J. J. (1986). How not to lie with statistics: The correct way to summarize the benchmark results. Communications of the ACM, 29(3), 218-221

Garcia, G. J., Pomares, J., & Torres, F. (2009). Automatic robotic tasks in unstructured environments using an image path tracker. Control Engineering Practice, 17(5), 597-608.

Gitelson, A. A., Gritz, Y., & Merzlyak, M. N. (2003). Relationships between leaf chlorophyll content and spectral reflectance and algorithms for non-destructive chlorophyll assessment in higher plant leaves. Journal of Plant Physiology, 160(3), 271-282.

HonYong, W., QiXin, C., Masateru, N., & JianYue, B. (1999). Image processing and robotic techniques in plug seedling production. Transactions of the Chinese Society of Agricultural Machinery, 30, 57-62.

Hu, X., He, Y., Pereira, A. G., & Gómez, A. H. (2005). Nondestructive Determination Method of Fruit Quantity Detection Based on Vis/NIR Spectroscopy Technique. 27th Annual International Conference of the Engineering in Medicine and Biology Society (IEEE-EMBS 2005), 1956-1959

Juckett, R. (2010). RGB color space conversion - Linear transformation of color. Retrieved 14/8/2010, from http://www.ryanjuckett.com

Kang, S. P., & Sabarez, H. T. (2009). Simple colour image segmentation of bicolour food products for quality measurement. Journal of Food Engineering, 94, 21-25.

Kawollek, M., & Rath, T. (2007). Robotic Harvest of Cut Flowers Based on Image Processing by Using Gerbera jamesonii as Model Plant. In S. DePascale, G. S. Mugnozza, A. Maggio & E. Schettini (Eds.), Proceedings of the International Symposium on High Technology for Greenhouse System Management (Greensys 2007), 557-563

Kelly, R., Carelli, R., Nasisi, O., Kuchen, B., & Reyes, F. (2000). Stable Visual Servoing of Camera-in-Hand Robotic Systems. IEEE/ASME Transactions on Mechatronics, 5(1), 39 - 48.

Menesatti, P., Antonucci, F., Pallottino, F., Roccuzzo, G., Allegra, M., Stagno, F., et al. (2010). Estimation of plant nutritional status by Vis–NIR spectrophotometric analysis on orange leaves. Biosystems Engineering, 105(4), 448-454.

Miranda, C., Girard, T., & Lauri, P. E. (2007). Random sample estimates of tree mean for fruit size and colour in apple. Scientia Horticulturae, 112 (1), 33-41.

Nicola¨ı, B. M., Beullens, K., Bobelyn, E., Peirs, A., Saeys, W., Theron, K. I., et al. (2007). Nondestructive measurement of fruit and vegetable quality by means of NIR spectroscopy: A review. Postharvest Biology and Technology, 46(2), 99-118.

OceanControls. (2010). Retrieved 12/04/2010, from www.oceancontrols.com.au

Omar, A. F. B., & MatJafri, M. Z. B. (2009). Optical Sensor in the Measurement of Fruits Quality: A Review on an Innovative Approach. International Journal of Computer and Electrical Engineering, 1(5), 557-561.

Otte, C., Schwanke, J., & Jensch, P. (1996). Automatic micropropagation of plants. Optics in Agriculture, Forestry, and Biological Processing II, Proc. SPIE 2907, 80-87.

Parallax Home. (2010). Retrieved 05/07/2010, from www.Parallax.com

Paz, P., S´anchez, M. ı.-T., P´erez-Mar´ın, D., Guerrerob, J. e.-E., & Garrido-Varob, A. (2009). Evaluating NIR instruments for quantitative and qualitative assessment of intact apple quality. Journal of the Science of Food and Agriculture, 89(5), 781-790.

Raja, A. S., & Sankaranarayanan, K. (2006). Use of RGB Color Sensor in Colorimeter for better clinical measurements of blood Glucose. BIME Journal 6(1), 23 - 28.

Reyes, J., & Chiang, L. (2009). Location And Classification Of Moving Fruits In Real Time With A Single Colour Camera. Chilean Journal Of Agricultural Research, 69, 179-187.

Scarfe, A. J., Flemmer, R. C., Bakker, H. H., & Flemmer, C. L. (2009). Development of An Autonomous Kiwifruit Picking Robot. 4th International Conference on Autonomous Robots and Agents, 380-384.

Slaughter, D. C., Giles, D. K., & Downey, D. (2008). Autonomous robotic weed control systems: A review. Computers and Electronics in Agriculture, 61(1), 63-78.

Sogaard, H. T., & Lund, I. (2007). Application accuracy of a machine vision-controlled robotic micro-dosing system. biosystems engineering, 96(3), 315-322.

Sony Professional. (2010). Retrieved 05/06/2009, from www.pro.sony.eu

van Henten, E. J., Hemming, J., van Tuijl, B. A. J., Kornet, J. G., Meuleman, J., van Os, E. A., et al. (2002). An autonomous robot for harvesting cucumbers in greenhouses. [Article]. Autonomous Robots, 13(3), 241-258.

Yadav, S. P., Ibaraki, Y., & Gupta, S. D. (2010). Estimation of the chlorophyll content of micropropagated potato plants using RGB based image analysis. Plant Cell Tissue and Organ Culture, 100(2), 183-188.

Yam, K. L., & Papadakis, S. E. (2004). A simple digital imaging method for measuring and analyzing color of food surfaces. Journal of Food Engineering, 61, 137-142.

Part 2

Control and Modeling

9

CPG Implementations for Robot Locomotion: Analysis and Design

Jose Hugo Barron-Zambrano and Cesar Torres-Huitzil
Information Technology Laboratory, CINVESTAV-IPN
Mexico

1. Introduction

The ability to efficiently move in a complex environment is a key property of animals. It is central to their survival, i.e. to avoid predators, to look for food, and to find mates for reproduction (Ijspeert, 2008). Nature has found different solutions for the problem of legged locomotion. For example, the vertebrate animals have a spinal column and one or two pairs of limbs that are used for walking. Arthropoda animals are characterized by a segmented body that is covered by a jointed external skeleton (exoskeleton), with paired jointed limbs on each segment and they can have a high number of limbs (Carbone & Ceccarelli, 2005). The biological mechanisms underlaying locomotion have therefore been extensively studied by neurobiologists, and in recent years there has been an increase in the use of computer simulations for testing and investigating models of locomotor circuits based on neurobiological observations (Ijspeert, 2001). However, the mechanisms generating the complex motion patterns performed by animals are still not well understood (Manoonpong, 2007).

Animal locomotion, for instance, requires multi-dimensional coordinated rhythmic patterns that need to be correctly tuned so as to satisfy multiple constraints: the capacity to generate forward motion, with low energy, without falling over, while adapting to possibly complex terrain (uneven ground, obstacles), and while allowing the modulation of speed and direction (Ijspeert & Crespi, 2007). In vertebrate animals, an essential building block of the locomotion controller is the Central Pattern Generator (CPG) located in the spinal cord. The CPG is a neural circuit capable of producing coordinated patterns of rhythmic activity in open loop, i.e. without any rhythmic inputs from sensory feedback or from higher control centers (Delcomyn, 1980; Grillner, 1985). Interestingly, very simple input signals are sufficient to modulate the produced patterns. Furthermore, CPG can adapt to various environments by changing the periodic rhythmic patterns. For instance, the cats and horses are able to change their locomotor patterns depending on the situation.

This relevance of locomotion both for biology and for robotics has led to multiple interesting interactions between the two fields. The interactions have mainly been in one direction, with robotics taking inspiration from biology in terms of morphologies, modes of locomotion, and/or control mechanisms. In particular, many robot structures are directly inspired by animal morphologies, from snake robots, quadruped robots, to humanoid robots. Increasingly, robotics is now providing something back to biology, with robots being used as scientific tools to test biological hypotheses (Ijspeert, 2008). Researchers have studied the CPGs for decades and some principles have been derived to model their functionality and structure. CPGs have been proposed as a mechanism to generate an efficient control strategy for legged robots based on biological locomotion principles. Locomotion in legged robots is much more complex than wheeled robots, since the formers have between twelve or twenty degrees of freedom, which must be rhythmically coordinated to produce specific gaits. The design of locomotion control systems of legged robots is a challenge that has been partially solved. To develop bio-inspired solutions, detailed analyses of candidate biological neural systems are essential as in the case of legged locomotion.

Models of CPGs have been used to control a variety of different types of robots and different modes of locomotion. For instance, CPG models have been used with hexapod and octopod robots inspired by insect locomotion, quadruped robots inspired by vertebrates, such as horse, biped robots inspired by humans and other kind of robots inspired by reptiles, such as snakes, as it will be discussed in the section 2. CPGs could been modeled with different levels of abstraction, these aspects will be presented in section 3. Additionally, CPGs present several interesting properties including distributed control, the ability to deal with redundancies, fast control loops, and allowing modulation of locomotion by simple control signals. These properties, when transferred to mathematical models, make CPGs interesting building blocks for locomotion controllers in robots (Fujii et al., 2002; Ijspeert, 2008), as shown in section 4.

In spite of its importance, the CPG-approach presents several disadvantages. One of the main drawbacks of CPGs is related with the learning methodologies to generate the rhythmic signals (Zielinska, 1996). For that purpose, a method to tune a CPG using genetic algorithms (GA) is proposed, whose details are provided in section 5. Moreover, a methodology for designing CPGs to solve a particular locomotor problem is yet missing. Other problem is that animals rarely perform steady-state locomotion for long time, and tend to superpose, and switch between, multiple motor behaviors. A remaining open challenge is therefore to design control architectures capable of exhibiting such richness motor skills.

Engineering CPG-based control systems has been difficult since the simulation of even rather simple neural network models requires a significant computational power that exceeds the capacity of general embedded microprocessors. As a result, CPG dedicated hardware implementations, both analog and digital, have received more attention (Nakada, 2003). On one hand, analog circuits have been already proposed, being computation and power efficient but they usually lack flexibility and dynamics and they involve large design cycles. On the other hand, Field-Programmable Gate Array (FPGA) substrate might provide real-time response, low power consumption and concurrent processing solutions for low-level locomotion issues (Barron-Zambrano et al., 2010b). So, a brief analysis of CPG implementations is presented in section 6. Also, the integration of environment information might allow to produce behaviors by means of selection and adaption of lower substrates. Under this scenario, legged robots may reconfigure their locomotion control strategies on-line according to sensory information so as to effectively handle dynamic changes in the real world. Yet, the FPGA-based approach fits well since it could handle efficiently higher perception functions for CPG parameter adaptation, and to provide the neural processing and coordination of a custom and adaptive CPG hardware module. For those reasons, section 7 presents a hardware implementation for quadruped locomotion control based on FPGA. The experimental results of hardware implementation and the future work are shown in sections 8 and 9, respectively. Finally, the conclusions of this work are presented in section 10.

2. Locomotion solution methods

In the literature, there are different approaches to the design of locomotion control systems such as: trajectory based methods, heuristic based methods and biologically inspired methods.

2.1 Trajectory based methods

In trajectory based methods, to move a leg in a desired trajectory, the joint angles between limbs are calculated in advance, by using a mathematical model that incorporates both robot and environment parameters, to produce a sequence of actions specifically scheduled. But these methods are not providing any methodology to design a controller. It is only a method to prove the stability of the motion. Therefore, the trajectories have to be designed by trial-and-error or from animal locomotion recording data (Jalal et al., 2009). The most popular stabilization criteria are the Center of Mass (CoM), Center of Pressure (CoP) and Zero Moment Point (ZMP). The gait is stable when one of these criteria remains inside the support polygon (the convex hull formed by the contact points of the feet with the ground).
These methods can generate either static or dynamic stability. The first one prevents the robot from falling down by keeping the CoM inside the support polygon by adjusting the body posture very slowly. Thus minimizing the dynamic effects and allowing the robot to pause at any moment of the gait without falling down. Using this criterion will generally lead to more power consumption since the robot has to adjust its posture so that the CoM is always inside the support polygon. The second one relies on keeping the ZMP or CoP inside the support polygon and this is a necessary and sufficient condition to achieve stability. Dynamic balance is particularly relevant during the single support phase, which means that the robot is standing in only one foot. This generally leads to more fast and reliable walking gaits (Picado et al., 2008).

2.2 Heuristic based methods

The major problem with the ZMP or CoP methods are the complexity of the equations used to compute the robot dynamic. With such a level of complexity directly exposed to the robot programmer, it becomes very difficult to have a global view over the robot behavior. This is because heuristics methods were developed to hide the complexity of the problem and make it more accessible.
Heuristic methods have strong similarities with the trajectory based approach. Joint trajectories are also pre-computed from an external optimization process, only the generation strategy differs. This time, heuristic or evolutionary based techniques are used (e.g. genetic algorithms) to design the desired trajectories (Lathion, 2006).
The most successful approach of this methods is called Virtual Model Control (VMC) (Picado et al., 2008). It is a motion control framework that uses simulations of virtual components to generate desired joint torques. These joint torques create the same effect that the virtual components would have created, thereby creating the illusion that the simulated components are connected to the real robot. Such components can include simple springs, dampers, dash pots, masses, latches, bearings, non-linear potential and dissipative fields, or any other imaginable component. The generated joint torques create the same effect that the virtual components would create if they were in fact connected to the real robot (Pratt et al., 2001). This heuristic makes the design of the controller much easier. First, it is necessary to place some virtual components to maintain an upright posture and ensure stability. Virtual

model control has been used to control dynamic walking bipedal robots (Pratt et al., 2001) and an agile 3D hexapod in simulation (Torres, 1996).

2.3 Biologically inspired methods

This approach uses CPGs which are supposed to take an important role in animal locomotion. By coupling the neural oscillators signals when they are stimulated through some input, they are able to synchronize their frequencies. In the field of artificial intelligence and robotics, it is possible to build structures that are similar to the neural oscillators found in animals by the definition of a mathematical model (Picado et al., 2008). The term central indicates that sensory feedback (from the peripheral nervous system) is not needed for generating the rhythms. In other words, the CPG has the ability to produce a periodic output from a non-periodic input signal. CPG is a proven fact which exists in animals and its task is to generate the rhythmic movements almost independent of central nervous system. There are a lot of CPG methods and different robotic application have been proposed by different authors (Lee et al., 2007; Loeb, 2001; Nakada, 2003). However, most of the time these CPGs are designed for a specific application and there are very few methodologies to modulate the shape of the rhythmic signals, in particular for on-line trajectory generation and the CPG internal parameters are usually tuned by trial and error methods.

2.4 Approaches analysis

Many different solutions have been experimented to achieve stable robot locomotion, from trajectory based methods to biologically inspired approaches. Each method presents different advantages as well as disadvantages. Next, a features summary of each approach under different requirements such as: robot knowledge, perturbation problems, design methodology, physical implementation and control scheme is presented.

- *Robot knowledge*: the major drawback of the trajectory based and heuristic based methods is that they both required an exact knowledge of the robot model. The CPG approach does not require an exact knowledge of the robot model.
- *Perturbation problems*: the CPG will recover smoothly and quickly after a perturbation, the ZMP needs external on-line control to deal with perturbations and the VMC is robust against small perturbation.
- *Design methodology*: the ZMP and VMC have well defined methodology to prove stability and intuitive ways of describing the controller. In the CPG approach, the number of parameters needed to describe the CPG is usually large and the stability is not guaranteed as in trajectory and heuristic based methods.
- *Physical implementation*: the ZMP is easy to implement in real robots. In the VMC, the control is fast and can be efficiently done on-line. Meanwhile, the CPG is able to generate new trajectories for various speeds, without the necessity to be trained again.
- *Control scheme*: In the ZMP and VMC, the control is more centralized in a black box connected to the robot. But the CPG will be distributed over the whole robot body, as a virtual spinal cord, this property allows to the robot to continue with the gait generation if some part of the control is missing.

The review shown that the CPG approach presents advantages, over the ZMP and VMC approaches, in the requirements of robot knowledge, perturbation problems, physical implementation and control scheme. For that reasons, more locomotion controllers for legged robots based on CPG are using currently.

3. Locomotion modeling based on CPG

The activity of the locomotor organs (legs, wings, etc.) generates a propulsive force that leads to locomotion. Each piece of a locomotor organ is rhythmically controlled by a CPG generating the rhythm. These neural networks, controlling each individual organ, interact with others so as to coordinate, in a distributed manner, the locomotor action of a complete specie. Moreover, they also adopt their activity to the surrounding environment through sensory feedback and higher level signals from the central nervous system.

CPGs are often modeled as neurons that have mutually coupled excitatory and inhibitory neurons, following regular interconnection structures. CPGs automatically generate complex control signals for the coordination of muscles during rhythmic movements, such as walking, running, swimming and flying (Delcomyn, 1980; Hooper, 2000; Kimura, Shimoyama & Miura, 2003).

The locomotion patterns can usually be modulated by some parameters, which offers the possibility to smoothly modify the gait (e.g. increase frequency and/or amplitude) or even to induce gait transitions. In CPG design, there are some common assumptions: the nonlinear oscillators are often assumed to be identical, the stepping movements of each limb are controlled by a single oscillator, while interlimb coordination is provided by the connections between the oscillators (Arena et al., 2004). Moreover, the sensory inputs from lower-level central nervous system and signals from higher-level central nervous system can modulate the activity of CPGs. The lower level signals are used for slow movements (low frequency response). Higher level signals are use to produce intelligent behavior and faster movements (higher frequency response). A comprehensive review of utilizing CPGs in robotics can be found in (Ijspeert, 2008), and an interesting overview of the different oscillators utilized for robotics purposes is given in (Buchli et al., 2006).

The vertebrate locomotor system is organized hierarchically where the CPGs are responsible for producing the basic rhythmic patterns, and that higher-level centers (the motor cortex, cerebellum, and basal ganglia) are responsible for modulating these patterns according to environmental conditions. Such a distributed organization presents several interesting features: (i) It reduces time delays in the motor control loop (rhythms are coordinated with mechanical movements using short feedback loops through the spinal cord). (ii) It dramatically reduces the dimensionality of the descending control signals. Indeed the control signals in general do not need to specify muscle activity but only modulate CPG activity. (iii) It therefore significantly reduces the necessary bandwidth between the higher-level centers and the spinal cord (Ijspeert, 2008; Loeb, 2001).

CPG have been modeled with several levels of abstraction as for example: biophysical models, connectionist models, abstract systems of coupled oscillators. In some cases, the CPG models have been coupled to biomechanical simulation of a body, in such case they are called neuromechanical models. Detailed biophysical models are constructed based on the Hodgkin-Huxley type of neuron models. That is, neuron models that compute how ion pumps and ion channels influence membrane potentials and the generation of action potentials. In particular, the cells excite themselves via fast feedback signals while they inhibit themselves and other populations via slower feedback signals. In some cases, the pacemaker properties of single neurons are investigated. While most models concentrate on the detailed dynamics of small circuits, some models address the dynamics of larger populations of neurons, for instance the generation of traveling waves in the complete lamprey swimming CPG (Arena, 2001). Connectionist models use simplified neuron models such as leaky-integrator neurons or integrate-and-fire neurons. This connectionist model has

shown that connectivity alone is sufficient to produce an oscillatory output. The goal of these models is on how rhythmic activity is generated by network properties and how different oscillatory neural circuits get synchronized via interneuron connections (Ijspeert, 2008).
Other extensively used oscillators include phase oscillators (Buchli & Ijspeert, 2004; Matsuoka, 1987). Most of the oscillators have a fixed waveform for a given frequency. In some cases, closed-form solutions or specific regimes, as for example phase-locked regimes, can be analytically derived but most systems are solved using numerical integration. Several neuromechanical models have been developed. The addition of a biomechanical model of the body and its interaction with the environment offers the possibility to study the effect of sensory feedback on the CPG activity.
Finally, oscillator models are based on mathematical models of coupled nonlinear oscillators to study population dynamics. In this case, an oscillator represents the activity of a complete oscillatory center (instead of a single neuron or a small circuit). The purpose of these models is not to explain rhythmogenesis (oscillatory mechanisms are assumed to exist) but to study how inter-oscillator couplings and differences of intrinsic frequencies affect the synchronization and the phase lags within a population of oscillatory centers. The motivation for this type of modeling comes from the fact that the dynamics of populations of oscillatory centers depend mainly on the type and topology of couplings rather than on the local mechanisms of rhythm generation, something that is well established in dynamical systems theory (Ijspeert & Crespi, 2007).

4. CPG-based locomotion design for quadruped robots

The locomotion control research field has been pretty active and has produced different approaches for legged robots. But the ability for robots to walk in an unknown environment is still much reduced at the time. Experiments and research work have addressed the feasibility of the design of locomotion control systems of legged robots taking inspiration from locomotion mechanisms in humans and animals. Legged locomotion is performed in rhythmic synchronized manner where a large number of degrees of freedom is involved so as to produce well-coordinated movements.
As described previously, one of the main drawbacks of CPGs is that there are few learning methodologies to generate the rhythmic signals and their parameters are usually tuned by trial and error methods. Learning and optimization algorithms can be used in different ways. The approaches can be divided into two categories: supervised learning and unsupervised learning. Supervised learning techniques can be applied when the desired rhythmic pattern that the CPG should produce is known. The desired pattern can then be used to define an explicit error function to be minimized. Such techniques can sometimes be used for designing CPGs, but they are restricted to situations where suitable patterns are available (e.g. they are obtained from kinematic measurements of animals). Some examples of these techniques are learning for vectors fields (Okada et al., 2002), gradient descent learning algorithms (Pribe et al., n.d.) and statistical learning algorithms (Nakanishi et al., 2004). Unsupervised learning techniques are used when the desired behavior of the CPG is not defined by a specific desired pattern (as in supervised learning), but by a high-level performance criterion, for instance, moving as fast as possible. Among unsupervised learning techniques, stochastic population-based optimization algorithms such as evolutionary algorithms have extensively been used to design CPG-like models (Ijspeert, 2008). The parameters that are optimized are usually the synaptic weights in fixed neural network architectures and coupling weights in systems of coupled oscillators.

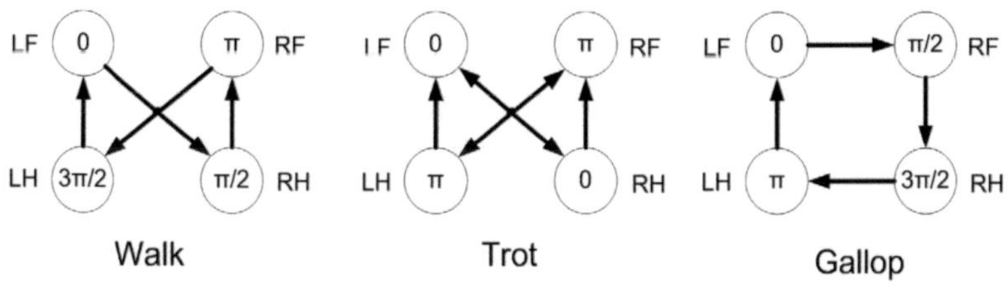

Fig. 1. Configurations of typical gait patterns in quadruped locomotion and their relative phases among the limbs.

4.1 Van Der Pol oscillator model

There are several models for neural oscillators to model the basic CPG to control a limb, such as the Amari-Hopfield model (Amari, 1988), Matsuoka model (Billard & Ijspeert, 2000) and Van Der Pol model (Van Der Pol B, 1928). For this work, the basic cell is modeled by a Van Der Pol (VDP) oscillator which is a relaxation oscillator governed by a second-order differential equation (equation 1):

$$\ddot{x} - \alpha(p^2 - x^2)\dot{x} + \omega^2 x = 0 \tag{1}$$

where x is the output signal from the oscillator, α, p and ω are the parameters that tune the properties of oscillators. In general, α affects the shape of the waveform, the amplitude of x depends largely on the parameter p. When the amplitude parameter p is fixed, the output frequency is highly dependent on the parameter ω. However, a variation of parameter p can slightly change the frequency of the signal, and α also can influence the output frequency.

4.2 Quadruped gaits

A simplified mathematical model of CPG-based locomotion consists of using one cell per limb and replacing each cell by a nonlinear oscillator. Thus, quadruped gaits are modeled by coupling four nonlinear oscillators, and by changing the connections among them, it is possible to reproduce rhythmic locomotion patterns. In rhythmic movements of animals, transition between these movements is often observed. As a typical example, horses choose different locomotive patterns in accordance with their needs, locomotive speeds or the rate of energy consumption. In addition, each gait pattern is characterized by relative phase among the limbs (Dagg, 1973). Figure 1 shows the typical horse gait pattern configurations and theirs relatives phases between the limbs. Here, *LF*, *LH*, *RF*, and *RH* stand for left forelimb, left hindlimb, right forelimb, and right hindlimb. In the rest of this work, *LF*, *RF*, *RH* and *LH* will be refer as *X*1, *X*2, *X*3 and *X*4 , respectively.

The mutual interaction among the VDP oscillators in the network produces a gait. By changing the coupling weights, it is possible to generate different gaits. The dynamics of the ith coupled oscillator in the network is given by:

$$\ddot{x}_i + \alpha(p_i^2 - x_{ai}^2)\dot{x}_i - \omega^2 x_{ai} = 0 \tag{2}$$

For $i = 1, 2, 3, 4$, where x_i is the output signal from oscillator i, x_{ai} denotes the coupling contribution of its neighbors given by the equation 3:

$$x_{ai} = \sum_j w_{ij} x_j \tag{3}$$

where w_{ij} is the coupling weight that represents the strength of ith oscillator over the jth oscillator. The generation of the respective gaits depends on the values of the oscillators parameters and the connection wights among oscillators.
The analysis of the behavior reported in (Barron-Zambrano & Torres-Huitzil, 2011) shows that *the sign of the coupling weight determines the phase difference. For inhibitory connections, the phase lag is around of* $360/n$ *to particular values of* w_{ij}*, where* n *is the number of coupled oscillators. For excitatory connections, the phase difference is equal to zero. Finally, the matrix built for the connection weights,* w_{ij}*, might be symmetrical and regular.*

5. GA-based approach for gait learning

To efficiently search for CPG parameters a genetic algorithm for oscillator parameters estimation was used. A genetic algorithm is a powerful optimization approach that can work in large-dimensional, nonlinear, and, often, largely unknown parameter spaces. In the simplest form of gait evolution, functional forward locomotion is the only goal and no sensor inputs are used. Gait learning is a form of locomotion learning, but it might be considered a somewhat more difficult problem in legged robots. The search space is represented by the genome defining the controller representation (or controller and morphology representation). Each evolvable parameter of the genome defines a dimension of the search space, and the fitness landscape is then given by the manifold defined by the fitness of each point in the search space.
The tuned method is divided into two stages and the robot controller was represented by a string of real numbers that describe oscillators and their connectivity. The first one estimates the oscillator parameters and the second one calculates the values of w_{ij} to produce the locomotion pattern. The two stages work in sequential way.
The first stage estimates the parameters $[\alpha, p, \omega]$ to generate a specific wave in frequency and amplitude. Here, the oscillators parameters correspond to the gene and represent the organization of an individual n in a group of population N. The individual, n, is represented by the concatenation of parameters in order shown in equation 4, where each parameter was represented by real numbers of 32-bit:

$$n = [\alpha p \omega] \quad (4)$$

Each individual n is evaluated according to the next fitness function which is minimized by the GA:

$$fitness_p = abs((A_d - A_p) * (f_d - DFT(S_o))) \quad (5)$$

where A_d is the desired amplitude of the signal, f_d is the desired frequency, A_p is the amplitude of the generated pattern, S_o is the pattern generated by the oscillator with the individual n and DFT is the Discrete Fourier Transform. Simulation of S_o is generated by around of 50 seconds by each individual, n, and only the last 5 seconds are considered for the DFT computation. This is to ensure a stable signal, in frequency and amplitude, and to reduce the amount of data to be processed. Figure 2 shows a block diagram of this stage.
The GA ranks the individuals according the fitness function. As result of ranking, the best individuals remain in the next generation without modification. For this stage, 2 individuals remain without modification and go to the next generation. By using the crossover operation, other individuals in the next generation are created by combining two

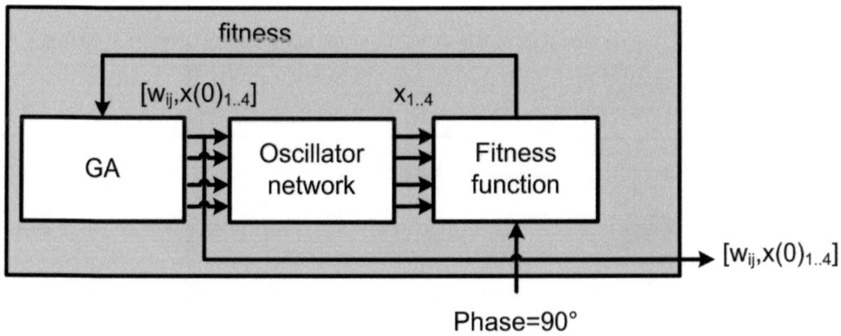

Fig. 2. Block diagram of first stage. It estimates the oscillator parameter values, $[\alpha, p, \omega]$, to produce a signal with specific frequency and amplitude.

individuals. Furthermore, the remaining individuals are randomly modified by mutation. The probabilities of crossover and mutation are set to be 0.8 and 0.2, respectively.
The second stage performs the synchronization among the oscillator outputs. The generation of different gaits needs patterns with specific phase amount the output signals, see figure 1. The stage computes the value of w_{ij}. To produce a desired phase is not enough only with connection weights, it is necessary to calculate the initial conditions of each oscillator, $x_{1..4}$, too. For that purpose, in the second stage each individual, n, has five values in order to generate a specific phase in the network. The values in the gene are organized from the right to left. The first four values are the initial values for each oscillator, $x_{1..4}$. And the fifth value codes the connection weight, w. The equation 6 shows the presentation of n in this stage.

$$n = [w x_4 x_3 x_2 x_1] \tag{6}$$

where the initial values and connection weight were represented by real numbers.
The GA only needs estimate one value of w for one gait and by changing the algebraic sign it is possible to generate the rest of the gaits. The fitness function is based on the phase among the patterns. To estimate the phase among the signals, the correlation was used. The oscillator labeled as LF ($X1$) is the reference signal, phase equal to zero, and the other signals are shifted with respect to this signal. The fitness function to evaluate the individual is given by equation 7. Here, the GA minimizes the value of the fitness function.

$$fitness_{\omega} = abs(90 - phase(x_1, x_3)) + abs(90 - phase(x_3, x_2)) + abs(90 - phase(x_2, x_4)) \tag{7}$$

where $phase_(x_i, x_j)$ is the phase between the ith and jth oscillator. The simulation of $x_{1..4}$ is generated by around of 50 seconds by each n and only the last 10 seconds are considered for the phase estimation. Figure 2 shows a block diagram of the second stage.

6. Related work of hardware implementations

Nowadays, many works have aimed to develop robust legged locomotion controllers (Fujii et al., 2002; Fukuoka et al., 2003; Susanne & Tilden, 1998). The design CPG-based control systems has been difficult given that the simulation of even rather simple neural network models requires a computational power that exceeds the capacity of general processors. For that reason, CPG dedicated hardware implementations, both analog and digital, have received more attention (Nakada, 2003). On one hand, CPGs have been implemented using

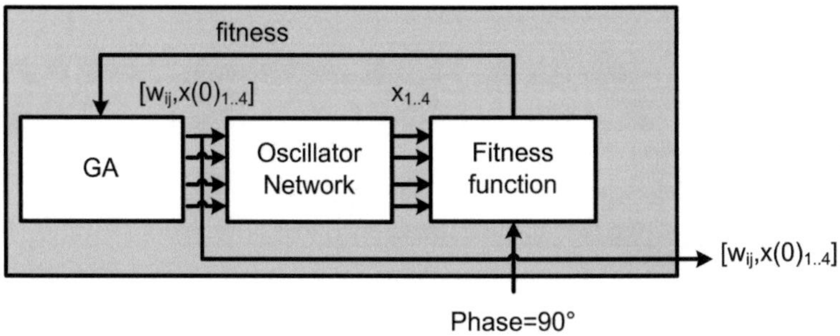

Fig. 3. Block diagram of second stage. It finds the values of coupling, w_{ij} and the initial values of $x_{1..4}$.

microprocessors providing high accuracy and flexibility but those systems consume high power and occupy a large area restricting their utility in embedded applications. On the other hand, analog circuits have been already proposed, being computation and power efficient but they usually lack flexibility and dynamics and they involve large design cycles.

Relatively few works have focused on adopting the hardware technology to fully practical embedded implementations. Examples of this adaptation using analog circuit are presented by (Kier et al., 2006; Lee et al., 2007; Lewis et al., 2001; Nakada, 2003). In the first one, Nakada et al present a neuromorphic analog CMOS controller for interlimb coordination in quadruped locomotion. Authors propose a CPG controller with analog CMOS circuits. It is capable of producing various rhythmic patterns and changing these patterns promptly. Lewis at al constructed an adaptive CPG in an analog VLSI (*Very Large Scale Integration*) chip, and have used the chip to control a running robot leg. They show that adaptation based on sensory feedback permits a stable gait even in an under actuated condition: the leg can be driven using a hip actuator alone while the knee is purely passive. Their chip has short-term on board memory devices that allow the continuous, real-time adaptation of both center-of-stride and stride amplitude. In addition, they make use of integrate-and-fire neurons for the output motor neurons. Finally, the authors report that their abstraction is at a higher level than other reported work, which lends itself to easier implementation of on-chip learning. Kier et al present a new implementation of an artificial CPG that can be switched between multiple output patterns via brief transient inputs. The CPG is based on continuous-time recurrent neural networks (CTRNNs) that contain multiple embedded limit cycles. The authors designed and tested a four-neuron CPG chip in AMI's 1.5 μm CMOS process, where each neuron on the chip implements the CTRNN model and is fully programmable. The authors report the measured results from the chip agree well with simulation results, making it possible to develop multi-pattern CPGs using off-line simulations without being concerned with implementation details. Lee at al report a feasibility study of a central pattern generator-based analog controller for an autonomous robot. The operation of a neuronal circuit formed of electronic neurons based on Hindmarsh-Rose neuron dynamics and first order chemical synapses modeled. The controller is based on a standard CMOS process with 2V supply voltage. Simulated results show that the CPG circuit with coordinate controller and command neuron is viable to build adaptive analog controller for autonomous biomimetic underwater robots. Finally, a biologically inspired swimming machine is presented by Arena in (Arena, 2001). The system used to generate locomotion is an analog array of nonlinear systems known as Cellular Neural Networks (CNN). The CNN is defined as a two-dimensional array of $M \times N$ identical cells arranged in a rectangular grid, where each cell was defined as the nonlinear first order circuit. The author reports that the application to service robot further underlines that CNNs can be used as powerful devices to implement biologically inspired locomotion and swimming.

Another solution to implement the CPG is using FPGAs as the hardware platform. The CPG is programmed in either VHDL or Verilog and download it onto the FPGA. This approach provide the hardware efficiency with software flexibility. In (Torres-Huitzil, 2008), Torres and Girau present an implementation of a CPG module based on the Amari-Hopfield structure into pure FPGA hardware. In their work, they have chosen to attached the CPG implementation to a Microblaze soft-core processor running uclinux. By doing this, they achieve the computation speed from having the CPG running the mathematical operations in the FPGA hardware making it possible to achieve some degree of parallelization in the calculations and then having the soft-core CPU free to handle other high level control algorithms and this could change the control parameters on the CPG implementation. Barron et al in (Barron-Zambrano et al., 2010b) present a FPGA implementation of a controller, based on CPG, to generate adaptive gait patterns for quadruped robots. The proposed implementation is based on an specific digital module for CPGs attached to a soft-core processor so as to provide an integrated and flexible embedded system. Experimental results show that the proposed implementation is able to generate suitable gait patterns, such as walking, trotting, and galloping.

Other CPGs implementations are using microprocessor. It gives the designer more freedom, because it is not constrained by the difficulty of designing e.g. a differential analog circuits. These implementations using a scheme where the control is distributed trough the robot body. The first implementation example of these is presented by (Inagaki et al., 2006). In that work, Inagaki et al proposed a method to control gait generation and walking speed control for an autonomous decentralized multi-legged robot using CPGs. The robot locomotion control is composed by subsystems. Each subsystem controls a leg and has a microcomputer. The subsystems are connected mechanically and communicate with neighbors. Each microcomputer calculates the dynamics of two oscillators, sends and receives the dynamic results from neighbor subsystems. The gait generation and the walking speed control are achieved by controlling the virtual energy of the oscillators (Hamiltonian). A real robot experiment showed the relationship to the Hamiltonian, the actual energy consumption and the walking speed. The effectiveness of the proposed method was verified. The proposed controller can be generalized as a wave pattern controller, especially for robots that have homogeneous components. In (Crespi & Ijspeert, 2006), Crespi and Ijspeert proposed an amphibious snake robot designed for both serpentine locomotion (crawling) and swimming. It is controlled by an on-board CPG inspired by those found in vertebrates. The AmphiBot II robot is designed to be modular: it is constructed out of several identical segments, named elements. Each element contains three printed circuits (a power board, a proportional-derivative motor controller and a small water detector) connected with a flat cable, a DC motor with an integrated incremental encoder, a set of gears. The elements are connected (both mechanically and electrically) using a compliant connection piece fixed to the output axis. The main contribution of this article is a detailed characterization of how the CPG parameters (i.e., amplitude, frequency and wavelength) influence the locomotion speed of the robot.

The related works present control schemes for robot locomotion based on biological systems under different implementation platforms. The schemes compute the walk pattern in a parallel way through the specialized either systems or coupled subsystems. The CPG control can be implemented by different approaches (analog circuits, reconfigurable hardware and digital processors). The analog circuit implementations present a good performance between power and energy consumption, but they have a large design cycle. For that, these approaches are useful in robots with either limited storage of energy or data processing with real time constraints. In the second approach is possible to implement CPG control with real time constraint sacrificing the energy performance. These implementations have smaller design cycle than the analog circuit implementations. The approach is ideal to be used in the early robot design stages. The last approach presents a high flexibility because it is not constrained by the difficulty of designing. The design cycle is shorter but it has the worst energy performance. Furthermore, those systems occupy a large area restricting their utility in embedded applications.

The analyzed implementations shown several common features. The first one is that they are reconfigurable. The second one, they are capable to adapt oneself to different environments. The most of the implementations presents a scheme where the control is distributed trough the robot body or they are implemented through the interaction of basic elements. These similar features are seen in animal locomotion tasks.

7. CPG hardware implementation for a quadruped robot

In this section, we describe the architecture of the CPG controller for interlimb coordination in quadruped locomotion. First, the design considerations for the implementation are presented. Next, the basic Van Der Pol Oscillator that constitute a part of the CPG network is given. Finally, the architecture of the complete system is described.

7.1 Design considerations

The Van Der Pol oscillator is suitable for CPG implementation as a digital circuit but two main factors for an efficient and flexible FPGA-based implementation should be taken into account: a) *arithmetic representation*, CPG computations when implemented in general microprocessor-based systems use floating point arithmetic. An approach for embedded implementations is the use of 2s complement fixed point representation with a dedicated wordlength that better matches the FPGA computational resources and that saves further silicon area at the cost of precision, and b) *efficiency and flexibility*, embedded hard processor cores or configurable soft processors developed by FPGA vendors add the software programmability of optimized processors to the fine grain parallelism of custom logic on a single chip (Torres-Huitzil, 2008). In the field of neural processing, several applications mix real-time or low-power constraints with a need for flexibility, so that FPGAs appear as a well-fitted implementation solution.

Most of the previous hardware implementation of CPGs are capable of generating sustained oscillations similar to the biological CPGs, however, quite a few have addressed the problem of embedding several gaits and performing transitions between them. One important design consideration in this work, is that the FPGA-based implementation should be a *platform well suited to explore reconfigurable behavior and dynamics*, i.e., the platform can be switched between multiple output patterns through the application of external inputs.

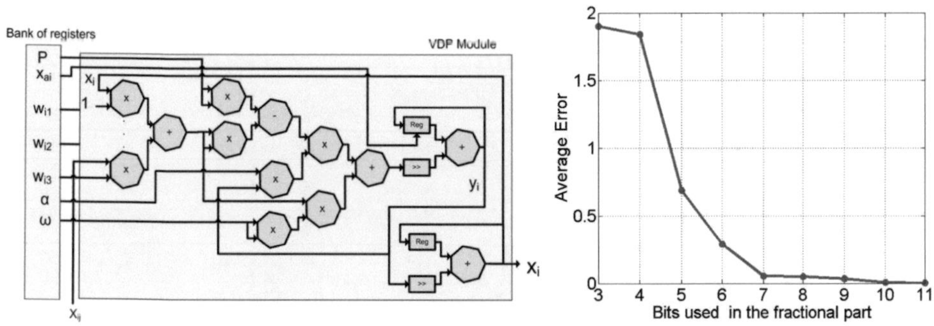

(a) (b)

Fig. 4. (a) Digital hardware architecture for the Van Der Pol oscillator and (b) average error as a function of the bit precision used in the basic blocks.

7.2 Module of Van Der Pol oscillator

From the analysis of Van Der Pol equation, Eq. 2, three basic operations were identified: addition, subtraction and multiplication. Thus, one block for each operation was implemented with 2's complement fixed-point arithmetic representation. Figure 4(a) shows a block diagram for the hardware implementation of the discretized VDP equation. In the first stage, the value of x_{ai} (equation 3) is calculated: this value depends on the x_i-neighbors and the coupling weight values. This stage uses four multipliers and one adder. The square values of p, x_{ai} and ω are calculated in the second stage, it uses three multipliers. In the third stage, the values of $\alpha * y_i$ and $p^2 - x_{ai}$ are calculated, one multiplier and a subtracter are used. The fourth stage computes the values of $\alpha * y_i * (p^2 - x_{ai})$ and $w^2 * x_{ai}$. This stage uses two multipliers. For the integration stage, the numerical method of Euler was implemented by using two shift registers and two adders. The integration factor is implemented by a shift register, which shifts six positions the values of $\dot{y}_i$ and $\dot{x}_i$ to provide an integration factor of 1/64. The block labeled as *Reg* stands for accumulators that hold the internal state of the VPD oscillators. Finally, the values y_i and x_i are obtained.

The size word for each block was 18-bit fixed point representation with 11-bit for the integer part and 7-bit for the fractional part. Figure 4(b) shows the amplitude average error using different precisions for the fractional part. The errors were obtained from the hardware implementation. Plot shows that the average error decreases as the resolution of the input variables is incremented. This reduction is not linear, and the graphic shows a point where such reduction is not significant. Seven bits were chosen as a good compromise between the fractional part representation and its average error.

7.3 Quadruped gait network architecture

In the CPG model for quadruped locomotion all basic VDP oscillators are interconnected through the connection weights (wij). In order to overcome the partial lack of flexibility of the CPG digital architecture, it has been attached as a specialized coprocessor to a microblaze processor following an embedded system design approach so as to provide a high level interface layer for application development. A bank of registers is used to provide communication channels to an embedded processor. The bank has twenty-three registers and it receives the input parameters from microblaze, α, p^2, ω^2, w_{ij} and the initial values of each oscillator. The architecture sends output data to specific FPGA pins. Figure 5 shows

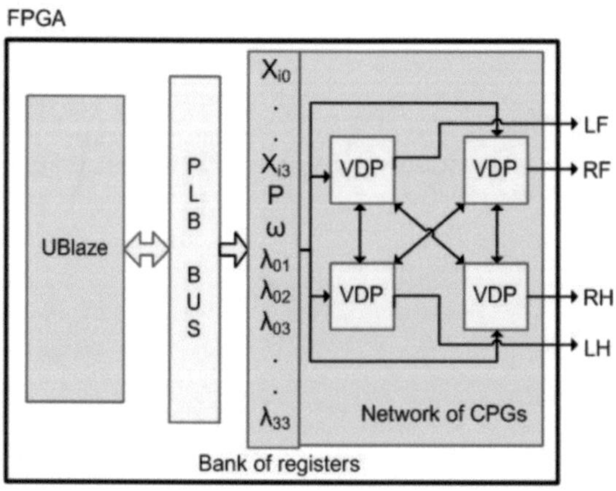

Fig. 5. Complete architecture for embedded implementation of a CPG-based quadruped locomotion controller.

a simplified block diagram of the VPD network interfacing scheme to the bank registers and microblaze processor.

8. Experimental results

8.1 GA-based method results

The proposed GA-based method was implemented in Matlab and its results were tested through hardware implementation reported in (Barron-Zambrano et al., 2010b). For the implementation of the GA, the specialized toolbox of Matlab was used. The parameters used to estimate the variables of the oscillator were: a population of 30 individuals and 200 generations. The probabilities of crossover and mutation are set to be 0.8 and 0.2 respectively. The method has an average error with the desired frequency of ± 0.03 Hz and with the desired amplitude of ± 0.15. Simulations show the state of the parameters in different generations, the first plot shows the simulations in the 2nd population generation. The generated signal in the early generations is periodic but it has low value of frequency and amplitude respect to the actual desired value. The second plot shows a signal with periodic behavior, 7th generation, but with a different frequency with respect to the desired frequency. Finally, the last plot shows the pattern with the desired frequency, 1 Hz, and amplitude, 2.5. The final parameters $[\alpha, p, \omega]$ estimated for the method are $[0.705, 0.956, 4.531]$.
Figure 6 shows the behavior of the fitness values at each generation for the oscillator parameters, plot 6(a), and for the network weights and initial conditions of $x_{1..4}$, plot 6(b). In both cases, the fitness function values decrease suddenly in the early stages of tuning. Also, plots show that the method needs a reduced number of generations to tune the oscillator and the network.
In a second step, the estimation of w_{ih} and the initial values of $x_{1..4}$ is computed. The parameters used for this step were: a population of 40 individuals and 300 generations. The probabilities of crossover and mutation are set to be 0.5 and 0.2 respectively. In the test, the desired pattern is a walk gait. This pattern has a phase of 90 degrees among the signal in fixed order, $X1 - X3 - X2 - X4$. Only, the walk matrix was estimated by the method. It is possible to generate the trot and gallop gait with the walk matrix and only changing the initial condition of $x_{1..4}$, but it is impossible to switch from one gait to another when the gait

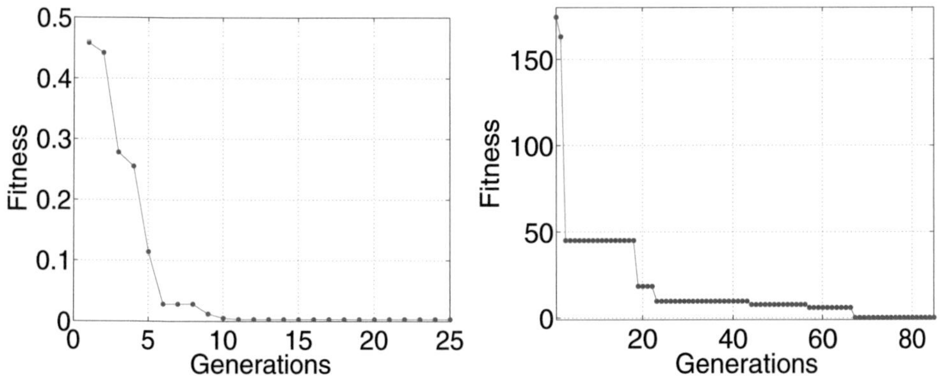

(a) (b)

Fig. 6. (a) Fitness convergence plots for the first stage, oscillator parameters and (b) for the second stage, network weights and initial conditions of $x_{1..4}$.

Gait	Walk	Trot	Gallop
Matrix	$\begin{pmatrix} 1.0 & -0.26 & -0.26 & -0.26 \\ -0.26 & 1.0 & -0.26 & -0.26 \\ -0.26 & -0.26 & 1.0 & -0.26 \\ -0.26 & -0.26 & -0.26 & 1.0 \end{pmatrix}$	$\begin{pmatrix} 1.0 & -0.26 & 0.26 & -0.26 \\ -0.26 & 1.0 & -0.26 & 0.26 \\ 0.26 & -0.26 & 1.0 & -0.26 \\ -0.26 & 0.26 & -0.26 & 1.0 \end{pmatrix}$	$\begin{pmatrix} 1.0 & 0.26 & -0.26 & -0.26 \\ -0.26 & 1.0 & 0.26 & -0.26 \\ -0.26 & -0.26 & 1.0 & 0.26 \\ 0.26 & -0.26 & -0.26 & 1.0 \end{pmatrix}$

Table 1. Weight matrix to configure the CPG network

is stable. Then, the remaining matrices, trot and gallop, were calculated using the behavior analysis presented in (Barron-Zambrano & Torres-Huitzil, 2011). The analysis shows that the combination of inhibitory and excitatory connections is able to modify the phase among the oscillators. Thus, it is possible to generate different gaits from walk matrix. To change the connections from inhibitory to excitatory type, and vice-versa, is only necessary to change the algebraic sign of the values in the gait matrix.

In the trot gait, the front limbs have the reverse phase and the limbs on a diagonal line move synchronously. From that, to generate the trot gait is necessary to change the algebraic sign in the matrix of weights for limbs on a diagonal line. The gallop has similarities with the walk gait, the phase is equal to 90 degrees but the order is different, $X1 - X2 - X3 - X4$. Therefore, the design of a new matrix for this gait is necessary. The gallop matrix is obtained only by changing the algebraic sign in the one connections off-diagonal, this produces a phase equal to 90 degrees among the limbs. Table 1 shows the matrix weights for the three basic gaits, walk, trot and gallop with parameters α, p and ω equal to $[1.12, 1.05, 4.59]$ and initial conditions $x_{1..4} = [2.44, 4.28, -1.84, 2.71]$.

Figure 7 shows the evolution of network weights and the initials condition of x. The plot 7(a) shows the pattern generated when the second stage starts. It shows a synchronized pattern where all signals have the same phase. The evolution of population shows how the signals start to have different phases, figure 7(b). Figure 7(c) shows that the method found the correct weight and the initial values of $x_{1..4}$ to generate the walk gait pattern. Finally, figure 7(d) and 7(e) show the pattern generated by the trot and gallop gaits with matrices estimated from the walk gait matrix.

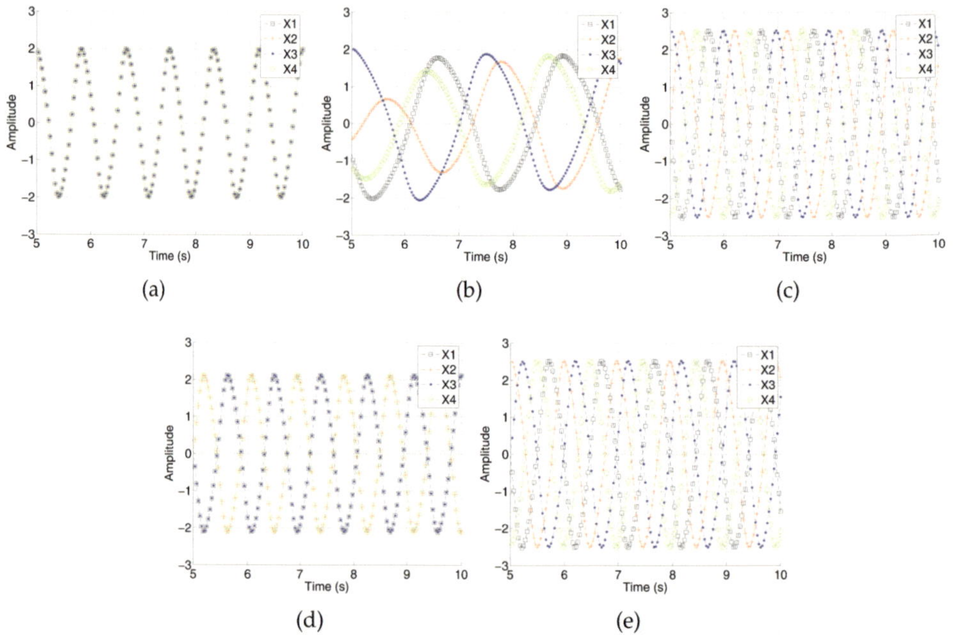

Fig. 7. (a)-(c)Evolution of pattern, in different generations, generated by the walk gait. (d)-(e) Pattern generated by the trot and gallop gait with matrices estimated from the walk gait matrix.

8.2 Physical implementation

The CPG digital architecture has been modeled using the Very High Speed Integrated Circuits Hardware Description Language (VHDL), and synthesized using the ISE Foundation and EDK tools from Xilinx targeted to the Virtex-4 FPGA device. In the hardware implementation test, a C-based application was developed on the microblaze embedded processor to set the values of the parameters in the CPG module. The implementation was validated in two ways. The first one, the results were sent to the host computer through serial connection to visualize the waveforms generated by the module. Then, the hardware waveforms were compared with the software waveforms. In the second way, results were sent to digital-analog converter (DAC) and the output signal from DAC was visualized on a oscilloscope. Figure 8 shows, the periodic rhythmic patterns corresponding to the gaits (walk, trot, gallop) generated by hardware implementation. The values of the weight matrix to configure the CPG network are shown in table 1. The initial value, $x_{1..4}$, the parameter values, $[\alpha, p, \omega]$, and the weight value, w_{ij}, were calculated automatically with a GA implementation.
The last CPG-hardware test was done through *Webots* robot simulator software. In this simulation the network with 8 VPD modules was used. Figure 9 shows, the walking locomotion pattern, in the bioloid robot simulator, and the periodic oscillations for the different joints, limbs and knees, produced by the hardware implementation. Figures 9(a)- 9(h) show the specific bioloid joint positions corresponding to the time labels shown in the walk time graph. Figures 9(i)- 9(k) show the phase relationship between the knee and limb joints in the right forelimb and hindlimb during walking. Finally, in figure 9(l) a time graph to show transition between walking and trotting is presented. Locomotions patterns for trot and gallop, and the transitions between them, were also tested.

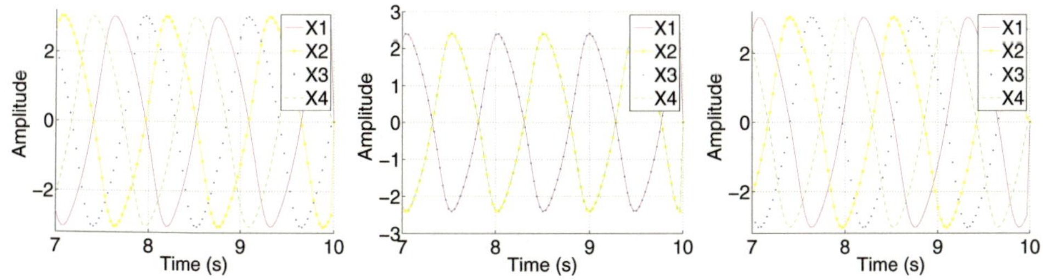

(a) Walk (b) Trot (c) Gallop

Fig. 8. (a)-(c) Three basic gaits for quadruped locomotion

9. Future work

In general, the discussion of future work will be focused in three goals: (a) to scale up the present approach to legged robots with several degrees of freedom to generate complex rhythmic movements and behavior, (b) integrate visual perception information to adapt the locomotion control in unknown environment, (c) incorporate the feedback from the robot body to improve the generation of patterns.

9.1 Network scalability

Nowadays, in robotics, the decentralization of control is an approach increasingly used in the design stage. In this approach, the robot control consists of group of modules where each module process the information in a local way. Then, the information of each module is sent to neighbor modules to produce a global result. One of the remarkable features of modular robots control is its scalability. This ability is important to achieve the best configuration required for doing the task, it is also useful in self-repair by separating faulty modules and replacing them with other modules (Murata & Kurokawa, 2007). Self-organization and distributed intelligence characterized by distribution of processing, decentralization of control and sensing tasks and modularization of components, have become essential control paradigms (Mutambara & Durrant-Whyte, 1994).

In the hardware architecture context, the scalability can be defined as the ability to increase the architecture throughput without a complete re-design, using additional hardware, in order to meet a user need. In other words, the architecture must be able to adapt oneself to robots with more degrees of freedom only adding the necessary modules. The first example of hardware scalability is presented in (Barron-Zambrano et al., 2010a). In this work, the scalability of a CPG, which control a quadruped robot with 1 DOF per limb, was the necessity to control a quadruped robot with two degrees of freedom per limb. In this case, the locomotion control system will be scalable a network with eight VDP oscillators as suggested in most works reported in the literature (Fujii et al., 2002; Kimura, Fukuoka, Hada & Takase, 2003).

9.2 Visual perception information

Locomotion and perception have been treated as separate problems in the field of robotics. Under this paradigm, one first solves the vision problem of recovering the three-dimensional geometry of the scene. This information is then passed to a planning system that has access to an explicit model of the robot (Lewis & Simo, 1999). This solution is computationally intense and too slow for real-time control using moderate power CPUs. Furthermore, this approach

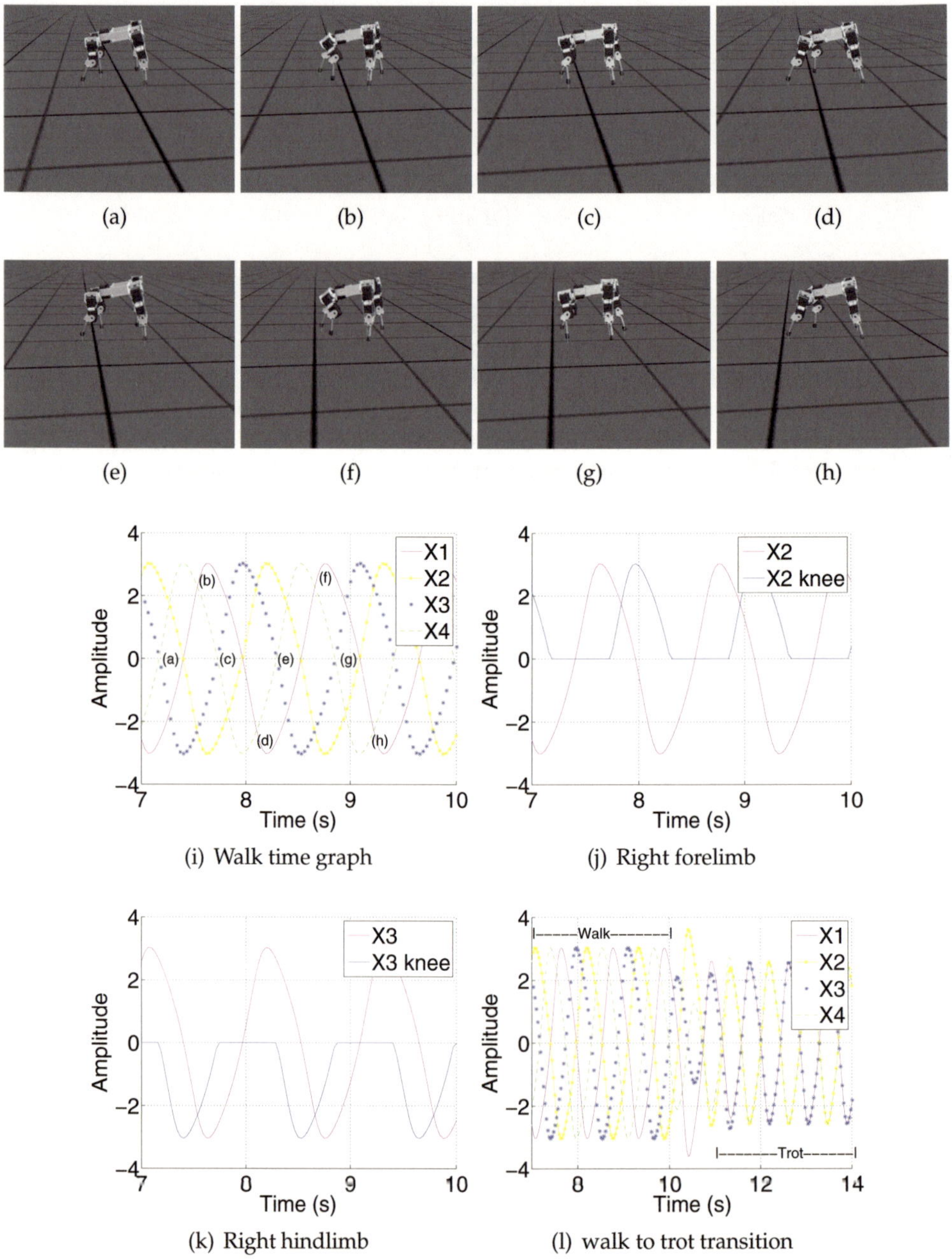

Fig. 9. (a)-(h) Walking locomotion patter snapshots for the bioloid robot simulator. (i) Walk time graph for different joints. (j)-(k) Oscillation patterns for the right forelimbs, hindlimbs and knees. (l) Waveforms during transition between walking and trotting.

does not exploit the fact that the walking machine will be presented with a similar situation again and again.

Recently, approaches consider to eliminate the intermediate explicit model and consider creating a direct coupling of perception to action, with the mapping being adaptive and based on experience. Also, continuous visual input is not necessary for accurate stepping. Not all visual samples have the same potential for control limb movements. Samples taken when the foot to be controlled is in stance are by far more effective in modulating gait. It has been suggested that during stepping visual information is used during the stance phase in a feed forward manner to plan and initiate changes in the swing limb trajectory (Hollands & Marple-Horvat, 1996; Patla et al., 1996).

Taken together, this may indicate that gait is modulated at discrete intervals. This modulation may be a program that depends on a brief sampling of the visual environment to instantiate it (c.f. (Patla Aftab E., 1991)). This hypothesis is intriguing because it implies that after a brief sample it is not necessary to store an internal representation of the world that needs to be shifted and updated during movement. This shifting and updating is problematic for both neural and traditional robotics models (Lewis & Simo, 1999).

9.3 Feedback from the robot body

Studies of mechanisms of adaptive behavior generally focus on neurons and circuits. But adaptive behavior also depends on interactions among the nervous system, body and environment: sensory preprocessing and motor post-processing filter inputs to and outputs from the nervous system; co-evolution and co-development of nervous system and periphery create matching and complementarity between them; body structure creates constraints and opportunities for neural control; and continuous feedback between nervous system, body and environment are essential for normal behavior. This broader view of adaptive behavior has been a major underpinning of ecological psychology and has influenced behavior-based robotics. (Chiel & Beer, 1997).

Recent work in the field of autonomous robotics has emphasized that intelligent behavior is an emergent property of an agent embedded in an environment with which it must continuously interact. The goal is to integrate information of the robot status with the CPG. The robot will be equipped with different sensors such as bumpers and accelerometers. These sensor signals could be added as coupling terms in the differential equation in the oscillator model. This can be able to simplify the control, allowing them to traverse irregular terrain and making them robust to failures.

10. Conclusions

The quadruped locomotion principles were studied and analyzed. It is feasible to generate locomotion patterns by coupling the neural oscillators signals that are similar to the neural oscillators found in animals by the definition of a mathematical model. The GA based approach takes advantage that the fitness function works directly with the oscillator and the network. It makes possible consider other restriction as the power consumption and it does not need knowledge of the robot dynamic to generate the pattern gait. The GA based approach uses small population, some tens individuals, and limited generations, some hundreds generations, ideal to be processed on computers with reduced resources. The applications only estimate the matrix for walk gait. Furthermore, it is possible to build the matrix for gallop and trot gaits using the walk gait matrix. This reduces the number of operations necessary to estimate the matrix for each gait

The hardware implementation exploits the distributed processing able to carry out in FPGA devices. The presented examples show that the measured waveforms from the FPGA-based implementation agree with the numerical simulations. The architecture of the elemental Van Der Pol oscillator was designed and attached as a co-processor to microblaze processor. The implementation provides flexibility to generate different rhythmic patterns, at runtime, suitable for adaptable locomotion and the implementation is scalable to larger networks. The microblaze allows to propose an strategy for both generation and control of the gaits, and it is suitable to explore the design with dynamic reconfiguration in the FPGA. The coordination of joint of a legged robot could be accomplished by a simple CPG-based network extremely small suitable for special-purpose digital implementation. Also, the implementation takes advantage of its scalability and reconfigurability and it can be used in robots with different numbers of limbs.

11. References

Amari, S.-I. (1988). Characteristics of random nets of analog neuron-like elements, pp. 55–69.

Arena, P. (2001). A mechatronic lamprey controlled by analog circuits, *MED'01 9th Mediterranean Conference on Control and Automation*, IEEE.

Arena, P., Fortuna, L., Frasca, M. & Sicurella, G. (2004). An adaptive, self-organizing dynamical system for hierarchical control of bio-inspired locomotion, *IEEE Transactions on Systems, Man, and Cybernetics, Part B* 34(4): 1823–1837.

Barron-Zambrano, J. H. & Torres-Huitzil, C. (2011). Two-phase GA parameter tunning method of CPGs for quadruped gaits, *International Joint Conference on Neural Networks*, pp. 1767–1774.

Barron-Zambrano, J. H., Torres-Huitzil, C. & Girau, B. (2010a). FPGA-based circuit for central pattern generator in quadruped locomotion, *Australian Journal of Intelligent Information Processing Systems* 12(2).

Barron-Zambrano, J. H., Torres-Huitzil, C. & Girau, B. (2010b). Hardware implementation of a CPG-based locomotion control for quadruped robots, *International Conference on Artificial Neural Networks*, pp. 276–285.

Billard, A. & Ijspeert, A. J. (2000). Biologically inspired neural controllers for motor control in a quadruped robot., *in a IEEE-INNS-ENNS International Joint Conference on Neural Networks. Volume VI*, IEEE Computer Society, pp. 637–641.

Buchli, J. & Ijspeert, A. J. (2004). Distributed central pattern generator model for robotics application based on phase sensitivity analysis, *In Proceedings BioADIT 2004*, Springer Verlag, pp. 333–349.

Buchli, J., Righetti, L. & Ijspeert, A. J. (2006). Engineering entrainment and adaptation in limit cycle systems: From biological inspiration to applications in robotics, *Biol. Cybern.* 95: 645–664.

Carbone, G. & Ceccarelli, M. (2005). *Legged Robotic Systems, Cutting Edge Robotics*, Vedran Kordic, Aleksandar Lazinica and Munir Merdan.

Chiel, H. J. & Beer, R. D. (1997). The brain has a body: adaptive behavior emerges from interactions of nervous system, body and environment, *Trends in Neurosciences* 20(12): 553 – 557.

Crespi, A. & Ijspeert, A. J. (2006). AmphiBot II: An Amphibious Snake Robot that Crawls and Swims using a Central Pattern Generator, *Proceedings of the 9th International Conference on Climbing and Walking Robots (CLAWAR 2006)*, pp. 19–27.

Dagg, A. (1973). Gaits in mammals, *Mammal Review* 3(4): 6135–154.

Delcomyn, F. (1980). Neural basis of rhythmic behavior in animals, *Science* 210: 492–498.

Fujii, A., Saito, N., Nakahira, K., Ishiguro, A. & Eggenberger, P. (2002). Generation of an adaptive controller CPG for a quadruped robot with neuromodulation mechanism, *IEEE/RSJ international conference on intelligent robots and systems* pp. 2619–2624.

Fukuoka, Y., Kimura, H., Hada, Y. & Takase, K. (2003). Adaptive dynamic walking of a quadruped robot 'tekken' on irregular terrain using a neural system model, *ICRA*, pp. 2037–2042.

Grillner, S. (1985). Neural control of vertebrate locomotion - central mechanisms and reflex interaction with special reference to the cat, *Feedback and motor control in invertebrates and vertebrates* pp. 35–56.

Hollands, M. A. & Marple-Horvat, D. E. (1996). Visually guided stepping under conditions of step cycle-related denial of visual information, *Experimental Brain Research* 109: 343–356.

Hooper, S. L. (2000). Central pattern generator, *Current Biology* 10(2): 176–177.

Ijspeert, A. (2001). A connectionist central pattern generator for the aquatic and terrestrial gaits of a simulated salamander, *Biological Cybernetics* 84(5): 331–348.

Ijspeert, A. (2008). Central pattern generators for locomotion control in animals and robots: A review, *Neural Networks* 21(4): 642–653.

Ijspeert, A. J. & Crespi, A. (2007). Online trajectory generation in an amphibious snake robot using a lamprey-like central pattern generator model, *Proceedings of the 2007 IEEE International Conference on Robotics and Automation (ICRA 2007)*, pp. 262–268.

Inagaki, S., Yuasa, H., Suzuki, T. & Arai, T. (2006). Wave CPG model for autonomous decentralized multi-legged robot: Gait generation and walking speed control, *Robotics and Autonomous Systems* 54(2): 118 – 126. Intelligent Autonomous Systems.

Jalal, A., Behzad, M. & Fariba, B. (2009). Modeling gait using CPG (central pattern generator) and neural network, *Proceedings of the 2009 joint COST 2101 and 2102 international conference on Biometric ID management and multimodal communication*, Springer-Verlag, Berlin, Heidelberg, pp. 130–137.

Kier, R. J., Ames, J. C., Beer, R. D. & Harrison, R. R. (2006). Design and implementation of multipattern generators in analog vlsi.

Kimura, H., Fukuoka, Y., Hada, Y. & Takase, K. (2003). Adaptive dynamic walking of a quadruped robot on irregular terrain using a neural system model, *in* R. Jarvis & A. Zelinsky (eds), *Robotics Research*, Vol. 6 of *Springer Tracts in Advanced Robotics*, Springer Berlin / Heidelberg, pp. 147–160.

Kimura, H., Shimoyama, I. & Miura, H. (2003). Dynamics in the dynamic walk of a quadruped robot, *Irregular Terrain Based on Biological Concepts. International Journal of Robotics Research 22(3Ű4):187Ű202*, MIT Press, pp. 187–202.

Lathion, C. (2006). Computer science master project, biped locomotion on the hoap2 robot, *Biologically Inspired Robotics Group* .

Lee, Y. J., Lee, J., Kim, K. K., Kim, Y.-B. & Ayers, J. (2007). Low power cmos electronic central pattern generator design for a biomimetic underwater robot, *Neurocomputing* 71: 284–296.

Lewis, M. A., Hartmann, M. J., Etienne-Cummings, R. & Cohen, A. H. (2001). Control of a robot leg with an adaptive VLSI CPG chip, *Neurocomputing* 38-40: 1409 – 1421.

Lewis, M. A. & Simo, L. S. (1999). Elegant stepping: A model of visually triggered gait adaptation.

Loeb, G. (2001). Learning from the spinal cord, *The Journal of Physiology* 533(1): 111–117.

Manoonpong, P. (2007). *Neural Preprocessing and Control of Reactive Walking Machines: Towards Versatile Artificial Perception-Action Systems (Cognitive Technologies)*, Springer-Verlag New York, Inc., Secaucus, NJ, USA.

Matsuoka, K. (1987). Mechanisms of frequency and pattern control in the neural rhythm generators, *Biological Cybernetics* 56(5): 345–353.

Murata, S., K. K. & Kurokawa, H. (2007). Toward a scalable modular robotic system, *IEEE Robotics & Automation Magazine* pp. 56–63.

Mutambara, A. G. O. & Durrant-Whyte, H. F. (1994). Modular scalable robot control, *Proceedings of 1994 IEEE International Conference on MFI 94 Multisensor Fusion and Integration for Intelligent Systems* pp. 121–127.

Nakada, K. (2003). An analog cmos central pattern generator for interlimb coordination in quadruped locomotion, *IEEE Tran. on Neural Networks* 14(5): 1356–1365.

Nakanishi, J., Morimoto, J., Endo, G., Cheng, G., Schaal, S. & Kawato, M. (2004). Learning from demonstration and adaptation of biped locomotion, *Robotics and Autonomous Systems* 47: 79–91.

Okada, M., Tatani, K. & Nakamura, Y. (2002). *Polynomial design of the nonlinear dynamics for the brain-like information processing of whole body motion*, Vol. 2, pp. 1410—1415 vol.2.

Patla, A. E., Adkin, A., Martin, C., Holden, R. & Prentice, S. (1996). Characteristics of voluntary visual sampling of the environment for safe locomotion over different terrains, *Experimental Brain Research* 112: 513–522.

Patla Aftab E., Stephen D., R. C. N. J. (1991). Visual control of locomotion: Strategies for changing direction and for going over obstacles., *Experimental Psychology: Human Perception and Performance* 17: 603–634.

Picado, H., Lau, N., Reis, L. P. & Gestal, M. (2008). Biped locomotion methodologies applied to humanoid robotics.

Pratt, J. E., Chew, C.-M., Torres, A., Dilworth, P. & Pratt, G. A. (2001). Virtual model control: An intuitive approach for bipedal locomotion, *I. J. Robotic Res.* 20(2): 129–143.

Pribe, C., Grossberg, S. & Cohen, M. A. (n.d.).

Susanne, S. & Tilden, M. W. (1998). Controller for a four legged walking machine, *Neuromorphic Systems Engineering Silicon from Neurobiology, World Scientific, Singapore*, pp. 138–148.

Torres, A. L. (1996). Virtual model control of a hexapod walking robot, *Technical report*.

Torres-Huitzil, C, Girau B. (2008). Implementation of Central Pattern Generator in an FPGA-Based Embedded System, *ICANN* (2): 179-187.

Van Der Pol B, V. D. M. J. (1928). The heartbeat considered as a relaxation oscillation, and an electrical model of the heart, 6: 763–775.

Zielinska, T. (1996). Coupled oscillators utilised as gait rhythm generators of a two-legged walking machine, *Biological Cybernetics* 74(3): 263–273.

10

Tracking Control in an Upper Arm Exoskeleton with Differential Flatness

E. Y. Veslin[1], M. Dutra[2], J. Slama[2], O. Lengerke[2,3] and M. J. M. Tavera[2]
[1]*Universidad de Boyacá,*
[2]*Universidade Federal do Rio de Janeiro,*
[3]*Universidad Autónoma de Bucaramanga*
[1,3]*Colombia*
[2]*Brazil*

1. Introduction

The Exoskeleton is classified inside of a wider technology known as Wearable Robots, that group describes a robotic field who studies the interaction between the human body and the robotics. In those systems, a mechatronics structure is attached to different members of the human body, and while the wearer commands the mechanical system using physical signals, like the electronic stimulation produced by the orders of the brain, or the movements of the body generated by the muscles; the mechanical system do the hard work, like carrying heavier objects or helping the movement of handicaps members of the body. Since its conception in the sixties (Croshaw, 1969) the bibliography speaks about two models of wearable robots: the prosthetics and the orthotics systems (Fig. 1). The first group replaces the lost body members, while the second one assists to the body movements or, in other cases, extends the body capabilities.

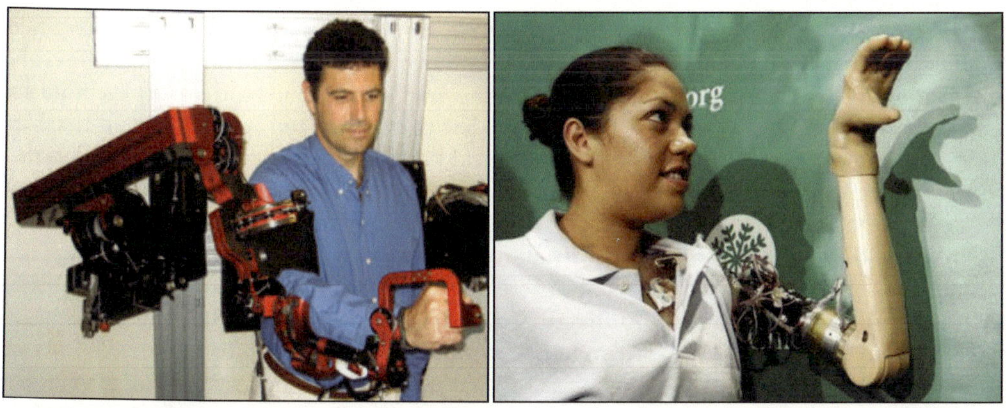

Fig. 1. (Left), orthotic robot of upper parts (Perry et al., 2007). (Right), prosthetic robot of upper parts (Rehabilitation Institute of Chicago, 2006)

The exoskeleton take part of the orthotics wearable robots field, it's a robotic structure (actually could be considered a mechatronics structure due to the integration of different

electronics equipments like sensors, actuators and an intelligent controller that are integrated into the mechanical system) fixed to the human body that follows along the movements of the wearer. At the present time are presented several models of exoskeletons: those that are attached to the legs was called lower part exoskeletons (Kazerooni, 2006) and also exists those that are attached to the arms, that are denominated upper part exoskeletons (Perry et al., 2007). In both structures, its applications goes from the medical camp to military applications; with functions like help handicap patients assisting on the body member's movement recuperation (Gopura & Kiguchi, 2007) or helping to the soldiers to carry heavier loads (Kazerooni, 2005). Also, exists prototypes that integrates both systems (Sankai, 2006), those mechanisms like Kazerooni's structure, used to help the wearer to extends his body capabilities in order to carry weights that are beyond his normal possibilities, or help to patients to made normal movements that a sane body could develop.
Our work are focussed into the study of the upper arm exoskeleton and a application of differential flatness in order to control the system that moves in a determinate path (in this case a path generated by a polynomial). This document will describe its consecution, initiating with a morphological analysis that leads to a mathematical model and posterior to a cinematic simulation. This model is useful to the consecution of the dynamic model, this study was based on the mechanical structure and the arm movements that were selected to made. This dynamic model will be applied into the problem of trajectory tracking, in this part, the arm is controlled using differential flatness, this part was considered our contribution, therefore, we will discuss about its applications, benefits and problems founded. All the analysis was supported by a code made in Matlab® and gives the necessaries simulation to this work, in this way, graphics of the model behaviours was characterized and captured in order to detail the different parts of this work.

2. Forward kinematics

The first problem to solve is the characterization of a human arm model than could be implemented for an exoskeleton design. In this part, forward kinematics plays an important role; offering a mathematical model that allows the prediction of the system's behaviour, giving the necessary preliminary information about the structure and movement capabilities. Case of modelling the human arm for the study of the human motion (Maurel & Thalman, 1999), (Klopčar & Lenarčič, 2005), or in a implementation study of a rehabilitation system in an arm (Culmer et al., 2005), or the design of a master arm for a man-machine interface (Lee et al., 1999), All needs of the forward kinematical analysis in order to initiate a system's modelling.
The forward kinematic study initiate with a morphological analysis in the part of the body that will be attached to the mechanical structure. At is says, the exoskeleton is a system that accompany the arm on their movements; therefore the design of the structure must not had interference with it. In order to do this, the morphological analysis gives the information about movements, lengths of the members, number of segments and its respective masses. Another justification is that the biological inspirations permits the creation of mechanical systems that are more compacted and reliable and also, more energy-efficient (Kazerooni, 2006).
In the nature, the biological designs of the living organisms allows the adaptation with the environment; the evolution mechanisms have made transformation into the body as the circumstances rules, for this reason the diversity of organic designs and also, the extinction

of those species that didn't could adapt to its environment. In conclusion is possibly to say that exist optimizations mechanisms that transforms our structure in function of a goal: survival.

The human arm is a result of this optimization, following the bones that conforms it (Fig. 2) its seen that the member are divided into three segments: the arm, the forearm and the hand, united among them by three joints: the shoulder that unites the upper limb to the rest of the body, the elbow that unites the arm with the forearm and finally the wrist that unites the forearm with the hand. The mechanical model of the exoskeleton responds to this configuration, conserving the systems general characteristics, this process of make a copy of the nature design, is called bioimitation (Pons, 2008).

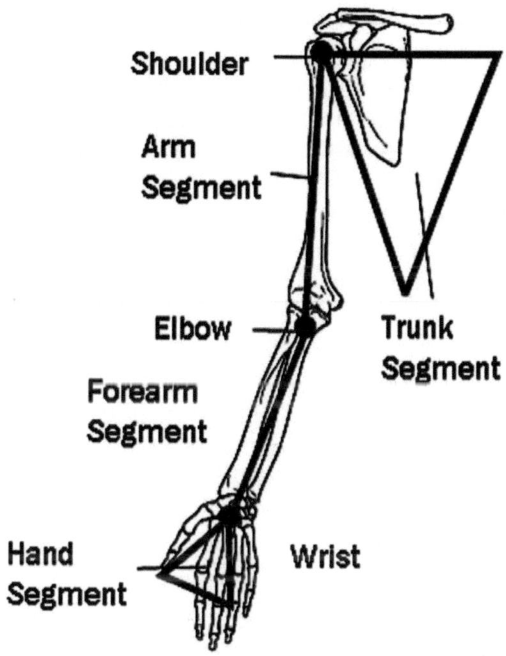

Fig. 2. Human arm structure (Pons, 2008)

Initially, for this study the hand and fingers movements will not be considered, so, it could be reduced to a single segment (Pons, 2008), in this way the kinematics is constructed with three segments, the arm, the forearm and the hand, also exists other considerations that could help into the study of the system (Rocon & Pons, 2005):

1. The mechanical behaviour of the arm is independent of the rest of the body.
2. All the components of each segment, including the bones and the soft parts (muscles, skin, etc.) take part of the same rigid body.
3. The deformations of the soft part will not affect significantly the mechanical properties of the entire segment.

The lengths of the segments are considered constants and depend of the height of the individual (*H*), this numerical relationship are detailed into Table 1, who also related another parameters of the human arm, like the position of the centre of mass of each segment.

Body Segment	Length, L	Centre of Mass (% of L)	
		Proximal	Distal
Upper Arm	0.186 H	0.436	0.564
Forearm	0.146 H	0.43	057
Hand	0.108 H	0.506	0.494

Table 1. Human arm Anthropometric data (Pons, 2008)

Each joint has its proper movements, which depending of the movement could it be: uni, bi or multiaxial (Gowitzke & Milner, 2000), it means that the joints allows movements in one, two or three degrees of freedom and planes generated by the rotation of the segments. For this study all the joints could be considered ideals (Rocon & Pons, 2005).

The movements of the arm are showed into Fig. 3. The first joint, the elbow, is compared with a spherical joint that allows movements in three degrees of freedom (multiaxial), its movements are defined as: flexion-extension, abduction-adduction, and rotation. The second joint, the elbow is an uniaxial joint, its movement is compared with a hinge, nevertheless, the bone of the forearm generates a movement of rotation around its own axis, that not affects the position of the three spatial coordinates, but if it rotation. This rotation could be located on the elbow joint or into the wrist joint. On early studies this rotation was considered on the elbow but later in order to make a simplification in the numerical operations it was transported to the wrist (Veslin, 2010).

Some research studies the movement of the arm in the plane (Gerald et al., 1997); others articles analyse the movement of the forearm for applications of trembling reduction (Rocon & Pons, 2007). The kinematic analysis on this work will be focused on the movement of the three joints of the elbow, and specifically to the posterior control of the hand position using a mechanical structure supported by the arm, for this reason is necessary to know its capabilities based in an analysis of the workspace.

Knowing the structure of the arm, the distance of the segments and the degrees of freedom of its joints, the mathematical model could be developed, in this part the concepts of morphology previously studied are combined with the classical robotic concepts, in this case using the model of homogenous transformation matrix (Siciliano et al., 2009). Equations (1-3) descript the rotations R around the three axis X, Y and Z, while (4) descript the translations T in whatever axis.

$$R_z(\beta)=\begin{bmatrix} \cos\beta & -\text{sen}\beta & 0 & 0 \\ \text{sen}\beta & \cos\beta & 0 & 0 \\ 0 & 0 & 1 & 0 \\ 0 & 0 & 0 & 1 \end{bmatrix} \quad (1)$$

$$R_y(\alpha)=\begin{bmatrix} \cos\alpha & 0 & \text{sen}\alpha & 0 \\ 0 & 1 & 0 & 0 \\ -\text{sen}\alpha & 0 & \cos\alpha & 0 \\ 0 & 0 & 0 & 1 \end{bmatrix} \quad (2)$$

$$R_x(\gamma)=\begin{bmatrix} 1 & 0 & 0 & 0 \\ 0 & \cos\gamma & \text{-sen}\gamma & 0 \\ 0 & \text{sen}\gamma & \cos\gamma & 0 \\ 0 & 0 & 0 & 1 \end{bmatrix} \quad (3)$$

$$T(d)=\begin{bmatrix} 1 & 0 & 0 & d_x \\ 0 & 1 & 0 & d_y \\ 0 & 0 & 1 & d_z \\ 0 & 0 & 0 & 1 \end{bmatrix} \quad (4)$$

In (4) is necessary to define in which, or what axis the translation movement will be effectuated. The cinematic model obeys to a kinematics open chain, because only exists one sequence of joints together, connecting the two endings of the chain (Siciliano et al., 2009). Fig. 3 shows the arm structure from the shoulder to the hand, each movement is associated to a Latin symbol and an orientation respects to the inertial axis. In this way, all the movements of flexion-extension occur into the Y axis, the rotation movements on the bone into the Z axis, and the others into the X axis. Its seen that in the shoulder exists a special case where three rotations intercepts in of one point without the existence of any translation, this kind of configuration it's called spherical wrist. All the translations referents to the segments lengths (l1,l2 and l3) happens in the Z axis.

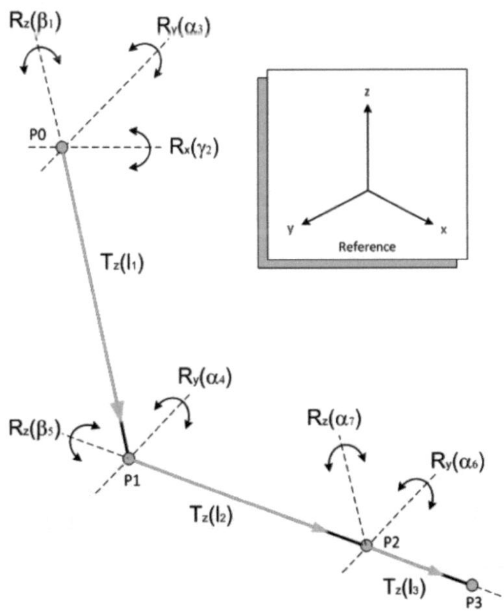

Fig. 3. Description of arm transformations

The four reference Points P0, P1, P2 and P3 will define the path a across the system. The set of equations (5) describes the number and kind of transformations given from the origin in the shoulder (P0) to the each reference point. In this way, the matrix result *T* will define the position en XYZ and the orientation respects to the origin in each final of segment, being T01 the path from the origin to the point P1, T02 the path from P0 to P2 and T03 the path from P0 to P3.

$$
\begin{aligned}
&T_0^1(\beta_1,\alpha_2,\gamma_3)=R_y(\alpha_2)R_x(\gamma_3)R_z(\beta_1)T_z(l_1)\\
&T_0^2(\beta_1,\alpha_2,\gamma_3,\alpha_4,\beta_5)=T_0^1R_y(\alpha_4)R_z(\beta_5)T_z(l_2)\\
&T_0^3(\beta_1,\alpha_2,\gamma_3,\alpha_4,\beta_5,\alpha_6,\gamma_7)=T_0^2R_y(\alpha_6)R_x(\gamma_7)
\end{aligned}
\tag{5}
$$

The order of multiplication of each matrix is given according to an observation analysis; the orientation due to the consecutive operations on the rotation matrix could make variations into the final response, affecting both the visualization and the position of a point of the system. It's important to be clear in what will be the order of the final equation (5), this gives an analysis of what happened into the arm with the variation of each joint. This analysis is compatible to both arms, only changes the direction of the rotation, remembering that according to the established parameters of the homogeneous transformation, every rotation made counter the clock is considered positive.

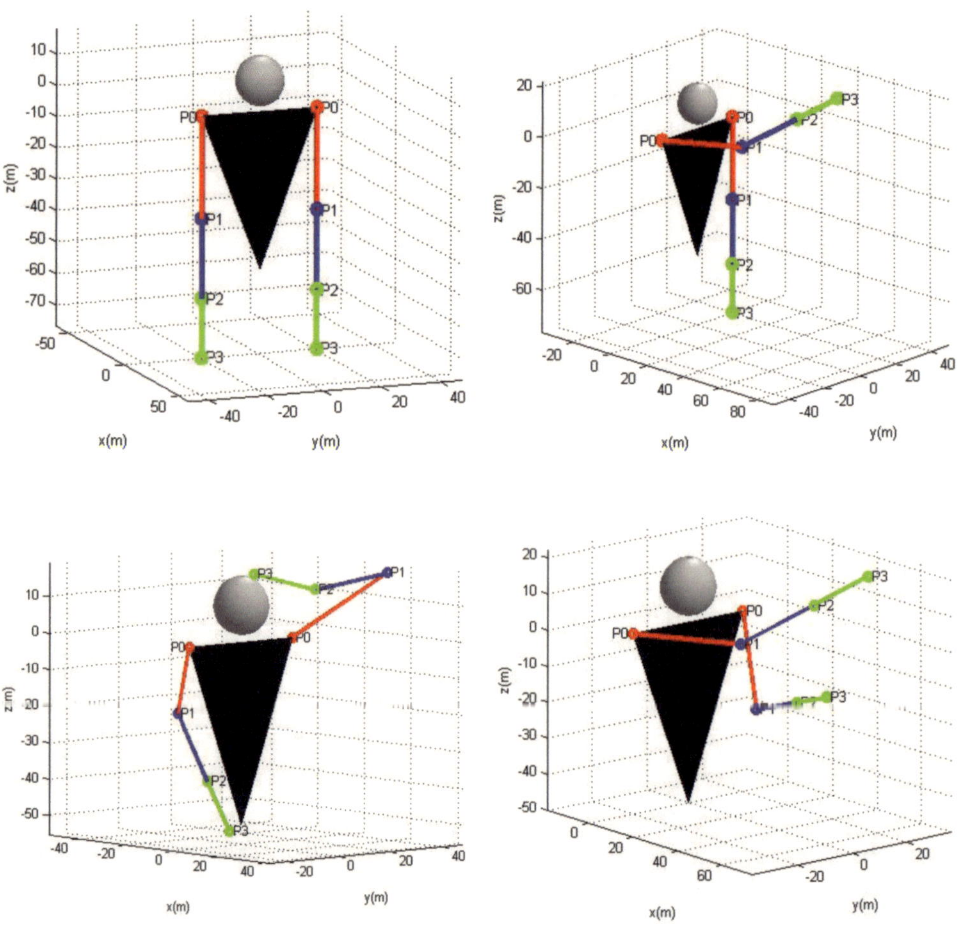

Fig. 4. Different positions of the arm

This mathematical model was implemented in MATLAB®, and the results graphics Fig. 4, describes different positions of the arm structure, as the anatomical position with all the rotations in 0°. The three segments represent the arm divisions, the triangle the body and the sphere the human head. All the graphics was made for an individual of 1.75 m of height.

The model obtained of the kinematical analysis give to the researcher a map of the system behaviour, the restrictions on the joint movements and the extensions of the segments of the arm can detail the workspace (Fig. 5), a graphical analysis that gives an understanding of the system capabilities. Analysing the workspace of the human arm, is possible to obtain information for the design and control of the mechanism, and also, gives an understanding of the movement properties. In the case of the human arm, this study allows the planning and programming of rehabilitation process comparing the workspace of a normal patient, with a person with pathology deficient, this differences could be used for the diagnosis study (Lenarčič & Umek, 1994).

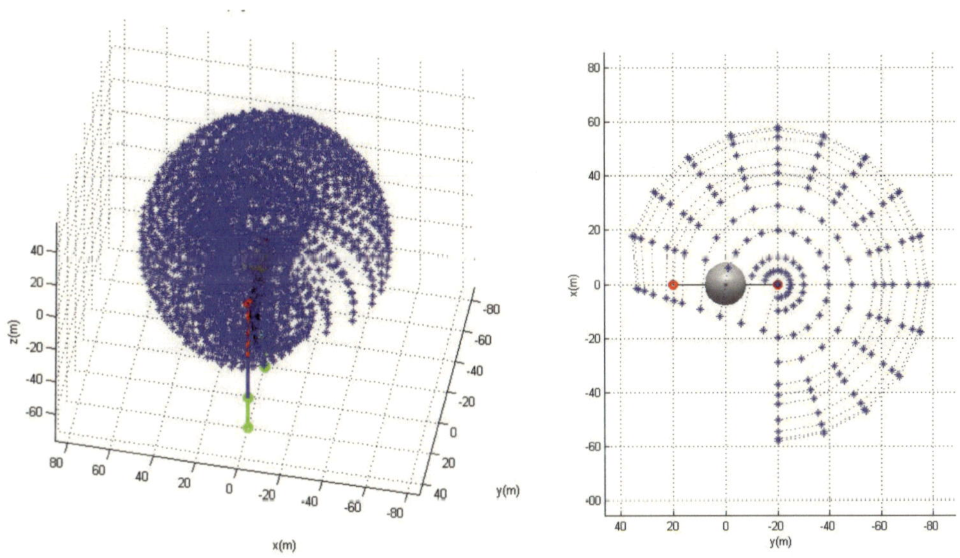

Fig. 5. Arm Workspace

In an exoskeleton analysis, exist two workspaces in interaction: the human arm movement and the mechanical system movement (Pons, 2008), both generates a range of positions in the space that intercepts, this interception will be the plausible movement generated by the wearer of the exoskeleton. In the case of the human arm, the graphics was obtained using the three movements of the shoulder and the flections of the elbow. The arm was moved with movements of flexion-extension and, finally, the rotation of the elbow. The Fig. 5 shows the path made by the right arm around the shoulder, the figure describes a form like a circumference with radio equal to the distance of the arm (l1) and forearm (l2), all the dots points the wrist position for each movement. Exists a space that the arm cannot reach, this space correspond to the sector near to the back, where is impossible to reach due to the physics restrictions for the body. Is important to say that in ideal conditions, the workspace in both arms is the same.

The workspace is defined by the mobility of the articulations, these being from the human arm or the mechanical system that support the arm. Therefore, the workspace of a wearer that carries an exoskeleton could be affected by the mechanical restrictions that the system imposes, for this reason, better mechanical designs allows the performance of movements more obedient and with a wider rank.

3. Dynamics

While the kinematics analysis gives an approach of the system, the dynamic model generates the necessary information about the forces needed to umbalnce the segments an initiate the movement, and posteriorly the behavior of the system across the differents possible positions that the arm could reach.

Refers to other autors (Rosen et al. 2005), the conception of upper arm exoskeleton approach to the behavior of an human arm due to the morphologycal design of the mechanism. In some designs, the weight of the sytems doesn't influentiate in the dynamical analysis, and all the intertial forces are related to the human arm physiollogy (Carignan et al., 2005), in this case the center of mass is realted to the extenstion of the arm, and its mass depends of the organical structure (Table 1 and Table 2). In other cases, the mechanical system is more heavier than the human arm (Croshaw, 1969), in these models the structure must be included into the analysis, this cause that parameters like intertial center and estructure change drastically. Nevertheless, with a adecuate code this changes could be easilly modifcated, because this information could be parametrized or previouslly changed to the simulation code.

Figure 6 illustrated the model of the arm, C1 and C2 are the possition of the center of mass on the arm segments, the rotations consdired into the study are the shoulder movements and the flexion-extension of the foream. The hand movements doesn´t be considered into this part of the study.

The initial evaluation are focused into the minimal forces necessaries to move an human arm, in order to do that, is necessary to known its composition . In this way, parameters like mass, length and position of the inertial center that depends of the wearer stature, are introduced into the model and simulated to obtain a further analysis.

Body Segment	Percent respect to the body weight m(%)
Upper Arm	2.70
Forearm	1.60
Hand	0.66

Table 2. Mass of the segments body (Tözerem 2000)

To obtain the body mass, exists a relation with its height, this relation is called Body Mass Index (BMI) established by Adolphe Quelet in the XIX century, that evaluates the health conditions of the person using a number, a discussion of its validation it's not an interest in this work, but this model gives a mathematical relation (6) that is useful to introduce into the code of the model, high values of BMI evidences complication in the health, a BDI of 24 considerate that the person have good health conditions, and this number will be used to obtain the body mass that finally gives the segments mass due to Table 2.

$$\text{BMI} = \text{mass}/\text{weight}^2 \quad (6)$$

To do the simulation, the arm is modeled using a Lagrangian conception, where the human arm is considered a robotic manipulator. With the kinematics model, the information of the position of the ends of segments and inertial centers could be added into the model. The final result is a nonlinear mechanical model that simulates the arm behavior in different movements.

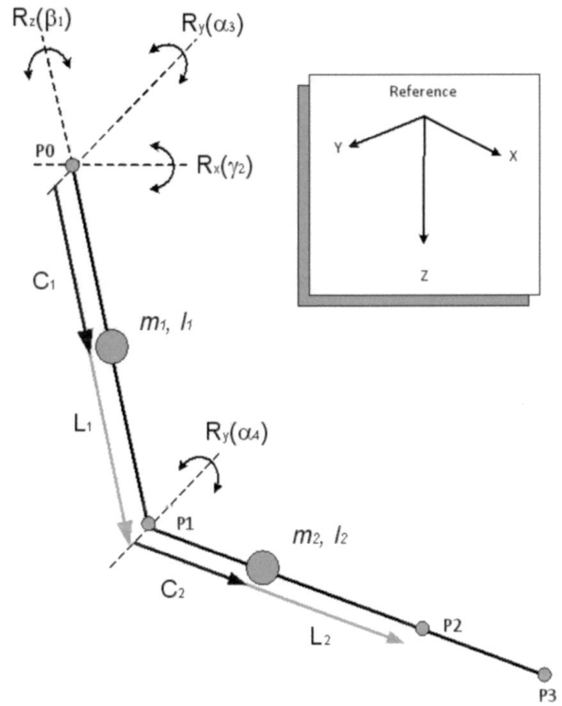

Fig. 6. Human arm model

For the evaluation of the forces that involves the movement of the system, a PD controller is use to move the arm from an initial position to a desired position. Movements like lift the arm while the forearm is static or move only the forearm could be simulated. Initiating in different positions and with various wearer statures, the code, gives the necessary information about the maximal and minimal forces involved, the influence of the Coriollis and centrifugal forces and the gravitational forces into the arm, and also plausible reductions into the model to reduce the computational time in the simulation.
In previously works (Veslin 2010), the dynamical model was obtained and finally condensed into a form described in (7), in this equation, the joints are considered ideals so the mathematical model ignore the friction forces, also, the perturbations that could exist into the movement doesn't be introduced because we don´t have a way to quantify it, and finally is considered that the system is fully actuated, its means that each movement have its proper force actuating into it.

$$M(q)\ddot{q} + C(q,\dot{q}) + K(q) = \tau \quad (7)$$

The mass of the segments was concentrated in its geometrical center, the form of the arm was approximated into a cylinder (Admiral et al. 2004, Soechting et al., 1995) this gives a mathematical relationship and evades the use of tables or interpolations to find the inertial center of the arm. Was demonstrated (Veslin, 2010) that the presence of the inertial tensor doesn´t influence significantly the system behaviour, this consideration gives a reduction of the model equation and converts the tensor on a single equation.
Each movement was controlled using a feedback rule (8), it gives to the system the force needed to move the segments to de desired position, given by q_{nd}, its equivalent to a PD controller, its applications are seen in industrial robots and lower parts exoskeletons (Carignan, et al. 2005).

$$\tau_n = -\beta\dot{q}_n - \gamma(q_n - q_{nd}) + W_n(q_n) \tag{8}$$

The Gain β involve the arm velocities and the energy dissipation, in oscillatory movements it controls its amplitude and the duration of the movement, γ gives a force that depends of the distance between the actual position of the segment q_n and the desired position, in this way the equation reduce the force while the arm is reaching the goal. W_n are the forces that the actuators have to support when the segments are in static equilibrium. The implementation of these controllers adds a problem called overshoot (a high input that occurs in a small portion of time) into the beginning of the simulation (Figure 6).

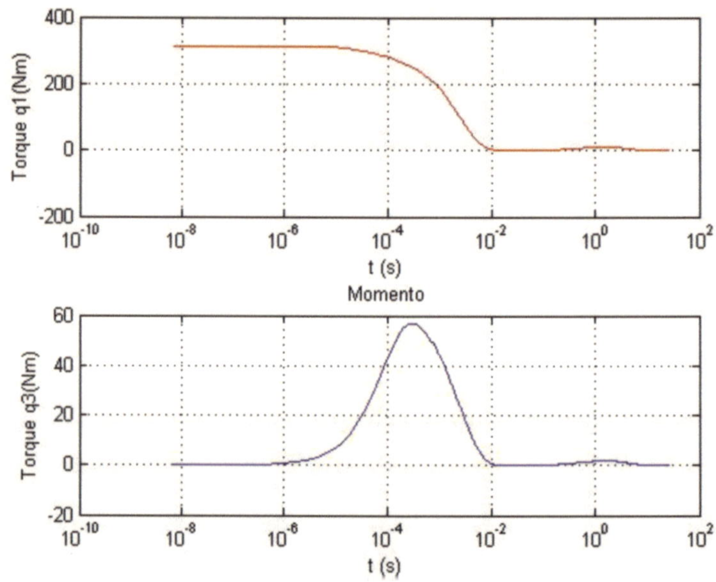

Fig. 7. Presence of overshoots at the beginning of the simulation

In order to resolve this problem, a trajectory generator is added to the system control, this generator increments the position of the segment considering the initial an finally positions, in this way is possible to create trajectory that descript a desired behavior applying interpolation methods, like cubical, quadratic or another equation. Considering that the model initiate and finish with zero speed, the form that could be implemented into the model is a cubical equation (Figure 7), which is implemented into (8) replacing the term q_{nd}.

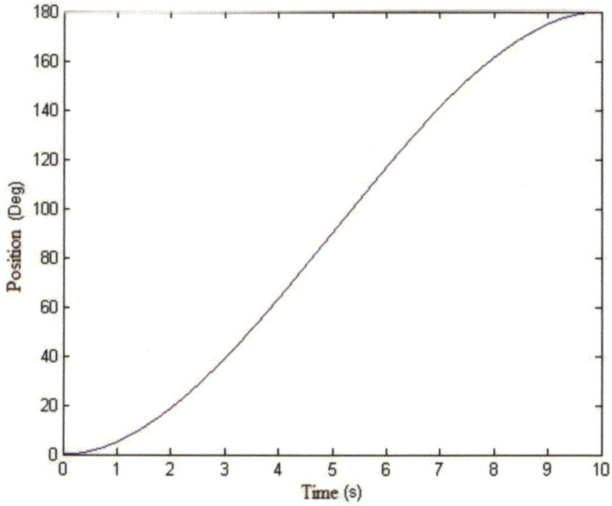

Fig. 8. Variation of the angular position in a cubical way

The movement is generated around the movement determined by the trajectory equation, the Figures 8 and 9, shows a common displacement between two points in the shoulder and elbow respectively, this movement is called plane movement because only actuate in one degree of freedom of the shoulder. The dot line represents the movement interpolated by the cubic equation, and the solid line is the movement generated by the degrees of freedom of the system that are actuated, moving according by the path and controlled by equation 8.

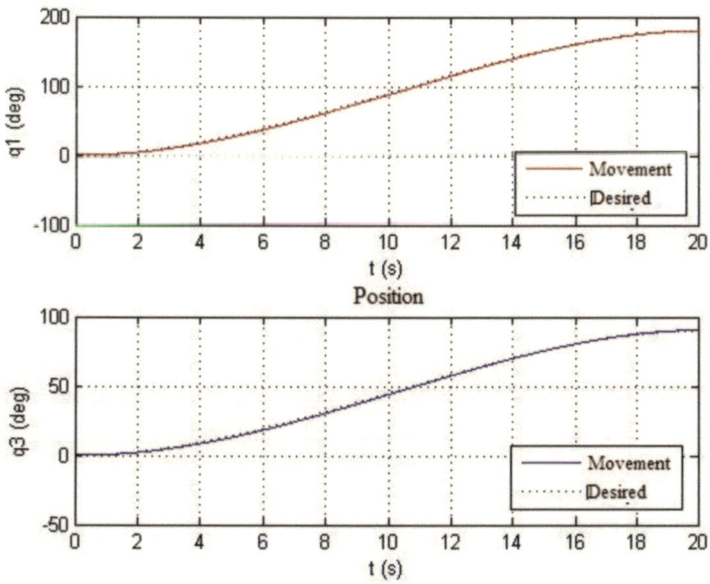

Fig. 9. Comparison between the movement generated by the controller into the shoulder and the elbow respects the desired path

The absolute error (Fig. 10) between the movements executed and the path is evaluated, the mayor displacement happens in the middle of the movement around the 8 seconds, being major in the shoulder, presenting a displacement of 2.5 degrees and 1.4 degrees into the elbow. The final position also shows a displacement, the error in this position is near to 0,5 degrees in both movements.

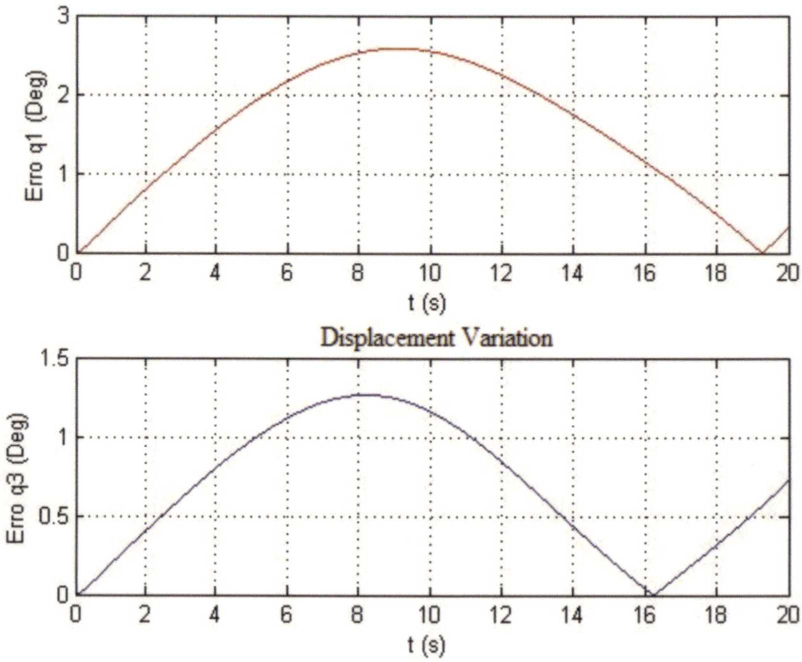

Fig. 10. Difference between the shoulder and elbow movement with the desired path

To evaluate the controller, different movements were generated and the result is showed in Fig. 11, the forearm final position is evaluated because it gives the location of the wrist, which is the object of control. The graphic evaluates the final position of the forearm initiating its movement from 0° to 140° during a elevation movement of the arm (that moves from 0° to 180°), the mayor displacements happens when the final position of the forearm is 90°. The internal forces between the segments cause this variations, the forearm in this position generates the biggest moment to the arm in movement. The controller assure that them doesn't influencing moving the arm to another position, the moments caused by the mass segments and the energy induced by the velocity is controlled by the model proposed, with a minimal error as is illustrated that doesn't overcome the 0.3° in the worst case.

Therefore, this error could be reduced according with the variation of the parameter γ in the controller, Fig. 12 and 13 evaluates different errors in the final position with different values in γ in a range of 100 Nm/rad to 600 Nm/rad, into ascending (Fig. 12) and descending (Fig. 13) movements. Bigger values into the constants reduces the difference in the displacement generated by the internal forces of the arm, the controller supplies the necessary force and consecutively the error of displacements change according to the variation of the constant.

For this graphics the forearm was conserved into a position of 0° while the arm was moving in a range of 0° to 180°.

The maximal forces generated by the controller are illustrated in Fig. 14. In this graphics two movements are compared, in the red line the movements begins from the 0° and moves into different positions in a range of 0° to 180° with increments of 10°, always executed in a time of 20 seconds. The blue line effectuates the same movement in a time that is proportional with the space travelled, in other words, faster than the first move. This movement generates bigger forces because effectuate the same movement with less time, the augmentation of the velocities into the generalized coordinate generates an augmentation of the Coriollis forces between the segments, generating in the controller a bigger force to counteract the impulse generated by the movement and the mass of the arm.

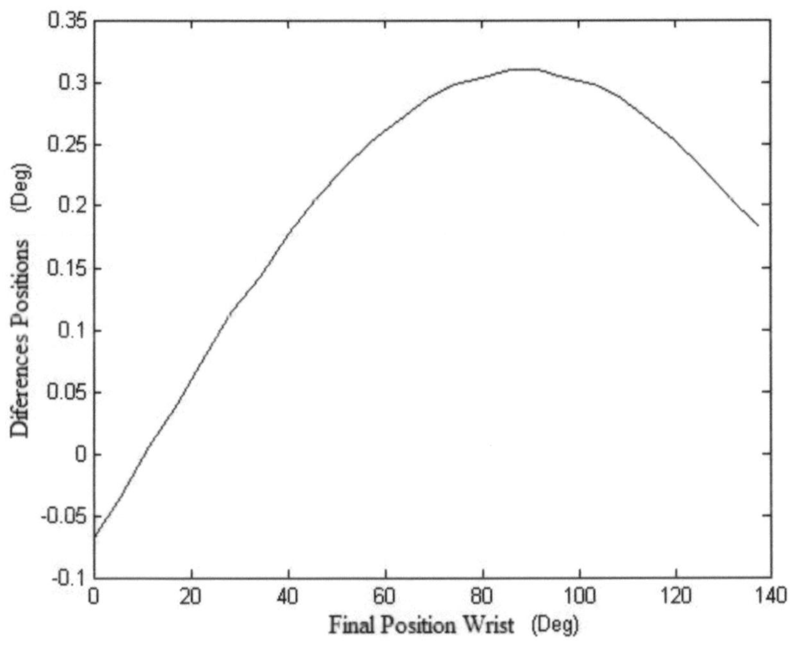

Fig. 11. Deviation of the final position of the wrist depending of the position of the forearm while the arm is rotating into flexion movement

Also this forces changes depending of the height in the individual, as is show in (6) and Table 2, the mass on the arms differ depending of the stature of the individual and his BMI, in this case, the maximal forces in the arm (which actuator have to support the weight of the complete system) are illustrated in Fig 16, where in a range of heights from 1.5 m to 2m, the forces necessaries to arise the arm increases in almost 10 Nm.

Movements out of the plane, with a rotation of two degrees of freedom in the shoulder are illustrated in Fig 16, the shoulder was commanded to move the arm in flexion and abduction, while the forearm rotates from the rest position to 90°, (1) illustrates the movement according with the generalized coordinates, while (2), (3) y (4) illustrates the behaviour on the system in the space.

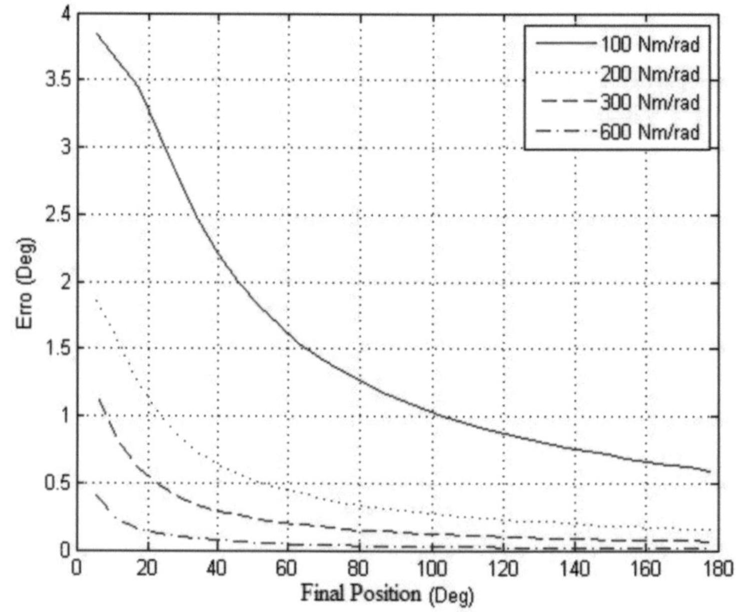

Fig. 12. Reduction of the error according to different values of γ into the controller in a ascending movement

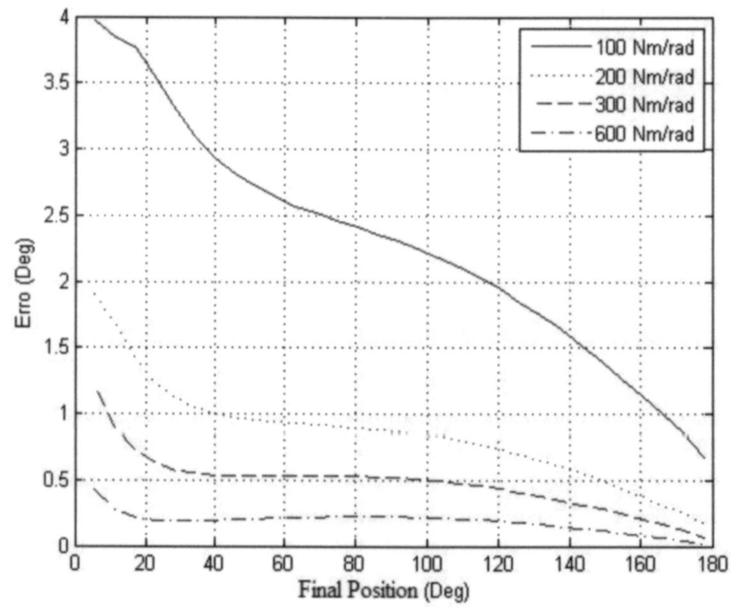

Fig. 13. Reduction of the error according to different values of γ into the controller in a descending movement

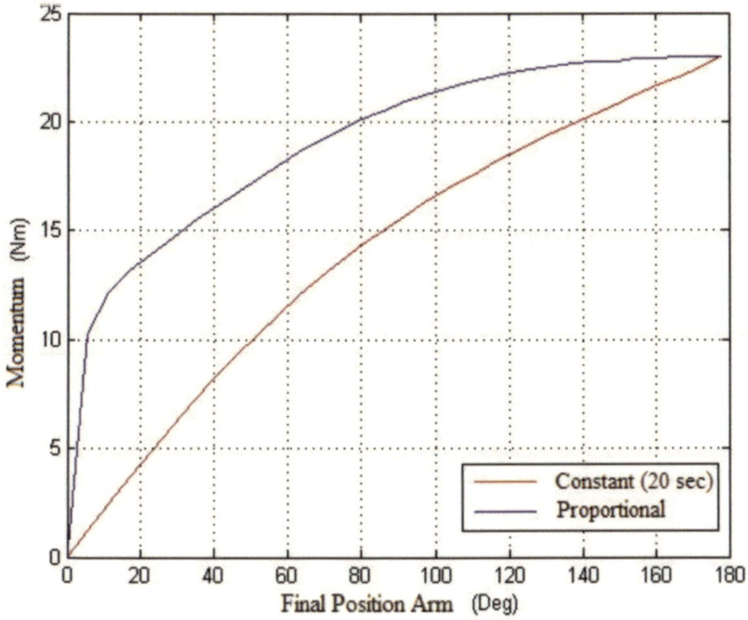

Fig. 14. Variation into the maximal forces for movements in the arm with different velocities

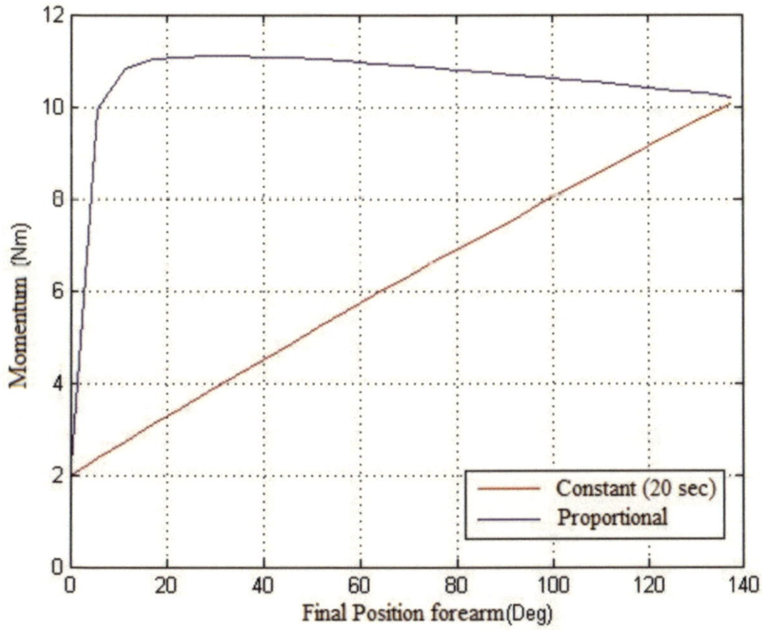

Fig. 15. Variation into the maximal forces for movements in the forearm arm with different velocities

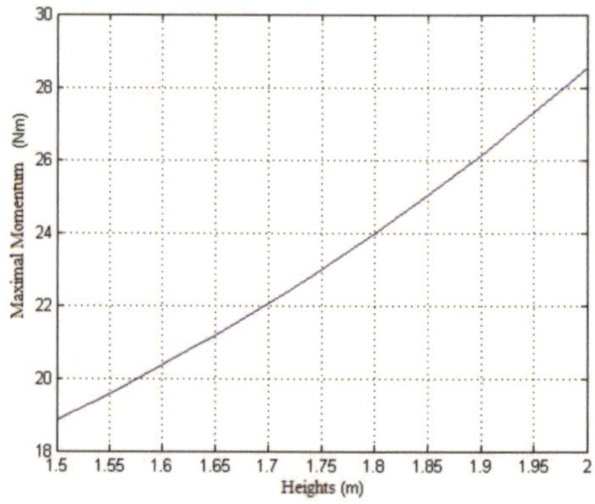

Fig. 16a. Variation into the maximal forces with different individual heights

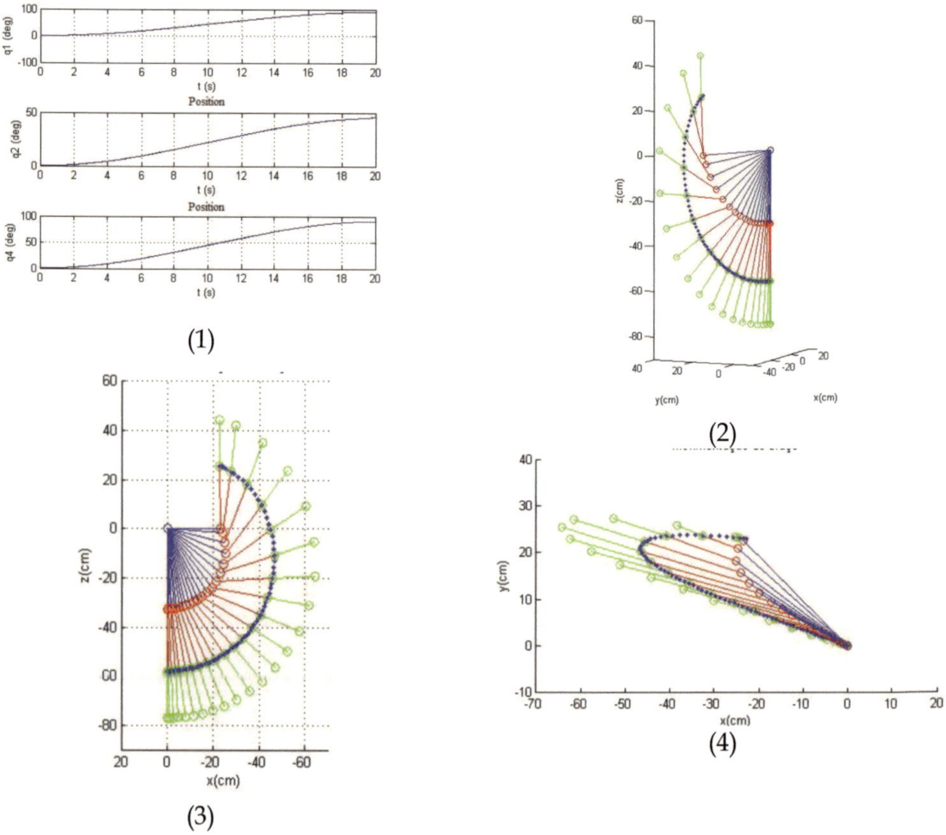

Fig. 16b. Movement in the generalized coordinates generated by the controller (1), trajectory in the space described by the arm (2), projection in the transversal plane (3) and the sagittal plane (4)

The influence of the gravitational forces in the arm model was evident when was founded an increase in the maximal forces effectuated by each generalized coordinate, this is caused by the increasing of the mass in the segments and also to the position of the forearm respects to the arm, meanwhile and augmentation on the velocities also increase the value on the applied forces to effectuate the movement. The implementation of the PD controller with a compensation of the gravitational forces in the arm position allows to made the previously forces analysis, the constants have an important role in the system behaviour, adding stabilities and reducing the error in the positioning of the arm wrist.

4. Differential flatness

Introduced by Fliess (Fliess et al., 1994), the differential flatness is a concept of the dynamic control, where a system is assumed to be flat, if it have a number of outputs known as flat outputs (9), such that them and its derivatives, could describe the system inputs (that have the same number of outputs) and the states of the system (10-11).

$$y = h(x, u, \dot{u}, \dots, u^{r}) \tag{9}$$

$$x = f\left(y, y, \dots \dot{,} u^{(q)}\right) \tag{10}$$

$$u = g\left(y, y, \dots \dot{,} u^{(q)}\right) \tag{11}$$

This property called is called by Fliess diffeomorphism, and is very useful in the control of systems that are needed to achieve a desired behaviour. In order to do that, the flat outputs are parameterized into functions, and posterior introduced into the system; the result is a non linear equation system that is transformed into a polynomial, so, the solution of the problem is reduced significantly because is not necessary the use of integration or numerical solutions.

The focus of this analysis is to assume than a system is differential flat, work that isn't easy to achieve, in fact, not all the systems have this property, and is considered that be flat it's a characteristic of the system itself . Many systems are defined as flat and some authors have studied its implications (Murray et al., 1995). Like Murray´s text, these works are focused into the demonstration of the flat property and its application, in systems like mobile robots and cars with n trailers.

The manipulator trajectory tracking, consist in determinate the necessary inputs (forces) in each system's generalized coordinate, in order to accomplish that the model moves between one point to another across a defined path. In this kind of problems the first step consist in the specification of a sequel of points in the manipulator space work, this points will become desires positions across the path, and posterior are integrated using an interpolation function (which is typically in a polynomial form). Actually, different techniques exist for the trajectory planning: (i) when is given each final and ending position, talk about a point to point motion, (ii) when is working with a finite point sequence, talk about motion through a sequence of points. Both techniques give a time function that describes the desired behavior (Siciliano, 2009, van Nieuwstadt, 1997a).

The second step is the implementation of this function into the system's dynamical model and verifies that effectively is given a trajectory tracking. In this part, is necessary to take a decision, if the problem is going to be worked into the operational space, or in each joint.

Due to his generality, the first solution, could give singularity problems and redundancy caused by the number of degrees of freedom in each joint, and the nonlinear effects could made it difficult to predict, meanwhile, a joint position path parameterization shows to be furthermore suitable, because this study allows to work in each element, considering factors like the restrictions movements and degrees of freedom on each joint.
Everything joint parameterized trajectory is defined by a function *q(t)*. This function is the source of a joint desired position collections for the manipulators movement across the time. Then, using different control techniques: like the PD control with gravity compensation, IDC (Inverse Dynamic Control) or adaptive control, is possible accomplish a manipulator movement on the defined path with the minimal deviation (Marguitu, 2009, Spong, 1992, Wang, 2009).
It was demonstrated that all robotic manipulator (if the upper arm exoskeleton is asumed of this form) could be flat if the system is fully actuated (Lévine, 2009); it means, that all its generalized coordinates have an actuator. In this case, from the model of the system (7), assuming that all the outputs are the position of the generalized coordinates q, and the inputs are the forces that moves each coordinate τ, the systems is differential flat if the position of each generalized coordinate and its derivatives (the velocity and the acceleration) are the flat outputs. It must be realized that all the systems outputs didn't are flat, so it's important that those be differentiated from the normal ones by the letter Z (Van Nieuwstadt, 1997b). Thus the equation (7), remembering that the joints are considered ideals, can be represented as:

$$M(z)\ddot{z} + C(z,\dot{z}) = \tau \tag{12}$$

The tracking trajectories problem doesn't present complications in a flat system; as the inputs and its states are defined in function of the flat outputs, is possible to create paths that vary in the time in order to define an behavior on its outputs, and through them and its respective derivatives, determine which are the inputs needed by the system to generate an answer that approach to the desired one by the path. This path could be polynomial or a function of C^∞ type as the trigonometric or the exponentials (Rotella & Zambettakis, 2008). The application of the flat theories allows the transformation of the differential equation system (12) into a polynomial of order n, making a simplification into the problem solution and transforming it to a completely algebraic system.
The next step is to applying the differential flat theories into the model; the general objective is control the behavior of the system in function of a desired action. This behavior will be parameterized into a trajectory. With knowledge of the flat outputs of the system, these trajectories could be easily implemented. The model in (12) shows that is possible to interpret the systems inputs knowing its flat outputs behaviors until the second derivative, therefore, is necessary add this functions and its derivatives into the model. By example, if is desired that the arm that is working in a plane movement rise starting from the origin, with a gradual variation of the positioning, it could be commanded a behavior using a cubic function in the generalized coordinates of the shoulder and the elbow, this is:

$$z_1(t) = 0{,}0094t^2 - 0{,}0006t^3 \tag{13}$$

$$z_4(t) = 0{,}0157t^2 - 0{,}001t^3 \tag{14}$$

The equation (13) leads the arm from the origin positioned in 0° to a position of 18°, on equation (14) the forearm moves from 0° to 30° (Fig. 4), both segments presents null velocities at the beginning and at the end, executing the translation in a time of 10 seconds. The equations (13) and (14) and its respective derivatives are introduced in the dynamic model (12). To determine the inputs τ only is necessary to solve the system in a given time. The response provided by the Fig. 17 is a set of outputs that are applied into the system and must allow to displace it according to the behavior implemented in equations (13) and (14), for each segment.
This response is evaluated into the dynamic model (8), the system response is verified in Fig. 18, where the trajectory made by the generalized coordinates is superposed on the desired path define by the cubic equations. Due to the proximity in each equation, Fig. 19 evaluates the error between them, showing an oscillatory trend of error variation, the maximum difference is the order of 10^{-2}.

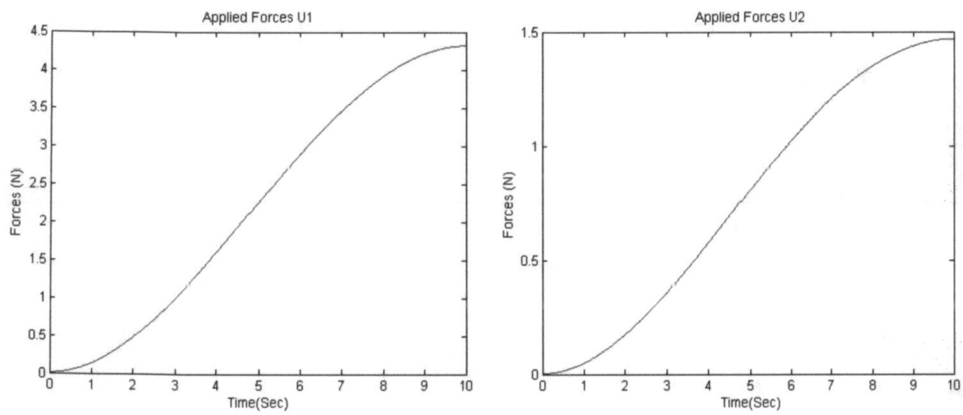

Fig. 17a. Set of forces obtained through equation (8) in each segment

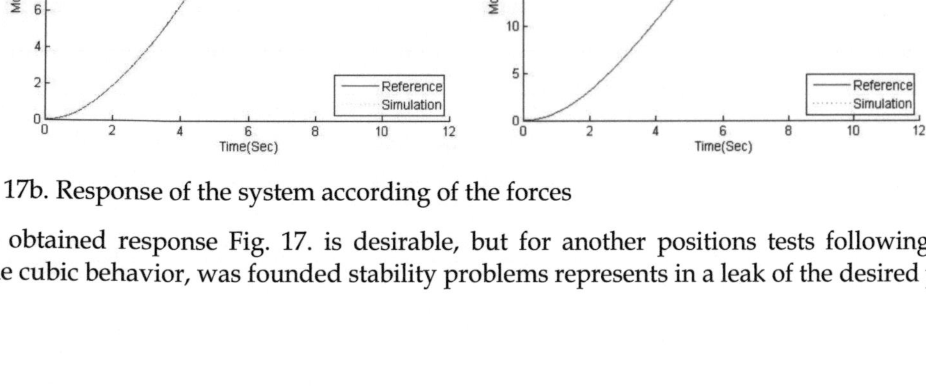

Fig. 17b. Response of the system according of the forces

The obtained response Fig. 17. is desirable, but for another positions tests following the same cubic behavior, was founded stability problems represents in a leak of the desired path system, this is evidenced in Fig. 19, where is tried to move both segments from 0° to 90° position.

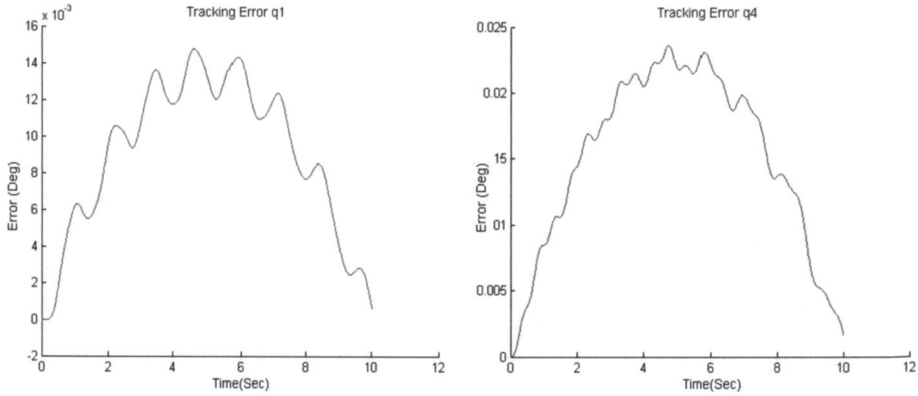

Fig. 18. Tracking error in each generalized coordinate

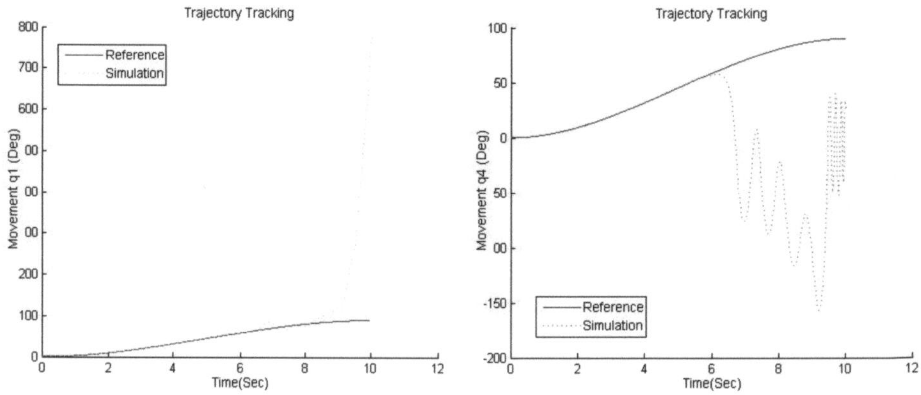

Fig. 19. Trajectory deviation for each generalized coordinate in higher movements

This doesn't mean that the flat systems theories doesn't can be applied to certain positions, only that, for these solutions , flatness not guarantee the tracking and the necessary stability if applied in open loop, case of the examples seen before. It is necessary then, to change the focus, in order to take advantage of the benefits that flatness offers. This scheme implementing a controller that feedback information from real estate and compare it with the desired output, generated from this difference a control action that finally allows the system track the trajectory desired with a minimum error, this is made with the implementation of a controller into the system.

Called two degrees of freedom control (Fig. 20), any closed loop system contains a controller and a trajectories generator. The controller modifies the output in function of existing error between the actual output and nominated desired output (Van Nieuwstadt, 1997b).

This system had actually three degrees of freedom, such system contains a higher level that generate the output depending of the desired paths (output trajectory). An intermediate level generates inputs based on these outputs, and finally the lower level stabilizes the system around the nominal trajectory. The uncertainties were variations caused by the gravitational, friction forces, and nonlinear in the models.

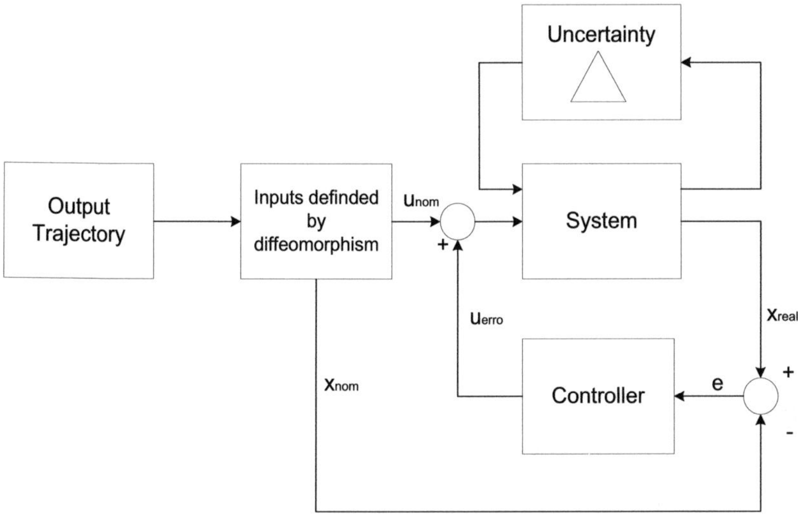

Fig. 20. Two degree of Freedom Control System (Van Nieuswtadt, 1997b)

The solution gives an input trajectory that could be used into the real model. This assumption works in theory, however some interferences, could separate the real behaviour of the desired, thus, is necessary to apply a PD feedback control into each generalized coordinate, with this, is possibly to enforce the manipulator to describe a desired performance.
For the controller in Fig. 20 is established a feedback law (Franch & Agrawal, 2003):

$$\tau = \tau + k_p(z_n - x_n) + k_d(\dot{z}_n - \dot{x_n}) \tag{15}$$

Equation (15) ensures that the nominal state X_n follow the desired path Z_n over a correct choice of k_p and k_d constants. k_p generates an action that modifies the inputs from the flatness system control in a proportional form. The derivative constant k_d, removes variations (oscillations) caused by the corrections in the proportional part. In Fig. 19 are visible the jumps in the position ranging from 100° to 800° in a small period of time, so hopefully abrupt change in the action of proportional control and consequently oscillations in the model. If each segment has its own control loop, with the constant k_p=25 and k_d=0,5 for the segment of the arm and k_p=2,5 and k_d=0,5 to the forearm. The results of the implementation of these values can be seen in Fig. 21 Obtained paths follow very near to the desired paths. Fig. 22 shows the performance of the arm according the inputs.
With this control, is easier to move the system using others trajectories, any position that is within the manipulator workspace, can be parameterized by a polynomial for then of differentially flatness systems theory, appropriate requirements. For example, is required that the arm makes a movement oscillating from a cosine function while the forearm elevate to 90° position following the same methodology of the previous path behavior, is obtained the behavior described in Fig 23.

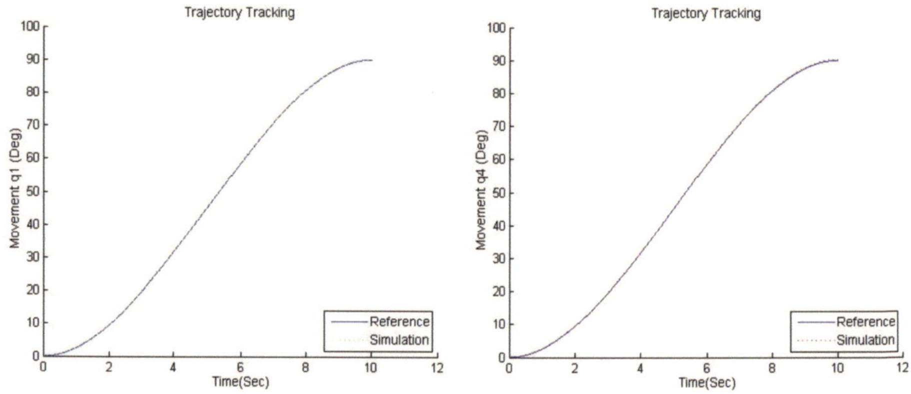

Fig. 21.Trajectory tracking using the controller

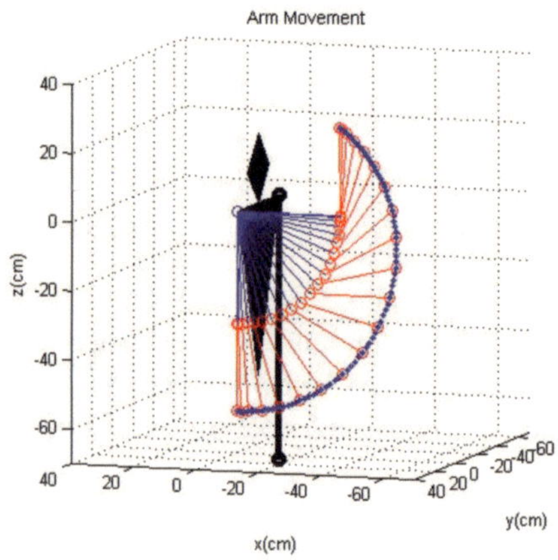

Fig. 22. System behavior using the inputs defined by flatness

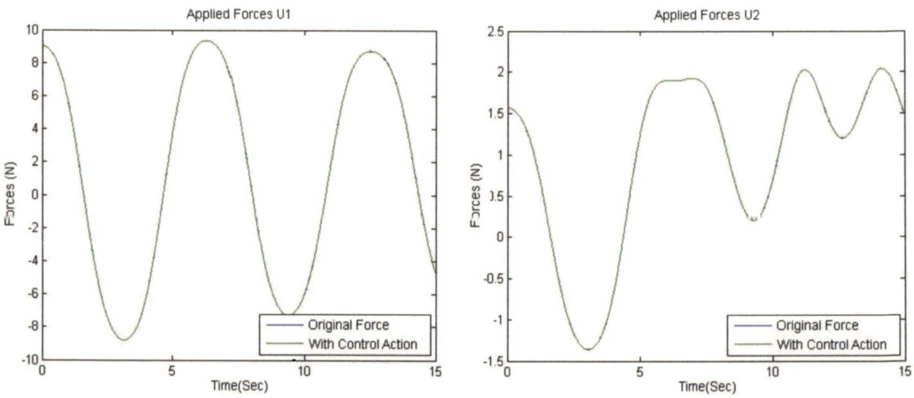

Fig. 23. Force behaviour obtained according with the cosine function implemented in the arm

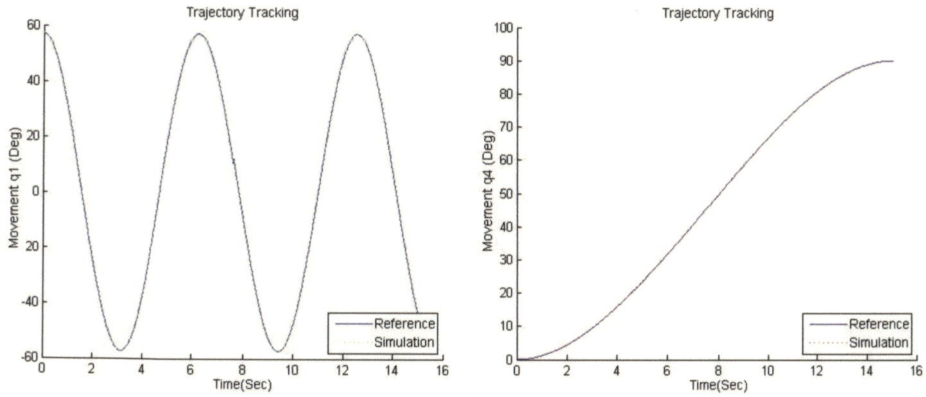

Fig. 24. System behaviour using the desired trajectories

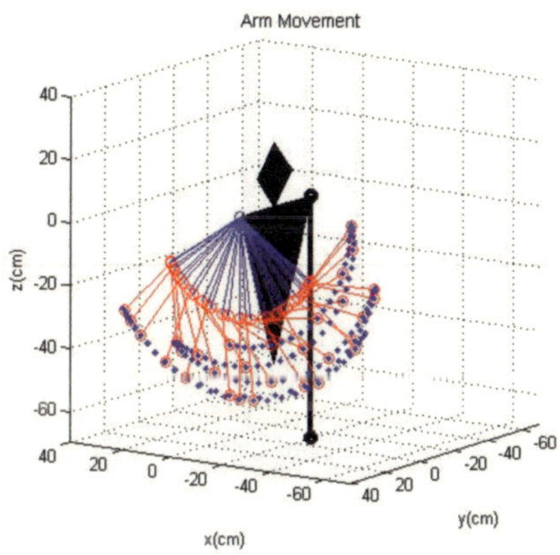

Fig. 25. System behaviour using the entries defined by differential flatness

The concept is also applicable to systems with more degrees of freedom, although the fact that the equations of motion are more complex it is possible to obtain path tracking and behaviors. But this implementation presents difficulties, inputs equations in function of four outputs are extensive and prevent the possibility of achieving an acceptable result for extensive times. Precisely this error, forced to resort to the use of a controller. Open loop system presented difficulties during the follow-up. The controller makes modifications imperceptibles on the inputs, but enough so that there is a stable trajectory. Fig. 26 visualize the behavior of one of these desired trajectories for a system that controls the four generalized coordinates.

The ability to follow-up on trajectories using differential flatness systems on robotics manipulators depends on the quality of input values. Was founded that small changes in these settings offer better performance, proving that the resolution of inputs is an important part of the results.

The tests carried out previously, the models were analyzed over a time period in increments of 0.01 seconds. But testing with greater increases do not provide appropriate behavior, as it may be identified in Fig. 27, this makes it difficult to determine the values of the constants of the controller or even worse, that do not exist. This restriction to use short times make it difficult for the calculation of inputs due to the high consumption of computing resources, therefore, in to reconfigure a strategy for solution of this situation.

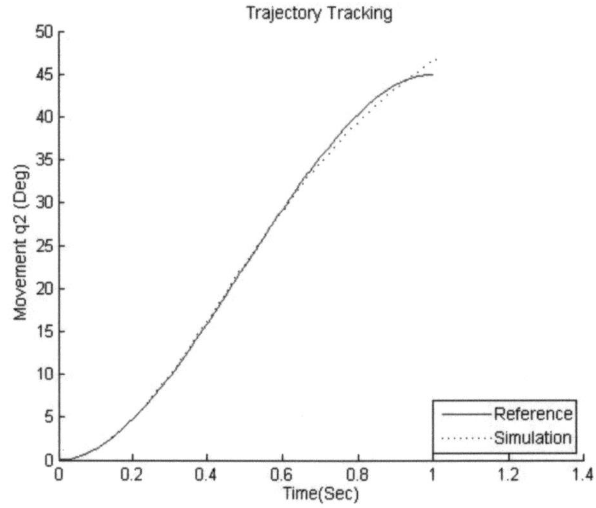

Fig. 26. Behavior of system for each trajectory, the dash line describes the actual movement while continuous line describes the desired trajectory

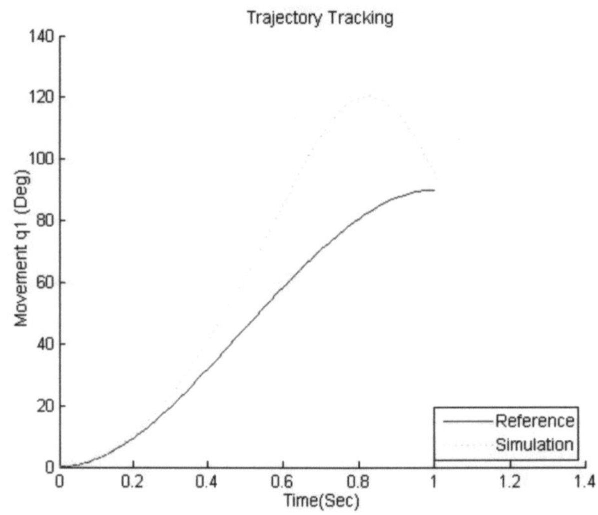

Fig. 27. Behavior model with higher time period, are evidence a difficulty in the follow-up to the desired output

5. Conclusions

This study gives to us the initial phases into the consecution of a trajectory control for an upper arm exoskeleton. Whit the forward kinematics study is possibly to generate a fist view on the system, this mathematical model achieved by means of a morphological study of the human arm gives a first view of the system behaviour, being this workspace a tool to predict the system capabilities in function of the mechanical restrictions and also the human movements. This one is an important implementation of the code made.

Therefore, the dynamical analysis provides the information about the system requirements in function of the stature of the wearer, the influences of the mechanical model was not considered into the dynamics, the movements was made considered the human arm as the objective to make an actuation, and the prototype as the element that actuated on the arm. Every consideration on the physical model of the system must be considered in the dynamical design. This dynamical analysis generates a mathematical model that is implementing into the problem of the trajectory tracking on the system. It was demonstrated that the arm is differential flat, and with this assumption, was possibly to generate any behaviour to be implemented into the system, it means a path in each generalized coordinate. In order to ensure a correct tracking, a PD controller was added into each generalized coordinate, the simulations shows that the model of the upper arm get a closely approach to the desired path, and with that is possibly the imposition of behaviours more specific, only defined by polynomial functions.

6. References

Carignan, C. & Liszka, M. (2005). Design of an Arm Exoskeleton with Scapula Motion for Shoulder Rehabilitation. *Proceedings of 12th International Conference of Advanced Robotics 2005*, pp. 524-531.

Croshaw, P.F. (1969). Hardiman I Arm Test - Hardiman I Prototype Project. *General Electric Co.* Shcenectady NY. Specialty Materials Handling Products Operation.

Culmer, P., Jackson, A., Levesley, M.C., Savage, J., Richardson, R., Cozens, J.A. & Bhatka, BB. (2005) An Admitance Control Scheme for a Robotic Upper-Limb Stroke Rehabilitation System. *Proceedings of 27th Annual International Conference of the IEEE Engineering in Medicine and Biology Society. IEEE Engineering in Medicine and Biology Society. Conference.* pp. 5081-5084.

Fliess, M. ; Lévine, J. ; Martin, P. & Rouchon, P. (1994). Flatness and defect of nonlinear systems : Introduction, theory and examples. *International Journal of Control.* Vol 61, No. 6, pp. 1327-1361.

Franch, J. & Agrawal, S.K. (2003). Design of differentially Flat Planar Space Robots : A Stepforward in their Planning and Control, *Proceedings of the IEEE/RSJ Interntational Conference on Intelligence Robots and Systems,* pp. 3053-3058.

Gerald, L.G. ; Song, Q. ; Almeida, G.L. ; Hong, D. & Corcos, D. (1997). Directional Control of Planar Human Arm Movement. *The Journal of Neurophysiology.* Vol 78, No. 6 (December 1997), pp. 2985-2998.

Gopura, R.A.R.C. & Kiguchi, K. (2007). Development of an Exoskeleton Robot for Human Wirst and Forearm Motion Assist. *Proceeding of the Second International Conference on Industrial and Information Systems, ICIIS 2007,* pp. 535-540, August 2007. Sri Lanka.

Gowitzke, B & Milner, M. (2000). *Scientific Base of Human Movements.* Ed. Paidotribo. ISBN : 84-8019-418-9. Barcelona, Spain.

Kazerooni, H. (2005). Exoskeletons for Human Power Augmentation. *Proceedings of the IEEE/RSJ International Conference on Intelligent Robots and Systems,* pp. 3120-3125. August 2005.

Kazerooni, H. (2006). The Berkeley Lower Extremity Exoskeleton Project. In : *Experimental Robotics* Vol IX. Ang, M.H.; Khatib, O. pp. 291-301, Springer-Verlag Berlin Heidelberg.

Klopčar, N. & Lenarčič, J. (2005) Kinematic Model for Determination of Human Arm Recheable Workspace. *Meccanica.* Vol 40. No. 2. Pp 203-219. Springer.

Lee, S., Lee, J., Kim, M. & Lee, C. (1999) A New Masterarm for Man- Machine Interface. *Proceedings of 1999 IEEE International Conference on Systems, Man and Cybernetics.* Vol 4. pp 1038-1043.

Lenarčič, J. & Umek, A. (1994). Simple Model of Humarn Arm Reachable Workspace. *IEEE Transactions on Systems, Man and Cybernetic.* Vol 24, No 8, pp. 1239-1246.

Lévine, J. (2009). *Analysys and Control of Non Linear Systems : A Flatness Based Approach.* Springer-Verlag Berlin Heilderberg, ISBN 978-3-642-00838-2, Germany.

Marguitu, D. B. (2009). *Mechanics and Robot Analysis with MATLAB.* Springer-Verlag London Limited. ISBN 978-1-84800-390-3.London, England. pp. 183-204.

Maurel, W & Thalman, D. (1999) A Case Study on Human Arm Upper Limb Modelling for Dynamic Simulation. *Computer methods in biomechanics and biomedical engineering,* Vol. 2, No. 1. pp. 65 – 82.

Murray, R. M.; Rathinam, M. & Sluis, W. (1995). Differential Flatness of Mechanical Control Systems : A catalog of Prototype Systems. *ASME International Mechanical Engineer.*

Perry, J.C. ; Rosen, J. & Burns, S. (2007). Upper Limb Powered Exoskeleton Design. *IEEE/ASME Transactions on Mechatronics,* Vol 12, No. 4, pp. 408-417.

Pons, J.L. (2008). *Wearable Robots : Biomechatronic Exokeletons,* John Willey & Sons, Ltd, ISBN 978-0-470-5194-4, England.

Rocon, E. & Pons, J.L. (2005). Case study : Study of tremor charcateristics based on a biomechanical model of the upper limb. In : Pons, J.L. (2008) *Wearable Robots : Biomechatronic Exokeletons,* John Willey & Sons, Ltd, ISBN 978-0-470-5194-4, England.

Rosen, J. ; Perry, J. C. ; Manning, N. ; Burns, S. & Hannaford, S. B. (2005). The Human Arm Kinematics and Dynamics during daily activities – Towar to a 7 DOF Upper Limb Powered Exoskeleton. *Proceedings fo the 12th Internatinal Conference on Advance Robotics,* pp. 532-539.

Rotella F. & Zambettakis, I. (2008) *Comande des Systèmes par Platitude* Available: http://www.techniquesingenieur.fr/book/s7450/commande-des-systemes-par-platitude.html.Site Access: August 2008, pp. 5-6.

Sankai, Y. (2006). Leading Edge of Cybernics : Robot Suit HAL. *Proceedings of SICE-ICASE International Joint Conference,* October 2006. Busan, Korea.

Siciliano, B. ; Sciavicco, L. ;Villani, H. & Oriolo, G. (2009). *Robotics : Modeling Planning and Control,* Springer-Verlag London Limited, ISBN 978-1-84628-641-4, England.

Spong, M.W. (1992). On the Robust Control of Robot Manipulators, *Transactions on Automatic Control of Robot Manipulators,* Vol 37 No. pp. 1782-1786.

Tözerem, A. (2000). *Human Body Dynamics : Classical Mechanis and Human Movements,* Springer Verlag Berlin Heidelberg. Germany.

Van Nieuwstadt, M.J. (1997a). Trajectory Generation for Nonlinear Control Systems, Technical Report CS 96-011 del Department of Mechanical Engineering, California Institute of Technology, Pasadena, California, pp. 54-57.

Van Nieuwstadt, M.J. (1997b). *Trajectory Generation of Non Linear Control Systems,* Ph. D. dissertation, University of California at Berkeley, Berkeley, California, USA.

Veslin E. (2010). *Implementação dos Sistemas diferencialmente planos para o contrôle de um manipulador robótico tipo braço humano.* M.Sc. Dissertation, Universidade Federal do Rio de Janeiro, Rio de Janeiro, RJ, Brazil.

Wang H. & Xie, Y. (2009). Adaptative Inverse Dynamics Control of Robot with Uncertain Kinematics and Dynamics, *Automatica 45,* pp. 2114-2119.

11

Real-Time Control in Robotic Systems

Alex Simpkins
University of Washington
USA

1. Introduction

Robotic systems are beginning to interact with humans and are increasing in dimensionality, control complexity, task difficulty, and dynamic capability. It is therefore becoming increasingly imperative that control be performed in real-time (i.e. on or faster than the timescale of the system dynamics), intelligently making decisions on demand, and then executing them in a dependable fashion. In addition, high-dimensional bio-inspired systems are becoming more common (Butterfab & et al., 2001; Movellan et al., 2005; Simpkins et al., 2011; Todorov et al., 2010; Vandeweghe et al., 2004; Wilkinson et al., 2003), and are based upon a different strategy for control (and therefore control system design) than traditional engineered systems - coordinated, compliant, and often coupled control rather than orthogonal, stiff, and independent control of joints. As a result, issues related to real-time capability, more critical now than ever, must be addressed with a coherent, measured approach.

In order to deal with this challenge appropriately, a number of sub-problems must be addressed and will be discussed in this chapter. These are timing, dimensionality, computational capabilities, system bandwidth, and safety. There are fields devoted to each problem independently, and the background of each will be given as each one is discussed. Rarely does a researcher or practitioner in one field work in any of the other fields, let alone all of them. However, ignoring any of these sub-problems will likely lead to a failure of a complex real-time system. Therefore, familiarization with these issues and methods will help the researcher and practitioner design, implement, and troubleshoot a real-time system most effectively. This chapter will present current solution methods for each component of the problem and show how they can be integrated together effectively.

1.1 Key points and contributions

- A rare integrative approach between real-time control theory and practical hardware issues which are closely linked with the functionality of these theories.
- A novel means of semi-redundant integrated systems to provide improved safety measures for human-robot interaction.
- Strategies which effectively deal with high dimensional systems and complex control strategies in real-time.
- The concept of real-time is different in different fields. Here we discuss a variety, how they relate to each other, and how each can enhance and integrate with the others.

- This unifies and expands upon approaches in the fields of real-time control systems, control engineering, mechatronics, dynamics, robotics, embedded systems, and computer science to address new design challenges.

1.2 Chapter organization

The rest of this chapter is organized as follows. Section 2 breaks down the real-time problem into sub-problems, and Section 3 presents how each problem is addressed. Each sub-problem is discussed in a way which exposes how they are all related, why it is important to address all aspects of the real-time problem, and how this can be done. Section 5 presents a set of results for a real-time system which has been designed with the methods presented in this chapter, demonstrating its effectiveness. Section 6 closes the chapter with a review, discussion of the problem and approach, and suggests further directions in research.

2. The sub-problems

By defining the sub-problems we can begin to address the overall issue of real-time control in robotic systems. The timing problem is centered around what is defined as real-time for a particular application. For example, an energy-producing chemical process may occur on the order of several hours, or the coordination of joints required for locomotion may occur on the order of milliseconds. Microprocessors are digital devices, and commands to the actuators are updated periodically. Thus timing issues become guaranteeing that sample time is short enough, that delays in the system are minimal (or properly addressed), that samples are not missed (or if sample time is variable, this is accounted for), and that information bottlenecks are avoided or part of the control strategy. Timing leads to issues of computational capability required versus what is available. This defines what dimensionality and control complexity is possible, and limits the overall system bandwidth. Traditionally designers have considered processing power versus requirements as an afterthought, with little concern over the criticality. This is due to the fact that, for simple problems, most modern processors are more than powerful enough. In complex systems, however, this quickly becomes a poor assumption. When one considers the actual realities of controlling a fifty degree of freedom (DOF) system, for example, where each DOF possesses multiple sensors and actuators, the required information flow bandwidth becomes large quickly. A system with such limitations can effectively lose control of some or all DOFs under dynamic load, though it may be guaranteed to be controllable in theory from a dynamical systems perspective. When considering these issues, and that in the future robots will be interacting directly with humans more often in everyday activities, safety becomes a major consideration. A robot which can exert large forces, is not back-drivable, and/or possesses significant mass is inherently unsafe for human interaction. Making the robot small and weak is not the only or best solution as this still does not guarantee that the system will not be dangerous. There are ways to minimize risk of human injury in the case of loss of control, even during a catastrophic failure of the entire primary control system.

3. The solution approach

Though there are a number of sub-goals for approaching real-time control in robotic systems, an overall and most significant approach is that the various design aspects of the system must be constrained such that each solution does not violate another's constraints. For example, when setting sample rate for the control algorithm, one possible constraint could be that this must not cause a need to communicate data more rapidly than the data bandwidth allows.

3.1 Timing issues

Timing problems can be mitigated by ensuring that traditional real-time control issues as well as new issues arising in complex robotics are addressed. If sample time is unavoidably significantly variable, some mathematical assumptions in discrete control do not hold (Nilsson, 1998; Wittenmark et al., 1995), and various equations in the dynamics will have variation over ones with the precise sample time assumption. However, a constant delay can be modeled, following (Wittenmark et al., 1995), as a standard state-space system,

$$\frac{dx(t)}{dt} = Ax(t) + Bu(t-\tau) \tag{1}$$

$$y(t) = Cx(t) \tag{2}$$

then the sampled-data system becomes

$$x(kh+h) = \Phi x(kh) + \Gamma_0 u(kh) + \Gamma_1 u(kh-h) \tag{3}$$

$$y(kh) = Cx(kh) \tag{4}$$

where

$$\Phi(h) = e^{Ah} \tag{5}$$

$$\Gamma_0(h,\tau) = \int_0^{h-\tau} e^{As} ds B \tag{6}$$

$$\Gamma_1(h,\tau) = e^{A(h-\tau)} \int_0^{\tau} e^{As} ds B, \tag{7}$$

and delays larger than a sample h can be dealt with fairly simply, by adding the relation

$$\tau = (d-1)h + \tau', \ \ 0 < \tau' \leq h \tag{8}$$

with an integer d. Now our system becomes, after replacing τ by τ' in Γ_0 and Γ_1,

$$x(kh+h) = \Phi x(kh) + \Gamma_0 u(kh - dh + h) + \Gamma_1 u(kh - dh). \tag{9}$$

Variable delays can be dealt with by simply computing time varying system matrices. Then the model described in Equation 3 is still valid. As long as the variation in sampling is less than a delay in duration, the model described is still valid, as the control signal does not vary with the delay. If the delay varies longer than a sample period, then a different model may be derived which causes the control signal to not be held constant, but rather to vary along with the delay. This allows the control signal to be scaled appropriately, and mitigates instabilities. Excessively large variations in sampling period can lead to unavoidable (or rather 'very difficult to avoid') instability due to violation of controllability guarantees. In other words, when the system finally processes the delayed sample, too much time may have passed to react properly, and a large error may already have been incurred from the previous control action. In the case that this type of problem is likely, it is possible to create robust controls, which guarantee certain bounded behaviors even in the event of large errors due to sample time variability, which may be lumped into disturbances. Another way to effectively alter

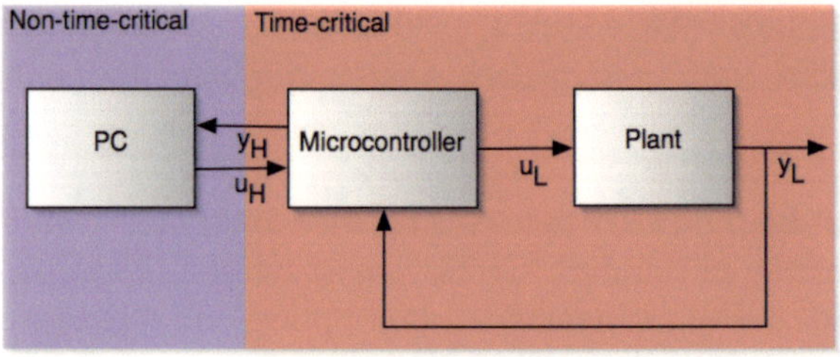

Fig. 1. For complex control systems, it is effective to split time-critical aspects of control from non-critical aspects. In this way a complex algorithm can be computed in near real-time, but may not be guaranteed, while the critical aspects which require such guarantees can be designed at the low level to handle variable updates to its reference. L stands for 'Low level' while H stands for 'High level.' y represents output, and u represents control input.

control parameters to match the delay is to include the delay as a parameter of the control equation (Åström & Wittenmark, 1990).

Thus we can attempt to model delays by measurement, and then compute an appropriate control, but that does not tell us everything that might happen, and how do we, in the context of complex systems, measure for every possible permutation? Additionally, it may not be trivial to make measurements of some features needing to be measured! That leads us to the next section regarding computational capabilities.

3.2 Computational capability issues

Computational capabilities required depend on required overall update rates, complexity of the control algorithm, number of peripherals involved and dimensionality of the system (thus the size of data passed around during code execution). There are a number of ways to ensure code will be processed on time and that computational requirements and capability are matched. With highly complex control algorithms required for artificial intelligence and complex coordinated movements, a very effective approach to ensuring real-time capability is to split the time-critical aspects of control from the computationally intensive aspects, as in Figure 1. Then a hierarchical but highly interconnected framework is formulated. This parallels the human brain's sensorimotor system structure, which can be thought of, most definitely, as the most intricate and advanced control system on the planet. Biological systems are also excellent at solving the types of control problems now beginning to be addressed by roboticists. The high level would consist of powerful personal computers or clusters of computers which are excellent for massive floating point computations but are limited for time-critical applications, while the low level would consist of embedded systems which are excellent at time-critical applications but less adept at complex calculations. This design approach has been applied and is detailed in (Simpkins et al., 2011). The challenge then becomes communication between the levels. There are a number of communication strategies and protocols which address this issue well. Additionally, the high level operating system must be carefully selected. Most common operating systems are not real-time.

GPU parallel processing is a powerful tool which is significantly under-applied in control applications at this time. It is possible to perform complex algorithms rapidly on thousands of processors in parallel on such devices, and the current state-of-the-art simulators are beginning to take advantage of such capabilities (such as introduced in the appendix of (Erez et al., 2011), which takes advantage of such an engine for speed). They are low cost (especially compared to cluster computing) and highly effective. This approach is useful not only to run single-thread algorithms, but also to run multiple model predictive algorithms in parallel. In this way, several potential control decisions or trajectories can be explored very rapidly, and the one with the best outcome selected.

Measuring execution time for all the computations required, including communication delays, and delays from peripherals is an important ability to have. This is not a trivial problem, however.

3.2.1 Worst case execution time

Worst Case Execution Time (WCET) is the time it takes to execute the longest branch of code, not including hardware peripherals. This is important to at least have a reliable estimate of, since a designer needs to know whether a task can finish in a set amount of time for real-time systems. Traditionally, with fixed code and no caches, this was not as difficult a problem as it is in the context of more advanced (often multicore) processors, infinite loops, complex pipelines, and dynamic code length.

The traditional methods of measuring WCET comprise an entire field and will be mentioned here fairly briefly, as the main contribution is when and how to use these tools in conjunction with all the other subproblems that need to be addressed (The interested reader is referred to (Ermedahl, 2003; Hansen et al., 2009; Wilhelm et al., 2008). This field is moderately mature, though advances continue steadily. Timing analysis is the process of determining estimates or bounds on execution times. (Wilhelm et al., 2008) provides an excellent overview of timing analysis and the WCET problem in general, as well as tools which are available.

There essentially are two approaches for WCET prediction. These are static and dynamic (Grob, 2001) approaches. Static prediction approaches measure small pieces of code which do not change, do not have recursion, and are otherwise limited (Puschner & Nossal, 1998), while dynamic approaches (sometimes called measurement-based methods) measure an entire execution pathway, running on a simulator or the actual hardware for a set of (probably limited) inputs.

The static method may not provide realistic estimates when code snippets are combined together, and scheduling complex algorithms which combine many snippets together may break down the estimates. In addition, some processors have speculative components, where the processor determines an expected path, begins to load items in memory, and performs some processing ahead of time in anticipation. However, these speculations can of course be mistaken, and then all the predicted data must be deleted and the proper execution performed, which is actually slower than just waiting then branching when the call happens, without prediction.

Modern processors provide means of measuring code performance in a variety of ways via peripherals built into the chip, but some of these methods are imprecise at best. They do give a measure of how many cycles a task takes, however performing the measurement itself takes some computation. This violates the basic principle of measurement - that the action of performing the measurement should have such a small effect on the system measured that no significant change occurs in the process, but it does give a great deal of information, and the effect is fairly small. These measures can be used to create an estimate of best and worst case execution times. In practice this is very helpful, though difficult or impossible to provide hard guarantees in all cases.

Two effective approaches which can be used on complex real-time robotic control systems, of the techniques discussed are either to split the time-critical components from the non-critical components, or to use a measurement technique which will not necessarily guarantee perfect timing, but may provide fairly good estimates on bounds of timing. In safety-critical applications it is suggested that having a low level which has timing guarantees is more appropriate than bounded estimates, in combination with a watchdog (discussed in Section 3.5). In some experimental robotic systems it may be sufficient to simply design controllers which deal with temporal variability without instability, but write code which minimizes delays. This latter method may be effective for rapid prototyping of new designs and testing experimental theories, while the hierarchical structure with some components real-time may be a more effective general strategy, as long as communication between levels is designed carefully, such as in (Simpkins et al., 2011). Standard timing analysis tools should be used at the programming stage to test algorithms and ensure their real-time performance. Even in the best cases, there are always limitations, and in complex high dimensional systems we must consider carefully what is the maximum capability available, and how much is required.

3.3 Dimensionality and bandwidth

The dimensionality of a system may technically be very high, but can be reduced in practice depending on the task. Or alternatively, simple coordination can be performed at a low level (akin to reflexes in biological systems, which are thought to be processed in the brain stem or lower levels, depending), leaving higher levels free to coordinate a few degrees of freedom (DOF) in more complex patterns. This reduces and efficiently distributes the processing load. For example, though human beings have over twenty-two DOFs in their hands, during the majority of grasping and manipulation tasks, they tend to use as few as two DOFs for a variety of common tasks. The individual DOFs are coupled together into movement 'synergies,' which makes the computational task significantly simpler.

3.3.1 Advantages of synergies and evidence to support their uses

Biological systems are almost unilaterally agreed to be capable of superior complex manipulation tasks in terms of complexity and variety of skills. Interestingly enough, several studies have been performed of humans undergoing complex manipulation tasks, and dimensionality of the movements have been explored using principle components analysis and canonical correlations analysis. See (Anderson, 2003) and (Mardia et al., 1992) for good discussions of these techniques. The results essentially state that biological systems use synergistic movements to achieve goals. It is likely that there are a number of reasons for this, but one advantage of synergistic control is that the number of individual dimensions controlled is reduced significantly over the total state space. Of the many DOFs available, many manipulation tasks require less than *three* (Santello et al., 1998). This has led some researchers to claim that this is a maximum for control but it appears that there is a sliding dimensionality. However dynamic complex movements require more training than lower dimensional, less dynamic ones. One theory is that when learning a new skill, humans go through stages of learning, where initially many dimensions are frozen, then gradually are unfrozen as learning takes place(Bernstein, 1967). One possible explanation for this strategy is that, in order to free bandwidth as much as possible, coordinated movement is encoded at as low a level as possible, freeing up the high level to devote bandwidth to complexity of the strategy, rather than simply giving individual joint commands. In fact, the manifold of

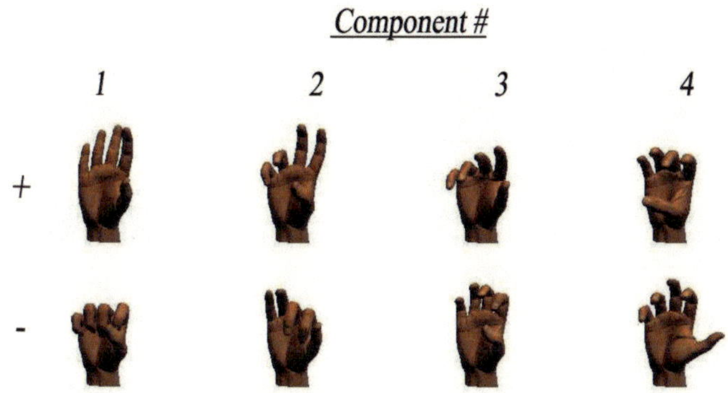

Fig. 2. First four principle components of postural movement data for humans undergoing a 'variability maximizing experiment,' scaled appropriately for joint angles and applied such that the size of the component for a joint determines its deviation from a neutral posture. It is interesting to note that the components, which represent the dimensions that were most significant in terms of accounting for the variability in the data, and the direction of variability, comprise certain basic components of manipulation - gripping with all fingers, using a wavelike pattern of all fingers or using the first few primary fingers and thumb, precision manipulation with all fingers in fine patterns, and finally the thumb dominating the movement. This may suggest that combining these synergies, or using subsets of fingers, many manipulations can be performed.

the subspace is quite dynamic, and changes depending on the task. In other words, joints are coupled together in a fairly optimal fashion for each task(Todorov & Ghahramani, 2003; 2004). In fact, the author has performed a series of studies attempting to maximize the dimensionality of hand movement control, and there is still a notion of synergies(Simpkins, 2009). The maximum dimensionality is approximately ten DOFs. The obvious advantage is that the less individual trajectories that must be computed, or more generally, control decisions to be individually made, the faster those decisions can happen, and the more complexity can go into computing what exactly those decisions will be. So couple joints together when possible into synergies such as Figure 2 and compute a global trajectory for them to follow, such as open vs. closed hand (See Figure 3), and this can be done quite quickly and efficiently computationally. A number of approaches can be taken to build control systems upon this general methodology, but optimal control has been found to effectively model much biological movement control data, and provides an intuitive means of building control systems. Optimal control (Stengel, 1986) is a methodology which makes decisions based upon a notion of optimality for a task. This notion comes in the form of a cost (or reward, hereafter referred to only as cost for simplicity) function. Behaviors are created by specifying the dynamics of the system as constraints, then actions are chosen which minimize the cost computed by the cost function. There is evidence that the sensorimotor system ignores task-irrelevant deviations (Todorov, 2004; Todorov & Jordan, 2003), which tends to reduce dynamic load and uses less system bandwidth to compute controls than actively controlling everything individually.

Bandwidth is a term which is used in many fields for different purposes. In control theory it relates the ability to track a reference signal (Franklin & et al., 1994); in communications and

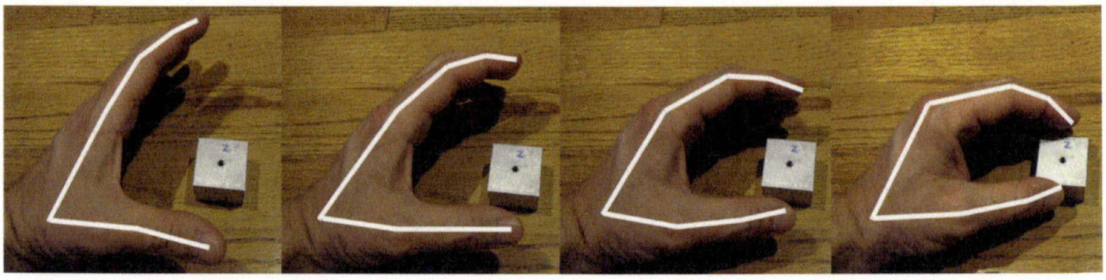

Fig. 3. Images of a hand gripping an object. Though there are many degrees of freedom in the hand, it is clear that there is a low dimensional synergy in this case. We can control the entire hand from this synergy for this task (open-close). In that way, more complex algorithms can be computed in real-time if needed. A sliding complexity (in terms of dimensionality and dynamics) can be used to guarantee real-time control even for a complex system.

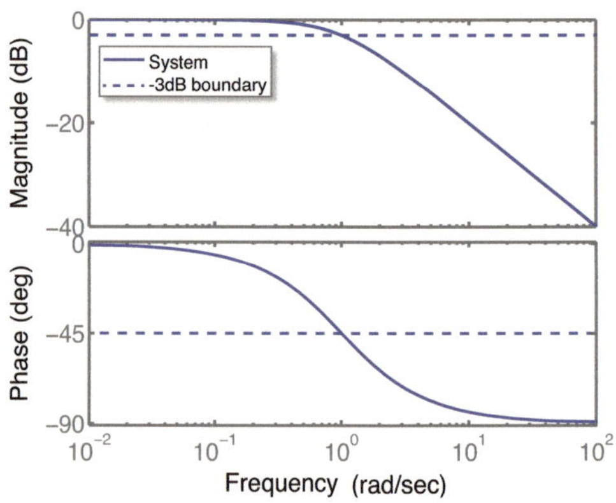

Fig. 4. The bandwidth cutoff for a system tracking a sinusoidal input is considered to be the point at which the amplitude of the output drops below -3dB, along with a corresponding phase shift. This is the bode plot for $G(s) = 1/(s+1)$.

networking it refers to how much data throughput is possible. However, the general concept is very similar in all instances. We will use as the following as the definition of bandwidth:
Definition: ***Bandwidth*** *shall be defined as the rate at which a system, open-loop or closed-loop, may successfully track a reference. The boundary between successful tracking and unsuccessful tracking is the point at which the amplitude of the output drops below -3 Decibels (dB) relative to the input signal in the frequency domain, such as Figure 4.*

The bandwidth requirement for a system is also reduced by synergistic movement control. It is also significant that a designer consider not what the maximum possible bandwidth of the overall system can be, but more what the maximum bandwidth *should* be. In other words, what is required (along with a factor of safety) to succeed at all tasks the system is designed to accomplish? Too high of a (dynamic) mechanical bandwidth may lead to instability, high forces, and other problems affecting system safety. Too high of a data bandwidth leads to higher probabilities of lost samples, delays, and related issues. It is also possible to have a sliding bandwidth which varies depending on computational complexity versus dynamic timing of a particular task. In fact, this appears to be closer to what biological systems do -

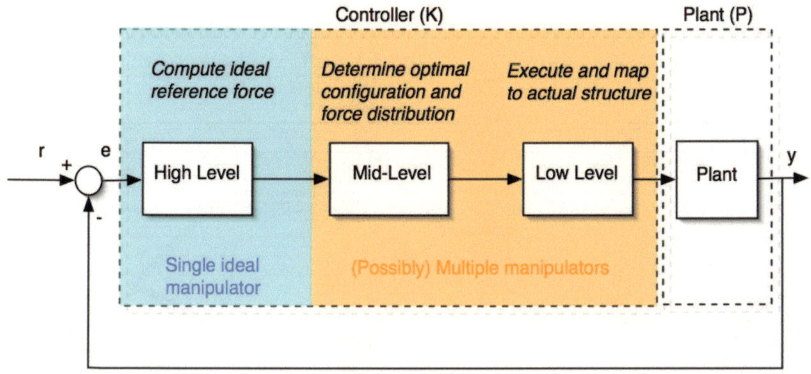

Fig. 5. Block diagram of hierarchical control scheme. This approach breaks a single complex problem into several smaller and more tractable problems.

use some subset of features of the complex system for any given task, but have the capability for a variety of tasks (Todorov & Jordan, 2002).

3.4 Control complexity

As more intelligent control is applied, the computations required to carry out these algorithms tend to increase. For example, the number of mathematical operations per second to run model-predictive nonlinear stochastic optimal control is far higher than proportional integral derivative control.

An approach which is promising is to break the one complex problem down into simpler components with a hierarchical control approach (See Figure 5). Then different components of the hierarchy, as discussed in Section 3.2, can address either time-critical or computationally intensive aspects of the controller. This will be explored through results from a system developed by the author (in collaboration with Michael S. Kelley and Emanuel Todorov) for complex manipulation and locomotion (Figure 6(a)).

Finally, some aspects of the control can be computed or adapted offline or on a slow timescale. This is parallel to some learning processes in humans. The resulting solutions (such as pre-computed trajectories) can be accessed from memory rapidly when required.

3.4.1 Model and approach

The basis for this approach is presented in (Simpkins & Todorov, 2011) for the robot in (Simpkins et al., 2011).

The goal of these robotic fingers was to develop a robotic system which parallels certain key aspects of biological systems for manipulation (and similarly for locomotion). The reason to do so is that most robots are not designed to behave in a similar fashion to biological systems. They are often solving different problems, such as how to follow a trajectory in order to weld two parts together or to cut away material in a prescribed pattern with a milling bit. In these standard cases it is very important for the robot to track a specific trajectory regardless of outside disturbances from the world. Biological systems such as a human being walking across a room, or selecting a key from a keychain full of keys in his or her pocket are solving a completely different problem. They require a bidirectional information flow with the world at an actuator level - many movements are more like a dance with the outside world as one

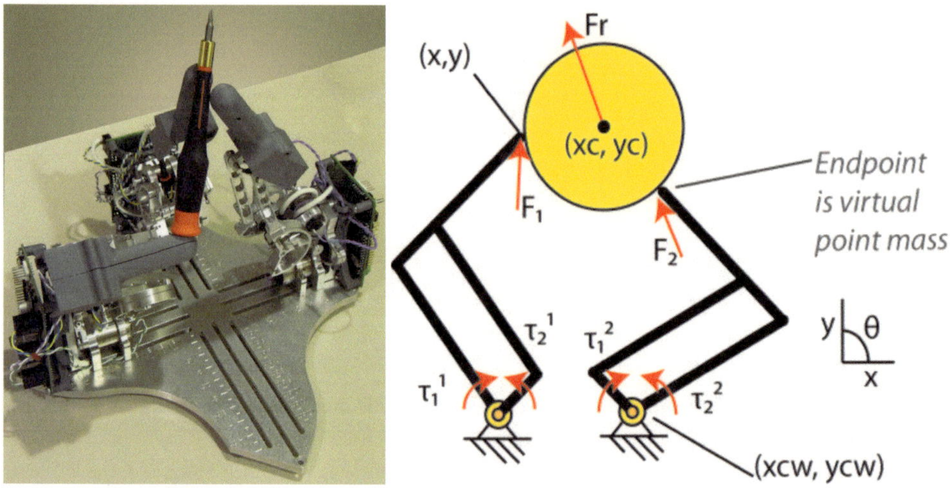

(a) (b)

Fig. 6. (a) A modular bio-mimetic robotic system developed by the author at the University of California, San Diego and the University of Washington in the Movement Control Laboratory. (b) A simplified 2-dimensional representation of the robot and object, used in the algorithm described in this section.

partner, and the human being as the other. The human acts on the keys by applying forces to the keychain, the keychain changes state, and by nature of the contacts with the human's hand, alters the state of the human, who gains information about the state of the keys. This also helps simplify the control problem in various ways, as the human hand conforms to the shape of the key, rather than the hand attempting to move *in spite of* the key.
Creating a robot which interacts with the world in this fashion involves solving a number of new design challenges without compromising on any one specification. For example, in order to dynamically manipulate objects, a robotic finger must be able to move rapidly, be completely compliant and back-drivable, communicate at a high rate with a host processor, be capable of implementing complex control in real-time, be sensitive, measure appropriate variables (such as force, velocity, and position), and it must be sufficiently robust. This is to the author's knowledge the first design which solves all these constraints and incorporates modularity (each finger is completely self-contained and has wireless capabilities, with provisions to be attached to any base with the appropriate hole pattern). Controlling such a device to perform dynamic manipulation in the way previously described is a new type of control problem, and requires a significantly new approach.
The goal is to apply forces to an object to cause it to behave in a desired fashion. This can be tracking a trajectory, or any other goal of control. An important note is that the two problems (manipulation and locomotion) are related in that we either have a small object relative to manipulators such as a hand, or the ground, an object of infinite size (in either case, it is often that one side of the problem is significantly larger than the other side - manipulators or object manipulated), as in Figure 7.
Forces are applied to the object by manipulators, described as linkages, as shown in Figure 6(b), which are a simplified model of the physical robots developed by the author, shown in Figure 6(a) and described in detail in (Simpkins et al., 2011). Torques (τ_i) are applied to either

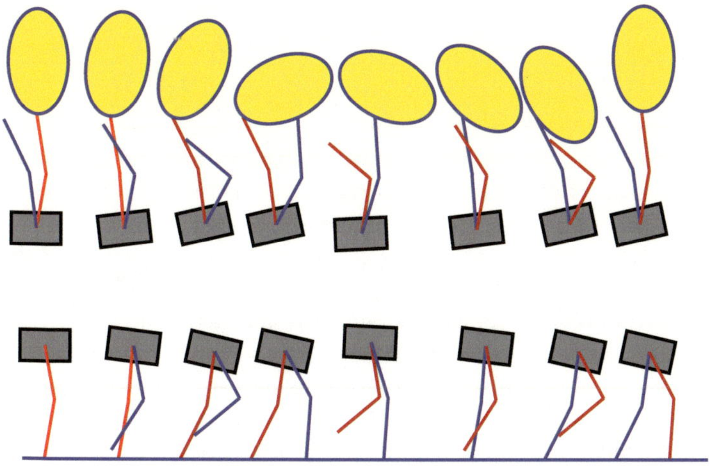

Fig. 7. Image sequences showing manipulation of an object (top) and locomotion (bottom). Note the parallels between the two. In fact, the two can be thought of as essentially the same problem - this is a significant insight.

of the two links shown for a particular manipulator in order to generate an output force at the endpoint. This is similar to human joints in terms of the way tendons pull on joints to cause an output force. However, here for design simplicity and robustness, the complex inner musculoskeletal structure is reduced to a four bar linkage model. The motors which produce the torque are coupled to the joints through a cable drive system which has ultra-low friction and little if any backlash - essential components for back-drivability. The cables act as the tendons would in biological systems in the sense of a pull-pull system, as well as providing some spring characteristics. As much mass as possible is kept at the base, and decoupled from the output, which is another reason for the linkage design. This helps keep inertia small relative to the objects being manipulated, and relative to the output force capabilities of the finger. This is significant because the finger needs to be capable of dynamic movement (rapid accelerations and decelerations), and must avoid wasting most of the force output capability of the motors in overcoming the structure's inertia (imagine going through your day with an extra 10kg of weight strapped to your wrists; it would indeed be very difficult to perform simple tasks because nearly all of your strength would be used to overcome the weight and inertia of the additional mass). Large inertia also is very detrimental to back-drivability.

The concept of these devices is not to attempt to duplicate the structure of a biological system. Rather, the concept is to capture the capabilities which are significant for developing controllers that solve the problems of manipulation and locomotion similarly to biological systems. Each finger is capable of producing a few Newtons of peak output force in each axis, and requires a fraction of that to be backdriven (frictional and inertial loads are small), enough for fine dynamic manipulation. Each can perform closed loop control at over 1kHz, communicating between all fingers and a higher level processor (such as a personal computer) all relevant data (position, velocity, force, actuator states, etc).

Consider for simplicity the manipulation or locomotion problem within a 2D context. Objects are represented by center of mass, position and angle, mass (m), inertia (J), and a boundary ($d()$), which may be described by splines, allowing any shape to be represented, convex or concave. The sum of forces on the object (F_x, F_y, and M) are given by

$$\begin{Bmatrix} \sum F_{x,o} \\ \sum F_{y,o} \\ \sum M_o \end{Bmatrix} = \begin{bmatrix} 1 & 0 \\ 0 & 1 \\ -d_y(\theta) & d_x(\theta) \end{bmatrix} \begin{Bmatrix} \sum_i f_{x_i} \\ \sum_i f_{y_i} \end{Bmatrix} - \begin{bmatrix} m_o & 0 & 0 \\ 0 & m_o & 0 \\ 0 & 0 & J \end{bmatrix} \begin{Bmatrix} a_{x,o} \\ (a_{y,o} + g) \\ \ddot{\theta}_o \end{Bmatrix} \tag{10}$$

Where f_i representing the force applied by manipulator i at the surface location relative to the object center of mass as a function of angle (θ) relative to horizontal in the global coordinate system ($d(\theta)$), g representing gravitational acceleration, a representing acceleration of either manipulator i's endpoint or the manipulated object, depending on the subscript, and $()_o$ representing 'with respect to the object being manipulated.' Time derivatives are represented by an over-dot, for example, the time derivative of x has the notation $\dot{x}$.

Contact dynamics can be complex to simulate accurately, and can bring about bottlenecks for otherwise lightning fast algorithms (important in real-time contexts). Though with the rigid body assumption it is true that accurate contact dynamics appear difficult to 'get right,' real objects are not infinitely rigid. This provides a means by which contacts are smooth, soft, and damped. The reader is encouraged to consider his or her hands as an example- note the flexibility of the tissue. In general we, as human beings, do very little manipulation of rigid objects with rigid objects. When was the last time you attempted to pick up and manipulate a coffee cup or other fairly rigid body while wearing rigid steel armor which has no padding over your hand? It is unlikely that you ever have, or ever will. In fact, it is important to have damped interactions to avoid damaging the manipulators and objects - consider shoes, which prevent joint and limb damage due to shocks over time. This makes the problem significantly simpler, and we can make strong assumptions which do not necessarily reproduce contact dynamics of rigid bodies perfectly, but instead assume that objects interacting with each other do so with damping. In addition, the concepts we are presenting here do not necessarily depend upon the contact algorithm, being a modular method, and so the physics and contact model can easily be switched for others, even on the fly. And it makes sense to have a more flexible strategy for simulating dynamic interactions than making one assumption and basing the entire methodology upon this such that the algorithm becomes 'brittle' to simple changes, or becomes computationally infeasible due to complexity.

Given this discussion we create a constraint for the form of the algorithm presented here (Simpkins & Todorov, 2011). We assume that there is no slip during contacts, effectively creating a sticky contact algorithm. The advantage here is that it allows a constraint on acceleration of the manipulators relative to the object,

$$\begin{Bmatrix} a_{x,i} \\ a_{y,i} \end{Bmatrix} = \begin{bmatrix} 1 & 0 & d_x(\theta_i) \\ 0 & 1 & d_y(\theta_i) \end{bmatrix} \begin{Bmatrix} a_{x,o} \\ a_{y,o} \\ \ddot{\theta}_o \end{Bmatrix} \tag{11}$$

where a represents the acceleration of the object o or point mass i, and $\ddot{\theta}_o$ is the angular acceleration of the object. This does create an instantaneous change of velocity when a contact occurs, which can be modeled as an impulsive force that causes the change in velocity which is required. There is also an assumption that the point masses are small enough in mass relative to the object that their contribution to the momentum of the system when in contact is insignificant.

When not in contact, the point masses experience accelerations due to a force which is applied by the robot fingers. The entire robot finger assembly is modeled as a point mass with external forces applied to the point mass in order to cause accelerations. This assumption is made because the rotor inertias of the motors dominate the overall finger inertia. It is also makes the system more simple to model and control than one with complex nonlinear dynamics, but this would be possible as well and is a trivial extension. The affine point mass dynamics for mass i are given by

$$\begin{Bmatrix} a_{x,i} \\ a_{y,i} \\ 1 \end{Bmatrix} = \begin{bmatrix} \frac{1}{m_i} & 0 & 0 \\ 0 & \frac{1}{m_i} & -g \\ 0 & 0 & 1 \end{bmatrix} \begin{Bmatrix} b_{x,i} \\ b_{y,i} \\ 1 \end{Bmatrix} \tag{12}$$

and the object dynamics, after rearranging the terms in Equation 10 are given by

$$\begin{Bmatrix} \sum_i f_{x,i} \\ \sum_i f_{y,i} \\ -\sum_i f_{x,i} d_y(\theta_{i,p}) + \sum_i f_{y,i} d_x(\theta_{i,p}) \end{Bmatrix} - \begin{Bmatrix} m_o a_{x,o} \\ m_o a_{y,o} \\ J\ddot{\theta}_o \end{Bmatrix} = \begin{Bmatrix} 0 \\ m_o g \\ 0 \end{Bmatrix} \tag{13}$$

and the end effector dynamics when in contact with the object can be similarly derived to give

$$\begin{Bmatrix} f_{x,i} \\ f_{y,i} \end{Bmatrix} + m_i \begin{Bmatrix} a_{x,o} \\ a_{y,o} \end{Bmatrix} + m_i \begin{Bmatrix} \ddot{\theta}_o d_x(\theta_i) \\ \ddot{\theta}_o d_y(\theta_i) \end{Bmatrix} = \begin{Bmatrix} b_{x,i} \\ b_{y,i} + m_i g \end{Bmatrix}. \tag{14}$$

We combine the object and end effector dynamics into a single equation, which allows us to solve for the unknown forces and accelerations in one operation

$$Aw = b, \tag{15}$$

where A is given by

$$A = \begin{bmatrix} 1\,0 & -m_o & 0 & 0 \\ 0\,1 & 0 & -m_o & 0 \\ 0\,0 & 0 & 0 & -J \\ I\,\mathbf{0} & m_i & 0 & m_i\ddot{\theta}_o d_x(\theta_i) \\ \mathbf{0}\,I & 0 & m_i & m_i\ddot{\theta}_o d_y(\theta_i) \end{bmatrix} \tag{16}$$

w is given by,

$$w = \left\{ f_{(x,i)}, f_{(y,i)}, a_{(x,o)}, a_{(y,o)}, \ddot{\theta}_o \right\}^T \tag{17}$$

and b is given by

$$b = \left\{ 0, g, 0, b_{(x,i)}, b_{(y,i)} + m_i g \right\}^T. \tag{18}$$

Then, we can compute a solution, or an approximation to the solution quite quickly using the solution method of choice. One method that works well is simply to use the pseudoinverse of A, which will yield the forces the point masses apply on the object and thus the forces on the point masses in response, and the acceleration of the object.

$$w = A^+ b \tag{19}$$

The A matrix changes dimensions as needed depending on which manipulator is in contact, on the fly. This is how we can easily compute an approximation to the contact forces and resulting acceleration.
The overall problem, now that the physics are laid out, can be presented as the following:

Compute the optimal forces and contact locations to track the reference if in contact, and if not, compute the trajectories for those manipulators not in contact in order to move them to new contact points, or into a useful position (if, for example, the object moves out of the particular manipulator's workspace).
These two trajectory classes form a pattern, which could be referred to as a gait for manipulation or locomotion.

There are several sub-problems to consider in order to not only solve the above overall problem, but to do so in a useful way - for example, manipulate objects without dropping them, make a robot walk without falling over. Additionally, it should be noted that most manipulation problems involve redundancy - usually one has more fingers than one needs to for any given task, so how does one decide what to do with each finger? There are many solutions (Simpkins & Todorov, 2011). Here we will state that this can be formulated within the optimal control framework, and solved using the method described below. In that way the solution which 'best' solves the problem at hand is selected from the set. It remains to further describe what 'best' means, and how the set of possible actions is generated.

It is possible to formulate this problem within the optimal control framework as a single level problem, however it is nonlinear, high dimensional, and stochastic. We are interested in real-time solutions, and it is difficult to solve these problems with an unprescribed length of time and computational capabilities, let alone limited resources and time. There is also support for the notion that the human brain solves such problems in a hierarchical but interconnected fashion. Thus we draw from nature to address this problem - break the one complex problem down into several more simple and tractable problems, organizing challenging aspects of the problems into groups which sum to the solution of the more difficult problem.

The three levels are the high level, which assumes one can apply an ideal force to the object directly at its center of mass, a mid-level, which takes as input the requirement for the ideal force, and decides how to break the force up into individual forces to apply with the manipulators, and where these forces should be applied, and finally a low level which deals with mapping endpoint forces to apply with manipulators to joint torques. The mid-level also, in computing optimal configurations, weighs out the difference between remaining in the same contact posture and breaking contact to create a new grip.

3.4.2 High-level

Another advantage of this approach is that the high level can employ more complex algorithms, as it will be lower dimensional overall. In fact, predicting individual trajectories of the manipulators would be prohibitively computationally expensive, however implemented in this way it is not.

To implement this high level controller, one can implement any desired strategy: optimal (Stengel, 1986), robust (Zhou & Doyle, 1998), risk-sensitive, model-predictive (Erez et al., 2011; Li & Todorov, 2004), classical (Franklin & et al., 1994), and adaptive (Krstic et al., 1995), to list a small subset of possibilities.

In an opposing notion, it is simpler to implement classical control using ideal forces, and the lower level controls may be implemented similarly as, for example, Proportional Derivative (PD) control which, in its discrete form, is given by, with ΔT the time increment, k the current time-step, K_d the differential gain and K_p the proportional gain, e the error, y the output state feedback signal, and finally r the reference,

$$u_k^{PD} = K_p(e_k) + \frac{K_d}{\Delta T}(e_k - e_{k-1}) \tag{20}$$

where the error is given by

$$e_k = r_k - y_k \tag{21}$$

3.4.3 Mid-level

Given the required force to apply to the object, the mid-level splits this force into (given the number of manipulators) what forces each manipulator should apply and where they should apply them. This level deals also with workspace limitations, what to do if the object moves outside the manipulator workspace, and contact breaking is handled as well, all in one equation.

This is done differently, depending on which of the manipulators are in contact and which are not. Given a particular contact grouping, the overall ideal force is distributed among the in-contact fingers with a fast quadratic constrained optimization (forces constrained to only have a negative surface normal component).

The forces (f_x) acting upon the object (at location specified by $d(\theta)$) are in the x, y, and angle θ coordinates, and are linear forces and the resulting moments, given in matrix form (at timestep k) , for manipulator i acting on the object as

$$f_{t_i,k} = \begin{bmatrix} f_{x_i} & f_{y_i}d_x(\theta_i) - f_{x_i}d_y(\theta_i) \\ f_{y_i}d_x(\theta_i) - f_{x_i}d_y(\theta_i) & f_{y_i} \end{bmatrix} \tag{22}$$

The objective $J(f_i)_k$ is to minimize the error between the ideal force and the sum of total forces acting on the object, which tends to keep any force from being excessively large,

$$J(f_i)_k = \left[[f_{r,k} - \sum_i f_{t_i,k}]^2 + \sum_{(x,y),i} f_{i,k}^2 \right] \tag{23}$$

We can pose this as a constrained optimization problem by stating (with n representing the normal to the surface of the object at the point of contact),

$$\left(f_i^{opt} = \underset{f_i}{\text{argmin}} [J(f_i)_k], s.t. [f \bullet n < 0,] \right) \tag{24}$$

So given the manipulators in contact, we can extremely quickly determine a set of forces which sum (as closely as possible) to an equivalent of the ideal desired force.

The behavior of the fingers that are not in contact (and the contact breaking behavior) can be encoded in a field equation which combines several different elements to create a pattern of action. This is easier to build from scratch or data than a cost function, since it is more direct and actions can be easily shaped. This is similar to directly constructing a policy - imagine a virtual force acting on a finger, so when you hold a coffee cup in your hand and rotate it continuously in one direction, there is a moment where you feel an increasing urge to change your grip (in such a way that you do not drop your cup). A virtual force field translates well to constructing these types of behaviors from observations. In the case presented here it is desirable to change grip when a manipulator is nearing the boundary of its workspace (i.e.

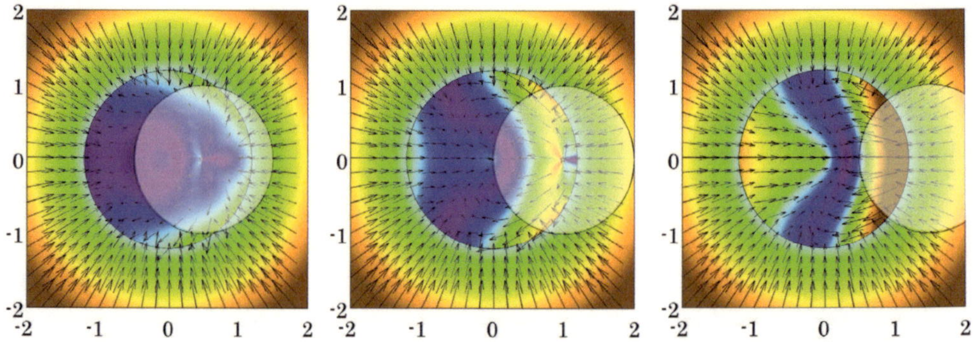

Fig. 8. Virtual force field rendering. Here one sees the forces acting on the manipulator endpoint. Note how the object (white) affects the field as it enters the workspace - the further the object is from the center of the workspace, the narrower the contact region becomes. This is a fast way of generating trajectories for the endpoint, and is very flexible for a variety of behaviors or adapting behaviors, as it is simply a matter of altering parameters of an equation.

apply a force which pulls the finger away from the object near the boundary of the workspace), and find a new contact which is closer to the middle of the workspace, in order to maximize potential for varied actions. This force should fall off near the line between the center of the workspace and the center of the object.

$$F_e = K_e \frac{(x - x_c)}{1 + ||x - x_w||^2} \tag{25}$$

Equation 25 pulls the manipulator toward the object, but falls off the further the object is from the center of the workspace - in other words if the object is too far away, the term which pulls the manipulator towards the center of the workspace (Equation 26, the spring element, dominates (here x_w represents the center of the workspace, x_c represents the center of the object, x the endpoint of the manipulator, K_e and K_s gains, and a is a constant determining the fall-off of equation 26 as a function of distance from the object, surface normal n,

$$F_s = K_s n ||x - x_w|| e^{-a(x - x_c)^2} \tag{26}$$

and finally F_t the total virtual force field equation.

$$F_t = F_e + F_s \tag{27}$$

Outside of the workspace, biological system tendons and joint assemblies act as springs, pulling toward the center of the workspace, as in Equation 28.

$$F_t = K_k(x - x_w) \tag{28}$$

These equations can be visualized as in Figure 8. Here it is clear that as the object enters the workspace, the virtual force field acting on the manipulator endpoint correspondingly alters. In the case just described we use (for clarity) circular objects and workspaces, but different shapes can be used by replacing the simple subtraction by a function of any shape.

3.4.4 Low-level

There are advantages to working in endpoint coordinates (Simpkins & Todorov, 2011), one of the most notable being the flexibility in actual structures the control is solving the problem for. In addition, the more challenging parts of the problem are solved for dynamics which are linear.

We here implement a bounded Newton's method (Ferziger, 1998) to compute the optimal feasible torques to apply which map as closely as possible to the desired forces at the endpoints. Each state is initialized with the current state rather than randomly in order to be near the goal initially, and this is quite effective and efficient in practice. When implemented on a physical system, accuracy is further improved by implementing a low level PD controller to track these joint torques more closely, and deal with issues of friction and un-modeled dynamics of the structure or system.

3.5 Safety and real-time control

Human and robot interaction can pose a number of dangers. This is especially true when the humans are interacting with dynamic or powerful robots. Not only do the robots present danger to humans directly, but the unpredictable nature of human behavior is such that they can introduce disturbances into the system which are difficult to account for ahead of time. There are a number of ways to mitigate hazards. One approach is to make the entire robotic system passive - based on inertia, back-drivability, and brakes. Then the human applies all the energy to the system. Though this is powerful, the human-robot system can still create accelerations which can cause injury to themselves or others, and the system is no longer autonomous. The human *must* be in the loop for the system to act. Making the entire robot small and weak, or to mechanically decouple the motors from the human is another approach. This may be effective in some cases, but has limitations as then the robot cannot perform many tasks. In the mechanical coupling case it may be difficult to keep disconnected components from moving unsafely.

In all cases it is beneficial to have a secondary monitoring system which is connected with all the necessary sensors and can control either braking or power to actuators. This 'watchdog' system performs checks on the control system to make sure it does not exceed safety specifications, that all components operate properly, and that the high level control has not crashed. If anything does fail checks, the watchdog can disable the power and stop the system safely. A small processor is all that is required to perform these duties, requiring minimal space and power. The overall architecture for this watchdog approach will be discussed, along with demonstrations of a robot with such a system integrated into the design.

3.5.1 Failure modes of a standard system

Consider a standard engineered control system, as shown in Figure 9(a). The system consists of a mechanical system, sensors, actuators, power input/supply, and some sort of computer system.

Case 1: Sensor failure

Assume this is a feedback system, such as a robotic arm used for high energy haptics. The robot is aware of its current position by a series of position sensors coupled to key joints. This allows the robot to follow a reference trajectory, as in the image sequence in Figure 12. Now consider what happens if one sensor suddenly discontinues to function (Figure 9(b)), or begins

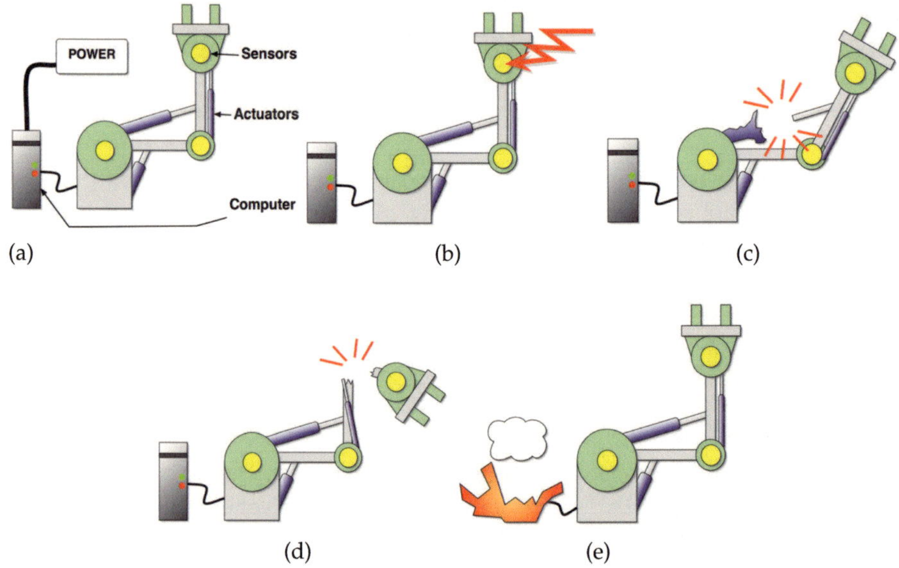

Fig. 9. (a) A typical engineered control system. It consists of a mechanical structure, sensors, actuators, power input/supply, and some computer system. (b)-(e) Some examples of failures of the typical system, and how they can be classified into one of the four cases (power supply omitted for clarity). (b) Case 1 : sensor failure - a sensor stops functioning properly. (c) Case 2 : actuator failure - in this case an actuator breaks, but it could be that it has an internal electrical failure. (d) Case 3 : mechanical failure - here we see that a main mechanical component has dislocated completely. (e) Case 4 : Computer failure - here it seems that the computer has been destroyed, but this type of failure also can be a software crash.

to function abnormally (a wire breaks, it wears out, shock destroys it, a bad ground, etc). The robot control algorithm assumes it is still being fed sensor information, and acts accordingly. This will likely cause a disastrous acceleration until a boundary is struck with maximum force, and the user may be injured.

In the case of a haptic system, it is better to have a motor turn off rather than accelerate uncontrollably, however how does one integrate available information to react properly? There actually is information available to tell us that there is a hardware failure, even if the sensor failure itself is literally undetectable.

Case 2: Actuator failure

There are a number of ways an actuator can fail (Figure 9(c)). An electrical actuator may fail such that it becomes passive, unresponsive, sluggish, or lacking in power output. It may also lock or jam due to some component failure. The internal circuit of the actuator may fail such that it shorts (which tends to make the actuator act as a viscous brake) or has an open circuit, which leads to a non-responsive, passive actuator. A hydraulic actuator may lock, or it may leak fluid which leads to pressure loss and either stiffness or loss of power, depending on where the leak occurs. It may also suffer a compressor failure (which may lead to a total failure of the actuator) or component wear (which may lead to inefficiency or lack of functionality as well). Pneumatic actuators may experience a lock-up, but due to gas compressibility this is less common. Often the failure of pneumatics lead to rapid functionality loss as the gas

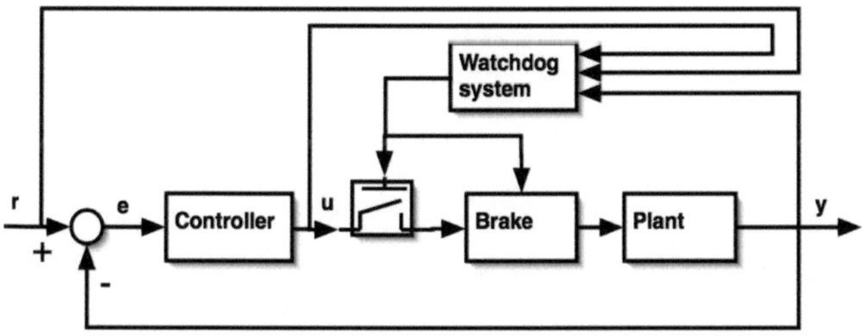

Fig. 10. One possible safety configuration for a watchdog system. In this system there is a simple microprocessor overseeing all safety-critical functions of the main device - sensors, actuators, and boundaries. If anything is incorrect, it can disable the motors and enable a brake directly.

pressure loss may be more rapid than for a fluid-based system. Many pneumatic, electrical, and hydraulic systems include integrated sensors which may fail, as previously discussed in Case 1.

Case 3: Mechanical failure

A mechanical failure is defined by the author as a failure in one or more mechanical aspects of the system. This can be a component breaking or yielding (deforming such that it does not return to its original shape), a mechanical jam, or a physical disconnection which is undesirable, as in Figure 9(d).

Case 4: Computer or electronic failure

A computer failure may occur through a software crash, electrical short or disconnection, a software bug or error, memory corruption, or computational insufficiency, as in Figure 9(e). Any of these can lead to the control system not acting as designed.

Achieving the goal regardless

Ultimately, many failures can be detected and 'handled' in such a way that potentially devastating effects are mitigated. It may be convenient to include as a safety device a redundant low level microprocessor-based 'watchdog' which is independent of the standard system, as in Figure 10. It is connected to all the sensor information and actuators, and has a programmed set of boundaries to maintain. Its job is to oversee all the components of the system and ensure they are functioning. The watchdog and main computer system ping each other frequently in a simple way such as with a pulse train. If there is a failure in either system component (main or watchdog) the system, depending on its function, may be shut down. The watchdog may be connected to the actuators such that it may disable the power output and engage some braking action if desired. It is also generally a high bandwidth system (since it is not performing many computations). If the system is mission-critical, such as in a military application or in a flight control or other situation where it may be dangerous for the system to be suddenly disabled entirely, it may be more effective to determine the safest course of action which still achieves the objective. In the case of a walking robot, where a Case 1-4 failure has occurred in the leg, it may still be possible to walk. Consider humans with an injured joint - they can often offload much of the dynamic load to anther leg or joint. It is simply a matter of an adaptive control algorithm which recomputes a solution for a different system configuration. Consider sending a robot into a situation where it will be gradually melting - a radioactive disaster or extreme conditions. The robot needs to be capable of adapting to loss of actuators, sensors, and some computational capabilities. An algorithm such as discussed

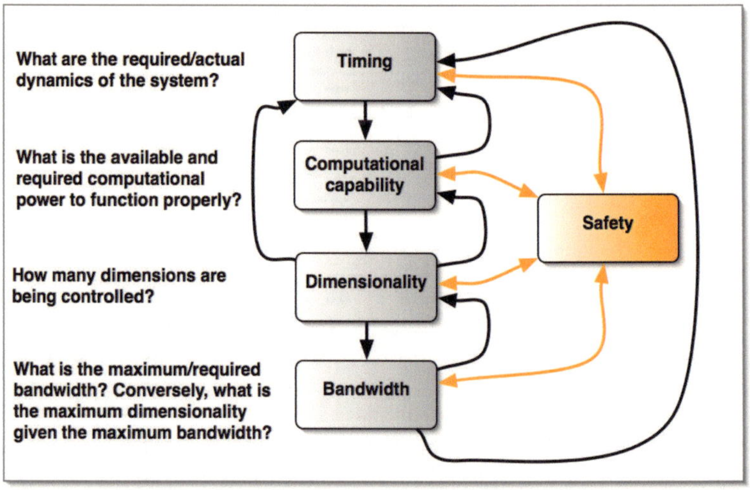

Fig. 11. Suggested flowchart for real-time system design. Note that safety is to be part of each stage of the design process. In addition, bandwidth, dimensionality, and timing are somewhat interdependent, which is the reason that feedback is incorporated in this chart. The first thing to start with as a question is 'what are the dynamics of interest?' This leads to the computations outlined in Section 3.1. Second, one must determine just how many degrees of freedom the system needs to perform all its tasks (Section 3.3), and how complex the control algorithms need to be (and can be, depending upon hardware available). This determines dimensionality, which affects the timing. Bandwidth limitations impose constraints on both dimensionality and timing (slower timing allows for higher dimensionality, and vice versa for faster timing).

in Section 3.4.1 can easily handle these sorts of challenges. Identification of the failure is a challenge. This can be achieved through the integration of the watchdog. A watchdog can be used to compare the measurements with expected outputs - for example a sensor with a zero voltage output which measures velocity of a motor normally, while a motor is being commanded to a high velocity, and is currently not drawing very much current (meaning it probably is not stalled), and has significant back-emf being generated can be a flag for a sensor or actuator failure. All the components of a system work together in unison, and the dynamics of a system express themselves through all the sensors and actuators. Expected values can be integrated into the watchdog in the form of a classifier which can output error states very quickly and efficiently, if well-trained and carefully coded. The watchdog and computer compare error signals, and can eliminate (from the control algorithm), freeze, or otherwise remove from or adjust the dynamics as much as possible to minimize failure impact. Then the control algorithm is simply updated with a new number of actuators or in the case of something similar to locking an injured joint, a new number of degrees of freedom, and it continues to solve the problem in the optimal fashion given the new system. Since everything is solved online this is possible.

4. Putting it all together

Methods have been presented which, when combined, effectively deal with the real-time control problem for complex robotic systems. Since some of the components of the design

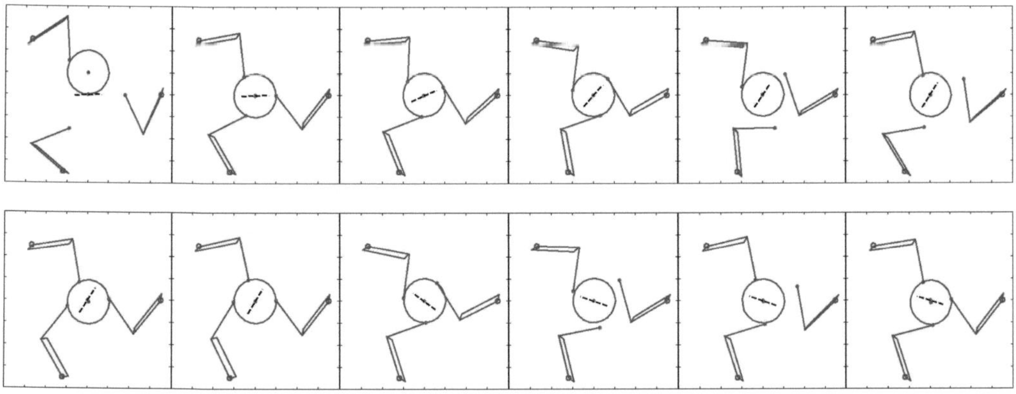

Fig. 12. Animation sequence of three manipulators tracking a rotating stationary trajectory (Frames advance starting from top-left in sequence to the right in the same fashion as normal english text). Note the multiple contact breaks in order to allow the object to be continuously rotated.

depend upon others, Figure 11 suggests a sequence which is typically effective. Each component is interdependent with each other component, however an effective starting point is to define the problem, or class of problems first. This gives a sense of the requirements for timing. Then the designer can compute required bandwidth and dimensionality of the system (as well as constraining the design by the available computational power), or generate something like the sensorimotor system of a human, with a high mechanical dimensionality, but limitations on overall bandwidth and dimensionality of control, leading to optimal actions for a particular task. At each stage of the process, safety engineers or the primary designer can design checks such as a secondary safety system and impose safety-based constraints.

Real-time control itself, in order to be most effective, should not be decoupled from the design of the actual system. Otherwise the system may be unnecessarily limited due to designed-in bottlenecks. However, being aware of the concepts presented in this chapter, a control engineer can effectively maximize the capabilities of a system which already exists and must merely be controlled.

5. Results and application

We begin with a description of the problem, originally described in (Simpkins et al., 2011), and discussed previously in Section 3.4.1. In order to study biological systems, and how they interact with their environment through manipulation and locomotion, it is useful to have a robotic device which mimics certain key aspects of biology (those systems capable of manipulation and dextrous locomotion), such as output-to-actuator backdrivability[1], low endpoint inertia, high dynamic capabilities, and sensitivity. After carefully determining the requirements for manipulation and locomotion, the author and collaborators drew upon biological systems in terms of timing and dynamic capability to define requirements. In addition, there was a desire to create a modular system which could ultimately be assembled quickly into a variety of configurations. There are a variety of requirements determined,

[1] Application of force at the output can about a change of state at the input of a back-drivable system.

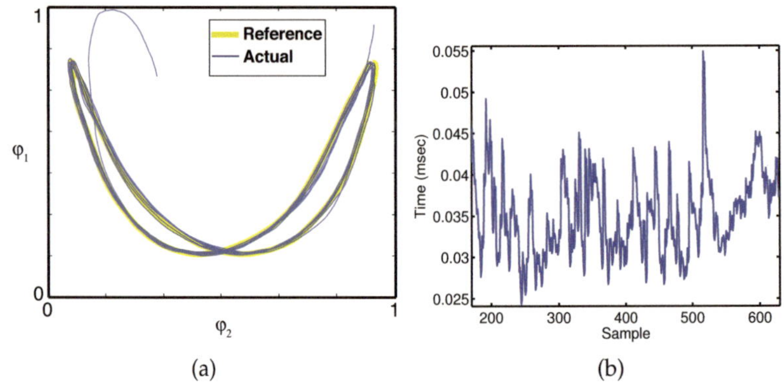

Fig. 13. (a) The robot, controlled hierarchically from a PC running matlab and communicating over bluetooth, with a local low level PID controller, is capable of tracking a cyclical motion very rapidly. The timing of the loop for this plot is approximately 1msec, with some variability in timing. (b) In this plot, the loop (which here includes contact dynamics and a more complex simulation) was deliberately executed in a non-optimal fashion in order to vary the timing of the sampling and updates, for demonstration purposes. When such variability is handled properly, as described in this chapter, the system can be built to remain stable.

and a summary is a 1kHz maximum bandwidth for the control of five fingers through one bluetooth or USB connection, ability to apply approximately 1N of endpoint force, minimal friction and endpoint inertia, resolution above one-tenth of a degree, and measures of force in addition to position. These requirements pose a challenge as communicating position, force, and current for each joint of each finger for five fingers at 1kHz becomes 9000 variables per second, and at 12-bits of resolution, for five fingers, this becomes 540kbps. If the robot is receiving, for example, no more than three variables for command of either position, velocity, or force output, then the most data must be sent from the low to the higher levels, as the serial interface used can simultaneously send and receive data.

There have been a few biologically-based robots recently developed, however they all suffer from some limitation imposed by a lack of an integrative real-time approach such as discussed in this chapter. It is not unusual for a well-designed hand mechanism to have a maximum bandwidth of one-tenth of a Hz to move from open-to-closed. Consider attempting most daily tasks where every motion could be performed in no shorter time than ten seconds. Dynamic manipulation would indeed be impossible! We omit the discussion of the design itself here, shown in Figure 6(a), instead focusing on how to maximize the real-time capabilities.

Figure 12 demonstrates a simulation based upon the real-time concepts discussed in previous sections. This is an implementation of a real time hierarchical optimal control of a number of manipulators tracking a reference trajectory, and can be applied to the physical robots. In this case the high level is a simple proportional-derivative controller tracking a 'rotate continuously' command. It is clear that the algorithm is capable of multiple contact breaks required in order to allow continuous rotation.[2]

[2] Further results and animations can be found at the author's home page, http://casimpkinsjr.radiantdolphinpress.com/pages/papersv2.html. In addition, this is the location for all the author's publications, with most available for download.

Fig. 14. Image sequence of a bio-mimetic robotic finger designed by the author undergoing real-time control as described in this chapter. The finger is tracking a cyclical shape in 3-Dimensional space at approximately a 10Hz frequency. The robot is controlled wirelessly, through bluetooth, and is communicating 9x12bit values for position, velocity, and coil current at 1kHz, and assuming it is receiving the same amount of data as it is sending (as a worst-case), required bandwidth is 9x12x1000x2 = 216,000 bits per second, or 216kbps. The bandwidth of the connection is 2e6bps, or 2Mbps, which is nearly 10 times greater. In this case the algorithm controlling the finger is not computing any contacts, and so executes in an average of less than 1msec at the PC side.

This simulation runs entirely in real-time, with no offline components. With code executing in Matlab R2008b for Macintosh upon an Intel Core 2 Duo processor at 2.33GHz with 2GB of RAM, each loop takes on average 30msec, as measured from within Matlab (See Matlab help on tic-toc), which is fast enough for real-time manipulation. Human beings have an update rate perceptually of approximately 30Hz, and dynamically of only a few Hz (Nigg & Herzog, 1994). Additionally, it would be straightforward to optimize this code to execute far faster by implementing it in C, for example.
The fully implemented control algorithm including hardware executes in approximately the same time, as the hardware is provided the command computed by the simulation, and the actual object states are fed back into the system to adjust for the difference between simulation and reality. Essentially including the hardware in the loop in parallel with the model adds a model reference control element to the hierarchical optimal control strategy. The majority of the processor time is spent on the contact dynamics computations, and could be further improved to minimize some calculations without changing the algorithm significantly. Therefore when some manipulators are not in contact (as in Figure 14), computation time is reduced dramatically. The 30Hz or even a 200Hz sample rate are well within the 1kHz bandwidth of the robot manipulator and wireless communication system. It is notable that, in Figure 13(b), the variability in timing is less than one sample, and so the state space representations described in Section 3.1 hold. In the case that the manipulator is tracking a trajectory, as in Figure 13(a), we see that this hierarchical strategy is capable of smooth rapid control of the manipulator. If the current Matlab implementation were to be used, and the system were required to perform manipulation at a faster rate than 30Hz, the concepts in Section 3.3 could be more explicitly included in the control strategy of Section 3.4.1 by adding a cost on the number of simultaneous manipulator control inputs. A lower dimensionality means the algorithm will execute faster and be capable of a higher bandwidth, while still being able to increase dimensionality on demand.

6. Conclusion

Modern robotic systems are becoming more and more complex, and requiring far more sophisticated controllers to be designed and implemented than ever before in order to deal with the demand for them to perform more complex tasks. These changes have raised a new challenge for engineers and roboticists. This chapter has presented a first step toward a unifying theory which combines and expands upon modern real-time design control techniques. It broke the problem down into sub-problems which each have a significant body of literature, while describing the fact that these seemingly separate fields are in fact intertwined, are part of a whole, and complement each other. By being at least aware of this method of organizing the problem, and the potential pitfalls associated with the sub-problems, it is possible to improve the end result of a design by making revisions and requirements which are more appropriate far earlier in the process than otherwise possible.
A number of results were presented for the control strategy discussed, as well as an implementation on a real dynamic bio-mimetic robot. Currently, many of the Matlab functions associated with this robot are being implemented in C and made available as MEX files which Matlab can still make use of. This should improve the execution time significantly of the simulation, as well as all other aspects of the computations. These techniques will be part of the development of new skills for these dynamic manipulators. The two dimensional representation has been extended to three dimensions, and will be part of a series of papers to be released soon by the author.
These new challenges require a different overall strategy for problem solving than ever before in engineered systems, but have the potential to truly revolutionize all aspects of human experience. By developing and refining methodologies, we will transition from individual designs which take years of iterative slow improvements with many dead ends to clear paths with fewer roadblocks, and new tools to illuminate the way.

7. References

Anderson, T. W. (2003). *An Introduction to Multivariate Statistical Analysis*, John Wiley and Sons.

Åström, K. & Wittenmark, B. (1990). *Computer controlled systems - theory and design*, 2nd edn, Prentice Hall, Englewood cliffs, NJ.

Bernstein, N. (1967). *The coordination and regulation of movements*, Pergamon, London.

Butterfab, J. & et al. (2001). Dlr-hand ii: Next generation of a dextrous robot hand, *Proc. of the IEEE International Conference on Robotics and Automation* pp. 109–114.

Erez, T., Todorov, E. & Tassa, Y. (2011). Fast model predictive control for complex robotic behavior, *Robotics Science and Systems (RSS'11)* .

Ermedahl, A. (2003). *A modular tool architecture for worst-case execution time analysis*, PhD thesis, Uppsala University.

Ferziger, J. (1998). *Numerical Methods for Engineering Application*, 2nd edition edn, John Wiley and Sons, New York, NY.

Franklin, G. F. & et al. (1994). *Feedback control of dynamic systems*, 3rd edn, Addison Wesley, Reading, MA.

Grob, H.-G. (2001). A prediction system for evolutionary testability applied to dynamic execution time analysis, *Information and Software Technology* (43): 855–862.

Hansen, J., Hissam, S. & Moreno, G. (2009). Statistical-based wcet estimation and validation, *9th International Workshop on Worst-Case Execution Time (WCET) Analysis* .

Krstic, M., Kanellakopoulos, I. & Kokotovic, P. (1995). *Nonlinear and Adaptive Control Design*, John Wiley and Sons.

Li, W. & Todorov, E. (2004). Iterative linear quadratic regulator design for nonlinear bio-logical movement systems, *Proc. of the 1st International Conference on Informatics in Control, Automation and Robotics* **1**: 222–229.

Mardia, K., Kent, J. & Bibby, J. (1992). *Multivariate Analysis*, Academic Press.

Movellan, J., Tanaka, F., Fortenberry, B. & Aisaka, K. (2005). The rubi/qrio project: Origins, principles, and first steps, *IEEE International Conference on Development and Learning* **1**: 80 – 86.

Nigg, B. & Herzog, W. (1994). *Biomechanics*, John Wiley and Sons, New York.

Nilsson, J. (1998). *Real-time control systems with delays*, PhD thesis, Dept. of automatic control, Lund institute of technology, Sweden.

Puschner, P. & Nossal, R. (1998). Testing the results of static worst-case execution-time analysis, *Proceedings of the 19th IEEE Real-Time Systems Symposium (RTSS98) Madrid.*

Santello, M., Flanders, M. & Soechting, J. (1998). Postural hand synergies for tool use, *Journal of Neuroscience* (18): 10105–15.

Simpkins, A. (2009). *Exploratory studies of sensorimotor control and learning with system identification and stochastic optimal control*, PhD thesis, University of California, San Diego, La Jolla, CA.

Simpkins, A., Kelley, M. & Todorov, E. (2011). Modular bio-mimetic robots that can interact with the world the way we do, *Proceedings of the IEEE International Conference on Robotics and Automation* .

Simpkins, A. & Todorov, E. (2011). Complex object manipulation with hierarchical optimal control, *IEEE Symposium of Adaptive Dynamic Programming and Reinforcement Learning, IEEE Computer Society, Paris, France.*

Stengel, R. (1986). *Stochastic Optimal Control: Theory and Application*, John Wiley and Sons.

Todorov, E. (2004). Optimality principles in sensorimotor control, *Nature Neuroscience* **7**: 907–915.

Todorov, E. & Ghahramani, Z. (2003). Unsupervised learning of sensory-motor primitives, IEEE, *In Proceedings of the 25th Annual International Conference of the IEEE Engineering in Medicine and Biology Society.*

Todorov, E. & Ghahramani, Z. (2004). Analysis of synergies underlying complex hand manipulation, IEEE, *Proceedings of the 26th Annual International Conference of the IEEE EMBS, San Francisco, CA, USA.*

Todorov, E., Hu, C., Simpkins, A. & Movellan, J. (2010). Identification and control of a pneumatic robot, *Proc. of the fourth IEEE RAS / EMBS International Conference on Biomedical Robotics and Biomechatronics* .

Todorov, E. & Jordan, M. (2002). Optimal feedback control as a theory of motor coordination, *Nature Neuroscience* **5**(11): 1226–1235.

Todorov, E. & Jordan, M. (2003). A minimal intervention principle for coordinated movement, *Advances in Neural Information Processing Systems* **15**: 27–34.

Vandeweghe, J. M., M.Rogers, M.Weissert & Matsuoka, Y. (2004). The act hand: Design of the skeletal structure, *Proc. of the IEEE International Conference on Robotics and Automation.*

Wilhelm, R., Engblom, J., Ermedahl, A., Holsti, N., Thesing, S., Whalley, D., Bernat, G., Ferdinand, C., Heckmann, R., Mitra, T., Mueller, F., Puaut, I., Puschner, P., Staschulat,

J. & Stenström, P. S. (2008). The worst-case execution time problem - overview of methods and survey of tools, *ACM Transactions on Embedded Computing Systems* **7**(3).

Wilkinson, D. D., Vandeweghe, J. M. & Matsuoka, Y. (2003). An extensor mechanism for an anatomical robotic hand, *Proc. of the IEEE International Conference on Robotics and Automation*, Vol. 1, pp. 238 – 243.

Wittenmark, B., Nilsson, J. & Torngren, M. (1995). Timing problems in real-time control systems, *In Proceedings of the 1995 American Control Conference. Seattle, WA* .

Zhou, K. & Doyle, J. (1998). *Essentials of Robust Control*, Prentice Hall, Upper Saddle River, NJ.

12

Robot Learning from Demonstration Using Predictive Sequence Learning

Erik Billing, Thomas Hellström and Lars-Erik Janlert
Department of Computing Science, Umeå University
Sweden

1. Introduction

In this chapter, the prediction algorithm Predictive Sequence Learning (PSL) is presented and evaluated in a robot *Learning from Demonstration* (LFD) setting. PSL generates hypotheses from a sequence of sensory-motor events. Generated hypotheses can be used as a semi-reactive controller for robots. PSL has previously been used as a method for LFD (Billing et al., 2010; 2011) but suffered from combinatorial explosion when applied to data with many dimensions, such as high dimensional sensor and motor data. A new version of PSL, referred to as *Fuzzy Predictive Sequence Learning* (FPSL), is presented and evaluated in this chapter. FPSL is implemented as a Fuzzy Logic rule base and works on a continuous state space, in contrast to the discrete state space used in the original design of PSL. The evaluation of FPSL shows a significant performance improvement in comparison to the discrete version of the algorithm. Applied to an LFD task in a simulated apartment environment, the robot is able to learn to navigate to a specific location, starting from an unknown position in the apartment.

Learning from Demonstration is a well-established technique for teaching robots new behaviors. One of the greatest challenges in LFD is to implement a learning algorithm that allows the robot pupil to generalize a sequence of actions demonstrated by the teacher such that the robot is able to perform the desired behavior in a dynamic environment. A behavior may be any complex sequence of actions executed in relation to sensor data (Billing & Hellström, 2010).

The LFD problem is often formulated as four questions, *what-to-imitate*, *how-to-imitate*, *when-to-imitate* and *who-to-imitate* which leads up to the larger problem of how to evaluate an imitation (Alissandrakis et al., 2002). Large parts of the literature approach the learning problem by trying to find the common features within a set of demonstrations of the same behavior. A skill is generalized by exploiting statistical regularities among the demonstrations (e.g. Calinon, 2009). This is reflected in the *what-to-imitate* question, originally introduced in a classical work by Nehaniv & Dautenhahn (2000) and is in a longer form described as:

> An action or sequence of actions is a successful component of imitation of a particular action if it achieves the same subgoal as that action. An entire sequence of actions is successful if it successively achieves each of a sequence of abstracted subgoals.

The problem is difficult since a certain behavior can be imitated on many different abstraction levels. Byrne & Russon (1998) identified two levels; the *action-level imitation* copying the surface of the behavior and a *program-level imitation* copying the structure of the behavior. A third level, the *effect-level imitation*, was introduced by Nehaniv & Dautenhahn (2001) in order to better describe imitation between agents with dissimilar body structures. Demiris & Hayes (1997) proposed three slightly different levels: *1) basic imitation* with strong similarities to the notion of action-level imitation, 2) *functional imitation* that best corresponds to effect-level imitation and 3) *abstract imitation* that represents coordination based on the presumed internal state of the agent rather than the observed behavior. Demiris and Hayes give the example of making a sad face when someone is crying.

The necessity to consider the level of imitation in LFD becomes apparent when considering two demonstrations that look very different considered as sequences of data, but that we as humans still interpret as examples of the same behavior since they achieve similar results on an abstract level. This would correspond to a functional or program-level imitation. In these situations it is very difficult to find similarities between the demonstrations without providing high level knowledge about the behavior, often leading to specialized systems directed to LDF in limited domains.

A related problem is that two demonstrations of the same behavior may not have the same length. If one demonstration takes longer time than another, they can not be directly compared in order to find common features. Researchers have therefore used techniques to determine the temporal alignment of demonstrations. One common technique is *dynamic time warping* (Myers & Rabiner, 1981), that can be used to compensate for temporal differences in the data.

Behaviors can be demonstrated to a robot in many different ways. Argall et al. (2009) outline four types of demonstrations: A direct recording of sensor stimuli, joint angles, etc., is referred to as an *identity record mapping*. In this case, the robot is often used during the demonstration and controlled via teleoperation or by physically moving the robot's limbs (kinestetic teaching). An external observation, e.g. a video recording of the teacher, is called a *non-identity record mapping*. This type of demonstrations poses a difficult sensing problem of detecting how the teacher has moved, but also allows much more flexible demonstration setting. The teacher may have a body identical to that of the pupil (*identity embodiment*) or a body with a different structure (*non-identity embodiment*). In the latter case, the demonstration has to be transformed into corresponding actions using the body of the pupil, a difficult problem known as the *correspondence problem* (Nehaniv & Dautenhahn, 2001). In this work we focus on LFD via teleoperation. Sensor data and motor commands are in this setting recorded while a human teacher demonstrates the desired behavior by tele-operating the robot, producing demonstrations with identity in both record mapping and embodiment.

1.1 Metric of imitation

Successful imitation requires that relevant features of the demonstration are selected at a suitable imitation level and processed into a generalized representation of the behavior. The process is difficult to implement in a robot since it is often far from obvious which imitation level that is optimal in a specific situation, and the relevance of features may consequently vary significantly from one learning situation to another. This problem has been formalized as a *metric of imitation*, defined as a weighted sum over all strategy-dependent metrics on all imitation levels (Billard et al., 2003).

The metric of imitation was originally demonstrated on a manipulation task with a humanoid robot (Billard et al., 2003). With focus on the correspondence problem, Alissandrakis et al. (2005) propose a similar approach to imitation of manipulation tasks. The what-to-imitate problem is approached by maximizing trajectory agreements of manipulated objects, using several different metrics. Some metrics encoded absolute trajectories while other metrics encoded relative object displacement and the relevant aspects of the behavior were in this way extracted as the common features in the demonstration set. Calinon et al. (2007) developed this approach by encoding the demonstration set using a mixture of Gaussian/Bernoulli distributions. The Gaussian Mixture Model approach is attractive since the behavior is divided into several distributions with different covariance, and different metrics can in this way be selected for different parts of the demonstrated behavior. More recently, similar encoding strategies have been evaluated for learning of a robot navigation task (de Rengervé et al., 2010).

1.2 Behavior primitives as a basis for imitation

Another common approach to LFD is to map the demonstration onto a set of pre-programmed or previously learned primitives controllers (Billing & Hellström, 2010). The approach has strong connections to *behavior-based architectures* (Arkin, 1998; Matarić, 1997; Matarić & Marjanovic, 1993) and earlier reactive approaches (e.g. Brooks, 1986; 1991). When introducing behavior primitives, the LFD process can be divided into three tasks (Billing & Hellström, 2010):

1. *Behavior segmentation* where a demonstration is divided into smaller segments.
2. *Behavior recognition* where each segment is associated with a primitive controller.
3. *Behavior coordination*, referring to identification of rules or switching conditions for how the primitives are to be combined.

Behavior segmentation and recognition can be seen as one way to approach the what-to-imitate problem, whereas behavior coordination is part of how-to-imitate. The approach represents one way of introducing good bias in learning and solve the generalization problem by relying on previous behavioral knowledge. While there are many domain specific solutions to these three subproblems, they appear very difficult to solve in the general case. Specifically, behavior recognition poses the problem of mapping a sequence of observations to a set of controllers to which the input is unknown. Again, the need to introduce a metric of imitation appears.

Part of the problem to find a general solution to these problems may lie in a vague definition of *behavior* (Matarić, 1997). The notion of behavior is strongly connected to the purpose of executed actions and a definition of goal. Nicolescu (2003) identified two major types of goals:

Maintenance goals: A specific condition has to be maintained for a time interval.

Achievement goals: A specific condition has to be reached.

The use of behavior primitives as a basis for imitation has many connections to biology (e.g. Matarić, 2002) and specifically the mirror system (Brass et al., 2000; Gallese et al., 1996; Rizzolatti et al., 1988; Rizzolatti & Craighero, 2004). While the role of the mirror system is still highly debated, several groups of researchers propose computational models where perception and action are tightly interweaved. Among the most prominent examples are the *HAMMER* architecture (Demiris & Hayes, 2002; Demiris, 1999; Demiris & Johnson, 2003) and the *MOSAIC* architecture (Haruno et al., 2001; Wolpert & Kawato, 1998). Both these architectures implement a set of modules, where each module is an inverse model (controller) paired with a forward model (predictor). The inverse and forward models are trained together such that the forward model can predict sensor data in response to the actions produced by the inverse model. The inverse model is tuned to execute a certain behavior when the forward model produces good predictions. The prediction error is used to compute a bottom-up signal for each module. Based on the bottom-up signal, a top-down responsibility signal or confidence value is computed and propagated to each module. The output of the system is a combination of the actions produced by each inverse model, proportional to their current responsibility. The responsibility signal also controls the learning rate of each module, such that modules are only updated when their responsibility is high. In this way, modules are tuned to a specific behavior or parts of a behavior. Since the prediction error of the forward model is used as a measure of how well the specific module fits present circumstances, it can be seen as a metric of imitation that is learnt together with the controller. The architecture can be composed into a hierarchical system where modules are organized in layers, with the lowest layer interacting with sensors and actuators. The bottom-up signal constitutes sensor input for the layer above and actions produced by higher levels constitutes the top-down responsibility signal.

One motivation for this architecture lies in an efficient division of labor between different parts of the system. Each module can be said to operate with a specific temporal resolution. Modules at the bottom layer are given the highest resolution while modules higher up in the hierarchy have decreasing temporal resolution. State variables that change slowly compared to a specific module's resolution are ignored by that module and are instead handled by modules higher up in the hierarchy. Slowly changing states that lead to high responsibility for the module is referred to as the module's *context*. In a similar fashion, variables that change fast in comparison to the temporal resolution are handled lower in the hierarchy. This allows each module to implement a controller where the behavior depends on relatively recent states. Long temporal dependencies are modeled by switching between modules, which removes the requirement for each model to capture these dependencies. Furthermore, updates of a single behavior or parts of a behavior will only require updates of a few modules and will not propagate changes to other modules. See Billing (2009) for a longer discussion on these aspects of hierarchical architectures.

The HAMMER and MOSAIC architectures make few restrictions on what kind of controllers each module should implement. We argue however, that modules should be *semi-reactive*, meaning that action selection and predictions of sensor events should be based on recent sensor and motor events. Strictly reactive modules are not desirable since each module must be able to model any dependency shorter than the temporal resolution of modules in the layer directly above.

The division of behavior into modules is however also producing a number of drawbacks. The possibility for the system to share knowledge between behaviors is limited. Moreover, the system has to combine actions produced by different modules, which may be difficult in cases when more than one module receives high responsibility.

One architecture with similarities to HAMMER and MOSAIC able to share knowledge between different behaviors is *RNNPB* (Tani et al., 2004). *RNNPB* is a recurrent neural network with parametric bias (PB). Both input and output layer of the network contains sensor and motor nodes as well as nodes with recurrent connections. In addition, the input layer is given a set of extra nodes, representing the PB vector. The network is trained to minimize prediction error, both by back-propagation and by changing the PB vector. The PB vector is however updated slowly, such that it organizes into what could be seen as a context layer for the rest of the network. In addition to giving the network the ability to represent different behaviors that share knowledge, the PB vector can be used for behavior recognition.

Another architecture known as *Brain Emulating Cognition and Control Architecture* (BECCA) (Rohrer & Hulet, 2006) heavily influenced our early work on the PSL algorithm. The focus of BECCA is to capture the discrete episodic nature of many types of human motor behavior, without introducing a priori knowledge into the system. BECCA was presented as a very general reinforcement learning system, applicable to many types of learning and control problems. One of the core elements of BECCA is the temporal difference (TD) algorithm *Sequence Learning* (SL) (Rohrer, 2007). SL builds sequences of passed events which is used to predict future events, and can in contrast to other TD algorithms base its predictions on many previous states.

Inspired by BECCA and specifically SL, we developed the PSL algorithm as a method for LFD (Billing et al., 2010; 2011). PSL has many interesting properties seen as a learning algorithm for robots. It is model free, meaning that it introduces very few assumptions into learning and does not need any task specific configuration. PSL can be seen as a variable-order Markov model. Starting out from a reactive (first order) model, PSL estimates transition probabilities between discrete sensor and motor states. For states that do not show Markov property, the order is increased and PSL models the transition probability based on several passed events. In this way, PSL will progressively gain memory for parts of the behavior that cannot be modeled in a reactive way. In theory, there is no limitation to the order of the state and hence the length of the memory, but PSL is in practice unable to capture long temporal dependencies due to combinatorial explosion.

PSL has been evaluated both as a controller (Billing et al., 2011) and as a method for behavior recognition (Billing et al., 2010). Even though the evaluation overall generated good results, PSL is subject to combinatorial explosion both when the number of sensors and actuators increase, and when the demonstrated behavior requires modeling of long temporal dependencies. PSL can however efficiently model short temporal dependencies in a semi-reactive way and should thus be a good platform for implementing a hierarchical system similar to the HAMMER and MOSAIC architectures.

In this chapter, we present and evaluate a new version of PSL based on Fuzzy Logic. While keeping the core idea of the original PSL algorithm, the new version can handle continuous and multi dimensional data in a better way. To distinguish between the two, the new fuzzy version of the algorithm is denoted FPSL, whereas the previous discrete version is denoted DPSL. A detailed description of FPSL is given in Section 2. An evaluation with comparisons between the two algorithms is presented in Section 3, followed by a discussion and conclusions in section 4.

2. Predictive Sequence Learning

FPSL builds fuzzy rules, referred to as *hypotheses* h, describing temporal dependencies between a sensory-motor event e_{t+1} and a sequence of passed events $\left(e_{t-|h|+1}, e_{t-|h|+2}, \ldots, e_t\right)$, defined up until current time t.

$$h : \left(\Upsilon_{t-|h|+1} \text{ } is \text{ } E^h_{|h|-1} \wedge \Upsilon_{t-|h|+2} \text{ } is \text{ } E^h_{|h|-2} \wedge \ldots \wedge \Upsilon_t \text{ } is \text{ } E^h_0\right) \overset{C}{\Rightarrow} \Upsilon_{t+1} \text{ } is \text{ } \bar{E}^h \tag{1}$$

Υ_i is the event variable and $E^h(e)$ is a fuzzy membership function returning a membership value for a specific e. The right hand side $\bar{E}^h$ is a membership function comprising expected events at time $t+1$. $|h|$ denotes the length of h, i.e., the number of left-hand-side conditions of the rule. Both E and $\bar{E}$ are implemented as standard cone membership functions with base width ε (e.g. Klir & Yuan, 1995).

A set of hypotheses can be used to compute a prediction $\hat{e}_{t+1}$ given a sequence of passed sensory-motor events η, defined up to the current time t:

$$\eta = (e_1, e_2, \ldots, e_t) \tag{2}$$

The process of matching hypothesis to data is described in Section 2.1. The PSL learning process, where hypotheses are generated from a sequence of data, is described in Section 2.2. Finally, a discussion about parameters and configuration is found in Section 2.3.

2.1 Matching hypotheses

Given a sequence of sensory-motor events $\eta = (e_1, e_2, \ldots, e_t)$, a match $\alpha_t(h)$ of the rule is given by:

$$\alpha_t(h) : \bigwedge_{i=0}^{|h|-1} E_i^h(e_{t-i}) \tag{3}$$

where $\wedge$ is implemented as a *min*-function.
Hypotheses are grouped in fuzzy sets C whose membership value $C(h)$ describes the confidence of h at time t:

$$C(h) = \frac{\sum_{k=t^h}^{t} \alpha_k(h) \bar{E}^h(e_{k+1})}{\sum_{k=t^h}^{t} \alpha_k(h)} \tag{4}$$

.

t^h is the creation time of h or 1 if h existed prior to training. Each C represents a *context* and can be used to implement a specific behavior or part of a behavior. The *responsibility signal* $\lambda_t(C)$ is used to control which behavior that is active at a specific time. The combined confidence value $\tilde{C}_t(h)$ is a weighted sum over all C:

$$\tilde{C}_t(h) = \frac{\sum_C C(h)\lambda_t(C)}{\sum_C \lambda_t(C)} \tag{5}$$

.

$\tilde{C}_t$ can be seen as a fuzzy set representing the active context at time t. Hypotheses contribute to a prediction in proportion to their membership in $\tilde{C}$ and the *match set* $\hat{M}$. $\hat{M}$ is defined in three steps. First, the best matching hypotheses for each $\bar{E}$ is selected:

$$M = \left\{ h \mid \alpha(h) \geq \alpha(h') \; for\; all \left\{ h' \mid \bar{E}^{h'} = \bar{E}^h \right\} \right\} \tag{6}$$

.

The longest $h \in M$ for each RHS is selected:

$$\tilde{M} = \left\{ h \mid |h| \geq |h'| \; for\; all \left\{ h' \in M \mid \bar{E}^{h'} = \bar{E}^h \right\} \right\} \tag{7}$$

.

Finally, the match set $\hat{M}$ is defined as:

$$\hat{M}(h) = \begin{cases} \alpha(h)\tilde{C}(h) & h \in \tilde{M} \\ 0 & otherwise \end{cases} \tag{8}$$

.

The aggregated prediction $\hat{E}(e_{t+1})$ is computed using the Larsen method (e.g. Fullér, 1995):

$$\hat{E}\left(e_{t+1}\right) : \bigvee_{h} \bar{E}_h\left(e_{t+1}\right) \hat{M}\left(h\right) \tag{9}$$

$\hat{E}$ is converted to crisp values using a squared *center of sum* defuzzification:

$$\hat{e} = \frac{\sum\limits_{e} e\hat{E}\left(e\right)^2}{\sum\limits_{e} \hat{E}\left(e\right)^2} \tag{10}$$

The amount of entropy in $\hat{M}$ also bears information about how reliable a specific prediction is, referred to as the *trust* $\hat{c}$:

$$\hat{c}\left(\hat{M}\right) = \begin{cases} 0 & \hat{M} = \varnothing \\ \exp\left[\sum\limits_{h} \hat{M}\left(h\right) \log_2\left(\hat{M}\left(h\right)\right)\right] & otherwise \end{cases} \tag{11}$$

The trust is important since it allows a controller to evaluate when to rely on PSL, and when to choose an alternate control method. The proportion of time steps in η for which $\hat{c} > 0$ and PSL is able to produce a prediction is referred to as the *coverage* $\phi\left(\eta\right)$:

$$\phi\left(\eta\right) = \frac{\sum\limits_{i=1}^{t} \begin{cases} 1 & \hat{c}_i > 0 \\ 0 & otherwise \end{cases}}{t} \tag{12}$$

2.2 Generating hypotheses

Hypotheses can be generated from a sequence of sensory-motor events η. During training, PSL continuously makes predictions and creates new hypotheses when no matching hypothesis produces the correct prediction $\bar{E}$. The exact training procedure is described in Algorithm 0.1.

For example, consider the event sequence $\eta = abccabccabcc$. Let $t = 1$. PSL will search for a hypothesis with a body matching a. Initially, the context set C is empty and consequently PSL will create a new hypothesis $(a) \Rightarrow b$ which is added to C with confidence 1, denoted $C\left(a \Rightarrow b\right) = 1$. The same procedure will be executed at $t = 2$ and $t = 3$ such that $C\left((b) \Rightarrow c\right) = 1$ and $C\left((c) \Rightarrow c\right) = 1$. At $t = 4$, PSL will find a matching hypothesis $(c) \Rightarrow c$ producing the wrong prediction c. Consequently, a new hypothesis $(c) \Rightarrow a$ is created and confidences are updated such that $C\left((c) \Rightarrow c\right) = 0.5$ and $C\left((c) \Rightarrow a\right) = 1$. The new hypothesis receives a higher confidence since confidence values are calculated from the creation time of the hypothesis (Equation 4). The predictions at $t = 5$ and $t = 6$ will be correct and no new hypotheses are created. At $t = 7$, both $(c) \Rightarrow a$ and $(c) \Rightarrow c$ will contribute to the prediction $\hat{E}$. Since the confidence of $(c) \Rightarrow a$ is higher than that of $(c) \Rightarrow c$, $\hat{E}$ will defuzzify towards a, producing the wrong prediction (Equation 10). As a result, PSL creates a new hypothesis $(b, c) \Rightarrow c$. Similarly, $(c, c) \Rightarrow a$ will be created at $t = 8$. PSL is now able to predict all elements in the sequence perfectly and no new hypotheses are created.

Source code from the implementation used in the present work is available online (Billing, 2011).

Algorithm 0.1 Predictive Sequence Learning (PSL)

Require: $\psi = (e_1, e_2, \ldots, e_T)$ where T denotes the length of the training set
Require: $\hat{\alpha}$ as the precision constant, see text

let $t \leftarrow 1$
let $\eta = (e_1, e_2, \ldots, e_t)$
let $C \leftarrow \varnothing$
let $\hat{E}$ as Eq. 9
if $\hat{E}(e_{t+1}) < \hat{\alpha}$ **then**
    let $h_{new} = CreateHypothesis(\eta, C)$ as defined by Algorithm 0.2
    $C(h_{new}) \leftarrow 1$
end if
Update confidences $C(h)$ as defined by Equation 4
set $t = t + 1$
if t<T **then**
    goto 2
end if

Algorithm 0.2 CreateHypothesis

Require: $\eta = (e_1, e_2, \ldots, e_t)$
Require: $C : h \rightarrow [0, 1]$
Require: α as defined by Eq. 3

let $\hat{M}(h)$ as Eq. 8
let $\bar{M} = \left\{ h \mid \bar{E}^h(e_{t+1}) \geq \hat{\alpha} \wedge \hat{M}(h) > 0 \right\}$ where $\hat{\alpha}$ is the precision constant, see Section 2.3
if $\bar{M} = \varnothing$ **then**
    let E^* be a new membership function with center e_t and base ε
    return $h_{new} : (\Upsilon_t \; is \; E^*) \Rightarrow \Upsilon_{t+1} \; is \; \bar{E}$
else
    let $\bar{h} \in \bar{M}$
    if $C(\bar{h}) = 1$ **then**
        return *null*
    else
        let E^* be a new membership function with center $e_{t-|\bar{h}|}$ and base ε
        return $h_{new} : \left(\Upsilon_{t-|\bar{h}|} \; is \; E^*, \Upsilon_{t-|\bar{h}|+1} \; is \; E^{\bar{h}}_{|\bar{h}|-1}, \ldots, \Upsilon_t \; is \; E^{\bar{h}}_0 \right) \Rightarrow \Upsilon_{t+1} \; is \; \bar{E}$
    end if
end if

2.3 Parameters and task specific configuration

A clear description of parameters is important for any learning algorithm. Parameters always introduce the risk that the learning algorithm is tuned towards the evaluated task, producing better results than it would in the general case. We have therefore strived towards limiting the number of parameters of PSL. The original design of PSL was completely parameter free, with the exception that continuous data was discretized using some discretization method. The version of PSL proposed here can be seen as a generalization of the original algorithm (Billing et al., 2011) where the width ε of the membership function E determines the discretization resolution. In addition, a second parameter is introduced, referred to as the *precision constant* $\hat{\alpha}$. $\hat{\alpha}$ is in fuzzy logic terminology an *α-cut*, i.e., thresholds over the fuzzy membership function in the interval $[0, 1]$ (Klir & Yuan, 1995).

ε controls how generously FPSL matches hypotheses. A high ε makes the algorithm crisp but typically increases the precision of predictions when a match is found. Contrary, a low ε reduces the risk that FPSL reaches unobserved states at the cost of a decreased prediction performance. The high value of ε can be compared to a fine resolution data discretization for the previous version of PSL.

$\hat{\alpha}$ is only used during learning, controlling how exact a specific $\bar{E}$ has to be before a new hypothesis with a different $\bar{E}$ is created. A large $\hat{\alpha}$ reduces prediction error but typically results in more hypotheses being created during learning.

Both ε and $\hat{\alpha}$ controls the tolerance to random variations in the data and can be decided based on how exact we desire that FPSL should model the data. Small ε in combination with large $\hat{\alpha}$ will result in a model that closely fits the training data, typically producing small prediction errors but also a low coverage.

3. Evaluation

Two tests were performed to evaluate the performance of FPSL and compare it to the previous version. A simulated Robosoft Kompai robot (Robosoft, 2011) was used in the Microsoft RDS simulation environment (Microsoft, 2011). The 270 degree laser scanner of the Kompai was used as sensor data and the robot was controlled by setting linear and angular speeds.

Demonstrations were performed via tele-operation using a joypad, while sensor and motor data were recorded with a temporal resolution of 20 Hz. The dimensionality of the laser scanner was reduced to 20 dimensions using an average filter. Angular and linear speeds were however fed directly into PSL.

The first test (Section 3.1) was designed to compare FPSL and DPSL as prediction algorithms, using sensor data from the simulated robot. The second test (Section 3.2) demonstrates the use of FPSL as a method for LFD.

3.1 Sensor prediction

The two versions of PSL were compared using a series of tests of prediction performance. Even though DPSL and FPSL are similar in many ways, a comparison is not trivial since DPSL works on discrete data whereas FPSL uses continuous data. Prediction performance of DPSL will hence depend on how the data is discretized while the performance of FPSL depends on the parameters ε and $\hat{\alpha}$.

To capture the prediction performance of the two algorithms using different configurations, a series of tests were designed. 10 discretization levels were chosen, ranging from a fine resolution where DPSL could only produce predictions on a few samples in the test set, to a low resolution where DPSL rarely met unobserved states. Laser readings were discretized

Fig. 1. Simulation Environment (Microsoft, 2011) used for evaluations. Blue stars and yellow dots represent starting positions used for demonstrations and test runs, respectively. The green area marked with a *G* represents the target position. The white area under star 10 is the robot.

over 0.8 m for the finest resolution, up to 8 m for the lowest resolution. Motor data was discretized over 0.06m/s for the finest resolution up to 0.6 m/s for the lowest resolution.
Similarly, 10 ε values were chosen, corresponding to a cone base ranging from 0.8 m to 8 m for laser data, and 0.06 m/s up to 0.6 m/s for motor data. $\hat{\alpha}$ was given a constant value of 0.9, corresponding to a error tolerance of 10% of ε.

10 data files were used, each containing a demonstration where the teacher directed the robot from a position in the apartment to the TV, see Figure 1. A rotating comparison was used, where PSL was tested on one demonstration at a time and the other nine demonstrations were used as training data. Prediction performance was measured in meters on laser range data.

3.1.1 Results

The results from the evaluation are illustrated in Figure 2. While the ε value of FPSL cannot directly be compared to the discretization level used for DPSL, the two parameters have similar effect on coverage. Prediction error is only calculated on the proportion of the data for which prediction are produced, and consequently, prediction error increases with coverage.

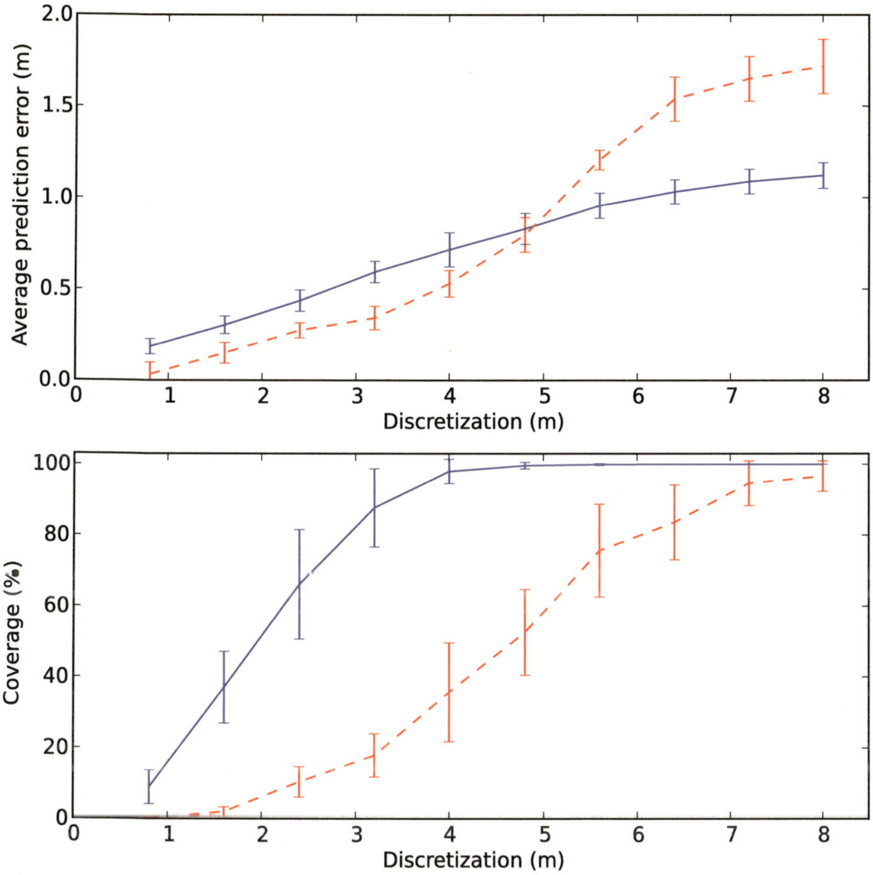

Fig. 2. Results from the prediction evaluation (Section 3.1). Upper plot shows prediction errors for FPSL (solid line) and DPSL (dashed line). Lower plot shows coverage , i.e. the proportion of samples for which the algorithm generated predictions, see Equation 12. Vertical bars represent standard deviation.

3.2 Target reaching

This evaluation can be seen as a continuation of previous tests with DPSL using a Khepera robot (Billing et al., 2011). The evaluation is here performed in a more complex environment, using a robot with much larger sensor dimensionality. Initial tests showed that DPSL has sever problems to handle the increased sensor dimensionality when used as a controller. A discretization resolution of about 2 m appeared necessary in order to produce satisfying discrimination ability. Even with this relatively low resolution, the 20 dimensional data produced a very large state space causing DPSL to frequently reach unrecognized states. DPSL could control the robot after intense training in parts of the test environment, but could

not compete with FPSL in a more realistic setting. We therefore chose to not make a direct controller comparison, but rather show the behavior of FPSL in an applied and reproducible setting.
FPSL was trained on 10 demonstrations showing how to get to the TV from various places in the apartment, see Figure 1. The performance of FPSL as a method for LFD was evaluated by testing how often the robot reached the target position in front of the TV starting out from 10 different positions than the ones used during training. FPSL controlled the robot by continuously predicting the next sensory-motor event based on the sequence of passed events. The motor part of the predicted element was sent to the robot controller. A standard reactive obstacle avoidance controller was used as fallback in cases where FPSL did not find any match with observed data. The task was considered successfully executed if the target position was reached without hitting any walls or obstacles. The experiment was repeated ten times, producing a total of 100 test runs.

3.2.1 Results

FPSL successfully reached the target position in front of the TV (the green area in Figure 1) in 79% of the test runs. In 68 runs, it stopped in front of the TV as demonstrated, but in 11 runs it failed to stop even though it reached the target position. The distribution over the 10 starting positions illustrated in Figure 3.

4. Discussion

Applied as a robot controller, PSL is a semi-reactive generative model that produces both actions and expected observations, based on recent sensory-motor events. We believe that this approach to robot learning has great potential since the behavior can be learnt progressively and previous knowledge contributes to the interpretation of new events. It is also general in the sense that very little domain specific knowledge is introduced. Memories are stored as sequences of sensory-motor events that in principle can represent any behavior. While PSL efficiently can represent behaviors with short temporal dependencies, it is subject to combinatorial explosion when the behavior requires representations over longer time spans. We argue that the gradually extending memory of PSL, from being purely reactive to containing representations over longer time when needed, provides a good bias in learning. It will however make learning of behaviors that do require long temporal dependencies slow. The fuzzy version of PSL presented in this work does not directly provide a solution to this problem, but is one step towards integrating PSL in a hierarchical structure as discussed in Section 1.2.
The seeds to FPSL came from the observation that a lot of training was required in order to cover the state space of DPSL with satisfying resolution. A better tradeoff between high precision in prediction and coverage would make PSL a more competitive alternative for real world LFD scenarios. Without sacrificing the strong attributes of the original PSL algorithm, such as the model free design, few parameters and progressively growing representations, FPSL was designed.
Expressing PSL with Fuzzy Logic is in many ways a natural extension and generalization of the discrete algorithm. By using a discrete uniform membership function E and a *max* operator for defuzzification, FPSL becomes very similar to DPSL. Even though the processing of continuous values does add significant processing requirements in comparison to DPSL, FPSL can still be efficiently implemented as a fuzzy rule controller.

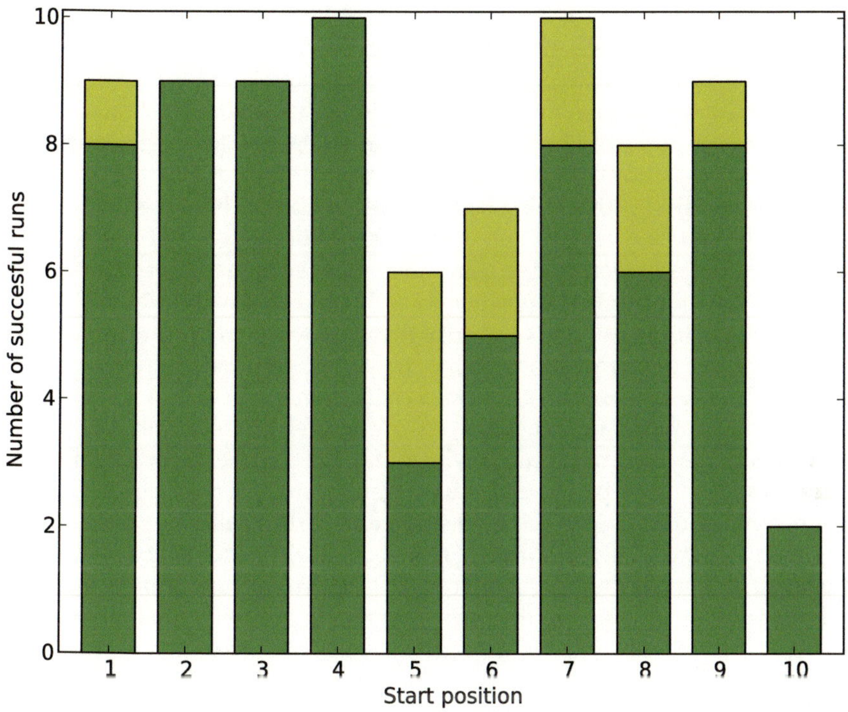

Fig. 3. Results from test runs in simulated environment (Section 3.2). Each bar corresponds to one starting position, see Figure 1. The green part of the bar represents number of successful runs where the robot reached and stopped at the target position in front of the TV. The yellow part represents runs when the robot successfully reached the target, but did not stop. The test was executed 10 times from each starting position.

The evaluation shows that FPSL produces significantly smaller prediction errors in relation to the coverage than DPSL (Section 3.1). This was expected since FPSL can be trained to produce small prediction errors by keeping a high precision constant $\hat{\alpha}$, while the coverage is still kept high by using a large ε. In contrast, when using DPSL, one must choose between a small prediction error with low coverage or a high coverage at the price of an increased prediction error. As can be seen in Figure 2, FPSL is also affected by the precision/coverage tradeoff, but not nearly as much as DPSL. Furthermore, the number of generated hypotheses will increase with $\hat{\alpha}$, which also has a positive effect on coverage for multidimensional data.

While FPSL performs much better than DPSL on large and multidimensional state spaces, it should not be seen as a general solution to the dimensionality problem. The increased number of hypotheses results in increased processing and memory requirements. Furthermore, FPSL is still not able to ignore uncorrelated dimensions in data, making it subject to the curse of dimensionality. One potential solution is to modify the size of membership functions in relation to the information content of the dimension. However, initial tests did not produce satisfying results and further experiments in this direction were postponed to future work.

We found the results from the controller evaluation very promising (Section 3.2). The application environment has been scaled up significantly in comparison to previous work (Billing et al., 2011) and we are now able to perform learning in a fairly realistic setting. When observing these results one should remember that PSL does not solve a spatial task. There is no position sensor or internal map of the environment and the robot is still able to navigate from almost any position in the environment, to a specific target location. The goal is better described as an attractor in a dynamic system, where the robot in interaction with the environment finally reaches a stable state in front of the TV (Figure 1).

An interesting observation is that it is often difficult to predict how well PSL will be able to handle a specific situation. For example, starting position 6 was not passed during any demonstration, but PSL still managed to control the robot such that it reached the target in 7 out of 10 test runs. On the other hand, position 5 and 10 produced worse results than expected. Even though these starting positions were spatially close to several positions passed during the demonstrations, the directions at which the robot reached these positions were different, producing different laser scans, and PSL could consequently not find a suitable match. In some of the cases, inappropriate matches were found and the robot turned in the wrong direction. In other cases, no match at all was found causing the robot to fall back on the reactive controller for a longer period and usually getting stuck in a corner.

The amount of training data used in this evaluation was fairly small. Only one demonstration from each starting position was performed. One reason why FPSL is able to solve the task despite the small amount of training is that all data potentially contribute to every action selection, independently of where in the demonstration it originally took place. Techniques that represent the whole behavior as a sequence with variations, typically require more training since information from the beginning of the demonstration does not contribute to action selection in other parts of the behavior. PSL does not relies on common features within a set of demonstrations and consequently does not require that demonstrations are compared or temporally aligned, see Section 1. In its current design, PSL is of course unable to perform a program-level imitation since it always rely on sensory-motor events, but it does not suffer from a large diversity in the demonstration set as long as the recent sensory-motor events bear necessary information to select a suitable action.

4.1 Conclusions and future work

In this chapter, we show that PSL can be used as a method for LFD, in a fairly realistic setting. The move from a discrete state space used in previous work to the continuous state space appears to have a positive effect on generalization ability and prediction performance, especially on multi-dimensional data. The next step is to conduct experiments with the physical Kompai robot (Robosoft, 2010) in an attempt to verify the results in the real world.

The fuzzy version of PSL proposed here, and specifically the introduction of context sets C, should be seen as one step towards integrating PSL in a hierarchical architecture. The higher level controller may be another instance of PSL working on a lower temporal resolution, or a completely different control system interacting with PSL by changing the responsibility $\lambda_t(C)$ for each context (Equation 5). For an actual interaction to take place, PSL also has to feed information upwards, to higher level controllers. In previous work on behavior recognition (Billing et al., 2010), we have shown that PSL can be used to compute a bottom-up signal providing information about how well each context corresponds to present circumstances. While this has not been the focus of this chapter, we intend to evaluate these aspects of PSL in future work.

5. References

Alissandrakis, A., Nehaniv, C. L. & Dautenhahn, K. (2002). Imitation With ALICE: Learning to Imitate Corresponding Actions Across Dissimilar Embodiments, *IEEE Transactions on Systems, Man and Cybernetics, Part A: Systems and Humans* 32: 482–496.

Alissandrakis, A., Nehaniv, C. L., Dautenhahn, K. & Saunders, J. (2005). An Approach for Programming Robots by Demonstration: Generalization Across Different Initial Configurations of Manipulated Objects, *Proceedings of 2005 International Symposium on Computational Intelligence in Robotics and Automation*, Ieee, Espoo, Finland, pp. 61–66.

Argall, B. D., Chernova, S., Veloso, M. & Browning, B. (2009). A survey of robot learning from demonstration, *Robotics and Autonomous Systems* 57(5): 469–483.

Arkin, R. C. (1998). *Behaviour-Based Robotics*, MIT Press.

Billard, A., Epars, Y., Cheng, G. & Schaal, S. (2003). Discovering imitation strategies through categorization of multi-dimensional data, *Proceedings of the 2003 IEEE/RSJ Intl. Conference on Intelligent Robots and Systems*, Vol. 3, Las Vegas, Nevada, pp. 2398–2403 vol.3.

Billing, E. A. (2009). *Cognition Reversed - Robot Learning from Demonstration*, Lic. thesis, UmeÃ¥ University, Department of Computing Science, UmeÃ¥, Sweden.

Billing, E. A. (2011). www.cognitionreversed.com.

Billing, E. A. & Hellström, T. (2010). A Formalism for Learning from Demonstration, *Paladyn: Journal of Behavioral Robotics* 1(1): 1–13.

Billing, E. A., Hellström, T. & Janlert, L. E. (2010). Behavior Recognition for Learning from Demonstration, *Proceedings of IEEE International Conference on Robotics and Automation*, Anchorage, Alaska.

Billing, E. A., Hellström, T. & Janlert, L. E. (2011). Predictive learning from demonstration, *in* J. Filipe, A. Fred & B. Sharp (eds), *Agents and Artificial Intelligence*, Springer Verlag, Berlin, pp. 186–200.

Brass, M., Bekkering, H., Wohlschläger, A. & Prinz, W. (2000). Compatibility between observed and executed finger movements: comparing symbolic, spatial, and imitative cues., *Brain and cognition* 44(2): 124–43.

Brooks, R. A. (1986). A Robust Layered Control System For A Mobile Robot, *IEEE Journal of Robotics and Automation* 2(1): 14–23.

Brooks, R. A. (1991). New Approaches to Robotics, *Science* 253(13): 1227–1232.

Byrne, R. W. & Russon, A. E. (1998). Learning by Imitation: a Hierarchical Approach, *The Journal of Behavioral and Brain Sciences* 16(3).

Calinon, S. (2009). *Robot Programming by Demonstration - A Probabilistic Approach*, EFPL Press.

Calinon, S., Guenter, F. & Billard, A. (2007). On Learning, Representing and Generalizing a Task in a Humanoid Robot, *IEEE Transactions on Systems, Man and Cybernetics, Part B. Special issue on robot learning by observation, demonstration and imitation* 37(2): 286–298.

de Rengervé, A., D'halluin, F., Andry, P., Gaussier, P. & Billard, A. (2010). A study of two complementary encoding strategies based on learning by demonstration for autonomous navigation task, *Proceedings of the Tenth International Conference on Epigenetic Robotics*, Lund, Sweden.

Demiris, J. & Hayes, G. (1997). Do robots ape?, *Proceedings of the AAAI Fall Symposium on Socially Intelligent Agents*, pp. 28–31.

Demiris, J. & Hayes, G. R. (2002). *Imitation as a dual-route process featuring predictive and learning components: a biologically plausible computational model*, MIT Press, pp. 327–361.

Demiris, Y. (1999). *Movement Imitation Mechanisms in Robots and Humans*, PhD thesis, University of Edinburgh.

Demiris, Y. & Johnson, M. (2003). Distributed, predictive perception of actions: a biologically inspired robotics architecture for imitation and learning, *Connection Science* 15(4): 231–243.

Fullér, R. (1995). *Neural Fuzzy Systems*, Abo Akademi University.

Gallese, V., Fadiga, L., Fogassi, L. & Rizzolatti, G. (1996). Action recognition in the premotor cortex, *Brain* 119(2): 593–609.

Haruno, M., Wolpert, D. M. & Kawato, M. M. (2001). MOSAIC Model for Sensorimotor Learning and Control, *Neural Comput.* 13(10): 2201–2220.

Klir, G. J. & Yuan, B. (1995). *Fuzzy Sets and Fuzzy Logic: Theory and Applications*, Prentice Hall.

Matarić, M. J. (1997). Behavior-Based Control: Examples from Navigation, Learning, and Group Behavior, *Journal of Experimental and Theoretical Artificial Intelligence* 9(2-3): 323–336.

Matarić, M. J. (2002). *Sensory-motor primitives as a basis for imitation: linking perception to action and biology to robotics*, MIT Press, pp. 391–422.

Matarić, M. J. & Marjanovic, M. J. (1993). Synthesizing Complex Behaviors by Composing Simple Primitives, *Proceedings of the European Conference on Artificial Life*, Vol. 2, Brussels, Belgium, pp. 698–707.

Microsoft (2011). Microsoft Robotic Developer Studio. URL: *http://www.microsoft.com/robotics/*

Myers, B. C. S. & Rabiner, L. R. (1981). A Comparative Study of Several Dynamic Time-Warping, *The Bell System Technical Journal* 60(7): 1389–1409.

Nehaniv, C. L. & Dautenhahn, K. (2000). *Of hummingbirds and helicopters: An algebraic framework for interdisciplinary studies of imitation and its applications*, Vol. 24, World Scientific Press, pp. 136–161.

Nehaniv, C. L. & Dautenhahn, K. (2001). Like Me? - Measures of Correspondence and Imitation, *Cybernetics and Systems* 32: 11–51.

Nicolescu, M. (2003). *A Framework for Learning from Demonstration, Generalization and Practice in Human-Robot Domains*, PhD thesis, University of Southern California.

Rizzolatti, G., Camarda, R., Fogassi, L., Gentilucci, M., Luppino, G. & Matelli, M. (1988). Functional organization of inferior area 6 in the macaque monkey. II. Area F5 and the control of distal movements., *Experimental brain research. Experimentelle Hirnforschung. Expérimentation cérébrale* 71(3): 491–507.

Rizzolatti, G. & Craighero, L. (2004). The Mirror-Neuron System, *Annual Review of Neuroscience* 27: 169–192.

Robosoft (2010). www.robosoft.com.

Robosoft (2011). Kompai Robot.

Rohrer, B. (2007). S-Learning: A Biomimetic Algorithm for Learning, Memory, and Control in Robots, Kohala Coast, Hawaii, pp. 148 – 151.

Rohrer, B. & Hulet, S. (2006). BECCA - A Brain Emulating Cognition and Control Architecture, *Technical report*, Cybernetic Systems Integration Department, Univeristy of Sandria National Laboratories, Alberquerque, NM, USA.

Tani, J., Ito, M. & Sugita, Y. (2004). Self-Organization of Distributedly Represented Multiple Behavior Schemata in a Mirror System : Reviews of Robot Experiments Using RNNPB, *Neural Networks* 17: 1273–1289.

Wolpert, D. M. & Kawato, M. (1998). Multiple paired forward and inverse models for motor control.

13

Biarticular Actuation of Robotic Systems

Jan Babič
"Jožef Stefan" Institute
Slovenia

1. Introduction

In biology, biarticular muscles are muscles that are passing two joints. One of the biarticular muscles that is passing the knee and the ankle joints is the Gastrocnemius muscle (see Fig. 1). Functions of the biarticular muscles during human movement have been studied extensively by many researchers but these functions still remain unclear. One of such functions is the transportation of mechanical energy from proximal to distal points (Ingen Schenau, 1989). It is believed that this transportation causes an effective transformation of rotational motion of body segments into translation of the body centre of gravity. The mechanism by which this is achieved is the timely activation of biarticular muscles before the end of the explosive phase of the movement. The activation of the biarticular muscle prior to the end of the explosive phase of the motion enables the transportation of the power generated by the proximal joint extensors from the proximal joint to the distal joint. This transfer of mechanical energy can be explained using the following example of the gastrocnemius muscle activation. During the push-off phase of the human jump, the knee joint is rapidly extended as a result of the positive work done by the knee extensor muscles. If the biarticular gastrocnemius muscle contracts isometrically (its length does not change), the additional mechanical work is done at the ankle joint because of the gastrocnemius muscle, which contributes no mechanical work by itself. A part of the energy generated by the knee extensors appears as mechanical work at the ankle joint and the height of the jump is significantly increased. This is because, as the jump proceeds and the knee straightens, the angular position changes of the knee have progressively less effect on vertical velocity of the jumper centre of gravity. By gastrocnemius muscle activation, a rapid extension of the foot is produced. This extension has a greater effect on the vertical velocity than the extension of the almost straightened knee. The energy is more effectively translated into vertical velocity and a greater height of the jump is achieved. However, the timing of the gastrocnemius muscle activation is critical to obtain a maximum effect. This was demonstrated by an articulated physical model of the vertical jump by Bobbert et al (1986).
Besides biarticularity, the gastrocnemius muscle has one more interesting feature. It is connected to the foot by an elastic tendon (see Fig. 1). The elasticity in the muscle fibres and tendons plays an important role in enhancing the effectiveness and the efficiency of human performance. An enhanced performance of human motion has been most effectively demonstrated for jumping and running (Cavagna, 1970; Bobbert et al., 1996; Shorten, 1985; Hobara, 2008). An important feature of elastic tissues is the ability to store elastic energy when stretched and to recoil this energy afterwards as a mechanical work (Asmussen and

Bonde-Petersen, 1974). Beside this feature, oscillatory movements performed at the natural frequency of muscle-tendon complex could maximize the performance. A countermovement vertical jump can be treated as one period of oscillatory movement and from this point of view the natural frequency, as well as parameters that define the natural frequency, can be determined.

Lower extremities of today's humanoid robots are mostly serial mechanisms with simple rotational joints that are driven directly or indirectly by electrical servo drives. Such design of humanoid robot mechanism allows only rotational motion in joints to occur. This means that translations of the robot's centre of gravity are solely a result of the transformation of rotations in joints into translations of the robot centre of gravity. Especially in ballistic movements such as fast running or jumping where the robot centre of gravity is to be accelerated from low or zero velocity to a velocity as high as possible, this transformation is handicapped. The transfer of the angular motion of the lower extremity segments to the desired translational motion of the robot centre of gravity is less effective the more the joints are extended. When the joint is fully extended, the effect of this joint on the translational motion of the robot centre of gravity in a certain direction equals zero. Besides, the motion of the segments should decelerate to zero prior to the full extension to prevent a possible damaging hyperextension. Where relatively large segments which may contain considerable amounts of rotational energy are involved, high power is necessary to decelerate the angular motion.

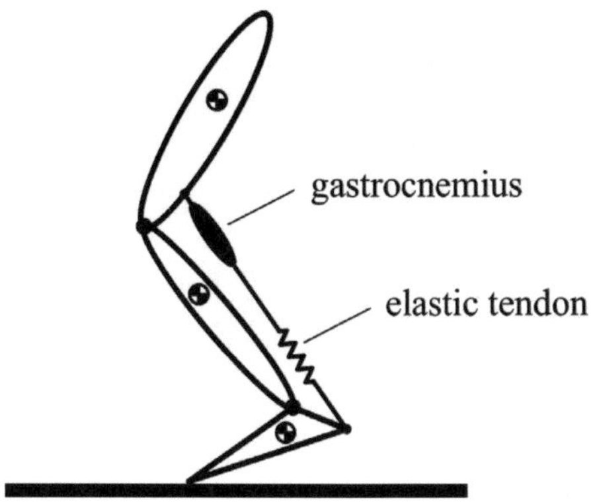

Fig. 1. Biarticular muscle gastrocnemius passing the knee and the ankle joints. It acts as a knee flexor and ankle extensor. Gastrocnemius muscle is connected to the foot by an elastic tendon

The purpose of this chapter is, first, to make an extensive review of the research of the role of biarticular muscles in motion of the humans and to present this biomechanical body of knowledge from an engineering perspective, and second, to describe the research results where biarticular actuation was used to achieve energy efficient and high performance motion of a humanoid robotic leg. The chapter is based mostly on the previous journal publications of the author and on the yet unpublished research results.

2. Role of biarticular muscles in human jump

Vertical jumping is a complex task requiring quick and harmonized coordination of jumper's body segments, first for the push-off, then for the flight and lastly for the landing.

The prime criterion for vertical jump efficiency is the height of the jump that depends on the speed of the jumper's centre of gravity (COG) in the moment when the feet detach from the ground. Besides maintaining the balance, the task of the muscles during the push-off phase of the jump is to accelerate the body's COG up in the vertical direction to the extended body position. During the push-off phase of the jump, the jumper's centre of gravity must be above the supporting polygon that is formed by the feet (Babič et al., 2001). In contrast to the humans, today's humanoid robots are mostly unable to perform fast movements such as the vertical jump. They can mostly perform only slow and statically stable movements that do not imitate the human motion. Besides, these slow and statically stable movements are energy inefficient. With the understanding of the anatomy and the biomechanics of the human body, one can find out that, beside the shape, majority of today's humanoid robots and human bodies do not have a lot of common properties. To achieve a better imitation of the human motion and ability to perform fast movements such as the vertical jump or running, other properties and particularities, beside the shape of the body, should be considered in the design of the humanoid robot.

In this section we will first analyse the human vertical jump and show that for each and every subject there exists an optimal triceps surae muscle-tendon complex stiffness that ensures the maximal possible height of the vertical jump. We define the influence of the m. gastrocnemius activation timing and the m. gastrocnemius and Achilles tendon stiffness on the height of the vertical jump and establish the methodology for analysis and evaluation of the vertical jump. We monitored kinematics, dynamics and m. gastrocnemius electrical activity during the maximum height countermovement jump of human subjects and measured viscoelastic properties of the m. gastrocnemius and Achilles tendon using the free-vibration technique. Based on the findings of the biomechanical study of the human vertical jump we performed a simulation study of the humanoid robot vertical jump. As a result we propose a new human inspired structure of the lower extremity mechanism by which a humanoid robot would be able to efficiently perform fast movements such as running and jumping.

2.1 Biorobotic model of vertical jump

Biorobotic model of the vertical jump consists of the dynamic model of the musculoskeletal system and of the mathematical model used for the motion control of the model. The results of the modelling are differential equations and a diagram for simulation and optimization of the vertical jump.

2.1.1 Dynamic model of musculoskeletal system

Vertical jump is an example of a movement that can be satisfactorily observed and analysed in just a sagittal plane. Therefore we built a model of the musculoskeletal system in a two dimensional space of the sagittal plane. Because both lower extremities perform the same movement during the vertical jump, we joined both extremities in one extremity with three rigid body segments. Trunk, head and upper extremities were modelled as one rigid body with a common COG, mass and moment of inertia. The model of the musculoskeletal system is therefore a planar model composed of four segments that represent the foot, shank, thigh and trunk together with the head and both upper extremities. Segments of the model are joined together by frictionless rotational hinges whose axes are perpendicular to the sagittal plane. The contact between the foot and the ground is modelled as a rotational joint between the tip of the foot and the ground. A model, whose foot is connected to the ground by a rotational joint, is applicable only for the push-off and landing phases of the vertical jump and is not applicable for the flight. As the motion of the COG during the flight is simply described and depends only on the speed vector of the COG just prior to the take-off, this simplification does not present any limitations.

Fig. 2 shows the planar model of the musculoskeletal system, composed of four rigid bodies that represent the foot, shank, thigh and trunk together with the head and both upper extremities. The origin of the base coordinate system is in the centre of the virtual joint that connects the foot with the ground.

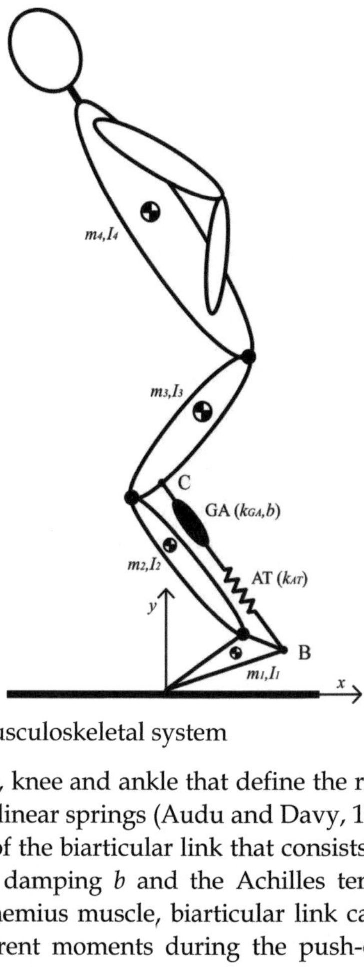

Fig. 2. Planar model of the musculoskeletal system

Passive constraints in the hip, knee and ankle that define the range of motion of these joints were modelled as simple nonlinear springs (Audu and Davy, 1985; Davy and Audu, 1987).

Fig. 2 also shows the model of the biarticular link that consists of the gastrocnemius muscle (GA) with stiffness k_{GA} and damping b and the Achilles tendon (AT) with stiffness k_{AT}. Contrary to the real gastrocnemius muscle, biarticular link cannot contract. It can only be enabled or disabled at different moments during the push-off phase of the jump. High pennation angle of the gastrocnemius muscle fibres suggest that the predominant role of the m. gastrocnemius is not in its contraction but in its ability to bear high forces and to enable the energy transportation from the knee to the ankle (Bogert et al., 1989; Legreneur et al., 1997). Therefore our simulated biarticular link that cannot contract and can only be enabled or disabled is considered as an appropriate model for this study.

Insertions of the biarticular link on the foot (B) and thigh (C) have been determined from the muscle data provided by Brand et al. (1982) and Delp (1990).

Vector of the force in the biarticular link *f* is

$$f = k \cdot (\boldsymbol{BC} - \boldsymbol{BC}_0) - b \cdot \dot{\boldsymbol{BC}}, \quad (1)$$

where k represents the stiffness of the m. gastrocnemius and Achilles tendon connected in series, $\boldsymbol{BC}$ is the vector between the insertions of the biarticular link on the foot and thigh. $\boldsymbol{BC}_0$ is the vector $\boldsymbol{BC}$ in the moment of the gastrocnemius muscle activation. Force in the biarticular link *f* causes a torque in the ankle joint

$$Q_{bl2} = -\left\| \boldsymbol{r}_B \times f \right\|, \quad (2)$$

where $\boldsymbol{r}_B$ is the moment arm vector from the centre of the ankle joint to the insertion of the biarticular link on the foot and a torque in the knee joint

$$Q_{bl3} = \left\| \boldsymbol{r}_C \times f \right\|, \quad (3)$$

where $\boldsymbol{r}_C$ is the moment arm vector from the centre of the knee joint to the insertion of the biarticular link on the thigh.
Motion of the musculoskeletal system is written with a system of dynamic equations of motion

$$\mathbf{H}(q)\ddot{q} + h(q,\dot{q}) + G(q) = Q_{mov} + Q_{pas} + Q_{bl}, \quad (4)$$

where $\boldsymbol{Q}_{pas}$ is the vector of joint torques caused by the passive constraints in the joints and $\boldsymbol{Q}_{bl}$ is the vector of joint torques caused by the biarticular link. $\boldsymbol{Q}_{mov}$ is the vector of joint torques caused by muscles and represents the input to the direct dynamic model of the musculoskeletal system. The output from the direct dynamic model is the vector of joint displacements $\boldsymbol{q}$. We determined parameters of (4) $\mathbf{H}(q), h(q,\dot{q}), G(q)$ using the equations published by Asada and Slotine (1986). Simulation diagram of the direct dynamic model of the musculoskeletal system is shown in the shaded rectangle of the Fig. 3.

2.1.2 Motion control

Motion controller of the musculoskeletal system was designed to meet the following four requirements:

1. Perpendicular projection of the body's COG on the ground coincides with the virtual joint that connects the foot with the ground during the entire push-off phase of the vertical jump. Therefore balance of the jumper and verticality of the jump is assured. Equation that describes this requirement is

$$x_T^d(t) = 0, \quad (5)$$

   where $x_T^d(t)$ is the distance between the desired perpendicular projection of the body's COG on the ground from the origin of the base coordinate system in time t.
2. Motion controller assures the desired vertical movement of the body's COG relative to the ankle $y_{TA}^d(t)$. By controlling the movement of the body's COG relative to the ankle, we excluded the influence of the biarticular link on the motion $y_{TA}^d(t)$. Therefore parameters of the biarticular link can be varied and optimized for a certain desired motion of the body's COG relative to the ankle.

3. Motion controller assures a constant angle of the foot relative to the ground q_1 before the biarticular link activation occurs. Thus the number of degrees of freedom of the model remains constant during the push-off phase of the vertical jump
4. In the moment, when the biarticular link activates, motion controller sets the torque in the ankle joint Q_2 to zero and thus enable a free motion of the foot relative to the ground. By setting the torque in the ankle joint to zero, the motion in the ankle joint is only a function of the motion in the knee joint that is transmitted to the ankle by the biarticular link.

Motion controller that considers the requirement (5) and enables the desired vertical motion of the COG relative to the ankle $y_{TA}^d(t)$ in the base coordinate system is

$$\ddot{x}_T^c = \begin{bmatrix} 0 \\ \ddot{y}_{TA}^d + \ddot{y}_A \end{bmatrix} + k_p\left(\begin{bmatrix} 0 \\ y_{TA}^d + y_A \end{bmatrix} - x_T\right) + k_d\left(\begin{bmatrix} 0 \\ \dot{y}_{TA}^d + \dot{y}_A \end{bmatrix} - \dot{x}_T\right), \tag{6}$$

where k_p and k_d are coefficients of the PD controller, $\ddot{x}_T^c$ is the vector of the control acceleration of the COG in the base coordinate system, y_A is the current height of the ankle joint relative to the ground, $\dot{x}_T, x_T$ are the vectors of the current speed and position of the COG in the base coordinate system.

The relation between the vector of the control speed of the COG in the base coordinate system $\dot{x}_T^c$ and the vector of the control angular velocities in the joints $\dot{q}_c$ is

$$\dot{x}_T^c = \mathbf{J}_T \dot{q}_c, \tag{7}$$

where $\mathbf{J}_T$ is the Jacobian matrix of the COG speed in the base coordinate system. Equation (7) represents an under determined system of equations. From the requirement that the motion controller assures a constant angle of the foot relative to the ground q_1 before the biarticular link activation occurs, follows the condition

$$\dot{q}_{c1} = 0. \tag{8}$$

An additional condition that abolishes the under determination of (7) is the relationship of the knee and hip joint angles

$$\dot{q}_{c4} = n \cdot \dot{q}_{c3}, \tag{9}$$

where n is the coefficient that describes the relationship.

By substitution of (8) and (9) into (7) we get a relation between the vector of the control speed of the COG in the base coordinate system $\dot{x}_T^c$ and the vector of the control angular velocities in the ankle and knee joints $\dot{q}'_c$

$$\dot{x}_T^c = \mathbf{J}'_T \dot{q}'_c, \tag{10}$$

where $\mathbf{J}'_T$ is a new Jacobian matrix of the centre of gravity speed in the base coordinate system. Differentiation of (10) with respect to time yields

$$\ddot{q}'_c = \begin{bmatrix} \ddot{q}_{c2} \\ \ddot{q}_{c3} \end{bmatrix} = \mathbf{J}_T'^{-1}(\ddot{x}_T^c - \dot{\mathbf{J}}'_T \dot{q}'), \tag{11}$$

where

$$\dot{q}' = \begin{bmatrix} \dot{q}_2 \\ \dot{q}_3 \end{bmatrix}. \tag{12}$$

On the basis of conditions (8) and (9) and relation (11) we define control angular accelerations in all four joints

$$\ddot{q}_c = \begin{bmatrix} 0 \\ \ddot{q}_{c2} \\ \ddot{q}_{c3} \\ n\ddot{q}_{c3} \end{bmatrix}. \tag{13}$$

By substitution of (13) into a system of dynamic equations of motion (4) we get control torques in joints Q_{mov} that we need to control the model of the musculoskeletal system

$$Q_{mov} = \mathbf{H}(q)\ddot{q}_c + h(q,\dot{q}) + G(q). \tag{14}$$

Direct dynamic model of the musculoskeletal system together with the motion controller compose the biorobotic model of the vertical jump. Simulation diagram for the simulation of the vertical jump is shown in Fig. 3. Inputs into the simulation diagram are the desired trajectory of COG relative to the ankle y_{TA}^d and a signal for biarticular link activation a. Output from the simulation diagram is the vector of body's COG position x_T.

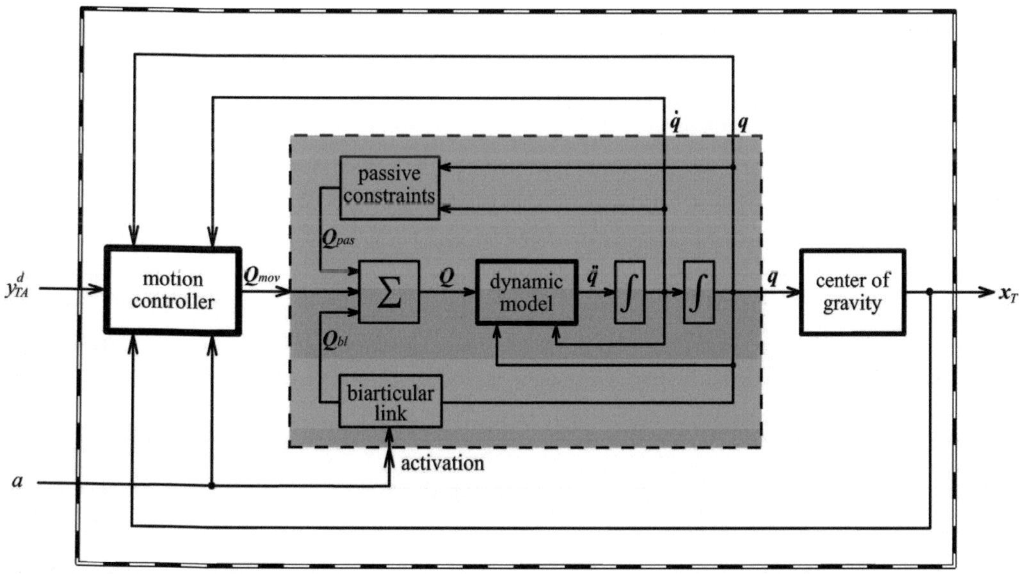

Fig. 3. Simulation diagram for the simulation of the vertical jump. Inputs into the simulation diagram are the desired trajectory of the COG relative to the ankle and a signal for biarticular link activation. Output from the simulation diagram is the vector of jumper's COG position

2.2 Biomechanical analysis of human vertical jump

2.2.1 Subjects and experimental protocol

Ten trained male subjects (age 26 ± 4 years, height 180.3 ± 6.56 cm, body mass 77.1 ± 7.24 kg) participated in the study. Informed consent was obtained from all of them. The protocol of the study was approved by the National Ethics Committee of the Republic of Slovenia. The experiments comply with the current laws of the Republic of Slovenia.
After a warm-up procedure and three practice jumps, subjects subsequently performed four countermovement vertical jumps. They were instructed to keep their hands on the hips and to jump as high as possible. At the beginning of each vertical jump, subjects stood on their toes and in the erected position. During each jump, position of anatomical landmarks over epicondylus lateralis and fifth metatarsophalangeal joint were monitored, ground reaction forces were measured and electromyogram of m. gastrocnemius was recorded. After the jumping, we determined viscoelastic properties of the triceps surae muscle tendon complex of all subjects. Details on methods and procedures are provided in the following sections.

2.2.2 Anthropometric measurements

Segmental anthropometric parameters, such as masses, moments of inertia about the transverse axes, lengths and locations of the centres of gravity were estimated using regression equations (Zatsiorsky and Seluyanov 1983; Leva 1996). Position of the first metatarsophalangeal joint and the Achilles tendon insertion on the calcaneus were determined by palpation for each subject. Insertion of the m. gastrocnemius on femur was determined using the muscle data collected by Brand et al. (1982) and Delp (1990).

2.2.3 Kinematics and dynamics

To determine the motion of the body's COG during the vertical jump we measured the vertical component of the ground reaction force caused by the subject during the jump. Subjects performed vertical jumps on a force platform (Kistler 9281CA) that is capable to measure ground reaction forces with the frequency of 1000 Hz. We zeroed the force platform just prior to the vertical jump when the subject was standing still in the erected position. Thus we enabled the precise determination of the subject's body mass. Body mass of the subject standing on the force platform is therefore equal to the quotient between the negative ground reaction force during the flight phase of the jump and the ground acceleration. The vertical position of the body's COG relative to the ground in time t $y_T^m(t)$ was obtained by double integrating with respect to time the vertical ground reaction force in time t $F(t)$

$$y_T^m(t) = \frac{1}{m}\int_0^t\int_0^t F(t)dtdt + y_T^m(0) \tag{15}$$

where m is the body mass of the subject and $y_T^m(0)$ is the initial height of the body's COG relative to the ground. To determine the vertical position of the body's COG relative to the ankle in time t $y_{TA}^m(t)$ we measured the motion of the ankle during the vertical jump by means of the contactless motion capture system (eMotion Smart). The vertical position of the body's COG relative to the ankle is therefore

$$y_{TA}^{m}(t) = y_{T}^{m}(t) - y_{A}^{m}(t) \quad (16)$$

where $y_{A}^{m}(t)$ is the vertical position of the ankle in time t.

2.2.4 Electromyography

The activity of the m. gastrocnemius was recorded using a pair of surface electrodes put over the medial head of the m. gastrocnemius. Analogue EMG signals were amplified and filtered with a band-pass filter with cut off frequencies at 1 Hz and 200 Hz. The signals were then digitalized with 1000 Hz sampling frequency and full-wave rectified. To reduce the variability of sampled EMG signal, the signal was then smoothed with a digital low-pass Butterworth filter. Finally the EMG signal was normalized with respect to the maximum value attained during the vertical jump.

2.2.5 Measurements of muscle-tendon viscoelastic properties

Triceps surae muscle-tendon complex viscoelastic properties of both legs were measured for each subject using the free-vibration method described by Babič and Lenarcič (2004). The measurement device and the procedure have been designed in such a manner that as few human body segments move as possible during the measurement. Thus the measurement uncertainty due to the approximation of the properties of the human body segments was minimized. The results of the measurements are the elastic stiffness k_{GA} and viscosity b of the m. gastrocnemius and the elastic stiffness k_{AT} of the Achilles tendon.

2.2.6 Treatment of data

For the purposes of analysis and optimization of the vertical jump we adjusted the biomechanical model of the musculoskeletal system with segmental anthropometric parameters, such as masses, moments of inertia about the transverse axes, lengths and locations of the centers of gravity of each subject. Parameters of the biarticular link, such as insertion of the m. gastrocnemius on femur, insertion of the Achilles tendon on calcaneus, elastic stiffness and viscosity were adjusted to match the measured parameters of each subject.

To simulate the vertical jump of the individual subject we used the measured trajectory of the body's COG as the input into the biomechanical model of the vertical jump. Biarticular link activation that is also an input into the biomechanical model of the vertical jump was determined from the EMG signal of the m. gastrocnemius. The moment of biarticular link activation was determined as the moment when the rectified, normalized and filtered EMG signal of the m. gastrocnemius increased to 95% of its maximum value. After the activation, the biarticular link remains active during the entire push-off phase of the jump.

2.2.7 Results

To determine the optimal timing of the biarticular link activation that results in the highest vertical jump, a series of the countermovement vertical jump simulations have been performed for each subject. Each simulation was performed with a different timing of the biarticular link activation.

All subjects activated their m. gastrocnemius slightly before the optimal moment, determined by means of simulations. In average, the difference between the optimal and measured knee angle when the m. gastrocnemius was activated was 6.4 ± 2.22°. Because the dynamic model of the musculoskeletal system does not include the monoarticular muscle soleus, the measured heights of the jumps were higher than the jump heights determined with the simulations for 4.3 ±1.12 % in average. The primary motive for omitting the m. soleus from the modelling is that we wanted to control the motion of the body's COG relative to the ankle so that the parameters of the biarticular link could be varied and optimized for a certain measured motion of the subject's COG relative to the ankle. If the dynamic model of the musculoskeletal system would include the m. soleus, the motion of the ankle would be fully determined by the force of the m. soleus and we would not be able to control it with regard to the desired body's COG relative to the ankle. Moreover if the dynamic model of the musculoskeletal system would include the m. soleus, force produced by the m. soleus would be another input into the biomechanical model of the vertical jump and we would have to accurately measure the force produced by the m. soleus of subjects performing the vertical jump. An additional cause for the differences between the measured heights of the jumps and the jump heights determined by means of simulations can be the simplified model of the foot that we modelled as one rigid body. The arch of the foot is linked up by elastic ligaments that can store elastic energy when deformed and later reutilize it as the mechanical work (Alexander, 1988). Ker et al. (1987) measured the deformation of the foot during running and determined the amount of the energy stored during the deformation. They showed that the elasticity of the foot significantly contribute to the efficiency of the human movement. To be able to compare the measurements and the simulation results, we corrected the simulation results by adding the contribution of the m. soleus to the height of the vertical jump. Such corrected heights of the vertical jumps at the optimal moments of m. gastrocnemius activation are insignificantly larger than the measured heights of the vertical jumps for all subjects. In average the height difference is only 1.6 ± 0.74 cm.

To determine the optimal stiffness of the Achilles tendon regarding to the height of the vertical jump, a series of the countermovement vertical jump simulations have been performed, each with a different stiffness of the Achilles tendon. Optimal timing of the biarticular link has been determined for each stiffness of the Achilles tendon as described in the previous paragraph. The measured values of the Achilles tendon stiffness for all subjects were always higher than the optimal values determined by means of simulations. By considering the elastic properties of the arch of the foot, we can assume that the optimal values of the Achilles tendon stiffness would increase and therefore the differences between the optimal and measured values would decrease.

Results of the measurements, simulations and optimizations of the human vertical jumps are presented in Fig. 4. Subjects are arranged with regard to the ratio between the optimal stiffness of the Achilles tendon determined by means of simulations and the measured stiffness of the Achilles tendon. If we want to evaluate the contribution of the viscoelastic properties to the height of the jump, ranking of the subjects with regard to the height of the jump is not appropriate because the main parameter that influences the height of the vertical jump is the power generated by muscles during the push-off phase of the jump. The elasticity of the Achilles tendon has the secondary influence on the height of the jump. Results show that for the same power generated by an individual subject during the push-off phase of the jump, the height of the vertical jump varies with the Achilles tendon stiffness. Therefore the appropriate criterion for ranking the subjects have to consider the elasticity of the Achilles tendon.

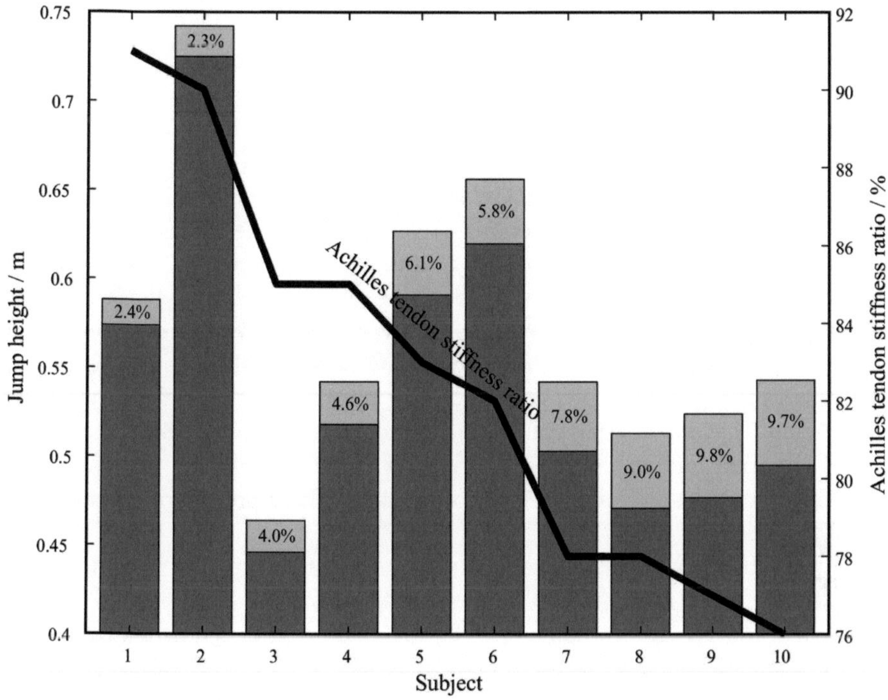

Fig. 4. Results of the measurements, simulations and optimizations of the vertical jumps. Whole bars represent the maximal heights of the vertical jumps achieved with the optimal values of the Achilles tendon stiffness and determined by means of simulations. The dark shaded parts of the bars represent the measured heights of the vertical jumps while the light shaded tops represent the differences between the maximal and measured heights of the vertical jumps. The differences are also shown as percentages relative to the measured heights. Bold line represents the ratio between the optimal stiffness of the Achilles tendon determined by means of simulations and the measured stiffness of the Achilles tendon for each subject

3. Biarticular legged robot

In the past, several research groups developed and studied jumping robots but most of these were simple mechanisms not similar to humans. Probably the best-known hopping robots were designed by Raibert and his team (Raibert, 1986). They developed different hopping robots, all with telescopic legs and with a steady-state control algorithm. Later, De Man et al. developed a trajectory generation strategy based on the angular momentum theorem which was implemented on a model with articulated legs (De Man et al., 1996). Hyon et al. developed a one-legged hopping robot with a structure based on the hind-limb model of a dog (Hyon et al., 2003). Recently, Iida et al. investigated a model of biped robot which makes use of minimum control and elastic passive joints inspired by the structures of biological systems (Iida et al., 2009).

We built an efficient dynamic model of the humanoid robot and designed and built a human inspired half sized humanoid robotic mechanism that is capable of performing vertical jump and long jump. We describe the complete design process of the real robotic system. Besides, we describe the controller of the robot and show the results of the vertical jump experiments performed by the hardware prototype of the robot.

3.1 Design and specifications

To perform explosive movements such as the vertical jump, very high torques in the robot's joints are required in addition with the low mass of the whole system. To achieve high joint torque with a lightweight electromotor available on the market today, high gear ratio is usually necessary. However, high gear ratio is not desired because we want to maintain the friction in the joints as low as possible. Besides, the back drivability of the joints, where the joints can be easily moved by an external force, cannot be achieved using gears with high ratio. A suitable method to achieve high torques and back drivability at the same time is to use a motor with low gear ratio and overload it for the short periods of time when the high torque is needed. This allows us to get a sufficient torque for a shorter time, that depends on the cooling situation and the desired overloaded torque. Explosive movements such as vertical jump usually last a very brief period of time. Therefore this overloading method seems appropriate for our task.
We used Maxon RE 40 DC servo drives with stall torque of 2.5 Nm and weight of only 480 g. The gear ratio we chose is 1:8, which ensures both low friction and back drivability of the joint. Using this motor/gear combination, the maximal joint torque is 20 Nm, which can, based on the simulations, be sufficient to perform a 20cm high vertical jump.
Each joint is activated by a servo drive mounted inside the proximal segment with regard to the joint. The largest part of the robot weight is in the trunk segment where, besides the motor that activates the hip joint, a computer, a motion controller, all power amplifiers and an inclinometer are installed. To have a heavy trunk is not desired from a control point of view. However, it can be shown that this kind of weight distribution can be beneficial for improvement of fast and explosive movements such as the vertical jump.
The biarticular actuation mechanism is attached between the thigh and heel as in the simulation model. It is realized by a stiff steel wire of diameter 1mm that is connected to the heel via a spring. The spring is interchangeable – springs of varying stiffness determined by measurements are available. At one end the steel wire is wound on a 10mm diameter grooved shaft that is directly connected to a flat servo motor (the spiral groove prevents the wire from knotting itself). This motor functions solely as a break: when the motor is not activated the biarticular link smoothly follows the motion of the robot (the force in the biarticular link is zero).

3.1.1 Construction

The real robotic system was constructed considering the CAD model built in I-DEAS. The real robotic system and its comparison with the CAD model is shown in Fig. 5. All mechanical parts are made of aluminium except the gears and axes that are made of titanium. The total weight of the robot is approximately 4.4 kg and its height is 972mm.
To better understand the biarticular actuation mechanism, consider a jumping motion starting from a squat position. Until the instant of activation, the biarticular actuation mechanism has no effect on the system, essentially acting as a passive prismatic joint. When the biarticular actuation mechanism is activated, the robot essentially becomes redundantly actuated by the biarticular force. Immediately after robot pushes off the ground, the

biarticular actuator is deactivated. This sequence must be executed over a short time in an optimal fashion to maximize jump performance.

Fig. 5. Real jumping robot

3.2 Dynamic modelling

To efficiently control the robot we determined the dynamic model of the robot. We approximated the robotic system as a planar articulated system composed of four rigid bodies connected together by three rotational joints. To define the dynamic model of the robot, all kinematic and dynamic properties of the robot segments have to be determined. We determined these parameters using I-DEAS modeling software. To increase the accuracy of the model, the weights that we obtained from the I-DEAS model were further adjusted by the measured weights of the manufactured parts.

The obtained properties of all four links including servo drives, computer, power amplifiers and wirings are shown in Table 1.

link	mass kg	length mm	COM position mm	Inertia $kgmm^2$
foot	0.399	93.5	70.7, -3.4	1057
shank	0.963	255	127.8, 10.5	7487
thigh	1.033	254	119.2, -12.0	7440
trunk	2.031	370	186.4, -12.0	20366

Table 1. Kinematic and dynamic properties of the robot links

The dynamic model of the robot was used in the simulation of the robotic system and for the control of the real system.

The dynamic model is a combination of two separate models, the model of the robot in contact with the ground and the model of the robot in the air. The contact between the tip of the foot and the ground is modeled as a rotational hinge joint between the foot tip and the ground. With the assumption that the foot tip does not slip and does not bounce, the robot has four degrees of freedom during stance. While in the air, the robot has two more degrees of freedom. The generalized coordinates that describe the state of the robot and its motion are denoted by q_i. The dynamic model of the robot is

$$H(\mathbf{q})\ddot{\mathbf{q}} + C(\mathbf{q},\dot{\mathbf{q}}) + g(\mathbf{q}) = \tau \,, \tag{17}$$

where **H**, **C** and **g** denote the inertia matrix, the vector of Coriolis and centrifugal forces and the vector of gravity forces, respectively. $\boldsymbol{\tau}$ is the vector of the joint torques and $\mathbf{q}$ is the vector of the joint positions. The dynamic equation does not consider the friction in the gears. For this purpose, an additional controller was implemented in the control strategy.

3.2.1 Description of model using SD/FAST

SD/FAST is a physically-based simulation tool for simulating the mechanical systems. Based on the short description of an articulated mechanical system of rigid bodies it derives the full nonlinear equations of motion for the given system. These equations can then be directly used in a C programming language and further compiled or linked with an arbitrary simulation or animation environment. The symbolic derivation of the equations provides one of the fastest possible simulations of the mechanical system.

		On the ground	In the air
segment	connected	joint type	joint type
foot	ground	rot	2 trans + 1 rot
shank	foot	rot	rot
thigh	shank	rot	rot
trunk	thigh	rot	rot
Number of DOFs		4 (q_3-q_6)	6 (q_1-q_6)

Table 2. Description of the jumping robot

When the system was completely described, SD/FAST generated the equations of motion for both models. To perform simulations and to control the robotic system, these equations were then used in the simulation environment.

3.3 Simulation and control

The equations of motion which were generated by SD/FAST have to be used together with a programming tool that provides the integration of the equations of motion. For this purpose we used Matlab/Simulink. The C code describing the equations of motion is used in S-function, that can be used directly as a Simulink block. We designed two S-functions, one for the inverse dynamics and one for the direct dynamics.

3.3.1 Inverse dynamics

Depending on the current state of the robot ($\mathbf{q},\dot{\mathbf{q}}$) the inverse dynamics block in Simulink defines the matrices **H**, **C** and **g** that form the dynamic equations of motion (17). The inverse dynamics block also defines the position of the COM and its Jacobian matrix, which is used

in the controller. The output of the inverse dynamics block depends on whether the robot is in the contact with the ground or in the air.
To get the requested output of the block, the following SD/FAST functions were used: *sdpos* for the position, *sdrel2cart* to calculate the Jacobian matrix and *sdmassmat, sdequivht, sdfrcmat* to calculate the **H**, **C** and **g** matrices..

3.3.2 Direct dynamics

The direct dynamic model performs the integration of the equations of motion (17) and returns the states of the robot $(\mathbf{q},\dot{\mathbf{q}})$ depending on the input torque τ. In our case where Matlab/Simulink performs the integration, we only have to define the derivatives of the states $(\mathbf{q},\dot{\mathbf{q}})$. To calculate the derivatives of the states and to apply the input torque, the following functions were used: *sdstate* to get states, *sdderiv* to get the derivatives of the states and *sdhinget* to set the torque.

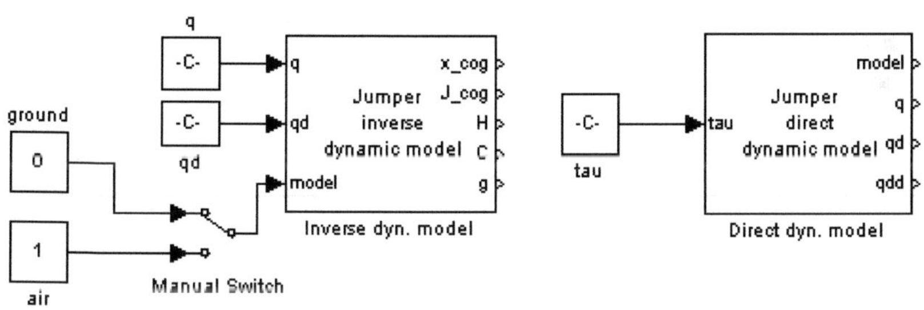

Fig. 6. Inverse and direct dynamic model blocks in simulation environment Simulink

Special attention should be paid to switching between the dynamic models. Usually all movements start with the foot touching the ground. In case of vertical jumping the switch occurs when the ground reaction forces exerted by the robot in the vertical direction change direction from downwards to upwards. In the moment of the switch all joint states have to maintains their values. Hence continuity and dynamic consistency is achieved.
On the other hand when the robot lands, a force impact occurs and the joint velocities change their values. The joint velocities after the impact have to be calculated with regard to the robot configuration and the joint velocities before the impact using following equation (Zheng & Hemami, 1985):

$$\Delta\dot{\mathbf{q}} = H_{\text{air}}^{-1}S^{\mathrm{T}}(SH_{\text{air}}^{-1}S^{\mathrm{T}})^{-1}\Delta\mathbf{v}, \tag{18}$$

where $\mathbf{H}_{\text{air}}$ represents the **H** matrix of the system in the air, $\Delta\mathbf{v}$ represents the velocity change in the moment of the impact and $\Delta\dot{\mathbf{q}}$ represents the resulting changes of the joint velocities at the impact. Matrix **S** defines the constraints of the impact and is equal to

$$S = \begin{bmatrix} 1 & 0 & 0 & 0 & 0 & 0 \\ 0 & 1 & 0 & 0 & 0 & 0 \end{bmatrix}. \tag{19}$$

To get the ground reaction force and the position of the toe, the following functions are used: *sdreac* to calculate the reaction force and *sdpos* to calculate the position.

3.3.3 Control of the system

To control the robot to perform the vertical jump, two separate controllers were used, one for the push-off phase of the jump and the other for the flight phase of the jump. Furthermore, not all degrees of freedom of the robot can be actuated. Both degrees of freedom that describe the position of the toe (q_1 and q_2) and the degree of freedom in the passive toe joint (q_3) are considered as free degrees of freedom that cannot be actuated by the robot. While the robot is in contact with the ground, position of the toe (q_1 and q_2) is fixed. In the free degrees of freedom the control torque should remain zero at all times. The system of dynamic equations that describe the robot can be divided into two parts, for actuated and inactivated joints separately,

$$\begin{bmatrix} H_{11} & H_{12} \\ H_{21} & H_{22} \end{bmatrix} \begin{bmatrix} \ddot{\mathbf{q}}_1 \\ \ddot{\mathbf{q}}_2 \end{bmatrix} + \begin{bmatrix} C_1 \\ C_2 \end{bmatrix} + \begin{bmatrix} g_1 \\ g_2 \end{bmatrix} = \begin{bmatrix} \tau_1 \\ \tau_2 \end{bmatrix}, \tag{20}$$

where τ_1 represents the degrees of freedom that cannot be actuated. This condition leads to

$$\begin{aligned} H_{11}\ddot{\mathbf{q}}_1 + H_{12}\ddot{\mathbf{q}}_2 + C_1 + g_1 &= 0, \\ H_{21}\ddot{\mathbf{q}}_1 + H_{22}\ddot{\mathbf{q}}_2 + C_2 + g_2 &= \tau_2. \end{aligned} \tag{21}$$

Accelerations in the uncontrollable joints $\ddot{\mathbf{q}}_1$ are unknown and can therefore be eliminated. Combining both equations (21) we get

$$\ddot{\mathbf{q}}_1 = -H_{11}^{-1}(H_{12}\ddot{\mathbf{q}}_2 + C_1 + g_1) \tag{22}$$

and

$$-H_{21}H_{11}^{-1}(H_{12}\ddot{\mathbf{q}}_2 + C_1 + g_1) + H_{22}\ddot{\mathbf{q}}_2 + C_2 + g_2 = \tau_2. \tag{23}$$

Finally we get

$$\begin{aligned} &\underbrace{(-H_{21}H_{11}^{-1}H_{12} + H_{22})}_{H_{new}}\ddot{\mathbf{q}}_2 + \\ &\underbrace{(-H_{21}H_{11}^{-1}C_1 + C_2)}_{C_{new}} + \\ &\underbrace{(-H_{21}H_{11}^{-1}g_1 + g_2)}_{g_{new}} = \tau_2. \end{aligned} \tag{24}$$

We have acquired a new dynamic model with three degrees of freedom. The other degrees of freedom cannot be controlled directly; however they can be controlled indirectly using other conditions such as COM and ZMP.

3.3.4 Vertical jump controller

To assure the verticality of the jump, the robot's COM has to move in the upward direction above the support polygon during the push-off phase of the jump. The second condition, which refers to the balance of the robot during the push-off phase, is the position of the zero moment point (ZMP). Both conditions are described in details in (Babič et al. 2006).

The vertical jump consists of a phase when the robot is in contact with the ground and a phase when the robot is in the air without any contact with the environment. To sufficiently control the vertical jump we used ground model of the robot in the former case and air model of the robot in the latter case. Switching between the models was performed by the controller on the basis of the foot switch.

In the push-off phase of the jump we controlled the position of the COM and ZMP in such a way that the robot performed a vertical jump. The task of the controller in the flight phase of the jump was to stop the motion in all joints and to prevent the mechanical damage by over extension of the joints. At this point we did not consider any strategies for the landing phase of the jump.

To control all joints we used the controller described by

$$\boldsymbol{H}_{new}\ddot{\mathbf{q}}_{2c} + \boldsymbol{C}_{new} + \boldsymbol{g}_{new} = \boldsymbol{\tau}_{2c}, \tag{25}$$

where $\boldsymbol{\tau}_{2c}$ is the control torque and $\ddot{\mathbf{q}}_{2c}$ is the control acceleration of all active degrees of freedom. $\ddot{\mathbf{q}}_{2c}$ is defined as

$$\ddot{\mathbf{q}}_{2c} = \ddot{\mathbf{q}}_{2d} + K_p\mathbf{e} + K_d\dot{\mathbf{e}}, \quad \mathbf{e} = \mathbf{q}_{2d} - \mathbf{q}_2. \tag{26}$$

Here, $\mathbf{q}_{2d}$ and $\mathbf{q}_2$ are the desired and the actual joint positions of the joints, respectively. The desired joint trajectory during the jump is defined with regard to the COM and ZMP as described in (Babič et al. 2006).

Fig. 7 schematically shows the complete system consisting of the controller and the direct dynamic model of the robot. The controller includes both ground and air controllers of the vertical jump. These two controllers are merged together as shown in Fig. 8.

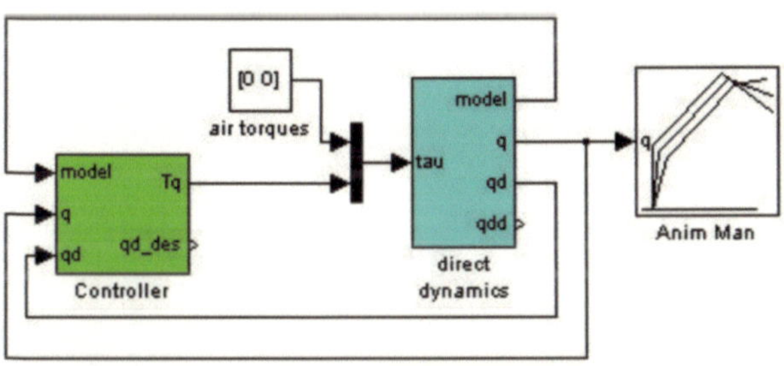

Fig. 7. Block diagram of robotic system consisting of controller and direct dynamic model of robot

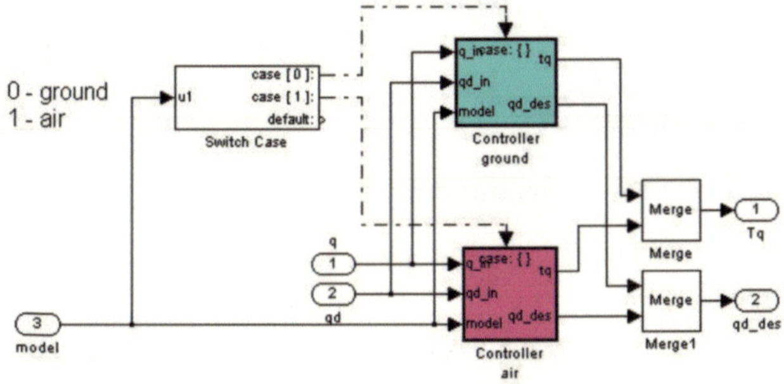

Fig. 8. Block diagram of vertical jump controller made of two separate controllers for push-off and flight phases of vertical jump

3.4 Vertical jump experiments

Based on the numerical results, we performed vertical jumping experiments with the hardware prototype and the controller described above. Joint torques obtained in the simulation study were used as the desired joint torques for the experiments. In the first case we performed squat vertical jumps for a conventional robot that does not have a biarticular link. In the second case we performed squat vertical jumps for a biarticular legged robot. In this latter case the robot was equipped with biarticular link that represents the gastrocnemius muscle in a human leg.

The robot performed jumps on a force platform that was used to measure ground reaction forces generated by the robot during the push-off and landing phases of the vertical jump. By double integration of the measured forces that are proportional to the acceleration of the robot's center of mass during the jump, the trajectory of the robot's center of mass is determined during the vertical jump experiments. In this way we obtained the height of the jump of the robot without the biarticular link (0.11m) and the height of the jump of the biarticular legged robot (0.13m). The biarticular legged robot jumped approximately 18% higher than the conventional robot.

A typical sequence of the vertical jump performed by the biarticular legged robot is shown in Fig. 9. Images in the first row show the robot between the beginning of the push-off phase of the jump and the moment when the robot reached the maximal height of the jump. In the second row of Fig. 9 the second part of the flight phase and the landing phase is shown.

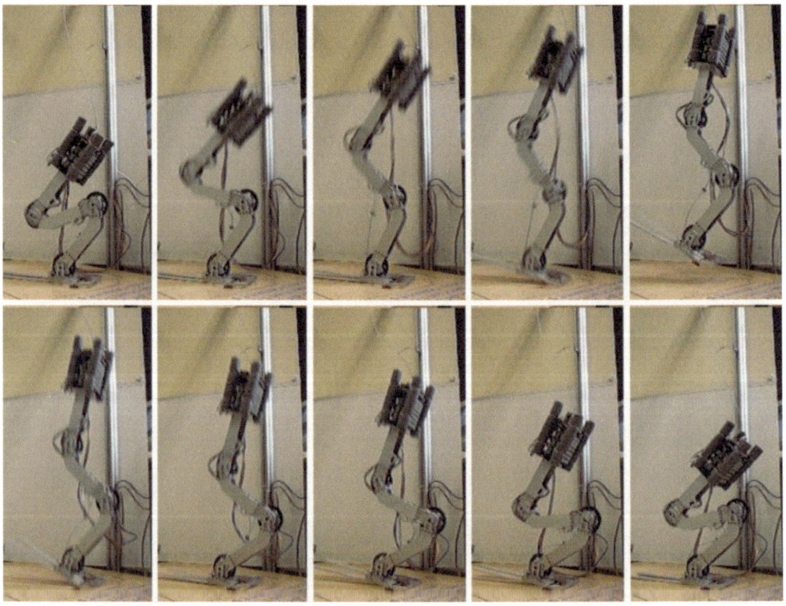

Fig. 9. A sequence of the biarticular legged robot performing a vertical jump experiment

4. Acknowledgement

This investigation was supported by the Slovenian Ministry of Education, Science and Sport. Thanks to Borut Lenart, Jožef Stefan Institute for designing the robot prototype and to Damir Omrčen, Jožef Stefan Institute for designing the vertical jump controller and for the help with the experiments.

5. References

Alexander, R.McN. (1988). *Elastic mechanisms in animal movement,* Cambridge University Press, Cambridge

Asada, H. & Slotine, J.J.E. (1986). *Robot analysis and control,* John Wiley and sons, New York

Asmussen, E. & Bonde-Petersen, F. (1974). Storage of elastic energy in skeletal muscles in man. *Acta Physiologica Scandinavica,* 91, 385-392

Audu, M.L. & Davy, D.T. (1985). The influence of muscle model complexity in musculoskeletal motion modeling. *Journal of Biomechanical Engineering,* 107, 147-157

Babič, J. & Lenarčič, J. (2004). In vivo determination of triceps surae muscle-tendon complex viscoelastic properties. *European Journal of Applied Physiology,* 92, 477-484

Babič, J., Omrčen, D. & Lenarčič, J. (2006). Balance and control of human inspired jumping robot, *Advances in Robot Kinematics,* J. Lenarčič, B. Roth (Eds.), pp. 147-156, Springer

Babič, J.; Karčnik, T. & Bajd, T. (2001). Stability analysis of four-point walking, *Gait & Posture,* 14, 56-60

Bobbert, M.F.; Gerritsen, K.G.M.; Litjens, M.C.A. & Soest, A.J. van (1996). Why is countermovement jump height greater than squat jump height? *Medicine & Science in Sports & Exercise,* 28, 1402-1412

Bobbert, M.F.; Hoed, E.; Schenau, G.J. van.; Sargeant, A.J. & Schreurs, A.W. (1986). A model to demonstrate the power transportating role of bi-articular muscles, *Journal of Physiology,* 387 24

Bogert, A.J. van den; Hartman, W.; Schamhardt, H.C. & Sauren, A.A. (1989). In vivo relationship between force, EMG and length change in the deep digital flexor muscle in the horse, In *Biomechanics XI-A,* G. de Groot; A.P. Hollander; P.A. Huijing; G.J. van Ingen Schenau (Eds.), pp. 68-74, Free University Press, Amsterdam

Brand, R.A.; Crowninshield, R.D.; Wittstock, C.E.; Pederson, D.R.; Clark, C.R. & Krieken, F.M. van (1982). A model of lower extremity muscular anatomy. *Journal of Biomechanics,* 104, 304-310

Cavagna, G.A. (1970). Elastic bounce of the body, *Journal of Applied Physiology,* 29, 279-282

Davy, D.T. & Audu, M.L. (1987). A dynamic optimization technique for predicting muscle forces in the swing phase of gait. *Journal of Biomechanics,* 20, 187-201

Delp, S.L. (1990). Surgery simulation: A computer graphics system to analyze and design musculoskeletal reconstructions of the lower limb, Doctoral Dissertation, Stanford University, Palo Alto

De Man, H., Lefeber, F., Daerden, F. & Faignet, E. (1996). Simulation of a new control algorithm for a one-legged hopping robot (using the multibody code mechanics motion), In *Proceedings International Workshop on Advanced Robotics and Intelligent Machines,* pp.1-13

Hyon, S., Emura, T. & Mita, T. (2003). Dynamics-based control of a one-legged hopping robot, *Journal of Systems and Control Engineering,* 217, 83–98

Hobara, H. (2008). Spring-like leg behavior and stiffness regulation in human movements, Coctoral Dissertation, Waseda University, Japan

Iida, F., Minekawa, Y., Rummel, J., Seyfarth, A. (2009). Toward a human-like biped robot with complient legs. *Robotics and Autonomous Systems,* 57, 139-144

Ker, R.F.; Bennett, M.B.; Bibby, S.R.; Kester, R.C. & Alexander, R.McN. (1987). The spring in the arch of the human foot. *Nature,* 325, 147-149

Legreneur, P.; Morlon, B. & Hoecke, J. van (1997). Joined effects of pennation angle and tendon compliance on fibre length in isometric contractions: A simulation study. *Archives of Physiology and Biochemistry*, 105, 450-455

Leva, P. de (1996). Adjustments to Zatsiorsky-Seluyanov's segment inertia parameters. *Journal of Biomechanics*, 29, 1223-1230

Raibert, M. (1986). *Legged Robots That Balance*, MIT Press.

Schenau, G.J. van. (1989). From rotation to translation: constraints of multi-joint movements and the unique action of bi-articular muscles, *Human Movement Science*, 8, 301-337

Shorten, M.R. (1985). Mechanical energy changes and elastic energy storage during treadmill running, In: *Biomechanics IXB*, D.A. Winter; R.W. Norman; R. Wells; K.C. Hayes; A. Patla (Eds.), pp. 65-70, Human Kinetics Publ, Champaign

Zatsiorsky, V. & Seluyanov, V. (1983). The mass and inertia characteristics of the main segments of the human body, In *Biomechanics VIII-B*, H. Matsui; K. Kobayashi (Eds.), pp. 1152-1159, Human Kinetics, Illinois

Zheng, Y.F. & Hemami, H. (1985). Mathematical modeling of a robot collision with its environment, *Journal of Robotic Systems*, 2, 289-307

14

Optimization and Synthesis of a Robot Fish Motion

Janis Viba, Semjons Cifanskis and Vladimirs Jakushevich
Riga Technical University
Latvia

1. Introduction

Inverse method algorithm for investigation of mechatronic systems in vibration technology is used for robotic systems motion control synthesis. The main difference of this method in comparison with simple analysis method is that before synthesis of a real system the optimal control task for abstract initial subsystem is solved [1 - 4]. As a result of calculations the optimal control law is found that allows to synthesize series of structural schemes for real systems based on initial subsystem. Is shown that near the optimal control excitation new structural schemes may be found in the medium of three kinds of strongly non-linear systems:

- systems with excitation as a time function;
- systems with excitation as a function of phase coordinates only;
- systems with both excitations mixed [2 – 4].

Two types of vibration devices are considered. The first one is a vibration translation machine with constant liquid or air flow excitation. The main idea is to find the optimal control law for variation of additional surface area of machine working head interacting with water or air medium. The criterion of optimization is the time required to move working head of the machine from initial position to end position. The second object of the theoretical study is a fin type propulsive device of robotic fish moving inside water. In that case the aim is to find optimal control law for variation of additional area of vibrating tail like horizontal pendulum which ensures maximal positive impulse of motion forces acting on the tail. Both problems have been solved by using the Pontryagin's maximum principle. It is shown that the optimal control action corresponds to the case of boundary values of area. One real prototype was investigated in linear water tank.

2. Translation motion system

First object with one degree of freedom x and constant water or air flow $\bar{V}_0$ excitation is investigated (Fig. 1., 2.). The system consists of a mass m with spring c and damper b. The main idea is to find the optimal control law for variation of additional area *S(t)* of vibrating mass m within limits (1.):

$$S_1 \le S(t) \le S_2, \qquad (1)$$

where S_1 - lower level of additional area of mass m; S_2 - upper level of additional area of mass m, t - time.
The criterion of optimization is the time T required to move object from initial position to end position.
Then the differential equation for large water velocity $|V0| \geq |\dot{x}|$ is (2):

$$m\,\ddot{x} = -c\,x - b\,\dot{x} - u(t) \cdot (V_0 + \dot{x})^2\,, \tag{2}$$

where $u(t) = S(t) \cdot k$, c – stiffness of a spring, b – damping coefficient, V_0 - constant velocity of water, S(t) – area variation, $u(t)$ - control action, k – constant.
It is required to determine the control action $u = u(t)$ for displacement of a system (2) from initial position $x(t_0)$ to end position $x(t_1)$ in minimal time (criterion K) K= T, if area $S(t)$ has the limit (1).

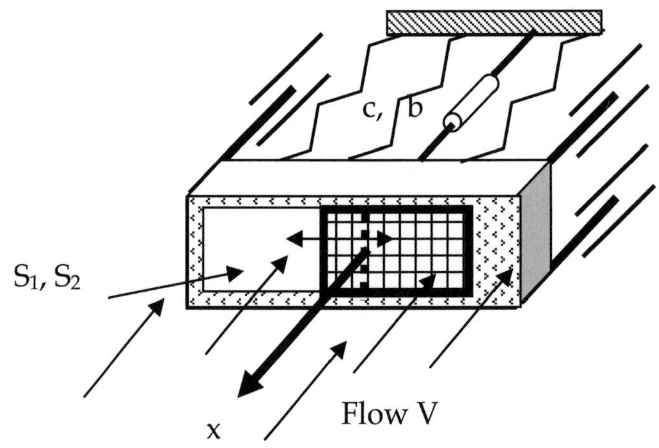

Fig. 1. Working head construction

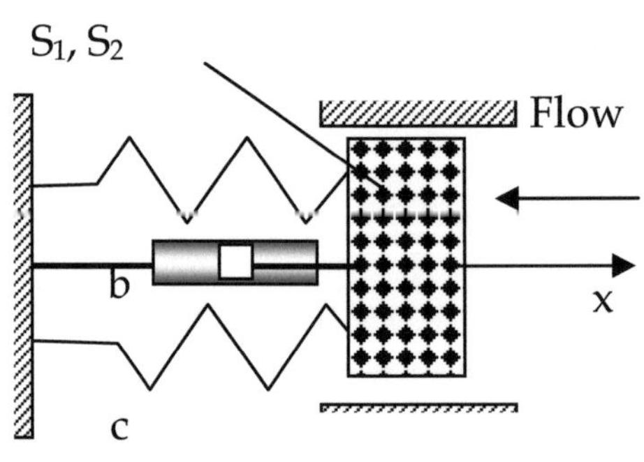

Fig. 2. Side-view of a mathematical model

2.1 Solution of optimal control problem

For solution of problem the Pontryagin's maximum principle may be used [5 - 13]. We have the high-speed problem $K = \int_{t0}^{t1} 1 \cdot dt$.

To assume $t0 = 0;\quad t1 = T,$ we have $K = T$.

From the system (2), we transform $x_1 = x;\quad \dot{x}_1 = x_2$ or

$$\dot{x}_1 = x_2;\quad \dot{x}_2 = \frac{1}{m}\cdot(-c\,x - b\,\dot{x} - u(t)\cdot(V_0 + \dot{x})^2)$$

and we have Hamiltonian (3):

$$H = \psi_0 + \psi_1 x_2 + \psi_2\left(\frac{1}{m}\cdot\left(-cx_1 - bx_2 - u(t)\cdot(V_0 + x_2)^2\right)\right), \tag{3}$$

here $H = \psi \cdot X$, where (4)

$$\psi = \begin{Bmatrix}\psi_0\\ \psi_1\\ \psi_2\end{Bmatrix};\; X = \begin{Bmatrix}0\\ \dot{x}_1\\ \dot{x}_2\end{Bmatrix}. \tag{4}$$

Scalar product of those vector functions ψ and X in any time (function H) must be maximal. To take this maximum, control action *u(t)* must be within limits $u(t) = u1;\quad u(t) = u2$, depending only on function ψ_2 sign (5):

$$\begin{aligned} &H = \max H,\\ &if\quad \psi_2\cdot(-u(t)\cdot(V_0 + x_2)^2) = \max. \end{aligned} \tag{5}$$

If $\psi_2 > 0,\quad u(t) = u1$ and if $\psi_2 < 0,\quad u(t) = u2$, where $u1 = S_1 \cdot k$ and $u2 = S_2 \cdot k$, see (1).
The examples of control action with one, two and three switch points are shown in Fig. 3. – Fig. 5.

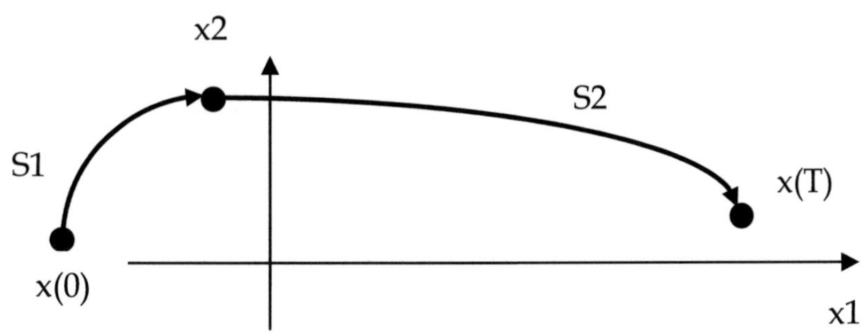

Fig. 3. Example of optimal control with one switch point

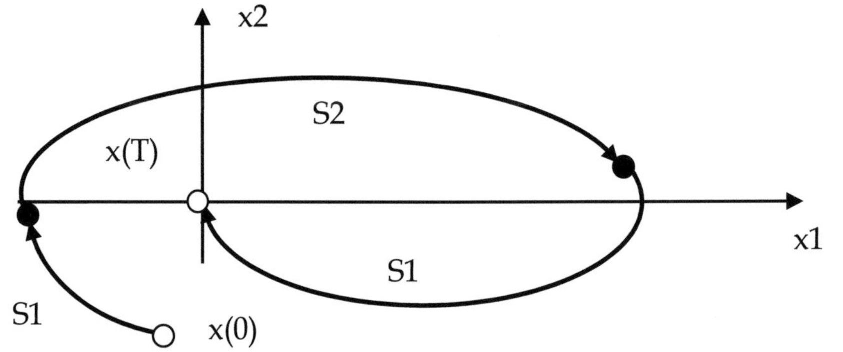

Fig. 4. Example of optimal control with two switch points in the system with damping

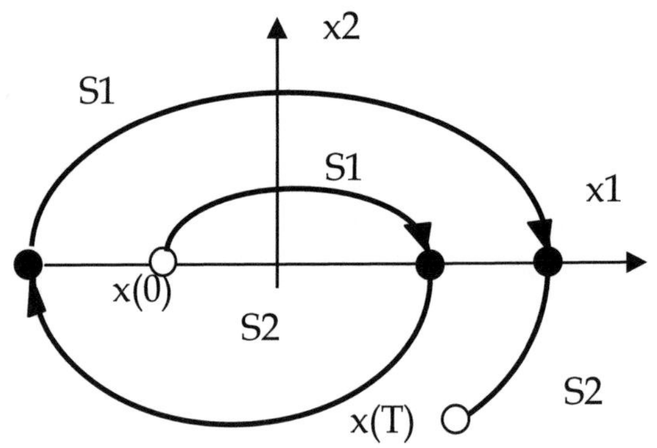

Fig. 5. Example of optimal control with three switch points in the task of system excitation

2.2 Theoretical synthesis of control action

For realizing of optimal control actions (in general case) system of one degree of freedom needs a feedback system with two adapters: one to measure a displacement and another one to measure a velocity. There is a simple case of control existing with only one adapter when directions of motion are changed (Fig. 5) [3]. It means that control action is like negative dry friction and switch points are along zero velocity line. In this case equation of motion is (6):

$$m \cdot \ddot{x} = -c \cdot x - b \cdot \dot{x} - \left[k \cdot (V0 + \dot{x})^2 \cdot A1 \cdot (0{,}5 - 0{,}5 \cdot \frac{\dot{x}}{|\dot{x}|}) \right] - \\ + \left[k \cdot (V0 + \dot{x})^2 \cdot A2 \cdot (0{,}5 + 0{,}5 \cdot \frac{\dot{x}}{|\dot{x}|}) \right], \quad (6)$$

where m – mass; c, b, k, V0 – constants. Examples of modeling are shown in Fig. 6. – Fig. 9.

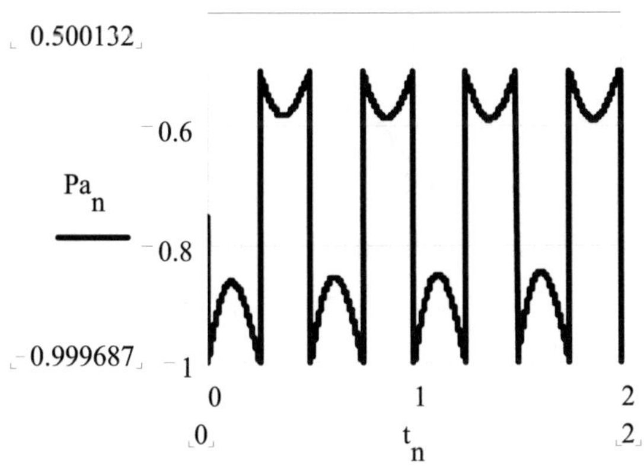

Fig. 6. Full control action in time domain

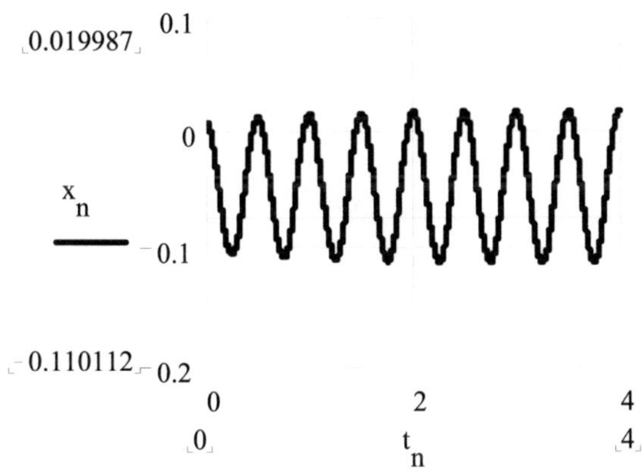

Fig. 7. Displacement in time domain

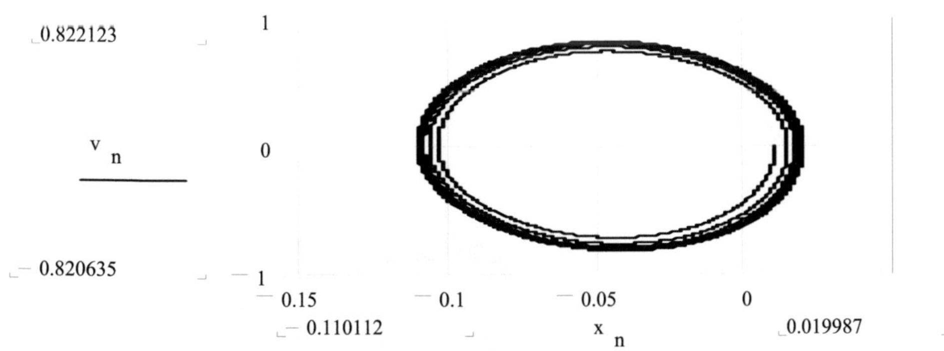

Fig. 8. Motion in phase plane starting from inside of limit cycle

A search for the case with more than one limit cycle was investigated in a very complicated system with linear and cubic restoring force, linear and cubic resistance force and dry friction. It was found that for a system with non-periodic excitation (like constant velocity of water or air flow) more than one limit cycles exist (Fig. 10., 11.).

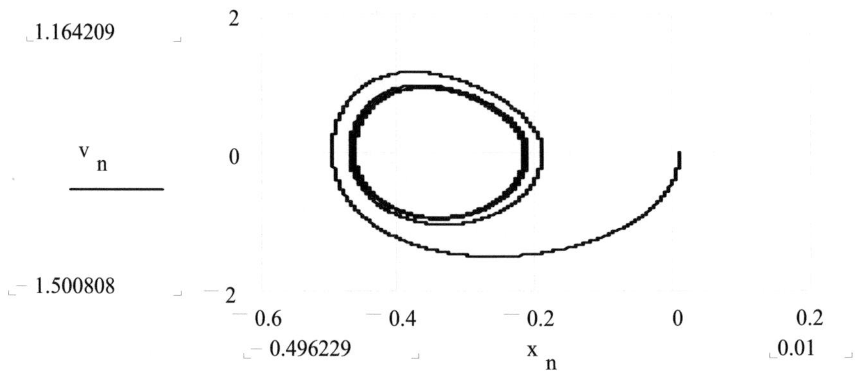

Fig. 9. Motion with initial conditions outside of limit cycle

$$\begin{bmatrix} x_0 \\ v_0 \end{bmatrix} := \begin{bmatrix} 0.09 \\ 0 \end{bmatrix}$$

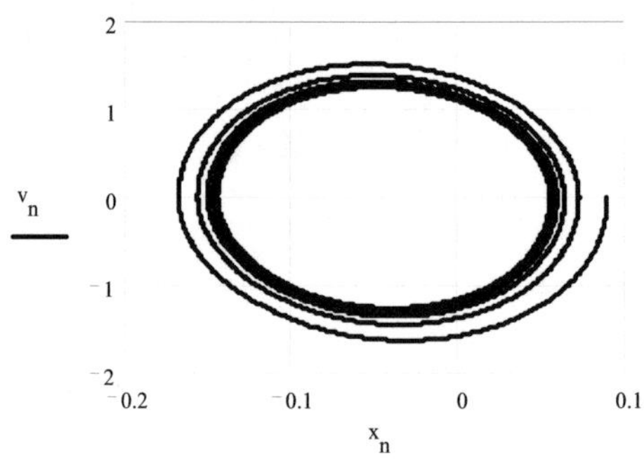

Fig. 10. Motion in phase plane for small limit cycle

$$\begin{bmatrix} x_0 \\ v_0 \end{bmatrix} := \begin{bmatrix} 9 \\ 0 \end{bmatrix}$$

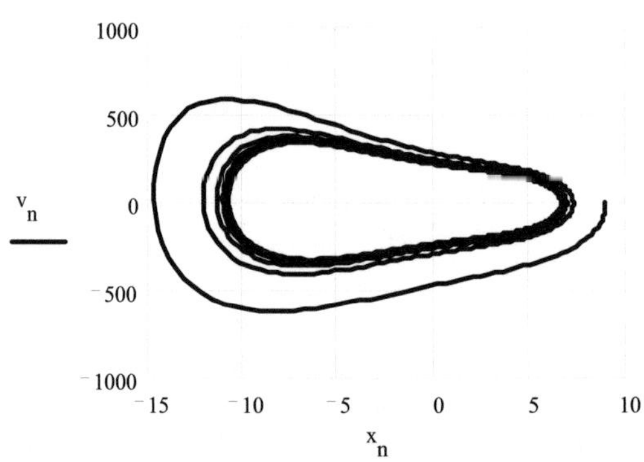

Fig. 11. Motion for large limit cycle

2.3 Synthesis of a real system with rotating blades

Scheme of a system includes main mass of the object, spring and viscous damper (Fig. 12.). Part of mass with blades does not rotate. Second part of main mass like blades rotates around fixed axis inside body. Blades have symmetric areas A1 and A2.

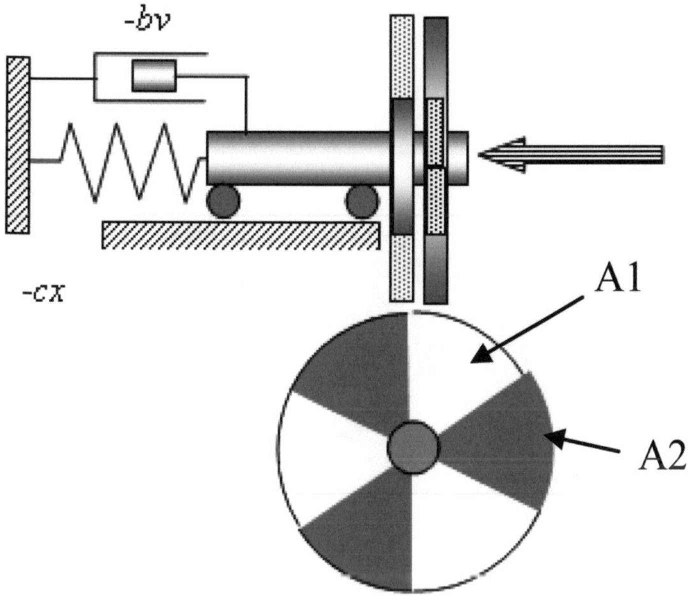

Fig. 12. System with rotating blades

Results of investigation are shown in Fig. 13. – Fig. 14.

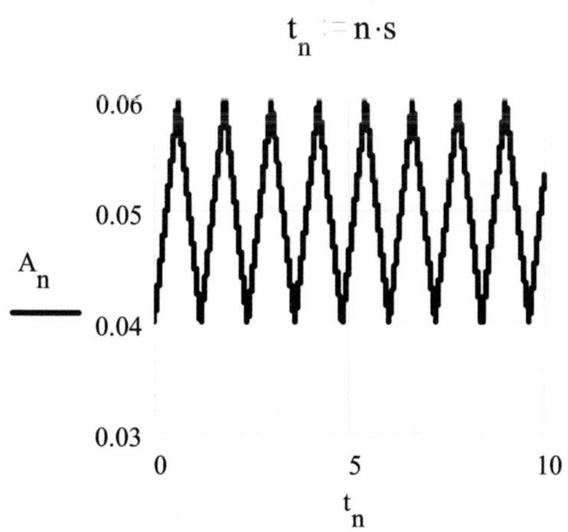

Fig. 13. Area change function in time domain when blades rotate with constant angular velocity ω

Investigation shows that system is very stable because of air excitation and damping forces depending from velocity squared.

$$\frac{\sqrt{\frac{c}{m}}}{\omega} = 1$$

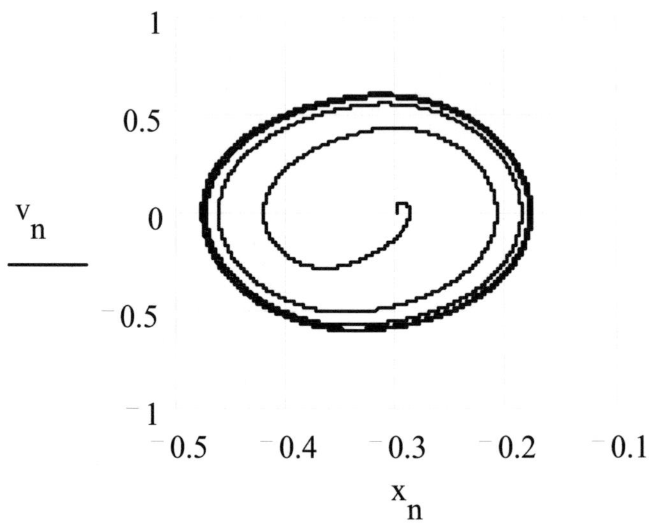

Fig. 14. Motion in phase plane for parameters close to the first resonance

2.4 Synthesis of real system inside tube with variable hole area - A

Adaptive control where analyzed by next area exchange functions f1(A) (Fig. 15, 16.):

$$f1(A) = A2 \cdot (1 + a \cdot \frac{v \cdot x}{|v \cdot x|}),$$

where A2, a – constants, v, x – velocity and coordinate of moving mass m2.

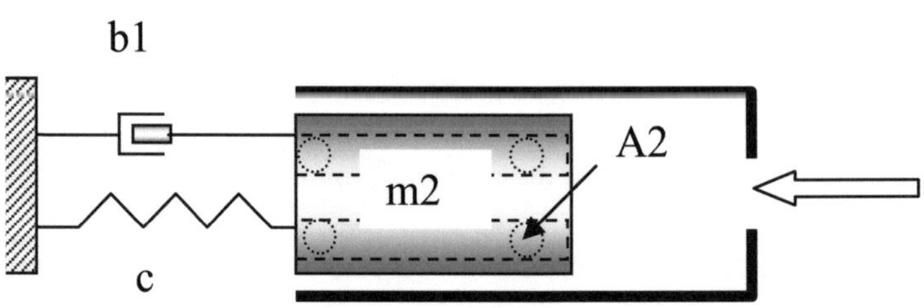

Fig. 15. Synthesis of system inside tube

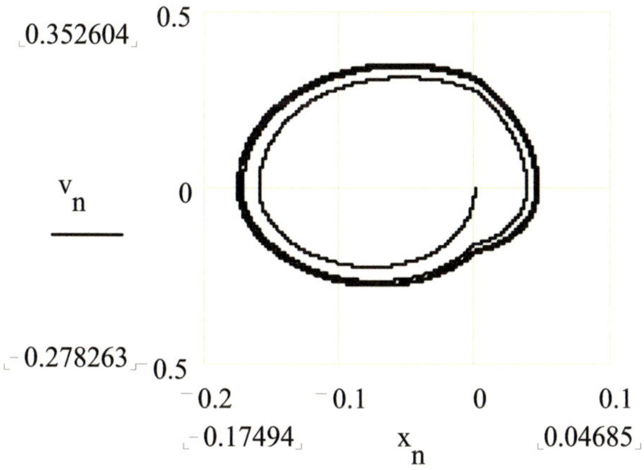

Fig. 16. Motion in phase plane

It is shown that adaptive systems are also very stable because of water or air excitation and damping forces depending from velocity squared.
In order to get values of parameters in equations, experiments were made in the wind tunnel. It should be noted that the air flow is similar to the water flow, with one main difference in the density. It is therefore convenient to perform synthesis algorithm experiments done with air flow in the wind tunnel. Experimental results obtained in the wind tunnel to some extent may be used to analyze water flow excitation in robot fish prototypes. Using "Armfield" subsonic wind tunnel several additional results were obtained. For example, drag forces, acting on different shapes of bodies, were measured and drag coefficients were calculated by formula $F_d = C_d \cdot S \cdot \frac{\rho \cdot V^2}{2}$, where ρ – density, V – flow velocity, S – specific area, C_d – drag coefficient. These coefficients depend on bodie's geometry, orientation relative to flow, and non-dimensional Reynolds number.

Fig. 17. Subsonic wind tunnel used for experiments at Mechanics Institute of Riga TU

All experiments were done at constant air flow velocity in the range of 10 - 20 m/s. This corresponds to *Re* value of about 40 000.
Drag coefficient for a plate perpendicular to the flow depends on plate's length and chord ratio. For standard plates C_d it is about 1.2. For infinitely long plates C_d reaches the value of 2. For a cone C_d = 0,5; for a cylinder along the flow - about 0,81; for both cylinder and cone in one direction - 0,32. All these forms can be used for robot fish main body form designing, (together with the tail) in real robot fish models (see part 3.4, where cone and cylinder were used). Depth control of prototypes was provided by two front control wings (see part 3.4).
Parameters of subsonic wind tunnel were as follows: length 2.98m, width 0.8m., height 1.83m. Variable speed motor drive unit downstream of the working section permits stepless control of airspeed between 0 and 26 ms-1. Experiments prove that air flow excitation is very efficient.

3. Investigation of a rotating system

The second object of the theoretical study is a fin type propulsive device of robotic fish moving inside water. The aim of the study is to find out optimal control law for variation of additional area of vibrating tail like horizontal pendulum, which ensures maximal positive impulse of motion forces components acting on a tail parallel to x axis (Fig.18).

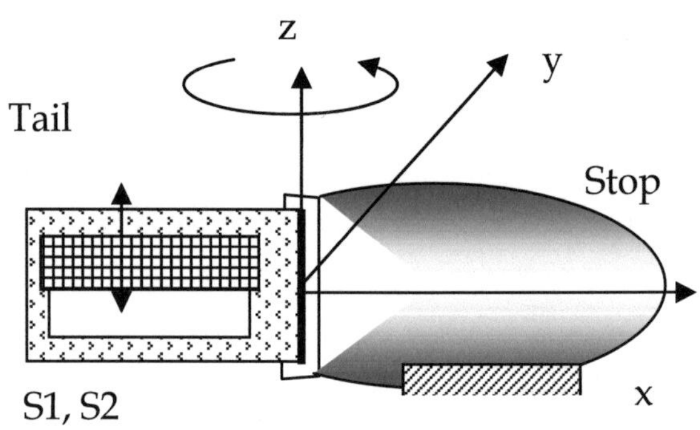

Fig. 18. Side-view of a robot fish tail like horizontal pendulum

Problem has been solved by using Pontryagin's maximum principle. It is shown that optimal control action corresponds to the case of boundary values of area.
Mathematical model of a system consists of rigid straight flat tail moving around pivot (Fig. 18.,19.). Area of the tail may be changed by control actions. For excitation of the motion a moment $M(t,\varphi,\omega)$ around pivot must be added, where $\omega = d\varphi/dt$ – angular velocity of a rigid tail. To improve performance of a system additional rotational spring with stiffness *c* may be added (Fig. 19.).

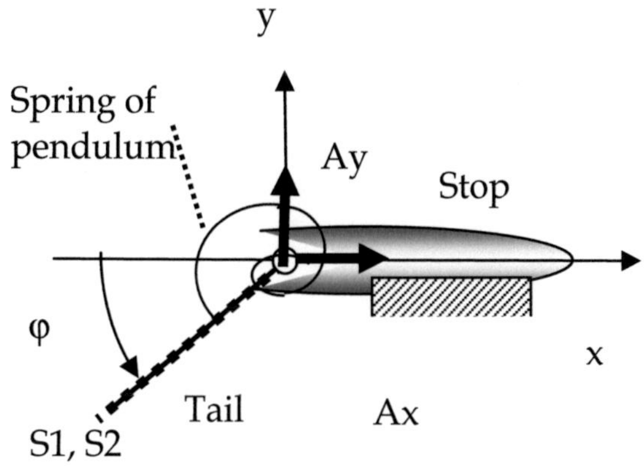

Fig. 19. Top-view of a robot fish tail like horizontal pendulum.
Simplified model with fixed pivot A

The principal task described in this report is to find optimal control law for variation of additional area *B(t)* of vibrating tail within limits (7.):

$$B_1 \le B(t) \le B_2, \tag{7}$$

where B_1 - lower level of additional area of a tail; B_2 - upper level of additional area of a tail, t - time.
The criterion of optimization is full reaction Ax^* impulse in pivot, acting to a hull (indirect reaction to a tail) which is required to move an object from one stop - initial position ($\varphi_{\max}, \dot{\varphi}=0$) to another stop - end position ($\varphi_{\min}, \dot{\varphi}=0$) in time T(8).

$$K = -\int_0^T Ax^* \cdot dt, \tag{8}$$

Differential equation of motion of the system with one degree of freedom (including a change of angular momentum around fixed point A) is (9):

$$J_A\ddot{\varphi} = M(t,\varphi,\dot{\varphi}) - c\cdot\varphi - k_t\cdot B\cdot sign(\dot{\varphi}\cdot\varphi)\cdot\int_0^L(\dot{\varphi}\cdot\xi)^2\cdot\xi\cdot d\xi. \tag{9}$$

Here J_A - moment of inertia of the tail around pivot point; $\ddot{\varphi}$ - angular acceleration of a rigid straight tail; $M(t,\varphi,\dot{\varphi})$ - excitation moment of the drive; c - angular stiffness; k_t - constant coefficient; B - area; $(\int_0^L(\dot{\varphi}\cdot\xi)^2\cdot\xi\cdot d\xi)$ - component of the moment of resistance force expressed as an integral along tail direction; $\dot{\varphi}$ - angular velocity; L - length of the tail.
From the principle of the motion of mass center C it follows (10):

$$m \cdot \left(\dot{\varphi}^2 \cdot \frac{L}{2} \cdot \cos\varphi + \ddot{\varphi} \cdot \frac{L}{2} \cdot \sin\varphi \right) = Ax - k_t \cdot B \cdot \sin(\varphi) \cdot sign(\varphi \cdot \dot{\varphi}) \cdot \left(\int_0^L (\dot{\varphi} \cdot \xi)^2 \cdot d\xi \right) \quad (10)$$

After integration from equations (9) and (10) we have (11, 12):

$$\ddot{\varphi} = \frac{1}{J_A} \cdot \left[M(t, \varphi, \dot{\varphi}) - c \cdot \varphi - k_t \cdot B \cdot sign(\dot{\varphi}) \cdot \dot{\varphi}^2 \cdot \frac{L^4}{4} \right]. \quad (11)$$

$$\begin{aligned} Ax = m \cdot \{ \dot{\varphi}^2 \cdot \frac{L}{2} \cdot \cos(\varphi) + \frac{1}{J_A} \cdot \left[M(t, \varphi, \dot{\varphi}) - c \cdot \varphi - k_t \cdot B \cdot sign(\dot{\varphi}) \cdot \dot{\varphi}^2 \cdot \frac{L^4}{4} \right] \cdot \frac{L}{2} \cdot \sin(\varphi) \} + \\ + k_t \cdot B \cdot \sin(\varphi) \cdot sign(\varphi \cdot \dot{\varphi}) \cdot \dot{\varphi}^2 \cdot \frac{L^3}{3}. \end{aligned} \quad (12)$$

Then the criterion of optimization (full impulse) is:

$$K = -\int_0^T \left(\begin{aligned} & m \cdot \{ \dot{\varphi}^2 \cdot \frac{L}{2} \cdot \cos(\varphi) + \frac{1}{J_A} \cdot \left[M(t, \varphi, \dot{\varphi}) - c \cdot \varphi - k_t \cdot B \cdot sign(\dot{\varphi}) \cdot \dot{\varphi}^2 \cdot \frac{L^4}{4} \right] \cdot \frac{L}{2} \cdot \sin(\varphi) \} + \\ & + k_t \cdot B \cdot \sin(\varphi) \cdot sign(\varphi \cdot \dot{\varphi}) \cdot \dot{\varphi}^2 \cdot \frac{L^3}{3} \end{aligned} \right) \cdot dt, \quad (13)$$

Equations (13) will be used to solve the optimization problem.

3.1 Task of optimization

Solution of optimal control problem for a system with one degree of freedom (11) by using the Pontryagin's maximum principle includes following steps [8 - 13]:

1. Formulation of a criterion of optimization (13):

$$K = -\int_0^T Ax \cdot dt .$$

2. Transformation of the equation (9) and equation (13) in the three first order equations with new variables φ0, φ1 and φ2 (phase coordinates):

$$\varphi 0 = (-1) \cdot \left[\begin{aligned} & m \cdot \{ \varphi 2^2 \cdot \frac{L}{2} \cdot \cos(\varphi 1) + \frac{1}{J_A} \cdot \left[M(t, \varphi 1, \varphi 2) - c \cdot \varphi 1 - k_t \cdot B \cdot sign(\varphi 2) \cdot \varphi 2^2 \cdot \frac{L^4}{4} \right] \cdot \frac{L}{2} \cdot \sin(\varphi 1) \} + \\ & + k_t \cdot B \cdot \sin(\varphi 1) \cdot sign(\varphi 1 \cdot \varphi 2) \cdot \varphi 2^2 \cdot \frac{L^3}{3}, \end{aligned} \right]$$

$$\begin{aligned} & \dot{\varphi} 1 = \varphi 2; \\ & \dot{\varphi} 2 = \frac{1}{J_A} \cdot \left[M(t, \varphi 1, \varphi 2) - c \cdot \varphi 1 - k_t \cdot B \cdot sign(\varphi 2) \cdot \varphi 2^2 \cdot \frac{L^4}{4}. \right] \end{aligned} \quad (14)$$

According to procedure of Pontryagin's maximum principle, Hamiltonian (*H*) is [9 -12]:

$$H=\psi_0\cdot(-1)\cdot\begin{bmatrix} m\cdot\}\varphi2^2\cdot\frac{L}{2}\cdot\cos(\varphi1)+\frac{1}{J_A}\cdot[M(t,\varphi1,\varphi2)-c\cdot\varphi1- \\ -k_t\cdot B\cdot sign(\varphi2)\cdot\varphi2^2\cdot\frac{L^4}{4}\bigg]\cdot\frac{L}{2}\cdot\sin(\varphi1)\})+ \\ +k_t\cdot B\cdot\sin(\varphi1)\cdot sign(\varphi1\cdot\varphi2)\cdot\varphi2^2\cdot\frac{L^3}{3}\end{bmatrix}+$$

$$+\psi_1\cdot\varphi2+\psi2\cdot\frac{1}{J_A}\cdot\left[M(t,\varphi1,\varphi2)-c\cdot\varphi1-k_t\cdot B\cdot sign(\varphi2)\cdot\varphi2^2\cdot\frac{L^4}{4}\right]. \tag{15}$$

here $H=\psi\cdot\Phi$, where (10, 11)

$$\psi=\begin{Bmatrix}\psi_0\\ \psi_1\\ \psi_2\end{Bmatrix}; \tag{16}$$

$$\Phi=\begin{Bmatrix}\Phi_0\\ \Phi_1\\ \Phi_2\end{Bmatrix}, \tag{17}$$

Φ_0,Φ_1,Φ_2 - right side of system equations (14).

For functions Φ_0,Φ_1,Φ_2 calculations are following differential equations [9 - 12]:

$$\dot{\Phi}_0=-\frac{\partial H}{\partial\phi_0};\ \dot{\Phi}_1=-\frac{\partial H}{\partial\phi_1};\ \dot{\Phi}_2=-\frac{\partial H}{\partial\phi_2}, \tag{18}$$

where left side is the derivation in time *t*: $\dot{\Phi}_{0,1,2}=\frac{d\Phi_{0,1,2}}{dt}$.

From equations (15) and (18) we get nonlinear system of differential equations to find functions ψ_0,ψ_1,ψ_2. Solution of such system is out of the scope of this report because it depends from unknown moment *M(t,φ,ω)*. But some conclusions and recommendations may be given from Hamiltonian if excitation moment *M(t,φ,ω)* does not depend from phase coordinates *φ = φ1, ω = φ2:*

$$M = M(t).$$

In this case scalar product of two last vector functions ψ and Φ in any time (Hamiltonian *H* [11]) must be maximal (supremum - in this linear B case) [8 – 12]. To have such maximum (supremum), control action *B(t)* must be within limits $B(t)=B_1;\ \ B(t)=B_2$, depending only from the *sign* of a function (19) or (20):

$$sign\left(\begin{array}{l}\psi_0\cdot\left[\frac{m}{J_A}\cdot\left[k_t\cdot 1\cdot sign(\varphi 2)\cdot\frac{L^5}{4}\right]\cdot\frac{1}{2}\cdot\sin(\varphi 1)+k_t\cdot 1\cdot\sin(\varphi 1)\cdot sign(\varphi 1\cdot\varphi 2)\cdot\frac{L^3}{3}\right]+\\ +\psi 2\cdot\frac{1}{J_A}\cdot\left[-k_t\cdot 1\cdot sign(\varphi 2)\cdot\frac{L^4}{4}\right]\end{array}\right)\geq 0, \quad (19)$$

$$B = B2;$$

$$sign\left(\begin{array}{l}\psi_0\cdot\left[\frac{m}{J_A}\cdot\left[k_t\cdot 1\cdot sign(\varphi 2)\cdot\frac{L^5}{4}\right]\cdot\frac{1}{2}\cdot\sin(\varphi 1)+k_t\cdot 1\cdot\sin(\varphi 1)\cdot sign(\varphi 1\cdot\varphi 2)\cdot\frac{L^3}{3}\right]+\\ +\psi 2\cdot\frac{1}{J_A}\cdot\left[-k_t\cdot 1\cdot sign(\varphi 2)\cdot\frac{L^4}{4}\right]\end{array}\right)\leq 0. \quad (20)$$

$$B = B1;$$

From inequalities (19) and (20) in real system synthesis following quasi-optimal control action may be recommended (21):

$$B=\left[B2\cdot(0,5-0,5\cdot sign(\varphi 1\cdot\varphi 2))+B1\cdot(0,5+0,5\cdot sign(\varphi 1\cdot\varphi 2))\right],$$

or

$$B=\left[B2\cdot(0,5-0,5\cdot sign(\varphi\cdot\dot{\varphi}))+B1\cdot(0,5+0,5\cdot sign(\varphi\cdot\dot{\varphi})\right]. \quad (21)$$

3.2 Synthesis of mixed system with time-harmonic excitation and area adaptive control

In the case of time-harmonic excitation moment M in time domain is (see equations (9) and (11)):

$$M(t) = M0\cdot sin(k\cdot t).$$

Results of modeling are shown in Fig. 20. – 25. Comments about graphics are given under all Fig. 20. – 25.

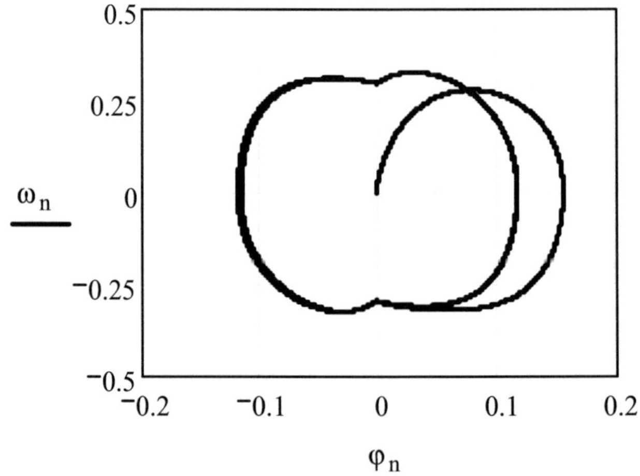

Fig. 20. Tail angular motion in the phase plane – angular velocity as function of angle

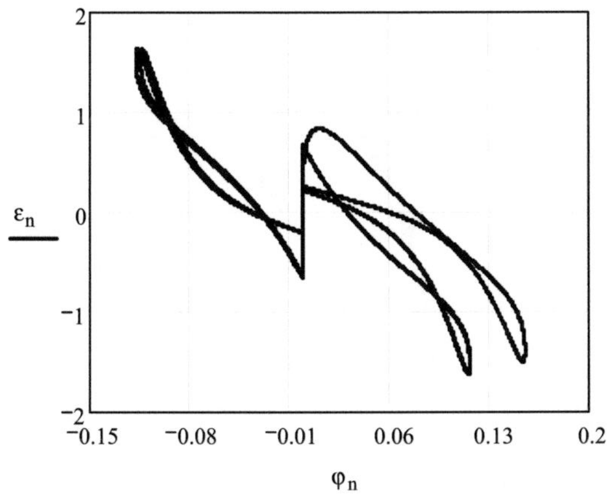

Fig. 21. Angular acceleration of the tail as function of angle. At the end of transition process graph is symmetric

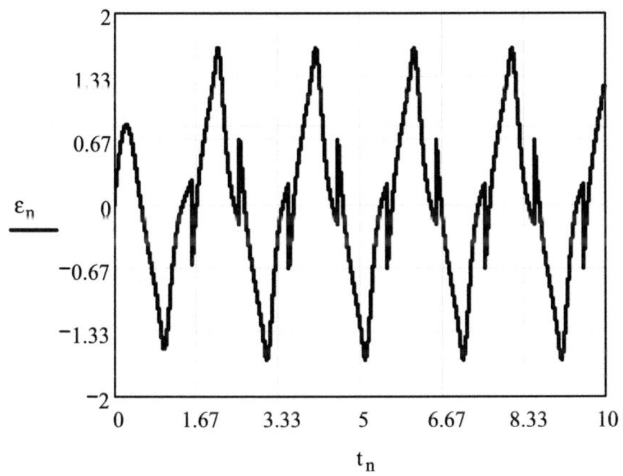

Fig. 22. Angular acceleration $\varepsilon = \ddot{\varphi}$ of a tail in time domain (see equation (9)). Practically angular acceleration of the tail reaches steady-state cycle after one oscillation

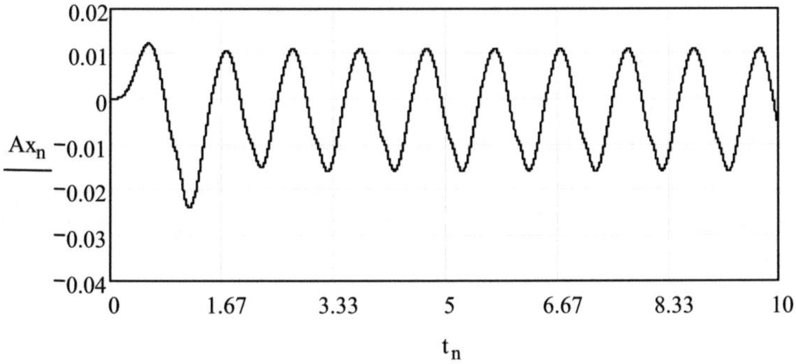

Fig. 23. Impulse *Ax(t)* in time domain (see equation (12)). Impulse is non-symmetric against zero level (non-symmetry is negative)

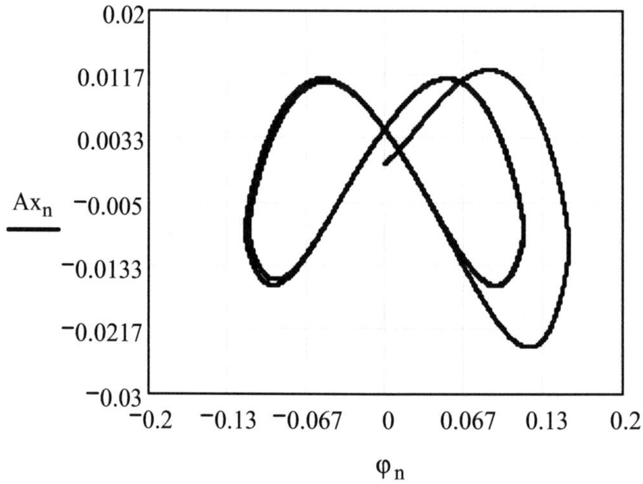

Fig. 24. Impulse *Ax(t)* as a function of angle φ.
Impulse is non-symmetric against zero level (non-symmetry is negative)

$$IAx_n := \sum_n Ax_n$$

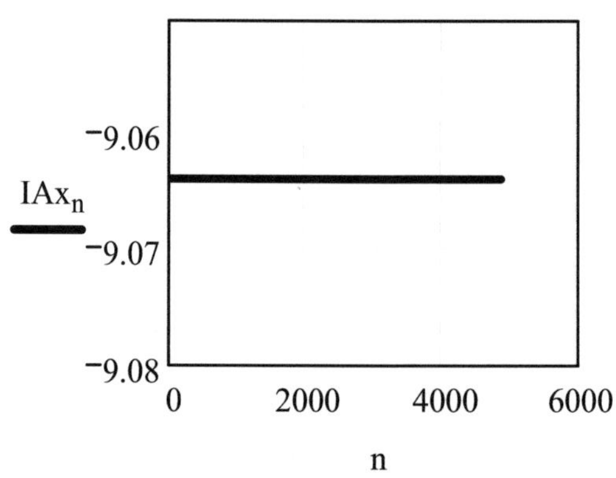

Fig. 25. Negative mean value of *Ax(t)* in time domain.
Force of pivot to push a hull (necessary for robotic fish motion)

3.3 Synthesis of a system with adaptive excitation and adaptive area control

In a case of adaptive excitation a moment *M* may be used as the function of angular velocity in the form [3. 4]:

$$M(t) = M0 \cdot sign(\omega).$$

Results of modeling are shown in Fig. 26. – 31. Comments about graphics are given under all Fig. 26. – 31.

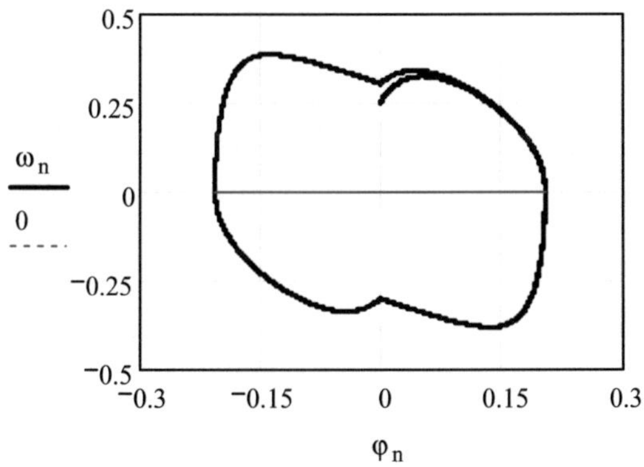

Fig. 26. Tail angular motion in phase plane - angular velocity as the function of angle. Due to large water resistance the transition process is very short. Adaptive excitation is more efficient than harmonic excitation (see Fig.20.)

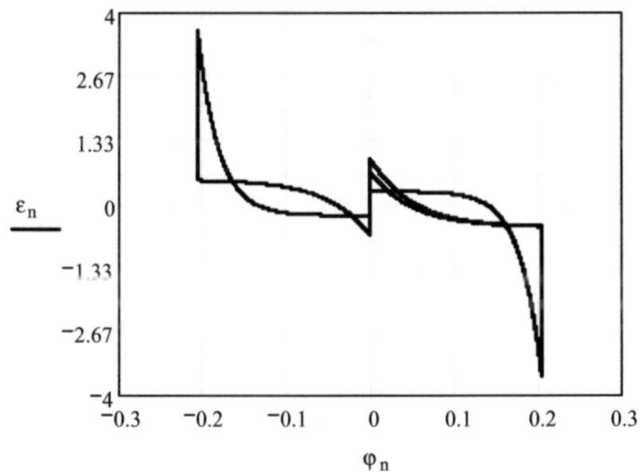

Fig. 27. Angular acceleration of a tail as the function of angle. At the end of transition process graph is more symmetric than the graph shown in Fig. 22

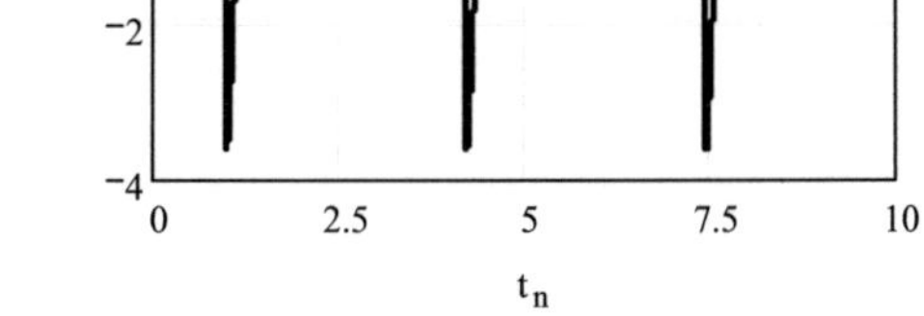

Fig. 28. Angular acceleration $\varepsilon = \ddot{\varphi}$ of a tail in time domain (see equation (9)). Typically angular acceleration of the tail reaches steady-state cycle after half oscillation

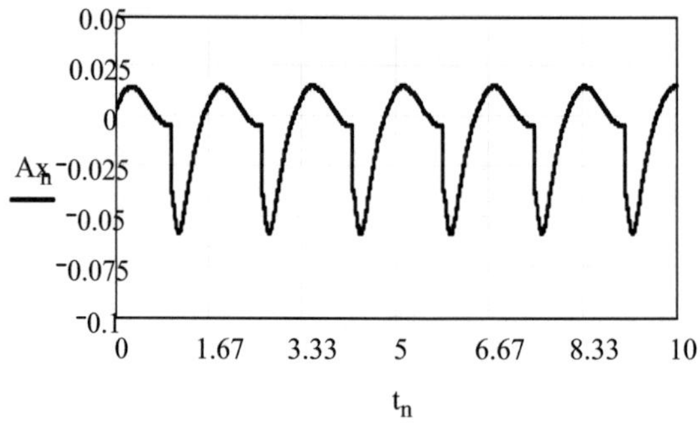

Fig. 29. Impulse *Ax(t)* in time domain (see equation (12)). Impulse is non-symmetric against zero level (non-symmetry is negative)

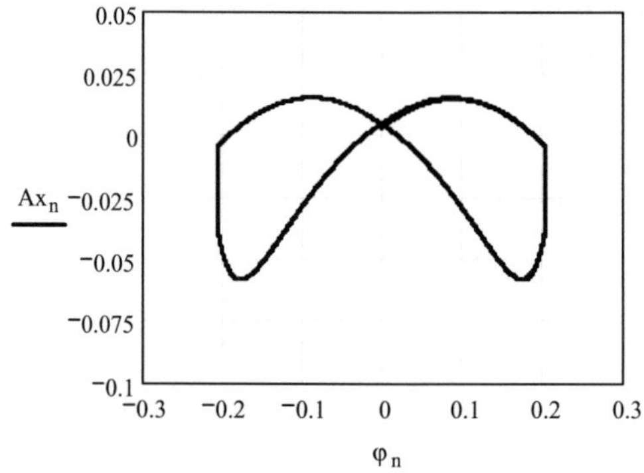

Fig. 30. Impulse *Ax(t)* as a function of angle φ

Impulse is non-symmetric against zero level (non-symmetry is negative), it means that criterion of optimization (8) is positive and force of pivot pushes fish hull to the right.

$$IAx_n := \sum_n Ax_n$$

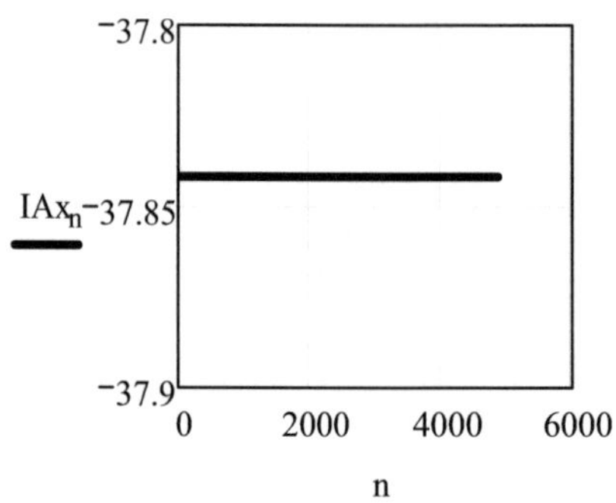

Fig. 31. Negative mean value of *Ax(t)* in time domain. Force of pivot to push fish hull to the right (in order to have robotic fish motions to the right). This graph shows that adaptive excitation of a moment is about four times more efficient than harmonic excitation (see Fig. 25.)

3.4 Robot fish model

A prototype of robot fish for experimental investigations was made (Fig. 32, 33). This prototype was investigated in a linear water tank with different aims, for example: - find maximal motion velocity depending on the power pack capacity; - find minimal propulsion force, depending on the system parameters.

Additionally this prototype was investigated in a large storage lake with autonomy power pack and distance control system, moving in real under water conditions with waves (Fig. 34.). The results of the theoretical and experimental investigation may be used for inventions of new robotic systems. The new ideas of synthesising robotic systems in Latvia can be found in [15 - 23].

Fig. 32. Top-view of prototype

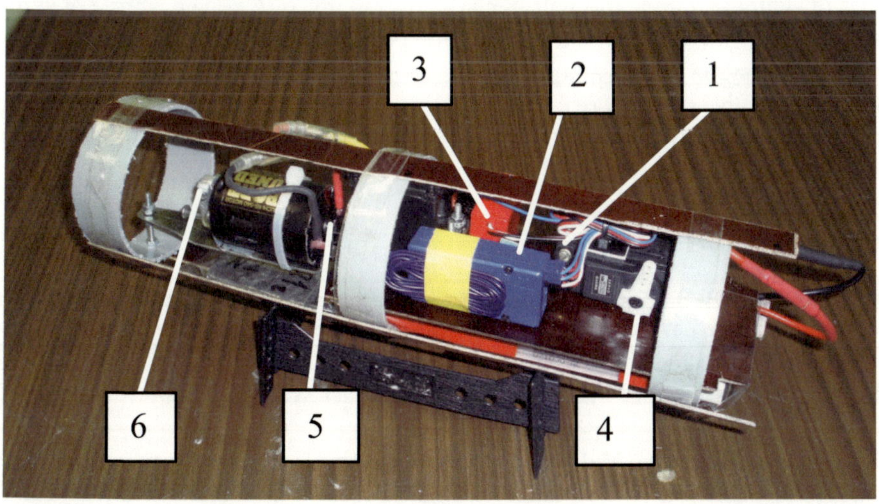

Fig. 33. Inside prototype there are: microcontroller 1; three actuators (for level 4, direction 5 and velocity 6 control); power supply 2 and radio signal detector 3)

Fig. 34. Prototype was investigated in a large storage lake with autonomy power pack and distance control system, moving in real under water conditions with waves

4. Conclusion

Motion of robotic systems vibration by simplified interaction with water or air flow can be described by rather simple equations for motion analysis. That allows to solve mathematical problem of area control optimization and to give information for new systems synthesis. Control (or change) of object area under water or in air allows to create very efficient mechatronic systems. For realization of such systems adapters and controllers must be used. For this reason very simple control action have solutions with use of sign functions. Examples of synthesis of real mechatronic systems are given. As one example of synthesis is a system with time-harmonic moment excitation of the tail in the pivot. The second example of synthesis is a system with adaptive force moment excitation as the function of phase coordinates. In both systems area change (from maximal to minimal values) has control action as the function of phase coordinates. It is shown that real controlled systems vibration motion is very stable.

5. References

[1] E. Lavendelis: Synthesis of optimal vibro machines. Zinatne, Riga, (1970). (in Russian).
[2] E. Lavendelis and J. Viba: Individuality of Optimal Synthesis of vibro impact systems. Book: Vibrotechnics". Kaunas. Vol. 3 (20), (1973). (in Russian).
[3] J. Viba: Optimization and synthesis of vibro impact systems. Zinatne, Riga, (1988). (in Russian).
[4] E. Lavendelis and J. Viba: Methods of optimal synthesis of strongly non - linear (impact) systems. Scientific Proceedings of Riga Technical University. Mechanics. Volume 24. Riga (2007).
[5] http://en.wikipedia.org/wiki/Lev_Pontryagin. (2008).

[6] http://en.wikipedia.org/wiki/Pontryagin%27s_maximum_principle.(February (2008).
[7] http://en.wikipedia.org/wiki/Hamiltonian_%28control_theory%29. (2007).
[8] V.G. Boltyanskii, R.V. Gamkrelidze and L. S. Pontryagin: On the Theory of Optimum Processes (In Russian), *Dokl. AN SSSR*, 110, No. 1, 7-10 (1956).
[9] L. S. Pontryagin, V.G. Boltyanskii, R.V.Gamkrelidze and E.F. Mischenko: (Fizmatgiz). *The Mathematical Theory of Optimal Processes*, Moscow (1961).
[10] L.S. Pontryagin, V.G. Boltyanskii, R.V. Gamkrelidze, and E.F. Mishchenko: *The Mathematical Theory of Optimal Processes*, Wiley-Interscience, New York. (1962).
[11]V. G. Boltyanskii: *Mathematical Methods of Optimal Control*, Nauka, Moscow. (1966).
[12] V.G. Boltyanskii: Mathematical methods of optimal control: Authorized translation form the Russian. Translator K.N. Trirogoff. Balskrishnan-Neustadt series. New York. Holt, Rinehart and Winston. (1971).
[13] E.B. Lee and L. Markus: *Foundations of Optimal Control Theory*, Moscow: Nauka, (1972). (in Russian). 14. Sonneborn, L., and F. Van Vleck (1965): *The Bang-Bang Principle for Linear Control Systems*, SIAM J. Control 2, 151-159.
[14] http://en.wikipedia.org/wiki/Bang-bang_control. (2008).
[15] Patent LV 13928, Republic of Latvia, Int.Cl. B25 D9/00, B25 D11/00. Method for control of operation condition of one-mass vibromachine on elastic suspension / J. Viba, V. Beresnevich, M. Eiduks, L. Shtals, E. Kovals, G. Kulikovskis. - Applied on 03.03.2009, application P-09-38; published on 20.09.2009 // Patenti un preču zīmes, 2009, No. 9, p. 1209 - 1210.
[16] Patent LV 14033, Republic of Latvia, Int.Cl. B63 H 1/00. Method for tractive force forming in fin propulsive device / J. Viba, V. Beresnevich, S. Cifanskis, G. Kulikovskis, M. Kruusmaa, W. Megill, J.-G. Fontaine, P. Fiorini. - Applied on 04.09.2009, application P-09-151; published on 20.02.2010 // Patenti un preču zīmes, 2010, No. 2, p. 203.
[17] Patent LV 14034, Republic of Latvia, Int.Cl. B63 H1/00. Hydrodynamic fin-type vibration propulsive device / S. Cifanskis, J. Viba, V. Jakushevich, O. Kononova, J.-G. Fontaine, W. Megill. - Applied on 29.09.2009, application P-09-160; published 20.02.2010 // Patenti un preču zīmes, 2010, No. 2, p. 203.
[18] Patent LV 14054, Republic of Latvia, Int.Cl. B63 H1/00. Fin of hydrodynamic vibration propulsive device / S. Cifanskis, J. Viba, V. Jakushevich, V. Beresnevich, O. Kononova, G. Kulikovskis. - Applied on 15.10.2009, application P-09-174; published 20.03.2010 // Patenti un preču zīmes, 2010, No. 3, p. 428.
[19] Patent LV 14055, Republic of Latvia, Int.Cl. B63 H1/00. Method for control of operating condition of hydrodynamic fin-type vibration propulsive device / J. Viba, S. Cifanskis, V. Jakushevich, O. Kononova, J.-G. Fontaine, E. Kovals. - Applied on 02.11.2009, application P-09-191; published 20.03.2010 // Patenti un preču zīmes, 2010, No. 3, p. 428.
[20] Patent LV 14076, Republic of Latvia, Int.Cl. B63 H1/00. Hydrodynamic fin-type vibration propulsive device / S. Cifanskis, J. Viba, V. Jakushevich, O. Kononova, M. Kruusmaa, G. Kulikovskis. - Applied on 24.11.2009, application P-09-205; published 20.04.2010 // Patenti un preču zīmes, 2010, No. 4, p. 631.

[21] Patent LV 14077, Republic of Latvia, Int.Cl. B63 H 1/00. Method for adjustment of operation condition of fin-type vibration propulsive device / S. Cifanskis, V. Beresnevich, J. Viba, V. Yakushevich, J.-G. Fontaine, G. Kulikovskis. - Applied on 10.12.2009, application P-09-219; published on 20.04.2010 // Patenti un preču zīmes, 2010, No. 4, p. 631.

[22] Patent LV 14175, Republic of Latvia, Int.Cl. B63 H 1/00. Method for adjustment of operation condition of fin-type vibration propulsive device / S. Cifanskis, J. Viba, V.Beresnevich, V. Jakushevich, M. Listak, T. Salumäe. - Applied on 28.04.2010, application P-10-63; published 20.10.2010 // Patenti un preču zīmes, 2010, No. 10, p. 1515 - 1516.

[23] Patent LV 14237, Republic of Latvia, Int.Cl. B63 H1/00. Hydrodynamic fin-type vibration propulsive device / S. Cifanskis, J. Viba, V. Beresnevich, V. Jakushevich, J.-G. Fontaine, W. Megill. - Applied on 07.10.2010, application P-10-141; published 20.02.2011 // Patenti un preču zīmes, 2011, No. 2, p. 188.

15

Modeling of Elastic Robot Joints with Nonlinear Damping and Hysteresis

Michael Ruderman
Institute of Control Theory and Systems Engineering, TU-Dortmund
Germany

1. Introduction

Elastic robot joints gain in importance since the nowadays robotics tends to the lightweight structures. The lightweight metals and composite materials deployed not only in the links but also in the joint assemblies affect the overall stiffness of robotic system. The elastic joints provide the loaded robot motion with additional compliance and can lead to significant control errors and vibrations in the joint as well as operational space. A better understanding of the compliant joint behavior can help not only to analyze and simulate robotic systems but also to improve their control performance.
Elastic robot joints are often denoted as flexible joints or compliant joints as well. The former modeling approaches aimed to describe the dynamic behavior of elastic robot joints lead back to Spong (1987). Spong extended the general motion equation of a rigid robotic manipulator to the case of joint elasticity captured by a linear connecting spring. Remember that the general motion equation derived either from Lagrange or Newton-Euler formalism for a class of rigid robotic manipulators (Sciavicco & Siciliano (2000)) can be expressed as

$$\mathbf{M}(\mathbf{q})\ddot{\mathbf{q}} + \mathbf{C}(\mathbf{q}, \dot{\mathbf{q}})\dot{\mathbf{q}} + \mathbf{G}(\mathbf{q}) = \mathbf{u}\,. \tag{1}$$

Here, $\mathbf{q} \in \mathbb{R}^n$ is the vector of Lagrangian (also joint) coordinates and $\mathbf{u} \in \mathbb{R}^n$ is the vector of generalized input forces. The configuration dependent matrixes $\mathbf{M} \in \mathbb{R}^{n\times n}$, $\mathbf{C} \in \mathbb{R}^{n\times n}$, and $\mathbf{G} \in \mathbb{R}^n$ constitute the inertia, Coriolis-centrifugal, and gravity terms correspondingly. The model introduced by Spong (1987) has been widely used in the later works which deal with a compliance control (Zollo et al. (2005)) and passivity-based impedance control (Ott et al. (2008)) of robots with joint elasticities. Despite a good acceptance for the control applications this modeling strategy misses the damping factor related to the connecting spring. In this regard, the approach proposed by Ferretti et al. (2004) which provides two masses connected by a linear spring and damper arranged in parallel is more consistent with the physical joint structure. Also the inverse dynamic model used for vibration control of elastic joint robots (Thummel et al. (2005)) incorporates a torsional spring with both stiffness and damping characteristics. From a slightly diversing point of view the modeling of elastic robot joints was significantly influenced by the studies of harmonic drive gear transmission performed in the end-nineties and last decade. The harmonic drives are widely spread in the robotic systems due to their compact size, high reduction ratios, high torque capacity, and low (nearly zero) backlash. However, a specific mechanical assembly provides the harmonic drive gears with additional elasticities which are strongly coupled with frictional effects. A structure-oriented dissipative model of the harmonic drive torque transmission was proposed by Taghirad & Bélanger (1998). The model composes multiple component-related frictional elements and an additional structural damping attributed to the flexspline. Another notable modeling issue provided by Dhaouadi et al. (2003) presents the torsional torque in harmonic drives as an integro-differential hysteresis function of both angular displacement and angular velocity across the flexspline. Later, Tjahjowidodo et al. (2006) describe the dynamics in harmonic drives using nonlinear stiffness characteristics combined with distributed Maxwell-slip elements that capture the hysteresis behavior. A complex phenomenological model of elastic robot joints with coupled hysteresis, friction and backlash nonlinearities was proposed by Ruderman et al. (2009). However, realizing the complexity of decomposing and identifying the single nonlinearities a simplified nonlinear dynamic model was later introduced in Ruderman et al. (2010).

The leitmotif provided in this Chapter is to incorporate a combined physical as well as phenomenological view when modeling the joint transmission with elasticities. Unlike the classical approaches which rather operate with linear stiffness and damping elements arranged either in series or in parallel the structure-oriented effects are emphasized here. The inherently nonlinear compliance and damping of elastic robot joints are tackled from the cause-and-effect point of view. It is important to note that such approaches can rapidly increase the number of free parameters to be identified, so that the model complexity has to be guarded carefully. The Chapter is organized as follows. Section 2 introduces the robot joint topology which captures equally the dynamic behavior of both rigid and elastic revolute joints. Three closed subsystems are described in terms of their eigenbehavior and feedforward and feedback interactions within the overall joint structure. Section 3 provides the reader with description of the developed joint model capable to capture two main nonlinearities acting in the joint transmission. First, the dynamic joint friction is addressed. Secondly, the nonlinear stiffness combined with hysteresis map is described in detail. An experimental case study provided in Section 4 shows some characteristical observations obtained on a laboratory setup of the joint with elasticities and gives some identification results in this relation. The main conclusions are derived in Section 5.

2. Robot joint topology

Before analyzing the dynamic behavior of an elastic robot joint an underlying joint topology has to be assumed first. Different topologies of robotic joints are conceivable starting from the more simple linear structures as used e.g. by Ferretti et al. (2004), Zollo et al. (2005), Albu-Schaffer et al. (2008) and migrating towards the more complex nonlinear approaches as proposed e.g. by Taghirad & Bélanger (1998), Dhaouadi et al. (2003), Tjahjowidodo et al. (2006), particularly in context of harmonic drive gear transmissions. Here it is important at which level of detail the robot joint transmission could be described. Mostly it is conditioned by the available knowledge about the mechanical joint structure and the accessibility of system measurements required for the identification. In majority of applications, the actuator measurements such as the angular position and velocity as well as active motor current are available only prior to the gear transmission. In more advanced but also more cost-intensive applications like DLR lightweight robots (see e.g. by Albu-Schaffer et al. (2008)), the angular motion and torque measurements are available at both actuator and load side of the joint transmission. Although those hardware solutions are often not technically or (and) economically profitable, a prototypic laboratory measurement performed on the load side of robotic joints can yield adequate data sufficient for analysis and identification. Here, one can think about highly accurate static as well as dynamic measurements of the joint output position performed by means of the laser interferometry or laser triangulation. The load torques can also be determined externally either by applying an appropriate accelerometer or by using locked-load mechanisms equipped by the torque (load) cells. However, the latter solution is restricted to the quasi-static experiments with a constrained output motion.

Now let us consider the topology of an elastic robot joint as shown in Fig. 1. In terms of the input-output behavior the proposed structure does not substantially differ from a simple fourth-order linear dynamic model of two connected masses. An external exciting torque u constitutes the input value and the relative position of the second moving mass θ constitutes the output value we are interested in. Going into the input-output representation,

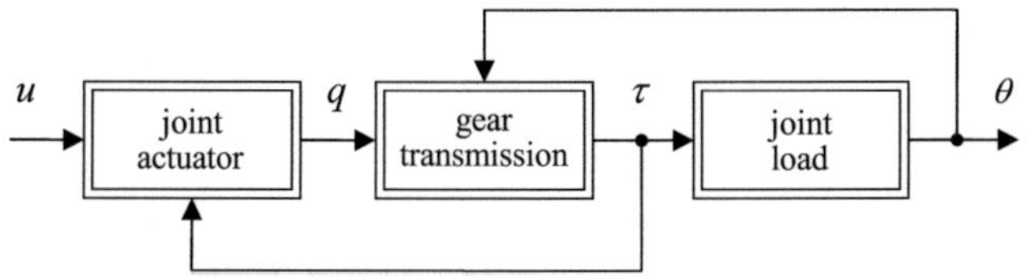

Fig. 1. Topology of elastic robot joint

let us subdivide the robotic joint into three close subsystems connected by the feedforward and feedback actions realized by appropriate physical states. The joint actuator loaded by the feedback torque τ provides the output angular displacement q of a rigid shaft. This value is an inherent determinant of the relative motion entering the gear transmission and mostly measurable prior to that one. Since the gear transmission captures the intrinsic joint elasticity, the angular output displacement of the joint load constitutes the second feedback state. Assuming that the gear with elasticities behaves like a torsion spring with a certain stiffness capacity its output value represents the transmitted torque which drives the joint load. When cutting free the forward and feedback paths the joint model decomposes into three stand-alone submodels, each one describing the specified physical subsystem. Note that from energy conversion point of view we obtain three dissipative mappings with two different sets of the input values. The first input set constitutes the forward propagation of the energy fed to the system. The second input set represents the system reaction with a negative or positive energy feedback depending on the instantaneous operation state. The superposition of the feed-in and reactive joint torque provides the internal angular motion $(u - \tau) \mapsto q$ damped by the friction. The constrained relative displacement across the gear transmission provides the elastic tension in the joint $(q - \theta) \mapsto \tau$ damped by its structure. Finally, the transmitted torque provides the output relative motion $\tau \mapsto \theta$ whose damping depends on the topology of subsequent load.

The actual joint topology represents a general case which covers both elastic and rigid transmission. The actuator submodel can be equally used for a rigid joint modeling, where τ will appear as an input disturbance fed back directly from the joint load and acting upon the overall torque balance. The stand-alone gear transmission modeling is however meaningful only when the relative displacement occurs, i.e. $q \neq \theta$. When considering a rigid robotic joint the transmission submodel will describe a particular case with an unlimited stiffness. At this, each infinitesimal change in q will lead to an immediate excitation of the load part, so that the eigendynamics of the joint transmission disappears.

The following analytical example of two equivalent linear joint models explains the upper mentioned ideas in more detail. For instant, consider a simple motion problem of two connected masses m and M with two damping factors d and D, once with a rigid and once with an elastic joint. Assume that the last one is represented by the spring with the stiffness K. The first case can be described by the simple differential equation

$$(m + M)\,\ddot{q} + (d + D)\,\dot{q} = u\,. \tag{2}$$

It is evident that the system (2) provides the single dynamic state, and the mass and damping factors appear conjointly whereas $q = \theta$. The more sophisticated second case with elasticity requires two differential equations

$$\begin{aligned} m\,\ddot{q} + d\,\dot{q} + K(q - \theta) &= u\,, \\ M\,\ddot{\theta} + D\,\dot{\theta} - K(q - \theta) &= 0 \end{aligned} \tag{3}$$

which constitute the fourth-order dynamic system. For both systems, let us analyze the step response in time domain and the amplitude response in frequency domain shown in Fig. 2 (a) and (b). Here, $\mathrm{H}_1(s)$ denotes the transfer function $\mathrm{H}(s) = \dot{\theta}(s)/U(s)$ of Eq. (2), and $\mathrm{H}_2(s)$ as well as $\mathrm{H}_3(s)$ denote the one of Eq. (3). At this, the stiffness K is set to 1e3 for $\mathrm{H}_2(s)$, and to 1e8 for $\mathrm{H}_3(s)$. The case described by $\mathrm{H}_3(s)$ should approximate a rigid system at which the stiffness increases towards unlimited. Note that the residual parameters of Eqs. (2) and (3) remain constant. It is easy to recognize that the step response of $\mathrm{H}_3(s)$ coincides well with those one of the absolute rigid joint $\mathrm{H}_1(s)$. When analyzing the frequency response function it can be seen that the resonance peak of $\mathrm{H}_3(s)$ is shifted far to the right comparing to $\mathrm{H}_2(s)$. Up to the resonance range the frequency response function $\mathrm{H}_3(s)$ coincides exactly with $\mathrm{H}_1(s)$. Note that the shifted resonance of $\mathrm{H}_3(s)$ provides the same peak value as $\mathrm{H}_2(s)$. Hence, during an exceptional excitation exactly at the resonance frequency the oscillatory output will arise again. However, from the practical point of view such a high-frequently excitation appears as hardly realizable in a mechanical system. Even so, when approaching the resonance range, the high decrease of the frequency response function provides a high damping within the overall system response.

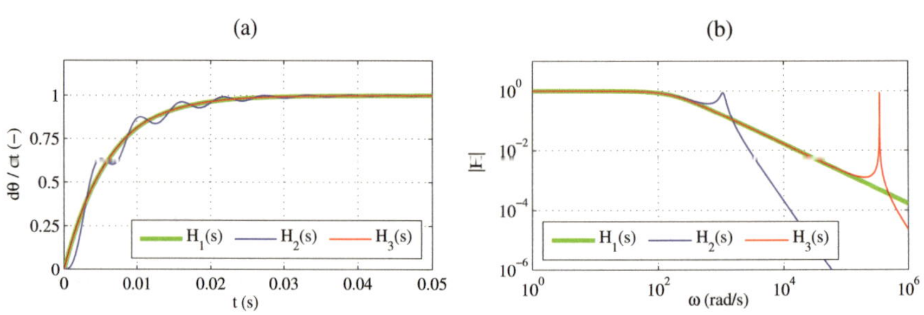

Fig. 2. Comparison of linear joint approaches; step response (a), frequency response (b)

The performed qualitative example demonstrates the universal character of the proposed joint topology. At this, the principal mechanisms conducting the joint dynamics are independent

from the complexity of a particular robotic system. Being conditional upon the gear and bearing structure an optimal level of detail for modeling can be determined using a top-down approach. Starting from the simplest rigid case with a single actuator damping, additional compliant and frictional elements can be included hierarchically in order to particularize the case-specific system behavior.
In the following, we consider the single modeling steps required to achieve an adequate description of each subsystem included in the topology of a nonlinear elastic joint.

2.1 Joint actuator

Considering the joint actuator which is driven by an electrical servo motor the angular velocity and angular position of the output shaft are determined by two input values. The first one is the input torque u which induces the angular acceleration of the rotor. The second one is a feedback load torque τ which antagonizes the controllable motion and thus behaves as a disturbance value to be accounted for. The conjoint moment of inertia m includes the rotor with shaft as well as the additional rotating elements such as encoders, breaks, and couplings. Mostly, it can be supposed that all rotating bodies are homogenous and axially symmetrical. This straightforward simplification allows us to deal with the concentrated model parameters and to consider the relative motion at this stage as the single body dynamics. The joint actuator can be described as follows

$$m\ddot{q} + f(\dot{q}) + \tau = u\,, \tag{4}$$

in which the input torque is assumed to be linear to the regulated motor current i. The fast current controls of the common servo motors operate at sampling rates about or larger than 10 kHz. Hence, the transient behavior of the current control loop can be easily neglected taking into account the time constants of the mechanical system part. The latter amount to several tens up to some hundred milliseconds. With an appropriate current regulation an electrical servo motor can be considered as a nearly reactionless power source which provides the joint actuator with the driving input torque $u(t) \cong k_m i(t)$.
The friction $f(\cdot)$ acting in the bearings is in large part load-independent. However, since a particular joint actuator can be mounted on the robotic manipulator with multiple degrees of freedom (DOF) its frictional characteristics can vary dependent on the actual robot configuration. Particularly, the orientation of the supported motor shaft to the gravity field can influence the breakaway and sliding friction properties to a certain degree. Aside of the dependency from the robotic configuration the most significant frictional variations, as well known, are due to the thermal effects. Both the external environment temperature and the internal temperature of contacting elements play a decisive role, whereas the last one increases usually with the operating time and intensity. However, the thermal frictional effects can be rather attributed to a slow time-variant process and a corresponding adaptivity of the model parameters. For reason of clarity an explicit temperature dependency is omitted here just as in most known approaches of modeling the robot dynamics. In the following, we also assume that no significant eccentricities are present in the system, so that almost no periodic torque ripples occur on a shaft revolution. Otherwise, the harmonic disturbances partially attributed to the position-dependent friction can be efficiently estimated and compensated as proposed e.g. by De Wit & Praly (2000).
The complexity of the obtained actuator model can differ in number of the free parameters to be identified, mostly dependent on the selected mapping of the friction behavior. The lumped

moment of inertia can be easily estimated by decoupled load using the large steps and (or) free run-out experiments. However, in doing so a linear damping has to be assumed. Since the motion is essentially damped by a nonlinear friction it appears as more reasonable to identify the actuator inertia together with the corresponding friction parameters. For these purposes more elaborated and specially designed experiments either in time or frequency domain can be required (see e.g. by Ruderman & Bertram (2011a)). Often, it is advantageous to identify the actuator friction together with the friction acting in the gear transmission. The friction effects in the actuator and gear assembly are strongly coupled with each other since no significant elasticities appear between both. When identifying the dynamic friction one has to keep in mind that the captured torque value, mostly computed from the measured motor current, represents a persistent interplay between the system inertia and frictional damping.

2.2 Gear transmission

The mechanical gear transmission can be considered as a passive transducer of the actuator motion to the output torque which drives the joint load. At this, the angular position of the joint load constitutes the feedback value which contains the signature of elasticities and backlash acting in the transmission system. Often, the gear transmission provides the main source of nonlinearities when analyzing the joint behavior, the reason for which several applications arrange the direct drives without gear reduction (see e.g. by Ruderman et al. (2010)). However, most nowadays robotic systems operate using the gear units which offer the transmission ratios from 30:1 up to 300:1. Apart from the classical spur gears the more technologically elaborated planetary and harmonic drive gears have been established in robotics for several years. Different gear types exhibit quite differing level of torsional compliance and mechanical play, also known as backlash.

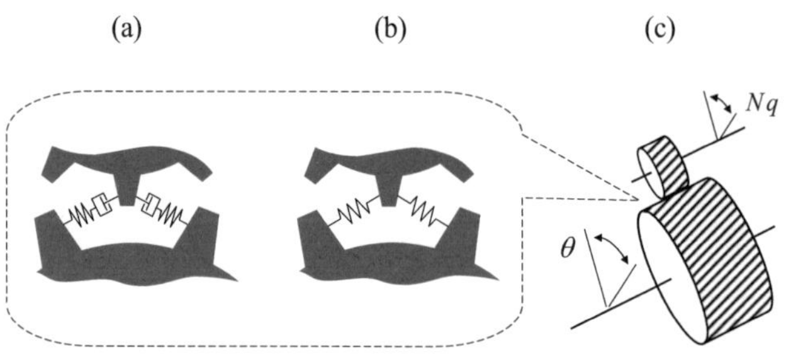

Fig. 3. Gear transmission with elasticities and backlash

An idealized rigid gear with a transmission ratio N provides the angular motion reduction

$$\theta = \frac{1}{N} q \tag{5}$$

and the corresponding torque gain

$$T = N \tau . \tag{6}$$

Due to the gear teeth meshing the disturbing torsional compliance and backlash can occur during a loaded motion. Assuming the rotationally symmetrical gear mechanisms, i.e. excluding any rotary cam structures, the cumulative effects of the gear teeth meshing can be represented as shown schematically in Fig. 3. Here in (a), the angular relative displacement

$$\delta = N q - \theta \tag{7}$$

is additionally subject to a backlash. However, the latter one can be neglected when the play size stays below the resolution of q and θ measurement. Thereafter, the case (b) represents a pure elastic deflection of the teeth meshing, whereat the corresponding compliance is not mandatory linear. In fact, it is rather expected that the teeth interaction exhibits a hardening stiffness property. Due to internal frictional mechanisms within the teeth engagement area the overall torsional joint compliance can behave piecewise as elasto-plastic and thus give rise to substantial hysteresis effects. The backlash, even when marginal, is coupled with an internal teeth friction and thus provides a damped bedstop motion. Due to a mutual interaction between the mentioned nonlinear phenomena the resulting hysteresis is hardly decomposable in proper frictional, structural and backlash elements. Hence, it must be rather considered as a compound input-output nonlinearity, while keeping in mind the nature of its internal mechanisms.

The overall hysteresis torque can be considered as a general nonlinear function

$$T(t) = h\big(\delta(t),\, \delta(0)\big) \tag{8}$$

of the relative displacement, and that under impact of the initial condition $\delta(0)$. The latter is poorly known when the absolute angular position is captured neither on the input nor output side of the joint, that is the most typical case in the robotic praxis. In this regard, it could be necessary first to saturate the transmission load in order to determine a proper hysteresis state which offers a well known memory effect. Here, the hysteresis memory manifests itself in the transmitted torque value which depends not only on the recent relative displacement between the gear input and output but equally on the history of the previous values. Thus, the dynamic hysteresis map has to replace the static stiffness characteristic curve of the gear transmission. However, the last one remains a still suitable approximation sufficient for numerous practical applications. Well understood, the hysteretic torque transmission includes an inherent damping which is characterized by the energy dissipation on a closed load-release cycle. The enclosed hysteresis loop area provides a measure of the corresponding structural losses, where the damping ratio is both amplitude- an frequency-dependent. Thus, the structural hysteresis damping differs from the linear viscous one and can lead to the limit cycles and multiple equilibrium states.

The following numerical example demonstrates the damping characteristics of the joint transmission in a more illustrative way. For instance, the single mass with one DOF is connected to the ground, once using a linear spring with linear viscous damping, and once using a nonlinear hysteretic spring. At this, the linear stiffness is selected so as to coincide with the average value of the nonlinear stiffness map included in the hysteresis model. When exciting the system by the Dirac impulse (at time t=1) the eigenmotion behaves as a damped oscillation shown in Fig. 4. The linear case (a) provides a typical second-order oscillatory response which is dying out towards zero equilibrium state, having an exponential enveloping function. In case (b), the nonlinear hysteretic spring is higher damped at the beginning, though does not provide a final equilibrium state. Instead of that, the motion enters a stable limit cycle up from a certain amplitude. The last one indicates the hysteresis cancelation close to zero displacement that is however case-specific and can vary dependent

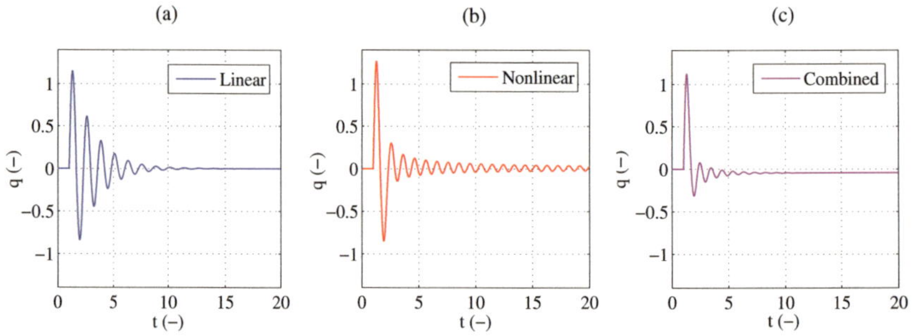

Fig. 4. Damped response of the excited eigenmotion when applying linear and nonlinear hysteretic spring

on the selected hysteresis map. From a physical point of view an autonomous mechanical system will surely converge to a static idle state due to additional damping mechanisms not necessarily captured by the hysteresis. However, theoretically seen the hysteresis damping does not guarantee the full decay of all eigenmotions. An interesting and more realistic case (c) represents a combination of the linear and nonlinear hysteretic damping. Due to an additional velocity-dependent damping term the eigenmotion does not stay inside of a limit cycle and converges towards a non-zero idle state, thus providing the system response with memory. The characteristical nonzero equilibrium states denote the forces remaining in the gear transmission after the input and output loads drop out.

2.3 Joint load

The robot links which connect the single joints within the kinematic chain of manipulator constitute the moving bodies with additional inertia and elasticities. They may generate lightly damped vibrational modes, which reduce the robot accuracy in tracking tasks according to Zollo et al. (2005).

The overall multi-body manipulator dynamics (1) provides the interaction between the actuated joints and passive links. Recall that Eq. (1) captures a rigid robot structure with no elasticities at all. However in general case, the contribution of the gravity and Coriolis-centrifugal forces has to be taken into account independent of the considered rigid or elastic manipulator. For reason of traceability, we cut free the kinematic chain and consider the single robot joint with the following link as depicted in Fig. 5. At this, the joint link can be represented either as the concentrated L or distributed $L_1, \dots, L_n$ mass. Here, no additional inertia and Coriolis-centrifugal terms fed back from the following joints and links contribute to the overall load balance. Note that the recent approach constitutes a strong simplification of the link dynamics so that the additional degree of freedom $\theta_L = \theta - \alpha$ or $\theta_L = \theta - \sum \alpha_i$ for case (b) is used to capture the vibrational disturbances only. It means that in terms of the distributed impact of gravity the angular link deflection α is assumed to be small enough, so that $\sin(\theta) \approx \sin(\theta - \alpha)$ and

$$G = l\,g\,\sin(\theta)\,. \tag{9}$$

Here, g denotes the gravity constant and l is the lever of COG (center of gravity). The transmitted torque τ drives the joint output with inertia M and bearing friction which can be captured in a simplified way using Coulomb and viscous friction $F_c\,\mathrm{sgn}(\dot{\theta}) + d_f\,\dot{\theta}$ only.

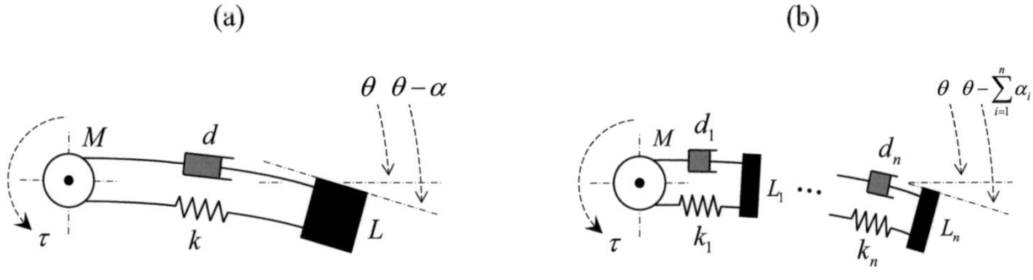

Fig. 5. Concentrated (a) and distributed (b) joint load

For reasons of clarity let us analyze here the case of a concentrated mass shown in Fig. 5 (a). At this point it is important to say that the load model with distributed masses shown in Fig. 5 (b) can behave more accurately and capture more complex vibration modes. At the same time, the identification of distributed mass-stiffness-damper parameters can rapidly exceed the given technical conditions of the system measurement. Besides, the achieved gain in accuracy of capturing the reactive load related to vibrations is not necessarily high. Recall that the primary objective here is to determine the reactive torque fed back from the oscillating elastic link and thus to improve the prediction of θ. An accurate computation of the link end-position due to link elasticities is surely also an important task in robotics. However, this falls beyond the scope of the current work whose aim is to describe the robot joint transmission with nonlinearities. Introducing the state vector $\mathbf{x} = (\theta\ \dot{\theta}\ \theta_L\ \dot{\theta}_L)^T$ and the vector of nonlinearities $\mathbf{h} = (\sin(\theta)\ \mathrm{sgn}(\dot{\theta})\ 0\ 0)^T$ one can obtain the state-space model of the concentrated joint load given by

$$\dot{\mathbf{x}} = \begin{pmatrix} 0 & 1 & 0 & 0 \\ -k/M & -(d_f+d)/M & k/M & d/M \\ 0 & 0 & 0 & 1 \\ k/L & d/L & -k/L & -d/L \end{pmatrix} \mathbf{x} + \begin{pmatrix} 0 & 0 & 0 & 0 \\ -lg/M & -F_c/M & 0 & 0 \\ 0 & 0 & 0 & 0 \\ 0 & 0 & 0 & 0 \end{pmatrix} \mathbf{h} + \begin{pmatrix} 0 \\ N/M \\ 0 \\ 0 \end{pmatrix} \tau . \tag{10}$$

The 4×4 matrices in Eq. (10) constitute the system matrix of a linear part and the coupling matrix of nonlinearities correspondingly. For all physically reasonable parameter values the system proves to be controllable whereas the input torque τ occurs as always lagged due to preceding dynamics of the joint actuator and gear transmission.

The following numerical example demonstrates the disturbing link vibrations during the torque step whose unavoidable lag is approximated by a first-order low pass filter with cut-off frequency 1000 rad/s. The parameter values are selected so as to provide the relation between both inertia $M/L = 0.1$ and damping $d_f/d = 10000$. Further, two relative stiffness values $1k$ and $10k$ are considered. Since the gravity term is not directly involved into vibrational characteristics it can be omitted at this stage. In the same manner the Coulomb friction nonlinearity which constitutes a constant torque disturbance during an unidirectional motion is also excluded from the computation. The response of the simulated, insofar linear, joint load is shown in Fig. 6. Two stiffness values differing in order of magnitude lead to the oscillating link deflection depicted in Fig. 6 (b) and (c). Besides the differing eigenfrequencies the principal shape of the enveloping function appears as quite similar for both stiffness values.

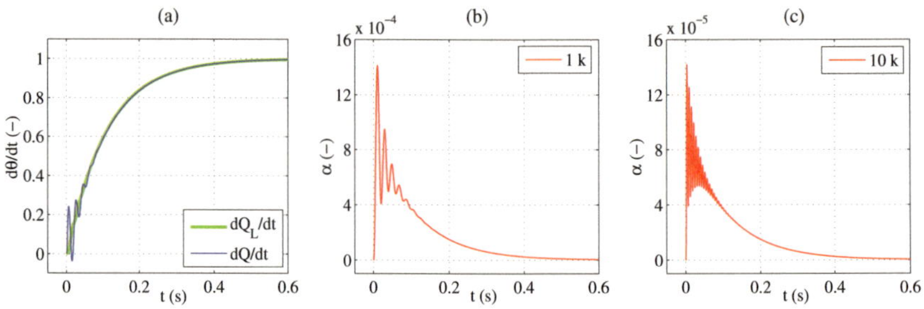

Fig. 6. Response of the joint load with elasticities to the lagged step of the input torque

However, most crucial is the fact that the amplitudes differ in the same order of magnitude, so that a lower stiffness provides a definitely higher level of link vibrations. Considering the correspondent velocity response, once in the joint output and once in the link deflection coordinates depicted in Fig. 6 (a) the impact of the link vibrations becomes evident. Here, the output joint velocity $\dot{\theta}$ is subject to the substantial oscillations at the beginning of relative motion. As a consequence, the oscillating behavior of the joint load will be propagated back to the joint transmission, thus increasing the complexity of overall joint behavior.

3. Nonlinear joint model

The main focus of the current Chapter relates to the grey-box modeling of elastic robot joints with nonlinearities. With respect to the proposed joint topology shown in Fig. 1 that means to derive a particular structure and to find a suitable set of equations to capture the transmission behavior for a class of robotic joints with elasticities. At this, the use of freely parameterizable sub-models and functions allows to describe the system dynamics for different types of gears and bearings involved in the joint assembly. The class of the robot joints aimed to be described here is characterized by a rotary-to-rotary transmission of the actuated relative motion with no position-dependent changes in the joint behavior. That is the input drive torque is transmitted to the joint output by means of the axially symmetric gear mechanisms. Note that the latter, including all bearings and couplings, allow one degree of freedom only. Several types of the gears like harmonic drives, planetary, and diverse subtypes of spur gears fulfill these assumptions, though offering quite differing mechanical principles of the torque transmission.

3.1 Joint structure

The proposed single elastic robot joint with nonlinear damping and hysteresis is shown in Fig. 7. The input drive torque u accelerates the combined actuator mass m, where the friction f constitutes the overall coupled friction of the actuator and gear input. The mechanical joint interface arranges the observation of angular motion, once prior $(q, \dot{q})$ and once beyond $(\theta, \dot{\theta})$ the gear transmission. That is the input and output joint axes are assumed to be rigid outside the employable state measurements. Note that the output joint axis does not necessarily coincide with some physical rotary shaft coming out from the mechanical joint assembly. Just as well it can be a virtual axis of a rotary motion for which the interface will indicate the measuring points. More simply, the input motion interface coincides with the rotary actuator (motor) shaft to which the angular position/velocity measurement is directly applied in the most cases. The nominal gear transmission ratio is denoted by N. The main transmission element with elasticities can be represented by a nonlinear rotary spring with hysteresis Γ.

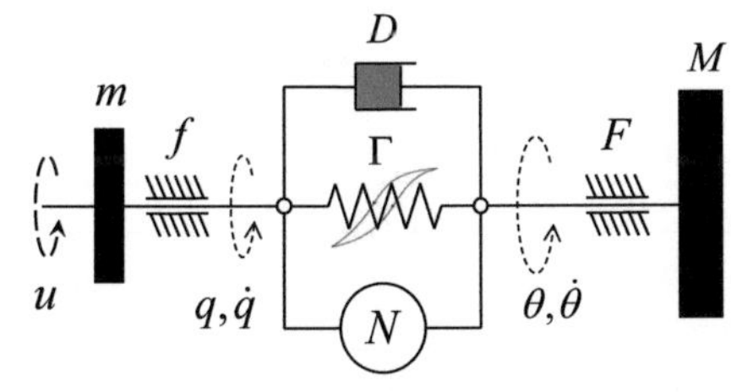

Fig. 7. Elastic robot joint with nonlinear damping and hysteresis

Apart from the hardening stiffness and hysteresis property, the rotary spring can also include the backlash nonlinearities as mentioned afore in Section 2.2. Note that the complexity degree of nonlinearities involved in the transmitting rotary spring can vary. This depends on the required model accuracy as well as significance of each particular phenomenon observable when measuring the joint behavior.

Since the hysteresis map captures the frequency-independent structural damping, an additional linear damping term D is connected in parallel to the rotary spring. The introduced linear damping comprises all viscous (velocity-dependent) dissipative effects arising from the internal teeth and slider interactions as mentioned afore in Section 2.2. Here, one must be aware that the velocity-dependent damping map can ditto possess the nonlinear characteristics. However, for reasons of clarity and more straightforward identification the simple linear damping is further assumed. As shown previously in Fig. 4 (c) the superposition of both damping mechanisms provides the system dynamics with multiple non-zero equilibrium states. Apart from the curvature and area of the underlying hysteresis, which determine the dissipative map, the magnitude of the linear damping affects a faster settling of the oscillatory joint motion.

The moving output joint part is mandatory supported by some rotary-type bearing, so that the output friction torque F has to be included. A proper decomposition of the input and output joint friction appears as one of the most challenging identification tasks when modeling the joint with elasticities. When no sufficient input and output torque measurements are available, certain assumptions about the friction distribution across the joint transmission have to be made. One feasible way to decompose a coupled joint friction is to identify first the overall friction behavior under certain conditions, at which the impact of joint elasticities is either negligible or it constitutes a constant bias term. Here, one can think about the drive experiments under reduced (decoupled) or stationary handled joint load. The steady-state velocity drive experiments can constitute the basis for such an identification procedure. Once the overall joint friction is identified it can be decomposed by introducing the weighting functions and finding the appropriate weighting rates within the overall joint dynamics. However, this heuristic method requires a good prior knowledge about the acting joint elasticities. That is the nonlinear spring and damping models have to be identified afore.

3.2 Dynamic friction

The modeling of dynamic friction is one of the crucial tasks in robotics since the joint friction constitutes not only the stabilizing damping but also can lead to large errors in the control. There is a huge number of works dedicated to modeling and compensating the frictional effects in robotic joints and machine tools (see Armstrong-Helouvry et al. (1994), Bona & Indri (2005), Al-Bender & Swevers (2008) for overview).

Generally, the existent approaches can be subdivided into static and dynamic mapping of kinetic friction. Note that in terms of kinetics the static friction signifies the friction forces acting during a relative motion at constant or slightly changing velocities and not the sticking forces of stationary bodies. Here, the classical constant Coulomb friction and the linear velocity-dependent viscous friction can be involved to describe the frictional forces between two contacting surfaces which slide upon each other. Moreover, the well-known Stribeck effect can be involved to describe the nonlinear transition between the break-away friction at zero velocity and linear viscous friction in a more accurate way. The most dynamic friction models capture both pre-sliding and sliding regimes of the friction and provide the smooth transition through zero velocity without discontinuities typical for static friction models. The most advanced dynamic friction models are also capable of describing such significant frictional phenomena as position-dependent pre-sliding hysteresis and frictional lag in the transient behavior. An extensive comparative analysis of several dynamic friction models, among Dahl, LuGre, Leuven, and GMS one, can be found in Al-Bender & Swevers (2008) and Armstrong-Helouvry & Chen (2008).

Apart from the pre-sliding frictional mechanisms, the main differences between the static and dynamic friction modeling can be explained when visualized in the velocity-force (w, F) coordinates as shown in Fig. 8. The static friction map provides a discontinuity at zero velocity which can lead to significant observation errors and control problems since the system operates with frequent motion reversals. However, when the steady-state operation modes are predominant the static map can be accurate enough to capture the main frictional phenomena. Further, it is easy to recognize that the closer the relative velocity is to zero the larger is the impact of nonlinear Stribeck effect. The peak value at zero velocity indicates the maximal force, also called break-away, required to bear the system from the sticking to the sliding state of continuous motion. Comparing the Stribeck static map with the dynamic

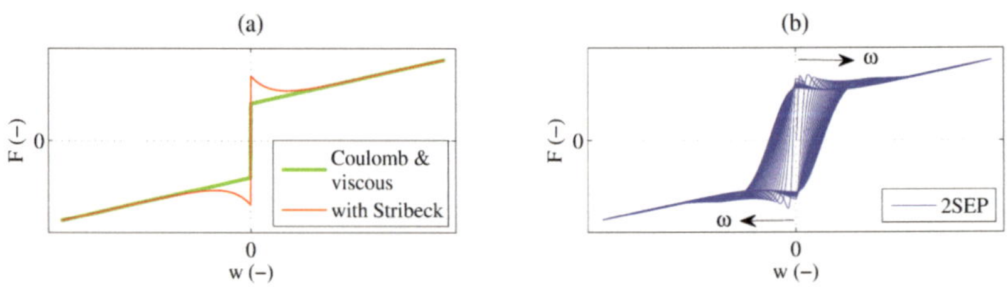

Fig. 8. Static (a) and dynamic (b) friction map in velocity-force coordinates

one shown in Fig. 8 (b) it can be seen that the principal shape of the velocity-force curves remain the same. However, the transient frictional behavior when passing through zero velocity is quite different dependent on the ongoing input frequency ω. Here, the overall friction response to a cyclic input velocity with an increasing frequency is shown. The higher is the ongoing frequency when changing the motion direction the higher is the stretch of the resulting frictional hysteresis. This phenomenon is also known as the frictional lag (see by Al-Bender & Swevers (2008)) since the friction force lags behind the changing relative velocity. Roughly speaking, the frictional lag occurs due to the time required to modify either the hydro-dynamic lubricant properties during the viscous, or the adhesion contact properties during the dry sliding. With an increasing relative velocity after transient response the friction

force is attracted towards the static characteristic curve which represents the steady-state friction behavior. Depicted in Fig. 8 (b), the dynamic friction response is obtained using 2SEP (two-state with elasto-plasticity) model which is briefly described in the rest of this Section.
The 2SEP dynamic friction model (Ruderman & Bertram (2010), Ruderman & Bertram (2011a)) maintains the well-established properties of dynamic friction which can be summarized as i) Stribeck effect at steady-state velocity-dependent sliding, ii) position-dependent pre-sliding hysteresis, and iii) friction lag at transient response. The model describes the dynamic friction behavior using a linear combination of the pre-sliding and transient friction state that converges towards the velocity-dependent steady-state. At this, one independent and one dependent state variables are intuitively related to such frictional phenomena as the pre-sliding hysteresis and friction lag at transient behavior. The weighted superposition of both yields the total friction force as

$$F = A\, z_1 + B\, |w|\, z_2 \,. \tag{11}$$

Note that the instantaneous friction value is conducted coevally by the relative displacement and relative velocity without switching functions and (or) thresholds to separate the pre-sliding and sliding regimes. The pre-sliding friction behavior, predominantly captured by z_1, is described using the Modified Maxwell-Slip (MMS) approximation (Ruderman & Bertram (2011b)). The MMS structure can be imagined similar to a single Maxwell-slip element, but with a pivotal difference of representing the applied connecting spring. In MMS model, the irreversible nonlinear spring exhibits a saturating elasto-plastic deflection until a motion reversal occurs. Assuming an exponentially decreasing stiffness by an increasing relative displacement after reversal, the overall hysteretic state converges towards the constant tangential force. It is assumed that the deflection of elasto-plastic asperity contacts saturates at the Coulomb friction level, so that the weighting factor A can be often set to one. A saturated friction state characterizes the plastic sliding at which the deflected asperities slip upon each other and provide an approximately constant value of the tangential force at non-zero velocity. The state dynamic is defined as a first-order nonlinear differential equation

$$\dot{z}_1 = |\Omega|\, w\, K \exp\big(-K|q_r|\big)\,, \tag{12}$$

with

$$\Omega = \mathrm{sgn}(w)\, F_c - F\,, \tag{13}$$

when the velocity sign changes. The average stiffness capacity Ω of interacting asperities memorizes the last motion reversal, at which the asperity junctions are released and reloaded again in the opposite direction. The integrated relative displacement q_r is reset to zero whenever the motion direction changes. Solely two parameters, the Coulomb friction F_c and the initial stiffness K, determine the overall hysteresis map of the pre-sliding friction.
The transient friction state

$$\dot{z}_2 = \frac{S(w) - F}{|S(w)|}\,, \tag{14}$$

behaves as a first-order time delay element that transfers the velocity-dependent dynamic friction and attracts it towards the steady-state $S(w) = s(w) + \sigma\, w$. The factor σ constitutes the linear viscous friction term and the nonlinear characteristic curve

$$s(w) = \mathrm{sgn}(w)\Big(F_c + (F_s - F_c)\exp\big(-|\frac{w}{V_s}|^{\delta}\big)\Big) \tag{15}$$

describes the well-known Stribeck effect. The velocity-dependent Stribeck map is upper bounded by the static friction (also called stiction) force F_s and low bounded by the Coulomb friction force. The exponential factors V_s and δ denote the Stribeck velocity and the Stribeck shape factor correspondingly. In Eq. (11), the attraction gain B determines how fast the overall friction force converges towards the steady-state. The attraction gain is additionally subject to the velocity magnitude. On the one hand, this is done in order to reduce the impact of the transient behavior in vicinity to zero velocity, at which the pre-sliding frictional mechanisms predominate. On the other hand, the velocity-dependent attraction gain ensures the steady-state convergence as $|w|$ increases.

Overall seven free parameters are required to describe the dynamic friction using 2SEP model, where five of them are already spent to support the standard Stribeck effect. Here, it is important to say that other dynamic friction models which capture the basic frictional phenomena can be applied in equal manner when describing the overall joint behavior. At this, the most significant aspects by the choice of a particular friction model are the ease of implementation, generality, as well as practicability in terms of the identification to be done under the circumstances of a real operation mode.

3.3 Nonlinear stiffness with hysteresis

In the simplest case the robot joint stiffness can be represented by a linear function of the relative angular displacement across the joint transmission. This somehow archaic but still widely used and, in particular, robust approach relates the joint torque transmission to a classical spring behavior. Nevertheless, most compliant mechanical structures are well known to exhibit a kind of the hardening properties. That is an increasing relative displacement or better to say deflection of the transmitting elements causes an increase of the overall joint stiffness, and thus leads to higher torque rates. Hence, the joint stiffness which is the inverse compliance of all embedded transmitting elements appears as variable, depending on the ongoing joint load. In order to account for variable stiffness properties several manufacturers of gears and components as well as some works published by researchers suggest to use a piecewise linear stiffness approximation with multiple characteristical segments. An alternative to describe the variable stiffness characteristics is to use the polynomial functions whose coefficients have to be fitted from the accurate laboratory measurements.

Since a nonlinear transmission spring is coupled with corresponding hysteresis effects as mentioned in Section 2.2 a suitable hysteresis map has to be also incorporated. A huge number of available hysteresis approaches (see e.g. Bertotti & Mayergoyz (2006)), originated from different domains of natural and technical science, allow a certain freedom by choosing an appropriate model. The well established Bouc-Wen (Bouc (1967), Wen (1976)) hysteresis model as well as its numerous differential extensions (see survey provided by Ismail et al. (2009)) appears to be particulary suitable for those purposes. The Bouc-Wen hysteresis model, which originated from the structural mechanics, uses a first-order non-linear differential equation that relates the relative displacement to the restoring force in a hysteretic way. At this, the shape of the displacement-force hysteresis is controlled by a compact set of free parameters to be determined in each particular case. Another advantage of a Bouc-Wen-like hysteresis modeling is the possibility to incorporate more complex nonlinear stiffness characteristics, as suggested by Ruderman et al. (2010) and shown in the following in more detail.

The restoring torque arising in a hysteretic spring is subdivided into an elastic and plastic (irreversible) part related to each other by a weighting factor. When using the polynomials $k_1, \ldots, k_i$ for describing the characteristic stiffness curve the total restoring torque can be expressed as a superposition

$$\Gamma(\delta, x) = \sum_i \left(\mu\, k_i \,\mathrm{sgn}(\delta)\, |\delta|^i + (1-\mu)\, k_i \,\mathrm{sgn}(x)\, |x|^i\right). \tag{16}$$

At this, the weighting coefficient $0 < \mu < 1$ provides a relationship between the purely elastic ($\mu = 1$) and purely hysteretic, further also as plastic, ($\mu = 0$) deflection. The dynamic state variable x which captures the elasto-plastic deflection, therefore responsible for arising hysteresis, is described by

$$\dot{x} = A\,\dot{\delta} - \beta\,|\dot{\delta}|\,|x|^{n-1}x - \gamma\,\dot{\delta}\,|x|^n. \tag{17}$$

The amplitude and shape of hysteresis are determined by the parameters A, β, and γ. The factor $n \geq 1$ provides the smoothness of transition between an elastic and plastic response. For an interested reader a more extensive parameter analysis of Bouc-Wen-like differential

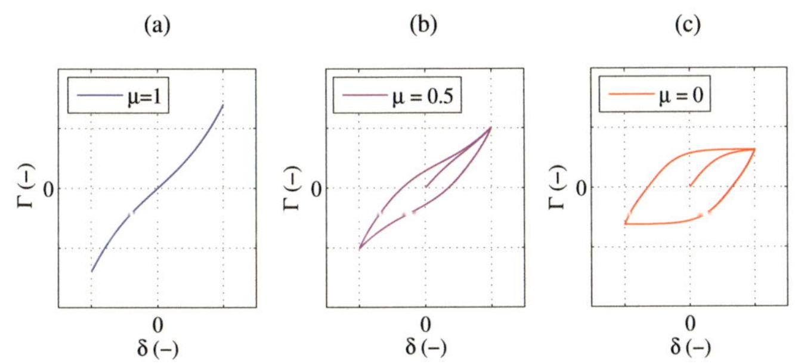

Fig. 9. Hysteretic restoring torque for purely elastic (a), elasto-plastic (b), and plastic (c) torsional deflection

hysteresis models can be found by Ma et al. (2004). In Fig. 9, an example of the restoring torque on a closed load-release cycle is shown dependent on the factor μ. At this, the residual model parameters have been held by the same values. It is evident that no hysteresis losses occur for μ=1, and the case represented in Fig. 9 (a) provides a simple nonlinear characteristic stiffness curve with hardening properties. Though the larger a plastic contribution to the overall restoring torque is the larger is the hysteresis area in Fig. 9 (b) and (c), and the larger is the corresponding structural damping acting in the joint transmission.

With respect to the joint structure derived in Section 3.1 the overall transmitted joint torque is given by

$$T = \Gamma(\delta) + D\,\dot{\delta}. \tag{18}$$

Note that the transmitted joint torque already includes the nominal gear ratio N captured by δ. When replacing the nonlinear function $\Gamma(\cdot)$ by a linear spring described by $K\delta$ a simple joint transmission model with linear stiffness and damping can be achieved again. However, when accounting for hysteresis motion losses the use of a hysteresis stiffness map becomes indispensable and has to be included in the overall joint dynamics.

4. Experimental case study

In the following, let us analyze the experimental observations together with some exemplary identification results obtained on the laboratory setup of an elastic revolute joint. The deployed laboratory setup depicted in Fig. 10 has been designed and constructed at the Institute of Control Theory and Systems Engineering of Technical University Dortmund and is primarily intended for investigating the single joint behavior as well as for the control experiments. The testbed contains a standard BLDC motor with rated power 400 W and idling speed 3100 rpm. The servo drive is equipped by a 15-bit resolver and is energized by a PWM electronic unit with onboard current control loop of 11 kHz rate. Through a torsionally stiff

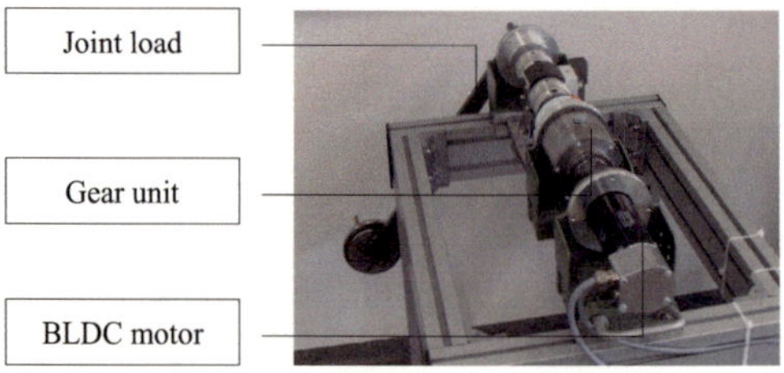

Fig. 10. Laboratory setup of elastic revolute joint

compensating coupling the servo drive is connected to the lightweight harmonic drive gear set (160:1) embedded in a rigid casting. On the output of gear transmission a high-precise absolute encoder with 20-bit resolution is mounted. Using the fixture with couplings a contactless torque cell is applied to measure the output torque for both bounded and free joint motion. Behind the torque measurement the output shaft is equipped with a mechanical interface in order to attach the link rod with adjustable additional weights. Note that the torsional deflection is determined as a difference between the output joint angle and input joint angle captured on the motor shaft.

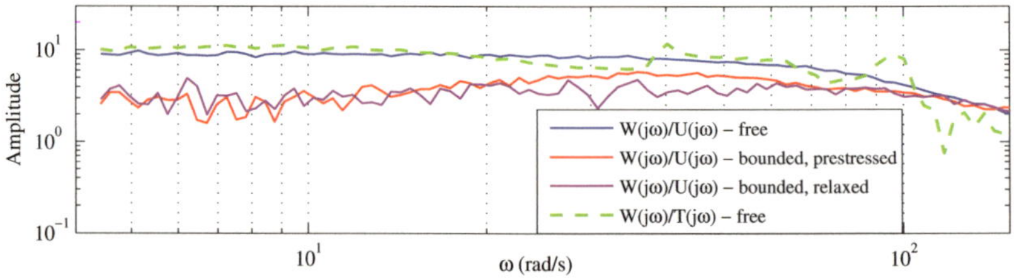

Fig. 11. Frequency response functions measured at different excitation conditions

Before looking on the identification results let us analyze first the obtained frequency response functions (FRF)s of the joint actuator. Four FRFs depicted in Fig. 11 are measured at the different excitation conditions. The transfer from the input torque $U(j\omega)$ to the angular velocity $W(j\omega)$ is realized once for a free and once for a bounded motion for which the output joint shaft was locked. As a controlled input excitation the short impulse of 5 A amplitude and 1.5 ms duration is applied. The bounded motion response is obtained once for a relaxed and once for a prestressed joint state. The latter one means that before applying the input impulse the motor shaft was turned manually in the same motion direction until the elastic transmission becomes fully twisted but does not rotate back due to the friction. The bounded

prestressed FRF exhibits a slight resonance increase due to the raised joint elasticity comparing to the bounded relaxed and the free FRF. At the same time, the high structural and frictional damping impedes the occurrence of significant resonances which could be related to the joint elasticities. The transfer from the joint torque $T(j\omega)$ to the angular velocity $W(j\omega)$ illustrates a backward propagation of reactive restoring torque coming from the joint load. The torque excitation is realized by an external shock applied mechanically to the load part of the joint. The $W(j\omega)/T(j\omega)$ FRF coincides qualitatively with the $W(j\omega)/U(j\omega)$ FRF of free motion despite a high level of process and measurement noise.
Further, let us examine the identification of the joint actuator including the nonlinear friction and the total inertia of the motor and gear assembly. Using the available motor current and joint torque as the inputs and the angular actuator velocity as the output the identification problem can be easily solved in the least-squares sense. The system is excited by a down-chirp signal with 30–1 Hz frequency range. The measured and identified velocity response shown in Fig. 12 exhibits some resonant and anti-resonant phases related to the overall system dynamics.

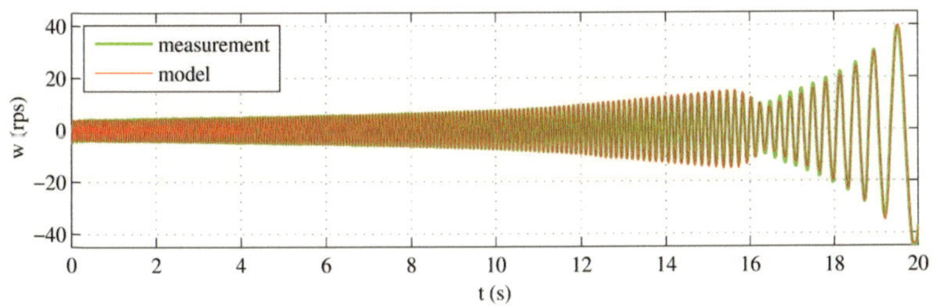

Fig. 12. Measured and identified velocity response to the down-chirp excitation

In the similar manner, the nonlinear joint transmission can be identified using the joint torque and torsion measurements obtained from the same experiment. The time series of the measured and identified joint torque is depicted in Fig. 13 (a). It is easy to recognize that both curves coincide fairly well with each other, thus proving the transmission model to be capable of describing the dynamic joint behavior. Note that the visible discrepancies occur rather at low frequencies at $t > 19$ s. Here, the joint output vibrations which come from the non-smoothly moving load part influence the torque and angular measurements in a non-congruent way. The corresponding torsion-torque hysteresis depicted in Fig. 13 (b) confirms ditto a good agreement between the measured and modeled response, and that for a large set of the nested hysteresis loops.

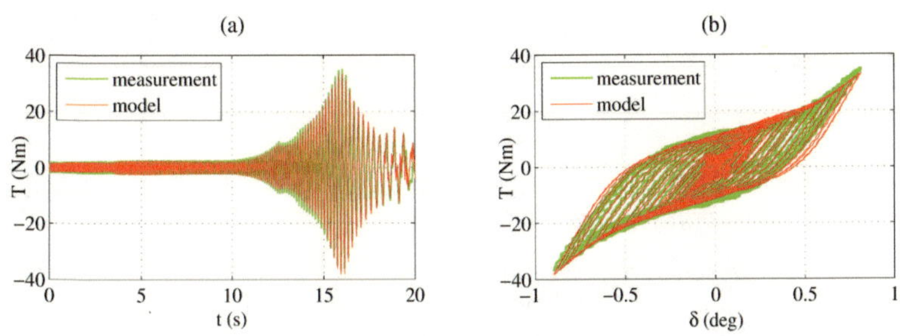

Fig. 13. Measured and identified joint torque, (a) time series, (b) torsion-torque hysteresis

Finally, the diagrams provided in Fig. 14 visualize some identification results related to the joint load. The gravity term can be determined in a quite simple way using the quasi-static motion experiments. The joint link is driven slowly across the overall operation space by using the lowest possible constant input torque. The obtained measurements of the joint torque and output angular position are used to fit the gravity as shown in Fig. 14 (a). The measured and predicted velocity response of the identified joint load is shown in Fig. 14 (b) for the applied step excitation. The peak velocity of about 60 deg/s is in the upper operation range of the experimental joint setup. A substantial velocity decrease occurs due to an increasing impact of gravity. Note that the depicted (measured) joint torque constitutes the total torque balance behind the gear affected simultaneously by the transmitted actuator torque and reactive torque of the joint load. A clearly visible oscillation pattern indicates the impact of the output joint shaft and link vibrations.

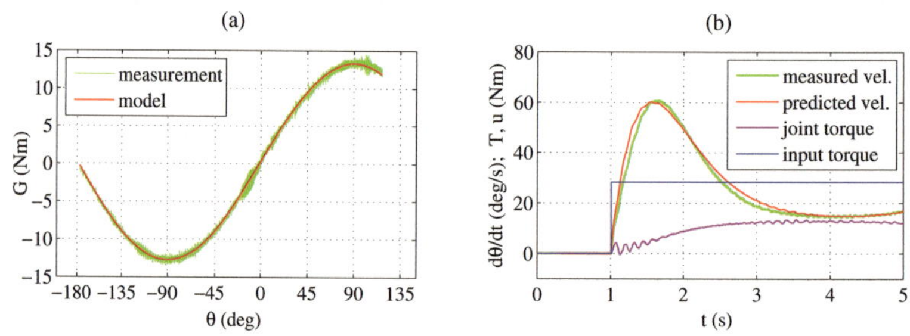

Fig. 14. Measured and identified gravity torque (a), measured and predicted velocity response of the joint load together with the measured applied and transmitted torque (b)

5. Conclusions

The presented methodology aims for analyzing and modeling the behavior of elastic robot joints with nonlinearities such as hysteresis and friction which affect the stiffness and damping characteristics. Independent of the rigid or elastic joint consideration a general topology of revolute robot joints has been demonstrated as a starting position for a subsequent, more detailed system modeling. The defined forward and feedback paths allow to decompose the robot joint into the close physical subsystems with appropriate eigenbehavior. This permits to cut free the entire joint, and to analyze and identify its subsystems stand-alone, under the terms of fully or partially measurable states. For a large class of rotary joint transmissions a sophisticated but nevertheless manageable model has been proposed. In addition to numerical examples an experimental case study has been shown to prove the proposed methods.

6. References

Al-Bender, F. & Swevers, J. (2008). Characterization of friction force dynamics, *IEEE Control Systems Magazine* 28(6): 64–81.

Albu-Schaffer, A., Eiberger, O., Grebenstein, M., Haddadin, S., Ott, C., Wimbock, T., Wolf, S. & Hirzinger, G. (2008). Soft robotics, *IEEE Robotics Automation Magazine* 15(3): 20–30.

Armstrong-Helouvry, B. & Chen, Q. (2008). The Z-properties chart, *IEEE Control Systems Magazine* 28(5): 79–89.

Armstrong-Helouvry, B., Dupont, P. & de Wit, C. C. (1994). A survey of modeling, analysis tools and compensation methods for the control of machines with friction, *Automatica* 30: 1083–1138.

Bertotti, G. & Mayergoyz, I. D. (2006). *The Science of Hysteresis 1–3*, first edn, Academic Press.

Bona, B. & Indri, M. (2005). Friction compensation in robotics: an overview, *Proc. 44th IEEE Conference on Decision and Control, and the European Control Conferenc*, Seville, Spain, pp. 4360–4367.

Bouc, R. (1967). Forced vibration of mechanical systems with hysteresis, *Proc. 4th Conference on Nonlinear Oscillations*, Prague, Czechoslovakia.

De Wit, C. C. & Praly, L. (2000). Adaptive eccentricity compensation, *IEEE Transactions on Control Systems Technology* 8(5): 757–766.

Dhaouadi, R., Ghorbel, F. H. & Gandhi, P. S. (2003). A new dynamic model of hysteresis in harmonic drives, *IEEE Transactions on Industrial Electronics* 50(6): 1165–1171.

Ferretti, G., Magnani, G. & Rocco, P. (2004). Impedance control for elastic joints industrial manipulators, *IEEE Transactions on Robotics and Automation* 20(3): 488–498.

Ismail, M., Ikhouane, F. & Rodellar, J. (2009). The hysteresis Bouc-Wen model, a survey, *Archives of Computational Methods in Engineering* 16(2): 161–188.

Ma, F., Zhang, H., Bockstedte, A., Foliente, G. C. & Paevere, P. (2004). Parameter analysis of the differential model of hysteresis, *ASME Journal of Applied Mechanics* 71(3): 342–349.

Ott, C., Albu-Schaeffer, A., Kugi, A. & Hirzinger, G. (2008). On the passivity-based impedance control of flexible joint robots, *IEEE Transactions on Robotics* 24(2): 416–429.

Ruderman, M. & Bertram, T. (2010). Friction model with elasto-plasticity for advanced control applications, *Proc. IEEE/ASME International Conference on Advanced Intelligent Mechatronics (AIM2010)*, Montreal, Canada, pp. 914–919.

Ruderman, M. & Bertram, T. (2011a). FRF based identification of dynamic friction using two-state friction model with elasto-plasticity, *Proc. IEEE International Conference on Mechatronics (ICM 2011)*, Istanbul, Turkey, pp. 230–235.

Ruderman, M. & Bertram, T. (2011b). Modified Maxwell-slip model of presliding friction, *Proc. 18th IFAC World Congress*, Milan, Italy, pp. 10764–10769.

Ruderman, M., Bertram, T. & Aranovskiy, S. (2010). Nonlinear dynamic of actuators with elasticities and friction, *Proc. IFAC 5th Symposium on Mechatronic Systems*, MA, USA, pp. 255–260.

Ruderman, M., Hoffmann, F. & Bertram, T. (2009). Modeling and identification of elastic robot joints with hysteresis and backlash, *IEEE Transactions on Industrial Electronics* 56(10): 3840–3847.

Sciavicco, L. & Siciliano, B. (2000). *Modelling and Control of Robot Manipulators*, second edn, Springer, Berlin, Germany.

Spong, M. W. (1987). Modeling and control of elastic joint robots, *Journal of Dynamic Systems, Measurement, and Control* 109(4): 310–319.

Taghirad, H. D. & Bélanger, P. R. (1998). Modeling and parameter identification of harmonic drive systems, *Journal of Dynamic Systems, Measurement, and Control* 120(4): 439–444.

Thummel, M., Otter, M. & Bals, J. (2005). Vibration control of elastic joint robots by inverse dynamics models, *Solid Mechanics and its Applications* 130: 343–354.

Tjahjowidodo, T., Al-Bender, F. & Brussel, H. V. (2006). Nonlinear modelling and identification of torsional behaviour in Harmonic Drives, *Proc. International Conference on Noise and Vibration Engineering (ISMA2006)*, Leuven, Belgium, pp. 2785–2796.

Wen, Y. K. (1976). Method for random vibration of hysteretic systems, *ASCE Journal of the Engineering Mechanics Division* 102(2): 249–263.

Zollo, L., Siciliano, B., Luca, A. D., Guglielmelli, E. & Dario, P. (2005). Compliance control for an anthropomorphic robot with elastic joints: Theory and experiments, *Journal of Dynamic Systems, Measurement, and Control* 127(3): 321–328.

16

Gravity-Independent Locomotion: Dynamics and Position-Based Control of Robots on Asteroid Surfaces

Marco Chacin[1] and Edward Tunstel[2]
[1]*Singularity University*
[2]*John Hopkins University Applied Physics Laboratory*
USA

1. Introduction

In recent years, the scientific community has had an increased interest in exploring the asteroids of the solar system (JAXA/ISAS, 2003; JHU/APL, 1996; NASA/JPL, 2007). Technological advances have enabled mankind for the first time to take a closer look at these small solar system objects through sensors and instruments of robotic deep space probes. However, most of these space probe missions have focused on the reconnaissance of the asteroids' surfaces and their compositional analysis from a distance. Little attention has been given to locomotion on their surfaces with a mobile robotic system, due to the challenging gravity conditions found in these small solar system bodies.

In small bodies like asteroids, the gravitational fields are substantially weaker than those of Earth or Mars, therefore the likelihood of a robot's unintentional collision with the surface while attempting a movement is substantially higher. In one of the latest missions, the Japanese Hayabusa spacecraft carried onboard a small robot named MINERVA (Yoshimitsu et al., 2001) to be deployed and used to explore the asteroid surface. The robot was designed with a single reaction wheel, located inside of it, to produce the necessary inertial reaction to move. But with this system the location of the robot when the motion is damped out is very challenging to predict or control. Subsequently, in order to maximize the scientific return from any given mission on an asteroid's surface, future missions must have the ability to conduct stable mobility and accurate positioning on the rough terrain.

In the robotics field, limbed locomotion is broadly recognized as superior in its capability to traverse terrains with irregularities such as obstacles, cliffs and slants. Exotic types of wheeled rovers (Bares et al., 1999; Wilcox & Jones, 2000) can only drive over obstacles of heights that are at best a fraction of the vehicle's body length. Thus, some terrains are not accessible to wheeled vehicles. Conversely, legged or limbed locomotion has the possibility to provoke minimum reactions on the asteroid surface that could push the robot with sufficient force to reach escape velocity and drift into space. It also facilitates achievement of desired goal configurations that deal with new complex situations, ensuring that a robot's behavior does not deviate from a stable condition.

In this chapter, the focus is on gravity-independent locomotion approaches, technologies and challenges of robotic mobility on asteroids. Recommendations and methods to perform

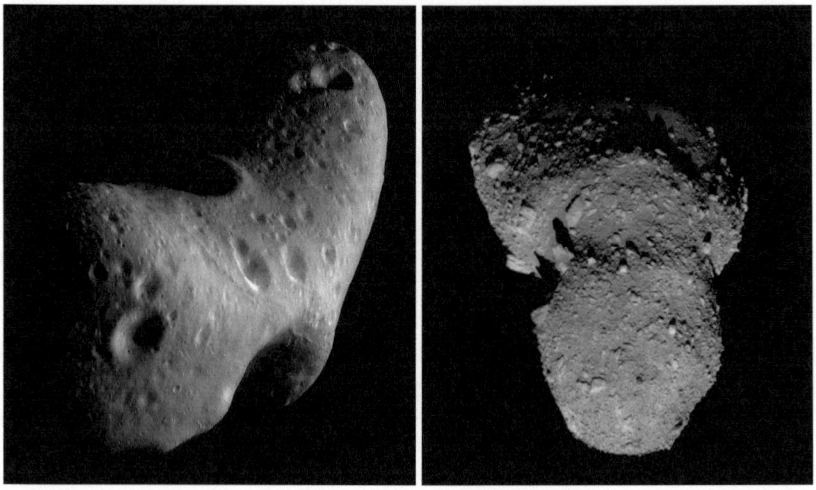

Fig. 1. Mosaic of Eros ©Applied Physics Lab/JHU and Asteroid 25143 Itokawa ©JAXA/ISAS

compliant motions during operations by a surface robot under microgravity conditions are presented. Taking into account the contact dynamics and the reaction force from its environment, a control scheme enables a limbed robot to move using the natural features of the environment and the friction of the surface for forward progress in any direction having contact only with the limbs' end tips.

2. Asteroid exploration

Orbiting between 2.1 to 3.2 AU[1] from the sun, primitive asteroids in the main asteroid belt represent key bodies to research on the early planetary system origin and evolution. Asteroids could provide clues about the birth and growth of our planetary system, but without any in-situ observation, we are not able to link measurements of asteroid material with a corresponding catalog of meteorite groups on Earth. The in-situ study of asteroids can lead to important scientific findings in the effort to map the main asteroid belt. Mapping the belt by spectral classes and knowledge about which region on Earth the meteorites have landed can provide key clues about the origin and evolution of our solar system, even including the geological history of our planet Earth (Fujiwara et al., 2006).

Asteroids' physical characteristics provide a very hostile environment distinguished by the absence of (almost any) gravity. The effects of the microgravity environment can be approximated for convenience as those on order of $10^{-6}g$ (Scheeres, 2004) (where g is the acceleration due to gravity on Earth). In such an environment, objects basically do not fall, but remain orbiting unless they reach the low escape velocity of the asteroid on order of 20 cm/s (Scheeres, 2004), as in the case of the asteroid 25143 Itokawa (Fig. 1, right). To attain stable mobility on these bodies, it is critical to consider the interaction forces between a robot and the asteroid's surface in such microgravity environments.

Relatively little attention from planetary scientists and planetary robotics engineers has been focused on surface mobility on asteroids. As a result, there exists some risk that premature conclusions about the feasibility of stable mobility on asteroid surfaces may be drawn without thorough consideration of all possible alternatives. However, it is clear that in order to increase

[1] AU stands for astronomical unit and is based on the mean distance from Earth to the Sun, 1.5×10^{8} [km].

any scientific return from a mission operating on an asteroid, movement on the surface would require a closer look at stability control against the forces interacting between bodies in a microgravity environment.

The following section presents recent developments in robotics which may pose the most feasible solutions to asteroid surface exploration challenges.

3. Mobility in microgravity

Weak gravitational fields (micro-g to milli-g) characteristic of asteroids make it difficult to achieve normal forces usually required for stable surface locomotion. Various approaches to mobility in weak gravity domains of asteroids are discussed below.

3.1 Rolling and hopping locomotion

Although all rovers sent to other planetary surfaces have been wheeled vehicles, wheels are not an obvious solution for asteroid surface mobility. However, at least one study has suggested wheels to be viable in certain cases for rovers with mass less than one kilogram. Baumgartner et al (Baumgartner et al., 1998) reported that the analysis of whether adequate tractive forces can be achieved for rolling mobility depends on the wheel-terrain interaction model employed. Traction sufficient for a small rover to traverse at a rate of 1 cm/sec was shown to be feasible via dynamic simulations revealing traction losses at the beginning of traverses. The simulations separately considered Coulomb friction alone and a combined Coulomb friction and adhesive force model.

Behar conducted computer-based dynamic simulation studies of hopping and wheeled vehicles concluding that both types have limited use for asteroid mobility (Behar et al., 1997). Hopping robots were deemed to be of limited utility due to complexity of both thruster control for accurate maneuvers and robot pose estimation. Wheeled robot utility was viewed as limited due to difficulty maintaining wheels on the surface when undesired surface reactions led to long periods of ballistic floating before touching down. To mitigate difficulties of maintaining wheel contact with asteroid terrain, Behar proposed a colony of small robots that would traverse asteroid surfaces while connected to a common physical net that was anchored to the asteroid. Robots connected to the anchored net would use wheeled locomotion to traverse along dedicated strands of the net. Another variation on rolling locomotion for asteroid mobility was proposed by Hokamoto and Ochi (Hokamoto & Ochi, 2001) based on a vehicle with a dodecahedron shape and 12 individually actuated prismatic leg joints oriented radially around the body to provide intermittent walking and rolling.

When complexities associated with accurate 3-axis thruster-based maneuvers are avoided and replaced by simpler propulsion mechanisms, ballistic hopping is perhaps the simplest means of mobility for reaching discrete patches of asteroid terrain. The aforementioned MINERVA vehicle was the only asteroid hopping rover fully developed for a space flight mission thus far. It was a 600 g vehicle designed for several asteroid days of autonomous operation involving ballistic hopping with variable hop speed and some control of hop direction depending on its attitude on the surface (Yoshimitsu et al., 2001). Other designs are in development or have been proposed as viable concepts. A 1.3 kg wheeled rover, at one time considered a payload for the Hayabusa mission on which MINERVA flew (Kawaguchi et al., 2003; Wilcox, 2000), was proposed with a novel mobility mechanism that also enables ballistic hopping. It further offered a capability to self-right in response to inevitable tumbling during landing from a hop or while traversing the surface at 1.5 mm/sec in low gravity (Tunstel, 1999).

Another hopping vehicle intended for asteroids is the Asteroid Surface Probe (Cottingham et al., 2009), an 8 kg battery-powered (100 hours) spherical body of 30 cm diameter that uses thrusters to hop. When stationary, the sphere opens up using 3 petals to expose a science instrument payload. The petals also provide the means to self-right the probe (Ebbets et al., 2007). Similar, in principle, is a 12 kg thruster-propelled ballistic free-flyer concept designed by the German Aerospace Center (DLR) as part of a European Space Agency study (Richter, 1998). Other hopping robots proposed for asteroid exploration include a pyramid-shaped, 533 g prototype with four single degree-of freedom (DOF) flippers at its base to enable hopping plus a lever arm for self-righting (Yoshida, 1999), and a spherical 1 kg robot with internal iron ball actuated by electro-magnets to induce hopping (Nakamura et al., 2000).

A recent study concluded that wheeled and hopping locomotion modes in low gravity are comparable in locomotion speed (Kubota et al., 2009). The study considered ideal conditions (e.g., flat terrain and no loss of contact between wheels and terrain). A similar conclusion regarding energy consumption was reached in a comparative study of wheeled and hopping rovers for Mars gravity (Schell et al., 2001).

3.2 Locomotion by crawling and climbing

Nature offers many existence proofs, in the form of animals and insects, for solutions capable of traversing rough terrain against forces of gravity. Limbed locomotion solutions are common among alternatives that could enable crawling or climbing to achieve viable mobility across asteroid surfaces.

Certain limbed mobility solutions for planetary rovers received limited consideration in the past for reasons of lower efficiency as compared to wheeled systems. Related arguments are less persuasive when dealing with the microgravity environment encountered on asteroids. The same holds when considering mobility on planetary surfaces of extreme topography that are impossible to access using conventional wheeled systems. On asteroids, a means to cling to the surface (Yoshida et al., 2002) would offer a critical capability for controlled motion and fine positioning. Limbs can also be beneficial as an active suspension that damps and prevents "bouncing" during traverse or upon landing after a hop. Crawling/climbing locomotion approaches without the use of limbs may also have merit for the asteroid domain. Such "limbless" approaches are briefly discussed below followed by promising technologies and concepts for clinging and gripping to surfaces.

3.2.1 Limbless crawling

A significant volume of engineering literature exists on research advances for snake-like or serpentine robots. A recent survey is provided in (Transeth et al., 2009). Robots have been developed to execute a variety of locomotion gaits inspired by snakes. Among them, the side-winding gait is perhaps best suited for effective locomotion under low surface friction conditions (Dalilsafaei, 2007) and especially effective for traversal of loose or slippery terrain (Hatton & Choset, 2010). The latest technologies for implementing side-winding gaits may provide viable solutions to local locomotion on asteroid surfaces. Relying largely on lateral friction forces, side-winding seems promising as a means to maintain surface contact but, in lieu of limbed mechanisms, may require additional means to adhere to the surface. Technologies such as dry adhesion are advancing for robotics applications in combination with mechanisms for climbing. Similar technologies can enhance the capability of a robotic side-winding mechanism. Discussed next are adhesive means of achieving secure surface contact while crawling or climbing based on examples of technology proposed for space and planetary rovers.

3.2.2 Enabling adhesion technologies

Dry adhesive and electrostatic adhesion approaches that permit walking or climbing robotic systems to stick to natural surfaces hold promise for gravity-independent locomotion. An example is the Automated Walking Inspection and Maintenance Robot (AWIMR), a concept intended for operation on the exterior of crewed space vehicles or structures in space rather than on planet or asteroid surfaces (Wagner & Lane, 2007). The AWIMR engineers established the feasibility of walking on such surfaces with the aid of prototype sticky feet, inspired by gecko feet, using dry adhesive polydimethylsiloxane for adhesion. The robot's sticky feet could walk on any clean, non-fragile surface (of the types found on space vehicle exteriors) and required a certain pull-off force. The AWIMR project also tested electrostatic means of sticking to surfaces, finding that greater shear forces were possible and that 2-3 kV was suitable for locomotion in this case (Wagner & Lane, 2007).

Bombardelli (Bombardelli et al., 2007) proposed artificial dry adhesives inspired by geckos and spiders for securing ballistically delivered microprobes to asteroid surfaces upon landing. The preliminary study suggests that multilevel conformal adhesive structures may be key to the performance of the microprobe attachment system for unknown asteroid terrain. The concept is motivated by the successful fabrication of several engineering prototypes of artificial reusable gecko adhesives. It is reported that the strongest such dry adhesive was recently fabricated using bundles of carbon nanotubes exhibiting four times the stickiness of natural gecko foot hairs (Bombardelli et al., 2007; Ge et al., 2007). Some researchers have found carbon nanotubes to be intrinsically brittle but express confidence in their near-term robustness for climbing robots (Menon et al., 2007).

Among the desirable characteristics of synthetic, gecko-like dry adhesion for enabling asteroid traversal by rovers is its effectiveness on many surface types (as its functional basis is van der Waals forces), its effectiveness in the vacuum of space, the fact that no additional energy is required to maintain an established grip on a surface, and their potential for mimicking the self-cleaning or dust resistant property of natural gecko footpads (Menon et al., 2007; Silva & Tenreiro, 2008). The applicability of this technology for space and planetary robotic vehicles that would walk or climb on in-space structures and terrestrial surfaces is highlighted in (Menon et al., 2007). What could be considered early phases of suitable asteroid robot designs are briefly described in that work. We next discuss examples of technologies that offer mechanical means for gripping with robot limbs or momentarily anchoring robot limbs to natural surfaces.

3.2.3 Limbs with gripping end-effectors

Limbed approaches employing gripping end-effectors as feet/hands can enable walking/climbing locomotion while maintaining contact with asteroid surfaces. The ability to grapple onto surfaces is key to gravity-independent locomotion allowing mobility in any orientation including steeply sloped natural terrain and upside down. Such "grapple-motion" capability enables natural surface traversal by clawing into regolith or forming grasping configurations against rough, hard surfaces of high friction.

During the past decade, prototypes of such limbed systems have been under development for planetary mobility and more recently focused on the problem of climbing steep terrain on Mars. A representative example of the state of the art for such applications is an 8 kg four-limbed planetary rover, LEMUR IIb, for which several types of climbing end-effectors have been investigated (Kennedy et al., 2005). The locomotion functionality for the LEMUR-class of robots evolved (kinematically) from 6-limbed robots for walking on space structures in orbit to 4-limbed free-climbing on steep terrain. Technologies addressed during the development of free-climbing capabilities for LEMUR IIb (e.g., gripping end-effectors, force control, and stability-based motion planning) should be useful for gravity-independent locomotion on asteroids as well.

Recent work at Tohoku University spearheaded limbed locomotion solutions and prototypes that are more specific to asteroid surface mobility and explored the feasibility of statically stable grapple-motion in microgravity (Chacin & Yoshida, 2005). The work is motivated by a desire to achieve finer and more deterministic control of robot motion and position. The focus thus far has been a 6-legged rover with 4 DOF per leg and spiked/gripping end-effectors for grasping the asteroid surface. Motion control complexities are handled using a behavior-based control approach in addition to bio-inspired central pattern generators for rhythmic motion and sensor-driven reflexes. Dynamic simulation results showed that static locomotion is feasible when grasping forces on the surface can be achieved (Chacin & Yoshida, 2005). A 2.5 kg prototype of the Tohoku asteroid rover was built using a piercing spike at the tip of each limb to serve as momentary anchors in soft regolith or as contact points of a static grip on hard surfaces when used in combination (Chacin et al., 2006). Crawling gaits feasible for locomotion in microgravity environments using this system are analyzed in (Chacin & Yoshida, 2006) for stability (in the sense that they hold the rover to the asteroid surface).

The next section focuses on this robotic system as an example of an asteroid mobility solution and control approach. It considers issues related to the microgravity environment and its effect on dynamics of robotic systems on asteroids.

4. Asteroid robot - ASTRO

In a future asteroid exploration mission (Chacin & Yoshida, 2005; 2006; Yoshida et al., 2002), it is expected to have a smart design of a robotic system that would allow scientists to benefit from more accurate instrument positioning capability on the asteroid's surface in microgravity. However, the engineering complexity of this task makes the design of an effective robot with stable locomotion difficult. A feasible robotic design for such a mission would be a small limbed robot deployed over the asteroid to crawl on its rough surface.

The asteroid robot ASTRO (Fig. 2) is a multi-limbed ambulatory locomotion system (Chacin et al., 2006) developed through the observation and imitation of clever solutions exhibited by biological systems. The robot is intended to use the natural features of the environment and the friction of the surface for omni-directional walking, having contact only at the limb end tips. This type of locomotion has the possibility to provoke minimum reactions on the asteroid surface thus avoiding excessive forces that could push the robot into space, or even the possibility to grasp the surface when some legs are controlled in a coordinated manner (Chacin & Yoshida, 2006).

During its development, the limbed robot concept was inspired from nature and other robots that have adopted the radially symmetric hexapod topology commonly used to obtain a platform complexity with a type of mechanism expected to perform robustly in unstructured environments, through the replication of walking gaits using six limbs like certain insects. A similar but much larger lunar rover concept is ATHLETE (Wilcox et al., 2007). However, while ASTRO and ATHLETE are kinematically similar, their distinguishable sizes and purposes enforce different sets of constraints to their designs. To eventually move on the Moon's

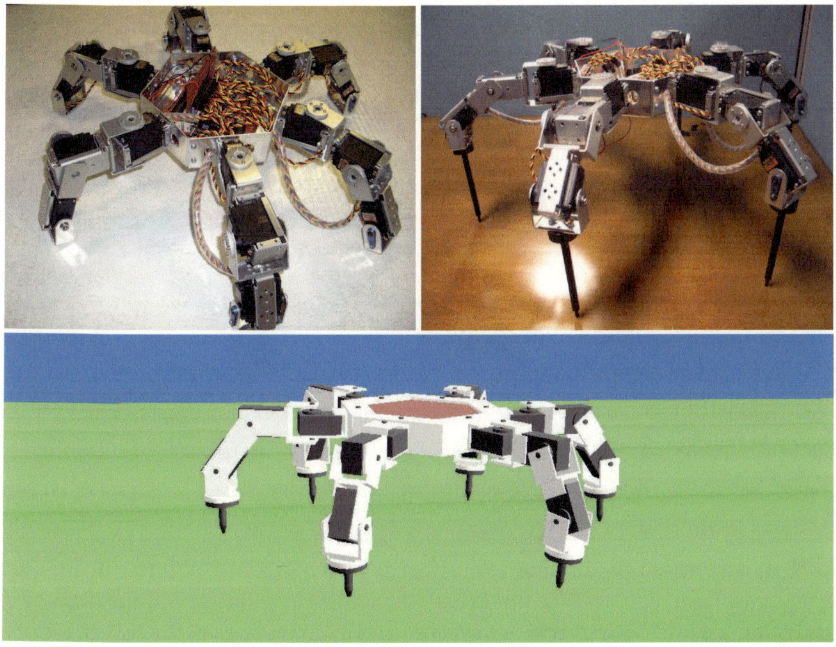

Fig. 2. Robot during early stages of development (top) and its simulated model (bottom).

surface, ATHLETE uses a combination of conventional rolling on wheels and quasi-static walking on wheels, and in ASTRO's case, the use of six limbs was decided based on the needs of a mission to an asteroid, where the microgravity environment would impose on the robot the challenging task of moving with high accuracy and still performing science operations in a stable manner. The main purpose of this form of motion is to avoid getting ejected from the surface. Therefore, six limbs would be better than two or four, given the possibility of using at least three limbs to grasp the surface to maintain the robot's attachment to the asteroid as a stable base while still using the remaining limbs for locomotion towards any direction or available for manipulation. Therefore, ASTRO is equipped with force/torque sensors on its end tips to provide feedback of the applied force to the surface.

ASTRO is expected to move on an asteroid surface with the capability of fine positioning over the surface to achieve science studies and mapping at several locations. As a result, the robot's design has been developed in a modular basis for all of its components, facilitating any normal changes in the adaptation process to accomplish a robust design for the demanding activities imposed.

Finally, because the scientific objectives of asteroid exploration missions are so ambitious, extensive prioritization on the development had to be made. Given that no robot designed by the current effort is intended to be tested in space, the space-environment requirements were given the lowest priority. This means that, the robot needs to be tested first in a $1g$ environment with as little information from its surroundings as possible.

4.1 General assumptions

Regular operations for limbed robots are characterized by multiple contacts between the limbs and the environment. In a microgravity environment, a planning algorithm can exploit this property by generating motions in contact, or more formally *compliant motions* (Borenstein,

1995). During the execution of a compliant motion, the trajectory of the end tip is reflected according to the sensed forces derived from the contacts.

In this context, the problem of specifying a compliant motion command is similar to the problem of planning using pure position control to orient the end tip. The compliant motion control (Klein & Briggs, 1980) allows specification of the forces and velocities to be maintained in the motion frame until the meeting of a set of termination conditions is detected.

Throughout the following analysis, to simplify the discussion, it is assumed that:

a_1) The object is a rigid body in contact with a rigid link of the robot;

a_2) Accurate models of the limbs and object are given;

a_3) Interference between limbs is ignored;

a_4) Each limb has only one frictional contact point at a fixed location;

a_5) The z direction of the contact point is always inward of the surface normal;

a_6) Contact points are known and the mass of each link of the robot is negligible;

a_7) Dynamic and static frictional coefficients are not distinguished between each other;

a_8) The motion is quasi-static[2] to suppress any dynamic effect.

Assumption a_4 allows us to consider only forces at the contact points. In this way, while executing a compliant command, the robot controller can interpret the sensed forces to automatically generate the corrective actions needed to comply with the task while preserving contact during motion.

Static friction occurs when the relative tangential velocity at a contact point is zero; otherwise, the friction is called dynamic friction. Assumptions a_7 and a_8 allow us to consider a "first-order" (or quasi-static) world where forces and velocities are related and also has static friction but no dynamic friction.

4.2 Dynamics model

Surface mobile robots have the same structure in terms of dynamics equations as free flying or floating robots which do not have a fixed point but have interaction with the ground. The model of the robot developed in this research (Chacin & Yoshida, 2005) consists of a central body with a hexagonal shape and six identical limbs (Inoue et al., 2001; 2002) symmetrically distributed around it.

Each limb has three links and three actuated revolute joints. Two of these joints are located in the junction of the leg/limb with the central body. The third joint is located at the knee connecting the upper and lower link, which results in each limb having three DOF. Considering six legs and the additional six DOF for central body translation and rotation, the system has a total of 24 DOF.

In this model, self-collision between limbs is ignored, meaning that links are allowed to cross each other, equal mass and length is assumed for all eighteen links, and that the joints are limited by internal mechanical stops. Any configuration of the robot can be defined by a set of parameters, the coordinates and orientation of the body, and the joint angles of each limb.

The dynamic motion of the free-flying multi-body system with the presence of the external forces $\boldsymbol{F}_{ex}$ is described as (Yoshida, 1997):

$$\boldsymbol{H}\begin{bmatrix}\ddot{\boldsymbol{x}}_b\\ \ddot{\boldsymbol{\phi}}\end{bmatrix}+\begin{bmatrix}\boldsymbol{c}_b\\ \boldsymbol{c}_m\end{bmatrix}=\begin{bmatrix}\boldsymbol{F}_b\\ \boldsymbol{\tau}\end{bmatrix}+\boldsymbol{J}^T\boldsymbol{F}_{ex} \tag{1}$$

[2] A quasi-static ("quasi-" (almost) "static") motion ensures that the system will go through a sequence of states that are infinitesimally close to equilibrium.

where,

$\boldsymbol{H}$: inertia matrix of the robot
$\boldsymbol{x}_b$: position/orientation of the base
$\boldsymbol{\phi}$: articulated joint angles
$\boldsymbol{c}_b, \boldsymbol{c}_m$: velocity/gravity dependent non-linear terms
$\boldsymbol{F}_b$: forces/moments directly applied on the base
$\boldsymbol{\tau}$: joint articulated torque
$\boldsymbol{J}^T$: Jacobian matrix
$\boldsymbol{F}_{ex}$: external forces/moments on the end points.

And the kinematic relationship around the end points is expressed as follows:

$$\dot{\boldsymbol{x}}_{ex} = \boldsymbol{J}_m \dot{\boldsymbol{\phi}} + \boldsymbol{J}_b \dot{\boldsymbol{x}}_b \tag{2}$$

$$\ddot{\boldsymbol{x}}_{ex} = \boldsymbol{J}_m \ddot{\boldsymbol{\phi}} + \dot{\boldsymbol{J}}_m \dot{\boldsymbol{\phi}} + \boldsymbol{J}_b \ddot{\boldsymbol{x}}_b + \dot{\boldsymbol{J}}_b \dot{\boldsymbol{x}}_b \tag{3}$$

where $\boldsymbol{J}_b$ and $\boldsymbol{J}_m$ denote the Jacobian of the base (main) body and the Jacobian of a given manipulator (limb) respectively.
The magnitude of the forces is determined by the friction of the contact surface. Let us consider the i-th limb of the robot. Vectors $\boldsymbol{p}_{ij} \in R^{3\times1}$, $\boldsymbol{f}_{ij} \in R^{3\times1}$ and $\boldsymbol{\tau}_{ij} \in R^{3\times1}$ denote the contact position, the contact force and the joint torque vectors, respectively; i and j express the i-th limb and the j-th contact point. Let $\boldsymbol{\tau}_i^j$ be the torque due to the contact force $\boldsymbol{f}_{ij}$. The relationship between the contact force and the joint torque is given by

$$\boldsymbol{\tau}_i^j = \boldsymbol{J}_{ij}^T \boldsymbol{f}_{ij} \tag{4}$$

where $\boldsymbol{J}_{ij}^T$ denotes the transpose of the Jacobian matrix that maps the contact force into the joint torque. Then, using the principle of superposition for the relationship between $\boldsymbol{\tau}_i^j$ and $\boldsymbol{f}_i$, we have:

$$\boldsymbol{\tau}_i = \sum_{j=1}^{k_i} \boldsymbol{\tau}_i^j \tag{5}$$

$$\boldsymbol{\tau}_i = \sum_{j=1}^{k_i} \boldsymbol{J}_{ij}^T \boldsymbol{f}_i \tag{6}$$

where, k_i is the number of contact points of the i-th limb. Since it is assumed that a limb can only have one contact point, Equation 6 can be rewritten in the following form.

$$\boldsymbol{\tau}_i = \boldsymbol{J}_i^T \boldsymbol{f}_i \tag{7}$$

where $\boldsymbol{\tau}_i = [\boldsymbol{\tau}_{i1}, \cdots, \boldsymbol{\tau}_{im}]^T$, $\boldsymbol{f}_i = [\boldsymbol{f}_{i1}^T, \cdots, \boldsymbol{f}_{im}^T]^T$ and $\boldsymbol{J}_i^T = [\boldsymbol{J}_{i1}^T, \cdots, \boldsymbol{J}_{im}^T]$.
Mathematically, both arms (upper limbs) and legs (lower limbs) are linked multi-body systems. Forces and moments yielded by the contact with the ground, or planetary surface, govern the motion of the system.

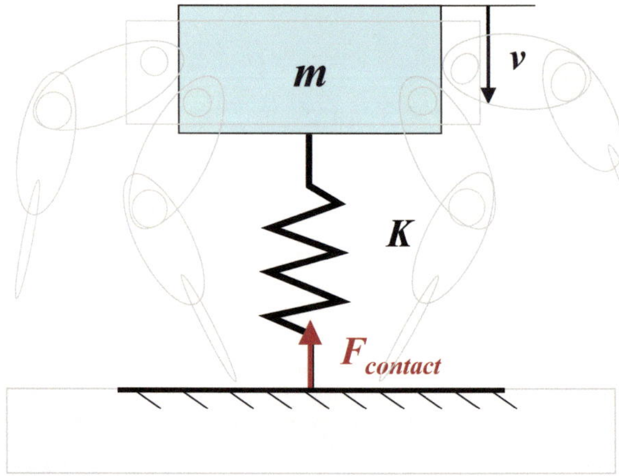

Fig. 3. Contact model.

4.3 Limb compliance and stiffness

Given an applied force on the end tip, any kinematic chain (on any given limb) experiences a deformation. This relation between the applied force and the corresponding deformation is defined as the "stiffness" of the kinematic chain. This stiffness can be related to two possibilities: the mechanical elasticity of the system and/or the dynamic behavior of the control system.
In general, for each joint there is a deflexion component defined by:

$$\tau = K\Delta\phi \tag{8}$$

where

$$K = \begin{bmatrix} k_1 & & 0 \\ & \ddots & \\ 0 & & k_N \end{bmatrix}. \tag{9}$$

The relation between the force f_i and the corresponding spatial deformation Δp is determined from the following relationship:

$$\Delta p = J\Delta\phi. \tag{10}$$

Solving Equation 8 for $\Delta\phi$,

$$\Delta\phi = K^{-1}\tau. \tag{11}$$

And substituting in Equation 10,

$$\Delta p = JK^{-1}\tau. \tag{12}$$

Finally, from Equation 7 and Equation 12,

$$\Delta p = JK^{-1}J^T f_i. \tag{13}$$

From this point the compliance matrix can be defined as:

$$C = JK^{-1}J^T. \tag{14}$$

The impact phase can be divided into two stages: compression and restitution. During the compression stage, the elastic energy is absorbed by the deformation of the contact surfaces

of the impacting bodies. In the restitution stage the elastic energy stored in the compression stage is released back to the bodies making the relative velocity greater than zero.
The robot can be modeled as a mass-spring system (Fig. 3) with a purely elastic contact at its end tips (Chacin & Yoshida, 2008). As a result, the following relation between the mass, the velocity and the force should hold:

$$2mv = F\Delta t. \tag{15}$$

Isolating the reaction force, we have:

$$F = \frac{2mv}{\Delta t}. \tag{16}$$

Now, considering the time of contact as a function of the mass of the system $\boldsymbol{m}$ and the stiffness coefficient of the limbs $\boldsymbol{K}$, we have:

$$\Delta t = \pi\sqrt{\frac{m}{K}}. \tag{17}$$

Finally, from Equation 16 and Equation 17,

$$F_{contact} = \frac{2}{\pi}\left(\sqrt{mK}\right)v \tag{18}$$

with

$$F_{contact} = \sum_{i=1}^{N} f_i \tag{19}$$

where N is the number of limbs in contact at landing.

4.4 Contact dynamics

In commonly used contact models (Brach, 1991; Keller, 1986), the relationship of momentum exchange and force-time product assumes infinitesimal impact. However, the infinitesimal impact between two single rigid bodies is a much idealized case. When modeling the ground (natural terrain), it is usually observed that as the stiffness coefficient is lowered, greater penetration in the ground occurs. The lower the damping constants, the longer the vibrations occur. However, a very high increment in stiffness or in damping constants in the limbs' model produces instabilities in simulations due to numerical issues that can be avoided by using a rigid limb, thus reducing the model order. The following discussion looks at how to determine the contact force F_{ex} (Der Stappen et al., 1999).
Where there is friction at a contact point, the friction force at a point acts tangential to the contact surface. We will denote the magnitude of the friction force at the i-th contact as f_i, an magnitude of the normal force as f_n. To specify the tangential acceleration and friction force completely in a three-dimensional system, we would also need to specify the direction of the acceleration and friction force in the tangent plane (Gilardi & Shraf, 2002; Yoshida, 1999).
Since the authors assume a model with a purely elastic deformation in the normal (z) direction of the contact point, and Coulomb friction in the tangential directions, we have the following general expressions from Fig. 4:

$$f_{ci-tg} = f_{ci}cos\theta \tag{20}$$

$$f_{ci-normal} = f_{ci}sin\theta \tag{21}$$

where $\boldsymbol{\theta}$ is the angle of the surface normal.

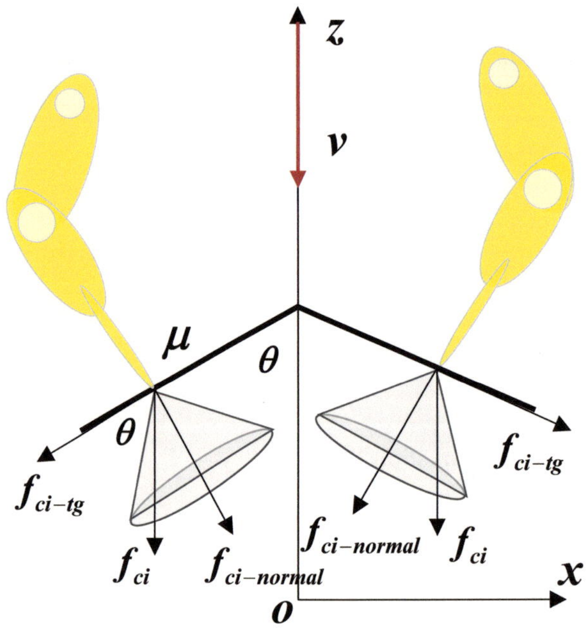

Fig. 4. Decomposition of the contact forces.

Next, the coefficient of friction can be denoted as:

$$\mu = \frac{f_{ci}}{f_{ci-normal}} > \frac{f_{ci}cos\theta}{f_{ci}sin\theta} \tag{22}$$

$$\mu > f_{ci}tan\theta. \tag{23}$$

Considering $F_{contact} = f_{ci}$ on the last expression, and using Equation 16 we have,

$$\mu > \frac{2mv}{\Delta t}tan\theta. \tag{24}$$

And substituting Equation 17,

$$\frac{tan\theta\mu}{2m}\left(\pi\sqrt{\frac{m}{K}}\right) > v. \tag{25}$$

Equation 25 shows that the considered contact stability will strictly depend on the approach velocity of the robot.

Quasi-static stability is a more general stability criterion than that used in the previous discussion. Under this condition, inertia forces are included but limb dynamics are not separately considered; the masses of the limbs are considered with that of the body. In the previous argument, quasi-static stability is partly assumed, provided all the normal components of the contact points' f forces are positive. Since the contact point cannot support a negative normal force (as shown in Fig. 4), the appearance of a negative force indicates that the given limb will lift and, since it cannot provide the required movement about the center of mass, the robot will jump in a microgravity environment like MINERVA (Yoshimitsu et al., 2001).

4.5 Motion control

If a robot is to successfully achieve walking with static stability, the control system must ensure that the behavior of the robot does not deviate from the following stable condition.

$$\sum_{i=1}^{m} f_i + m_0 g = 0 \tag{26}$$

$$\sum_{i=1}^{m} p_i \times f_i = 0 \tag{27}$$

To remain balanced, the robot must be able to apply forces with its end tips on the terrain that compensate for gravity without slipping. A necessary condition is that the robot's center of mass lies above the support polygon. But on an irregular surface, the support polygon does not always correspond to the base of the contact points. To compute the support polygon the contact interface (all contact points) is modeled as shown in (Chacin, 2007), with

$$\|f_n\| \geq 0 \tag{28}$$

$$\|f_t\| \leq \mu \|f_n\|. \tag{29}$$

The support polygon is the projection of these polyhedrons onto the global coordinate system. The body attitude also needs to be controlled in order for the robot to maintain static balance. Assuming that the robot is well balanced in the lateral plane, the body attitude at any given moment is an important determinant of the acceleration experienced by the center of mass (COM) in the sagittal plane. If the desired value of the acceleration output is known at all times throughout the gait cycle, then a corrective action can be generated which will maintain the robot in a stable state. If the body attitude needs to be increased or decreased depending upon the error signal, and assuming that we can determine the desired acceleration in the sagittal plane for all times throughout the gait cycle, we can implement a continuous control system.

4.6 Zero moment point and momentum of the system

In the walking robotics community, *Zero Moment Point* (ZMP), which was first introduced by (Vukobratovic et al., 1970), is a key concept to discuss the tip-over stability and gait control of a robot. Fig. 5 is a schematic drawing to explain ZMP. Point $\boldsymbol{O}$ is the center of gravity (COG) of the entire robot. Let vector $\boldsymbol{r}$ be a position vector from an arbitrary point on the ground P to the point $\boldsymbol{O}$, and vector $\boldsymbol{l}_i$ be a position vector from the point P to each ground contact point of the limbs. For this model, the following dynamic equilibria hold true

$$\dot{\boldsymbol{P}}_p = \sum \boldsymbol{f}_{exi} + M\boldsymbol{g} \tag{30}$$

$$\dot{\boldsymbol{L}}_p = \boldsymbol{n}_p + \boldsymbol{r} \times \dot{\boldsymbol{P}}_p + \sum (\boldsymbol{l}_i \times \boldsymbol{f}_{exi}) \tag{31}$$

where $\boldsymbol{P}_p$ and $\boldsymbol{L}_p$ are linear and angular momentum around point P, and M is the total mass of the robot. The ZMP is a position P at which the component of moment $\boldsymbol{n}_p$ around the horizontal axes, $\boldsymbol{n}_{px}$ and $\boldsymbol{n}_{py}$, becomes zero. The robot is stable, otherwise if the ZMP stays inside a polygon formed by ground contact points of the limbs. Otherwise the robot starts tipping over.

Gait generation and motion control algorithms for walking robots have been developed based on this concept. In addition, an advanced planning and control algorithm with a special attention to the kinetic momentum has been proposed recently (Kajita et al., 2003).

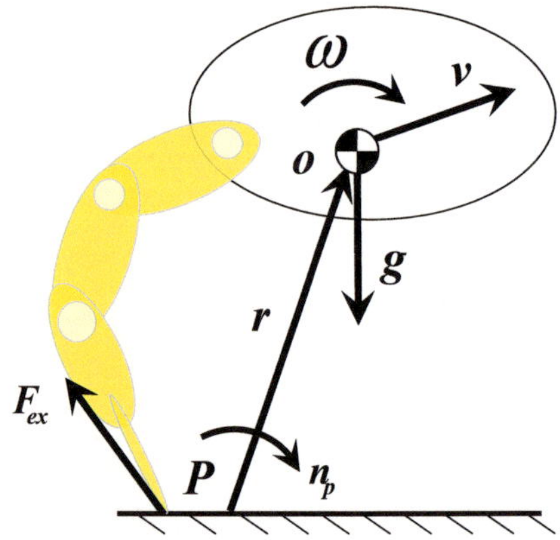

Fig. 5. A schematic force/moment model for an articulated surface robot

For a surface limbed robot, the gravity forces exerted on it can be neglected; the non-linear term in Equation 1 then becomes $c_b = \dot{H}_b \dot{x}_b + \dot{H}_{bm} \dot{\phi}_b$. Integrating its upper set with respect to time, we obtain the total momentum of the system as:

$$\mathcal{L} = \int J_b^T F_{ex} dt = H_b \dot{x}_b + H_{bm} \dot{\phi}. \tag{32}$$

Going back to Equation 2, and eliminating $\dot{\phi}$, we can obtain the following equation:

$$\mathcal{L} = (H_b - H_m J_s^{-1}) \dot{x}_b + H_m J_s^{-1} \dot{x}_{ex}. \tag{33}$$

In this way, if the system does depart from static stability, then the control system can identify this condition and bring the robot back to the statically stable condition.

4.7 Generalized control algorithm

Since walking is a continuous and cyclic process, we can consider two main types of control systems, a closed-loop control system and an event driven control system (Fig. 6). A general closed-loop control system for controlling the continuous process of the walking gait can be considered. However, since the positioning of the end tip itself can be modeled as a discrete process, we use an event driven control system to identify the existence of such states and modify the closed-loop control depending upon the current state of the system.
Given a motion command (a vector $\vec{X}$), the motion planning and control algorithm to move in the commanded direction and magnitude is carried out in the following way:

① Use the gait planner presented in (Chacin & Yoshida, 2006) to plan the complete set of limb motions to move in the desired direction.

② At time t, compute link positions and velocities, recursively from link 0 to n.

③ Set accelerations $\ddot{x}_b$ and $\ddot{\phi}$ to zero, and external forces F_b and F_{ex} to zero, then compute the inertial forces recursively from link n to 0. The resultant forces on the coordinates x_b and ϕ are equal to the non-linear forces c_b and c_m, respectively.

④ Plan the end point trajectory of each limb x_{exi} using the kinematics in Subsec. 4.6, so that ZMP satisfies the stability condition. Then obtain the end point velocity $\dot{x}_{ex}$ along the trajectory.

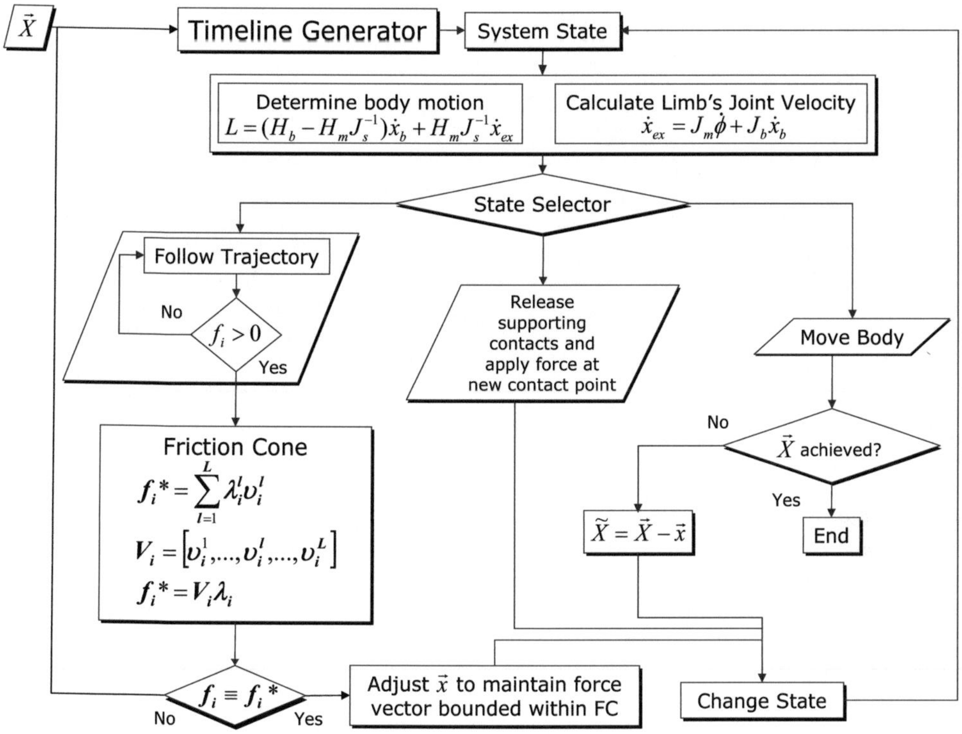

Fig. 6. Gait planner detail

⑤ Considering the friction cone estimation (Chacin, 2007) and the contact stability conditions shown by Equation 25, determine the robot base motion $\dot{x}_b$ by

$$\begin{bmatrix} \ddot{x}_b \\ \ddot{\phi} \end{bmatrix} = H^{-1}\left\{\begin{bmatrix} F_b \\ \tau \end{bmatrix} + J^T F_{ex} - \begin{bmatrix} c_b \\ c_m \end{bmatrix}\right\}$$

⑥ Calculate the joint velocity of the limbs $\dot{\phi}$ by Equation 2, using $\dot{x}_b$ and $\dot{x}_{ex}$ while considering Equation 29; change the state of the control system. If necessary, adjust $\vec{x}$ to maintain the force vectors bounded within the friction cones.

⑦ Adopt the new contact configuration to release the support contacts and apply a permissible contact force at the new contact points. Dynamic exploration can be applied to reduce the surface position/orientation uncertainties. Change the state of the control system.

⑧ Control the joints along with the solution from ⑥ to move the body. Verify if the goal position $\vec{X}$ has been reached; if it has not, then repeat.

One difference of this algorithm with respect to conventional ones is the consideration of momentum of the robot in ③. Without this step, the obtained joint motion may have errors from the originally planned end point trajectory, thus may not satisfy the stability condition. Conventionally, a feedback control may be employed to correct these errors. But using Equation 33 in ③, the error can be compensated in advance.

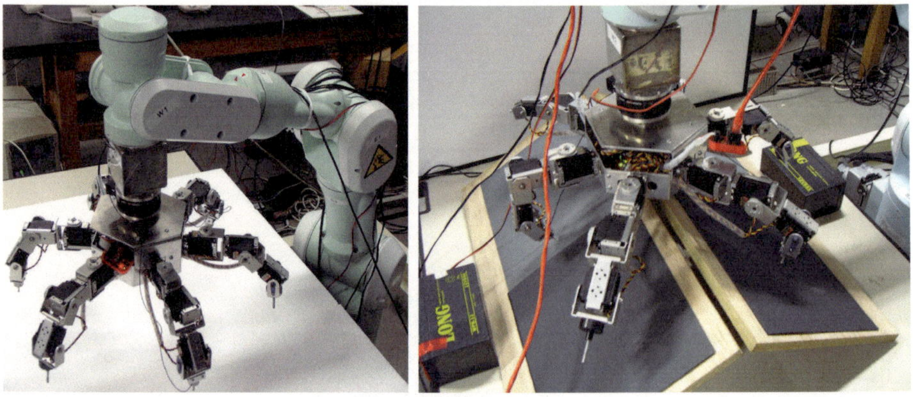

Fig. 7. The experimental setup: robot mounted on a PA10 manipulator (left) and over an uneven surface (right).

4.8 Emulating microgravity

To simulate the real dynamic conditions of a robot on an asteroid, the authors used a robust control system that could counterbalance Earth's gravity and leave the robot in an emulated state of microgravity. For this reason, ASTRO was mounted on the tip of a Mitsubishi PA10 manipulator arm (Fig. 7) controlled by a dynamics based model simulation equipped with an ATI Industrial Automation Gamma SI-32-2.5 Force/Torque sensor (ATI, 2005). This sensor allows the appropriate control of the manipulator arm to emulate the microgravity condition that asteroid 25143 Itokawa would impose on the robot by means of calculating its position, force, acceleration and velocity with high accuracy using real-time impedance and dynamic compliance control.

Experiments were performed to observe and analyze the robot's behavior under microgravity conditions, as it would approach the surface of an asteroid and perform motions on it. Gravity emulation is achieved through impedance control, also known as virtual compliance control, as shown in (Hirabayashi et al., 2000). The control algorithm is based upon the basic equation of motion shown in Equation 34.

$$[\boldsymbol{m}]\frac{d\overline{\boldsymbol{v}}}{dt} = \overline{\boldsymbol{q}} - [\boldsymbol{K}]\Delta\overline{\boldsymbol{x}} - [\boldsymbol{C}]\overline{\boldsymbol{v}} \tag{34}$$

where $\overline{q}$ is the external force and torque applied to the manipulator's end tip, $\Delta\overline{x}$ is the displacement of the end tip relative to the reference and $\overline{v}$ is its velocity. $[m] \in R^6$ is the virtual mass, $[K] \in R^6$ is the virtual spring constant and $[C] \in R^6$ is the damping constant. Equation 34 can be transformed into:

$$\overline{\boldsymbol{v}} = \frac{1}{[\boldsymbol{m}]}\int(\overline{\boldsymbol{q}} - [\boldsymbol{K}]\Delta\overline{\boldsymbol{x}} - [\boldsymbol{C}]\overline{\boldsymbol{v}})dt \tag{35}$$

which can be represented as the following finite differential equation:

$$\overline{\boldsymbol{v}}_n = \frac{\boldsymbol{\Delta t}}{[\boldsymbol{m}]}(\overline{\boldsymbol{q}}_{n-1} - [\boldsymbol{K}]\boldsymbol{\Delta}\overline{\boldsymbol{x}}_{n-1}) + ([\boldsymbol{I}] - \frac{\boldsymbol{\Delta t}}{[\boldsymbol{m}]}[\boldsymbol{C}])\overline{\boldsymbol{v}}_{n-1} \tag{36}$$

where $\boldsymbol{\Delta t}$ is the time step and $[\boldsymbol{I}]$ is the identity matrix. Equation 36 describes the velocity at the sampling time based upon values from the previous time step. Based on this equation,

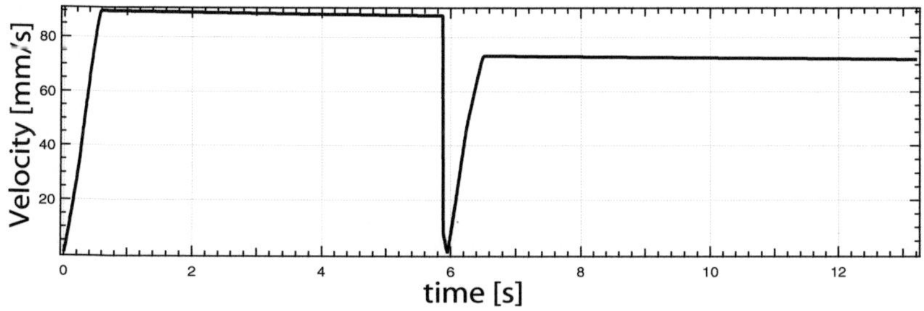

Fig. 8. Composite velocity profile of the manipulator's end tip.

and knowing the forces and torques as well as the displacement, the arm can be controlled in real-time using Equation 36 and Equation 1 to simulate the attached robot and to compensate the input of the F/T sensor on its tip. In short, the control policy uses the relationship between force and velocity to find the position and then calculate a velocity command as inputs to the system. The behavior of the system can be altered by changing the impedance constants $[m]$, $[K]$ and $[C]$ as defined in Subsec. 4.3 and Subsec. 4.4 or can be fine tuned in order to exhibit the desired properties.

4.9 Preliminary tests

To evaluate the performance of the manipulator's control strategy, several tests were designed to examine how the impedance constants $[m]$, $[K]$ and $[C]$ influence the behavior of the system. For simplicity, only four typical cases are described while keeping $[m]$ constant to damp out small mechanical vibrations on the system.

During the first case, the parameters were selected under the following constraint $K \gg C$. A force is applied on the manipulator's end tip, which is displaced before it starts swinging back and forth. A small reduction in the position is observed with time since the damping is not zero.

In the second case the condition changes to $K > C$. As in the first experiment, the end tip starts to swing back and forth when a force is applied, but since the damping constant $[C]$ has been increased, this swing motion dies out much faster as the end tip is back to its initial position without any further movement.

For the third case, the condition is switched to $K \ll C$. When a force is applied, the velocity increases but it is quickly reduced back to zero. The position also exhibits a rather small change where it remains when the velocity is damped out.

In the fourth case, the spring and damping constants are selected to satisfy $K < C$, but both parameters are chosen similar to what they are expected to be in a microgravity environment, where both $[K]$ and $[C]$ tend to zero[3]. It was observed that the manipulator's end tip velocity changes rapidly, and then it slowly decreases due to the small damping coefficient.

Several tests designed to verify the energy absorption and the velocity profile of the manipulator arm were performed. Typical experimental results are shown in Fig. 8.

The manipulator's end tip velocity changes after impacting the surface but remains within 10%-15% of its approaching phase yielding reasonable results that are in accordance with what is expected in microgravity conditions. At this point, if the virtual mass value is to be changed in any of the experiments, the behavior of the system would not change, but the force

[3] In reality $[C]$ is chosen to be 0.1 to avoid problems in the computational solution of Equation 36 and to satisfy the condition.

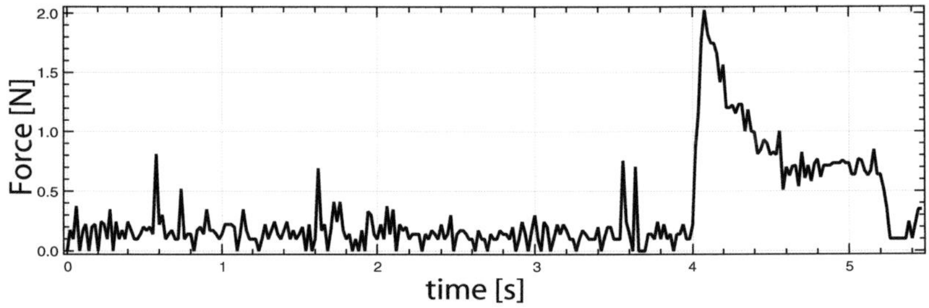

Fig. 9. Force profile during motion without control feedback.

required in order to get the same velocity and position displacement would have increased accordingly. Similar experiments have been carried out for rotational and multi-translational movements, but their description has been omitted to show these results as the basic principle behind the selection of proper parameter values.
After achieving a satisfactory response from the control system, ASTRO was attached to the end tip of the manipulator and was suspended above a table that serves as analog to the asteroid surface.

4.10 Experimental results

To demonstrate the concept and to verify the control strategy, further experiments were performed under the dynamic conditions shown in the previous section, and for comparison, motion was attempted first without control or feedback and then using the control algorithm described in Subsec. 4.7. In the first motion experiment, the robot was commanded to go forward over a rugged surface[4] with a known inclination angle, while keeping a stable and statically balanced position. The problem of control is simplified by giving the control system the geometry[5] of the environment and the gait is calculated off-line.
Although the robot successfully took three contact points and executed one gait motion, the selected contact points were not ideal, and the robot exhibited pronounced oscillating swaying motions in the lateral plane causing slip. This can be seen in the force data (Fig. 9). It was observed that after the gait missed the contact points the stability of the system was severely undermined. This instability at the end tips could cause the unpredictable behavior of the robot due to the escalating effect of the accumulative error. This is somewhat indicative of the dynamic behavior of uncontrolled hopping robots in similar microgravity environments.
Due to the contact instability of the previous attempt, the ability of the robot to use the information from the friction forces (friction cones) while walking on the surface was further examined (Fig. 11). The overall stability of the walking gait depends upon the timing of the motion from the gait planner. If the timing is incorrect, it will affect the stability of the robot during the walking gait. It is, therefore, desirable to study the relationship between the stability of the robot and the motion conditions. The stability of the robot can be examined by enabling the robot to walk continuously on the surface using the control strategy presented in Subsec. 4.7. This has the desired effect of reducing end tip momenta which are created when the robot translates or rotates an end tip, which can increase the instability of the contact.

[4] Sandpaper of different grit sizes was used in the experiments to simulate the potential roughness of an asteroid surface.

[5] The term geometry is used, when referring to robot control and gait planning, to mean that a robot is controlled based on fixed relationships between models of the robot and models of the world in which the robot operates.

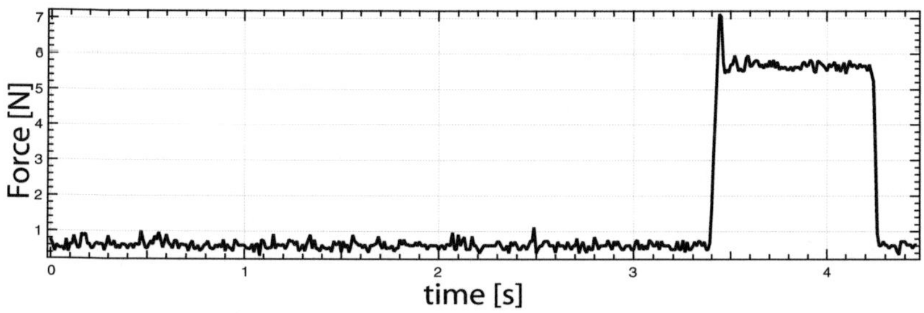

Fig. 10. Force profile during motion with control feedback.

The results of this experiment are shown in Fig. 10. It can be noted that the force values are higher, but more importantly the robot is resting balanced on the limbs in contact with the surface. With the feedback from the force sensors, the control system was able to determine an appropriate corrective measure for the robot to take in order to maintain the stability of the contact while keeping its attitude (expressed in roll-pitch-yaw angles) mostly unchanged during motion. For the case where the robot is under a known environment and the desired forces to be exerted on the surface are known in advance, any deviation from this state should cause an appropriate corrective action to be taken.

5. Challenges

A rich set of challenges are encountered during development and evaluation of prospective solutions for gravity-independent locomotion on asteroids. The experiments reported here are indicative of a few, but several additional key challenges deserve early attention by researchers. One of them is the mechanics of controlled ballistic hopping on rotating asteroids and in non-uniform gravity fields due to their irregularly shaped bodies. Bellerose et al (Bellerose et al., 2008; Bellerose & Scheeres, 2008) modeled the dynamics of hopping vehicles to enable hops covering designated distances by computing and controlling initial hop velocity. The model accounts for distance covered by residual bounces as the vehicle comes to rest (considering surface friction coefficient and restitution). A particularly challenging aspect to consider is that some asteroid shapes may have surface locations where a vehicle could stay in equilibrium, thus affecting vehicle dynamics on the surface (Bellerose et al., 2008; Bellerose & Scheeres, 2008). Conceivably, a hopping rover could be perturbed away from predicted ballistic trajectories by such equilibria. This can affect exploration objectives by constraining the total area that a rover can safely or reliably traverse to on an asteroid surface when stable and unstable equilibrium locations happen to coincide with surface regions of scientific interest. Purely hopping vehicles that operate primarily at the mercy of small body physics can have limited accessibility of such surface regions. Bellerose's model also provides insight into the effects of non-uniform gravity fields and how centripetal and Coriolis forces due to asteroid rotation may assist or hinder hop performance (Bellerose & Scheeres, 2008).
Another key challenge is achieving the ability to land after hopping in such a way as to avoid rebound. Control and robotics techniques can be used to address this challenge. One robot concept employs a spring and linear actuators with horizontal velocity control to achieve this (Shimoda et al., 2003), while other research is experimenting with active grappling of the surface upon landing (Chacin, 2007; Chacin & Yoshida, 2008; 2009). The related challenge, central to gravity-independent locomotion, is maintaining grip or temporary anchoring while controlling force for closure and compliance. The work presented herein and in (Chacin

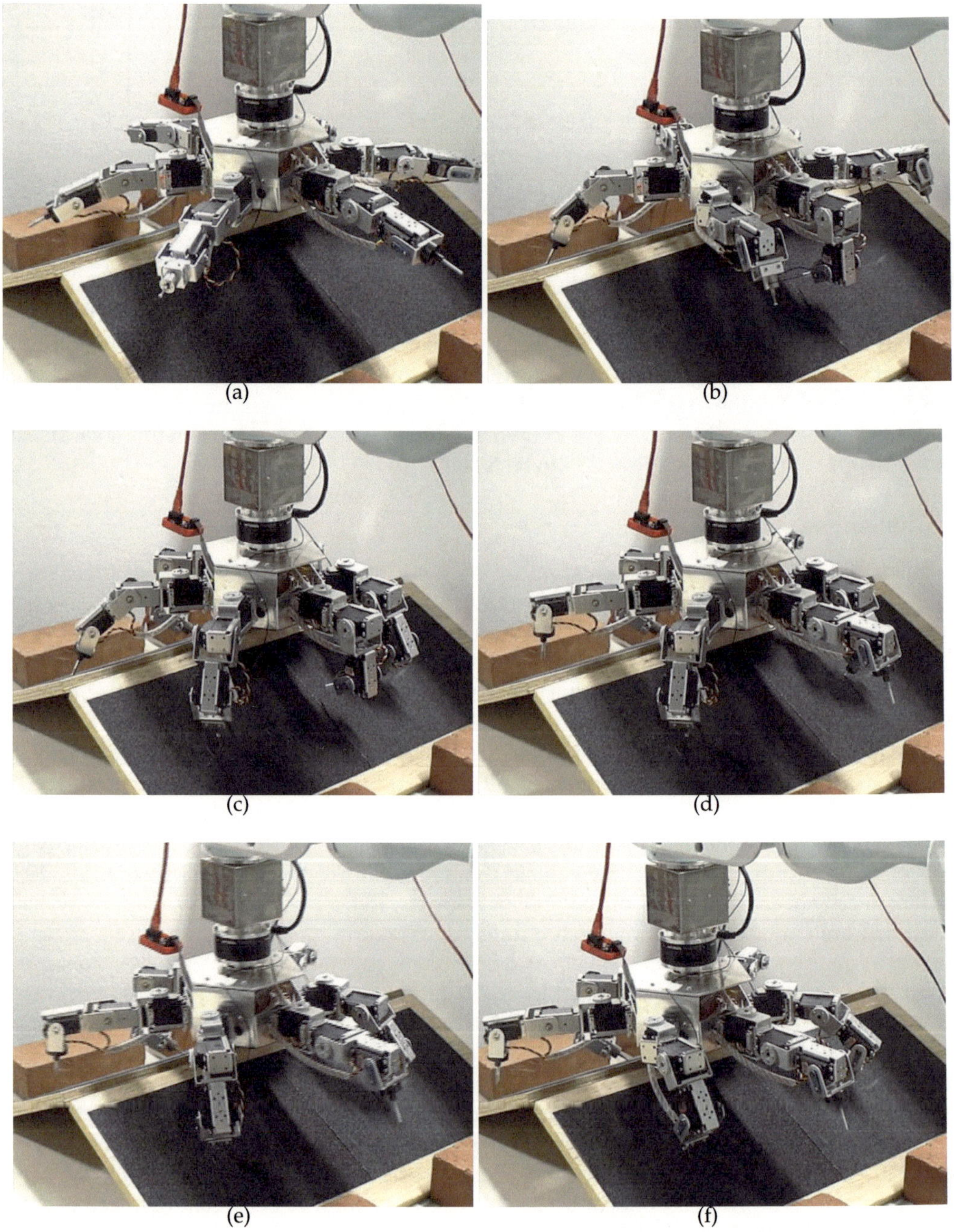

Fig. 11. Movement over inclined frictional surface with friction feedback gait control.

& Yoshida, 2009) examines the motion/force control and dynamic modeling germane to the problem of stable crawling and force closure needed to maintain contact/grip with an asteroid surface under microgravity conditions. Experiments with ASTRO reveal the utility of force feedback for maintaining contact during execution of compliant motion. Kennedy et al (Kennedy et al., 2005) address active force control to achieve anchoring associated with stable free-climbing motion control. Tactile sensing and related motion planning algorithms (Bretl et al., 2003) have been implemented on the LEMUR IIb robot.

The low gravity environment and its effect on surface vehicles present a key challenge for rover localization whether hopping, crawling, or climbing. Determining, updating and maintaining knowledge of rover position and orientation on an asteroid surface can be important for recording spatial context for surface science measurements and for certain mission concepts of operation. Localization approaches for hopping robots have been proposed with some reliance on range measurements to an orbiting or station-keeping mother spacecraft (Yoshimitsu et al., 2001) and via use of more general approaches such as particle filters (Martinez, 2004), Kalman Filters with landmark geo-referencing (Fiorini et al., 2005), and optical flow as well as visual odometry without continuous terrain feature tracking while tumbling (So et al., 2008; 2009). During local navigation across the terrain, existing localization approaches for rolling or walking robots may apply. These may be based on use of extended Kalman Filters on fused celestial sensor data and optical-flow measurements (Baumgartner et al., 1998).

Finally, a key challenge is the testing and verification of gravity-independent locomotion systems to ensure confidence in their technology readiness for asteroid missions. This is always a challenge for space systems and particularly those intended for operation in microgravity domains. The testbed described in the previous section and its means of emulating reduced gravity are representative solutions for addressing the challenge using relatively affordable technology. Other testbed approaches to emulating reduced gravity in the laboratory include the use of overhead gantry systems with frictionless air-bearing pulleys from which to suspend prototype rovers, and the use of prototype rovers on flat air-tables or mounted on a mobile base with integrated air bearings. Beyond the fundamental feasibility of controlled surface mobility in low gravity fields of asteroids, additional challenges of high relevance and importance remain to be addressed by advanced research and technology development.

6. Summary and conclusions

In this chapter, various approaches to gravity-independent locomotion on weak gravity surfaces of asteroids are discussed along with related technologies. Challenges are also described that affect planning and control of surface exploration robots that use hopping and rolling mechanisms and/or articulated limbs for the ground contact.

Given the focus on gravity-independent locomotion approaches, technologies, and challenges of robotic mobility on asteroids, an in-depth representative example of an asteroid mobility solution and control approach is provided. The control approach considers reaction and friction forces with the asteroid surface and is demonstrated using a prototype robot (ASTRO) and laboratory testbed that emulates microgravity. This example considered issues that most solutions must address related to the microgravity environment and its effect on dynamics of robotic systems on asteroids.

The research presents a planning and control structure for the locomotion of a multi-limbed robot under microgravity conditions. The algorithm can be considered as constructive proof

that there exists a solution that satisfies the force conditions for any system with friction. It works by reacting to the current locations of contact points and estimating the force-closure condition for stable motion. Such a mechanism is central in the control structure.
The control methods proposed in this research are useful to improve the operational performance and efficiency for robots capable of position-based controlled motion on an asteroid. They demonstrated that proper knowledge of the force cone interaction with the surface plays a significant role in the development of proper control procedures that can allow next-generation surface robots to gain proper mobility in a microgravity environment.

7. Acknowledgments

The research reported in this chapter was partly sponsored by the Japanese Ministry of Education, Culture, Sports, Science and Technology and conducted at the Space Robotics Laboratory in Tohoku University, Sendai, Japan. Partial support of the JHU/APL Civilian Space Business Area is acknowledged.

8. References

ATI Industrial Automation (2005). Force Torque Transducer TWE F/T Instalation and oparation manual. *Component data sheet*, ATI Industrial Automation. 2005.

Bares, J., Hebert, M., Kanade, T., Krotkov, E., Mitchell, T., Simmons, R. & Whittaker, W. (1999). Ambler: An Autonomous Rover for Planetary Exploration. *IEEE Computer*, pp. 18-26, Vol. 22, No 6, 1999.

Brach, R. M. (1991). Mechanical Impact Dynamics: Rigid Body Collisions. John Wiley & Sons. 1991.

Baumgartner, E.T., Wilcox, B.H., Welch, R.V. & Jones, R.M. (1998). Mobility performance of a small-body rover. *Proceedings 7th International Symposium on Robotics and Applications, World Automation Congress*, Anchorage, Alaska. 1998.

Behar, A., Bekey, G., Friedman, G., & Desai, R. (1997). Sub-kilogram intelligent tele-robots (SKIT) for asteroid exploration and exploitation. *Proceedings SSI/Princeton Conference on Space Manufacturing, Space Studies Institute*, pp. 65-83, May, Princeton, NJ. 1997.

Bellerose, J., Girard, A. & Scheeres, D.J. (2008). Dynamics and control of surface exploration robots on asteroids. *Proceedings 8th International Conference on Cooperative Control and Optimization*, pp. 135-150, January, Gainesville, FL. 2008.

Bellerose, J. & Scheeres, D.J. (2008). Dynamics and control for surface exploration of small bodies. *Proceedings AIAA/AAS Astrodynamics Specialist Conference and Exhibit*, Paper 6251, Honolulu, Hawaii, 2008.

Bombardelli, C., Broschart, M. & Menon, C. (2007). Bio-inspired landing and attachment system for miniaturised surface modules. *Proceedings 58th International Astronautical Congress* Hyderabad, India. 2007.

Borenstein, J. (1995). Control and Kinematic Design of Multi-degree-of-freedom Mobile Robots with Compliant Linkage. *IEEE Transactions on Robotics and Automation*, pp. 21-35, Vol. 11, No. 1, February, 1995.

Bretl, T. & Rock, S. & Latombe, J.C. (2003). Motion planning for a three-limbed climbing robot in vertical natural terrain. *IEEE International Conference on Robotics and Automation*, pp. 2947-2953, Taipei, Taiwan, 2003.

Chacin, M. & Yoshida, K. (2005). Multi-limbed rover for asteroid surface exploration using static locomotion. *Proceedings of the 8th International Symposium on Artificial Intelligence, Robotics and Automation in Space*, Munich, Germany. September 2005.

Chacin, M. & Yoshida, K. (2006). Stability and Adaptability Analysis for Legged Robots Intended for Asteroid Exploration. *Proceedings of the 2006 IEEE International Conference on Intelligent Robots and Systems (IROS2006)* Beijing, China. 2006.

Chacin, M. & Yoshida, K. (2006). Evolving Legged Rovers for Minor Body Exploration Missions. *Proceedings of the 1st IEEE / RAS-EMBS International Conference on Biomedical Robotics and Biomechatronics, BioRob2006* Pisa, Italy. 2006.

Chacin, M., Nagatani, K. & Yoshida, K. (2006). Next-Generation Rover Development for Asteroid Surface Exploration: System Description. *Proceedings of the 25th International Symposium on Space Technology and Science and 19th International Symposium on Space Flight Dynamics* Kanazawa, Japan. June 2006.

Chacin, M. (2007). Landing Stability and Motion Control of Multi-Limbed Robots for Asteroid Exploration Missions. *Ph.D. dissertation*, Dept. of Aerospace Engineering, Tohoku University, Tohoku, Japan, 2007.

Chacin, M. & Yoshida, K. (2008). A Microgravity Emulation Testbed for Asteroid Exploration Robots. Proceedings of International Symposium on Artificial Intelligence, Robotics, Automation in Space (i-SAIRAS08) Los Angeles, CA, February 2008.

Chacin, M. & Yoshida K., (2009). Motion control of multi-limbed robots for asteroid exploration missions, *Proceedings 2009 IEEE International Conference on Robotics and Automation*, Kobe, Japan, May 2009.

Cottingham, C.M., Deininger, W.D., Dissly, R.W., Epstein, K.W., Waller, D.M. & Scheeres, D.J. (2009). Asteroid Surface Probes: A low-cost approach for the in situ exploration of small solar system objects. *Proceedings IEEE Aerospace Conference*, Big Sky, MT, 2009.

Dalilsafaei, S. (2007). Dynamic analyze of snake robot. *Proceedings World Academy of Science, Engineering and Technology*, pp. 305-310, Issue 29, Berlin, Germany. May, 2007.

Der Stappen, A. V., Wentink, C. & Overmars, M. (1999). Computing form-closure configurations. *Proceedings of the IEEE International Conference on Robotics and Automation, ICRA* pp. 1837-1842, USA. 1999.

Ebbets, D., Reinert, R. & Dissly, R. (2007). Small landing probes for in-situ characterization of asteroids and comets. *textitProceedings 38th Lunar and Planetary Science Conference*, League City, TX. 2007.

Fiorini, P., Cosma, C. & Confente, M. (2005). Localization and sensing for hopping robots. *Autonomous Robots*, pp. 185-200, Vol. 18, 2005.

Fujiwara A., Kawaguchi J., Yeomans D. K. et al. (2006). The Rubble Pile Asteroid Itokawa as observed by Hayabusa. Report: Hayabusa at asteroid Itokawa. *Science*, pp. 1330-1334, Vol. 312, No. 5778, June, 2006.

Ge, L., Sethi, S., Ci, L., Ajayan, P.M. & Dhinojwala, A. (2007). Carbon nanotube-based synthetic gecko tapes. *Procedings National Academy of Sciences*, pp. 10792-10795, Vol. 104, No. 26, 2007.

Gilardi, G. & Shraf, I. (2002). Literature Survey of Contact Dynamics Modeling. *Mechanism and Machine Theory*, pp.1213-1239, Vol. 37, 2002.

Hatton, R.L. & Choset, H. (2010). Generating gaits for snake robots: annealed chain fitting and keyframe wave extraction. *Autonomous Robots*, pp. 271-281, Vol. 28, 2010.

Hirabayashi, H., Sugimoto, K., Enomoto, A. & Ishimaru, I. (2000). Robot Manipulation Using Virtual Compliance Control. *Journal of Robotics and Mechatronics*, pp. 567-575, Vol. 12, No.5, 2000.

Hokamoto, S. & Ochi, M. (2001). Dynamic behavior of a multi-legged planetary rover of isotropic shape. *Proceedings 6th International Symposium on Artificial Intelligence, Robotics, and Automation in Space*, June, St-Hubert, Quebec, Canada, 2001.

Inoue, K., Arai, T., Mae, Y., Takahashi, Y., Yoshida, H. & Koyachi, N. (2001). Mobile Manipulation of Limbed Robots -Proposal on Mechanism and Control. *Preprints of IFAC Workshop on Mobile Robot Technology*, pp. 104-109, 2001.

Inoue, K., Mae, Y., Arai, T. & Koyachi, N. (2002). Sensor-based Walking of Limb Mechanism on Rough Terrain. *Proceedings of the 5th International Conference on Climbing and Walking Robots (CLAWAR)*, 2002.

JAXA / Institute of Space and Astronautical Science (2003), Asteroid Explorer HAYABUSA, *http://hayabusa.jaxa.jp/*, 2003.

Johns Hopkins University Applied Physics Laboratory (1996), Near Earth Asteroid Rendezvous - Shoemaker Mission, *http://near.jhuapl.edu*, 1996.

Jones, R. et al. (1998). NASA/ISAS collaboration on the ISAS MUSES C asteroid sample return mission. *Proceedings of 3rd IAA International Conference on Low-Cost Planetary Missions*, Pasadena, CA. 1998.

Kajita, S., Kanehiro, F., Kaneko, K., Fujiwara, K., Harada, K., Yokoi, K. & Hirukawa, H. (2003). Resolved Momentum Control: Humanoid Motion Planning based on the Linear and Angular Momentum. *Proceedings of the IEEE/RSJ International Conference on Intelligent Robots and Systems (IROS'03)*, pp. 1644-1650, 2003.

Kawaguchi, J., Uesugi, K. & Fujiwara, A. (2003). The MUSES-C mission for the sample and returnits technology development status and readiness. *Acta Astronautica*, pp. 117-123, Vol. 52, 2003.

Keller, J. B. (1986). Impact With Friction. *ASME Journal of Applied Mechanics*, Vol. 53, No. 1, 1986.

Kennedy, B., Okon, A., Aghazarian, H., Badescu, M., Bao, X. & Bar-Cohen, Y. et al. (2005). LEMUR IIb: A robotic system for steep terrain access. *Proceedings 8th International Conference on Climbing and Walking Robots*, pp. 595-695, London, UK, 2005.

Klein, D.P.C. & Briggs, R. (1980). Use of compliance in the control of legged vehicles. *IEEE Transactions on Systems, Man and Cybernetics*, pp. 393-400, Vol. 10, No. 7, 1980.

Kubota, T., Takahashi, K., Shimoda, S., Yoshimitsu, T. & Nakatani, I. (2009). Locomotion mechanism of intelligent unmanned explorer for deep space exploration. *Intelligent Unmanned Systems: Theory and Applications*, pp. 11-26, Vol. 192, Springer, Berlin, 2009.

Martinez-Cantin, R. (2004). Bio-inspired multi-robot behavior for exploration in low gravity environments. *Proceedings 55th International Astronautical Congress*, Vancouver, Canada, 2004.

Menon, C., Murphy, M., Sitti, M. & Lan, N. (2007). Space exploration Towards bio-inspired climbing robots. *Bioinspiration and Robotics: Walking and Climbing Robots*, pp. 261-278, M.K. Habib, Ed., Vienna, Austria: I-Tech Education and Publishing, 2007.

Nakamura, Y., Shimoda, S. & Shoji, S. (2000). Mobility of a microgravity rover using internal electro-magnetic levitation. *Proceedings IEEE/RSJ International Conference on Intelligent Robots and Systems*, pp. 1639-1645, Takamatsu, Japan, 2000.

NASA / Jet Propulsion Laboratory (2007), Dawn Mission, *http://dawn.jpl.nasa.gov/*, September 2007.

Richter, L. (1998). Principles for robotic mobility on minor solar system bodies, *Robotics and Autonomous Systems*, pp. 117-124, Vol. 23, 1998.

Schell, S., Tretten, A., Burdick, J., Fuller, S.B. & Fiorini, P. (2001). Hopper on wheels: Evolving the hopping robot concept. *Proceedings International Conference on Field and Service Robotics*, pp. 379-384, Helsinki, Finland, 2001.

Scheeres, D. (2004). Dynamical Environment About Asteroid 25143 Itokawa. University of Michigan, Department of Aerospace Engineering, USA, 2004.

Scheeres, D., Broschart, S., Ostro, S.J. & Benner, L.A. (2004). The Dynamical Environment About Asteroid 25143 Itokawa. *Proceedings of the 24th International Symposium on Space Technology and Science*, pp. 456-461, Miyazaki, Japan, 2004.

Shimoda, S., Wingart, A., Takahashi, K., Kubota, T. & Nakatani, I. (2003). Microgravity hopping robot with controlled hopping and landing capability. *textitProceedings IEEE/RSJ Intl. Conference on Intelligent Robots and Systems*, pp. 2571-2576, October, Las Vegas, NV, 2003.

Silva, M.F. & Tenreiro Machado, J.A. (2008). New technologies for climbing robots adhesion to surfaces. *Proceedings International Workshop on New Trends in Science and Technology*, November, Ankara, Turkey, 2008.

So, E.W.Y., Yoshimitsu, T. & Kubota, T. (2008). Relative localization of a hopping rover on an asteroid surface using optical flow. *Proceedings Intl. Conf. on Instrumentation, Control, and Information Technology, SICE Annual Conference*, pp. 1727-1732, August, Tokyo, Japan. 2008.

So, E.W.Y., Yoshimitsu, T. & Kubota, T. (2009). Hopping odometry: Motion estimation with selective vision. *Proceedings IEEE/RSJ International Conference on Intelligent Robots and Systems*, pp. 3808-3813, October, St. Louis, MO. 2009.

Tunstel, E. (1999). Evolution of autonomous self-righting behaviors for articulated nanorovers. *Proceedings 5th International Symposium on Artificial Intelligence, Robotics & Automation in Space*, pp. 341-346, Noordwijk, The Netherlands, June, 1999.

Transeth, A.A., Pettersen, K.Y. & Liljebck, P. (2009). A survey on snake robot modeling and locomotion. *Robotica*, pp. 999-1015, Vol. 27, No. 7, December. 2009.

Vukobratovic, M., Frank, A. & Juricic, D. (1970). On the Stability of Biped Locomotion. *IEEE Transactions on Biomedical Engineering*, pp. 25-36, Vol.17, No.1, 1970.

Wagner, R. & Lane, H. (2007). Lessons learned on the AWIMR project. *Proceedings of the IEEE International Conference Robotics and Automation, Space Robotics Workshop*, Rome, Italy. 2007.

Wilcox, B.H. & Jones, R.M. (2000). The MUSES-CN nanorover mission and related technology. *IEEE Aerospace Conference*, Big Sky, MT, USA, pp. 287-295, 2000.

Wilcox, B.H. & Jones, R.M. (2001). A 1 kilogram asteroid exploration rover. *Proceedings IEEE International Conference on Robotics and Automation (ICRA)*, San Francisco, CA. 2001.

Wilcox, B., Litwin, T., Biesiadecki, J., Matthews, J., Heverly, M., Morrison, J., Townsend, J., Ahmad, N., Sirota, A. & Cooper, B. (2007). ATHLETE: A cargo handling and manipulation robot for the moon. *Journal of Field Robotics*, pp. 421-434, Volume 24, Issue 5, May 2007.

Yano, H., Kubota, T., Miyamoto, H. & Yoshida, K. et al. (2006). Touchdown of the Hayabusa Spacecraft at the Muses Sea on Itokawa. Report: Hayabusa at asteroid Itokawa. *Science*, Vol. 312, No. 312, June, 2006.

Yoshida, K. (1997). A General Formulation for Under-Actuated Manipulators. *Proceedings IEEE/RSJ Int. Conf. on Intelligent Robots and Systems*, pp.1651-1957, Grenoble, France, 1997.

Yoshida, K. (1999). Touch-Down Dynamics Simulation of MUSES-C with a Contact Friction Model. *Proceedings of 9th Workshop on Astrodynamics and Flight Mechanics* JAXA, Kanagawa, Japan. 1999.

Yoshida, K. (1999). The jumping tortoise: A robot design for locomotion on micro gravity surface. *Proceedings 5th International Symposium on Artificial Intelligence, Robotics & Automation in Space*, pp. 705-707, Noordwijk, The Netherlands, June, 1999.

Yoshida, K., Kubota, T., Sawai, S., Fujiwara, A.& Uo M. (2001). MUSES-C Touch-down Simulation on the Ground. *In AAS/AIAA Space Flight Mechanics Meeting*, pp. 481-490, AAS/AIAA, Santa Barbara, California, February, 2001.

Yoshida, K., Maruki, T. & Yano, H. (2001). A Novel Strategy for Asteroid Exploration with a Surface Robot. *Proceedings of the 3rd International Conference on Field and Service Robotics*, pp. 281-286, Finland, 2002.

Yoshimitsu, T., Kubota, T. & Akabane, S. et al. (2001). Autonomous navigation and observation on asteroid surface by hopping rover MINERVA. *Proceedings 6th International Symposium on Artificial Intelligence, Robotics & Automation in Space*, Quebec, Canada. 2001.

Kinematic and Inverse Dynamic Analysis of a C5 Joint Parallel Robot

Georges Fried[1], Karim Djouani[1] and Amir Fijany[2]
[1]*University of Paris Est Créteil LISSI-SCTIC Laboratory*
[2]*The Italian Institute of Technology*
[1]*France*
[2]*Italy*

1. Introduction

Parallel manipulators have been proposed to overcome accuracy problem in the end effector positioning, exhibited by serial manipulators(Stewart, 1965)(Reboulet, 1988)(Merlet, 2000). These parallel robots are primarly used in the applications for which the considered processes require a high degree of accuracy, high speeds or accelerations. Aircraft simulator (Stewart, 1965), machining tools (Neugebauer et al., 1998)(Poignet et al., 2002), and various other medical applications (Merlet, 2002)(Leroy et al., 2003)(Plitea et al., 2008) constitute some of the many possible applications of parallel robots.

The computation of the inverse dynamic model is essential for an effective robot control. In the field of parallel robots, many approaches have been developed for efficient computation of the inverse dynamics. The formalism of d'Alembert has been used to obtain an analytical expression of the dynamics model (Fichter, 1986)(Nakamura & Ghodoussi, 1989). The principle of virtual works has been applied in (Tsai, 2000) for solving the inverse dynamics of the Gough-Stewart platform and in (Zhu et al., 2005) for a Tau parallel robot. Lagrangian formalism is applied in (Leroy et al., 2003) for the dynamics modeling of a parallel robot used as a haptic interface for a surgical simulator. These various approaches do not seem effective for a robot dynamic control under the real time constraint. A better computational efficiency can be achieved by the development of approaches using recursive schemes, in particular, based on the Newton-Euler formulation. Gosselin (Gosselin, 1996) proposed an approach for the computation of the inverse dynamic model of planar and spatial parallel robots, in which all the masses and inertias are taken into account. This proposed method is difficult to generalize for all the parallel architectures. Dasgupda et al (Dasgupta & Choudhury, 1999) applied this method to several parallel manipulators. Khan (Khan, 2005) has developed a recursive algorithm for the inverse dynamics. This method is applied to a $3R$ planar parallel robot. Bi et al (Bi & Lang, 2006) use the Newton-Euler recursive scheme for the computation of the articular forces of a tripod system. Khalil et al (Khalil & Guegan, 2004) proposed a general method for the inverse and direct dynamic model computation of parallel robots, which is applied to several parallel manipulators (Khalil & Ibrahim, 2007).

Despite the large amount of contributions in this field, there is still a need for improving the computational effeciency of the inverse kinematic and dynamic model clculation for real-time control. In this paper, a parallel robot is considered as a multi robot system with

k serial robots (the k parallel robot segments) moving a common load (the mobile platform). The proposed approach uses the methodology developed by Khalil et al (Khalil & Guegan, 2004)(Khalil & Ibrahim, 2007).

In this paper, we first review the proposed approaches for computation of kinematics and inverse dynamics of of a C5 joint parallel robot. One of the interesting aspects of the parallel robots is that, unlike the serial robots, it is easy and efficient to directly compute the inverse of Jacobian matrix: However, there seems to be no proposed approach for direct computation of the Jacobian matrix. We propose a new method which allows the derivation of the Jacobian matrix in a factored form, i.e., as a product of two highly sparse matrices. This factored form enables a very fast computation and application of the Jacobian matrix, which is also needed for the inverse dynamics computation. Another issue for inverse dynamics computation is the determination of joint accelerations given the acceleration of the mobile platform. We propose a new scheme that, by using projection matrices, enablse a fast and direct calculation of the joint accelerations. Since this calculation is needed in any inverse dynamics formalism, our proposed new method improves the efficiency of the computation. For calculation of the inverse dynamics, we consider the formalism developped by Khalil et al (Khalil & Guegan, 2004)(Khalil & Ibrahim, 2007). In this approach, since the inverse of Jacobian is used, the calculation of the joint forces would require a linear system solution. We show that, by using our factorized form of the Jacobian, our proposed scheme not only eliminates the need for linear system solution but it also does not require the explicit computation of either Jacobian or its inverse. As a result, a significantly better efficiency in the computation can be achieved.

This paper is organized as follows. In Section 2, we present some preliminaries and the notation used in our appraoches. The C5 parallel robot is presented in Section 3. The proposed methodologies for computation of the inverse kinematics and the inverse Jacobian matrix are reviewed in Sections 4 and 5. The new methodology for derivation of the Jacobian matrix in a factored form is presented in Section 6. In Section 7, we present a fast and direct scheme for calculation of joint accelerations, given the desired acceleration of the mobile platform. The formalism for computation of the inverse dynamics, developed by Khalil et al (Khalil & Guegan, 2004)(Khalil & Ibrahim, 2007), is discussed in Section 8 and it is shown how the new scheme for calculation of joint accelerations as well as the use of factored form of the Jacobian matrix can significantly improve the computational efficiency. A simulation of the proposed scheme for computation of the inverse dynamics is provided in section 9 validating the proposed approach. Finally, some concluding remarks are presented in Section 10.

2. Preliminaries

In this section, the required notation for a serial chain are presented (see also Fig. 1).

2.1 System model and notations

2.1.1 Joint and link parameters

- O_j: Origin of frame j which is taken to be the center of jth joint
- $\boldsymbol{P}_j$: position vector from O_j to O_{j+1}
- N: number of bodies
- $\boldsymbol{Q}_j$, $\dot{\boldsymbol{Q}}_j$: position and velocity of the jth joint

2.1.2 Spatial quantities

- H_j: spatial-axis (map matrix) of joint j. For a joint with one rotational degree of freedom (DOF) around z-axis, H_j is given by:

$$H_j = \begin{bmatrix} 0 \\ 0 \\ 1 \\ 0 \\ 0 \\ 0 \end{bmatrix}$$

- $\boldsymbol{V}_j = \begin{bmatrix} \omega_j \\ v_j \end{bmatrix} \in \Re^6$: spatial velocity of point O_j
- $I_j \in \Re^{6\times 6}$: spatial inertia of body j

  The spatial inertia of body j about its center of mass,G_j, is denoted by I_{G_j} and is given by:

$$I_{G_j} = \begin{bmatrix} J_{G_j} & 0 \\ 0 & m_j\,U \end{bmatrix} \in \Re^{6\times 6}$$

  where J_{G_j} is the second moment of mass of link j about its center of mass and m_j is the mass of link j. The spatial inertia I_j can be calculated by:

$$I_j = \hat{\boldsymbol{S}}_j\, I_{G_j}\, \hat{\boldsymbol{S}}_j^{\boldsymbol{t}}$$

  Where $\boldsymbol{s}_j$ represents the position vector from O_j to G_j

- $\boldsymbol{V_{N+1}} \in \Re^6$: spatial velocity of the end effector

2.1.3 Global quantities

The following global quantities are defined for $j = N$ to 1

- $\dot{\boldsymbol{Q}} = Col\,(\dot{\boldsymbol{Q}}_j)$: vector of joint velocity
- $\mathcal{V} = Col\,(\,V_j\,) \subset \Re^{6N}$: global vector of spatial velocities
- $\mathcal{H} = Diag\,(\,H_j\,) \in \Re^{6N\times N}$: global matrix of spatial axis.
- $\mathcal{M}$: Symmetric positive definite (SPD) mass matrix

2.2 General notation

With any vector $\boldsymbol{V} = \left[\,V_x\; V_y\; V_z\,\right]^t$, a tensor $\tilde{V}$ can be associated whose representation in any frame is a skew symmetric matrix given by:

$$\tilde{V} = \begin{bmatrix} 0 & -V_z & V_y \\ V_z & 0 & -V_x \\ -V_y & V_x & 0 \end{bmatrix}$$

The tensor $\tilde{V}$ has the property that $\tilde{V} = -\tilde{V}^t$ and $\tilde{V}_1\boldsymbol{V_2} = \boldsymbol{V_1} \times \boldsymbol{V}_2$ i.e., it is the vector cross-product operator. A matrix $\hat{\boldsymbol{V}}$ associated to the vector $\boldsymbol{V}$ is defined as:

$$\hat{V} = \begin{bmatrix} U & \tilde{V} \\ 0 & U \end{bmatrix}$$

where U and 0 stand for unit and zero matrices of appropriate size.
In our derivation, we also make use of global matrices and vectors which lead to a compact representation of various factorizations. A bidiagonal block matrix $\mathcal{P} \in \Re^{6N \times 6N}$ is defined as:

$$\mathcal{P} = \begin{bmatrix} U & & & & \\ -\hat{P}_{N-1} & U & 0 & & \\ 0 & -\hat{P}_{N-2} & U & & \\ 0 & 0 & & & \\ \vdots & \vdots & & & \\ 0 & 0 & 0 & -\hat{P}_1 & U \end{bmatrix}$$

The inverse of matrix $\mathcal{P}$ is a block triangular matrix given by:

$$\mathcal{P}^{-1} = \begin{bmatrix} U & 0 & & \cdots & 0 \\ \hat{P}_{N,N-1} & U & 0 & \cdots & 0 \\ \hat{P}_{N,N-2} & \hat{P}_{N-1,N-2} & U & 0 & 0 \\ \vdots & \vdots & \vdots & \vdots & \vdots \\ \hat{P}_{N,1} & \hat{P}_{N-1,1} & \cdots & \hat{P}_{2,1} & U \end{bmatrix}$$

2.3 Equations of motion

In this section, we briefly review the equations of motion: velocity, force and acceleration propagation, for a serial chain of interconnected bodies.

2.3.1 Velocity propagation

The velocity propagation for a serial chain of interconnected bodies, shown in Fig. (1), is given by the following intrinsic equation:

$$\boldsymbol{V}_j - \hat{\boldsymbol{P}}^t_{j-1} \boldsymbol{V}_{j-1} = H_j \dot{\boldsymbol{Q}}_j \tag{1}$$

By using the matrix $\mathcal{P}$, Eq. (1) can be expressed in a global form as:

$$\mathcal{P}^t \boldsymbol{\mathcal{V}} = \mathcal{H} \dot{\boldsymbol{Q}} \tag{2}$$

thus:

$$\boldsymbol{\mathcal{V}} = \left(\mathcal{P}^t\right)^{-1} \mathcal{H} \dot{\boldsymbol{Q}} \tag{3}$$

The end effector spatial velocity $\boldsymbol{V}_{N+1}$ is obtained by the following relation:

$$\boldsymbol{V}_{N+1} - \hat{\boldsymbol{P}}^t_N \boldsymbol{V}_N = \boldsymbol{0} \tag{4}$$

thus:

$$\boldsymbol{V}_{N+1} = \hat{\boldsymbol{P}}^t_N \boldsymbol{V}_N \tag{5}$$

Let $\beta \in \Re^{6 \times 6N}$ be the matrix defined by $\beta = \left[\hat{\boldsymbol{P}}^t_N \ 0 \cdots 0 \right]$, Eq. (5) becomes:

$$\boldsymbol{V}_{N+1} = \beta \boldsymbol{\mathcal{V}} \tag{6}$$

Thus, inserting the expression of $\mathcal{V}$ from Eq. (3), we obtain:

$$\boldsymbol{V}_{N+1} = \beta \left(\mathcal{P}^t\right)^{-1} \mathcal{H}\dot{\mathcal{Q}} \tag{7}$$

Thus:

$$\mathcal{J} = \beta \left(\mathcal{P}^t\right)^{-1} \mathcal{H} \tag{8}$$

2.3.2 Acceleration and force propagation

The propagation of accelerations and forces among the links of serial chain are given by:

$$\dot{\boldsymbol{V}}_j = \hat{\boldsymbol{P}}^t_{j-1}\, \dot{\boldsymbol{V}}_{j-1} + H_j \ddot{\boldsymbol{Q}}_j \tag{9}$$

$$\boldsymbol{F}_j = I_j\, \dot{\boldsymbol{V}}_j + \hat{\boldsymbol{P}}_j\, \boldsymbol{F}_{j+1} \tag{10}$$

Eqs. (9)-(10) represent the simplified N-E algorithm (with nonlinear terms being excluded) for the serial chain (Luh et al., 1980).

The force $\boldsymbol{F}_j$ can be written, by using a rather unconventional decomposition of inter body force of the form (see, for example (Fijany et al., 1995) (Fijany et al., 1997), as:

$$\boldsymbol{F}_j = H_j\, \boldsymbol{F}_{T_j} + W_j\, \boldsymbol{F}_{S_j} \tag{11}$$

Where $\boldsymbol{F}_{S_j}$ represents the constraint force.

Complement to the Degrees Of Freedom (*DOF*), Degrees Of Constraint (*DOC*) are introduced ($DOC = 6 - DOF$).

The projection matrices H_j and W_j are taken to satisfy the following orthogonality conditions:

$$H^t_j\, W_j = W^t_j\, H_j = 0 \tag{12}$$

$$H_j H^t_j + W_j W^t_j = U \tag{13}$$

$$H^t_j H_j = W^t_j\, W_j = U \tag{14}$$

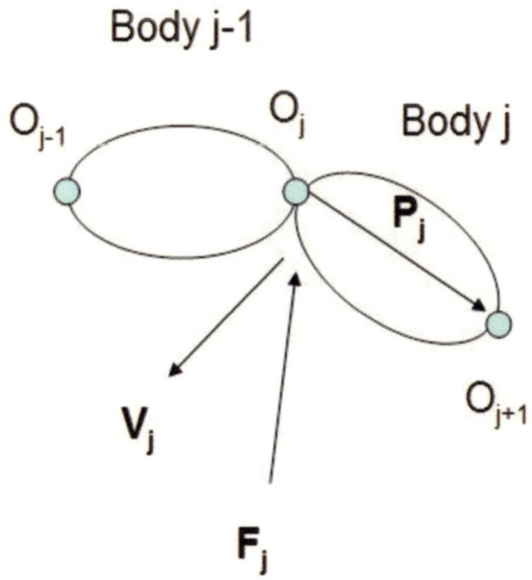

Fig. 1. Joint force and position vector of a serial chain.

3. C5 parallel robot

The C5 joint parallel robot (Dafaoui et al., 1998) consists of a static and a mobile part connected together by six actuated segments (Fig. 2 and 3). Each segment is connected to the static part at point A_i and linked to the mobile part through a C5 passive joint ($3DOF$) in rotation and $2DOF$ in translation) at point B_i. Each C5 joint consists of a spherical joint tied to two crossed sliding plates (Fig 4). Each segment is equipped with a ball and a screw linear actuator driven by a DC motor.

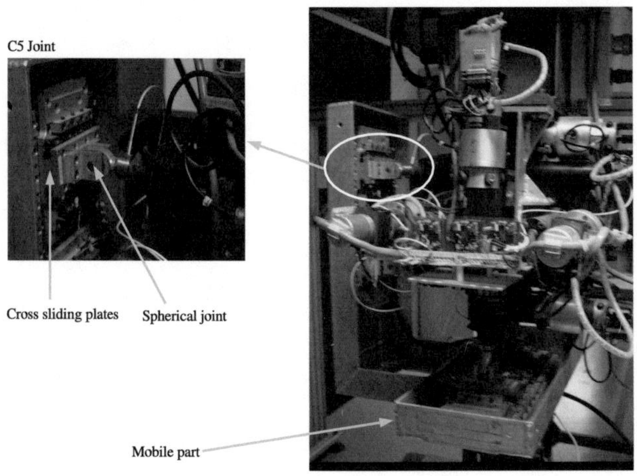

Fig. 2. C5 joint parallel robot

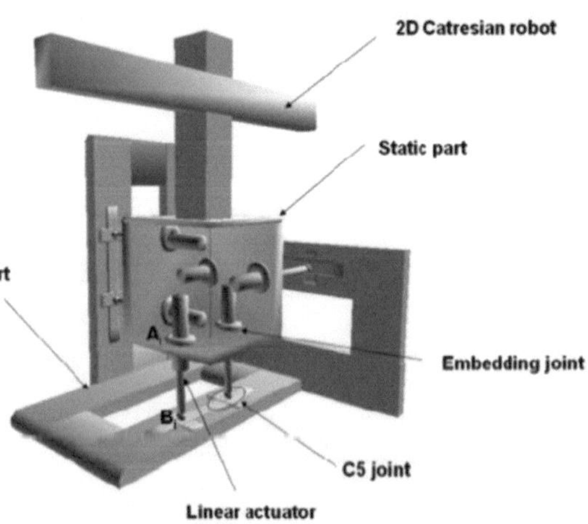

Fig. 3. C5 parallel robot representation.

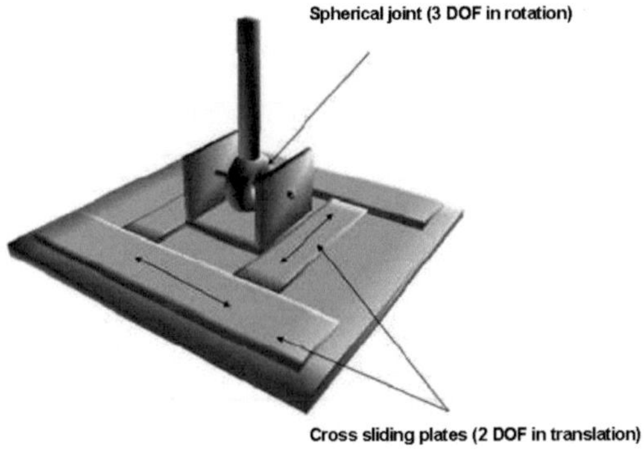

Fig. 4. Details of the C5 joint.

Following notations are used in the description of the parallel robot:

- R_b is the absolute frame, attached to to the fixed base: $R_b = (0, x, y, z)$.
- R_p is the mobile frame, attached to the mobile part: $R_p = (C, x_p, y_p, z_p)$.
- O is the origin of the absolute coordinate system.
- C is the origin of the mobile coordinate system, whose coordinates in the absolute frame are given by:

$$\boldsymbol{OC}_{/R_b} = \begin{bmatrix} x_c \; y_c \; z_c \end{bmatrix}^t$$

- A_i is the center of the joint between the segment i and the fixed base:

$$\boldsymbol{OA_i}_{/R_b} = \begin{bmatrix} a_i^x \; a_i^y \; a_i^z \end{bmatrix}^t$$

- B_i is the center of the rotational joint between the segment i and the mobile part:

$$\boldsymbol{CB_i}_{/R_p} = \begin{bmatrix} b_i^x \; b_i^y \; b_i^z \end{bmatrix}^t$$

- R is the rotation matrix, with elements r_{ij} (using the RPY formalism), expressing the orientation of the R_p coordinate system with respect to the R_b coordinate system. The expression for this matrix is given by:

$$R = \begin{bmatrix} r_{11} \; r_{12} \; r_{13} \\ r_{21} \; r_{22} \; r_{23} \\ r_{31} \; r_{32} \; r_{33} \end{bmatrix} \tag{15}$$

where:
$r_{11} = \cos\beta\cos\gamma$
$r_{12} = -\cos\beta\sin\gamma$
$r_{13} = \sin\beta$
$r_{21} = \sin\gamma\cos\alpha + \cos\gamma\sin\beta\sin\alpha$
$r_{22} = \cos\alpha\cos\gamma - \sin\alpha\sin\beta\sin\gamma$

$$
\begin{aligned}
r_{23} &= -\cos\beta\sin\alpha \\
r_{31} &= \sin\gamma\sin\alpha - \cos\gamma\sin\beta\cos\alpha \\
r_{32} &= \sin\alpha\cos\gamma + \cos\alpha\sin\beta\sin\gamma \\
r_{33} &= \cos\beta\cos\alpha
\end{aligned}
$$

- The rotation angles, α, β and γ , also called Roll Pitch and Yaw (RPY), describe the rotation of the mobile platform with respect to the fixed part. α is the rotation around x axis, β around y axis and γ around z axis.
- $\boldsymbol{X}$ is the task coordinate vector.

$$\boldsymbol{X} = \begin{bmatrix} \alpha\ \beta\ \gamma\ x_c\ y_c\ z_c \end{bmatrix}^t$$

In the following, the parallel robot is considered as six serial robots (the six segments) moving a common load (the mobile platform). According to our notation presented in the previous section, we define the following quantities:

- i is the segment index
- j is the body index
- O_{ij} is the jth joint center of the segment i
- $\boldsymbol{P}_{ij}$ is the position vector from O_{ij} to $O_{i,j+1}$
- $\boldsymbol{Q}_{ij}$ is the position of the jth joint of the segment i
- $\boldsymbol{Q}_i = \begin{bmatrix} \boldsymbol{Q}_{i2} \\ \boldsymbol{Q}_{i1} \end{bmatrix} \in \Re^6$ is the joint coordinate vector of the segment i
- $\dot{\boldsymbol{Q}}_i = \begin{bmatrix} \dot{\boldsymbol{Q}}_{i2} \\ \dot{\boldsymbol{Q}}_{i1} \end{bmatrix} \in \Re^6$ is the joint velocity vector of the segment i
- $\ddot{\boldsymbol{Q}}_i = \begin{bmatrix} \ddot{\boldsymbol{Q}}_{i2} \\ \ddot{\boldsymbol{Q}}_{i1} \end{bmatrix} \in \Re^6$ is the joint acceleration vector of the segment i
- H_{ij} is the spatial-axis of joint j^{th} joint of the segment i. For the C5 joint robot the projection matrices H_{ij} and W_{ij}, describe in the base frame, are given as:

$$
\begin{aligned}
H_{11} = H_{21} &= \begin{bmatrix} 0\,0\,0\,1\,0\,0 \end{bmatrix}^t \\
H_{31} = H_{41} &= \begin{bmatrix} 0\,0\,0\,0\,1\,0 \end{bmatrix}^t \\
H_{51} = H_{61} &= \begin{bmatrix} 0\,0\,0\,0\,0\,1 \end{bmatrix}^t
\end{aligned}
$$

$$H_{12} = H_{22} = \begin{bmatrix} 0\,0\,0\,0\,1 \\ 0\,0\,0\,1\,0 \\ 0\,0\,1\,0\,0 \\ 0\,0\,0\,0\,0 \\ 0\,1\,0\,0\,0 \\ 1\,0\,0\,0\,0 \end{bmatrix}$$

$$H_{32} = H_{42} = \begin{bmatrix} 0\,0\,0\,0\,1 \\ 0\,0\,0\,1\,0 \\ 0\,0\,1\,0\,0 \\ 0\,1\,0\,0\,0 \\ 0\,0\,0\,0\,0 \\ 1\,0\,0\,0\,0 \end{bmatrix}$$

$$H_{52} = H_{62} = \begin{bmatrix} 0\,0\,0\,0\,1 \\ 0\,0\,0\,1\,0 \\ 0\,0\,1\,0\,0 \\ 0\,1\,0\,0\,0 \\ 1\,0\,0\,0\,0 \\ 0\,0\,0\,0\,0 \end{bmatrix}$$

$$W_{12} = W_{22} = \begin{bmatrix} 0\,0\,0\,1\,0\,0 \end{bmatrix}^t$$
$$W_{32} = W_{42} = \begin{bmatrix} 0\,0\,0\,0\,1\,0 \end{bmatrix}^t$$
$$W_{52} = W_{62} = \begin{bmatrix} 0\,0\,0\,0\,0\,1 \end{bmatrix}^t$$

- $\boldsymbol{V}_{ij} = \begin{bmatrix} \omega_{ij} \\ v_{ij} \end{bmatrix} \in \Re^6$ is the spatial velocity of the link j for the segment i
- $I_{ij} \in \Re^{6\times 6}$: spatial inertia of body j for the segment i
- $\boldsymbol{\mathcal{V}}_i = Col\left(V_{ij}\right) \in \Re^{12}$: global vector of spatial velocities for the segment i
- $\mathcal{H}_i = Diag\left(H_{ij}\right) \in \Re^{12\times 6}$: global matrix of spatial axis for the leg i
- $\mathcal{M}_i$: Symmetric positive definite (SPD) mass matrix of the segment i

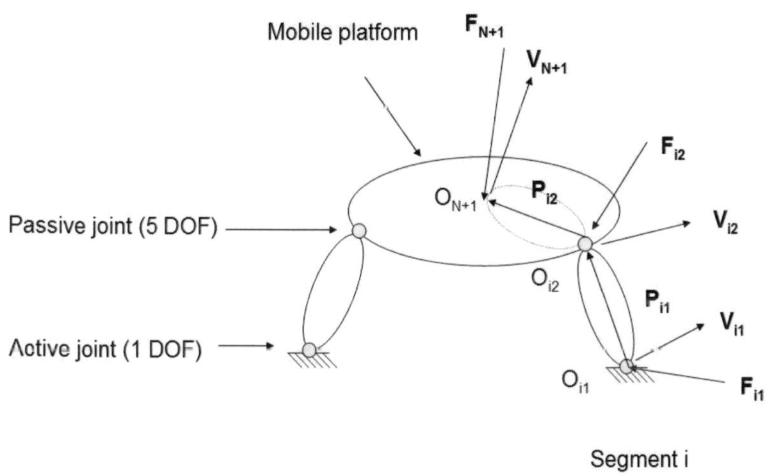

Fig. 5. Force and position vectors.

4. Inverse kinematic model

In this section, we briefly review the methodology used for the inverse kinematic model computation. More details can be found in (Dafaoui et al., 1998).
The inverse kinematic model relates the active joint variables $\boldsymbol{Q_a} = \left[Q_{61}\; Q_{51}\; Q_{41}\; Q_{31}\; Q_{21}\; Q_{11} \right]^t$ to the operational variables which define the position and the orientation of the end effector ($\boldsymbol{X}$). This relation is given by the following equation

(Dafaoui et al., 1998):

$$\begin{aligned}
Q_{11} &= x_c + \tfrac{r_{31}(z_c-L)+r_{21}y_c}{r_{11}} \\
Q_{21} &= x_c + \tfrac{r_{31}(z_c+L)+r_{21}y_c}{r_{11}} \\
Q_{31} &= y_c + \tfrac{r_{12}(x_c-L)+r_{32}z_c}{r_{22}} \\
Q_{41} &= y_c + \tfrac{r_{12}(x_c+L)+r_{32}z_c}{r_{22}} \\
Q_{51} &= z_c + \tfrac{r_{23}(y_c-L)+r_{13}x_c}{r_{33}} \\
Q_{61} &= z_c + \tfrac{r_{23}(y_c+L)+r_{13}x_c}{r_{33}}
\end{aligned} \tag{16}$$

where:

$$L = \tfrac{\|\boldsymbol{A_i A_{i+1}}\|}{2} \text{ for } i = 1, 3 \text{ and } 5$$

For the C5 joint parallel robot, the actuators are equidistant from point O (Fig. 6).

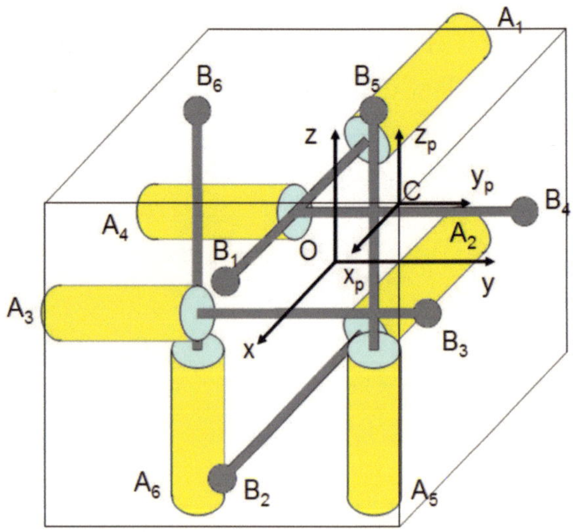

Fig. 6. The spatial arrangement of the C5 joint parallel robot segments.

5. Determination of the inverse Jacobian matrix

For parallel robots, the inverse Jacobian matrix computation ($\mathcal{J}^{-1}$) is obtained by the determination of the velocity of point B_i (Merlet, 2000)(Gosselin, 1996):

$$\dot{\boldsymbol{OB_i}} = \boldsymbol{v_{N+1}} + \boldsymbol{B_i C} \times \boldsymbol{\omega_{N+1}} \tag{17}$$

By using the following:

$$\dot{Q}_{i1} = \dot{\boldsymbol{OB_i}}\, \boldsymbol{n_i} \tag{18}$$

where $\boldsymbol{n_i}$ is the unit vector of the segment i, defined by:

$$\boldsymbol{n_i} = \frac{\boldsymbol{A_i B_i}}{Q_{i1}} \tag{19}$$

and inserting Eq. (17) into Eq. (18), we obtain the following expression:

$$\dot{Q}_{i1} = \boldsymbol{n_i v_{N+1}} + \boldsymbol{\omega_{N+1}} (\boldsymbol{n_i} \times \boldsymbol{B_i C}) \tag{20}$$

The $(6-i)^{th}$ row of the inverse Jacobian matrix is given as:

$$\left[(\boldsymbol{n_i} \times \boldsymbol{B_i C})^t \ \boldsymbol{n}_i^t \right] \tag{21}$$

The inverse Jacobian matrix of the C5 parallel robot is then given by the following relation:

$$\mathcal{J}^{-1} = \begin{bmatrix} (\boldsymbol{n_6} \times \boldsymbol{B_6 C})^t \ \boldsymbol{n}_6^t \\ (\boldsymbol{n_5} \times \boldsymbol{B_5 C})^t \ \boldsymbol{n}_5^t \\ (\boldsymbol{n_4} \times \boldsymbol{B_4 C})^t \ \boldsymbol{n}_4^t \\ (\boldsymbol{n_3} \times \boldsymbol{B_3 C})^t \ \boldsymbol{n}_3^t \\ (\boldsymbol{n_2} \times \boldsymbol{B_2 C})^t \ \boldsymbol{n}_2^t \\ (\boldsymbol{n_1} \times \boldsymbol{B_1 C})^t \ \boldsymbol{n}_1^t \end{bmatrix} = \begin{bmatrix} (\boldsymbol{n_6} \times \boldsymbol{P_{62}})^t \ \boldsymbol{n}_6^t \\ \vdots \\ (\boldsymbol{n_1} \times \boldsymbol{P_{12}})^t \ \boldsymbol{n}_1^t \end{bmatrix} \tag{22}$$

As stated before, for parallel robots, unlike the serial robots, the inverse of Jacobian matrix can be directly and efficiently obtained. In fact, the cost of computation of $\mathcal{J}^{-1}$ from Eq. (22) is $(18m + 30a)$ where m and a denote the cost of multiplication and addition, respectively.

6. Factorized expression of the Jacobian matrix

6.1 General approach

The differential kinematic model of a manipulator is defined by the relationship between the spatial velocity of the end effector and the vector of generalized coordinate velocities of the robot: $\boldsymbol{V_{N+1}} = \mathcal{J} \dot{\boldsymbol{Q}}_{\boldsymbol{a}}$, where $\mathcal{J}$ is the Jacobian matrix.

For parallel robots, it seems more efficient to compute the Jacobian matrix $\mathcal{J}$ by inverting the inverse Jacobian matrix $\mathcal{J}^{-1}$ (see for example (Khalil & Ibrahim, 2007)). In deriving the forward kinematic model of the C5 parallel robot, an analytical expression of the Jacobian matrix is presented in (Dafaoui et al., 1998). From a computational efficiency point of view, such a classical method, which is only applicable to the C5 parallel robot, is not well suited for real-time control.

Here, we present opur approach for direct and efficient computation of the Jacobian matrix (Fried et al., 2006). In this approach, an analytical expression of the Jacobian matrix is obtained in factorized form as a product of sparse matrices which achieves a much better computational efficiency.

In our approach, the parallel robot is considered as a multi-robot system, composed of serial robots (the segments) moving a common load (the mobile platform). A relationship between the Jacobian matrix of the parallel robot ($\mathcal{J}$) to the Jacobian matrix of each segment ($\mathcal{J}_i$) is first derived.

The principle of this approach consists of first computing the Jacobian matrix for each leg considered as an open serial chain. Secondly, the closing constraint is determined, allowing the computation of the parallel robot Jacobian matrix.

The Jacobian matrix $\mathcal{J}$ of the parallel robot is obtained by the closing constraint determination of the kinematic chain. This determination can be obtained by expressing the actuated joint velocity $\dot{\boldsymbol{Q}}_{\boldsymbol{a}}$ of the parallel robot in function of vectors $\dot{\boldsymbol{Q}}_{\boldsymbol{i}}$ associated to each segment i. Let the matrix Π_i be defined as:

$$\dot{\boldsymbol{Q}}_{\boldsymbol{i}} = \Pi_i \dot{\boldsymbol{Q}}_{\boldsymbol{a}} \tag{23}$$

Inserting Eq. (23) into Eq. (7), we obtain:

$$\boldsymbol{V_{N+1}} = \beta_i \left(\mathcal{P}_i^t\right)^{-1} \mathcal{H}_i \Pi_i \dot{\boldsymbol{Q}}_{\boldsymbol{a}} \tag{24}$$

Therefore, a factorized expression of the parallel robot Jacobian matrix is obtained as:

$$\mathcal{J} = \beta_i \left(\mathcal{P}_i^t\right)^{-1} \mathcal{H}_i \Pi_i \tag{25}$$

The matrices $\mathcal{J}$ and $\mathcal{J}_i$ are related by the following relationship:

$$\mathcal{J} = \mathcal{J}_i \Pi_i \tag{26}$$

The computation of matrix of Π_i depends on the considered parallel robot's structure. In the following, we present the computation of this matrix for the C5 parallel robot.

6.2 Application to the C5 parallel robot

Let $\boldsymbol{P_{i2}} = \left[\, x_i \; y_i \; z_i \,\right]^t$ denote the position vector from B_i to C:

$$\boldsymbol{P_{i2}} = \boldsymbol{B_i C}_{/R_b} = -Q_{i1}\, \boldsymbol{n_i} + \boldsymbol{A_i O} + \boldsymbol{OC} \tag{27}$$

The spatial arrangement of the segments (see Fig. 6) is as follows:

- The segments 1 and 2 are in the direction of the x-axis ($\boldsymbol{n_i} = \left[\,1\;0\;0\,\right]^t$ for $i = 1, 2$).
- The segments 3 and 4 are in the direction of the y-axis ($\boldsymbol{n_i} = \left[\,0\;1\;0\,\right]^t$ for $i = 3, 4$).
- The segments 5 and 6 are in the direction of the z-axis ($\boldsymbol{n_i} = \left[\,0\;0\;1\,\right]^t$ for $i = 5, 6$).

Thus, we deduce the following relations:

$$\begin{aligned} y_1 &= y_2 = y_c \\ z_3 &= z_4 = z_c \\ x_5 &= x_6 = x_c \end{aligned} \tag{28}$$

The global vector of articular coordinate velocity of the leg i is given by:

$$\dot{\boldsymbol{Q}}_{\boldsymbol{i}} = \left[\, \dot{w}_{p_i} \; \dot{u}_{p_i} \; \dot{\gamma}_{p_i} \; \dot{\beta}_{p_i} \; \dot{\alpha}_{p_i} \; \dot{Q}_{i1} \,\right]^t \tag{29}$$

where $\dot{u}_{p_i}$ and $\dot{w}_{p_i}$ are translation velocities due to the crossed sliding plates.

6.2.1 Determination of matrix Π_i

The matrix Π_i given in Eq. (23) is obtained as follows. We have:

$$\begin{bmatrix} \dot{w}_{p_i} \\ \dot{u}_{p_i} \\ \dot{\gamma}_{p_i} \\ \dot{\beta}_{p_i} \\ \dot{\alpha}_{p_i} \\ \dot{Q}_{i1} \end{bmatrix} = \Pi_i \begin{bmatrix} \dot{Q}_{61} \\ \dot{Q}_{51} \\ \dot{Q}_{41} \\ \dot{Q}_{31} \\ \dot{Q}_{21} \\ \dot{Q}_{11} \end{bmatrix} \tag{30}$$

The elements ${}^i\pi_{jk}$ of the matrix Π_i are computed by using Eq. (7). This equation is true for $i = 1$ to 6, thus:

$$\beta_i \left(\mathcal{P}_i^t\right)^{-1} \mathcal{H}_i \dot{\boldsymbol{Q}}_{\boldsymbol{i}} = \beta_j \left(\mathcal{P}_j^t\right)^{-1} \mathcal{H}_j \dot{\boldsymbol{Q}}_{\boldsymbol{j}} \tag{31}$$

for i and $j = 1$ to 6. From Eq. (31), we can show that for all $i, j = 1, \ldots, 6$, we have the following relations:

$$\begin{cases} \dot{\alpha}_{p_i} = \dot{\alpha}_{p_j} \\ \dot{\beta}_{p_i} = \dot{\beta}_{p_j} \\ \dot{\gamma}_{p_i} = \dot{\gamma}_{p_j} \end{cases} \tag{32}$$

After some manipulations on relation (31), we obtain:

- For $i = 1$ and $j = 2$

$$\dot{Q}_{11} = (z_2 - z_1)\,\dot{\beta}_{p_i} + (y_1 - y_2)\,\dot{\gamma}_{p_i} + \dot{Q}_{21} \tag{33}$$

- For $i = 3$ and $j = 4$:

$$\dot{Q}_{31} = (z_3 - z_4)\,\dot{\alpha}_{p_i} + (x_4 - x_3)\,\dot{\gamma}_{p_i} + \dot{Q}_{41} \tag{34}$$

- For $i = 5$ and $j = 6$:

$$\dot{Q}_{51} = (y_6 - y_5)\,\dot{\alpha}_{p_i} + (x_5 - x_6)\,\dot{\beta}_{p_i} + \dot{Q}_{61} \tag{35}$$

- For $i = 1$ and $j = 3$:

$$\dot{u}_{p_1} = (z_1 - z_3)\,\dot{\alpha}_{p_i} + (x_3 - x_1)\,\dot{\gamma}_{p_i} + \dot{Q}_{31} \tag{36}$$

- For $i = 1$ and $j = 5$:

$$\dot{w}_{p_1} = (y_5 - y_1)\,\dot{\alpha}_{p_i} + (x_1 - x_5)\,\dot{\beta}_{p_i} + \dot{Q}_{51} \tag{37}$$

From Eq. (28), we have $y_1 = y_2, z_3 = z_4$ and $x_5 = x_6$. Thus, the Eqs. (33, 34, and 35) can be written as:

$$\begin{aligned} \dot{Q}_{11} &= (z_2 - z_1)\,\dot{\beta}_{p_i} + \dot{Q}_{21} \\ \dot{Q}_{31} &= (x_4 - x_3)\,\dot{\gamma}_{p_i} + \dot{Q}_{41} \\ \dot{Q}_{51} &= (y_6 - y_5)\,\dot{\alpha}_{p_i} + \dot{Q}_{61} \end{aligned} \tag{38}$$

From Eqs. (30), (36), (37) and (38), the matrix Π_1 is computed as:

$$\Pi_1 = \begin{bmatrix} \frac{y_5-y_1}{y_5-y_6} & \frac{y_6-y_1}{y_6-y_5} & 0 & 0 & \frac{x_1-x_5}{z_1-z_2} & \frac{x_1-x_5}{z_2-z_1} \\ \frac{z_1-z_3}{y_5-y_6} & \frac{z_1-z_3}{y_6-y_5} & \frac{x_3-x_1}{x_3-x_4} & \frac{x_4-x_1}{x_4-x_3} & 0 & 0 \\ 0 & 0 & \frac{1}{x_3-x_4} & \frac{1}{x_4-x_3} & 0 & 0 \\ 0 & 0 & 0 & 0 & \frac{1}{z_1-z_2} & \frac{1}{z_2-z_1} \\ \frac{1}{y_5-y_6} & \frac{1}{y_6-y_5} & 0 & 0 & 0 & 0 \\ 0 & 0 & 0 & 0 & 0 & 1 \end{bmatrix} \tag{39}$$

The computational cost of explicit construction of the matrix Π_1 is $(17m + 28a)$ wherein the cost of division has been taken to be the same as multiplication.
Considering the matrix Π_1, the expression of the jacobian matrix is given by:

$$\mathcal{J} = \mathcal{J}_1\,\Pi_1 \tag{40}$$

With:

$$\mathcal{J}_1 = \beta_1\,\mathcal{P}_1^{-t}\,\mathcal{H}_1 = \begin{bmatrix} \hat{\boldsymbol{P}}_{12}^{\boldsymbol{t}} & 0 \end{bmatrix} \begin{bmatrix} U & \hat{\boldsymbol{P}}_{11}^{\boldsymbol{t}} \\ 0 & U \end{bmatrix} \begin{bmatrix} H_{12} & 0 \\ 0 & H_{11} \end{bmatrix}$$

Thus:

$$\mathcal{J}_1 = \begin{bmatrix} \hat{\boldsymbol{P}}_{12}^{\boldsymbol{t}} H_{12} & \hat{\boldsymbol{P}}_{12}^{\boldsymbol{t}} \hat{\boldsymbol{P}}_{11}^{\boldsymbol{t}} H_{11} \end{bmatrix} = \begin{bmatrix} 0 & 0 & 0 & 0 & 1 & 0 \\ 0 & 0 & 0 & 1 & 0 & 0 \\ 0 & 0 & 1 & 0 & 0 & 0 \\ 0 & 0 & -y_1 & z_1 & 0 & 1 \\ 0 & 1 & x_1 & 0 & -z_1 & 0 \\ 1 & 0 & 0 & -x_1 & y_1 & 0 \end{bmatrix} \tag{41}$$

Note the highly sparse structure of the matrix $\mathcal{J}_1$. In fact, if the matrix Π_1 is already computed then the computation of the matrix $\mathcal{J}_1$ does not require any operation. However, if the explicit computatiuon of $\mathcal{J}$ is needed it can be then computed as $\mathcal{J} = \mathcal{J}_1 \Pi_1$. Exploiting the sparse structure of matrices $\mathcal{J}_1$ and Π_1, this computation can be performed with a cost of $29m + 37a$.

7. Computation of joint accelerations of the segments

The conventional approach to calculate $\ddot{\boldsymbol{Q}}_i$ is based on time derivation of Eq. (7) as:

$$\ddot{\boldsymbol{Q}}_i = \mathcal{J}_i^{-1} \dot{\boldsymbol{V}}_{N+1} + \frac{d}{dt} \mathcal{J}_i^{-1} \boldsymbol{V}_{N+1} \tag{42}$$

Eq. (42) represents the second-order inverse kinematic model of the segment i.

In the following, we propose a new and more efficient approach for computation of $\ddot{\boldsymbol{Q}}_i$. From the propagation of acceleration given in Eq. (9), we can derive the following relations:

$$\dot{\boldsymbol{V}}_{N+1} = \hat{\boldsymbol{P}}_{i2}^t \dot{\boldsymbol{V}}_{i2} \tag{43}$$

$$\dot{\boldsymbol{V}}_{i2} = H_{i2} \ddot{\boldsymbol{Q}}_{i2} + \hat{\boldsymbol{P}}_{i1}^t \dot{\boldsymbol{V}}_{i1} + \frac{d}{dt} \hat{\boldsymbol{P}}_{i1}^t \boldsymbol{V}_{i1} \tag{44}$$

$$\dot{\boldsymbol{V}}_{i1} = H_{i1} \ddot{\boldsymbol{Q}}_{i1} \tag{45}$$

Considering the othogonality properties of the projection matrices H_{ij} and W_{ij} given in Eq. (12), by multiplying both sides of Eq. (44) by W_{i2}^t we get:

$$W_{i2}^t \dot{\boldsymbol{V}}_{i2} = \underbrace{W_{i2}^t H_{i2} \ddot{\boldsymbol{Q}}_{i2}}_{0} + W_{i2}^t \hat{\boldsymbol{P}}_{i1}^t \dot{\boldsymbol{V}}_{i1} + \underbrace{W_{i2}^t \frac{d}{dt} \hat{\boldsymbol{P}}_{i1}^t H_{i1}}_{0} \dot{\boldsymbol{Q}}_{i1} = W_{i2}^t \hat{\boldsymbol{P}}_{i1}^t \dot{\boldsymbol{V}}_{i1} \tag{46}$$

Note that the above projection indeed eliminates the term $\ddot{\boldsymbol{Q}}_{i2}$ from the equation. From Eqs. (45) and (46), we then obtain:

$$\ddot{\boldsymbol{Q}}_{i1} = \left(W_{i2}^t \hat{\boldsymbol{P}}_{i1}^t H_{i1} \right)^{-1} W_{i2}^t \dot{\boldsymbol{V}}_{i2} \tag{47}$$

Where the term defined by $\left(W_{i2}^t \hat{\boldsymbol{P}}_{i1}^t H_{i1} \right)^{-1}$ is a scalar.
Again, considering the properties given by Eq. (14), multiplying both sides of Eq. (44) by by H_{i2}^t we get:

$$\ddot{\boldsymbol{Q}}_{i2} = H_{i2}^t \dot{\boldsymbol{V}}_{i2} - H_{i2}^t \hat{\boldsymbol{P}}_{i1}^t \dot{\boldsymbol{V}}_{i1} - \underbrace{H_{i2}^t \frac{d}{dt} \hat{\boldsymbol{P}}_{i1}^t H_{i1}}_{0} \dot{\boldsymbol{Q}}_{i1} = H_{i2}^t \dot{\boldsymbol{V}}_{i2} - H_{i2}^t \hat{\boldsymbol{P}}_{i1}^t \dot{\boldsymbol{V}}_{i1} \tag{48}$$

Note that, $H_{i2}^t \frac{d}{dt} \hat{\boldsymbol{P}}_{i1}^t H_{i1} = \mathbf{0}$ since $\frac{d}{dt} \hat{\boldsymbol{P}}_{i1}^t$ is along H_{i1} and as can be seen from the description of H_{i1} and H_{i2} in section 3, $H_{i2}^t H_{i1} = \mathbf{0}$.
The joint accelerations of the segment i are then computed in four steps as follows:

1. Compute $\dot{\boldsymbol{V}}_{i2}$ from Eq. (43)
2. Compute $\ddot{\boldsymbol{Q}}_{i1}$ from Eq. (47)
3. Compute $\dot{\boldsymbol{V}}_{i1}$ from Eq. (45)
4. Compute $\ddot{\boldsymbol{Q}}_{i2}$ from Eq. (48)

Exploiting the sparse structure of matrices H_{i1}, H_{i2}, and W_{i2} as well as an appropriate scheme for projection of the equations, the cost of computing $\ddot{Q}_i = \begin{bmatrix} \ddot{Q}_{i2} \\ \ddot{Q}_{i1} \end{bmatrix}$ for each segment is of $(10m + 10a)$.

8. Inverse dynamic model

The inverse dynamic computation for the parallel robot consists of determination of the required active joints torques to achieve a desired acceleration of the mobile platform, which is needed for accurate control of the robot. In this section, we review the approch proposed by Khalil et al in (Khalil & Guegan, 2004) (Khalil & Ibrahim, 2007).
Our contibution is to show that by using the factorized expression of the Jacobian matrix, and the new formulation of acceleration joints, a significantly better computational efficiency can be achieved for the inverse dynamic calculation.

8.1 Computation of inverse dynamic model

The dynamical equation of motion for each segment is given by:

$$\mathcal{M}_i \ddot{Q}_i + C_i + G_i + \mathcal{J}_{B_i}^t F_{i2} = \Gamma_i \tag{49}$$

Where:

- F_{i2} is the force exterted to the mobile platform by the segment i (Fig. 5).
- $\mathcal{J}_{B_i}$ is the Jacobian matrix of the segment i, computed to the point B_i. The expression of $\mathcal{J}_{B_i}$ is given by:

$$\mathcal{J}_{B_i} = \hat{P}_{i2}^{-t} \mathcal{J}_i \tag{50}$$

- $C_i + G_i$ represents the contributation of the Coriolis, centrifugal, and gravitional terms.
- Γ_i represents the joint force vector of the segment i.

The contact forces exerted to the mobile platform by the segments, shown in Fig. 5, are computed from Eq. (49) as:

$$F_{i2} = -\mathcal{J}_{B_i}{}^t \left(\mathcal{M}_i \ddot{Q}_i + C_i + G_i\right) + \mathcal{J}_{B_i}{}^t \Gamma_i \tag{51}$$

The dynamic behavior of the mobile platform is given by the following relation:

$$F_{N+1} = \Lambda_C \dot{V}_{N+1} + (G_C + C_C) \tag{52}$$

Where:

- F_{N+1} is the spatial force applied at the point C, representing the contribution of the contact forces F_{iN} propagated to the point C:

$$F_{N+1} = \begin{bmatrix} n_{N+1} \\ f_{N+1} \end{bmatrix} = \sum_{i=1}^{6} \hat{P}_{i2}^t F_{i2} \tag{53}$$

- $\Lambda_C \in \Re^{6\times 6}$ is the spatial inertia matrix of the mobile platform:

$$\Lambda_C = \begin{bmatrix} I_C & m_C \widetilde{GC} \\ -m_C \widetilde{GC} & m_C U \end{bmatrix} \tag{54}$$

- m_C is the platform mass
- $I_C \in \Re^{3\times3}$ is the inertia tensor of the mobile platform expressed in the mobile platform center of mass and projected in the fixed frame R_b:

$$I_C = R\, I_{C_{/R_m}}\, R^t \tag{55}$$

- $\boldsymbol{C_C} \in \Re^6$ is the vector of Coriolis and centrifugal forces:

$$\boldsymbol{C_C} = \begin{bmatrix} \tilde{\omega}_{N+1}\, I_C\, \boldsymbol{\omega_{N+1}} \\ -m_C\, \tilde{\omega}_{N+1}\, \widetilde{GC}\, \boldsymbol{\omega_{N+1}} \end{bmatrix} \tag{56}$$

- $\boldsymbol{G_C} \in \Re^6$ is the vector of gravitational forces:

$$\boldsymbol{G_C} = \begin{bmatrix} -m_C\, \widetilde{GC} \\ -m_C\, U \end{bmatrix} \boldsymbol{g} \tag{57}$$

- $\boldsymbol{g}$ being the acceleration vector of gravity

Substituting (51) in (53), we obtain:

$$\boldsymbol{F_{N+1}} = \sum_{i=1}^{6} \boldsymbol{\hat{P}^t_{i2}} \left(\mathcal{J}_{B_i}^{-t}\, \boldsymbol{\Gamma_i} - \mathcal{J}_{B_i}^{-t} \left(\mathcal{M}_i \boldsymbol{\ddot{Q}_i} + \boldsymbol{C_i} + \boldsymbol{G_i} \right) \right) \tag{58}$$

The active joint forces vector is given by:

$$\boldsymbol{\Gamma} = \left[\Gamma_{61}\; \Gamma_{51}\; \Gamma_{41}\; \Gamma_{31}\; \Gamma_{21}\; \Gamma_{11} \right]^t$$

We have $\sum_{i=1}^{6} \boldsymbol{\hat{P}^t_{i2}}\, \mathcal{J}_{B_i}^{-t}\, \boldsymbol{\Gamma_i} = \mathcal{J}^{-1}\, \boldsymbol{\Gamma}$ where $\mathcal{J}^{-1}$ is the inverse Jacobian matrix of the parallel robot. Eq. (58) can be rewritten as:

$$\boldsymbol{F_{N+1}} = \mathcal{J}^{-t}\, \boldsymbol{\Gamma} - \sum_{i=1}^{6} \boldsymbol{\hat{P}^t_{i2}}\, \mathcal{J}_{B_i}^{-t} \left(\mathcal{M}_i \boldsymbol{\ddot{Q}_i} + \boldsymbol{C_i} + \boldsymbol{G_i} \right) \tag{59}$$

The inverse dynamic model, given by Kahlil et al in (Khalil & Ibrahim, 2007) is then expressed by:

$$\boldsymbol{\Gamma} = \mathcal{J}^t \left[\boldsymbol{F_{N+1}} + \sum_{i=1}^{6} \boldsymbol{\hat{P}^t_{i2}}\, \mathcal{J}_{B_i}^{-t} \left(\mathcal{M}_i \boldsymbol{\ddot{Q}_i} + \boldsymbol{C_i} + \boldsymbol{G_i} \right) \right] \tag{60}$$

8.2 Computational complexity analysis

The inverse dynamic model, given in Eq. (60), is computed in six steps:

1. Computation of joint accelerations from Eq. (42)
2. Computation of the vector defined as $\boldsymbol{T_i} = \mathcal{M}_i \boldsymbol{\ddot{Q}_i} + \boldsymbol{C_i} + \boldsymbol{G_i}$ with the recursive Newton-Euler algorithm (Cork, 1996).
3. Computation of the vector resulting from the propagation of forces exterted on the mobile platform by all the segments as $\boldsymbol{\Phi} = \sum_{i=1}^{6} \boldsymbol{\hat{P}^t_{i2}}\, \boldsymbol{T_i}$
4. Computation of $\boldsymbol{F_{N+1}}$ from Eq. (52)
5. Computation of the vector defined as $\boldsymbol{K} = \boldsymbol{F_{N+1}} + \boldsymbol{\Phi}$

6. Computation of the vector $\mathbf{\Gamma} = \mathcal{J}^t \boldsymbol{K}$

The computation of the last step, as discussed by Khalil et al in (Khalil & Guegan, 2004) (Khalil & Ibrahim, 2007), is performed by first computing $\mathcal{J}^{-t}$ from Eq. (22). As a result, the computation of $\mathbf{\Gamma}$ requires solving a linear system of dimension 6 as:

$$\mathcal{J}^{-t}\mathbf{\Gamma} = \boldsymbol{K} \tag{61}$$

The computation cost for the derivation of $\mathcal{J}^{-1}$, from Eq. (22), is $(18m + 30a)$. The cost of solving the linear system of dimension 6 in Eq. (61) is of $(116m + 95a)$ wherein the cost of division and multiplication are taken to be the same. Therefore, the total cost of Step 6 is of $(134m + 125a)$

Our contribution concerns the improvement of the efficiency of the inverse dynamic computational by using the factorized expression of the Jacobian matrix and the new formulation of the joint accelerations. Using these formulations, the computations can be then performed as follows:

Step 1 is performed with a cost of $(10m + 10a)$ for each segment by using the expression given in Eqs. (47) and (48). This represents a much better efficiency than that of using Eq. (42).

Steps 2, 3, 4 and 5 are the same as in Khalil et al in (Khalil & Guegan, 2004) (Khalil & Ibrahim, 2007) .

For computation of Step 6, the factorized expression of $\mathcal{J}$, given by Eq. (40) is used. To this end, we have:

$$\mathbf{\Gamma} = {\Pi_1}^t\, {\mathcal{J}_1}^t \boldsymbol{K} \tag{62}$$

Note that, the above equation not only eliminates the need for solving a linear system but it also does not require the explicit computation of neither $\mathcal{J}^{-1}$ nor $\mathcal{J}$. By exploiting the sparse structure of matrices ${\Pi_1}^t$ and $\mathcal{J}_1$, the cost of computation of the vector Γ from Eq. (62) is of $(65m + 67a)$, which represents a much better efficiency, almost half of the computations, compared by using Eq. (40).

9. Simulation of the inverse dynamic model

To validate our inverse dynamic model of the C5 joint parallel robot, a simulation under Matlab environment is presented in this section. The dynamic parameters used for the simulation are given in appendix.

The trajectory profile used for this study is given as follows:

- Fig. 7 shows cartesian trajectory of the mobile platform for a constant orientation ($\alpha = 10^o$, $\beta = 15^o$ and $\gamma = 20^o$).
- Fig. 8 and 9 show respectively the linear velocity and linear acceleration of the mobile platform along the given trajectory.
- The active joint forces are computed using inverse dynamic model given by Eq. (60). Fig. 10 shows the active joint force evolution along the trajectory.

  The simulation results show the feasibility of the proposed approach. The terms of Eq (60) have been computed, and especially, the joint accelerations and the Jacobian matrix in its factorized form, which represent our main contribution in this present paper.

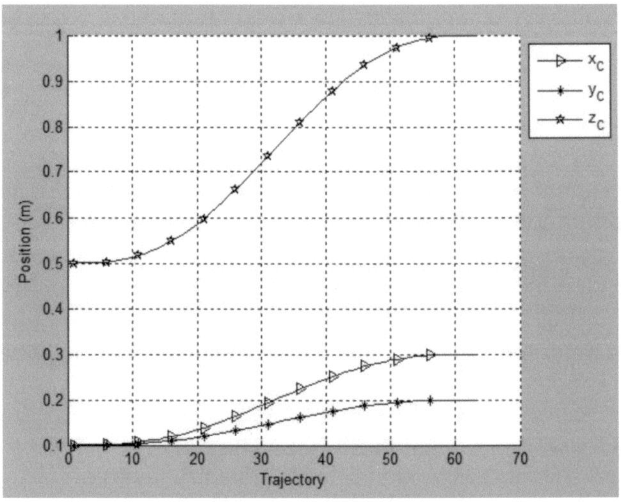

Fig. 7. Mobile platform cartesian trajectory.

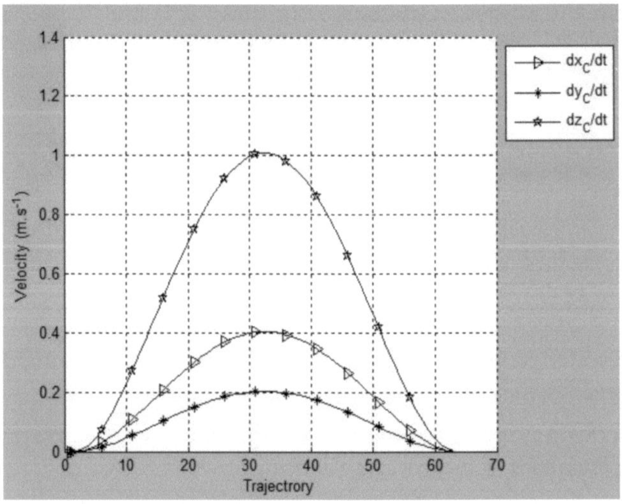

Fig. 8. Mobile platform linear velocity profile.

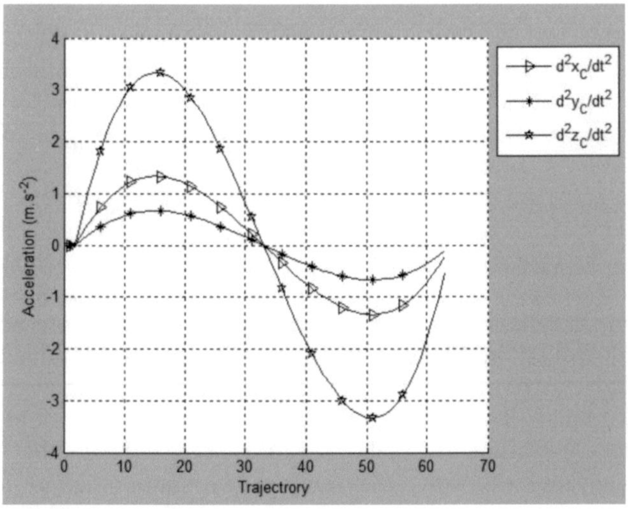

Fig. 9. Mobile platform linear acceleration profile.

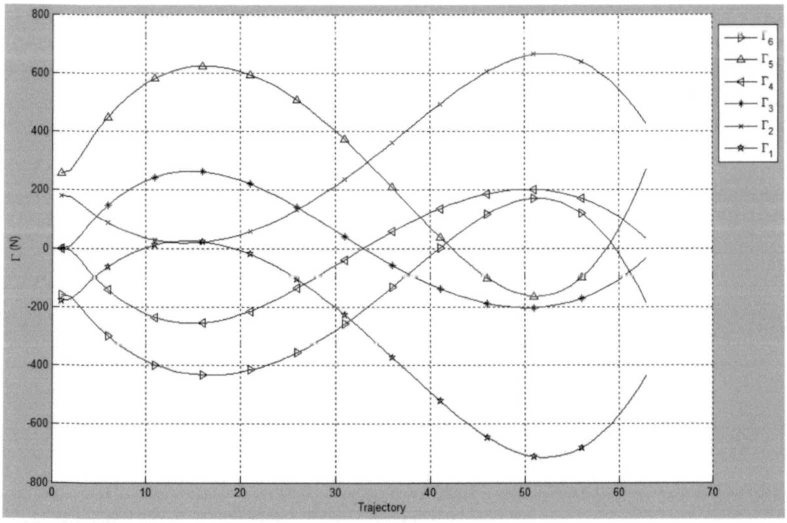

Fig. 10. Active joint forces Γ_i along the trajectory.

10. Conclusion

In this paper, we first presented a review of various proposed schemes for kinematics and inverse dynamics computation of the C5 parallel robot. We presented a new and efficient scheme for derivation of the Jacobian matrix in a factored form as a product two highly sparse matrices. We also presented a new scheme for fast and direct computation of the joint accelerations, given the desired acceleration of the mobile platform. We also reviewed the proposed scheme by Khalil et al (Khalil & Guegan, 2004)(Khalil & Ibrahim, 2007) for computation of the inverse dynamics of the C5 parallel robot. We showed that by using our new scheme for computation of the joint acceleration as well as the use of Jacobian in a factored form a much better efficiency in the computation can be achieved.
The proposed methodology by Khalil et al (Khalil & Guegan, 2004)(Khalil & Ibrahim, 2007) for computation of the inverse dynamics is very systematic and indeed has been applied to several parallel robots. However, we still believe that more efficient formulations can be derived by a further exploitation of specific structures of the parallel robots. We are currently investigating new approaches for the inverse dynamics computation of parallel robots. We are also extending our derivation of the Jacobian matrix in the factored form to other parallel robots.

11. Appendix: dynamic parameters

Body	Mass	Inertia
1	$2\,kg$	$J_{G_{i1}} = \begin{bmatrix} 0.0025 & 0 & 0 \\ 0 & 0.00125 + 0.1667\ (Q_{i1})^2 & 0 \\ 0 & 0 & 0.00125 + 0.1667\ (Q_{i1})^2 \end{bmatrix}$
2	$1.67\,kg$	$J_{G_{i2}} = \begin{bmatrix} 0.00167 & 0 & 0 \\ 0 & 0.0084 & 0 \\ 0 & 0 & 0.0084 \end{bmatrix}$

Table 1. Dynamic parameters of the segment $i = 1$ and 2

Body	Mass	Inertia
1	$2\,kg$	$J_{G_{i1}} = \begin{bmatrix} 0.00125 + 0.1667\ (Q_{i1})^2 & 0 & 0 \\ 0 & 0.0025 & 0 \\ 0 & 0 & 0.00125 + 0.1667\ (Q_{i1})^2 \end{bmatrix}$
2	$1.67\,kg$	$J_{G_{i2}} = \begin{bmatrix} 0.0084 & 0 & 0 \\ 0 & 0.00167 & 0 \\ 0 & 0 & 0.0084 \end{bmatrix}$

Table 2. Dynamic parameters of the segment $i = 3$ and 4

Body	Mass	Inertia
1	$2\,kg$	$J_{G_{i1}} = \begin{bmatrix} 0.00125 + 0.1667\ (Q_{i1})^2 & 0 & 0 \\ 0 & 0.00125 + 0.1667\ (Q_{i1})^2 & 0 \\ 0 & 0 & 0.0025 \end{bmatrix}$
2	$1.67\,kg$	$J_{G_{i2}} = \begin{bmatrix} 0.0084 & 0 & 0 \\ 0 & 0.0084 & 0 \\ 0 & 0 & 0.00167 \end{bmatrix}$

Table 3. Dynamic parameters of the segment $i = 5$ and 6

Body	Mass	Inertia
Mobile platform	$10\,kg$	$I_C = \begin{bmatrix} 0.375 & 0 & 0 \\ 0 & 0.1875 & 0 \\ 0 & 0 & 0.1875 \end{bmatrix}$

Table 4. Dynamic parameters of the mobile platform

12. References

Bi, Z. M. & Lang, S. Y. T. (2006). Kinematic and dynamic models of a tripod system with a passive leg, *IEEE/ASME Transactions on Mechatronics* Vol. 11(No. 1): 108–112.

Cork, P. I. (1996). A robotics toolbox for matlab, *IEEE Robotics and Automation Magazine* Vol. 3: 24–32.

Dafaoui, M., Amirat, Y., François, C. & Pontnau, J. (1998). Analysis and design of a six dof parallel robot. modeling, singular configurations and workspace, *IEEE Trans. Robotics and Automation* Vol. 14(No. 1): 78–92.

Dasgupta, B. & Choudhury, P. (1999). A general strategy based on the newton-euler approach for the dynamic formulation of parallel manipulators, *Mech. Mach. Theory* Vol. 4: 61–69.

Fichter, E. F. (1986). A stewart platform based manipulator: general theory and pratical construction, *International Journal of Robotics Research* Vol. 5(No. 2): 157–182.

Fijany, A., Djouani, K., Fried, G. & J. Pontnau, J. (1997). New factorization techniques and fast serial and parallel algorithms for operational space control of robot manipulators, *Proceedings of IFAC, 5th Symposium on Robot Control*, Nantes, France, pp. 813–820.

Fijany, A., Sharf, I. & Eleuterio, G. M. T. (1995). Parallel o(logn) algorithms for computation of manipulator forward dynamics, *IEEE Trans. Robotics and Automation* Vol. 11(No. 3): 389–400.

Fried, G., Djouani, K., Borojeni, D. & Iqbal, S. (2006). Jacobian matrix factorization and singularity analysis of a parallel robot, *WSEAS Transaction on Systems* Vol. 5/6: 1482–1489.

Gosselin, C. M. (1996). Parallel computational algorithms for the kinematics and dynamics of planar and spatial parallel manipulator, *Journal of Dynamic System, Measurement and Control* Vol. 118: 22–28.

Khalil, W. & Guegan, S. (2004). Inverse and direct dynamic modeling of gough-stewart robots, *IEEE Transactions on Robotics* Vol. 20(No. 4): 745–762.

Khalil, W. & Ibrahim, O. (2007). General solution for the dynamic modeling of parallel robots, *Journal of Intelligent and Robotic Systems* Vol. 49(No. 1): 19–37.

Khan, W. A. (2005). Recursive kinematics and inverse dynamics of a 3r planar parallel manipulator, *Journal of Dynamic System, Measurment and Control* Vol. 4(No. 4): 529–536.

Leroy, N., Kokosy, A. M. & Perruquetti, W. (2003). Dynamic modeling of a parallel robot. Application to a surgical simulator, *Proceedings of IEEE International Conference on Robotics and Automation*, Taipei, Taiwan, pp. 4330–4335.

Luh, J. Y. S., Walker, W. & Paul, R. P. C. (1980). On line computational scheme for mechanical manipulator, *ASME Journal of Dynamic System, Meas Control* Vol. 102: 69–76.

Merlet, J. P. (2000). *Parallel Robots*, Kluwer Academic Publishers.

Merlet, J. P. (2002). Optimal design for the micro parallel robot mips, *Proceedings of IEEE International Conference on Robotics and Automation, ICRA '02*, Washington DC, USA, pp. 1149–1154.

Nakamura, Y. & Ghodoussi, M. (1989). Dynamic computation of closed link robot mechanism with non redundant actuators, *IEEE Journal of Robotics and Automation* Vol. 5: 295–302.

Neugebauer, R., Schwaar, M. & F. Wieland, F. (1998). Accuracy of parallel structured of machine tools, *Proceedings of the International Seminar on Improving Tool Performance*, Spain, pp. 521–531.

Plitea, N., Pisla, D., Vaida, C., Gherman, B. & Pisla, A. (2008). Dynamic modeling of a parallel robot used in minimally invasive surgery, *Proceedings of EUCOMES 08*, Cassino, Italy, pp. 595–602.

Poignet, P., Gautier, M., Khalil, W. & Pham, M. T. (2002). Modeling, simulation and control of high speed machine tools using robotics formalism, *Journal of Mechatronics* Vol. 12: 461–487.

Reboulet, C. (1988). *Robots parallèles*, Eds Hermès, Paris.

Stewart, D. (1965). A plaform with 6 degrees of freedom, *Proceedings of the institute of mechanical engineers*, pp. 371–386.

Tsai, L. W. (2000). Solving the inverse dynamics of a gough-stewart manipulator by the principle of virtual work, *Journal of Mechanism Design* Vol. 122: 3–9.

Zhu, Z., Li, J., Gan, Z. & Zhang, H. (2005). Kinematic and dynamic modelling for real-time control of tau parallel robot, *Journal of Mechanism and Machine Theory* Vol. 40(No. 9): 1051–1067.

18

Utilizing the Functional Work Space Evaluation Tool for Assessing a System Design and Reconfiguration Alternatives

A. Djuric and R. J. Urbanic
University of Windsor
Canada

1. Introduction

1.1 Problem definition

The 'boundary of space' model representing all possible positions which may be occupied by a mechanism during its normal range of motion (for all positions and orientations) is called the work envelope. In the robotic domain, it is also known as the robot operating envelope or workspace (Nof, 1999). Several researchers have investigated workspace boundaries for different degrees of freedom (DOF), joint types and kinematic structures utilizing many approaches (Ceccarelli , 1996), (Ceccarelli & Vinciguerra, 1995), (Ceccarelli & Lanni, 2004), (Cebula & Zsombor-Murray, 2006), (Castelli, et al., 2008), (Yang, et al., 2008). A recent example utilizes a reconfigurable modeling approach, where the 2D and 3D boundary workspace is created by using a method identified as the Filtering Boundary Points (FBP) algorithm. This is developed fully in Djuric and ElMaraghy (2008).

However, this work envelope based work is limited as it does not contain relevant information regarding the relationships between the robot, or mechanisms within a system. This includes the general kinematic structures within the system, the location of the working part(s), tools, process parameters and other operation related parameters. Here an operation is defined as consisting of the travel path, manipulator/end-effector or working tool, tool and part location, and orientation, and any other related process related parameters. The work envelope provides essential boundary information, which is critical for safety and layout concerns, but the work envelope information does not by itself determine the feasibility of a desired configuration. The effect of orientation is not captured as well as the coupling related to operational parameters. Included in this are spatial occupancy concerns due to linking multiple kinematic chains, which is an issue with multi-tasking machine tools, reconfigurable machines, and manufacturing cells.

Multi-tasking machine tools can be considered CNC mill-lathe hybrids (Figure 1). These machines enable multiple tool engagement in multiple parts simultaneously. Each tool/ part/ spindle/ axis set combination follows its own unique programmed path. These machines minimize the number of manual operations, as well as reduce setup costs and potential quality issues. A single multi-tasking machine tool can completely machine complex parts from start to finish (Hedrick & Urbanic, 2004). Traditional computer numerical control (CNC) machines consist of multiple kinematic chains (two - one to support the tool and the other to support the fixture), and can be defined by standard families, i.e., 3 axis horizontal machine, 3 axis vertical, 4 axis horizontal and so forth. The valid kinematic chain loops need to be defined, but are static. It must be noted that depending on the machine type and configuration for multi-tasking machines, the kinematic chain loops can dynamically change, which is beyond the scope of this work, but needs to be considered when assessing the feasibility in time and space of these machine families.

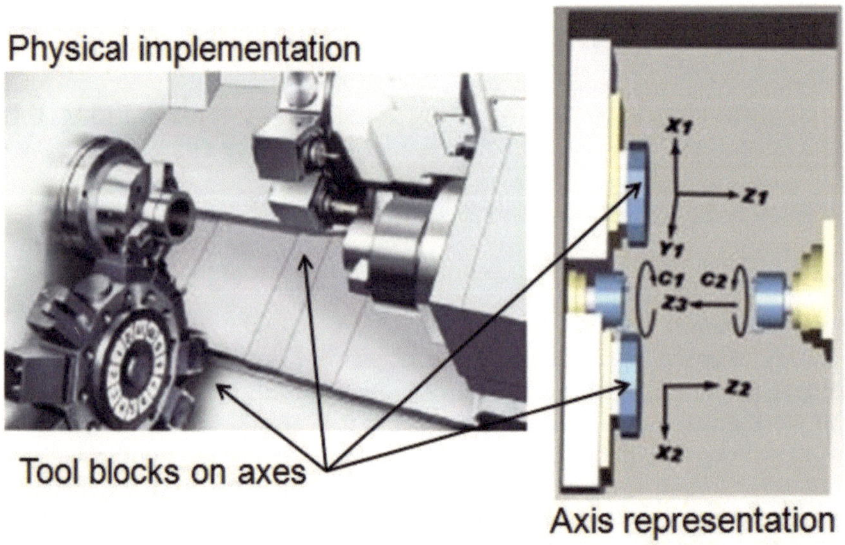

Fig. 1. Okuma LT300-MY (Okuma, 2002) and Programmable Axes (adapted from Hedrick & Urbanic, 2004)

In any manufacturing environment, responsiveness to unforeseen disturbances on the shop floor is highly desirable in order to increase throughput, reduce waste and improve resource utilization and productivity (ElMaraghy, 1993). Process improvement methods to increase shop floor responsiveness while minimizing capital investments have consistently been a topic for manufacturing research. This is presently being addressed through the concept of reconfigurable manufacturing (ElMaraghy, 2008). The key building blocks of a reconfigurable manufacturing system are scalability, adaptability, and transformability (Koren et al., 1999). This should reduce capital investments over the system life cycle, rapid changeovers, steeper launch curves, and minimal fluctuations in quality and this has great future potential (Anderson & Bunce, 2000). The reconfiguration concept is divided into two subsets for this work: extendibility (Figure 2) and reconfigurability (Figure 3).

Changing cutting tools, the tooling systems used to mount the tools into the machine, the end-effector or replacing an axis with another with different constraints is considered extendibility. For machining, this usually introduces shifts in the working boundaries allowing the same programs / travel paths to be used with appropriate offsets.

For robotic applications, this may not be the case. (Note: for systems where the end-effector requires specific orientations, it is assumed that the orientation remains constant for a given length offset.) These modifications are typically made in the tooling domain.

A reconfiguration consists of altering the machine topology, i.e. changing the DOF by adding or removing an axis. This may require re-programming, or depending on the reconfiguration, a totally different programming or process planning strategy may be required.

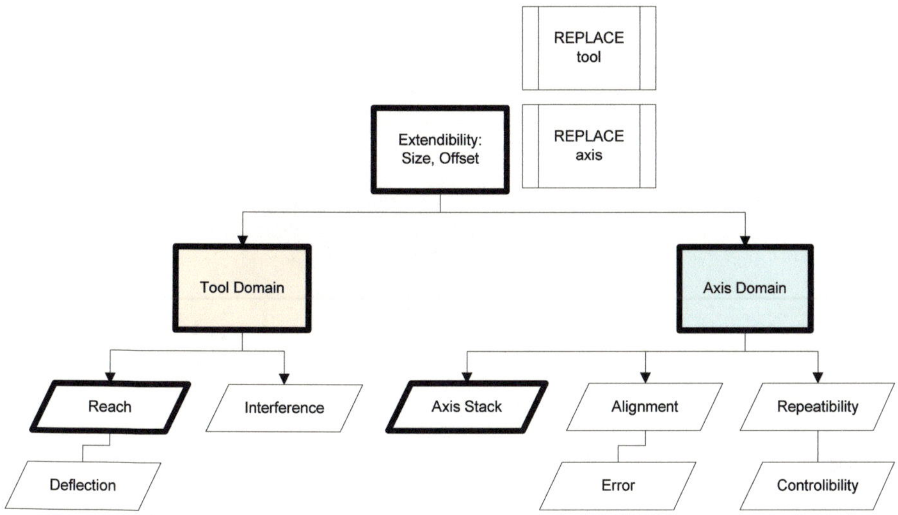

Fig. 2. Extendibility breakdown

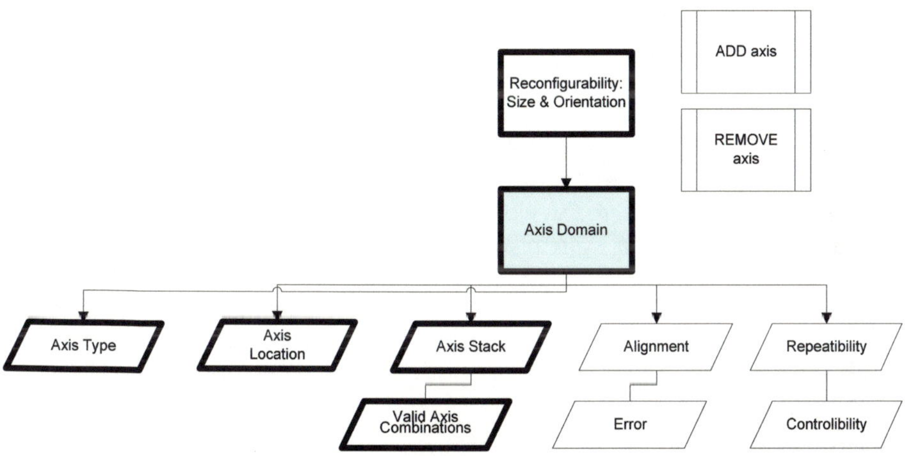

Fig. 3. Reconfigurability breakdown

The work envelope does not allow one to assess the operating feasibility of a machine, robot, or system configuration for a set of tasks. Another assessment parameter needs to be introduced to indicate the functional operating work space. Hence, the work window is introduced, and is defined as the valid functional space for a configuration to allow a kinematic structure (robot, machine tool, manufacturing cell, etc.) to follow a path for a set of conditions relating to the system configuration, tooling, fixture location and so forth (Djuric & Urbanic, 2009). A part may be located within the work envelope, but it may not be within the work window. In Figure 4, the encircled region cannot be reached for the 'normal to the base' orientation for the 6R Fanuc robot although the part is within the envelope. The items bounded by a bolded line in Figures 2 and 3 are discussed in this research.

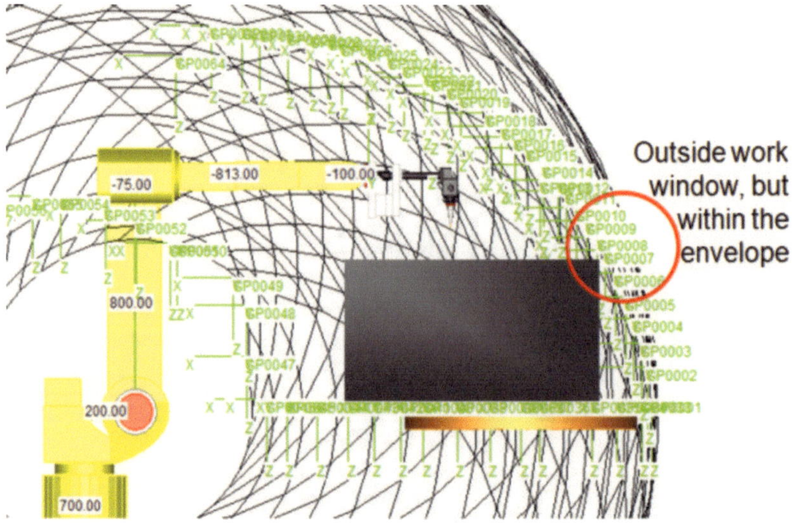

Fig. 4. Work envelope and work window for a 6R Fanuc robot (Djuric & Urbanic, 2009)

The actual work window conditions will vary based on the part, travel path tool parameters, geometry, machine kinematic composition and the application. For machining operations, the work window will vary for each unique orientation/ tooling /fixture combination. Structural reconfigurations, such as adding or removing an axis (DOF and type of joints) either virtually or physically, will also impact the work window. A methodology needs to be developed to define the work window for a configuration, and any potential reconfigurations. The assessment of system elements is necessary in order to examine the feasibility and practicality of a set of kinematic structures in a manufacturing scenario. This does not exist at this time. There is a direct relation between the kinematic structures, their manipulability and constraints, to the work window. Proper assessment of system elements will help in optimization of a system and help in deciding various parameters such as number of degrees of freedom, placement of machines, robots and other supporting mechanisms (in case of a work cell consisting of multiple kinematic chain elements) and so forth. The methodology needs to be applicable for the reconfigurable type of machinery. The goal of this work is to focus on foundational work window assessment approaches.

1.2 Research approach

This structure of the chapter is as follows. First, the assessment methodology for machine tool type mechanisms is presented. An appraisal of "primitive" or basic kinematic building blocks is performed and a breakdown of their characteristics is presented. These elemental building blocks are used to create the basic working boundaries using geometric modelling techniques and Boolean operations for more advanced kinematic structures.

A different solution approach is required for the selected industrial robot families (ABB and Fanuc), which is presented in detail in this work. However, the Boolean approach in order to couple the operating environment to the kinematic structure is common. The case study in this work in centred on a robotics work cell application. The feasibility of solution alternatives can then be assessed, as well as the determination of alternative options, which is explored in the case studies section. This research work is then summarized and concluded.

2. Methodology to define the work window

The approaches to determine the work window are different for machine tool type kinematic structures and 6 DOF robots at the base level due to their unique topologies. The machine tool type of devices is decomposed into elemental or primitive groups. The term primitive is utilized, as basic mathematical definitions are used to present the regions of interest. Combining these primitive elemental structures to characterize a selection of machine tools is then illustrated. The methodology to couple the fixture based and tooling based kinematic chains is then presented.

For the selected industrial robot families, an empirical and detailed analytical methodology to determine the workspace boundary for a desired orientation is presented, along with illustrative examples.

2.1 Decomposition analysis for machines consisting of elemental combinations

Machine types consisting of primitive elemental combinations are milling machines, lathes, routers, coordinate measuring machines (CMM's), and so forth. There are standard machine families, but there are also specialty combinations that are becoming more ubiquitous. There are several basic travel paths strategies that can be employed. The types and typical applications are as follows:

- 2 axis (X, Y) travel path motion (used for tool less profile machines such as water jet, plasma jet and laser cutting - this is trivial as the work space = work window and is a rectangular surface bounded by the axis stroke limits. This is presented as Case 4 in Table 1.)
- 2 axis (Z, X) travel path motion + rotary spindle (used for lathe machining - this is a subset of Case 4 in Table 1. The axes only move in Z and X, but are swept in a rotary profile)
- 2 axis (X, Y) + Z depth ('2 ½ axis') travel path motion (used for additive manufacturing machines - this is trivial as the work space = work window and is a surface bounded by the axis stroke limits, incremented by a specific Δz value. This is a subset of Case 6 in Table 1.)
- 3 axis travel path motion (used for standard milling machines - the work window is offset from the workspace by the length of the tool and is a solid, also bounded by the axis stroke limits. This is presented by Case 6 or Case 4 in combination with the Case 0, *T* in Table 1.)
- 3 axis travel path motion and 1 or 2 axis rotary positioning (used for standard 4 and 5 axis milling machines, and 5 axis CMMs. These are combinations of Cases 4-6 and 8-10 in Table 1.)
- 5 axis travel path motion and (used for 5 axis milling machines. These are combinations of Cases 4-6 and 8-10 in Table 1.)

More sophisticated machines, such as multi-tasking machines (Figure 1), consist of a combination of 3, 4, and 5 axis sub-machines on a lathe frame. The valid kinematic chain combinations must be understood for these machines. The common characteristic of these types of mechanisms is that they can be broken down into basic primitive elemental structures. One kinematic chain holds the tool, the other the part. These elemental structures are combined to create a closed kinematic loop. The reference planes reflect how the axes are physically stacked, and the datum locating points between these elemental structures, and are used to link relationships to define the usable three dimensional (3D) space. It is readily evident that the work envelope essentially consists the bounding stroke positions of the linear or translational axes. The process flow for determining the work window for these mechanisms is illustrated in Figure 5.

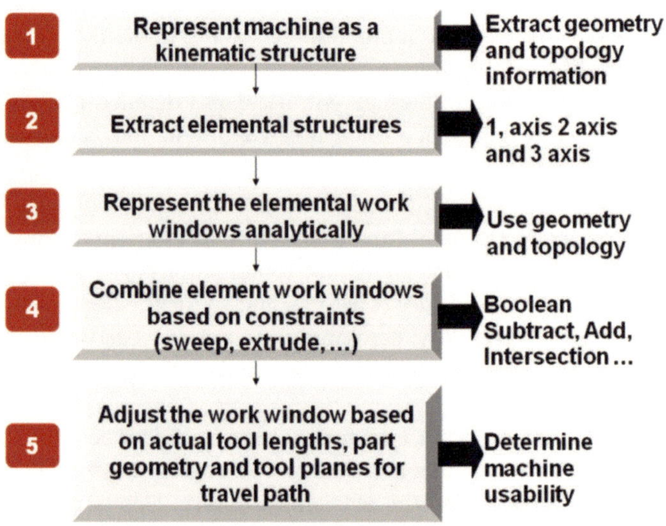

Fig. 5. Work window algorithm for mechanisms created from elemental kinematic chains, adapted from Djuric & Urbanic, 2009

To assist with steps 1 and 2 in the process flow, the workspace and window are analysed for a set of 2 and 3 DOF kinematic structures (a selected set presented in Table 1). The *R* represents a rotary axis, and the *T* a translational axis. For this work, it is assumed that axes are stacked orthogonally. The coded delineation represents the axis stack, i.e., *RT* indicates that a translational axis is orthogonally stacked at the end of a rotary axis (Case 2 in Table 1). When specific subscripts are included, for example Case 8 - $T_X R_B$, this is used to indicate an axis placed on top of another axis, in the order listed. The *A* axis rotates around the *X* translational axis, the *B* axis rotates around the *Y* translational axis, and the *C* axis rotates around the *Z* translational axis.

The workspace of the 2 DOF mechanisms may be a surface, curve or line. Depending on the axis combination and axis stack, the bounding conditions for an axis combination may be a segmented line - i.e., a void exists where the axis/part / fixture is located. The 3 DOF mechanisms usually have a solid workspace. Again, this depends of the joint type (rotational or translational) and their order in the kinematic chain. The revolve profile consists of the sweep volume of the rotary axis, part, and fixture profiles around the rotation axis.

Two combinations of vertical 3 axis milling machines are shown in Figure 6, which are comprised of Case 0, T and Case 4. In Figure 7, a 5 axis vertical milling machine which is comprised of Cases 5 and 6 is shown. In Figure 8, a horizontal 5 axis milling machine is illustrated, which is comprised of Cases 4 and 9. For this machine, the revolve profile is also displayed.

Case	Kinematic Structure	Label	Workspace	Work window	Comment
0, T		T	-	-	1 DOF basic element
0, R		R	-	-	1 DOF basic element
1		RR	Flat circular surface	Flat circular surface	Work space = Work window
2		RT	Cylindrical surface	Cylindrical surface	Work window = Work space - tool length
3		TR	Cylindrical surface	Cylindrical surface	Work window = Work space - tool length
4		T_XT_Y	Rectangular surface	Rectangular Surface	Work space = Work window
5		*RR-articulated*	*Spherical surface*	*Indeterminate*	*Point position also dependent on tool length / part fixture length*
6		$T_XT_YT_Z$	Cube solid	Cube solid	Work window = Work space offset by tool length
7		*RRR-articulated*	*Toroidal surface*	*Indeterminate*	*Point position also dependent on tool length*
8	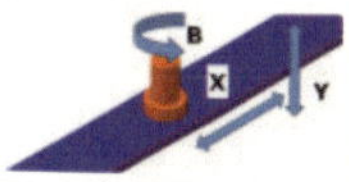	T_XR_B	**Line**	**Segmented or offset line - not completely defined until combined with other kinematic chains**	**Void in the region of the axis dependent on the revolve profile sweep radius and axis stack**
9		$T_XR_BR_A$	**Line**		**Void in the region of the axis dependent on the revolve profile sweep radii and axis stack**
10		$T_ZR_CR_A$	*Hemi--spherical + cylindrical surfaces*	*Indeterminate*	**Void in the region of the axis dependent on the sweep radii and tool length**

Table 1. Summary of selected elemental structures (adapted from Djuric & Urbanic (2009)

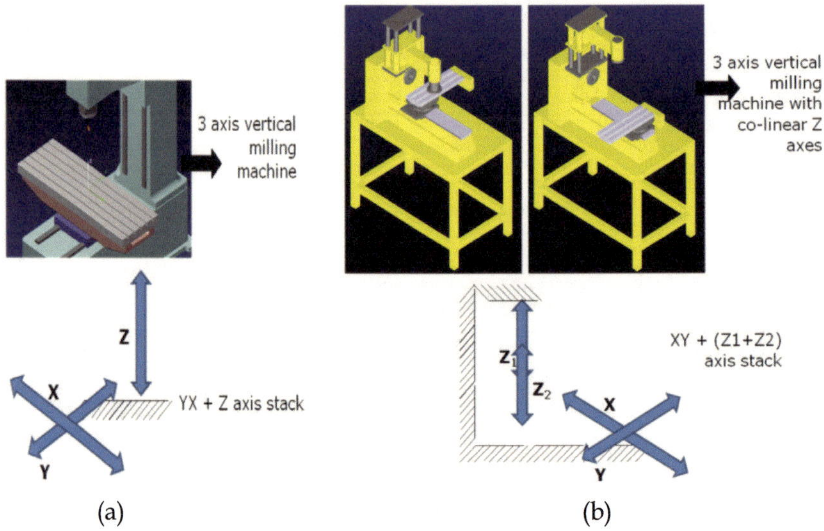

Fig. 6. (a) $T_Z + T_YT_X$ combination, (b) $(T_{Z1} + T_{Z2})+ T_XT_Y$ combination

For step 3 in the work window process flow, the analytical representation for Cases 1-7 is relatively simple, and related directly to the axis limits (all cases) and tool length (Cases 2, 3, 5 - 7). The complexity is introduced when coupling the kinematic chains for Cases 0, *R* and 0, *T*, and Cases 4, 5, and 6 with Cases 8-10. Reference planes need to be established to link the axes' grounds. Along with the axis stack order, the axes travel limits must be defined, as well as the revolve profiles, where appropriate. Set theory and Boolean operations are then performed to couple the fixture and tooling kinematic chains in relation to their bounding geometry.

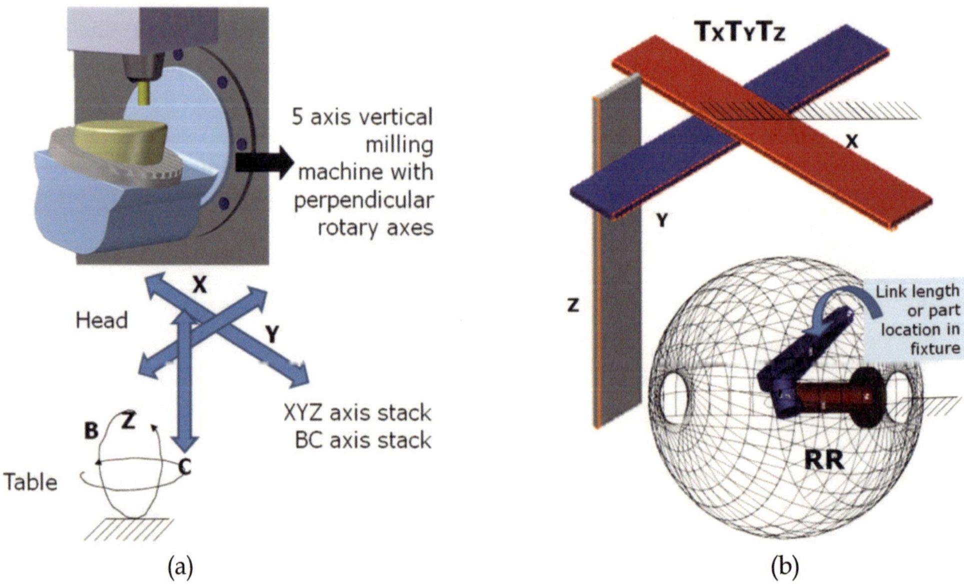

Fig. 7. (a) 5 axis milling machine and basic structure, (b) RR+ $T_XT_YT_Z$ combination

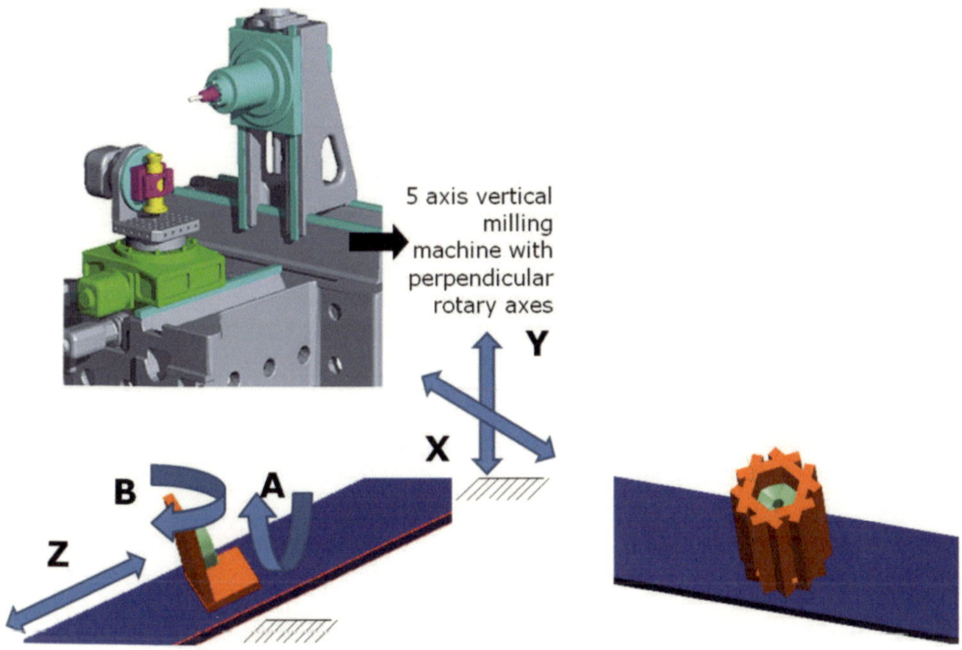

(a) (b)

Fig. 8. 5 axis milling machine and basic structure, $T_ZR_BR_A$+ T_XT_Y combination, (b) revolve profile around the B axis at the home position (centre) of X axis

The addition of the rotary axes immediately reduces the maximum part size that can be machined. This reduction of available work space (or volumetric void) depends on the centre of rotation for the rotary axis, its rotational limits, and the axis stack; hence, the importance of being able to generate the revolve profile. The inclusion of the fixture(s) and part(s) modifies the rotary profile for an axis stack located on the table. For an axis stack located on the end-effector, the tool length (cutting tool or CMM probe) needs to be taken into consideration. This needs to be defined in order to assess the optimal fixture(s) positioning, the maximum part size or number of parts to be located on a fixture (i.e., a tombstone configuration consisting of multiple parts), the maximum tool length and width parameters, and the retraction travel path. The table based revolve profile must be contained or be a volumetric space subset within the workspace region defined by the *X*, *Y*, *Z* travel limits. For mechanisms that contain Cases 8 and 9, initial volumetric assessments need to be performed at the *X* and *Y* travel limits with the *Z* axis fully retracted while considering the tool offset to ensure no basic crash conditions and the maximum part/fixture conditions. Then assessments need to be performed at the rotary axes limits to determine orientation feasibilities. A sample of Boolean operations for the region defined by the tool offset (Figure 9 (a)) and then an axis stack on the *Z* axis (Figure 9 (b)) are illustrated in Figure 9.

2.2 Industrial robot solution approach

To start, the Denavit-Hartenberg notation (i.e., DH parameters) is presented to provide the background for the subsequent analyses, as the DH parameters are commonly used in the

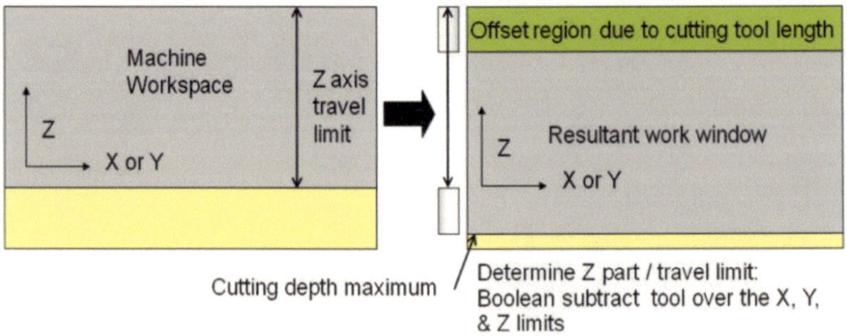

(a)

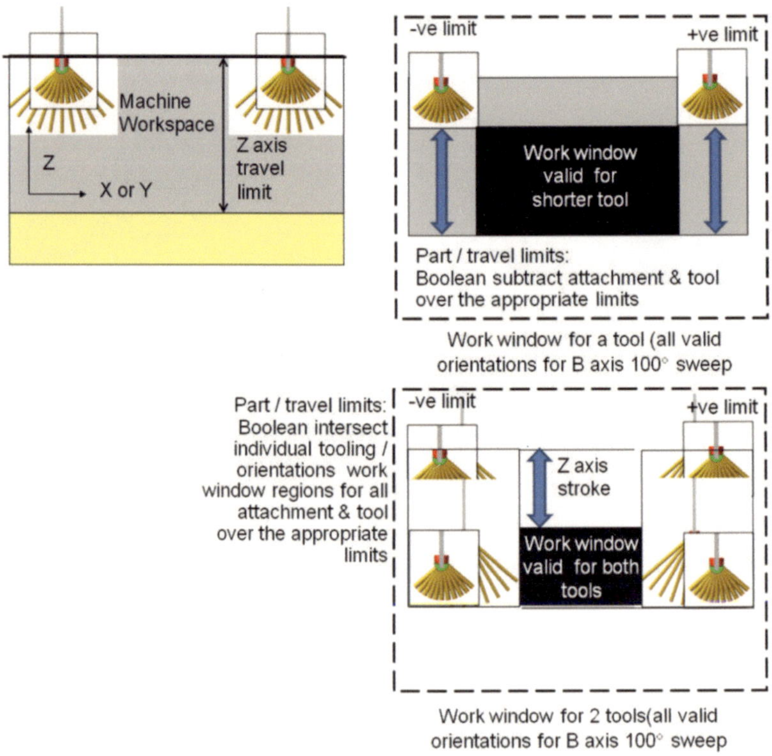

(b)

Fig. 9. Boolean operations for a Case 10 set of tools to establish valid operating region

robotics domain and provide a standard methodology to write the kinematic equations of a manipulator or end-effector. Each joint in a serial kinematic chain is assigned a coordinate frame. Using the DH notation (Denavit & Hartenberg , 1955), 4 parameters are needed to describe how a frame i relates to a previous frame $i-1$ as shown in Figure 10.

After assigning coordinate frames the four D-H parameters can be defined as following:

a_i - Link length is the distance along the common normal between the joint axes

α_i - Twist angle is the angle between the joint axes

θ_i - Joint angle is the angle between the links

d_i - Link offset is the displacement, along the joint axes between the links

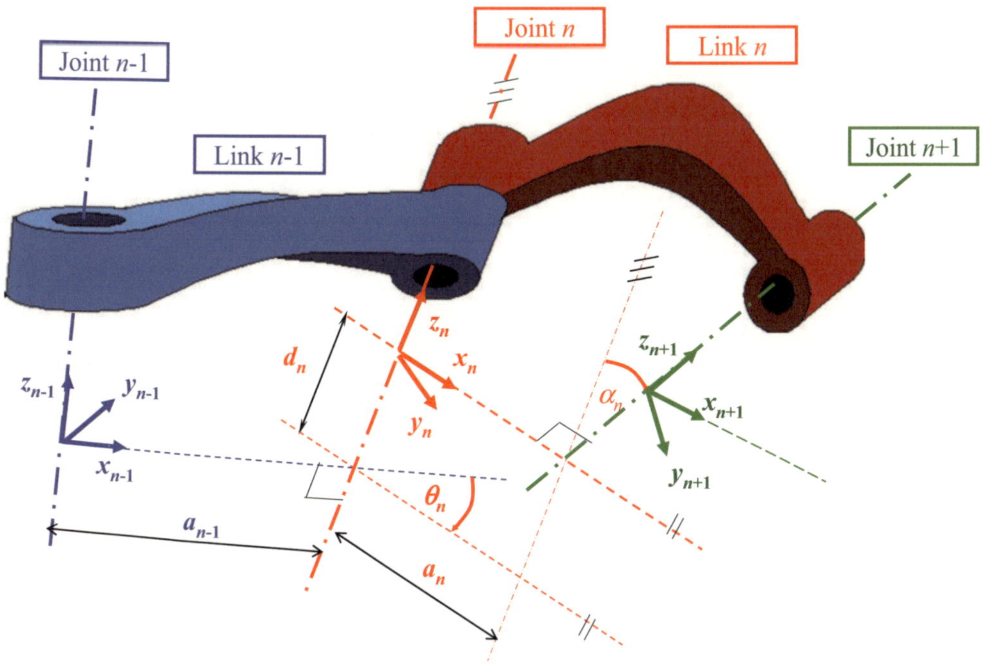

Fig. 10. Coordinate frames and D-H parameters

2.2.1 Empirical approach using a commercial robot simulation software package

The determination of open kinematic chain dexterity has been done using numerical and analytical methods. These methods are applied and compared using a six DOF manipulator, (Abdel-Malek & Yeh, 2000). The Jacobian formulation provides quantitative dexterity comparison between manipulators with different architectures. It was first applied to analyze serial robots and compared with parallel robots with the same DOF, (Pond & Carretro, 2011). Using the virtual angle analysis the dexterous workspace is generated to analyzed the orientation capability of a manipulator trough its equivalent mechanism (Dai & Shah, 2003). Placement of the manipulator has been done by implementing the Workspace towards the target points and subjecting it to a set of constraints to determine the reachability of the end-effector without using the inverse kinematics (Jingzhou Yang, et al., 2009).

Using the above information as a starting point, the methodology to generate the work window for robots using empirical equations follows. Two robots were selected with different kinematic structures (an ABB IRB 140 Robot and a Fanuc M16) and their work window was generated using the Workspace 5 robotic simulation software. A systematic manual valid point generation approach is used to develop the algorithm that supports both procedures. After this was completed, analytical equations were developed empirically and tested for their validity using different examples, an approach conceptually similar to that taken by Abdel-Malek & Yeh (2000). Then the methodology was expanded upon to reflect the problem being solved. By varying joint values to get the complete space representation with the desired orientation (in this example, the orientation is 'normal to the robot base' as this represents common pick and place operations associated with material handling), only four joints are used: Joint 1, Joint 2, Joint 3, and Joint 5. Joint 4 and Joint 6 remain constant. For this study the detailed kinematic information for the ABB IRB 140 Robot and Fanuc M16 robots is presented below. The ABB IRB 140 kinematic structure diagram is shown in Figure 11 and its D-H parameters are given in Table 2.

i	d_i	θ_i	a_i	α_i
1	d_1	$\theta_1 = 0^\circ$	a_1	-90°
2	0	$\theta_2 = -90^\circ$	a_2	0°
3	0	$\theta_3 = 180^\circ$	0	90°
4	d_4	$\theta_4 = 0^\circ$	0	-90°
5	0	$\theta_5 = 0^\circ$	0	90°
6	d_6	$\theta_6 = -90^\circ$	0	90°

Table 2. D-H parameters for the ABB IRB 140 Robot

The Fanuc M16 kinematic structure diagram is shown in Figure 12 and its D-H parameters are given in Table 3.

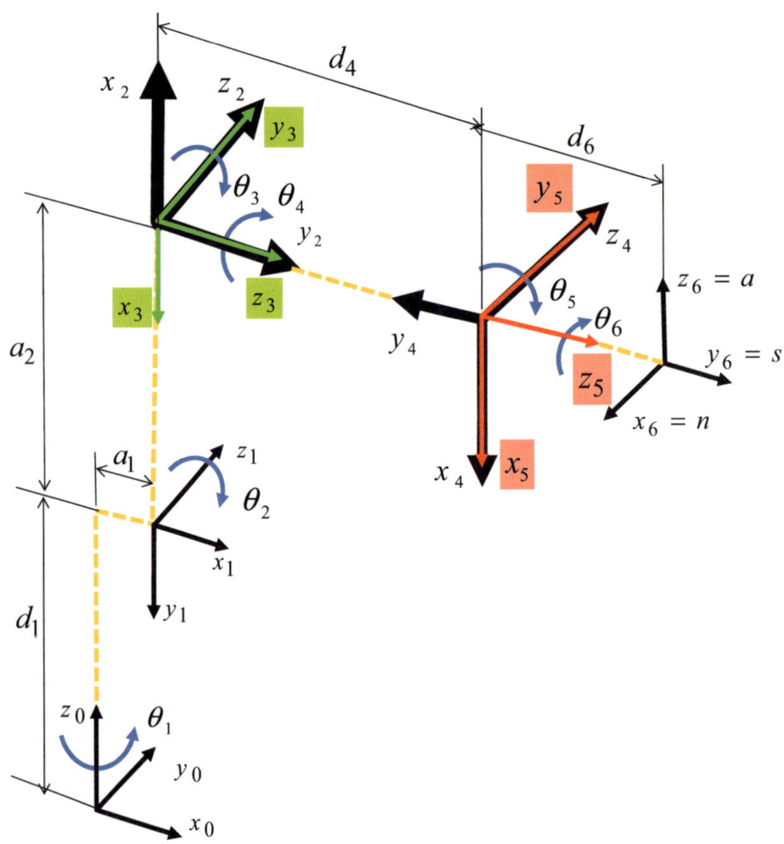

Fig. 11. Kinematic structure of the ABB IRB 140 robot

i	d_i	θ_i	a_i	α_i
1	d_1	$\theta_1 = 0^\circ$	a_1	-90°
2	0	$\theta_2 = -90^\circ$	a_2	180°
3	0	$\theta_3 = 180^\circ$	a_3	90°
4	d_4	$\theta_4 = 0^\circ$	0	-90°
5	0	$\theta_5 = 0^\circ$	0	90°
6	d_6	$\theta_6 = 180^\circ$	0	180°

Table 3. D-H parameters for the Fanuc M16 Robot

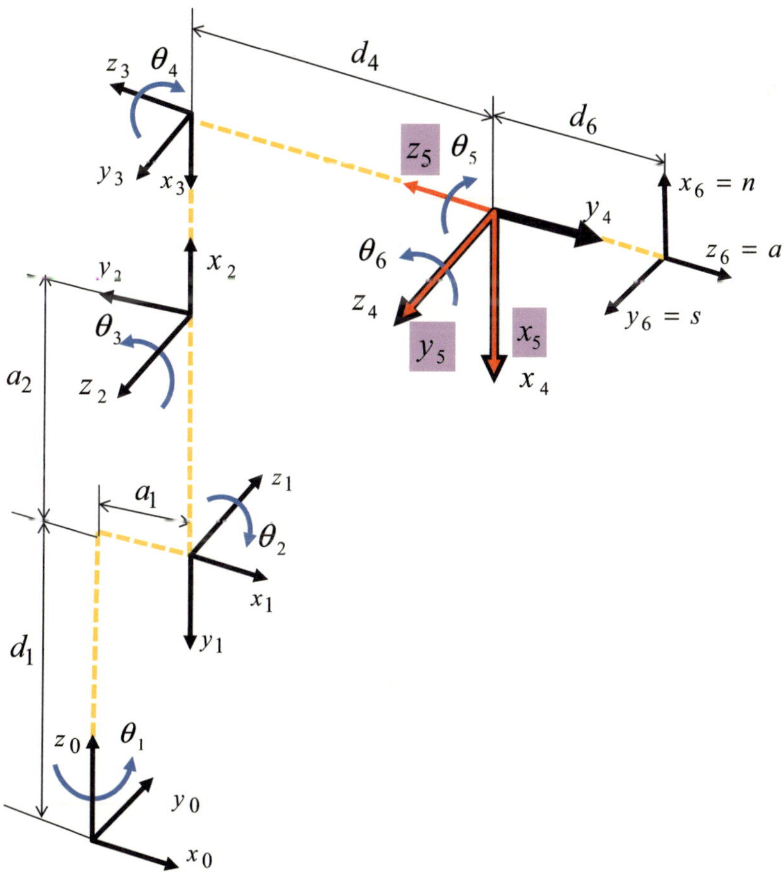

Fig. 12. Kinematic structure of the Fanuc M16 Robot

Using a commercial simulation and off-line programming software Workspace 5, the work window (subspace with 'normal to the base' orientation) is generated for the ABB IRB 140 robot. The black net represents the 2D workspace and the green coordinate frames represent the work window region. The 2D set of points for the ABB IRB 140 is presented in Figure 13. Similarly, the work window (subspace with 'normal to the base' orientation) is generated for the Fanuc M16 robot and this set of 2D points is presented in Figure 14.

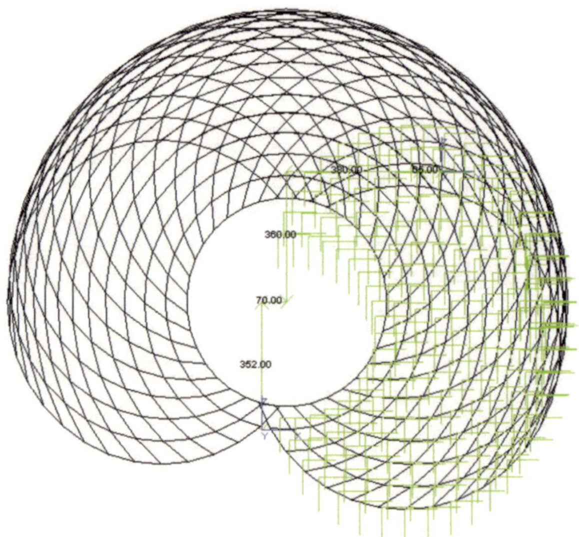

Fig. 13. Work window for the ABB IRB 140 Robot for the 90 degree (normal) orientation

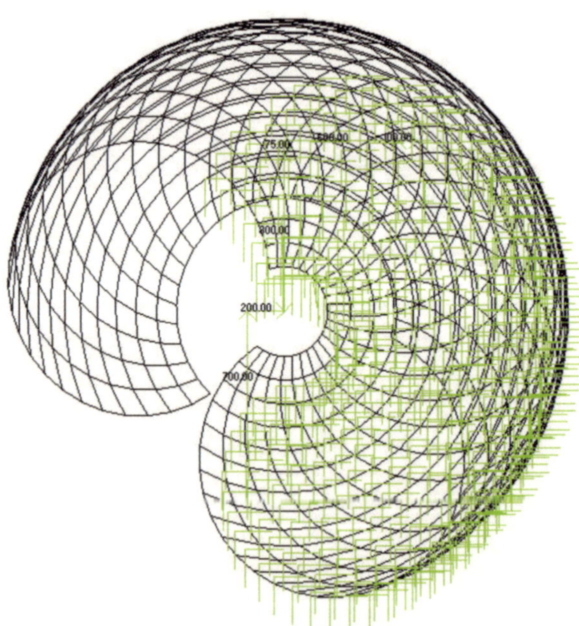

Fig. 14. Work window for the Fanuc M16 Robot for 90 degree (normal) orientation

Upon combining the procedures for the generation both work windows, an empirical formula for the joint five angle is developed, which is shown in equation (1).

$$\theta_5 = K\phi - (\theta_2 + K\theta_3) \tag{1}$$

The parameter K in Equation 1 is calculated using the formula below:

$$K = \cos\alpha_2 \tag{2}$$

Where α_2 is the Joint 2 twist angle.

The generalized algorithm for generating the work window for two different kinematic structures is given in Figure 15. This solution is valid for 6 DOF robots with rotational joints and the Fanuc and ABB robot kinematic structures. The outcome of the algorithm is either a 2D or 3D work window.

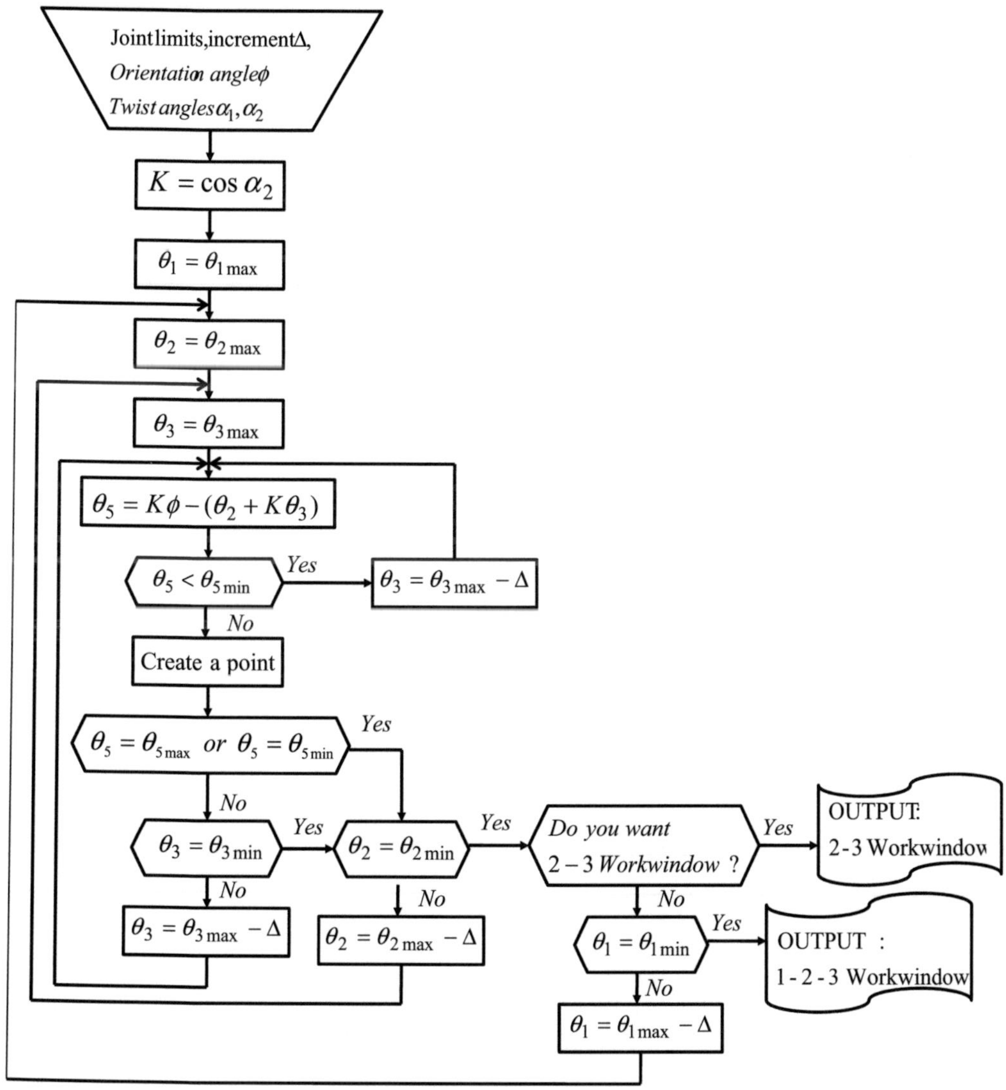

Fig. 15. Algorithm for work window generation

2.2.2 Generation of Work window using analytical equations

In the kinematic analysis of the manipulator position, there are two separate problems to be solved: the direct kinematics, and inverse kinematics. Assessing the direct kinematics involves solving the forward transformation equation to find the location of the hand in terms of the angles and displacements between the links. The angles and displacements between the links are called joint coordinates and are described with link variables, while the location of the hand in space is described using Cartesian coordinates. Inverse kinematics involves solving the inverse transformation equation to find the relationships between the links of the manipulator from the location of the hand in space. A serial link manipulator is a series of links, which connects the end-effector to the base, with each link connected to the next by an actuated joint. If a coordinate frame is attached to each link, the relationship between two links can be described with a homogeneous transformation matrix using D-H rules (Denavit & Hartenberg, 1955), and they are named ${}^{i-1}A_i$, where i is number of joints.

$$ {}^{i-1}A_i = \begin{bmatrix} \cos\theta_i & -\cos\alpha_i \sin\theta_i & \sin\alpha_i \sin\theta_i & a_i \cos\theta_i \\ \sin\theta_i & \cos\alpha_i \cos\theta_i & -\sin\alpha_i \cos\theta_i & a_i \sin\theta_i \\ 0 & \sin\alpha_i & \cos\alpha_i & d_i \\ 0 & 0 & 0 & 1 \end{bmatrix} \tag{3} $$

The robot can now be kinematically modeled by using the link transforms:

$$ {}^{0}A_n = {}^{0}A_1 \, {}^{1}A_2 \, {}^{2}A_3 \cdots {}^{i-1}A_i \cdots {}^{n-1}A_n \tag{4} $$

Where ${}^{0}A_n$ is the pose of the end-effector relative to base; ${}^{i-1}A_i$ is the link transform for the ith joint; and n is the number of links.

For the ABB IRB 140 robot, six homogeneous transformation matrices have been developed using Maple 12 symbolic manipulation software, Equation 2 and the D-H parameters from Table 2.

$$ {}^{0}A_1 = \begin{bmatrix} \cos\theta_1 & -\cos\alpha_1 \sin\theta_1 & \sin\alpha_1 \sin\theta_1 & a_1 \cos\theta_1 \\ \sin\theta_1 & \cos\alpha_1 \cos\theta_1 & -\sin\alpha_1 \cos\theta_1 & a_1 \sin\theta_1 \\ 0 & \sin\alpha_1 & \cos\alpha_1 & d_1 \\ 0 & 0 & 0 & 1 \end{bmatrix} = \begin{bmatrix} \cos\theta_1 & 0 & -\sin\theta_1 & a_1 \cos\theta_1 \\ \sin\theta_1 & 0 & \cos\theta_1 & a_1 \sin\theta_1 \\ 0 & -1 & 0 & d_1 \\ 0 & 0 & 0 & 1 \end{bmatrix} \tag{5} $$

$$ {}^{1}A_2 = \begin{bmatrix} \cos\theta_2 & -\cos\alpha_2 \sin\theta_2 & \sin\alpha_2 \sin\theta_2 & a_2 \cos\theta_2 \\ \sin\theta_2 & \cos\alpha_2 \cos\theta_2 & -\sin\alpha_2 \cos\theta_2 & a_2 \sin\theta_2 \\ 0 & \sin\alpha_2 & \cos\alpha_2 & d_2 \\ 0 & 0 & 0 & 1 \end{bmatrix} = \begin{bmatrix} \cos\theta_2 & -\sin\theta_2 & 0 & a_2 \cos\theta_2 \\ \sin\theta_2 & \cos\theta_2 & 0 & a_2 \sin\theta_2 \\ 0 & 0 & 1 & 0 \\ 0 & 0 & 0 & 1 \end{bmatrix} \tag{6} $$

$$ {}^{2}A_3 = \begin{bmatrix} \cos\theta_3 & -\cos\alpha_3\sin\theta_3 & \sin\alpha_3\sin\theta_3 & a_3\cos\theta_3 \\ \sin\theta_3 & \cos\alpha_3\cos\theta_3 & -\sin\alpha_3\cos\theta_3 & a_3\sin\theta_3 \\ 0 & \sin\alpha_3 & \cos\alpha_3 & d_3 \\ 0 & 0 & 0 & 1 \end{bmatrix} = \begin{bmatrix} \cos\theta_3 & 0 & \sin\theta_3 & 0 \\ \sin\theta_3 & 0 & -\cos\theta_3 & 0 \\ 0 & 1 & 0 & 0 \\ 0 & 0 & 0 & 1 \end{bmatrix} \tag{7} $$

$$ {}^{3}A_4 = \begin{bmatrix} \cos\theta_4 & -\cos\alpha_4\sin\theta_4 & \sin\alpha_4\sin\theta_4 & a_4\cos\theta_4 \\ \sin\theta_4 & \cos\alpha_4\cos\theta_4 & -\sin\alpha_4\cos\theta_4 & a_4\sin\theta_4 \\ 0 & \sin\alpha_4 & \cos\alpha_4 & d_4 \\ 0 & 0 & 0 & 1 \end{bmatrix} = \begin{bmatrix} \cos\theta_4 & 0 & -\sin\theta_4 & 0 \\ \sin\theta_4 & 0 & \cos\theta_4 & 0 \\ 0 & -1 & 0 & d_4 \\ 0 & 0 & 0 & 1 \end{bmatrix} \tag{8} $$

$$ {}^{4}A_5 = \begin{bmatrix} \cos\theta_5 & -\cos\alpha_5\sin\theta_5 & \sin\alpha_5\sin\theta_5 & a_5\cos\theta_5 \\ \sin\theta_5 & \cos\alpha_5\cos\theta_5 & -\sin\alpha_5\cos\theta_5 & a_5\sin\theta_5 \\ 0 & \sin\alpha_5 & \cos\alpha_5 & d_5 \\ 0 & 0 & 0 & 1 \end{bmatrix} = \begin{bmatrix} \cos\theta_5 & 0 & \sin\theta_5 & 0 \\ \sin\theta_5 & 0 & -\cos\theta_5 & 0 \\ 0 & 1 & 0 & 0 \\ 0 & 0 & 0 & 1 \end{bmatrix} \tag{9} $$

$$ {}^{5}A_6 = \begin{bmatrix} \cos\theta_6 & -\cos\alpha_6\sin\theta_6 & \sin\alpha_6\sin\theta_6 & a_6\cos\theta_6 \\ \sin\theta_6 & \cos\alpha_6\cos\theta_6 & -\sin\alpha_6\cos\theta_6 & a_6\sin\theta_6 \\ 0 & \sin\alpha_6 & \cos\alpha_6 & d_6 \\ 0 & 0 & 0 & 1 \end{bmatrix} = \begin{bmatrix} \cos\theta_6 & 0 & \sin\theta_6 & 0 \\ \sin\theta_6 & 0 & -\cos\theta_6 & 0 \\ 0 & 1 & 0 & d_6 \\ 0 & 0 & 0 & 1 \end{bmatrix} \tag{10} $$

The ABB IRB 140 robot can now be kinematically modeled by using Equations 4-10.

$$ {}^{0}A_6 = {}^{0}A_1\,{}^{1}A_2\,{}^{2}A_3\,{}^{3}A_4\,{}^{4}A_5\,{}^{5}A_6 \tag{11} $$

The pose matrix of the end-effector relative to the base is presented in Equation 12.

$$ {}^{0}A_6 = \begin{bmatrix} n_x & s_x & a_x & p_x \\ n_y & s_y & a_y & p_y \\ n_z & s_z & a_z & p_z \\ 0 & 0 & 0 & 1 \end{bmatrix} \tag{12} $$

The end-effector orientation is defined with the rotation matrix. The upper $3X3$ sub matrices of the homogeneous transformation matrices Equation 11, represents the rotational matrix, Equation 13. The graphical representation of the end-effector is shown in Figure16.

$$ {}^{0}R_6 = \begin{bmatrix} n_x & s_x & a_x \\ n_y & s_y & a_y \\ n_z & s_z & a_z \end{bmatrix} \tag{13} $$

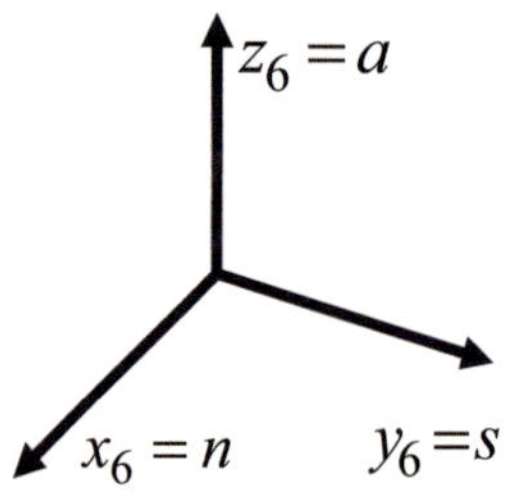

Fig. 16. The end-effector frame of the ABB IRB 140 robot

The normal vector n is along x_6 axis, the sliding vector s is along y_6 axis, and the approach vector a is along z_6 axis. The orientation of the end-effector, relative to the robot base frame, is defined with the three vectors: n, s, and a. Their projections onto the robot base frame are given with the rotational matrix ${}^0R_6$. For the ABB IRB 140 Robot the relation between end-effector and base frame, when the end-effector is normal to the robot base, is graphically shown in Figure 17. The orientation matrix for the given position is calculated (see Equation 13).

$$ {}^0R_6 = \begin{bmatrix} 0 & 0 & 1 \\ -1 & 0 & 0 \\ 0 & -1 & 0 \end{bmatrix} \quad (14) $$

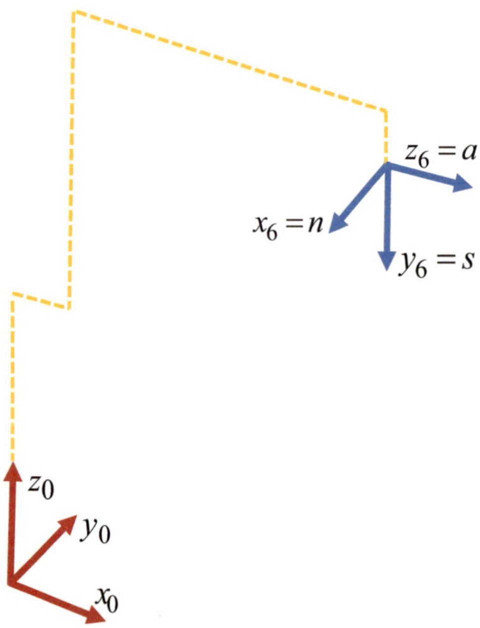

Fig. 17. The end-effector frame of the ABB IRB 140 robot

The calculation of the 2D end-effector orientation is dependent on the three joint angles: Joint 2, Joint 3 and Joint 5. The formula for Joint 5 is generated by assigning initial values for the Joint 1, Joint 4 and Joint 6 in the forward kinematic solution. The rotational matrix in that case is calculated and is shown in Equation 15.

$$
{}^{0}R_6 = \begin{bmatrix} 0 & \cos(\theta_2+\theta_3)\sin\theta_5+\sin(\theta_2+\theta_3)\cos\theta_5 & -\cos(\theta_2+\theta_3)\cos\theta_5+\sin(\theta_2+\theta_3)\sin\theta_5 \\ -1 & 0 & 0 \\ 0 & -\sin(\theta_2+\theta_3)\sin\theta_5+\cos(\theta_2+\theta_3)\cos\theta_5 & \cos(\theta_2+\theta_3)\sin\theta_5+\sin(\theta_2+\theta_3)\cos\theta_5 \end{bmatrix} \tag{15}
$$

By combining Equation 14 and 15 the formula for the Joint 5 angle is generated.

$$
\theta_5 = a\tan 2\left(\frac{\sin(\theta_2+\theta_3)}{-\cos(\theta_2+\theta_3)}\right) \tag{16}
$$

To be able to generate complete work window, the Joint 2 and Joint 3 angles must vary between their given limits for the desired increment value Δ and using the forward kinematic solution to generate the solution for Joint 5.

Equation 16 was evaluated using the commercial software Matlab. The results were compared to the forward kinematics calculated using the Maple 12 software and the Workspace 5 simulation and off-line programming software (Figure 18).

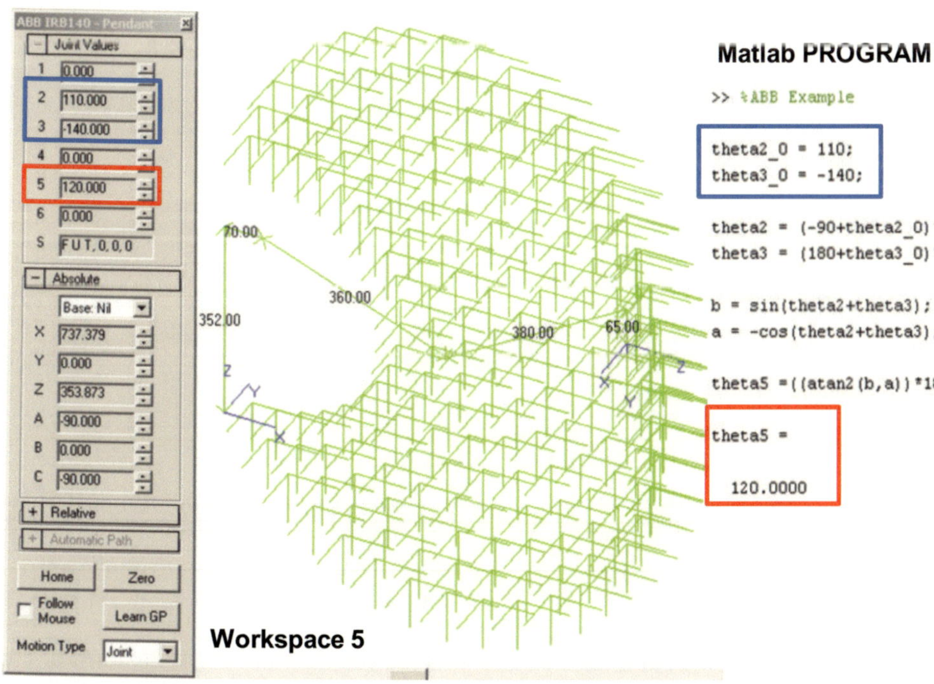

Fig. 18. Evaluation of the formula for the ABB IRB 140 robot

The forward kinematic calculations for the Fanuc M16 robot are done using Equations 2 and 3 and Table 3, which are shown in equations 17-22.

$$ {}^{0}A_1 = \begin{bmatrix} \cos\theta_1 & -\cos\alpha_1 \sin\theta_1 & \sin\alpha_1 \sin\theta_1 & a_1 \cos\theta_1 \\ \sin\theta_1 & \cos\alpha_1 \cos\theta_1 & -\sin\alpha_1 \cos\theta_1 & a_1 \sin\theta_1 \\ 0 & \sin\alpha_1 & \cos\alpha_1 & d_1 \\ 0 & 0 & 0 & 1 \end{bmatrix} = \begin{bmatrix} \cos\theta_1 & 0 & -\sin\theta_1 & a_1 \cos\theta_1 \\ \sin\theta_1 & 0 & \cos\theta_1 & a_1 \sin\theta_1 \\ 0 & -1 & 0 & d_1 \\ 0 & 0 & 0 & 1 \end{bmatrix} \quad (17) $$

$$ {}^{1}A_2 = \begin{bmatrix} \cos\theta_2 & -\cos\alpha_2 \sin\theta_2 & \sin\alpha_2 \sin\theta_2 & a_2 \cos\theta_2 \\ \sin\theta_2 & \cos\alpha_2 \cos\theta_2 & -\sin\alpha_2 \cos\theta_2 & a_2 \sin\theta_2 \\ 0 & \sin\alpha_2 & \cos\alpha_2 & d_2 \\ 0 & 0 & 0 & 1 \end{bmatrix} = \begin{bmatrix} \cos\theta_2 & \sin\theta_2 & 0 & a_2 \cos\theta_2 \\ \sin\theta_2 & -\cos\theta_2 & 0 & a_2 \sin\theta_2 \\ 0 & 0 & -1 & 0 \\ 0 & 0 & 0 & 1 \end{bmatrix} \quad (18) $$

$$ {}^{2}A_3 = \begin{bmatrix} \cos\theta_3 & -\cos\alpha_3 \sin\theta_3 & \sin\alpha_3 \sin\theta_3 & a_3 \cos\theta_3 \\ \sin\theta_3 & \cos\alpha_3 \cos\theta_3 & -\sin\alpha_3 \cos\theta_3 & a_3 \sin\theta_3 \\ 0 & \sin\alpha_3 & \cos\alpha_3 & d_3 \\ 0 & 0 & 0 & 1 \end{bmatrix} = \begin{bmatrix} \cos\theta_3 & 0 & \sin\theta_3 & a_3 \cos\theta_3 \\ \sin\theta_3 & 0 & -\cos\theta_3 & a_3 \sin\theta_3 \\ 0 & 1 & 0 & 0 \\ 0 & 0 & 0 & 1 \end{bmatrix} \quad (19) $$

$$ {}^{3}A_4 = \begin{bmatrix} \cos\theta_4 & -\cos\alpha_4 \sin\theta_4 & \sin\alpha_4 \sin\theta_4 & a_4 \cos\theta_4 \\ \sin\theta_4 & \cos\alpha_4 \cos\theta_4 & -\sin\alpha_4 \cos\theta_4 & a_4 \sin\theta_4 \\ 0 & \sin\alpha_4 & \cos\alpha_4 & d_4 \\ 0 & 0 & 0 & 1 \end{bmatrix} = \begin{bmatrix} \cos\theta_4 & 0 & -\sin\theta_4 & 0 \\ \sin\theta_4 & 0 & \cos\theta_4 & 0 \\ 0 & -1 & 0 & d_4 \\ 0 & 0 & 0 & 1 \end{bmatrix} \quad (20) $$

$$ {}^{4}A_5 = \begin{bmatrix} \cos\theta_5 & -\cos\alpha_5 \sin\theta_5 & \sin\alpha_5 \sin\theta_5 & a_5 \cos\theta_5 \\ \sin\theta_5 & \cos\alpha_5 \cos\theta_5 & -\sin\alpha_5 \cos\theta_5 & a_5 \sin\theta_5 \\ 0 & \sin\alpha_5 & \cos\alpha_5 & d_5 \\ 0 & 0 & 0 & 1 \end{bmatrix} = \begin{bmatrix} \cos\theta_5 & 0 & \sin\theta_5 & 0 \\ \sin\theta_5 & 0 & -\cos\theta_5 & 0 \\ 0 & 1 & 0 & 0 \\ 0 & 0 & 0 & 1 \end{bmatrix} \quad (21) $$

$$ {}^{5}A_6 = \begin{bmatrix} \cos\theta_6 & -\cos\alpha_6 \sin\theta_6 & \sin\alpha_6 \sin\theta_6 & a_6 \cos\theta_6 \\ \sin\theta_6 & \cos\alpha_6 \cos\theta_6 & -\sin\alpha_6 \cos\theta_6 & a_6 \sin\theta_6 \\ 0 & \sin\alpha_6 & \cos\alpha_6 & d_6 \\ 0 & 0 & 0 & 1 \end{bmatrix} = \begin{bmatrix} \cos\theta_6 & \sin\theta_6 & 0 & 0 \\ \sin\theta_6 & -\cos\theta_6 & 0 & 0 \\ 0 & 0 & -1 & d_6 \\ 0 & 0 & 0 & 1 \end{bmatrix} \quad (22) $$

Similarly to the previous calculations, the end-effector orientation matrix is found. The graphical representation of the end-effector is shown in Figure 19.
For the Fanuc M16 Robot the relationship between end-effector and base frame, when end-effector is normal to the robot base, is graphically shown in Figure 20. The orientation matrix for the given position is calculated using Equation 23.

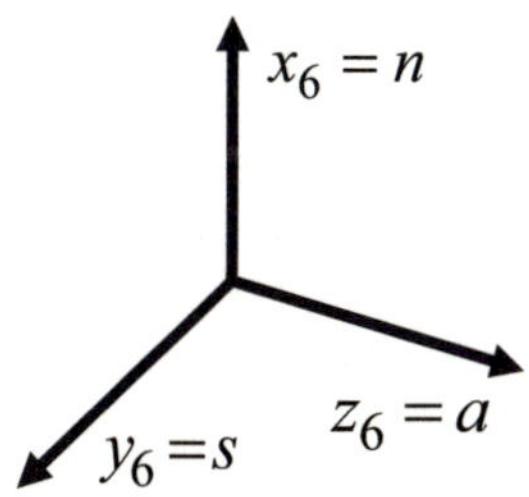

Fig. 19. The end-effector frame of the Fanuc M16 robot

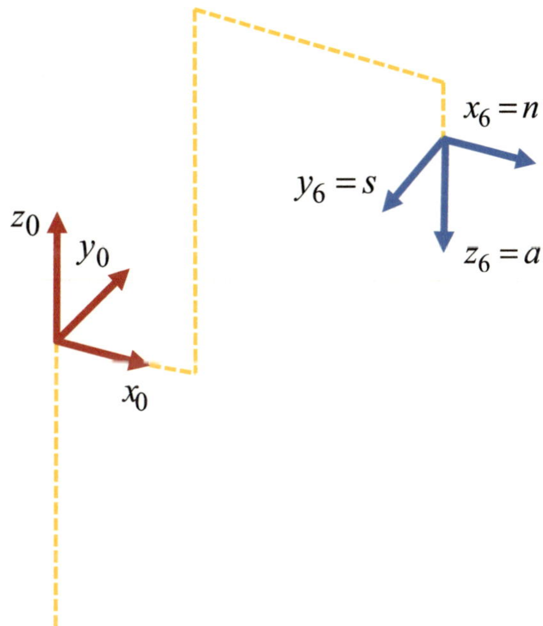

Fig. 20. The end-effector frame of the Fanuc M16 robot

$$ {}^{0}R_6 = \begin{bmatrix} 1 & 0 & 0 \\ 0 & -1 & 0 \\ 0 & 0 & -1 \end{bmatrix} \tag{23} $$

The rotational matrix in this case is calculated and is shown in Equation 24.

$$ {}^{0}R_6 = \begin{bmatrix} -\cos(\theta_2-\theta_3)\cos\theta_5 - \sin(\theta_2-\theta_3)\sin\theta_5 & 0 & -\cos(\theta_2-\theta_3)\sin\theta_5 + \sin(\theta_2-\theta_3)\cos\theta_5 \\ 0 & -1 & 0 \\ -\cos(\theta_2-\theta_3)\sin\theta_5 + \sin(\theta_2-\theta_3)\cos\theta_5 & 0 & \sin(\theta_2-\theta_3)\sin\theta_5 + \cos(\theta_2-\theta_3)\cos\theta_5 \end{bmatrix} \tag{24} $$

By combining Equations 23 and 24, the formula for Joint 5 angle is generated.

$$\theta_5 = a\tan 2\left(\frac{-\sin(\theta_2 - \theta_3)}{-\cos(\theta_2 - \theta_3)}\right) \tag{25}$$

To be able to generate the complete work window, again the Joint 2 and Joint 3 angles must be varied between their given limits for the desired increment value Δ and using the forward kinematic solution to generate the solution for the Joint 5.
As with the ABB IRB 140 robot, this resulting equation 25 is evaluated using Matlab, and compared with result assessments Maple 12 and Workspace 5 (Figure 21).

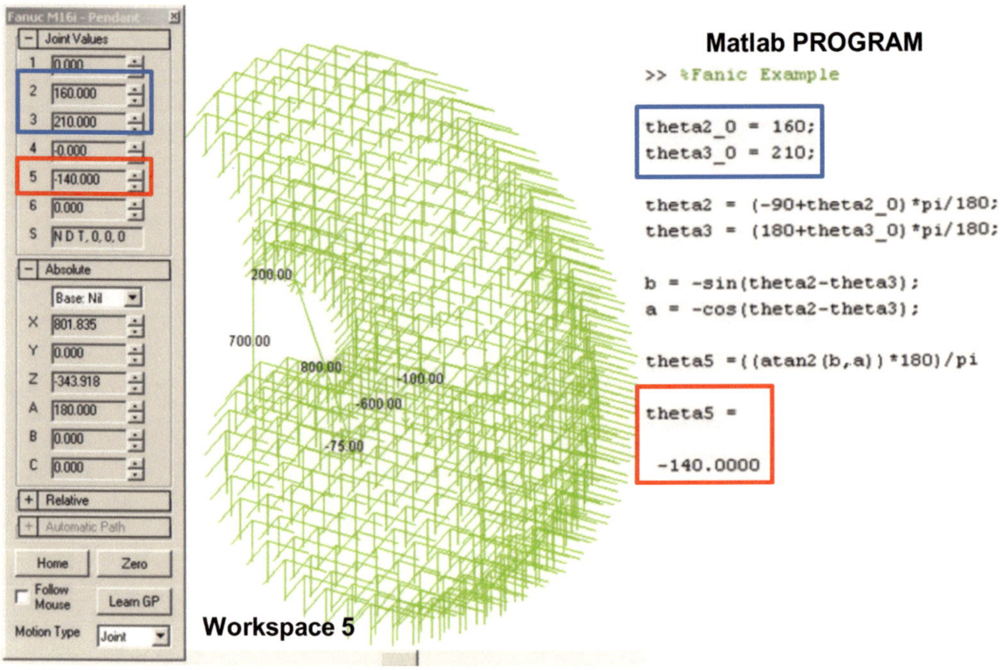

Fig. 21. Evaluation of the formula for Fanuc M16 Robot

Equations 16 and 25 can be unified into one general formula for calculating Joint 5 angle for either ABB IRB 140 robot or Fanuc M16 robot by using the parameter K in Equation 2. The unified formula is presented in Equation 26.

$$\theta_5 = a\tan 2\left(\frac{K\sin(\theta_2 + K\theta_3)}{-\cos(\theta_2 + K\theta_3)}\right) \tag{26}$$

This methodology can be applied to any similar kinematic structure and the unification procedure can be extended. The formula is valid for any desired orientation. The example using a 45 degree orientation for the ABB IRB 140 robot is illustrated in Figures 22 and 23. As with the machine tool example illustrated in Figure 9 (b), the Boolean intersection of the selected 45° orientation and the 90 ° normal to the base regions is the zone where both orientations are valid (Figure 23), and where the part should be located if both orientations are required for the process travel paths.

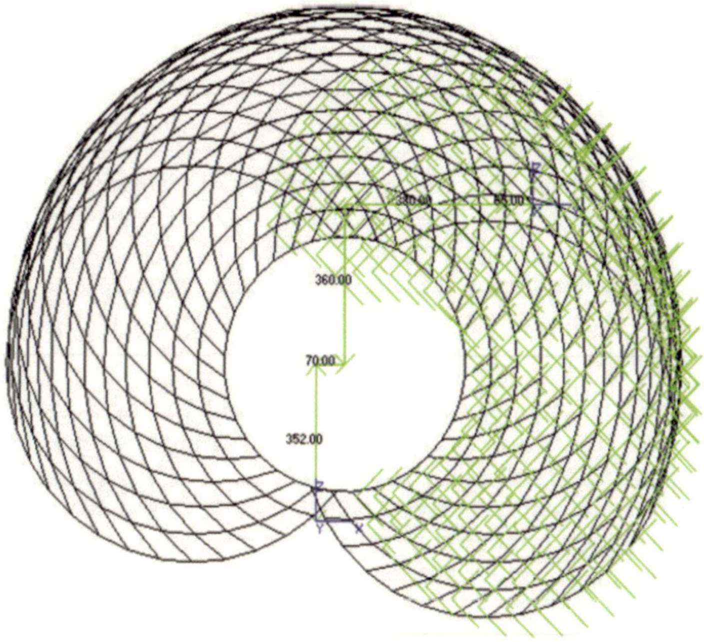

Fig. 22. Work window for the ABB IRB 140 robot for the 45 degree orientation

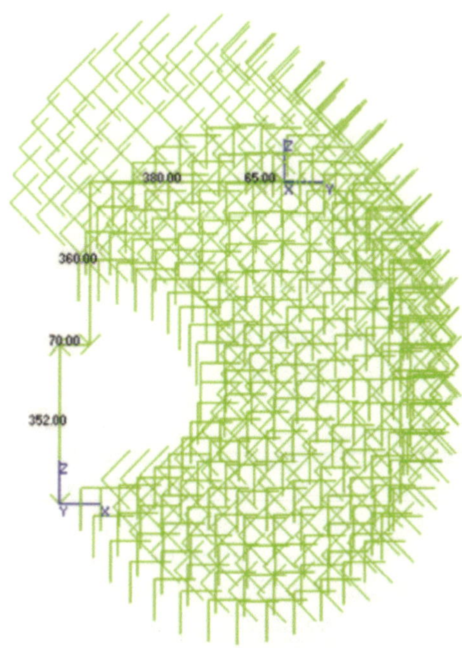

Fig. 23. Work window for the ABB IRB 140 robot showing the comparison between the 90 normal to the base orientation and the 45 degree orientation

To facilitate the 6 DOF work window calculation, empirical and analytical methods are used and validated. From the empirical study, it found out that Joint 2, Joint 3, and Joint 5 play the most important role. By varying Joint 2 and Joint 3 between their limits, Joint 5 needs to be calculated to satisfy the desired end-effector orientation angle. The study of a repeatable procedure led to the generation of the analytical method.
Two kinematically different robots are selected and their work windows are generated: the ABB IRB 140 and Fanuc M16 robots. The resulting work window space can then be assessed using Boolean operations to determine the valid working zone for multiple travel paths, orientations, or regions of overlap for a manufacturing cell. This is illustrated for the 90 degree normal to the base and selected 45 degree orientation example.

3. Robot cell case study

For the selected example, the work window is generated using the empirical formula (algorithm in Figure 11) and evaluated with analytical solution (Figure 24). Three 6DOF ABB robots are place in a work cell. The desired orientation is required for synchronous motion. The robots' placement will depend on the end-effector orientation. Using the work window for each robot, the layout can be done quickly, accurately and options can then be assessed. For example, the consideration that the robots may offset and sliding rails (or a 7th axis) added such that one robot could be repositioned to perform the other robot's tasks if a robot is non-performing for a reason. Reduced productivity would result, but production could still continue.

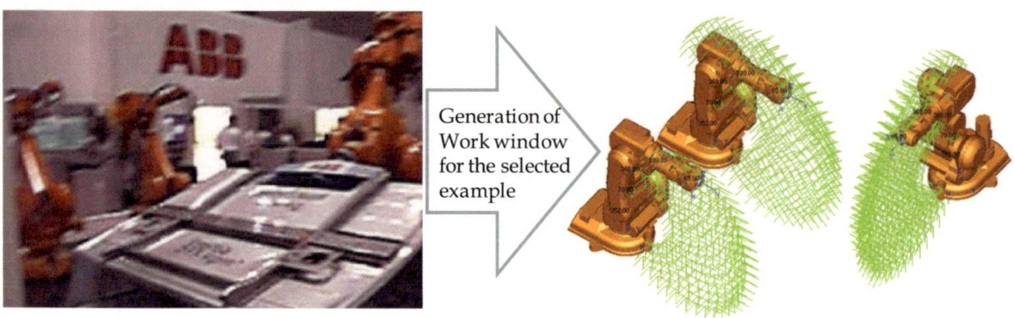

Fig. 24. Work window for the selected example

4. Conclusion

To conclude, a methodology to predetermine regions of feasible operation for multiple kinematic chain mechanisms and manufacturing cells is presented. The approach differs for mechanisms consisting of 1, 2, and 3 DOF subsets linked together via reference frames, and 6 DOF industrial robots. However, after the basic region of valid operation is determined, assessment of the part location and travel paths can be performed. With complex system configurations, it is not intuitive to define the region of task feasibility for the initial design and as well as design alternatives, as there is much coupling related to the kinematic structures and their manipulability, tooling, fixtures, part geometry and task requirements. Changing an operation / task set or a system reconfiguration can be executed virtually, and these methods employed to determine feasibility prior to physical set ups or modification in the manufacturing environment, which represents a time and cost savings.

5. References

Abdel-Malek, K. & Yeh, H. J. (2000). Local Dexterity Analysis for Open Kinematic Chains, *Mechanism and Machine Theory*, Vol. 35, pp. 131-154.

Anonymous - Okuma, 2002, Okuma Twin Star LT Series Turning Centres with Twin Spindles, *LT Series - (1) –300*, Okuma Co.

Castelli, G.; Ottaviano, E. & Ceccarelli, M. (2008). A Fairly General Algorithm to Evaluate Workspace Characteristics of Serial and Parallel Manipulators, *Mechanics Based Design of Structures and Machines*, Vol. 36, pp. 14-33

Cebula, A.J. & Zsombor-Murray, P.J. (2006). Formulation of the Workspace Equation for Wrist-Partitioned Spatial Manipulators, *Mechanisms and Machine Theory*, Vol 41, pp. 778-789

Ceccarelli, M. & Lanni, C. (2003). A Multi-objective Optimum Design of General 3R Manipulators for Prescribed Workspace Limits, *Mechanisms and Machine Theory*, Vol. 39, pp. 119-132

Ceccarelli, M. & Vinciguerra, A. (1995). On the Workspace of the General 4R Manipulators, *The International Journal of Robotics Research*, Vol.14, pp.152-160

Ceccarelli, M. (1996). A Formulation for the Workspace Boundary of General N-revolute Manipulators, *Mechanisms and Machine Theory*, Vol. 31, pp. 637-646

Dai, J. & Shah, P. (2003). Orientation Capability Planar Manipulators Using Virtual Joint Angle Analysis, *Mechanism and Machine Theory*, Vol. 38, pp. 241-252

Denavit J. & Hartenberg R. S. (1955). A Kinematic Notation for Lower-pair Mechanisms Based on Matrices, *Journal of Applied Mechanics*, Vol. 77, pp. 215-221.

Djuric, A. & Urbanic, R. J. (2009). A Methodology for Defining the Functional Space (Work Window) for a Machine Configuration, *3nd International Conference on Changeable, Agile, Reconfigurable and Virtual Production*, CD-ROM, Munich, October 5th-7th, 2009

Djuric, A. M. & ElMaraghy, W. H. (2008). Filtering Boundary Points of the Robot Workspace, *5th International Conference on Digital Enterprise Technology*. Nantes, France, October 2008

ElMaraghy, H. A. & ElMaraghy, W. H. (1993). Bridging the Gap Between Process Planning and Production Planning and Control, *CIRP Journal of Manufacturing Systems*, Vol. 22, No. 1, pp. 5-11

ElMaraghy, H. A. (2005). Flexible and Reconfigurable Manufacturing Systems Paradigms, *International Journal of Flexible Manufacturing Systems, Special Issue on Reconfigurable Manufacturing Systems*, Vol. 17, pp. 261-276

ElMaraghy, H. A. (2008). *Changeable and Reconfigurable Manufacturing Systems*, (editor) Springer-Verlag (Publisher), ISBN: 978-1-84882-066-1

Hedrick R. & Urbanic, R. J. (2007). A Methodology for Managing Change when Reconfiguring a Manufacturing System, *2nd International Conference on Changeable, Agile, Virtual and Reconfigurable Production (CARV)*, pp. 992-1001

Koren, Y.; Heisel, U.; Jovane, F.; Moriwaki, T.; Pritchow, G.; Van Brussel, H. & Ulsoy, A.G. (1999). Reconfigurable Manufacturing Systems, *CIRP Annals*, Vol. 48, No. 2.

Nof, S.Y. (1999). *Handbook of Industrial Robotics*, 2nd edn. New York: John Wiley & Sons.

Pond, G. & Carretro, J. (2011). Dexterity Measures and Their Use in Quantitative Dexterity Comparisons, *Meccanica*, Vol. 46, pp. 51-64

Yang, J.; Yu, W.; Kim, J. & Abdel-Malet, K. (2009). On the Placement of Open-Loop Robotic Manipulators for Reachability, *Mechanism and Machine Theory*, Vol. 44, pp. 671-684

Requirement Oriented Reconfiguration of Parallel Robotic Systems

Jan Schmitt, David Inkermann, Carsten Stechert,
Annika Raatz and Thomas Vietor
Technische Universität Braunschweig
Germany

1. Introduction

The development of industrial production is subject to a change, which is mainly caused by higher customer orientation. The possibility of an end-customer to configure nearly individual mass products results in sophisticated challenges for production. The diversity of variants is growing. In addition an increasing demand to offer a band of new products leads to shorter product life cycles. To be competitive a company has to provide the product at reasonable costs. Considering serial production lines in highly developed countries, a high degree of automation is necessary to gain economies of scale. These changes of the economic boundaries demand specific design methodologies for the production equipment. In particular reconfigurable robotic systems are promising alternatives to the purchase of new production equipment. In case of changed production organization, reconfigurable robotic systems reduce redesign efforts and do not require rescheduling of entire production lines. Major impact factors to the design of reconfigurable robots are shown in Figure 1. They are subdivided into product and production driven aspects.

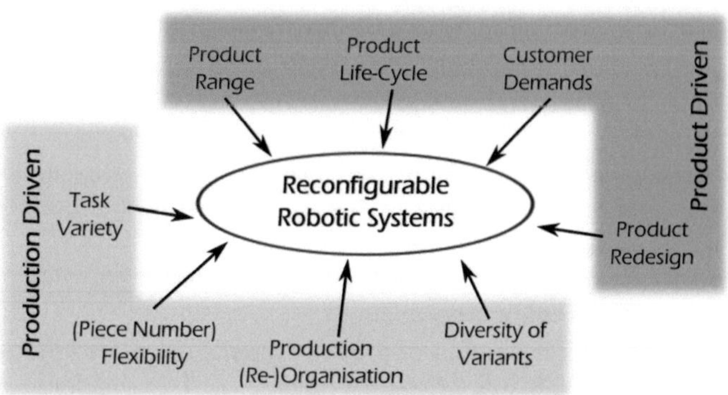

Fig. 1. Impact factors to reconfigurable robotic systems.

Regarding robotic systems, parallel robots, due to their high speed, acceleration and stiffness parameters have a huge potential to be reconfigurable systems. Several and entirely different reconfiguration strategies of parallel robotic systems can be derived based on these properties. Starting from the early design stage of requirement management, design considerations

and machine elements have to be considered in order to enable effective reconfiguration of the systems. Since development and design complexity for reconfigurable robotic systems increase necessary steps to be addressed are use case derivations, their segmentation and discretization as requirement spreads as well as the relation between the systems components and parameters of the robotic system, which are modified.
In this contribution general aspects of robotic reconfiguration approaches are shown and a methodological design procedure for reconfigurable systems is introduced. A brief literature review tends to determine the general idea of reconfiguration and the state of research of robotic reconfiguration by several approaches. As parallel robots constitute the focus in this chapter, some basics of this particular type of robots are shown. In order to develop reconfigurable robotic systems, the requirement management as the first step of the design process is significant. Therefore, paragraph 4 discusses the modeling of complex requirements and requirement spreads using SysML. In the following, static and dynamic reconfiguration approaches of parallel robots are presented and applied exemplary. Since the suitability of each reconfiguration strategy depends on the restrictions of the considered use cases, a systematic assessment of reconfiguration approaches is done in section 6. The chapter closes with a conclusion and a brief prospect to further research activities.

2. Literature review

The field of robotic research reconfiguration is often attended by modularization, standardization of interfaces or morphology on different mechatronic levels. In order to establish a consistent comprehension the term "'reconfiguration"' is defined, before representatives of different concepts are described briefly.

2.1 General idea and term definition

The general idea in developing reconfigurable robots is to enable a system to change its abilities to fulfil various tasks. These tasks are characterized by different requirements depending on the robots field of application. In this contribution parallel kinematic robots are focused on, where handling and assembly tasks are typical. Hence, a suitable definition for reconfiguration is given by Setchi (2004):

"'Reconfigurability is the ability to repeatedly change and rearrange the components of a system in a cost-effective way"'.

In literature a wide spread understanding of reconfigurable robots is existent (see Figure 2). In the following section different approaches and realization examples are introduced in order to distinguish the scope of this contribution.

2.2 Classification of reconfiguration approaches

Many approaches for the reconfiguration of robotic systems have been proposed in literature but are not yet established in industry. These approaches can be considered as either online or offline. While online reconfiguration is performed during the use of a robot without switching it off, offline reconfiguration demands a shut-down of the robot (see Figure 2). The following literature review helps to gain a deeper insight into the possibilities to realize reconfigurable robots that can adapt to external environments.

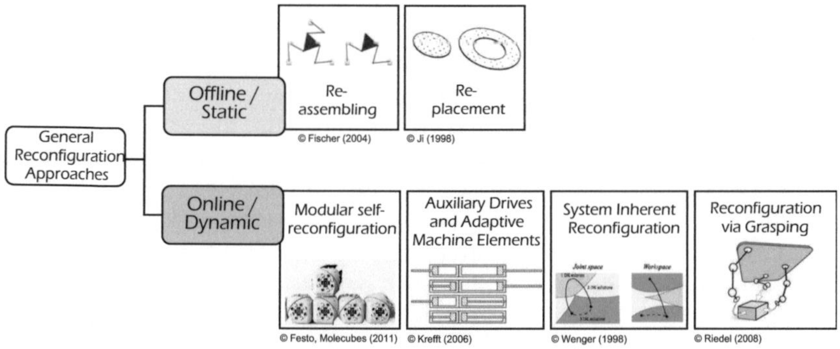

Fig. 2. General classification of reconfiguration approaches to adapted parallel robots to external environments.

2.2.1 Offline reconfiguration

The mechanical structure of fully parallel robotic systems is very suitable for offline reconfiguration since those structures are actuated in parallel and therefore consists of many consistent components. Components or sub-modules of the system are replaced or the same components are mounted in different positions and orientations. Thus, various morphologies of the kinematic structure can be arranged.

Reassembling and replacement. Various pieces of work focus on the design of modular kinematic structures in order to enable effective reconfiguration of robot systems [Krefft (2006), Fisher (2004), Wurst (2002)]. A modular robotic system consists of a set of standardized modules, such as actuators, passive joints, rigid links (connectors), mobile platforms, and end-effectors, that can be rapidly assembled into complete systems with various configurations. In Ji (1998) a reconfigurable platform manipulator based on a HEXAPOD kinematic structure is presented. Each module e.g. leg modules of the structure can be replaced and position and orientation of the joints on the mobile platform and the base can be varied, using different sets of pattened holes. Thus, different workspace dimensions are obtained. Yang et al. (2001) propose two types of robot modules in order to simply assemble three-legged parallel robots. Using fixed-dimension joint modules and variable dimension link modules that can be custom-designed rapidly, various kinematic structures can be facilitated.

In order to develop offline reconfigurable parallel robots numerous design approaches have been presented in literature. Since modular design is a research field in engineering design, specific methodologies for modular reconfigurable parallel kinematic machines (PKM) were introduced e.g. by Jovane (2002), Pritschow et al. (2000), and Koren et al. (1999). These approaches are mostly limited to concrete kinematic structures e.g. HEXAPOD robots, so transfer of constructive realizations is narrowed down.

2.2.2 Online reconfiguration

Online reconfiguration means to change the robots' properties without rearranging the mechanical parts. Hence, no shut-down of the system is essential, but particular hardware components as well as dedicated control functions are required.

Modular self-reconfiguration. Modular self-reconfigurable robotic systems have been topic of tradition in robotics science for years, a survey of Jantapremjit (2001) shows a variety of these reconfigurable robots. In general they are assembled by a set of identical modules with the same functional range. Each module has to carry actuation, sensing, control entities as well as energy supply, communication and offers several interfaces. In order to be self-reconfigurating the set of interlocked modules has to change configuration autonomously, which mostly imply the ability of locomotion for each module. In Yim (2007) the classification of modular self-reconfigurable robots by architecture and the way in which the units are reconfigured is provided. The control strategy of those robots can be centralized or distributed. Famous examples of this robot class are PolyBot G3 [Yim (2000)] or Molecubes [Molecubes (2011)].

Auxiliary drives and adaptive machine elements. In order to adapt the configuration of the kinematic structure during operation of the robot, passive components can be substituted by auxiliary drives or adaptive machine elements. Using these components, geometries of the kinematic structure e.g. strut length and degree of freedom (DoF) are affected, thus different configurations are achieved. Adaption of the strut length is carried out using auxiliary linear motion devices such as linear motors, hydraulic jacks, air cylinders, or lead screws driven by rotary actuators. An approach to select suitable components for reconfiguration of manufacturing systems is introduced by Lee (1997). However, a requirement oriented consideration of component parameters is not made. In Krefft (2006) a concept to vary the length of struts in discrete steps by means of air cylinders is introduced. This geometric reconfiguration results in workspace as well as stiffness variations. Approaches to affect systems DoF are presented e.g. by Theingi et al. (2007). Here, belt drives are used to couple joints and therefore reach new kinematic configurations as well as passing singularities within the workspace. O'Brien (2001) proposes a kinematic control of singularities applying passive joint brakes to affect the kinematic DoF.

Main difference of the described approaches is the impact to the robots' performance. Adding auxiliary drives, the performance is mostly influenced negatively. For instance system weight raises because of higher component weight. Hence, adaptive machine elements which offer basic function as well as additional ones are focus of this contribution. Other aspects such as the realization of redundant actuated mechanisms are not considered.

System inherent reconfiguration. Due to the fact that within the workspace of parallel robots singularities occur (see section 3), system inherent reconfiguration without demanding particular mechanical components is feasible. Several research works focus on passing through or avoiding singularities within the workspace [Budde (2010), Hesselbach et al. (2002), Wenger (1998)]. Passing actuator singularities, workspaces of different assembly modes can be combined to a larger workspace [Krefft (2006)]. Hesselbach et al. (2002) firstly introduced an approach to cross singularities using inertia of the Tool Center Point. A comprehensive procedure for the passing of singularities as well as the concerning control issues is described in Budde (2008).

Reconfiguration via grasping. The last approach to be considered in context of online reconfiguration of parallel robots is, that separate manipulators constituting an entire mechanism by grasping an object. In Riedel (2008) a reconfigurable parallel robot with an underactuated arm structure is presented. After grasping, the contact elements at the end of the underactuated arm mechanisms are connected to the object which forms a closed loop mechanism similar to the architecture of parallel manipulators. Each arm mechanism is a combination of a five-bar-linkage with a parallelogram arrangement, a revolute joint around the vertical axis and a spherical wrist joint [Riedel (2010)]. Consequently, different configurations of the entire system can be arranged to face specific requirements (workspace dimension, process forces, stiffness) of the current task.

Based on this brief literature review in the following section particular properties as well as design aspects of parallel robots are introduced. Subsequently, requirement management for reconfigurable robots is introduced in section 4. Further considerations will also focus on offline and online reconfiguration. Here, the use of adaptive machine elements as well as system inherent reconfigurations are considered.

3. Parallel robotic systems

Parallel robots constitute a special class of robotic systems mainly for industrial applications. The characteristic property of parallel robots, in comparison to their serial counterparts is the closed-loop kinematic chain referring to the kinematic structure. The constructive advantage of this robot architecture is the possibility to assemble the drive near the frame, which implicates very low moving masses. This specific design leads to several significant benefits in the robots' performance, such as high stiffness and high dynamic properties. Due to these aspects, parallel robots are employed for an increasing amount of handling and assembly tasks [Merlet (2001)]. The most famous example of a parallel kinematic structure is the Delta robot of Clavel (1991) (see Figure 3). Here, a four DoF Delta robot is shown, where three translational DoF are provided by the rotary drives mounted at the frame. The load transmission is carried out by the three close-loop chains, where the constant orientation of the end-effector platform is provided by the parallelogram structure in the mechanism. The middle axis offers an additional rotary DoF by the application of a length variable connecting shaft.

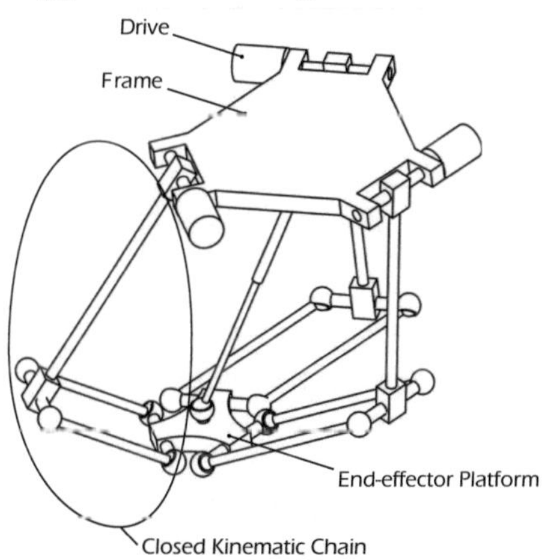

Fig. 3. Delta structure as famous example for parallel robotic systems.

The rather low market penetration of parallel robots (except the Delta kinematic) is reasoned by specific drawbacks. The unfavorable workspace-to-installation space ration in comparison to serial robots is the main aspect. Furthermore, the presence of diverse singularities within the workspace, which lead to a degeneration of the robots' manipulability. In addition, these singularities result in challenging tasks in control design. In Park (1998) singularities are classified as

1. configuration space singularities at the boundaries of configuration space
2. actuator singularities, where the DoF of the robot varies
   (a) degenerate types: links can be moved, even if the actuators are locked
   (b) nondegenerate types: internal forces are generated

3. end-effector singularities, in which the end-effector frame loses DoF of available motion.

Mathematically, the mentioned singularities occur, if one of the Jacobian Matrices $J_{A,B}$ with

$$J_A = \frac{\delta f}{\delta X} \vee J_B = \frac{\delta f}{\delta q} \quad (1)$$

where X represents the Tool Center Point (TCP) coordinates $X = (x, y, z)^T$ and q the actuator variables $q = (q_1, q_1, ..., q_n)^T$, becomes singular. Whereas, a configuration space singularity appears if,

$$det(J_A) \neq 0 \wedge det(J_B) = 0. \quad (2)$$

Actuation singularities are conditioned by

$$det(J_A) = 0 \wedge det(J_B) \neq 0. \quad (3)$$

The singularity analysis is hindered by the structural complexity of parallel mechanisms, which is attended by an extensive kinematic description. Due to reconfiguration approaches of parallel robots, singularity detection, avoidance or passing is especially focused in paragraph 5.2.

For the systematic design of parallel robotic systems different approaches have been proposed in literature e.g. [Frindt (2001)]. The parameter space approach proposed by Merlet (1997) considers requirements such as workspace and the articular velocities and leads to possible sets of robot geometries that satisfy these requirements. These first design solutions are sampled to identify the best compromise with regard to other requirements, i.e. accuracy. Following the cost function approach [Bourdreau (2001)] the robot geometries are arranged in the way that workspace dimensions are as closely congruent as possible with the prescribed as possible. Based on this demand the design problem becomes a multi objective optimization problem. Whereas these approaches consider only one use cases of the robotic system they lead to "'static"' solutions, which can hardly be adapted to other use cases. Therefore, the proposed approach can be seen as an addition to these approaches. It ensures the design of reconfigurable robotic systems in order to satisfy temporal changes of requirements which are caused by different use cases.

This brief overview of the properties of parallel robotic systems shows the benefits and drawbacks of these systems. With regard to the aforementioned aspects of reconfiguration parallel robots are suitable systems to be reconfigured, as their extensive dynamic, accuracy and stiffness parameters. Hence, they can cover a huge range of requirements. For this reason a specific methodical approach for the requirement management e.g. under consideration of the identification of requirement spreads is needed.

4. Requirement management for reconfigurable robotic systems

The development of reconfigurable robotic systems demands detailed analysis of the product surroundings and relevant use cases during each life-cycle phase (e.g. operation, maintenance). As a result of this analysis numerous requirements are refined to forecast claimed system properties. In order to enhance life-time and performance of the system several target values for one requirement have to be fulfilled. For instance, different use cases require different workspace dimensions. These varying values for one requirement we call *requirement spread* [Schmitt (2009)]. In order to identify spreading requirements and derive reconfiguration scenarios, a structured model of the expected life-cycle as well as the desired

system properties is needed. Nevertheless, a consistent model is essential to facilitate domain specific views on the gathered information and therefore enable an efficient treatment of the development task.
In this section an object oriented approach to work out a requirement model is introduced. In order to identify reconfiguration parameters a methodical procedure is presented.

4.1 Modeling requirements

The modeling of requirements is carried out in order to provide a functional, comprehensive, complete, operational, non redundant and minimal set of requirements, which can be used as a pattern for the whole development process [Roozenburg (1991)]. Analysing the requirements and their relations reconfigurations scenarios for the robotic system can be derived and estimated. Requirement management for reconfigurable robotic systems focuses on three aspects [Stechert (2010)], namely: Surroundings, structure, and relations. Based on these aspects an analysis can be arranged in order to indentify reconfiguration parameters and develop reconfiguration concepts.
Surroundings. As one of the first steps in the development process the surrounding of the robotic system has to be respected in order to recognize important requirements for the involved domains. Therefore, each life-cycle phase has to be taken into account including different scenarios (e.g. Anggreeni (2008), Brouwer (2008)) or use cases with all related actors, surrounding environment and possible disturbances. A systematic documentation helps to identify and to use gathered requirements and constraints. It further shows which requirements derive from what surrounding element.
Structure. In order to handle the gathered information in the following development steps, it has to be structured [Franke (1999)]. Requirements can be hierarchical structured such as goal, target, system requirement and subsystem requirement. Furthermore, they can be allocated to the respective domain and to the purpose in the development process. In progress of the development process requirements can be allocated to concerning subsystems. Assignment of certainty and change probability helps to focus on most relevant requirements during each development step.
Relations. Elements of complex models are related among one another. A basic classification for relations was presented in earlier work [Stechert (2009b)]. This classification helps to declare relations associated to development steps, granularity, support, direction, linking and quantifiability. Representing these relations within the requirement model helps designers to understand the influences of changes (e.g. change of rack radii), point out goal conflicts and highlight interfaces to other disciplines. However, the modeling of relations is the basis for the analysis of the requirement model, discussed in section 4.2.
Within the work of the Collaboration Research Centre (CRC) 562 "'Robotic Systems for Handling and Assembly - High Dynamic Parallel Structures with Adaptronic Components"' a SysML-based requirement model was developed [Stechert (2010)]. Since the Systems Modeling Language (SysML) uses parts of UML (Unified Modeling Language) it is a widely known notation in the fields of software development, electronic design and automation. In recent years it has become more and more popular in mechanical engineering e.g. Woelkl (2009). Using SysML a product can be modeled out of different viewpoints and on different levels of abstraction [Weilkiens (2008)]. However, within the introduced requirement model the above mentioned aspects are picked up in order to provide a coherent product model.
Major elements of the requirement model are "'Requirements"' and "'Surroundings"' [Stechert (2009a)]. The requirements are hierarchically distinguished into "'Goals"', "'Targets"', and

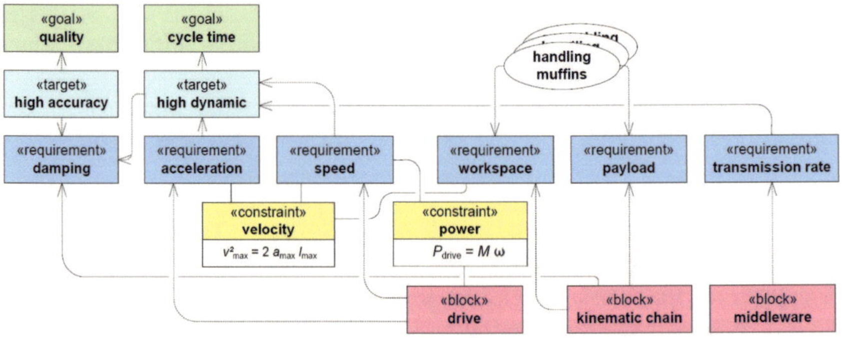

Fig. 4. Elements of an extended requirements diagramm based on SysML.

"'Technical Requirements"'. Within the element "'Surronding"' "'Product Environments"' (e.g. neighbor systems) and "'Use Cases"' are defined. These use cases are related to the product life-cycle and describe e.g. the handling of muffins within the use-phase of the robot. Figure 4 shows an excerpt of the hierarchical structure as well as different use cases. Pointing out the relations between use cases and requirements, several requirements are refined concerning their target values. For instance, the use case "'handling muffins"' correlates amongst others with the requirements "'workspace"' and "'payload"'. In addition, the requirements are linked to components of the robotic system (e.g. drive). Indicating constraints, correlation between components and requirements as well as requirements among each other are specified. In this way the requirement speed correlates with the component drive (constraint *power*).

4.2 Identification of reconfiguration parameters

Based on the introduced requirement model a systematic analysis can be carried out in order to identify meaningful scenarios and relevant requirement spreads for reconfiguration scenarios as well as their particular impact on the whole robotic system. By linking requirements which vary for different use cases with corresponding product parameters, reasonable reconfiguration concepts can be derived. Reconfiguration parameters can be systematically detected following the two steps shown in Figure 5. These two superordinated steps; *identify resonable reconfiguration scenarios* and *assess possible reconfiguration concepts*; constitute five working steps which are described below, following earlier work [Schmitt (2009)].

Initially, possible applications of the robotic systems and the corresponding use cases (e.g. handling muffins, PCB assembly) have to be considered. With regard to tasks (e.g. pick-and-place, feeding, high precision assembly) on the one hand and branches of trade (e.g. food industry, material processing, ICT) on the other hand, the identified use cases are structured. In addition, the reconfiguration of the robotic system is considered as a use case itself. Subsequently, relations between use cases and requirements are established. Using the relation "'refine"' each requirement is stated more precisely by an use case. As a result of this refinement a new sub requirement is created. For instance, the use case handling muffins refines the requirement workspace width to the exact value 500 mm. Assuming different use cases within the life-time of the robotic system it becomes evident, that diverse sub requirements for the same parent requirement occur.

Within the second step a hierarchical goal-system is developed. Based on the requirements considered before, traces are followed to superordinated targets and goals. Furthermore,

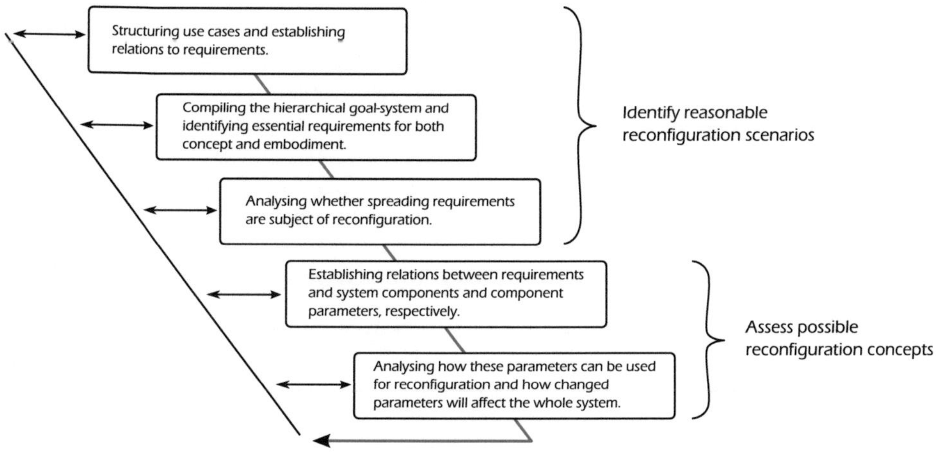

Fig. 5. Working steps for the identification of reconfiguration-relevant parameters.

the derived requirements have to be analysed to determine whether they are relevant for reconfiguration of the robot. First of all, concept-relevant requirements (e.g. DoF) are analysed. Based on the definition of concepts, embodiment related requirements have to be considered. For instance, different payloads and process loads lead to the relevant requirement operation forces, while accuracy leads to elastic deformation. Both are related by the structural stiffness of the kinematic structure.

The object of the third step is to estimate whether spreading requirements are subject of reconfiguration. It is obvious, that a reconfiguration is not probable between every use case. A robot used in the chemical industry first will not be sold to a company from the food industry, reconfigured and reused for pick-and-place tasks. However, a pick-and-place robot in the food industry might handle muffins first and convert to packaging of croissants in the same company later. Additionally the influence of requirement changes to other requirements have to be considered. The DoF at the gripper for instance might be subject of a wide spread. However, extending a fully parallel structure with an additional DoF causes a completely new control program. Therefore, it becomes obvious that not every requirement spread is a meaningful basis for reconfiguration concepts. Costs and engineering complexity are major limits. A detailed analysis of the goal-system leads to a lean set of essential reconfiguration-relevant requirements for robotic systems [Schmitt (2009)]:

- enabling object handling
- provide workspace
- enable operation forces
- provide degree of freedom
- assure high accuracy
- provide performance
- enable low lifecycle costs

These general requirements are specified by different use cases. For instance, the DoF is refined by the number of possible movements and the orientation of rotation (e.g. $\xi = \pm 25$).

In the fourth step, requirements are related to system components. System components can be hard- or software and are related to other components (e.g. information flow, geometric surfaces). For instance, the requirement workspace dimension is satisfied by the kinematic structure. More precisely the system component strut and its parameter strut length fulfill this requirement.

As a last step reconfiguration scenarios are developed and analyzed. Therefore, traces from a requirement to the involved parameters are followed and detected relations are described in a qualitative or - if possible - in a quantitative way. For instance, relation between strut length and workspace dimension can be described by kinematic problem, see section 3. Furthermore, it is necessary to determine the influence of the identified reconfiguration parameters to other requirements and detect possible goal-conflicts. It is obvious, that an extension of the strut length will lead to a higher moved mass, which might decrease the performance of the robotic system.

5. Reconfiguration approaches for parallel robotic systems

In Paragraph 2.1 general approaches for reconfigure robotic systems were introduced, in case of parallel robots for industrial applications three reconfiguration approaches are of major relevance. These different concepts are discussed in detailed, following the classification proposed by Krefft (2006):

- *Static reconfiguration (SR)* requires switching off the robot and rearranging machine parts such as drives, struts, or joints as well as complete sub-assemblies manually (see section 2.2.1).
- *Dynamic reconfiguration type 1 (DR I)* is carried out during operation of the system. Physical dimensions are not changed, but different configurations of the kinematic structure are used. Here singularities of type 1 have to be passed by the robot (see paragraph *system inherent reconfiguration*).
- *Dynamic reconfiguration type 2 (DR II)* is realized during operation of the system. The changing of kinematic properties is effected by the adjustment of geometric parameters such as strut length or modification of the DoF by joint couplings in order to avoid or replace singularities (see paragraph *auxiliary drives and adaptive machine elements*).

Table 1 illustrates the description of *SR, DR I* and *DR II* by characterizing the robot status and the configuration chance. The detailed discussion of the concepts is made in the following sections, whereas section 5.4 points out two case studies of reconfigurable parallel robots.

Reconfiguration Approach	Robot Status	Configuration Change
Static (SR)	off	manual rearrangement of components/ subassemblies
Dynamic Type I (DR I)	on	usage of different configuration by passing through singularities
Dymanic Type II (DR II)	on	adjustment of geometries by machine components

Table 1. Reconfiguration approaches for parallel robotic systems.

5.1 Static reconfiguration

According to the term of definition of static reconfiguration, the rearrangement of mechanical components is the objective to deal with. Based on a starting concept, the introduced methodology supports identification of reasonable reconfiguration modules (RM_i) of the parallel robot. Within this modularization it is possible to allocate several requirements to the system components. In this way it can be sorted out, which key property of the system is affected by what reconfiguration strategy. Figure 6 shows the $RM_{1...3}$ and their interrelation as well as the interfaces between single machine components of parallel robots.

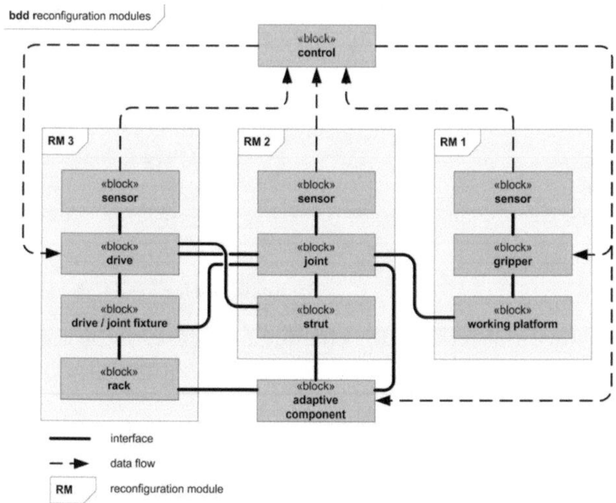

Fig. 6. SysML model for the modularization of the system components of a parallel robot.

The application of different grippers (RM_1) according to the requirement spread accomplishes the type of requirement, that result from the variation of the manufactured product. In this case, the workpiece size, geometries, surface or material properties are meant. The workspace or the required structural stiffness are only slightly influenced by the gripper. To integrate different kinds of grippers, e.g. mechanical or pneumatic ones, adequate constructive and control interfaces have to be designed. Furthermore, the changing manner of energy supply must be considered e.g. at the mobile platform. The change in mass and moments of inertia influences operation forces, accuracy and performance.
The strut/joint assembly (RM_2) can be reconfigured through the installation of different strut kits. Here, diverse strut length results in variations of the position and dimension of the workspace. In addition, changes in strut length affect performance and accuracy due to changes in moved masses and transmission ratio of the kinematic chain, thence possible acceleration, maximum speed and eigenfrequencies. In terms of a modular construction kit, the interfaces between modules have to correspond with one another.
In general the drives are the base of the characteristic kinematic chain of a parallel robot and represent the 3rd reconfiguration module (RM_3) together with the drive/joint fixture. They are assembled at or near the frame platform and are distinguishable between rotational and translational DoF. According to static reconfiguration, the specific properties such as weight or number and accessibility of the application points, must be observed during the design process for reconfigurability. Therefore, the frame can determine the ability for the static reconfiguration of parallel robots. If the frame provides a set of patterned holes to assemble the drives or base assembly points of the struts, the workspace dimension and position as well as stiffness can be influenced according to the identified requirement spread.
In order to show the benefits of static reconfiguration concerning the variation of the workspace position and dimension, a simple 2-DoF parallel mechanism is considered. The kinematic 2-R$\underline{P}$R structure consists of passive rotary joints (R) at the base and the end-effector platform and active prismatic joints ($\underline{P}$) (see Fig. 7).
The point of origin is fixed at base A and the Tool Centre Point (TCP) is defined by the vector $X = (x_p, y_p)^T$. The distance between base A and B is given by b. The vector

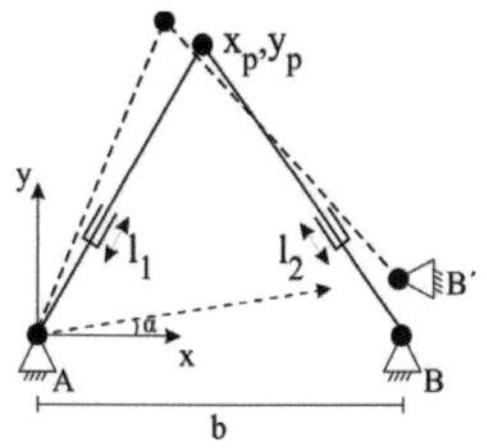

a) Kinematic structure of the RPR mechanism.

$$q_1 = \begin{bmatrix} x_p \\ y_p \end{bmatrix} = \sqrt{x_p^2 + y_p^2} \quad (4)$$

$$q_2 = \begin{bmatrix} cos(\alpha) \cdot (b - x_p) \\ sin(\alpha) \cdot b \cdot y_p \end{bmatrix} = \quad (5)$$

$$= \sqrt{cos(\alpha) \cdot (b - x_p))^2 + (y_p + b \cdot sin(\alpha))^2} \quad (6)$$

b) Relations for kinematic analysis.

Fig. 7. Kinematic structure and relation for kinematic analysis of the considered RPR mechanism.

$q = (q_1, q_2)^T$ includes the actuated strut variables, whereas each strut can be moved between the intervals $[q_{min}, q_{max}]$. The parameter α describes the angular displacement B of base B to A. For the kinematic analysis the solution of the relation between end-effector coordinates and drive parameter $(f(X, q) = 0)$ is essential. Here, the direct and inverse kinematic problem constitutes the motion behavior constrained by the geometric dimensions. The direct kinematic problem calculates the position and orientation of the TCP out of the given drive displacement. If the TCP coordinates are defined, the inverse kinematic problem assesses the actuator displacement. The relation given in Figure 7 b) is important for kinematic analysis.
In paragraph 3 the existence and kinematic influences of singularities of parallel robots were discussed briefly. In case of the shown planar 2-DoF parallel mechanism the configuration space singularities can be associated with the boundary of the respective workspace. Actuation singularities are existent for the condition

$$|q_1| + |q_2| = b. \quad (7)$$

Referring to the kinematic structure in Figure 7 a) a workspace analysis to point out several reconfiguration concepts was done using MATLAB. First, the variation of the interval $[q_{min}, q_{max}]$ can be associated as a change of strut kits. Here length range can be varied according to the boundary of each requirement in the chosen use case. The second reconfiguration concept is to reposition the assembly points. This can be shown by the variation of the parameter b in a way that the distance between A and B is in- or decreasable. Furthermore, the change α expresses the repositioning of B not only in x-, but also in y- direction to B. The following Figure 8 demonstrates the possibilities of workspace repositioning by changing the drive assembly points and varying strut length. In field (a) the range l with respect to $[q_{min} \dots q_{max}]$ of the actuated struts is changed. Illustration (b) shows the repositioning of the drive carrying struts in x-direction. In the kinematic model this is equivalent to the adjustment of parameter b. Furthermore in (c) the reconfiguration of the base assembly points in y-direction by varying the angel α between the drives is shown.

5.2 Dynamic reconfiguration type I

Dynamic reconfiguration of type I focuses on the passing through singularities during operation without adding specific components. Therefore, it also can be constituted as a system inherent ability of the parallel robot to reconfigure itself. Based on the works of *Budde* [Budde (2010), Budde (2007)] this reconfiguration approach is introduced in detail, using the TRIGLIDE structure as an example.
As pointed out before, workspace dimension is an essential system parameter, which has

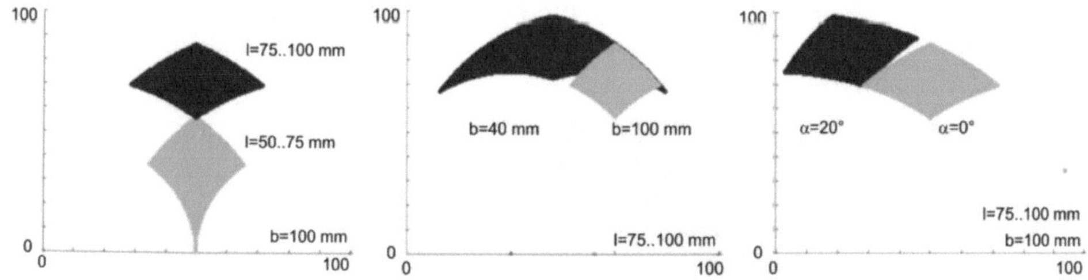

(a) (b) (c)

Fig. 8. Influence of the reconfiguration parameter on the position and dimension of the workspace: a) varied strut length l_1, l_2; b) varied frame diameter b and c) varied angle α at the frame connection level.

to be varied to fulfil different tasks. Since the main drawback of parallel mechanisms is the small workspace-to-installation-space ratio, dynamic reconfiguration type I facilitates the passing of singularties. Hence, several workspaces are combined during operation and the useable overall workspace is enlarged. The different workspaces are mathematically defined by different solutions of the direct kinematic problem (DKP) (assembly modes) or the indirect kinematic problem (IKP) (working modes). Each mode constitutes its own workspace. The workspaces of different working modes are divided by configuration space singularities, where assembly mode workspaces are distinguished by actuator singularities. In order to determine each of the possible configurations a vector **k** of binary configuration parameters (+1 or -1) has been established by Budde (2008) with

$$\mathbf{k} = (k_1, k_2, ...) = (\mathbf{k}_{IKP}, \mathbf{k}_{DKP}), \tag{8}$$

where k_{IKP} expresses the different solutions of the IKP and k_{DKP} denotes the different set of drive positions (solutions of the DKP). Each sprocket chain has two solutions of the IKP. Hence, a parallel mechanism with n_{SC} sprocket chains and n_{DKP} solutions of the DKP can have $2^{n_{SC}} \cdot n_{DKP}$ theoretical configurations, whereas not every solution is a real one and therefore relevant for the reconfiguration strategy. In order to use different workspaces or configurations generally (several) change-over configurations has to be captured to reconfigure the mechanism.

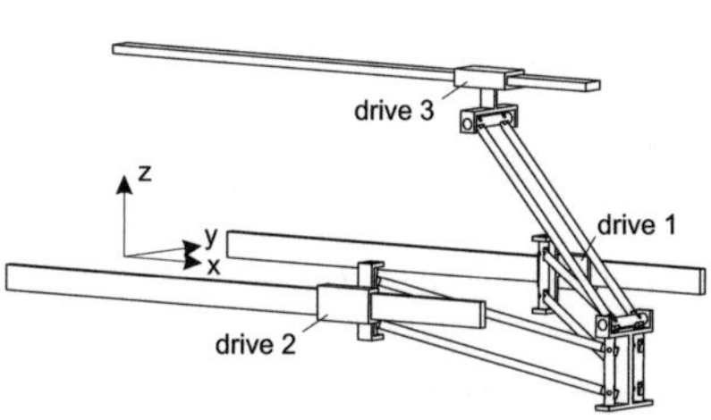

Fig. 9. Kinematic structure of the TRIGLIDE robot.

The described approach of DR I (passing through singularities to enlarge the workspace) should now be demonstrated using the example of the CRC 562 TRIGLIDE robot. The robot is based on the Linear-Delta structure. It consists of three linear driven kinematic chains, which guide the working platform. Each chain has a parallelogram structure in order to keep the end-effector platform in a constant orientation. This arrangement allows three DoF, additionally a fourth rotary axis can be mounted at the end-effector platform. Thus, a hybrid kinematic structure occurs. In Figure 9 the kinematic structure of the TRIGLIDE robot without the rotary axis is shown.

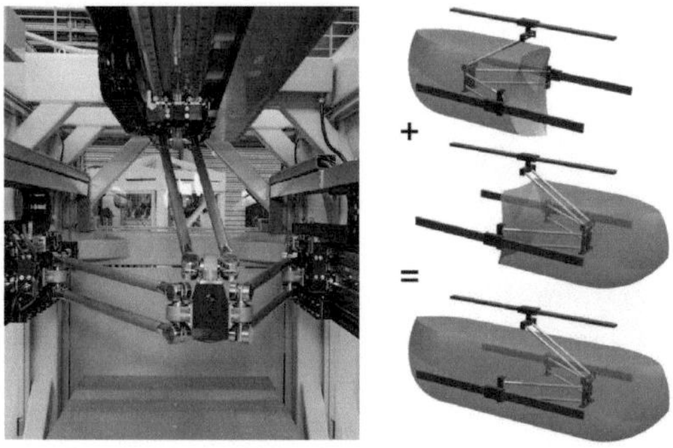

Fig. 10. Prototype of the TRIGLIDE robot (left); Usable working-configurations and resulting overall workspace (right).

In case of the TRIGLIDE robot the change between different configuration workspaces is accomplished by passing through singularities. Implemented strategies to go through the different singularity type are described in Budde (2010). In Figure 10 the prototype of the TRIGLIDE robot is shown as well as the desired main-workspaces and the resulting overall workspace.
As mentioned before the so called change-over configurations have to be captured in order to reach the main-workspaces, which are mostly capable to fulfil requirement spreads regarding to workspace. Several configurations are shown in Figure 11. The presented approach to reconfigure the TRIGLIDE robot leads to challenging tasks in control. The necessary steps and the adapted control strategy is explained in Maass (2006).
To complete the explanation of the named reconfiguration approaches for parallel robots, the next paragraph deals with *DR II* with a focus on adaptive machine elements.

5.3 Dynamic reconfiguration type II

Unlike the aforementioned reconfiguration strategies, the adjustment of geometric parameters is the subject of *DR II*. Here, the variation of strut length or the modification of kinematic DoF by joint couplings or lockings e.g. in order to avoid singularities or dislocate them are reasonable reconfiguration scenarios. Since this reconfiguration strategy demands no shut-down of the system, specific machine elements are required. In Schmitt (2010) an adaptive revolute joint is presented. Besides the basic functions of a passive joint, this revolute joint enables adaption of friction and stiffness. This functionality is based on the so-called quasi-statical clearance adjustment in the bearings using piezoelectric actuators Stechert (2007). This clearance modification between the inner and outer ring leads to an adaption of friction and stiffness (see Fig. 12 a)). Depending on the previous friction moment caused by the preload of the bearings the realized prototype enables continuous locking of rotary DoF by friction magnification of 440 percent [Inkermann (2011)]. In Figure 12 the first prototype of the adaptive revolute joint is illustrated.
In order to demonstrate a reconfiguration strategy of parallel mechanisms in the sense of *DR II* with an adaptive machine element a simple R̲RRRR̲-mechanism is further considered, following Schmitt (2010). The R̲RRRR̲-mechanism is a fully closed-loop planar parallel

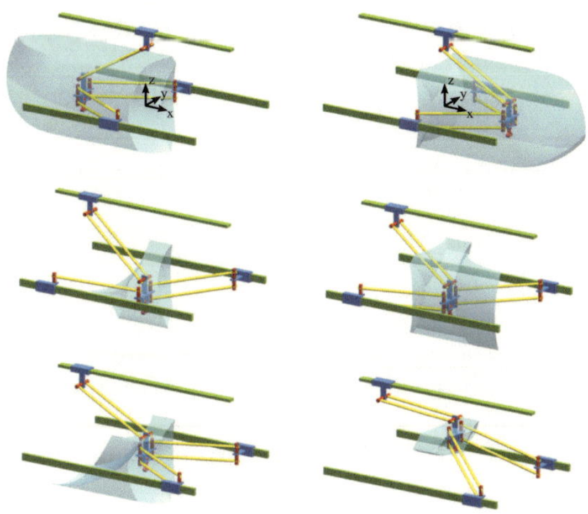

Fig. 11. Main-workspaces in working configuration (top); Selected changeover configurations (middle, buttom) of the TRIGLIDE robot.

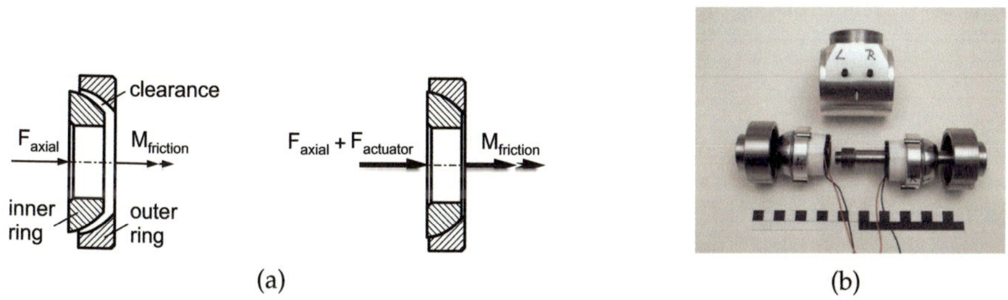

Fig. 12. Effect principle and realized prototype of the adaptive revolute joint used for dynamic reconfiguration type II.

manipulator with two DoF. The cranks at the base points A_1 and A_2 are actuated. The two passive rods L_{12} and L_{22} are coupled to each other by a passive revolute joint in the end-effector point C. The kinematic chain originating at base A_1 and the corresponding crank is connected by the described adaptive revolute joint, which is marked by ⊗ (see Figure 13).

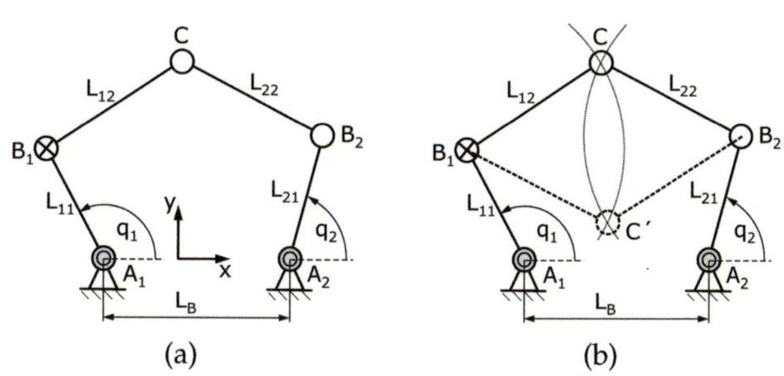

Fig. 13. Kinematic scheme and geometries of the RRRRR-mechanism (a) and solutions of the DKP (b).

The kinematic description of the mechanism with its two closed loop chains $i = 1,2$ can be geometrically done with the cartesian end-effector coordinates $\mathbf{X} = [x_c, y_c]^T$, the base coordinates $\mathbf{A} = [x_{Ai}, y_{Ai}]^T$, which can be derived by L_B, the actuator angles $\mathbf{q_i}$ and the given geometric parameters L_{ii} with respect to the base frame $\{0\}$. In eq. 9 the kinematic describtion of the mechanism is given, considering the parameters and variables shown in Figure 13 (a).

$$\mathbf{F} = \left[\begin{bmatrix} x_c \\ y_c \end{bmatrix} - \begin{bmatrix} x_{Ai} + cos(q_i) \cdot L_{i1} \\ y_{Ai} + sin(q_i) \cdot L_{i1} \end{bmatrix}\right]^2 - {L_{i2}}^2 = 0. \tag{9}$$

Expression 9 provides a system of equations which relate the end-effector coordinated $\mathbf{X}$ to the actuator coordinates $\mathbf{q_i}$. The equation for the direct kinematic problem (DKP) has two solutions, in case of the $\underline{R}RRR\underline{R}$-mechanism. They constitute to different configurations of the mechanism (see Figure 13 (b)). Changing these configurations an actuator singularity occurs, when moving from position C to C'. In general actuator singularities can occur in the workspace or separate workspaces into different areas.
In order to reconfigure the mechanism or to use these different workspaces adequate strategies have to be developed, in case of *DR II* using the adaptive revolute joint. Assembling the adaptive revolute joint in point B_1 (see Fig. 13 (a)) and blocking it entirely, it is possible to treat the mechanism as a $\underline{R}RRR$ 4-bar structure, with one actuated revolute joint at base point A_1. According to the theorem of GRASHOF different kinds of 4-bar mechanisms exist, depending on the geometric aspect ratio of the links [Kerle (2007)]:

$$\begin{aligned} l_{min} + l_{max} &< l' + l'' \quad \text{ability to turn around} \\ l_{min} + l_{max} &= l' + l'' \quad \text{ability to snap-through} \\ l_{min} + l_{max} &> l' + l'' \quad \text{no ability to turn around.} \end{aligned}$$

l_{min} and l_{max} are the lengths of the shortest and the longest link in the structure and l′, l″ are the lengths of the remaining links. In case of the shown 5-bar mechanism with the adaptive revolute joint, one link length can be adjusted according to the desired mechanism property. The resulting length L_R can be calculated by

$$\mathbf{L}_R = \sqrt{L_{12}^2 + L_{11}^2 - 2 \cdot L_{12} \cdot L_{11} \cdot cos(\Theta_1)}. \tag{10}$$

Beside the aforementioned possibility to reconfigure the robot, it is feasible to develop strategies to pass through an actuator singularity in an assured manner by blocking and releasing the mechanism at point B_1 by means of the adaptive revolute joint. This reconfiguration strategy is shown in Fig. 14.
The aspect ratios in the example are chosen by $L_{i1}/(L_{i2}, L_B) = 3/4$. If the mechanism is close to a singular position (Fig. 14.1), the adaptive revolute joint will be locked, ($\theta_1 = 40°$). The resulting 4-bar mechanism moves on a path, which is defined by a segment of a circle of radius L_R with center at A_1 through the singular position (Figure 14.2). Subsequently, the mechanism is in another configuration, which is not a singular position (Fig. 14.3) and the adaptive joint can be released.

5.4 Case studies

In the previous paragraphs different approaches to reconfigure parallel robotic systems were explained on a theoretical level. In order to give examples and fields of application two

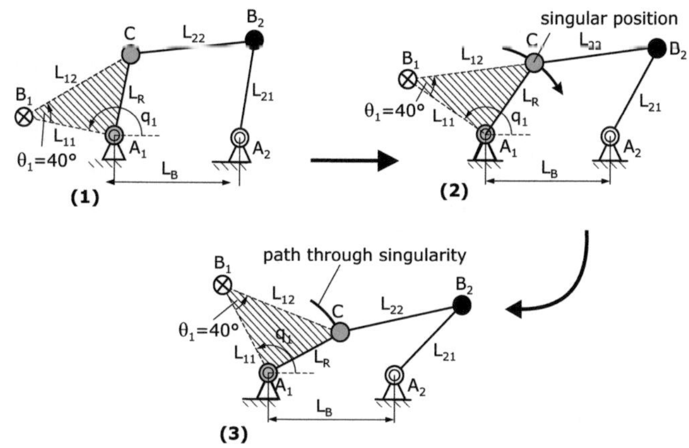

Fig. 14. Strategy of passing through singularity by blocking the adaptive joint.

case studies are explained and the benefits of reconfigurable robots are shown. The first one deals with the reconfiguration of a pick-and-place robot to an assembly robot, using the approach of static reconfiguration. Another case study is the classification of solar cells under consideration of their efficiency. This case study will illustrate an example of dynamic reconfiguration of type I.

5.4.1 Example I - reconfiguration of a "'pick-and-place"' robot into an assembly robot

In order to highlight the introduced procedure and the concept of static reconfiguration in this section case study is presented. Therefore, a "'pick-and-place"' robot should be reconfigured for an assembly task, while handling object and DoF of the system remain the same. To demonstrate the reconfiguration concept the kinematic structure described in section 5.1 is used. The use case assembly is characterised by a small workspace. However, higher process forces and accuracy are required. At the same time the performance (in this case numbers of assembly cycles per minute) compared with the "'pick-and-place"' task is not such important. The lifetime cost should be minimal. With regard to the introduced reconfiguration modules (see section 5.1), a change of the rack radius (RM_3) leads to a meaningful reconfiguration concept.

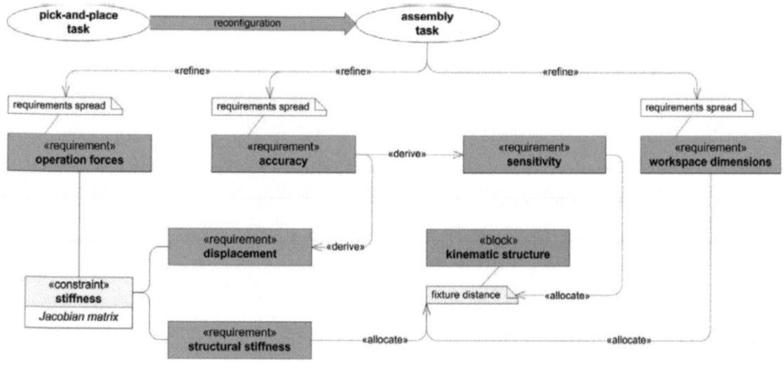

Fig. 15. Relation network for the static reconfiguration of a tow DoF parallel mechanism.

As can be seen in Figure 15 with the new use case the requirements *process forces*, *accuracy* and *workspace dimension* are refined. Since the workspace is directly influenced by the rack radius, the accuracy is influenced via the sensitivity. For increasing rack radius sensitivity in x-direction also increases and leads to higher accuracy in this direction. At the same time accuracy decreases in y-direction. In addition, the accuracy is influenced by the elastic deformation of the structure. The dimension of the deformation correlates with the value of the process forces and the stiffness of the structure. For the same stiffness higher process forces result in higher deformation and, therefore, decreased accuracy. This leads to a goal conflict which can be reduced with higher stiffness of the structure. At the same time the stiffness can be increased due to an adaption of the rack radius. Here, a higher rack radius results in a higher stiffness.

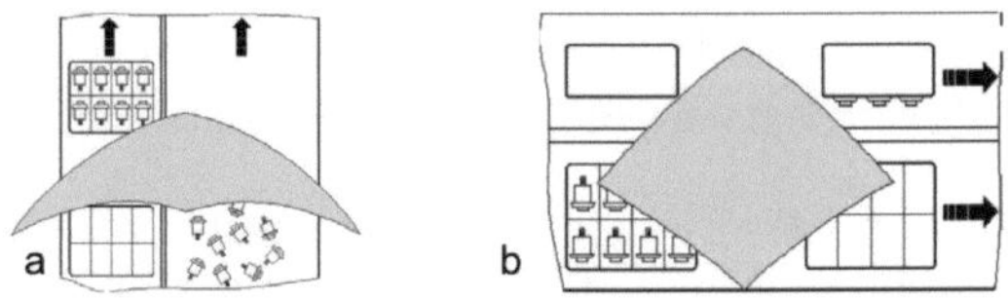

Fig. 16. Exemplary use cases for reconfiguration of a "'pick-and-place"' robot (a) into an assembly robot (b).

Static Reconfiguration from "'pick-and-place"' to assembly system in this case is carried out by increasing the rack radius. Figure 16 highlights the reconfiguration, showing the two use cases. In the first use case push-buttons are delivered via a first conveyer belt and placed into a container on a second belt. For this reason the workspace is elongated. After reconfiguration the same push-buttons are taken from a container and assembled into a casing. As the push-buttons provide a snap-in fastening operational forces in y-direction are needed and forces in x-direction might appear due to alignment deviations. Hence, the second configuration complies with the demanded changes.

5.4.2 Use case II - solar cell classification

The end-of-line process-step within the fabrication of crystalline solar cells is the classification of the cells according to their efficiency. Here, the "'flash-test"' is conducted, where each cell is lighted by a defined flash (time, brightness per area) comparable to the sun irradiation. The result determines the quality class of a cell. In order to sort the cells according to their class after the flash-test, handling devices are necessary. Due to the high number of classes, which are physically represented as square boxes, the handling robot has to provide a relatively huge workspace. Additionally, the requirement of cycle time about 1 sec. per "'pick-and-place"' operation is very challenging. This combination makes the application of parallel robots feasible, if it is possible to overcome the poor workspace-to-installation-space ratio. The process flow of the classification can be summarized as following:

1. Conveying the cells to the classifying facility via feed band
2. Determination of the efficiency value (flash test)
3. Define the right box/position of the robots' end-effector
4. Picking up the cell with the robot
5. Position the cell at the determined box according to the efficiency value

6. Drop the cell in the box
7. Pick up the next cell $\rightarrow$ return to process step 3

The values of the cell efficiency are subject to a distribution e.g. Gaussian, which means that a certain percentage rate (depends on the standard deviation σ) of the classes corresponding to the cell efficiency are distributed with a dedicated expectation value of μ. The remaining classes, which belong to

$$\mu \pm n_d \cdot \sigma, \tag{11}$$

where n_d constitutes the desired interval of σ.
Under this assumption the concept of DR I, the TRIGLIDE robot makes use of the two workspaces. The high frequented boxes to classify the solar cells are located in the first, the less ones in the second workspace. Due to the required time T_R to change the configuration, it is not efficient to vary workspace every cycle. Therefore, the proposed box distribution according to the expectation value μ is feasible. In Figure 17 (a) the workspaces with the corresponding intervals of the Gaussian distribution and in (b) the concept of the storage boxes are shown. The indices of the boxes c_{mno} describe the actual workspace (m), the column (n) and the row (o), so that the position is dedicated.

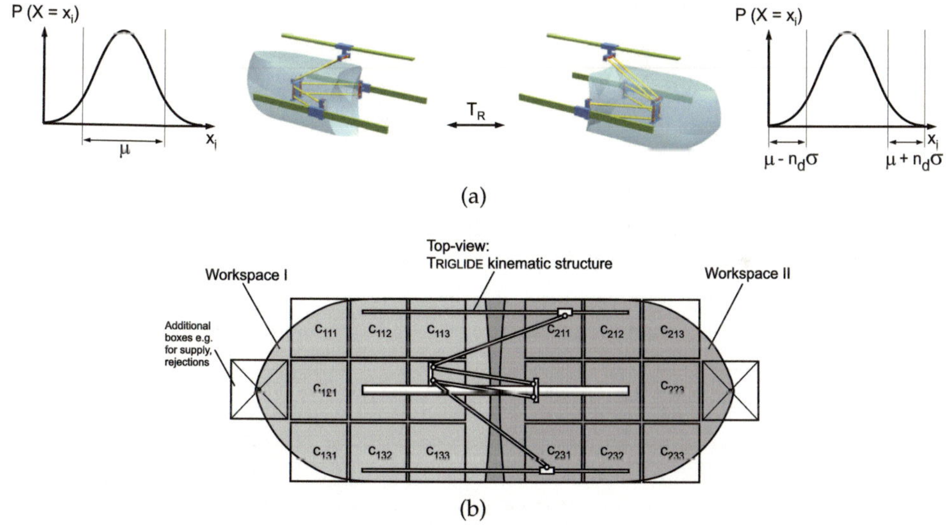

Fig. 17. Exemplarily probability distribution of supplied solar cells and used workspaces (a) and top view of a feasible conceptual design of a classification facility with the TRIGLIDE robot without periphery (b).

Since the introduced application examples highlight the advantages of reconfigurable parallel robotic systems to fulfil changing requirement, specific restrictions have to be considred in order to apply reasonable reconfiguration concepts. For this reason in the following section major aspects for the assessment and their interrelations to the presented static and dynamic reconfiguration strategies are proposed.

6. Assessment of reconfiguration concepts

As pointed out before the reconfiguration of robotic systems should not be considered as an end itself. It should rather follow an aim and should be subject of a systematic analysis. According to Steiner [(Steiner (1998), Steiner (1999))] changeability (reconfiguration) may not be cost efficient for systems which:

- are highly expedient, short life systems without needed product variety,
- are highly precendented systems in slowly changing markets and no customer variety,
- are insensitive to change over time, and
- are developed for ultrahigh performance markets with no performance loss allowables.

However, different use cases within the life time (e.g. handling muffins, packing croissants) and specific secondary functions (e.g. gripping principle) call for reconfiguration of robotic systems. Demands of high life-cycle times emphasize this need. In order to support the assessment of the before mentioned reconfiguration concepts a set of distinguishing aspects was derived. These aspects point out for which type of requirement spread a static or dynamic reconfiguration should be considered. Furthermore, the complexity of each reconfiguration approach is taken into account considering:

- *Duration time* characterizes the period needed to adapt the robotic systems performance, both manually and automatically.
- *Mechanical complexity* indicates the complexity of mechanical modifications and additional components, including costs to realize a reconfigurable robotic system.
- *Computational complexity* indicates the complexity of computational modifications and additional algorithms, including costs to realize a reconfigurable robotic system.
- *Parameter range* characterizes the potential of a reconfiguration approach to change several system properties e.g. workspace dimension in a wide range.
- *Robustness* indicates the vulnerability of a reconfiguration concept due to unforeseen operating conditions e.g. varying payload

The matrix in Figure 18 indicates which reconfiguration strategy is contributing to what reconfiguration-relevant requirement, see section 4.2. In applying the different approaches it is important to show up more detailed realizations. For this reason different reconfiguration concepts are stated following the introduced concepts. For instance, static reconfiguration can be realized by different strut kits and varying fixture distances. Relevant reconfiguration concepts are given for each approach.

While static reconfiguration concepts do not demand specific computational adaption and have high potencies to change several system properties such as workspace and accuracy, long time is needed to carry out the reconfiguration. The system has to be switched off, in order to rearrange several components. Thus, static reconfigurations approaches offer robust solutions since the configurations are changed manually. Regarding to dynamic reconfiguration concepts duration time is the essential advantage. However, to realize dynamic reconfiguration of a parallel robotic system specific components (DR II) have to developed or challenging control tasks have to be solved (DR I). Besides this in particular DR II features high potential to affect system properties.

This brief assessment of the introduced reconfiguration approaches should not be seen as a strict selecting tool. It rather helps to reason different concepts in early design phases of parallel robotic systems.

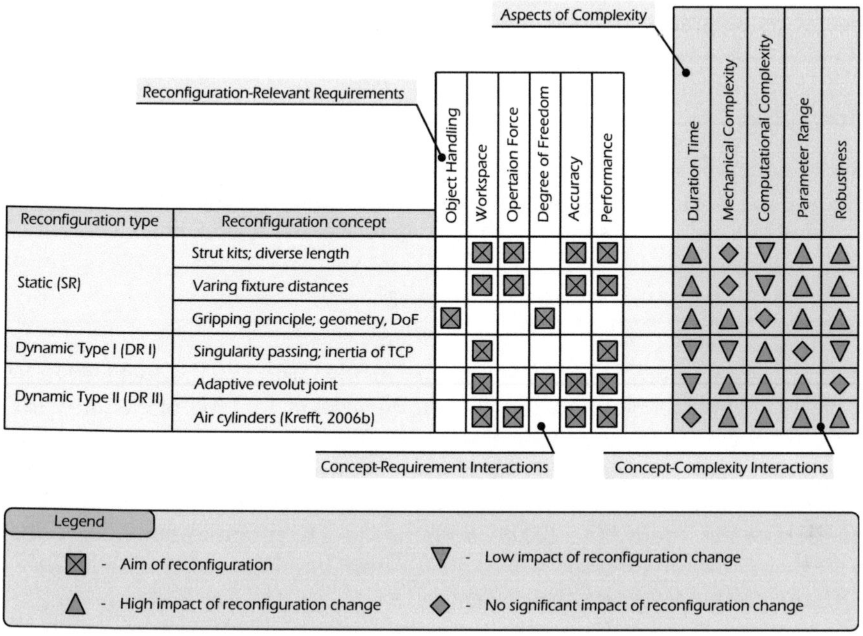

Fig. 18. Reconfiguration-approach-matrix.

7. Conclusion and further research

The need to increase flexibility of production systems via reconfiguration is mainly driven by the customers of the end product. Rising demands for individual products and shorter periods for product changes force the manufacturers to think about reconfigurable production equipment. In order to present ideas and strategies for reconfigurable robotic systems this contribution firstly shows the general meaning of reconfiguration and gives a brief literature review by classifying the most common types into offline and online approaches.

Such as parallel robots are in focus in this chapter, this special class of robotic systems was explained and the essential benefits as well as disadvantages, in particular the occurrence of singularities within the workspace and at its borders were introduced. Based on high potential of parallel robots to cover a huge range of requirements, an adequate requirements management referring to robotic reconfiguration has been introduced. Thereby, the identification of requirement spreads and the derivation of reconfiguration parameters were considered. The main part of the contribution examines the distinctive static and dynamic (type I and II) reconfiguration approaches of parallel robotic system. Each strategy is first discussed theoretically before two case studies are presented. In order to assess the different reconfiguration approaches, a matrix-based scheme is proposed, which serves the designer as a decision support.

Potential aims of further research activities should deal with the validation of the adapted development procedure with respect to reconfiguration issues. Further, regarding parallel robots, the proposed reconfiguration strategies should be surveyed to their transferability to other parallel structures. Another aspect, to be investigated, is the control implementation of the *SR* or *DR* concepts, in order to reduce efforts of changes. The treatment of an entirely changed kinematic structure after reassembling its mechanical components (*SR*) or the realization of passing through respectively to avoid singular positions (*DR*) are challenging research fields to deal with.

8. Acknowledgments

The authors gratefully thank the Deutsche Forschungsgemeinschaft (DFG) for supporting the CRC 562 "'Robotic Systems for Handling and Assembly"' and the Project "'Application-Oriented Analysis of High Dynamic Handling of Solar Cells with Parallel Robots"'.

9. References

Anggreeni, I., van der Voort, M.C. (2008). Classifying Scenarios in a Product Design Process: a study towards semi-automated scenario generation. In: Proc. of the 18th CIRP Design Conference - Design Synthesis, Enschede.

Boudreau, R. and C.M. Gosselin (2001). La synthèse d'une plate-forme de Gough-Stewart pour un espace atteignable prescrit. Mechanism and Machine Theory 36. pp. 327-342.

Brouwer, M., van der Voort, M.C. (2008). Scenarios as a Communication Tool in the Design Process: Examples from a Design Course. In: Proc. of the 18th CIRP Design Conference - Design Synthesis, Enschede.

Budde, C., Helm, M., Last, P., Raatz, A., Hesselbach, J. (2010). Configuration Switching for Workspace Enlargement In: Schütz, D., Wahl, F. (Ed.), Robotic Systems for Handling and Assembly, Springer-Verlag, Star Series, Berlin, pp. 175-189, ISBN 978-3-642-16784-3.

Budde, C., Last, P., Hesselbach, J. (2007). Development of a Triglide-Robot with Enlarged Workspace In: Proc. of the International Conference on Robotics and Automation, Rom, Italy, ISBN 978-1-4244-0602-9.

Budde, C. (2008). Wechsel der Konfiguration zur Arbeitsraumvergrösserung bei Parallelrobotern. Ph.D Thesis, TU Braunschweig, Vulkan, Essen, ISBN 978-3-8027-8747-8.

Clavel, R. (1991). Conception d'un robot parallel rapide a 4 degres de liberte, Ph.D. Thesis, EPFL, Lausanne.

Fisher, R., Podhorodeski, R.P., Nokleby, S.B. (2004). Design of a Reconfigurable Planar Parallel Manipulator. In: Journal of Robotic Systems Vol. 21, No. 12, (2004), pp. 665-675.

Franke, H.-J., Krusche, T. (1999). Design decisions derived from product requirements. In: Proc. of the 9th CIRP Design Conference - Integration of Process Knowledge into Design, Enschede, pp.371-382.

Frindt, M. (2001). Modulbasierte Synthese von Parallelstrukturen für Maschinen in der Produktionstechnik. Ph.D, TU Braunschweig, Vulkan, Essen, ISBN 978-3-8027-8659-4.

Hesselbach, J., Helm, M., Soetebier, S. (2002). Connecting Assembly Modes for Workspace Enlargement. In: Advances in Robotis, Kluwer Academic Publishers, 2002, pp. 347-356.

Inkermann, D., Stechert, C., Vietor, T. (2011). Einsatz multifunktionaler Werkstoffe für die Entwicklung einer adaptiven Gelenkachse. In: Proc. of 8. Paderborner Workshop Entwurf mechatronischer Systeme. HNI-Verlagsschriftenreihe Paderborn, 2011, pp. 231-244.

Jantapremjit, P., Austin, D. (2001). Design of a Modular Self-Reconfigurable Robot. In: Proc. of the 2001 Australien Conference on robotics and Automation. Syndney (Australia), 2001, pp. 38-43.

Ji, Z., Song, P. (1998). Design of a Reconfigurable Platform Manipulator. In: Journal of Robotic Systems Vol. 15, No. 6, 1998, pp. 341-346.

Jovane, F. et al. (2002). Design issues for reconfigurable PKMs. In: Proc. of the 4th Chemnitz Parallel Kinematic Seminar, Chemnitz, 2001, pp. 69-82.

Kerle, H., Pittschellis, R., Corves, B. (2007). Einführung in die Getriebelehre. Teubner Verlag, Wiesbaden, Germany, ISBN 978-3-8351-0070-1.

Koren, Y. et al. (1999). Reconfigurable Manufacturing Systems. Annals of the CRIP, Vol. 48, No. 2, 1999, pp. 527-540.

Krefft, M., Brüggemann, H., Herrmann, G. & Hesselbach, J. (2006). Reconfigurable Parallel Robots: Combining High Flexibility and Short Cycle Times. In: *Journal of Production Engineering*, Vol. XIII, No.1., pp. 109-112.

Lee, G.H. (1997). Reconfigurability Consideration Design of Components and Manufacturing Systems. In: The International Journal of Advanced Manufacturing Technology, Vol. 13, 1997, pp. 376-386.

Maass, J. et al. (2006). Control strategies for enlarging a spatial parallel robot's workspace by change of configuration. In Proc. of the 5th Chemnitz Parallel Kinematics Seminar, Chemnitz (Germany), 2006, pp. 515-530.

Merlet, J.P. (1997). DEMOCRAT: A DEsign MethOdology for the Conception of Robot with parallel ArchiTecture. In: Proceedings of the International Conference on Intelligent Robots and Systems, IROS, Grenoble France, pp. 1630-1636.

Merlet, J.-P. (2001). Parallel Robots - Solid Mechanisms and its applications, Vol. 74, ISBN 1402003854, Kluwer Academic Publishers, Dordrecht.

http://www.molecubes.org/

O'Brien, J.-F.; Wen, J.-T. (2001). Kinematic Control of Parallel Robots in the Presence of Unstable Singularities. In: Proceedings of IEEE International Conference on Robotics and Automation, Seoul, Korea, pp.354-359.

Park, F.C., Jin Wook Kim (1998). Manipulability and singularity analysis of multiple robot systems: a geometric approach, In: Proceedings of the IEEE International Conference on Robotics and Automation, Vol.2, pp.1032-1037.

Pritschow, G. (2000). Parallel Kinematic Machines (PKM) - Limitations and New Solutions. CIRP Anals-Manifacturing Technology. Vol. 49, No. 1, 2000, pp. 275-295.

Riedel, M., Nefzi, M., and Corves, B. (2010). Grasp Planning for a Reconfigurable Parallel Robot with an Underactuated Arm Structure. In: Mechanical Sciences, Vol. 1, 2010, pp. 33Ű42.

Riedel, M., Nefzi, M., Hüsing, M., and Corves, B. (2008). An Adjustable Gripper as a Reconfigurable Robot with a parallel Structure. In: Proc. of the 2nd Int. Workshop on Fundamental Issues and Future Research Directions for Parallel Mechanisms and Manipulators, 2008.

Roozenburg, N.F.M, Dorst, K., (1991). Some Guidelines for the Development of Performance Specifications in Product Design. In: Proc. of the 8th International Conference on Engineering Design ICED 91, pp.359-366.

Schmitt, J., Inkermann, D., Raatz, A., Hesselbach, J., Vietor, T. (2010). Dynamic Reconfiguration of Parallel Mechanisms. In: *EUCOMES - European Conference on*

Mechanism Design, Cluj-Napoca, Rumänien, Springer Berlin / Heidelberg, 2010, pp. 557-565, ISBN 978-90-481-9688-3.
Schmitt, J., Stechert, C., Raatz, A., Hesselbach, J., Vietor, T., Franke, H.-J. (2009). Reconfigurable Parallel Kinematic Structures for Spreading Requirements, In: Proc. of the 14th IASTED International Conference on Robotics and Applications (RA 2009), Cambridge.
Setchi, R-M. & Lagos, N. (2004). Reconfigurability and Reconfigurable Manufacturing Systems - State-of-the-art Review, In: *Proc. 2nd. IEEE International Conference on Industrial Informatics (INDIN)*, Berlin.
Stechert, C. (2010). Modellierung komplexer Anforderungen. Ph.D Thesis, TU Braunschweig, Verlag Dr. Hut, München, ISBN 978-3-86853-545-7.
Stechert, C., Franke, H.-J. (2009). Requirements Models for Collaborative Product Development. In: Proc. of the 19th CIRP Design Conference, Cranfield, pp.24-31.
Stechert, C., Franke, H.-J. (2009). Managing Requirements as the Core of Multi-Disciplinary Product Development. Cirp Journal of Manufacturing Science and Technology, Vol. 1, Is. 3, pp.153-158.
Stechert, C., Pavlovic, N., Franke, H.-J. (2007). Parallel Robots with Adaptronic Components - Design Through Different Knowledge Domains. In: Proc. of 12th IFToMM World Congress, Besancon, France, 2007.
Steiner, R. (1998). Systems Architectures and evolvability - definitions and perspectives. In: Proc. of the 8th Annu Symp. INCOSE, Vancouver, 1998.
Steiner, R. (1999). Enduring Systems Architectures. In: Proc. of the 10th Int. Symp. INCOSE, Brighton, 1999.
Theingi; Chen, I.M.; Angeles, J.; Chuan, Li. (2007). Management of Parallel-Manipulator Singularities Using Joint-Coupling. In: Advanced Robotics, Vol. 21, No.5-6, pp.583-600.
Weilkiens, T. (2008). Systems Engineering with SysML/UML - Modeling, Analysis, Design. Morgan Kaufmann. ISBN:9780123742742
Wenger, Ph., Chalbat, D. (1998). Workspace and Assembly Modes in Fully-Parallel Manipulators: A Descriptive Study. In: Proc. of the International Conference on Advances in Robotic Kinematics: Analysis and Control. Salzburg, 1998, pp. 117-126.
Woelkl, S., Shea, K. (2009). A Computational Product Model for Conceptual Design Using SysML In: Proc. of the ASME 2009 International Design Engineering Technical Conferences & Computers and Information in Engineering Conference IDETC/CIE 2009, San Diego.
Wurst, K.-H. & Peting, U. (2002). PKM Concept for Reconfigurable Machine Tools. Proc. of the 4th Chemnitz Parallel Kinematic Seminar, Chemnitz, 2002, pp. 683-696.
Yang, G., Chen, I-M., Lim, W.K., Yeo, S.H. (2001). Kinematic Design of Modular Reconfigurable In-Parallel Robots. In: Autonomous Robots No. 10, 2001, pp. 83Ű89.
Yim, M., et al. (2007). Modular Self-Reconfigurable Robot Systems - Challenges and Opportunities for the Future -. In: IEEE Robotics and Automation Magazine. March 2007, pp. 43-52
Yim, M.; Duff, D.G.; Roufas, K.D. (2000). "PolyBot: a modular reconfigurable robot," Robotics and Automation, 2000. In: Proc of ICRA '00., pp.514-520.

Part 3

Vision and Sensors

20

Real-Time Robotic Hand Control Using Hand Gestures

Jagdish Lal Raheja, Radhey Shyam, G. Arun Rajsekhar and P. Bhanu Prasad
Digital Systems Group, Central Electronics Engineering Research Institute (CEERI)/Council of Scientific & Industrial Research (CSIR), Pilani, Rajasthan India

1. Introduction

Interpretation of human gestures by a computer is used for human-machine interaction in the area of computer vision [3][28]. The main purpose of gesture recognition research is to identify a particular human gesture and convey information to the user pertaining to individual gesture. From the corpus of gestures, specific gesture of interest can be identified[30][36], and on the basis of that, specific command for execution of action can be given to robotic system[31]. Overall aim is to make the computer understand human body language [14], thereby bridging the gap between machine and human. Hand gesture recognition can be used to enhance human–computer interaction without depending on traditional input devices such as keyboard and mouse.

Hand gestures are extensively used for telerobotic control applications [20]. Robotic systems can be controlled naturally and intuitively with such telerobotic communication [38] [41]. A prominent benefit of such a system is that it presents a natural way to send geometrical information to the robot such as: left, right, etc. Robotic hand can be controlled remotely by hand gestures. Research is being carried out in this area for a long time. Several approaches have been developed for sensing hand movements and controlling robotic hand [9][13][22]. Glove based technique is a well-known means of recognizing hand gestures. It utilizes sensor attached mechanical glove devices that directly measure hand and/or arm joint angles and spatial position. Although glove-based gestural interfaces give more precision, it limits freedom as it requires users to wear cumbersome patch of devices. Jae-Ho Shin et al [14] used entropy analysis to extract hand region in complex background for hand gesture recognition system. Robot controlling is done by fusion of hand positioning and arm gestures [2][32] using data gloves. Although it gives more precision, it limits freedom due to necessity of wearing gloves. For capturing hand gestures correctly, proper light and camera angles are required. A technique to manage light source and view angles, in an efficient way for capturing hand gestures, is given in [26]. Reviews have been written on the subsets of hand postures and gesture recognition in [6]and [18]. The problem of visual hand recognition and tracking is quite challenging. Many early approaches used position markers or colored bands to make the problem of hand recognition easier, but due to their inconvenience, they cannot be considered as a natural interface for the robot control [24]. A 3D Matlab Kinematic model of a PUMA [1][4][16] robot, is used for executing actions by hand gesture[39]. It can be extended to any robotic system with a number of specific commands suitable to that system [17].

In this chapter, we have proposed a fast as well as automatic hand gesture detection and recognition system. Once a hand gesture is recognised, an appropriate command is sent to a robot. Once robot recieves a command, it does a pre-defined work and keeps doing until a new command arrives. A flowchart for overall controlling of a robot is given in figure 1.

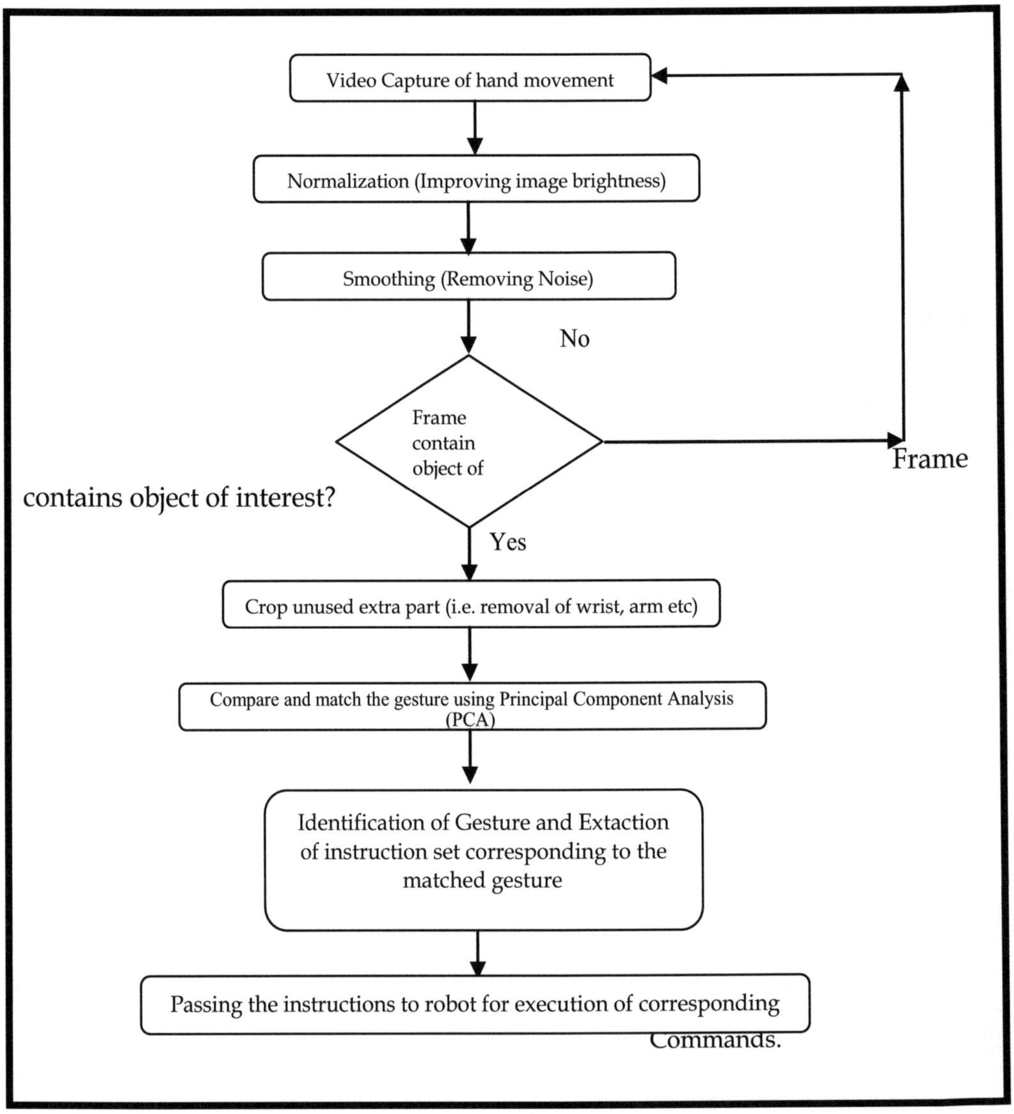

Fig. 1. Flow diagram for robot control using hand gestures

2. Methodology

The proposed methodology involves acquisition of live video from a camera for gesture identification. It acquire frames from live video stream in some given time interval[15]. In this case frame capture rate for gesture search is 3 frames per second. The overall proposed technique to acquire hand gestures and to control robotic system using hand gestures is divided into four subparts:

- Image acquisition to get hand gestures.
- Extracting hand gesture area from captured frame.
- Determining gesture by pattern matching using PCA algorithm.
- Generation of instructions corresponding to matched gesture, for specified robotic action.

2.1 Image acquisition to get hand gestures

First step is to capture the image from video stream. It is assumed that while capturing video, black background with proper light soure is used. To minimize number of light sources, arrangement of the light sources and camera is made as shown in figure 2.

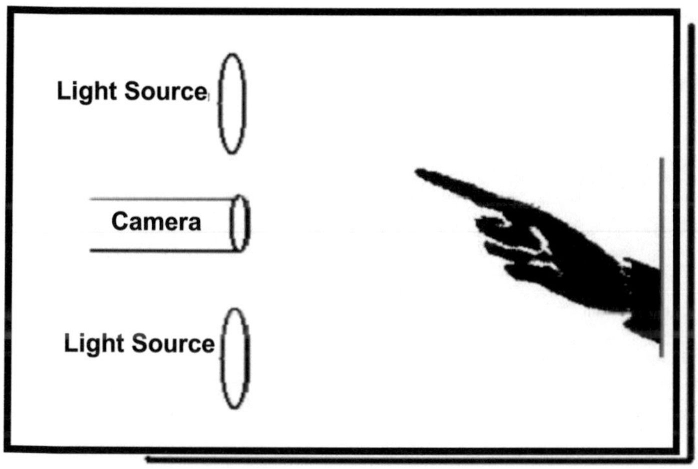

Fig. 2. Arrangement of light sources for image aquisition

From video stream one frame is captured every 1/3 of a second. Our aim is to identify the frame that contains hand gesture shown by human. For this we are searching the frame in which there is no or least movement. Required frame is identified by comparing three continuously captured frames. *Motion parameter* is determined for each frame by counting total pixels of mismatch. If *motion parameter* is less than the specified threshold value, it is considered as a frame having least movement of object i.e. the hand of the person.

Analysis of frames to find the frame of interest is carried out by converting the captured frame into a binary frame[23]. Since binary images have values of one or zero., differences of white pixels between newly captured frame and two previously captured frames are determined and they are added together to find *motion parameter*. XOR function is used to find the differences in white pixels [5][15].

If *frame1, frame2,* and, *frame3* are three matrices containing three frames captured in three continuous time slots respectively then:

$$fr1 = frame1\ XOR\ frame3 \quad (1)$$

$$fr2 = frame2\ XOR\ frame\ 3 \quad (2)$$

$$mismatch\_matrix = fr1\ OR\ fr2 \tag{3}$$

$$\text{Sum} = \sum_{i=1}^{r}\sum_{j=1}^{c} mismatch\_matrix[i][j] \tag{4}$$

$$\text{Motion parameter} = \frac{sum}{r * c} \tag{5}$$

Here r and c are the number of rows and columns in an image frame. Threshold value is set as 0.01. i.e. if total pixels of mismatch are less than 1% of total pixels in a frame, then it is considered as frame of interest. Required frame is forwarded for further processing and this module again starts searching for next frame of interest.

2.2 Extracting hand gesture area from frame of interest

It may happen that the frame, as shown in fig.3, with a gesture contains extra part along with required part of hand i.e. background objects, blank spaces etc. For better results in pattern matching, unused part must be removed. Therefore hand gesture is cropped out from the obtained frame. Cropping of hand gesture from the obtained frame contains three steps:

Fig. 3. Grayscale image

First step is to convert selected frame into HSV color space. As shown in fig.4, a HSV based skin filter was applied on the RGB format image to reduce lighting effects [8][1]. The thresholds for skin filter used in the experiment were as below:

$$\left.\begin{array}{c} 0.05 < H < 0.17 \\ 0.1 < S < 0.3 \\ 0.09 < V < 0.15 \end{array}\right\} \rightarrow \tag{6}$$

The resultant image was segmented to get a binary image of hand[37]. Binary images are bi-level images where each pixel is stored as a single bit (0 or 1). Smoothening was needed, as the output image had some jagged edges as clearly visible in figure 4(c). There can be some noise in the filtered images due to false detected skin pixels or some skin color objects (like wood) in background, it can generate few unwanted spots in the output image as shown in figure 4(e).

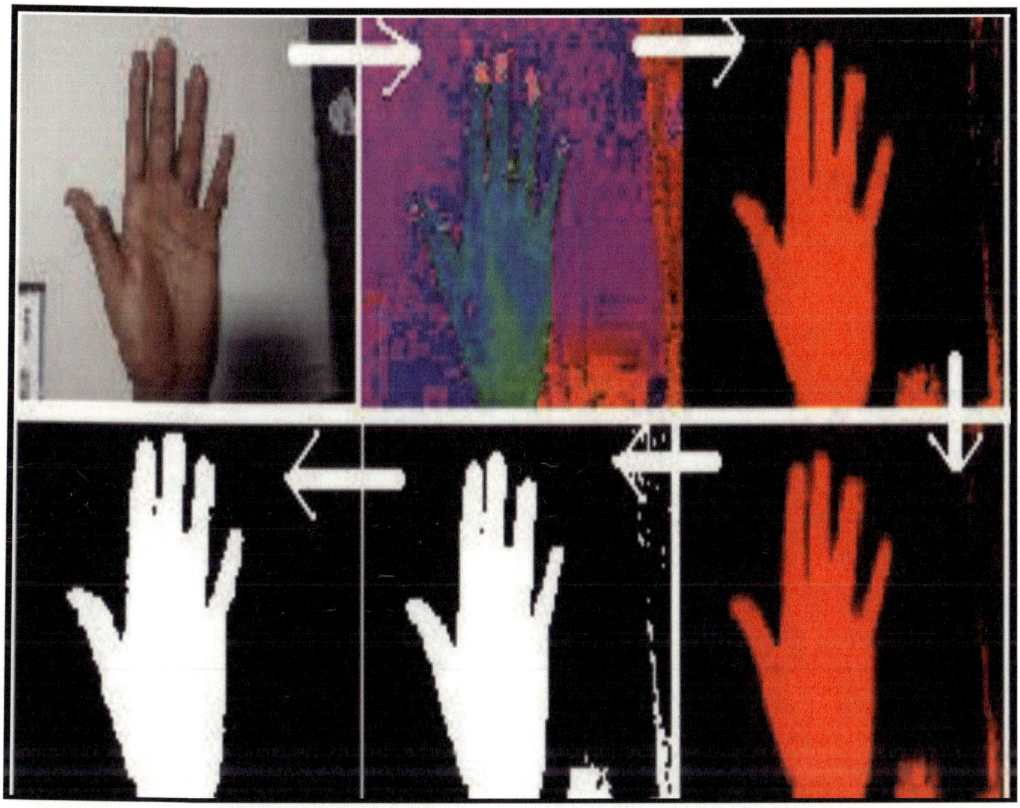

Fig. 4. Binary Image formation (a) Target image (b) HSV conversion (c) Filtered image (d) Smoothened image (e) Binary image (f) Biggest BLOB

To remove these errors, the biggest BLOB (Binary Linked Object) method was applied to the noisy image [24][25]. The error free image is shown in figure 5. The only limitation of this filter was that the BLOB for hand should be the biggest one. In this, masking background would be illuminated, so false detection is not possible from the background [7][10].

Second step is to extract object of interest from the frame. In our case, object of interest is the part of human hand showing gesture [29]. For this, extra part, other than the hand, is cropped out so that pattern matching can give more accurate results [12]. For cropping extra parts row and column numbers are determined, from where object of interest appears. This is done by searching from each side of binary image and moving forward until white pixels encountered are more than the offset value [25]. Experimental results show that offset value set to 1% of total width gives better result for noise compensation. If size of selected image is mXn then:

$$Hor\_offset = m/100 \tag{7}$$

$$Ver\_offset = n/100 \tag{8}$$

Min_col= minimum value of column number where total number of white pixels are more than *Hor_offset*.

Max_col= maximum value of column number where total number of white pixels are more than *Hor_offset*.

Min_row= minimum value of row number where total number of white pixels are more than *Ver_offset*.
Max_row= maximum value of row number where total number of white pixels are more than *Ver_offset*.

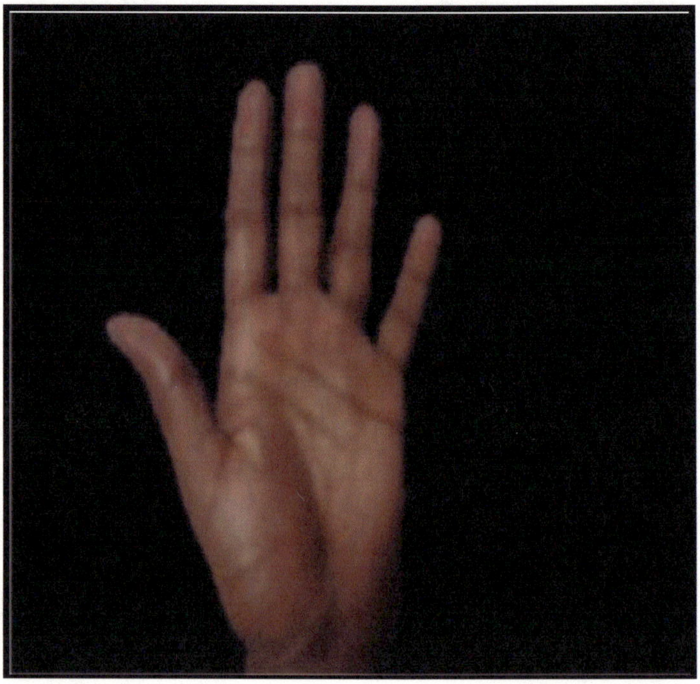

Fig. 5. Error free image

Figure 6 shows Min_row, Max_row, Min_col, and, Max_col for the selection of the boundary for cropping of hand.

Third step is to remove parts of hand not used in gesture i.e. removal of wrist, arm etc. As these extra parts are of variable length in image frame and pattern matching with gesture database gives unwanted results. Therefore, parts of hand before the wrist need to be cropped out.

Statistical analysis of hand shape shows that either we pose palm or fist, width is lowest at wrist and highest at the middle of palm. Therefore extra hand part can be cropped out from wrist by determining location where minimum width of vertical histogram is found. Figure 7.c and 7.d show global maxima and cropping points for hand gestures in figure 7.a and 7.b respectively.

Cropping point is calculated as:

Global Maxima = column number where height of histogram is highest (i.e. column number for global maxima as shown in figure 7).

Cropping point = column number where height of histogram is lowest in such a way that cropping point is in between first column and column of Global Maxima

If gesture is shown from opposite side (i.e. from other hand), then *Cropping point* is searched between column of *Global Maxima* and last column. Direction of the hand is detected using continuity analysis of object during hand gesture area determination. Continuity analysis shows that whether the object continues from the column of Global maxima to first column or to last column. i.e. whether extra hand is left side of palm or right side of palm.

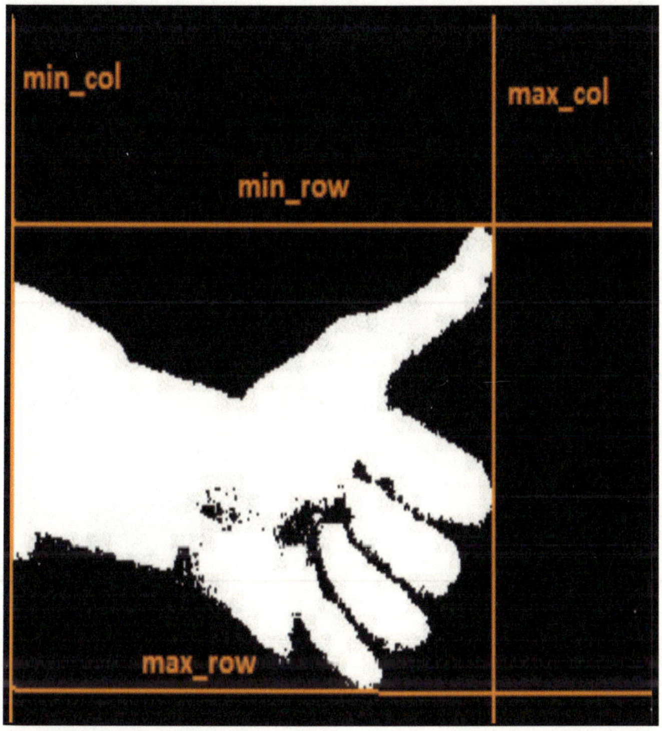

Fig. 6. Boundary selection for hand cropping

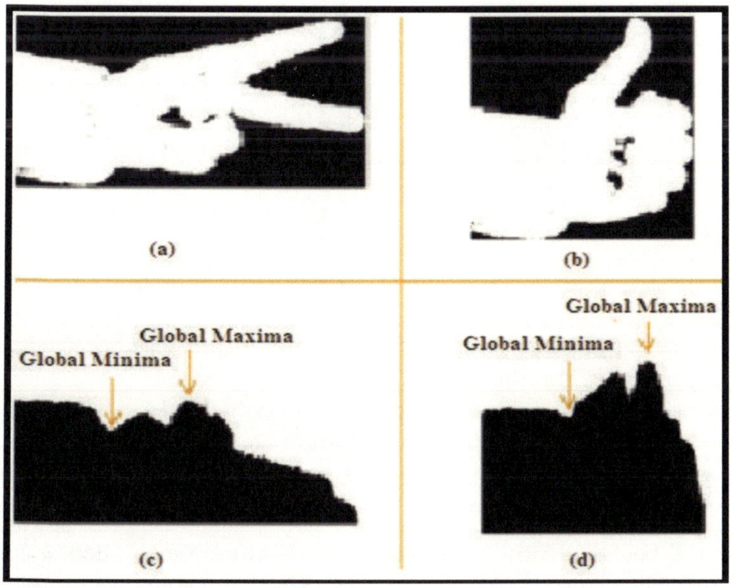

Fig. 7. Histogram showing white pixels in each column, with cropping point for hand gesture

2.3 Determining the gesture by resizing and pattern matching

Cropping results in a frame of 60x80 pixels, so that the gesture can be compared with the database. This particular size of 60x80 is chosen to preserve the aspect ratio of the original image, so that no distortion takes place.

Figure 8.b shows cropped hand gesture from figure 8.a. and figure 8.c. shows the result of scaling and fitting 8.b within 60X80 pixels. Figure 8.d represents final cropped and resized gesture of desired dimension.

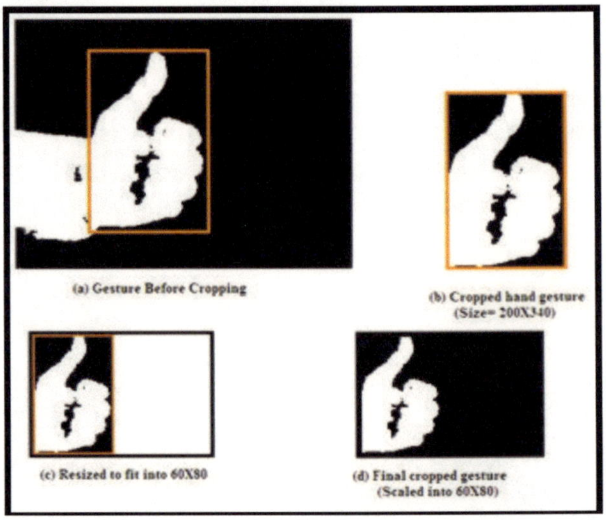

Fig. 8. Scaling cropped hand gesture to fit into desired dimension

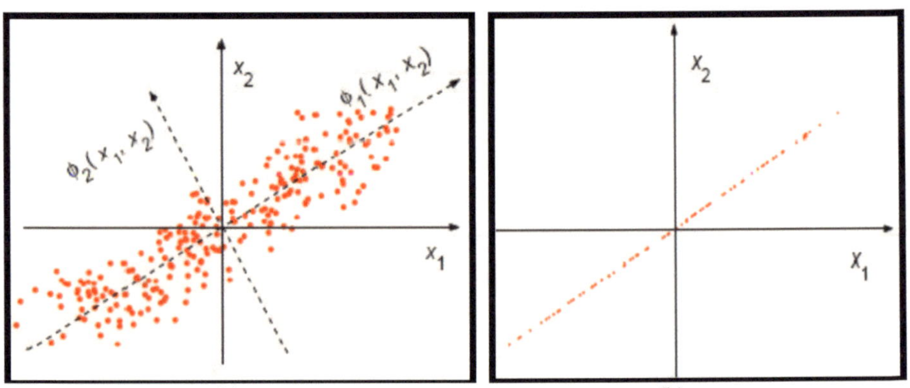

(a) (b)

Fig. 9. (a) PCA basis (b) PCA reduction to 1D

(a) The concept of PCA. Solid lines: the original basis; dashed lines: the PCA basis. The dots are selected at regularly spaced locations on a straight line rotated at 30degrees, and then perturbed by isotropic 2D Gaussian noise. (b) The projection (1D reconstruction) of the data using only the first principal component.

There are many different algorithms available for Gesture Recognition, such as principal Component Analysis, Linear Discriminate Analysis, Independent Component Analysis, Neural Networks and Genetic Algorithms.

Principal Component Analysis (PCA)[11] is used in this application to compare the acquired gesture with the gestures available in the database and recognize the intended gesture.

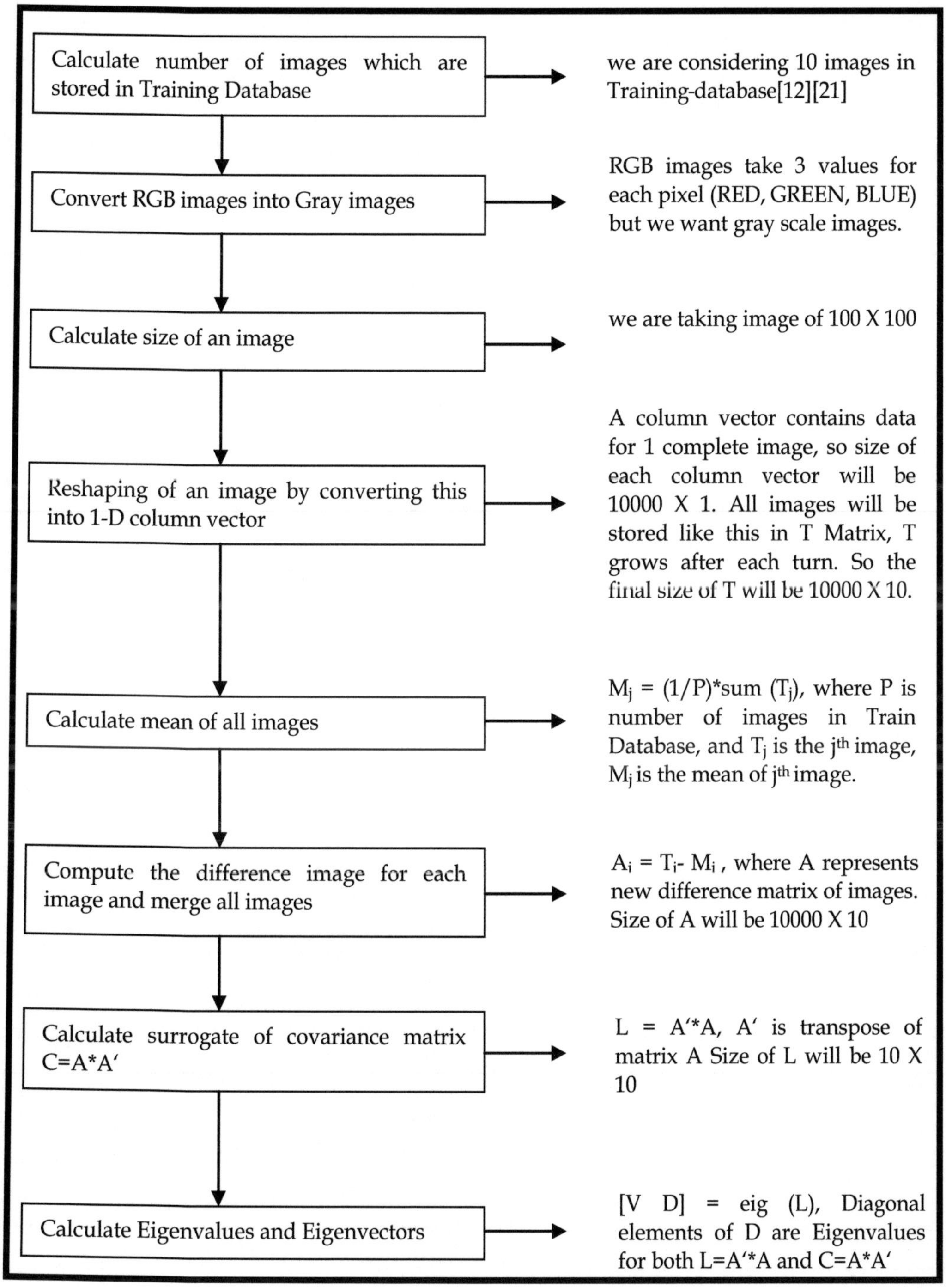

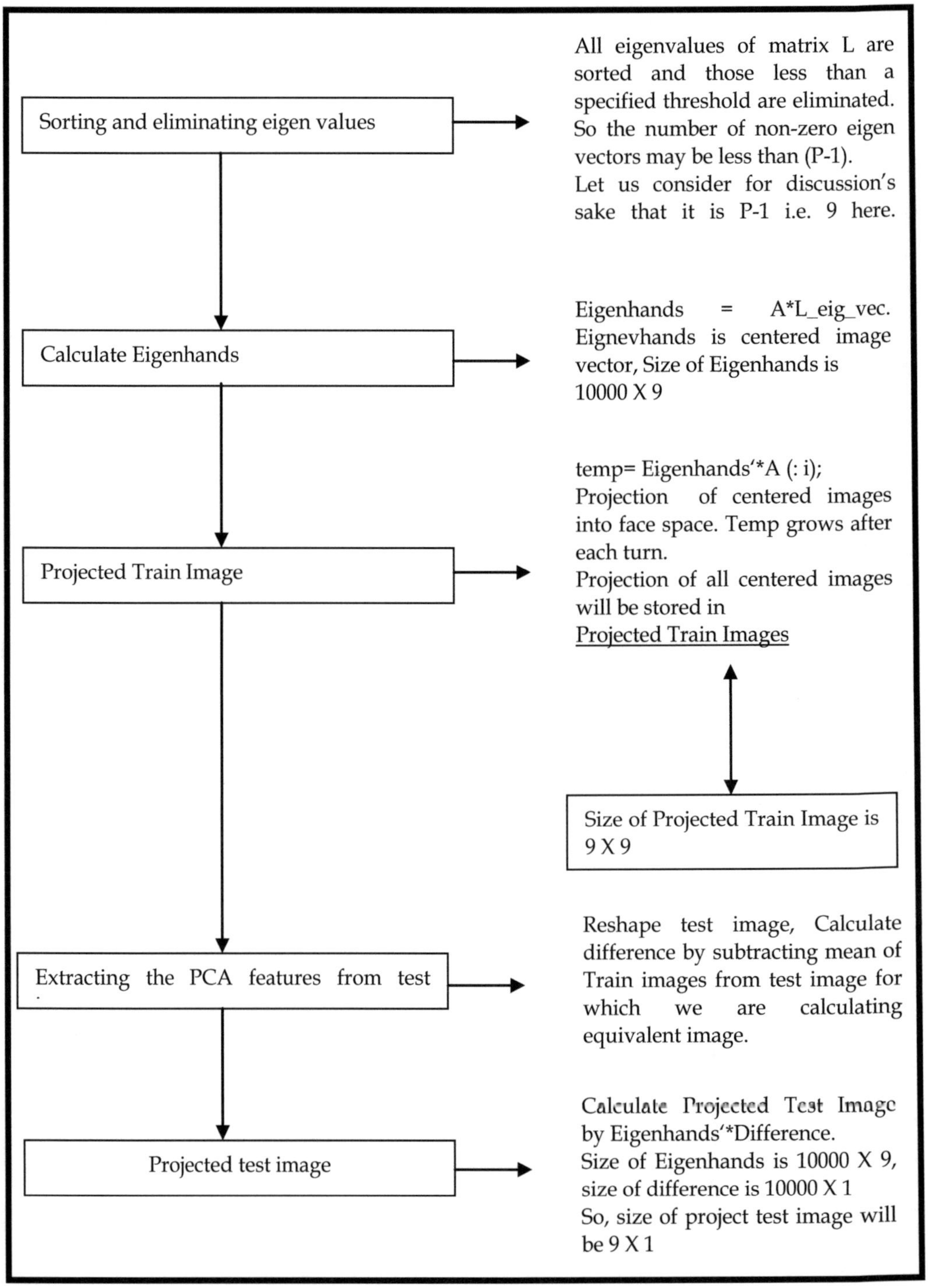

Fig. 10. PCA Algorithm steps for 10 Images

The concept of PCA is illustrated in Figure 9; The graph shows data that originally has two components along the x_1 and x_2 directions. This is now resolved along the Φ_1 and Φ_2 directions. Φ_1 corresponds to the direction of maximum variance and is chosen as the first principal component. In the 2D case currently being considered, the second principal component is then determined uniquely by orthogonality constraints; in a higher-dimensional space the selection process would continue, guided by the variances of the projections.

A comprehensive explanation of the above scheme for gesture recognition[19] within a database of ten images is given in figure 10. For simplicity of calculations and ease of understanding, each figure is taken to be of size of 100X100 pixels. This algorithms was tested on a PC platform.

2.4 Generation of control Instructions for a robot

Different functions corresponding to each meaningful hand gesture are written and stored in database for controlling the robotic Arm. For implementation of robotic arm control, PUMA robotic model has been chosen. Since this model was only available at the time of interface, hand gestures are used to control the various joints of the arm. However, these gestures can be used to control the robotic hand also. Altogether 10 different hand gestures are related to 10 different movements of robotic arm. Movement commands are written as a function in robot specific language. Whenever a gesture is matched with a meaningful gesture from the database, the instruction set corresponding to that gesture is identified and passed to robot for execution. In this way the robotic system can be controlled by hand gesture using live camera. Figure 11 shows movements of robotic hand corresponding to four different specific hand gestures.

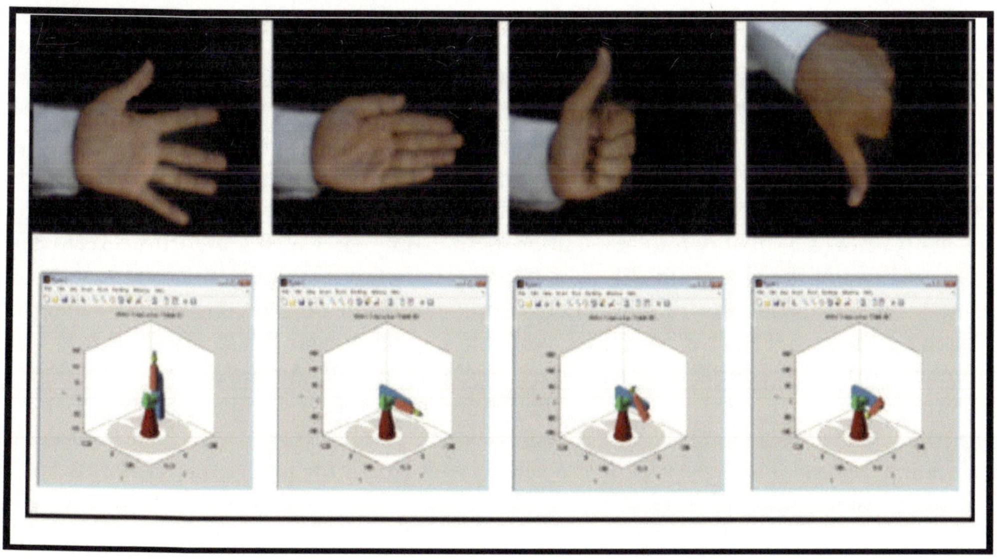

Fig. 11. Movements of PUMA 762 corresponding to some specific hand gestures

However, there are a total of 10 actionable gestures, and one background image for resetting purpose that have been identified as shown in fig. 12.

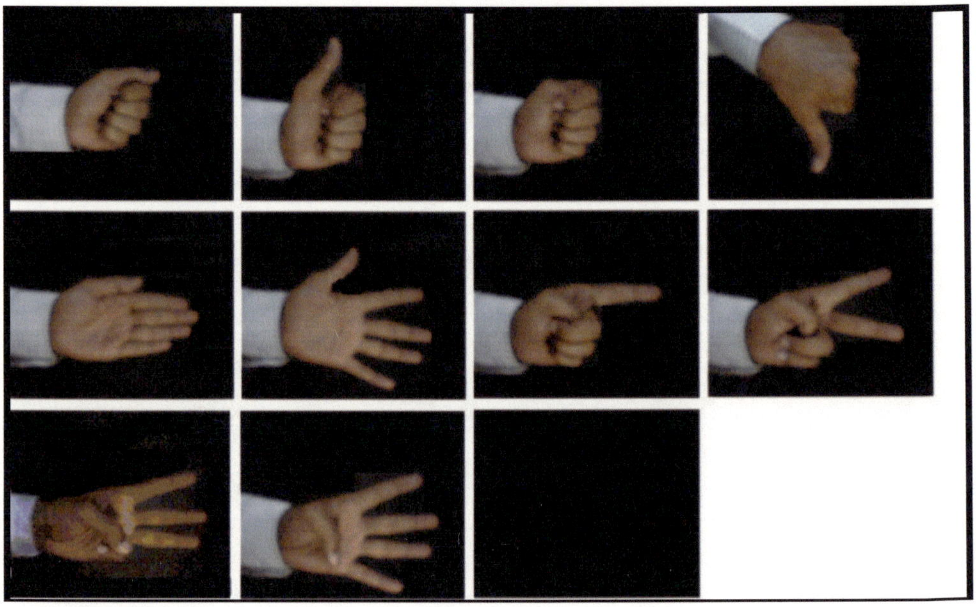

Fig. 12. Set of specified gestures used for robot control

3. Implementation on FPGA processor

We have implemented the algorithm on an FPGA, using a Database of 6 images. Each image size is 60X80 pixels. The reduction in number of images, image size and its resolution are due to FPGA memory constraints. The MATLAB code produces 6x6 projection vectors which are stored in the RAM of the FPGA using a JTAG cable [27]. After formation of the database, the system captures an image and projection vector of 6x1 is produced and that is given to the hardware for further processing. The hardware is responsible for calculating the Euclidian Distance [ED] with a dedicated block which we designate as ED block [34]. Then the minimum distance [MD] is found using a dedicated block named MD block. And the index corresponding to the minimum distance is returned to MATLAB to display the matched image from the Database [33]. The whole process is shown in fig. 13.
The main drawback of this implementation is that only part of the algorithm resides on FPGA and rest of the algorthim still runs on PC. Therefore it is not standalone system and requires interface with the PC[13]. FPGA implementation of PCA algorithm is a great challenge due to the limited resources available in commercial FPGA kits. But it is quite important to have a complete system on a FPGA chip [40]. As far as the technology is concerned, FPGA is most suited for implementation of real time image processing algorithm. It can be readily designed with custom parallel digital circuitry tailored for performing various imaging tasks making them well-suited for high speed real time vision processing applications[35].

4. Conclusion

Controlling a robot arm, in real time, through the hand gestures is a novel approach. The technique proposed here was tested under proper lighting conditions that we created in our laboratory. A gesture database consisting of binary images of size 60 X 80 pixels is pre-stored, so it takes less time and memory space during pattern recognition. Due to the use of cropped image of gesture, our database becomes more effective as one image is sufficient for one type of gesture presentation. So, neither we need to store more than one image for same gesture at different positions of image, nor have we to worry about positions of hand in front of camera. Experimental results show that the system can detect hand gestures with an accuracy of 95 % when it is kept still in front of the camera for 1 second.

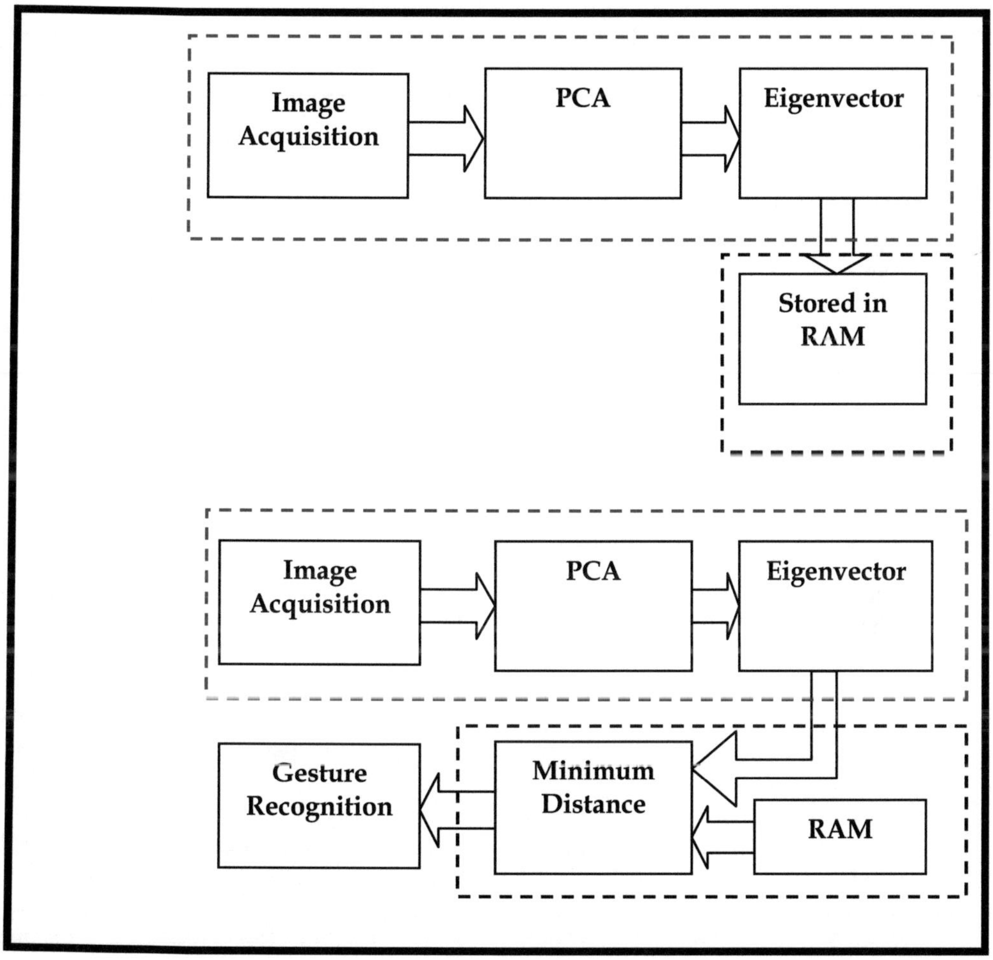

Fig. 13. Recognition process

5. Acknowledgment

This research is being carried out at Central Electronics Engineering Research Institute (CEERI), Pilani, India as part of a project "Supra Institutional Projects on Technology Development for Smart Systems". Authors would like to thank Director, CEERI for his active encouragement and support.

6. References

[1] Becerra V.M., Cage, C.N.J., Harwin, W.S., Sharkey, P.M. Hardware retrofit and computed torque control of Puma 560 robot updating an industrial manipulator. IEEE Control Systems Magazine, 24(5): 78-82. 2004.

[2] L. Brthes, P. Menezes, F. Lerasle, and J. Hayet. Face tracking and hand gesture recognition for human robot interaction. In International Conference on Robotics and Automation, pp. 1901–1906, 2004. New Orleans, Louisiana.

[3] Chan Wah Ng, Surendra Ranganath, Real-time gesture recognition system and application. Image and Vision computing (20): 993-1007, 2002.

[4] Corke, P.I. Operational Details of the Unimation Puma Servo System. Report, CSIRO Division of Manufacturing Technology. Australia, 1993.

[5] Crow, F. Summed-area tables for texture mapping. SIGGRAPH '84: Proceedings of the 11th annual conference on Computer graphics and interactive techniques. pp. 207-212.

[6] J. Davis, M. Shah. Recognizing hand gestures. In Proceedings of European Conference on Computer Vision, ECCV: 331-340, 1994. Stockholm, Sweden.

[7] Daggu Venkateshwar Rao, Shruti Patil, Naveen Anne Babu and V Muthukumar, Implementation and Evaluation of Image Processing Algorithms on Reconfigurable Architecture using C-based Hardware Descriptive Languages, International Journal of Theoretical and Applied Computer Sciences, Volume 1 Number 1 pp. 9–34, 2006.

[8] Dastur, J.; Khawaja, A.; Robotic Arm Actuation with 7 DOF Using Haar Classifier Gesture Recognition, Second International Conference on Computer Engineering and Applications (ICCEA), vol.1, no., pp.27-29, 19-21 March 2010.

[9] K G Derpains, A review of Vision-based Hand Gestures, Internal Report, Department of Computer Science. York University, February 2004.

[10] Didier Coquin, Eric Benoit, Hideyuki Sawada, and Bogdan Ionescu, Gestures Recognition Based on the Fusion of Hand Positioning and Arm Gestures, Journal of Robotics and Mechatronics Vol.18 No.6, 2006. pp. 751-759.

[11] Fang Zhong, David W. Capson, Derek C. Schuurman, Parallel Architecture for PCA Image Feature Detection using FPGA 978-1-4244-1643-1/08/2008 IEEE.

[12] Freeman W., Computer vision for television and games. In Recognition, Analysis, and Tracking of Faces and Gestures in Real-Time Systems, page 118, 1999. Copyright© MEITCA.

[13] Haessig D.,Hwang J., Gallagher S., Uhm M., Case-Study of a Xilinx System Generator Design Flow For Rapid Development of SDR Waveforms, Proceeding of the SDR 05 Technical Conference and Product Exposition. Copyright © 2005 SDR Forum.

[14] Jae-Ho Shin, Jong-Shill Lee, Se-Kee Kil, Dong-Fan Shen, Je-Goon Ryu, Eung-Hyuk Lee, Hong-Ki Min, Seung-Hong Hong, Hand Region Extraction and Gesture Recognition using entropy analysis, IJCSNS International Journal of Computer Science and Network Security, VOL.6 No.2A, 2006 pp. 216-222.

[15] Jagdish Raheja, Radhey Shyam and Umesh Kumar, Hand Gesture Capture and Recognition Technique for Real-time Video Stream, In The 13th IASTED International Conference on Artificial Intelligence and Soft Computing (ASC 2009), September 7 - 9, 2009 Palma de Mallorca, Spain.

[16] Katupitiya, J., Radajewski, R., Sanderson, J., Tordon, M. Implementation of PC Based Controller for a PUMA Robot. Proc. 4th IEEE Conf. on Mechatronics and Machine Vision in Practice. Australia, p.14-19. [doi:10.1109/MMVIP.1997.625229], 1997.

[17] R. Kjeldsen, J. Kinder. Visual hand gesture recognition for window system control, in International Workshop on Automatic Face and Gesture Recognition (IWAFGR), Zurich: pp. 184-188, 1995.

[18] Mohammad, Y.; Nishida, T.; Okada, S.; Unsupervised simultaneous learning of gestures, actions and their associations for Human-Robot Interaction, Intelligent Robots and Systems, 2009. IROS 2009. IEEE/RSJ International Conference on, vol., no., pp.2537-2544, 10-15 Oct. 2009.

[19] Moezzi and R. Jain, "ROBOGEST: Telepresence Using Hand Gestures". Technical report VCL-94-104, University of California, San Diego, 1994.

[20] M. Moore, A DSP-based real time image processing system. In the Proceedings of the 6th International conference on signal processing applications and technology, Boston MA, August 1995.

[21] Nolker, C.; Ritter, H.; Parametrized SOMs for hand posture reconstruction, Neural Networks, 2000. IJCNN 2000, Proceedings of the IEEE-INNS-ENNS International Joint Conference on , vol.4, no., pp.139-144 vol.4, 2000

[22] Nolker C., Ritter H., Visual Recognition of Continuous Hand Postures, IEEE Transactions on neural networks, Vol 13, No. 4, July 2002, pp. 983-994.

[23] K. Oka, Y. Sato and H. Koike. Real-Time Fingertip Tracking and Gesture Recognition. IEEE Computer Graphics and Applications: 64-71, 2002.

[24] Ozer, I. B., Lu, T., Wolf, W. Design of a Real Time Gesture Recognition System: High Performance through algorithms and software. IEEE Signal Processing Magazine, pp. 57-64, 2005.

[25] V.I. Pavlovic, R. Sharma, T.S. Huang. Visual interpretation of hand gestures for human-computer interaction, A Review, IEEE Transactions on Pattern Analysis and Machine Intelligence 19(7): 677-695, 1997.

[26] Quek, F. K. H., Mysliwiec, T., Zhao, M. (1995). Finger Mouse: A Free hand Pointing Computer Interface. Proceedings of International Workshop on Automatic Face and Gesture Recognition, pp.372 377, Zurich, Switzerland.

[27] A.K. Ratha, N.K. Jain, FPGA-based computing in computer vision, Computer Architecture for Machine Perception, CAMP '97. Proceedings Fourth IEEE International Workshop on, pp. 128–137, 20-22 Oct 1997.

[28] Ramamoorthy, N. Vaswani, S. Chaudhury, S. Banerjee, Recognition of Dynamic hand gestures, Pattern Recognition 36: 2069-2081, 2003.

[29] Sawah A.E., and et al., a framework for 3D hand tracking and gesture recognition using elements of genetic programming, 4th Canadian conference on Computer and robot vision, Montreal, Canada, 28-30 May, 2007, pp. 495-502.

[30] Sergios Theodoridis, Konstantinos Koutroumbas, Pattern Recognition, Elsevier Publication, Second Edition, 2003.

[31] Seyed Eghbal Ghobadi, Omar Edmond Loepprich, Farid Ahmadov Jens Bernshausen, Real Time Hand Based Robot Control Using Multimodal Images, IAENG International Journal of Computer Science, 35:4, IJCS_35_4_08, Nov 2008.

[32] Shin M. C., Tsap L. V., and Goldgof D. B., Gesture recognition using bezier curves for visualization navigation from registered 3-d data, Pattern Recognition, Vol. 37, Issue 5, May 2004, pp.1011–1024.

[33] N. Shirazi and J. Ballagh, "Put Hardware in the Loop with Xilinx System Generator for DSP", Xcell Journal online, May 2003, hitp://xvww.xilinix.coinjpoblication s/xcellonlince/xcell 47/xc svsen47.htm

[34] Stephen D.Brown, R.J. Francis, J.Rose, Z.G.Vranesic, Field Programmable Gate Arrays, 1992.

[35] Starner T., Pentland. A., Real-time American Sign Language recognition from video using hidden markov models. In SCV95, pages 265–270, Ames Street, Cambridge, 1995.

[36] Triesch J., Malsburg. C.V.D., Robotic gesture recognition. In Gesture Workshop, pages 233–244, Nara, Japan, 1997.

[37] Tomita, A., Ishii. R. J. Hand shape extraction from a sequence of digitized grayscale images. Proceedings of tweenth International Conference on Industrial Electronics, Control and Instrumentation, IECON '94, vol.3, pp.1925–1930, Bologna, Italy, 1994.

[38] Verma R., Dev A., Vision based Hand Gesture Recognition Using finite State Machines and Fuzzy Logic, International Conference on Ultra-Modern Telecommunications & Workshops, 12-14 Oct, 2009, pp. 1-6, in St. Petersburg, Russia.

[39] Walla Walla University, robotic lab PUMA 762. http://www.mathworks.com/matlabcentral/fileexchange/14932

[40] Xilinx Inc., System Generator for DSP, Users Guide http://wsssw.xilinx.con1i/products/softwa-e/svstzen/app docs/userLguide.htm

[41] Ying Wu, Thomas S Huang, Vision based Gesture. Recognition: A Review, Lecture Notes In Computer Science; Vol. 1739, Proceedings of the International Gesture Workshop on Gesture-Based Communication in Human-Computer Interaction, 1999.

21

Robotics Arm Visual Servo: Estimation of Arm-Space Kinematics Relations with Epipolar Geometry

Ebrahim Mattar
Intelligent Control & Robotics, Department of Electrical and Electronics Engineering, University of Bahrain Kingdom of Bahrain

1. Introduction

1.1 Visual servoing for robotics applications

Numerous advances in robotics have been inspired by reliable concepts of biological systems. Necessity for improvements has been recognized due to lack of sensory capabilities in robotic systems which make them unable to cope with challenges such as anonymous and changing workspace, undefined location, calibration errors, and different alternating concepts. Visual servoing aims to control a robotics system through artificial vision, in a way as to manipulate an environment, in a similar way to humans actions. It has always been found that, it is not a straightforward task to combine "Visual Information" with a "Arm Dynamic" controllers. This is due to different natures of descriptions which defines "Physical Parameters" within an arm controller loop. Studies have also revealed an option of using a trainable system for learning some complicated kinematics relating object features to robotics arm joint space. To achieve visual tracking, visual servoing and control, for accurate manipulation objectives without losing it from a robotics system, it is essential to relate a number of an object's geometrical features (object space) into a robotics system joint space (arm joint space). An object visual data, an play important role in such sense. Most robotics visual servo systems rely on object "features Jacobian", in addition to its inverse Jacobian. Object visual features inverse Jacobian is not easily put together and computed, hence to use such relation in a visual loops. A neural system have been used to approximate such relations, hence avoiding computing object's feature inverse Jacobian, even at singular Jacobian postures. Within this chapter, we shall be discussing and presenting an integration approach that combines "Visual Feedback" sensory data with a "6-DOF robotics Arm Controller". Visual servo is considered as a methodology to control movements of a robotics system using certain visual information to achieve a task. Visionary data is acquired from a camera that is mounted directly on a robot manipulator or on a mobile robot, in which case, motion of the robot induces camera motion. Differently, the camera can be fixed, so that can observe the robot motion. In this sense, visual servo control relies on techniques from image processing, computer vision control theory, kinematics, dynamic and real time computing.

Robotics visual servoing has been recently introduced by robotics, AI and control communities. This is due to the significant number of advantages over blind robotic systems. Researchers have also demonstrated that, VISUAL SERVOING is an effective and a robust framework to control robotics systems while relying on visual information as feedback. An image-based scheme task is said to be completely performed if a desired image is acquired by a robotic system. Numerous advances in robotics have been inspired by the biological systems. In reference to Fig. (1), visual servoing aims to control a robotics system through an artificial vision in a way as to manipulate an environment, comparable to humans actions. Intelligence-based visual control has also been introduced by research community as a way to supply robotics system even with more cognitive capabilities. A number of research on the field of intelligent visual robotics arm control have been introduced. Visual servoing has been classified as using visual data within a control loop, enabling visual-motor (hand-eye) coordination. There have been different structures of visual servo systems. However, the main two classes are; **P**osition-**B**ased **V**isual **S**ervo systems (**PBVS**), and the **I**mage-**B**ased **V**isual **S**ervo systems (**IBVS**). In this chapter, we shall concentrate on the second class, which is the Image-based visual servo systems.

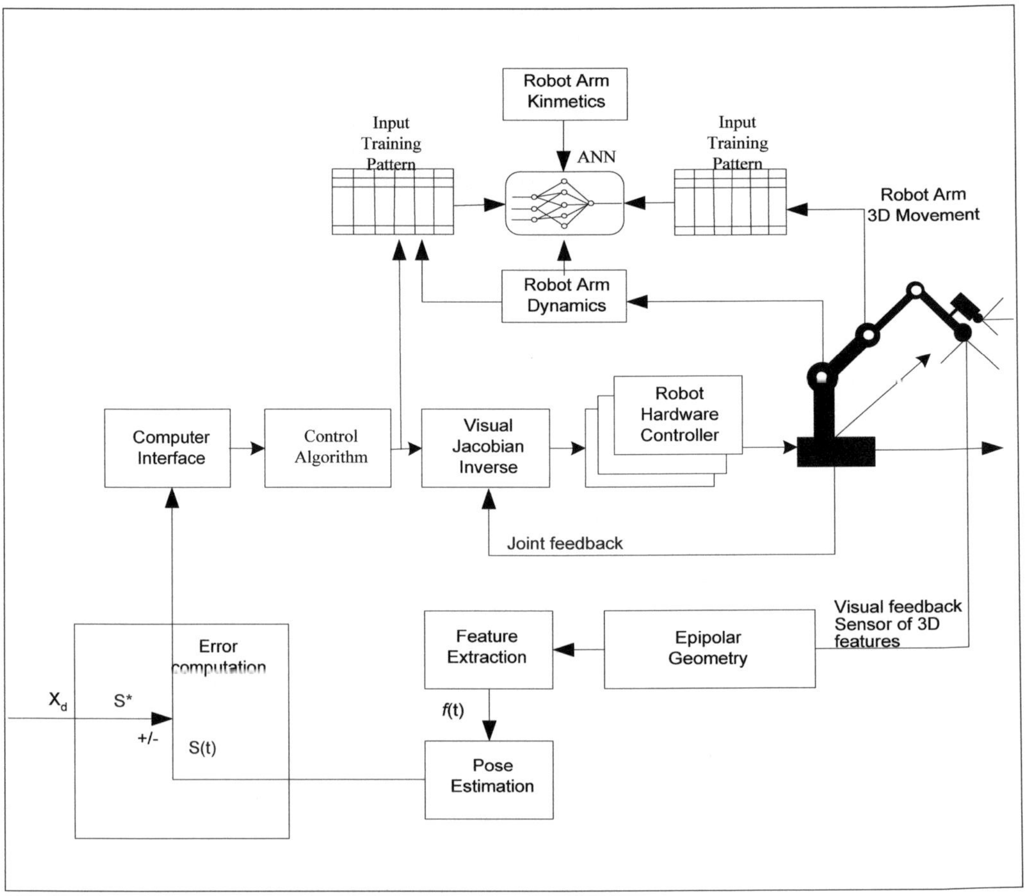

Fig. 1. Overall structure of an ANN based visual servoing

1.2 Literature surveys

EGT, Epipolar Geometry Toolbox, (Eleonora et al., 2004), was also built to grant MATLAB users with a broaden outline for a creation and visualization of multi-camera scenarios. Additionally, to be used for the manipulation of visual information, and the visual geometry. Functions provided, for both classes of vision sensing, the PINHOLE and PANORAMIC, include camera assignment and visualization, computation, and estimation of epipolar geometry entities. Visual servoing has been classified as using visual data within a control loop (Cisneros, 2004), thus enabling visual-motor (Hand-Eye) coordination.
Image Based Visual Servoing (**IBVS**) Using Takagi-Sugeno Fuzzy Neural Network Controller has been proposed by (Miao et. al, 2007). In their study, a T-S fuzzy neural controller based IBVS method was proposed. Eigenspace based image compression method is firstly explored which is chosen as the global feature transformation method. Inner structure, performance and training method of T-S neural network controller are discussed respectively. Besides that, the whole architecture of TS-FNNC was investigated.
An Image Based Visual Servoing using Takagi-Sugeno fuzzy neural network controller has been proposed by Miao, (Miao et. al, 2007). In this paper, a TAKAGI-SUGENO Fuzzy Neural Network Controller (TSFNNC) based Image Based Visual Servoing (IBVS) method was proposed. Firstly, an eigenspace based image compression method is explored, which is chosen as the global feature transformation method. After that, the inner structure, performance and training method of T-S neural network controller are discussed respectively. Besides, the whole architecture of the TS-FNNC is investigated.
Panwar and Sukavanam in (Panwar & Sukavanam 2007) have introduced Neural Network Based Controller for Visual Servoing of Robotic Hand Eye System. For Panwar and Sukavanam, in their paper a neural network based controller for robot positioning and tracking using direct monocular visual feedback is proposed. Visual information is provided using a camera mounted on the end-effector of a robotics manipulator. A PI kinematics controller is proposed to achieve motion control objective in an image plane. A Feed forward Neural Network (FFNN) is used to compensate for the robot dynamics. The FFNN computes the required torques to drive the robot manipulator to achieve desired tracking. The stability of combined PI kinematics and FFNN computed torque is proved by Lyapunov theory. Gracia and Perez-Vidal in (Gracia & Perez Vidal 2009), have presented a research framework through which a new control scheme for visual servoing that takes into account the delay introduced by image acquisition and image processing. The capabilities (steady-state errors, stability margins, step time response, etc.) of the proposed control scheme and of previous ones are analytically analyzed and compared. Several simulations and experimental results were provided to validate the analytical results and to illustrate the benefits of the proposed control scheme. In particular, it was shown that this novel control scheme clearly improves the performance of the previous ones.
Alessandro and researchers, as in (Alessandro et. al. 2007), in their research framework, they proposed an image-based visual servoing framework. Error signals are directly computed from image feature parameters, thus obtaining control schemes which do not need neither a 3-D model of the scene, nor a perfect knowledge of the camera calibration matrix. Value of the depth "Z" for each considered feature must be known. Thus they proposed a method to estimate on-line the value of Z for point features while the camera is moving through the scene, by using tools from nonlinear observer theory. By interpreting "Z" as a continuous unknown state with known dynamics, they build an estimator which asymptotically recovers the actual depth value for the selected feature.

In (Chen et. al. 2008), an adaptive visual servo regulation control for camera-in-hand configuration with a fixed camera extension was presented by Chen. An image-based regulation control of a robot manipulator with an uncalibrated vision system is discussed. To compensate for the unknown camera calibration parameters, a novel prediction error formulation is presented. To achieve the control objectives, a Lyapunov-based adaptive control strategy is employed. The control development for the camera-in-hand problem is presented in detail and a fixed-camera problem is included as an extension. Epipolar Geometry Toolbox as in (Gian et. al. 2004), was also created to grant MATLAB users with a broaden outline for a creation and visualization of multi-camera scenarios. In addition, to be used for the manipulation of visual information, and the visual geometry. Functions provided, for both class of vision sensing, the pinhole and panoramic, include camera assignment and visualization, computation, and estimation of epipolar geometry entities. Visual servoing has been classified as using visual data within a control loop, enabling visual-motor (hand-eye) coordination.

Image Based Visual Servoing Using Takagi-Sugeno Fuzzy Neural Network Controller has been proposed by (Miao et. al. 2007). In their study, a T-S fuzzy neural controller based IBVS method was proposed. Eigenspace based image compression method is firstly explored which is chosen as the global feature transformation method. Inner structure, performance and training method of T-S neural network controller are discussed respectively. Besides that, the whole architecture of TS-FNNC is investigated. For robotics arm visual servo, this issue has been formulated as a function of object feature Jacobian. Feature Jacobian is a complicated matrix to compute for real-time applications. For more feature points in space, the issue of computing inverse of such matrix is even more hard to achieve.

1.3 Chapter contribution

For robotics arm visual servo, this issue has been formulated as a function of object feature Jacobian. Feature Jacobian Matrix entries are complicated differential relations to be computed for real-time applications. For more feature points in space, the issue of computing inverse of such matrix is even more hard to achieve. In this respect, this chapter concentrates on approximating differential visual information relations relating an object movement in space to the object motion in camera space (which usually complicated relation), hence to joint space. This is known as the (object feature points). The proposed methodology will also discuss how a trained learning system can be used to achieve the needed approximation. The proposed methodology is entirely based on utilizing and merging of three MatLab Tool Boxes. Robotics Toolbox developed by Peter Corke (Corke, 2002), secondly is the Epipolar Geometry Toolbox (EGT) developed by Eleonora Alunno (Eleonora et. al. 2004), whereas the third is the ANN MatLab Tool Box.

This chapter is presenting a research framework which was oriented to develop a robotics visual servo system that relies on approximating complicated nonlinear visual servo kinematics. It concentrates on approximating differential visual changes (object features) relations relating objects movement in a world space to object motion in a camera space (usually time-consuming relations as expressed by time-varying Jacobian matrix), hence to a robotics arm joint space.

The research is also presenting how a trained Neural Network can be utilized to learn the needed approximation and inter-related mappings. The research whole concept is based on utilizing and mergence three fundamentals. The first is a robotics arm-object visual kinematics, the second is the Epipolar Geometry relating different object scenes during motion, and a learning artificial neural system. To validate the concept, the visual control loop algorithm developed by RIVES has been thus adopted to include a learning neural system. Results have indicated that, the proposed visual servoing methodology was able to produce a considerable accurate results.

1.4 Chapter organization

The chapter has been sub-divided into six main sections. In this respect, in section (1) we introduce the concept of robotics visual servo and related background, as related to visual servo. In section (2), we present a background and some related literatures for single scene via signal camera system. Double scene, as known as Epipolar Geometry is also presented in depth in section (3). Artificial Neural Net based **IBVS** is also presented in Section (4), whereas simulated of learning and training of an Artificial Neural Net is presented in section (5). Section (5) presents a case study and a simulation result of the proposed method, as compared to RIVES algorithm. Finally, Section (6) presents the chapter conclusions.

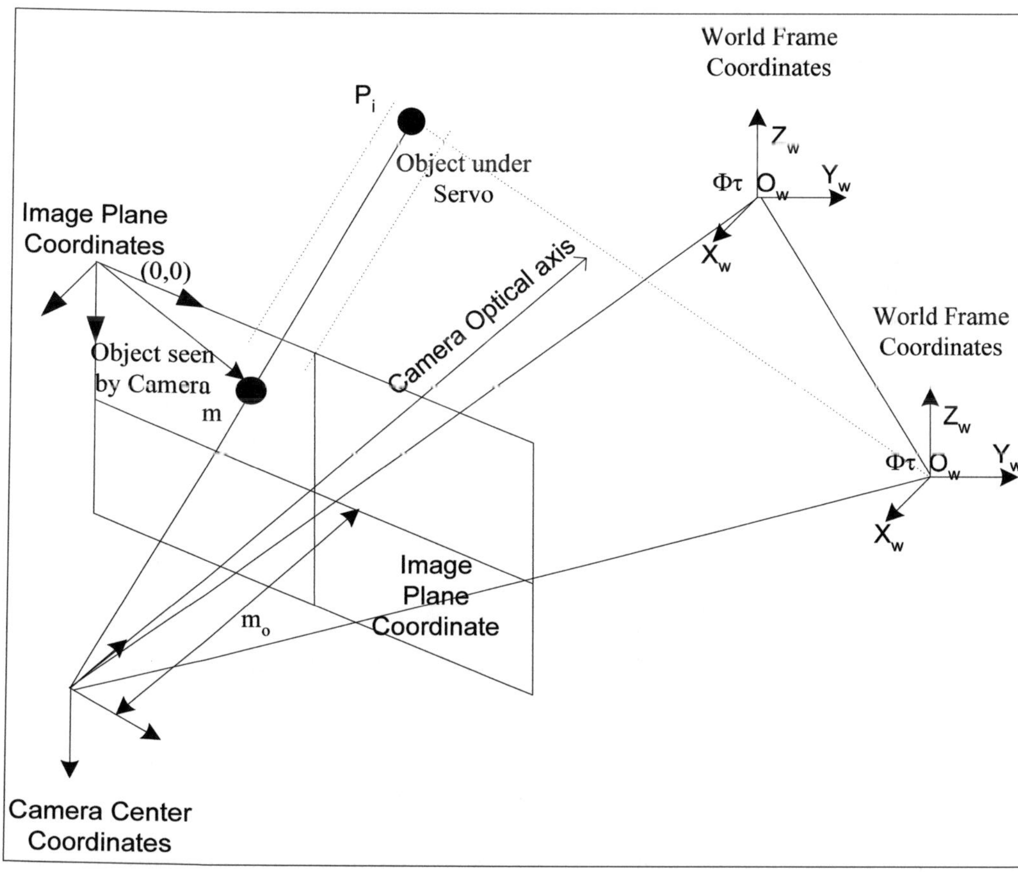

Fig. 2. Camera geometrical representation in a 3-D space

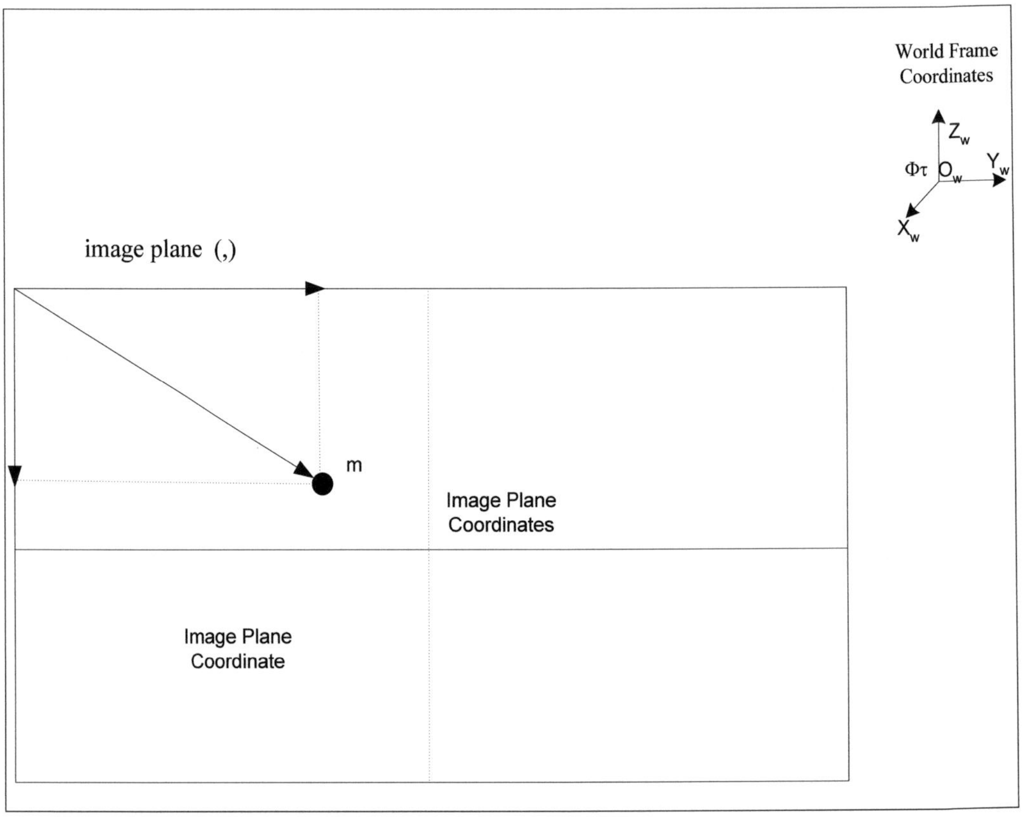

Fig. 3. Camera image plane and geometry

2. Single camera model: {object to camera plane kinmeatics}

Fig. (2) shows a typical camera geometrical representation in a space. Hence, to assemble a closed loop visual servo system, a loop is to be closed around the robotics arm system. In this study, this is a PUMA-560 robotics arm, with a Pinhole camera system. The camera image plane and associated geometry is shown in Fig. (3). For analyzing closed loop visual kinematics, we shall employ a Pinhole Camera Model for capturing object features. For whole representation, details of a Pinhole camera model in terms of image plane (ξ^a, ψ^a) location are expressed in terms (ξ, ψ, and ζ), as in Equ. (1). In reference to Fig. (2), we can express (ξ^a, ψ^a) locations as expressed in terms (ξ, Ψ, ξ):

$$\begin{cases} \xi^a = \phi^a\left(\dfrac{\xi}{\zeta}\right) \\ \psi^a = \phi^a\left(\dfrac{\psi}{\zeta}\right) \end{cases} \tag{1}$$

Both (ξ^a, ψ^a) location over an image plane is thus calculated by Equ. (1). For thin lenses (as the Pinhole camera model), camera geometry are thus represented by, (Gian et. al. 2004) :

$$\left(\frac{1}{\phi}\right)=\left(\frac{1}{\zeta^{a}}-\frac{1}{\zeta}\right) \tag{2}$$

In reference to (Gian et. al. 2004), using Craig Notation, denotes coordinate of point P in frame B. For translation case :

$$^{B}P = {}^{A}P + {}^{B}O_{A} \tag{3}$$

${}^{B}O_{A}$ is a coordinate of the origin O_A of frame "A" in a new coordinate system "B". Rotations are thus expressed :

$${}^{B}_{A}R=\left({}^{B}i_{A}\quad {}^{B}j_{A}\quad {}^{B}k_{A}\right)=\begin{pmatrix} {}^{A}i_{B}^{T} \\ {}^{A}j_{B}^{T} \\ {}^{A}k_{B}^{T} \end{pmatrix} \tag{4}$$

In Equ (4), ${}^{B}i_{A}$ is a frame axis coordinate of "A" in another coordinate "B". In this respect, for rigid transformation we have:

$${}^{B}P = {}^{B}_{A}R\,{}^{A}P$$

$${}^{B}P = {}^{B}_{A}R\,{}^{A}P + {}^{B}O_{A} \tag{5}$$

For more than single consecutive rigid transformations, (for example to frame "C"), i.e. form frames $A \rightarrow B \rightarrow C$, coordinate of point P in frame "C" can hence be expressed by:

$${}^{B}P = {}^{B}_{A}R\left({}^{B}_{A}R\,{}^{A}P + {}^{B}O_{A}\right) + {}^{C}O_{B}$$

$${}^{B}P = \left({}^{C}_{B}R\,{}^{B}_{A}R\,{}^{A}P\right) + \left({}^{C}_{B}R\,{}^{A}O_{A} + {}^{C}O_{A}\right) \tag{6}$$

For homogeneous coordinates, it looks very concise to express ${}^{B}P$ as :

$$\begin{pmatrix} {}^{B}P \\ 1 \end{pmatrix} = \begin{pmatrix} {}^{B}_{A}R & {}^{B}O_{A} \\ O^{T} & 1 \end{pmatrix}\begin{pmatrix} {}^{A}P \\ 1 \end{pmatrix} \tag{7}$$

$$\begin{pmatrix} {}^{C}P \\ 1 \end{pmatrix} = \begin{pmatrix} {}^{C}_{B}R & {}^{C}O_{B} \\ O^{T} & 1 \end{pmatrix}\begin{pmatrix} {}^{B}P \\ 1 \end{pmatrix} \tag{8}$$

$$\begin{pmatrix} {}^{C}P \\ 1 \end{pmatrix} = \begin{pmatrix} {}^{C}_{B}R & {}^{C}O_{B} \\ O^{T} & 1 \end{pmatrix}\begin{pmatrix} {}^{B}_{A}R & {}^{B}O_{A} \\ O^{T} & 1 \end{pmatrix}\begin{pmatrix} {}^{A}P \\ 1 \end{pmatrix} \tag{9}$$

We could express elements of a transformation (Γ) by writing :

$$\Gamma = \begin{pmatrix} \sigma_{11} & \sigma_{12} & \sigma_{13} & \sigma_{14} \\ \sigma_{21} & \sigma_{22} & \sigma_{23} & \sigma_{24} \\ \sigma_{31} & \sigma_{32} & \sigma_{33} & \sigma_{34} \\ \sigma_{41} & \sigma_{42} & \sigma_{43} & \sigma_{44} \end{pmatrix} \quad \Gamma = \begin{pmatrix} A & \Omega \\ O^T & 1 \end{pmatrix} \tag{10}$$

as becoming offline transformation. If $(A = R)$, i.e., a rotation matrix (Γ), once $R^T R = I$, then :

$$\Gamma = \begin{pmatrix} R & \Omega \\ O^T & 1 \end{pmatrix} \tag{11}$$

Euclidean transformation preserves parallel lines and angles, on the contrary, affine preserves parallel lines but not angles, Introducing a normalized image plane located at focal length $\phi = 1$. For this normalized image plane, pinhole (C) is mapped into the origin of an image plane using:

$$\hat{P} = (\hat{u} \quad \hat{v})^T \tag{12}$$

$\hat{C}$ and P are mapped to :

$$\hat{P} = \begin{pmatrix} \hat{u} \\ \hat{u} \\ 1 \end{pmatrix} = \frac{1}{\zeta^C}(I \quad 0)\begin{pmatrix} P^C \\ 1 \end{pmatrix} \tag{13}$$

$$\hat{P} = \frac{1}{\zeta^C}(I \quad 0)\begin{pmatrix} \xi^c \\ \psi^c \\ \zeta^c \\ 1 \end{pmatrix} \tag{14}$$

we also have :

$$\begin{pmatrix} u \\ v \\ 1 \end{pmatrix} = \frac{1}{\zeta^C}\begin{pmatrix} k\phi & 0 & a_0 \\ 0 & k\phi & v_0 \\ 0 & 0 & 1 \end{pmatrix}\begin{pmatrix} \xi^c \\ \psi^c \\ \zeta^c \end{pmatrix}$$

$$\begin{pmatrix} u \\ v \\ 1 \end{pmatrix} = \left(\frac{1}{\zeta^C}\right)\begin{pmatrix} k\phi & 0 & u_0 \\ 0 & k\phi & v_0 \\ 0 & 0 & 1 \end{pmatrix}(I \quad 0)\begin{pmatrix} \xi^c \\ \psi^c \\ \zeta^c \\ 1 \end{pmatrix} \tag{15}$$

Now once $\kappa = k\phi$ and $\beta = L\phi$, then we identify these parameters κ, β, u_o, v_o as intrinsic camera parameters. In fact, they do present an inner camera imaging parameters. In matrix notation, this can be expressed as :

$$\begin{pmatrix} u \\ v \\ 1 \end{pmatrix} = \left(\frac{1}{\zeta^C}\right) \begin{pmatrix} k\phi & 0 & u_0 \\ 0 & k\phi & v_0 \\ 0 & 0 & 1 \end{pmatrix} \begin{pmatrix} \xi^c \\ \psi^c \\ \zeta^c \\ 1 \end{pmatrix}$$

$$\begin{pmatrix} u \\ v \\ 1 \end{pmatrix} = \left(\frac{1}{\zeta^C}\right) \begin{pmatrix} \kappa & 0 & u_0 & 0 \\ 0 & \beta & v_0 & 0 \\ 0 & 0 & 1 & 0 \end{pmatrix} \begin{pmatrix} R & \Omega \\ O^T & 0 \end{pmatrix} \begin{pmatrix} \xi^c \\ \psi^c \\ \zeta^c \\ 1 \end{pmatrix} \tag{16}$$

Both (R) and (Ω) extrinsic camera parameters, do represent coordinate transformation between camera coordinate system and world coordinate system. Hence, any (u,v) point in camera image plan is evaluated via the following relation:

$$\begin{pmatrix} u \\ v \\ 1 \end{pmatrix} = \left(\frac{1}{\zeta^C}\right) M_1 M_2 p^w \qquad \begin{pmatrix} u \\ v \\ 1 \end{pmatrix} = \frac{1}{\zeta^C} M p^w \tag{17}$$

Here (M) in Equ (17) is referred to as a Camera Projection Matrix. We are given (1) a calibration rig, i.e., a reference object, to provide the world coordinate system, and (2) an image of the reference object. The problem is to solve (a) the projection matrix, and (b) the intrinsic and extrinsic parameters.

2.1 Computing a projection matrix

In a mathematical sense, we are given $\left(\xi_i^w \quad \psi_i^w \quad \zeta_i^w\right)$ and $(u_i \quad v_i)^T$ for i = (1 ...... n), we want to solve for M_1 and M_2:

$$\begin{pmatrix} u_i \\ v_i \\ 1 \end{pmatrix} = \frac{1}{\zeta^C} M_1 M_2 \begin{pmatrix} \xi_i^w \\ \psi_i^w \\ \zeta_i^w \\ 1 \end{pmatrix} \tag{18}$$

$$\begin{pmatrix} u_i \\ v_i \\ 1 \end{pmatrix} = \frac{1}{\zeta^C} M \begin{pmatrix} \xi_i^w \\ \psi_i^w \\ \zeta_i^w \\ 1 \end{pmatrix}$$

$$\zeta_i^c = \begin{bmatrix} u_i \\ v_i \\ 1 \end{bmatrix} = \begin{bmatrix} m_{11} & m_{12} & m_{13} & m_{14} \\ m_{21} & m_{22} & m_{23} & m_{24} \\ m_{31} & m_{32} & m_{33} & m_{34} \end{bmatrix} \begin{bmatrix} \xi_i^w \\ \psi_i^w \\ \zeta_i^w \\ 1 \end{bmatrix} \tag{19}$$

$$\zeta_i^c u_i = m_{11}\xi_i^w + m_{12}\psi_i^w + m_{13}\zeta_i^w + m_{14}$$

$$\zeta_i^c u_i = m_{21}\xi_i^w + m_{22}\psi_i^w + m_{23}\zeta_i^w + m_{24} \tag{20}$$

$$\zeta_i^c u_i = m_{31}\xi_i^w + m_{32}\psi_i^w + m_{33}\zeta_i^w + m_{34}$$

$$\zeta_i^c = \begin{bmatrix} u_i \\ v_i \\ 1 \end{bmatrix} = \begin{bmatrix} \sigma_{11} & \sigma_{12} & \sigma_{13} & \sigma_{14} \\ \sigma_{21} & \sigma_{22} & \sigma_{23} & \sigma_{24} \\ \sigma_{31} & \sigma_{32} & \sigma_{33} & \sigma_{34} \end{bmatrix} \begin{bmatrix} \xi_i^w \\ \psi_i^w \\ \zeta_i^w \\ 1 \end{bmatrix} \tag{21}$$

$$\zeta_i^c u_i = \sigma_{21}\xi_i^w + \sigma_{22}\psi_i^w + \sigma_{23}\zeta_i^w + \sigma_{24} \tag{22}$$

$$\zeta_i^c u_i = \sigma_{31}\xi_i^w + \sigma_{32}\psi_i^w + \sigma_{33}\zeta_i^w + \sigma_{34}$$

Obviously, we can let $\sigma_{34} = 1$. This will result in the projection matrix is scaled by σ_{34} . Once $KM = U$, $K \in \Re^{2n\times 11}$ is a matrix, a 11-D vector, and $U \in \Re^{2n-D}$ vector. A least square solution of equation $KM = U$ is thus expressed by:

$$M = K^{+}U$$

$$M = K^{T}K^{-1}K^{T}U \tag{23}$$

K^{+} is the pseudo inverse of matrix K, and m and m_{34} consist of the projection matrix M . We now turn to double scene analysis.

3. Double camera scene {epipolar geometry analysis}

In this section, we shall consider an image resulting from two camera views. For two perspective views of the same scene taken from two separate viewpoints O_1 and O_2, this is illustrated in Fig. (3). Also we shall assume that (m_1 and m_2) are representing two separate points in two the views. In other words, perspective projection through O_1 and O_2, of the same point X_w, in both image planes Λ_1 and Λ_2. In addition, by letting (c_1) and (c_2) be the optical centers of two scene, the projection E_1 (E_2) of one camera center O_1 (O_2) onto the image plane of the other camera frame Λ_2 (Λ_1) is the epipole geometry.

It is also possible to express such an epipole geometry in homogeneous coordinates in terms $\tilde{E}_1$ and $\tilde{E}_2$:

$$\tilde{E}_1 = \begin{pmatrix} E_{1x} & E_{1y} & 1 \end{pmatrix}^T \text{ and } \tilde{E}_2 = \begin{pmatrix} E_{2x} & E_{2y} & 1 \end{pmatrix}^T \quad (24)$$

One of the main parameters of an epipolar geometry is the fundamental Matrix H (which is $\Re \in 3\times3$). In reference to the H matrix, it conveys most of the information about the relative position and orientation (t, R) between the two different views. Moreover, the fundamental matrix H algebraically relates corresponding points in the two images through the Epipolar Constraint. For instant, let the case of two views of the same 3-D point X_w, both characterized by their relative position and orientation (t, R) and the internal, hence H is evaluated in terms of K_1 and K_2, representing extrinsic camera parameters, (Gian et al., 2004) :

$$H = K_2^{-T}(t)_x RK_1^{-1} \quad (25)$$

In such a similar case, a 3-D point (X_w) is projected onto two image planes, to points (m_2) and (m_1), as to constitute a conjugate pair. Given a point (m_1) in left image plane, its conjugate point in the right image is constrained to lie on the epipolar line of (m_1). The line is considered as the projection through C_2 of optical ray of m_1. All epipolar lines in one image plane pass through an epipole point.

This is a projection of conjugate optical centre: $\ddot{E}_1 = \ddot{P}_2 \begin{pmatrix} c_1 \\ 1 \end{pmatrix}$. Parametric equation of epipolar line of $\tilde{m}_1$ gives $\tilde{m}_2^T = \tilde{E}_2 + \lambda P_2 P_1^{-1} \tilde{m}_1$. In image coordinates this can be expressed as:

$$(u) = (m_2)_1 = \left(\frac{(\tilde{e}_2)_1 + \lambda(\tilde{v})_1}{(\tilde{e}_2)_3 + \lambda(\tilde{v})_3} \right) \quad (26)$$

$$(v) - (m_2)_2 - \left(\frac{(\tilde{e}_2)_2 + \lambda(\tilde{v})_2}{(\tilde{e}_2)_3 + \lambda(\tilde{v})_3} \right) \quad (27)$$

here $\tilde{v} = P_2 P_2^{-1} \tilde{m}_1$ is a projection operator extracting the (i^{th}) component from a vector. When (C_1) is in the focal plane of right camera, the right epipole is an infinity, and the epipolar lines form a bundle of parallel lines in the right image. Direction of each epipolar line is evaluated by derivative of parametric equations listed above with respect to (λ) :

$$\left(\frac{du}{d\lambda} \right) = \left(\frac{[\tilde{v}]_1 [\tilde{e}_2]_3 - [\tilde{v}]_3 [\tilde{e}_2]_1}{([\tilde{e}_2]_3 + \lambda[\tilde{v}]_3)^2} \right) \quad (28)$$

$$\left(\frac{dv}{d\lambda} \right) = \left(\frac{[\tilde{v}]_2 [\tilde{e}_2]_3 - [\tilde{v}]_3 [\tilde{e}_2]_2}{([\tilde{e}_2]_3 + \lambda[\tilde{v}]_3)^2} \right) \quad (29)$$

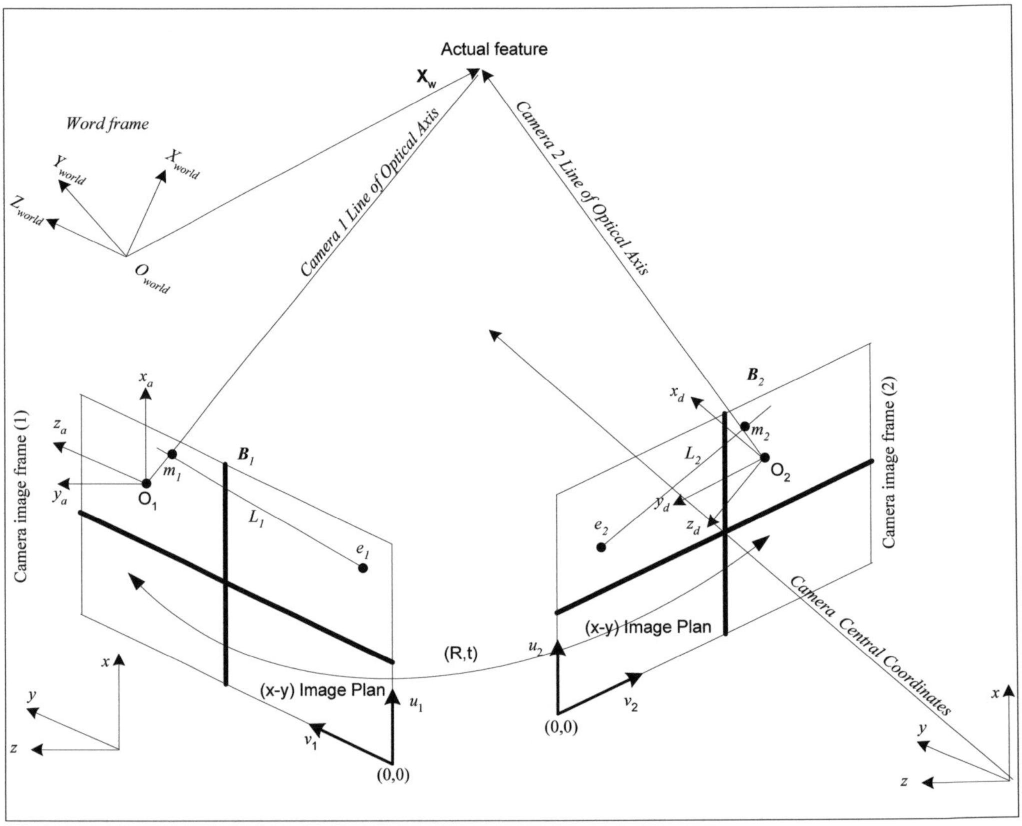

Fig 4. Camera image frame (Epipolar Geometry)

The epipole is projected to infinity once $\left(\tilde{E}_2\right)_3 = 0$. In such a case, direction of the epipolar lines in right image doesn't depend on any more. All epipolar lines becomes parallel to vector $\left(\left[\tilde{E}_2\right]_1 \ \left[\tilde{E}_2\right]_2\right)^T$. A very special occurrence is once both epipoles are at infinity. This happens once a line containing (C_1) and (C_2), the baseline, is contained in both focal planes, or the retinal planes are parallel and horizontal in each image as in Fig. (4). The right pictures plot the epipolar lines corresponding to the point marked in the left pictures. This procedure is called rectification. If cameras share the same focal plane the common retinal plane is constrained to be parallel to the baseline and epipolar lines are parallel.

4. Neural net based Image - Based Visual Servo control (ANN-IBVS)

Over the last section we have focused more in single and double camera scenes, i.e. representing the robot arm visual sensory input. In this section, we shall focus on "Image-Based Visual Servo" (IBVS) which uses locations of object features on image planes (epipolar) for direct visual feedback. For instant, while reconsidering Fig. (1), it is desired to move a robotics arm in such away that camera's view changes from (an initial) to (final) view, and feature vector from (ϕ_0) to (ϕ_d). Here (ϕ_0) may comprise coordinates of vertices, or areas of the object to be tracked. Implicit in (ϕ_d) is the robot is normal to, and centered over features of an object, at a desired distance. Elements of the task are thus specified in image space. For a robotics system with an end-effector mounted camera, viewpoint and features are functions of relative pose of the camera to the target, (${}^{c}\xi_{t}$). Such function is usually nonlinear and cross-coupled. A motion of end-effectors DOF results in complex motion of many features. For instant, a camera rotation can cause features to translate horizontally and vertically on the same image plane, as related via the following relationship :

$$\phi = f\left({}^{c}\xi_{t}\right) \tag{30}$$

Equ (30) is to be linearized. This is to be achieved around an operating point:

$$\delta\phi = {}^{f}J_{c}\left({}^{c}x_{t}\right)\delta\,{}^{c}x_{t} \tag{31}$$

$${}^{f}J_{c}\left({}^{c}x_{t}\right) = \left(\frac{\partial\phi}{\partial\,{}^{c}x_{t}}\right) \tag{32}$$

In Equ (32), ${}^{f}J_{c}\left({}^{c}x_{t}\right)$ is the Jacobian matrix, relating rate of change in robot arm pose to rate of change in feature space. Variously, this Jacobian is referred to as the feature Jacobian, image Jacobian, feature sensitivity matrix, or interaction matrix. Assume that the Jacobian is square and non-singular, then:

$${}^{c}\dot{x}_{t} = {}^{f}J_{c}\left({}^{c}x_{t}\right)^{-1}\dot{f} \tag{33}$$

from which a control law can be expressed by :

$${}^{c}\dot{x}_{t} = (K)\,{}^{f}J_{c}\left({}^{c}x_{t}\right)^{-1}\left(f_{d} - f(t)\right) \tag{34}$$

will tend to move the robotics arm towards desired feature vector. In Equ (34), K^{f} is a diagonal gain matrix, and (t) indicates a time varying quantity. Object posture rates ${}^{c}\dot{x}_{t}$ is converted to robot end-effector rates. A Jacobian , ${}^{f}J_{c}({}^{c}x_{t})$ as derived from relative pose between the end-effecter and camera, $({}^{c}x_{t})$ is used for that purpose. In this respect, a technique to determine a transformation between a robot's end-effector and the camera frame is given by Lenz and Tsai, as in (Lenz & Tsai. 1988). In a similar approach, an end-effector rates may be converted to manipulator joint rates using the manipulator's Jacobian (Croke, 1994), as follows:

$$\dot{\theta}_{t} = {}^{t6}J_{\theta}^{-1}(\theta)\,{}^{t6}\dot{x}_{c} \tag{35}$$

$\dot{\theta}_{t}$ represents the robot joint space rate. A complete closed loop equation can then be given by:

$$\dot{\theta}_{t} = K\,{}^{t6}J_{\theta}^{-1}(\theta)\,{}^{t6}J_{\theta}^{f}J_{c}^{-1}\,{}^{c}x_{t}\left(f_{d} - f(t)\right) \tag{36}$$

For achieving this task, an analytical expression of the error function is given by :

$$\phi = Z^{+}\phi_1 + \gamma\left(I_6 - Z^{+}Z\right)\frac{\partial \phi_2}{\partial X} \tag{37}$$

Here, $\gamma \in \Re^{+}$ and Z^{+} is pseudo inverse of the matrix Z, $Z \in \Re^{m \times n} = \Re(Z^{T}) = \Re(J_1^{T})$ and J is the Jacobian matrix of task function as $J = \left(\frac{\partial \phi}{\partial X}\right)$. Due to modeling errors, such a closed-loop system is relatively robust in a possible presence of image distortions and kinematics parameter variations of the Puma 560 kinematics. A number of researchers also have demonstrated good results in using this image-based approach for visual servoing. It is always reported that, the significant problem is computing or estimating the feature Jacobian, where a variety of approaches have been used (Croke, 1994). The proposed IBVS structure of Weiss (Weiss et. al., 1987 and Craig, 2004), controls robot joint angles directly using measured image features. Non-linearities include manipulator kinematics and dynamics as well as the perspective imaging model. Adaptive control was also proposed, since ${}^{f}J_{\theta}^{-1}({}^{c}\theta)$, is pose dependent, (Craig, 2004). In this study, changing relationship between robot posture and image feature change is learned during a motion via a learning neural system. The learning neural system accepts a weighted set of inputs (stimulus) and responds.

4.1 Visual mapping: Nonlinear function approximation ANN mapping

A layered feed-forward network consists of a certain number of layers, and each layer contains a certain number of units. There is an input layer, an output layer, and one or more hidden layers between the input and the output layer. Each unit receives its inputs directly from the previous layer (except for input units) and sends its output directly to units in the next layer. Unlike the Recurrent network, which contains feedback information, there are no connections from any of the units to the inputs of the previous layers nor to other units in the same layer, nor to units more than one layer ahead. Every unit only acts as an input to the immediate next layer. Obviously, this class of networks is easier to analyze theoretically than other general topologies because their outputs can be represented with explicit functions of the inputs and the weights.

In this research we focused on the use of Back-Propagation Algorithm as a learning method, where all associated mathematical used formulae are in reference to Fig. (5). The figure depicts a multi-layer artificial neural net (a four layer) being connected to form the entire network which learns using the Back-propagation learning algorithm. To train the network and measure how well it performs, an objective function must be defined to provide an unambiguous numerical rating of system performance. Selection of the objective function is very important because the function represents the design goals and decides what training algorithm can be taken. For this research frame work, a few basic cost functions have been investigated, where the sum of squares error function was used as defined by Equ. (38):

$$E = \frac{1}{NP}\sum_{p=1}^{P}\sum_{i=1}^{N}\left(t_{pi} - y_{pi}\right)^2 \tag{38}$$

where p indexes the patterns in the training set, i indexes the output nodes, and t_{pi} and y_{pi} are, respectively, the target hand joint space position and actual network output for the ith output unit on the pth pattern. An illustration of the layered network with an input layer, two hidden layers, and an output layer is shown in Fig. (5). In this network there are i

inputs, (m) hidden units, and (n) output units. The output of the jth hidden unit is obtained by first forming a weighted linear combination of the (i) input values, then adding a bias:

$$a_j = \left(\sum_{i=1}^{l} w^1_{ji} x_i + w^1_{j0} \right) \tag{39}$$

where $w^{(1)}_{ji}$ is a weight from input (i) to hidden unit (j) in the first layer. $w^{(1)}_{j0}$ is a bias for hidden unit j. If we are considering a bias term as being weights from an extra input $x_0 = 1$, Equ. (39) can be rewritten to the form of:

$$a_j = \left(\sum_{i=0}^{l} w^1_{ji} x_i \right) \tag{40}$$

The activation of hidden unit j then can be obtained by transforming the linear sum using a *nonlinear activation function* $g(x)$:

$$h_j = g(a_j) \tag{41}$$

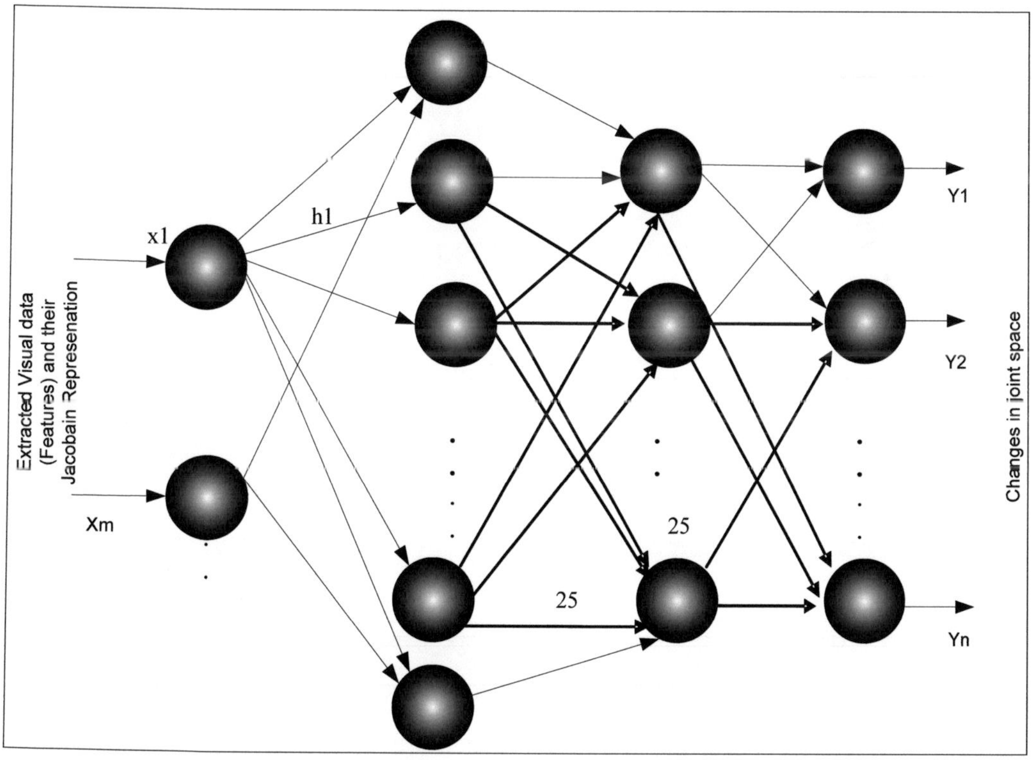

Fig. 5. Employed four layers artificial neural system

Outputs of the neural net is obtained by transforming the activation of the hidden units using a second layer of processing units. For each output unit k, first we get the linear combination of the output of the hidden units, as in Equ. (42):

$$a_k = \left(\sum_{j=1}^{m} w_{kj}^{2} h_j + w_{k0}^{2} \right) \tag{42}$$

absorbing the bias and expressing the above equation to:

$$a_k = \left(\sum_{j=0}^{m} w_{kj}^{2} h_j \right) \tag{43}$$

Applying the activation function $g_2(x)$ to Equ. (43), we can therefore get the kth output :

$$y_k = g_2(a_k) \tag{44}$$

Combining Equ. (40), Equ. (41), Equ. (43) and Equ. (44), we get a complete representation of the network as:

$$y_k = g_2\left(\sum_{j=0}^{m} w_{kj}^{2} g\left(\sum_{i=0}^{l} w_{ji}^{2} x_i \right) \right) \tag{45}$$

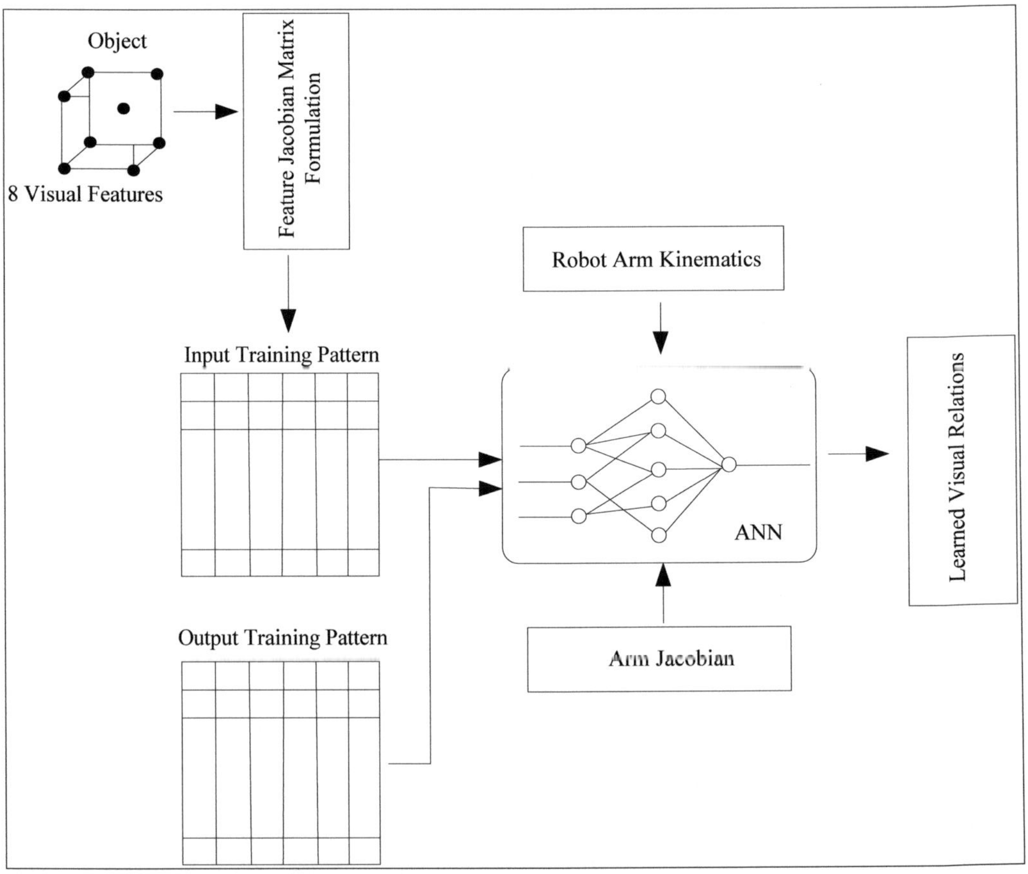

Fig. 6. Features based data gathering: Training patterns generations

The network of Fig. (5) is a synthesized ANN network with two hidden layers, which can be extended to have extra hidden layers easily, as long as we make the above transformation further. Input units do transform neural network signals to the next processing nodes. They are hypothetical units that produce outputs equal to their supposed inputs, hence no processing is done by these input units. Through this approach, the error of the network is propagated backward recursively through the entire network and all of the weights are adjusted so as to minimize the overall network error. The block diagram of the used learning neural network is illustrated in Fig. (6). The network learns the relationship between the previous changes in the joint angles $\Delta\Theta_{k-1}$, changes in the object posture Δu_a^c, and changes joint angles $\Delta\Theta_k$. This is done by executing some random displacements from the desired object position and orientation. The hand fingers is set up in the desired position and orientation to the object. Different Cartesian based trajectories are then defined and the inverse Jacobian were used to compute the associated joints displacement $\Theta_h(k)$. Different object postures with joint positions and differential changes in joint positions are the input-output patterns for training the employed neural network. During the learning epoch, weights of connections of neurons and biases are updated and changed, in such away that errors decrease to a value close to zero, which resulted in the learning curve that minimizes the defined objective function shown as will be further discussed later. It should be mentioned at this stage that the training process has indeed consumed nearly up to three hours, this is due to the large mount of training patterns to be presented to the neural network.

4.2 Artificial neural networks mapping: A biological inspiration

Animals are able to respond adaptively to changes in their external and internal environment and surroundings, and they use their nervous system to perform these behaviours. An appropriate model/simulation of a nervous system should be able to produce similar responses and behaviours in artificial systems. A nervous system is built by relatively simple; units, the neurons, so copying their behaviour and functionality should be the solution, (Pellionisz, 1989). In reality, human brain is a part of the central nervous system, it contains of the order of (10^{+10}) neurons. Each can activate in approximately 5ms and connects to the order of (10^{+4}) other neurons giving (10^{+14}) connections, (Shields & Casey, 2008). In reality, a typical neural net (with neurons) is shown in Fig. (5), it does resemble actual biological neuron, as they are made of:

- *Synapses*: Gap between adjacent neurons across which chemical signals are transmitted: (known as the input)
- *Dendrites*: Receive synaptic contacts from other neurons
- *Cell body /soma*: Metabolic centre of the neuron: processing
- *Axon*: Long narrow process that extends from body: (known as the output)

By emulation, ANN information transmission happens at the synapses, as shown in Fig. (5). Spikes travelling along the axon of the pre-synaptic neuron trigger the release of neurotransmitter substances at the synapse. The neurotransmitters cause excitation or inhibition in the dendrite of the post-synaptic neuron. The integration of the excitatory and inhibitory signals may produce spikes in the post-synaptic neuron. The contribution of the signals depends on the strength of the synaptic connection (Pellionisz, 1989). An Artificial Neural Network (ANN) is an information processing paradigm that is inspired by the way

biological nervous systems, such as the brain, process information. The key element of this paradigm is the novel structure of the information processing system. It is composed of a large number of highly interconnected processing elements (neurons) working in unison to solve specific problems. ANN system, like people, learn by example.
An ANN is configured for a specific application, such as pattern recognition or data classification, through a learning process. Learning in biological systems involves adjustments to the synaptic connections that exist between the neurons. This is true of ANN system as well (Aleksander & Morton, 1995). The four-layer feed-forward neural network with (n) input units, (m) output units and N units in the hidden layer, already shown in the Fig. (5), and as will be further discussed later. In reality, Fig. (5). exposes only one possible neural network architecture that will serve the purpose. In reference to the Fig. (5), every node is designed in such away to mimic its biological counterpart, the neuron. Interconnection of different neurons forms an entire grid of the used ANN that have the ability to learn and approximate the nonlinear visual kinematics relations. The used learning neural system composes of four layers. The {input}, {output} layers, and two {hidden layers}. If we denote (${}^{w}v_c$) and (${}^{w}\omega_c$) as the camera's linear and angular velocities with respect to the robot frame respectively, motion of the image feature point as a function of the camera velocity is obtained through the following matrix relation:

$$\dot{\gamma} = -\left(\frac{\alpha\lambda}{{}^{c}p_c}\right)\begin{pmatrix} 0 & 0 & \frac{{}^{c}p_x}{{}^{c}p_z} & \frac{{}^{c}p_x\,{}^{c}p_x}{{}^{c}p_z} & -\frac{{}^{c}p_x\,{}^{c}p_x}{{}^{c}p_z} & {}^{c}p_x \\ 1 & -1 & \frac{{}^{c}p_y}{{}^{c}p_z} & \frac{{}^{c}p_x\,{}^{c}p_x}{{}^{c}p_z} & -\frac{{}^{c}p_x\,{}^{c}p_x}{{}^{c}p_z} & -{}^{c}p_x \end{pmatrix}\begin{pmatrix} {}^{c}R_w & 0 \\ 0 & {}^{c}R_w \end{pmatrix}\begin{pmatrix} {}^{w}v_c \\ {}^{w}\omega_c \end{pmatrix} \tag{46}$$

Instead of using coordinates (${}^{x}P_c$) and (${}^{y}P_c$) for the object feature described in camera coordinate frame, which are a priori unknown, it is usual to replace them by coordinates (u) and (v) of the projection of such a feature point onto the image frame, as shown in Fig. (7).

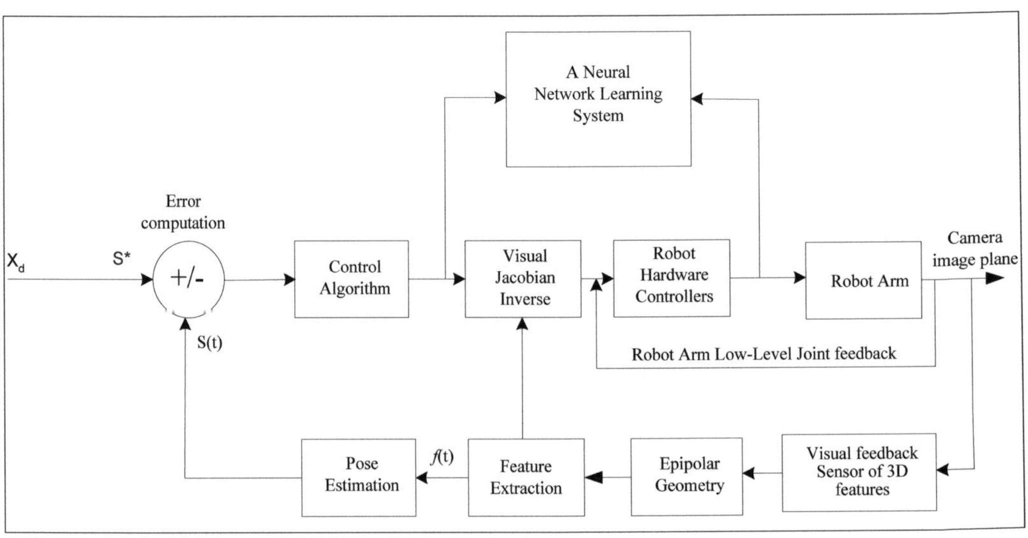

Fig. 7. Neural net based visual servo system

5. Simulation case study: Visual servoing with pin-hole camera for 6-DOF PUMA robotics arm

Visual servoing using a "pin-hole" camera for a 6-DOF robotics arm is simulated here. The system under study is a (PUMA) and integrated with a camera and ANN. Simulation block is shown Fig. (7). Over simulation, the task has been performed using 6-DOF-PUMA manipulator with 6 revolute joints and a camera that can provide position information of the robot gripper tip and a target (object) in robot workplace. The robot dynamics and direct kinematics are expressed by a set of equations of PUMA-560 robotics system, as documented by Craig, (Craig, 2004). Kinematics and dynamics equations are already well known in the literature, therefore. For a purpose of comparison, the used example is based on visual servoing system developed by RIVES, as in (Eleonora, 2004). The robotics arm system are has been servoing to follow an object that is moving in a 3-D working space. Object has been characterized by some like (8-features) marks, this has resulted in 24, $\Re\in^{8\times3}$ size, feature Jacobian matrix. This is visually shown in Fig. (7). An object 8-features will be mapped to the movement of the object in the camera image plane through defined geometries. Changes in features points, and the differentional changes in robot arm, constitute the data that will be used for training the ANN. The employed ANN architecture has already been discussed and presented in Fig. (5).

5.1 Training phase: visual training patterns generation

The foremost ambition of this visual servoing is to drive a 6-DOF robot arm, as simulated with Robot Toolbox (Corke , 2002), and equipped with a pin-hole camera, as simulated with Epipolar Geometry Toolbox, EGT (Gian et al., 2004), from a starting configuration toward a desired one using only image data provided during the robot motion. For the purpose of setting up the proposed method, RIVES algorithm has been run a number of time before hand. In each case, the arm was servoing with different object posture and a desired location in the working space. The EGT function to estimate the fundamental matrix H , given U_1 and U_2, for both scenes in which U_1 and U_2 are defined in terms of eight (ξ,ψ,ζ) feature points:

$$
U_1 = \begin{pmatrix} \xi_1^1 & \xi_1^2 & \cdots & \xi_1^8 \\ \psi_1^1 & \psi_1^2 & \cdots & \xi_1^8 \\ \zeta_1^1 & \zeta_1^2 & \cdots & \zeta_1^8 \end{pmatrix}
\quad \text{and} \quad
U_2 = \begin{pmatrix} \xi_2^1 & \xi_2^2 & \cdots & \xi_2^8 \\ \psi_2^1 & \psi_2^2 & \cdots & \xi_2^8 \\ \zeta_2^1 & \zeta_2^2 & \cdots & \zeta_2^8 \end{pmatrix} \tag{47}
$$

Large training patterns have been gathered and classified, therefore. Gathered patterns at various loop locations gave an inspiration to a feasible size of learning neural system. Four layers artificial neural system has been found a feasible architecture for that purpose. The net maps 24 (3×8 feature points) inputs characterizing object cartesian feature position and arm joint positions into the (six) differential changes in arm joints positions. The network is presented with some arm motion in various directions. Once the neural system has learned with presented patterns and required mapping, it is ready to be employed in the visual servo controller. Trained neural net was able to map nonlinear relations relating object movement to differentional changes in arm joint space. Object path of motion was defined and simulated via RIVES Algorithm, as given in (Gian et al., 2004), after such large number of running and patterns, it was apparent that the learning neural system was able to capture such nonlinear relations.

5.2 The execution phase

Execution starts primary while employing learned neural system within the robotics dynamic controller (which is mainly dependent on visual feature Jacobian). In reference to Fig. (7), visual servoing dictates the visual features extraction block. That was achieved by the use of the Epipolar Toolbox. For assessing the proposed visual servo algorithm, simulation of full arm dynamics has been achieved using kinematics and dynamic models for the Puma 560 arm. Robot Toolbox has been used for that purpose. In this respect, also Fig. (8) shows an "aerial view" of actual object "initial" posture and the "desired" posture. This is prior to visual servoing to take place. The figure also indicates some scene features. Over simulation, Fig. (9) shows an "aerial view" of the Robot arm-camera servoing, as

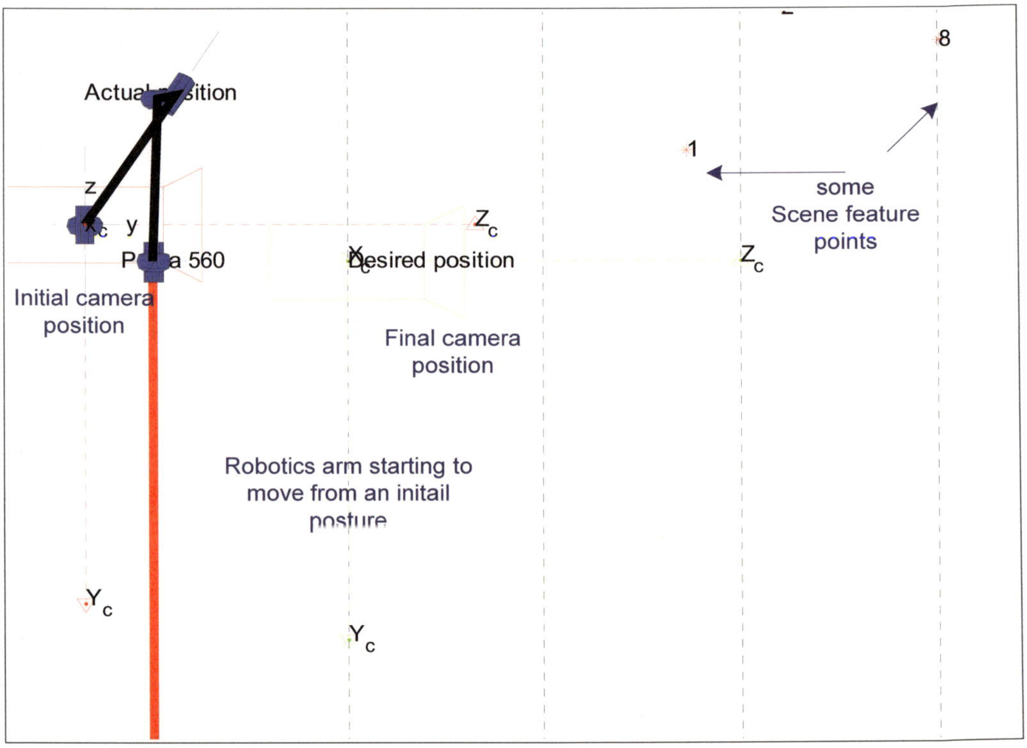

Fig. 8. Top view. Actual object position and desired position before the servoing

approaching towards a desired object posture. ANN was fed with defined patterns during arm movement. Epipolars have been used to evaluate visual features and the update during arm movement.

5.3 Object visual features migration

Fig. (10) shows the error between the RIVES Algorithm and the proposed ANN based visual servo for the first (60) iterations. Results suggest high accuracy of identical results, indicating that a learned neural system was able to servo the arm to desired posture. Difference in error was recorded within the range of (4×10^{-6}) for specific joint angles. Fig. (11) shows migration of the eight visual features as seen over the camera image plan. Just the once the Puma robot arm was moving, concentration of features are located towards an end within camera image plane. In Fig. (12), it is shown the object six dimensional movements. They indicate that they are approaching the zero reference. As an validation of the neural network ability to servo the robotics arm toward a defined object posture, Fig. (13) show that the trained ANN visual servo controller does approach zero level of movement. This is for different training patterns and for different arm postures in the 6-dimenstional space. Finally, Fig. (14) shows the error between RIVES computed joint space values and the proposed ANN controller computed joint space values. Results indicate excellent degree of accuracy while the visual servo controller approaching the target posture with nearly zero level of erroe for different training visual servo target postures.

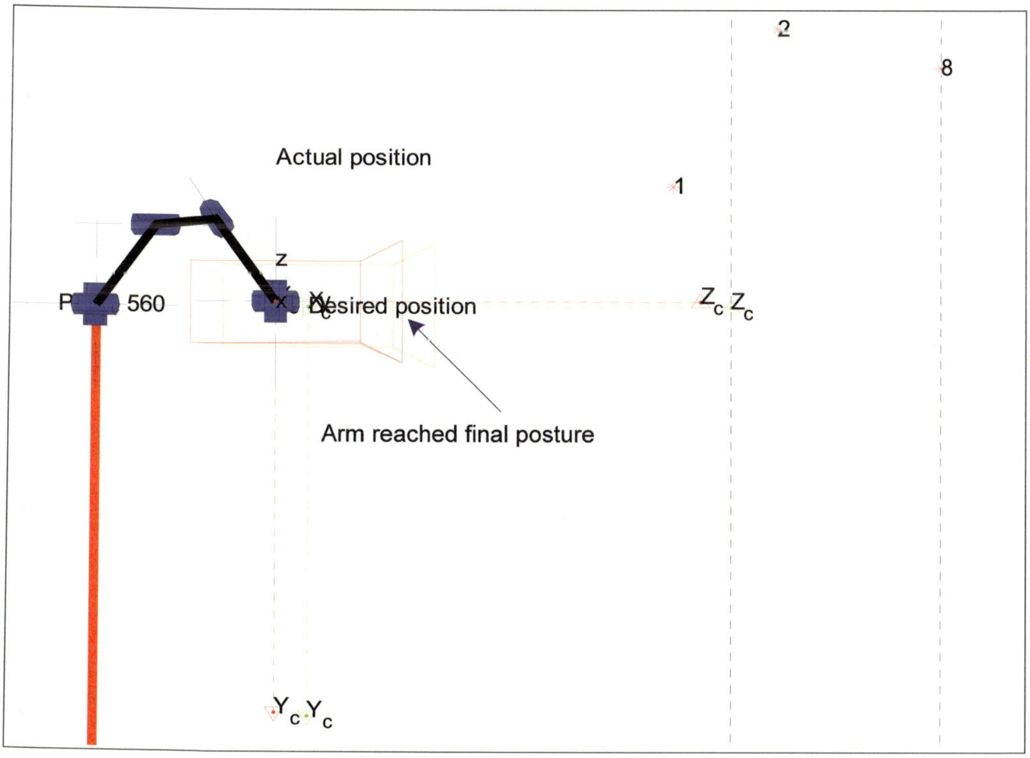

Fig. 9. Robot arm-camera system: Clear servoing towards a desired object posture

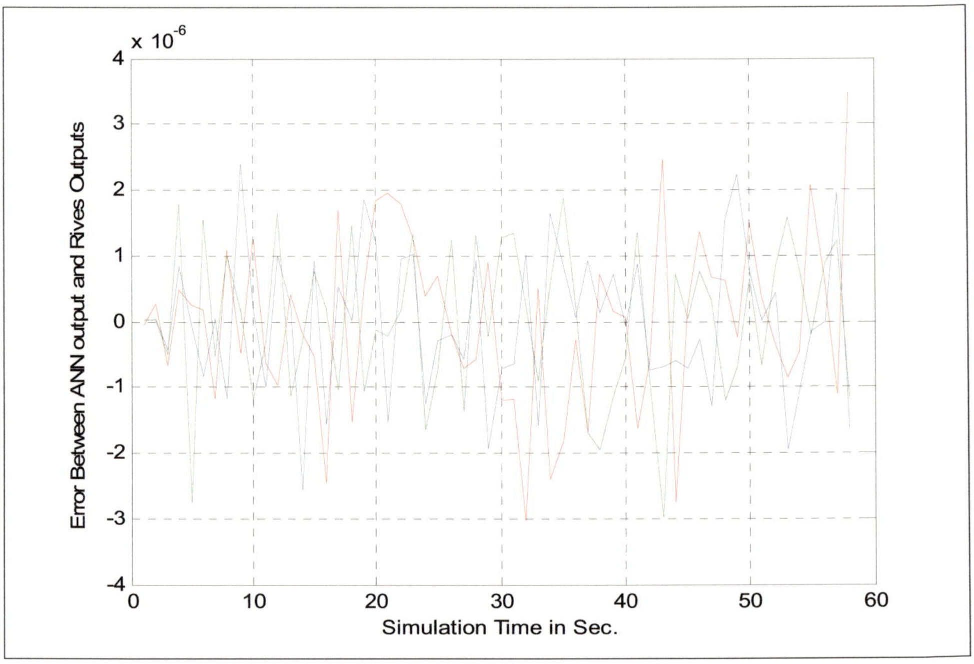

Fig. 10. Resulting errors. Use of proposed ANN based visual servo

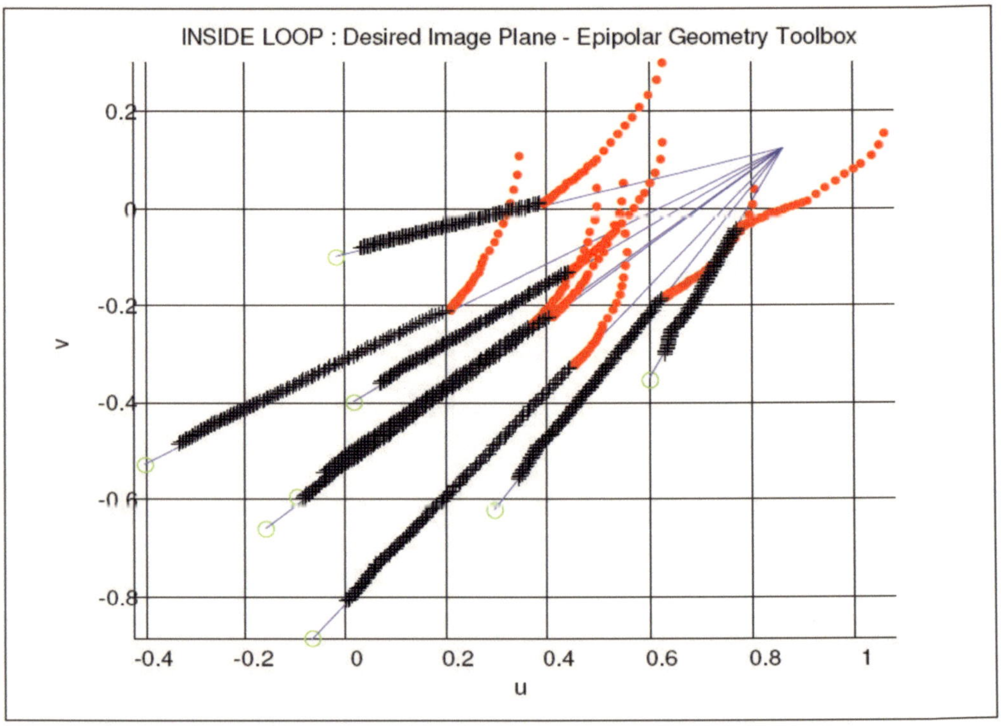

Fig. 11. Migration of eight visual features (as observed over the camera image plan)

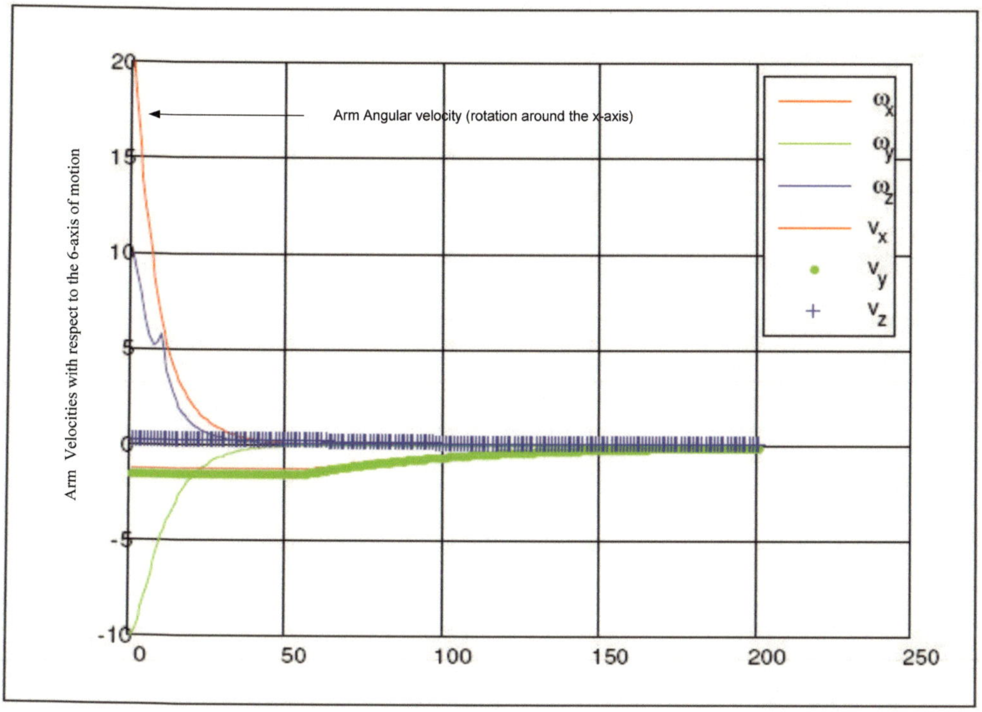

Fig. 12. Puma arm six dimensional movements

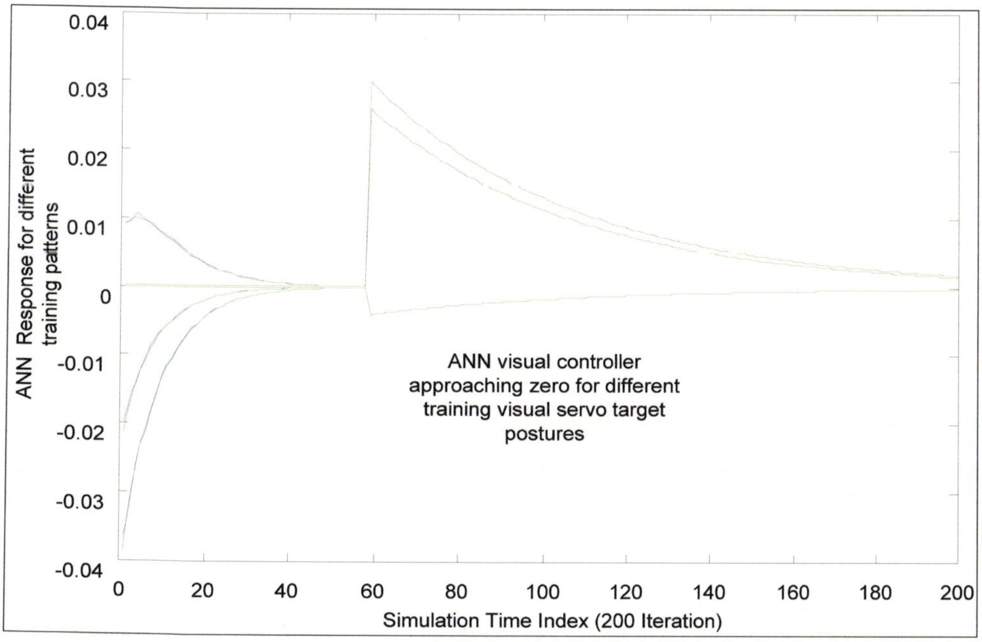

Fig. 13. ANN visual servo controller approaching zero value for different training visual servo target postures

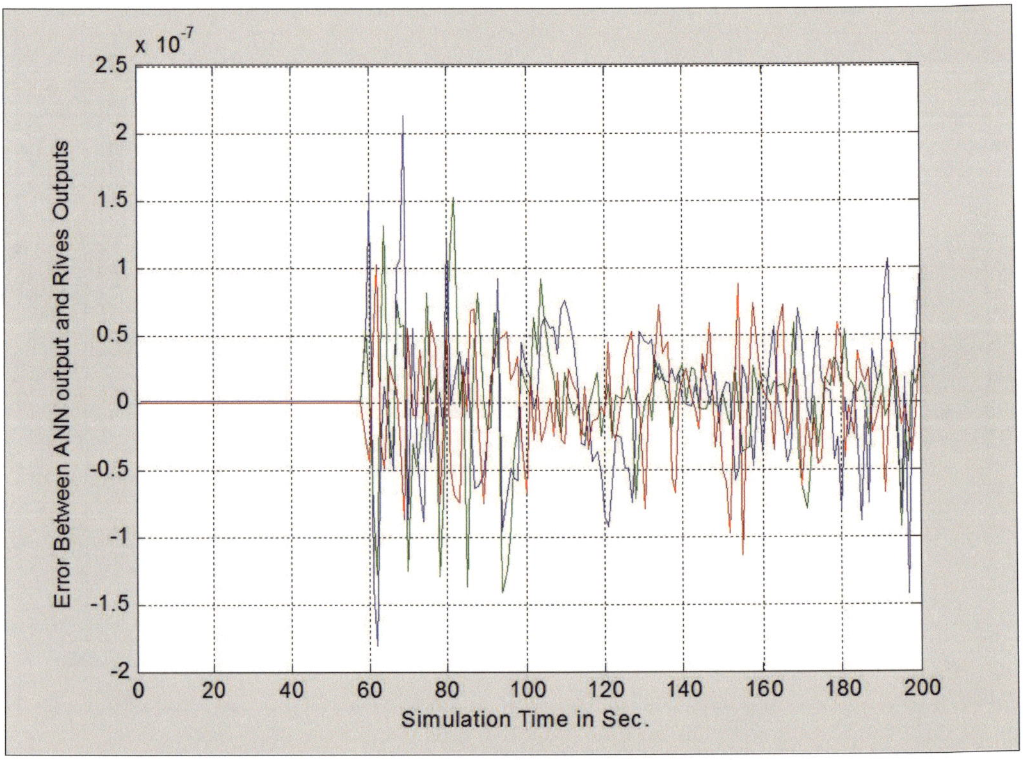

Fig. 14. Error of arm movements: ANN based controller and RIVES output difference ANN visual Servo

6. Conclusions

Servoing a robotics arm towards a moving object movement using visual information, is a research topic that has been presented and discussed by a number of researchers for the last twenty years. In this sense, the chapter has discussed a mechanism to learn kinematics and feature-based Jacobian relations, that are used for robotics arm visual servo system. In this respect, the concept introduced within this chapter was based on an employment and utilization of an artificial neural network system. The ANN was trained in such away to learn a mapping relating the " complicated kinematics" as relating changes in visual loop into arm joint space. Changes in a loop visual Jocobain depends heavily on a robotics arm 3-D posture, in addition, it depends on features associated with an object under visual servo (to be tracked). Results have shown that, trained neural network can be used to learn such complicated visual relations relating an object movement to an arm joint space movement. The proposed methodology has also resulted in a great deal of accuracy. The proposed methodology was applied to the well-know Image Based Visual Servoing, already discussed and presented by RIVES as documented in (Gian et al., 2004). Results have indicated a close degree of accuracy between the already published "RIVES Algorithm" results and the newly proposed "ANN Visual Servo Algorithm". This indicates ANN Visual Servo, as been based on some space learning mechanisms, can reduce the computation time.

7. References

Aleksander, I. & Morton, H. (1995), An Introduction to Neural Computing, *Textbook, International Edition 2nd Edition.*

Alessandro, D.; Giuseppe, L.; Paolo, O. & Giordano, R. (2007), On-Line Estimation of Feature Depth for Image-Based Visual Servoing Schemes, *IEEE International Conference on Robotics and Automation*, Italy, 1014

Chen, J.; Dawson, M.; Dixon, E. & Aman, B. (2008), Adaptive Visual Servo Regulation Control for Camera-in-Hand Configuration With a Fixed-Camera Extension, *Proceedings of the 46th IEEE Conference on Decision and Control*, CDC, 2008, pp. 2339-2344, New Orleans, LA, United States

Cisneros P. (2004), Intelligent Model Structures in Visual Servoing, *Ph.D. Thesis*, University of Manchester, Institute for Science and Technology

Corke, P. (2002), Robotics Toolbox for MatLab, *For Use with MATLAB, User Guide*, Vol. 1.

Craig, J. (2004), Introduction to Robotics: Mechanics and Control, Textbook, *International Edition, Prentice Hall*

Croke, P. (1994), High-Performance Visual Closed-Loop Robot Control, Thesis submitted in total fulfillment of the Requirements for the Degree of Doctor of Philosophy.

Eleonora, A.; Gian, M. & Domenico, P. (2004), Epipolar Geometry Toolbox, *For Use with MATLAB, User Guide*, Vol. 1.

Gian, M.; Eleonora, A. & Domenico, P. (2004), The Epipolar Geometry Toolbox (EGT) for MATLAB, *Technical Report*, 07-21-3-DII University of Siena, Siena, Italy

Gracia, L. & Perez-Vidal, C. (2009), A New Control Scheme for Visual Servoing, *International Journal of Control, Automation,and Systems*, Vol. 7, No. 5, pp. 764-776, DOI 10.1007/s12555-009-0509-9

Lenz, K. & Tsai, Y. (1988), Calibrating a Cartesian Robot with Eye-on-Hand Configuration Independent of Eye-To-Hand Relationship, *Proceedings Computer Vision and Pattern Recognition*, 1988 CVPR apos., Computer Society Conference, Vol.5, No. 9, pp. 67 – 75

Martinet, P. (2004), Applications In Visual Servoing, *IEEE-RSJ IROS'04 International Conference*, September 2004, Japan.

Miao, H.; Zengqi, S. & Fujii, M. (2007), Image Based Visual Servoing Using Takagi-Sugeno Fuzzy Neural Network Controller, *IEEE 22nd International Symposium on Intelligent Control, ISIC 2007*, Singapore, Vol. 1, No. 3, pp. 53 – 58

Miao, H.; Zengqi, S. & Masakazu, F. (2007), Image Based Visual Servoing Using Takagi-Sugeno Fuzzy Neural Network Controller, *22nd IEEE International Symposium on Intelligent Control*, Part of IEEE Multi-conference on Systems and Control Singapore, pp. 1-3

Panwar, V. & Sukavanam, N. (2007), Neural Network Based Controller for Visual Servoing of Robotic Hand Eye System, *Engineering Letters*, Vol. 14, No.1, EL_14_1_26, Advance Online Publication

Pellionisz, A. (1989), About the Geometry Intrinsic to Neural Nets, *International Joint Conference on Neural Nets*, Washington, D.C., Vol. 1, pp. 711-715

Shields, M. & Casey, C. (2008), A Theoretical Framework for Multiple Neural Network Systems, *Journal of Neurocomputing*, Vol. 71, No.7-9, pp. 1462-1476

Weiss, L.; Sanderson, A & Neuman, A. (1987), Dynamic Visual Servo Control of Robots: An adaptive Image-Based Approach , *IEEE Journal on Robotics and Automation*, Vol. 3, No.5, pp. 404–417.

22

Design and Construction of an Ultrasonic Sensor for the Material Identification in Robotic Agents

Juan José González España, Jovani Alberto Jiménez Builes and Jaime Alberto Guzmán Luna
Research group in Artificial Intelligence for Education, Universidad Nacional de Colombia Colombia

1. Introduction

Quality assurance in all industrial fields depends on having a suitable and robust inspection method which let to know the integrity or characteristics of the inspected sample. The testing can be destructive or not destructive. The former is more expensive and it does not guarantee to know the condition of the inspected sample. Though, it is used in sectors where the latter cannot be used due to the technical limitations. The second one is an ongoing research field where cheaper, faster and more reliable methods are searched to assess the quality of specific specimens.

Once a suitable reliability and quality are achieved by using a specific method, the next step is to reduce the duration and cost of the associated process. If it is paid attention to the fact that most of these processes require of a human operator, who is skilled in the labour of interpreting the data, it can be said that those previously mentioned factors (process duration and cost) can be reduced by improving the automation and autonomy of the associated processes. If a robotic system is used this can be achieved. However, most of the algorithms associated to those processes are computationally expensive and therefore the robots should have a high computational capacity which implies a platform of big size, reduced mobility, limited accessibility and/or considerable cost. Those constraints are decisive for some specific applications.

One important factor which should be considered to develop a low cost, small size and mobile robotic system is the design of the software and hardware of the sensor. If it is designed with a depth analysis of the context of the specific application it can be obtained a considerable reduction on the requirements and complexity of the robotic system.

The appropriated design of the hardware and software of the sensor depends on the proper selection of the signal pattern which is going to be used to characterize the condition of the inspected sample. For ultrasonic waves the patterns are changes on amplitude, frequency or phase on the received echo because of the properties of a specific target. Some of those properties are attenuation, acoustic impedance and speed of sound, among others.

Among the many applications of ultrasound, one of them which is important for the aerospace, automotive, food and oil industries, among others, is the material identification. Depict the fact that there are many ultrasonic sensors which let to identify the material of the sample being inspected, most of them are not appropriated to be implemented in a robot. The high computational capacity demanded by the algorithm of the sensor is the main constraint for the implementation of the ultrasonic identification of materials in a robot. In this book chapter is addressed this problem and it is solved by the Peniel method. Based on this method is designed and constructed a novel sensor which is implemented in two robots of the kit TEAC^{2}H-RI.

This book chapter is organized as follows: Section Two briefly reviews some approaches in Material Identification with ultrasonic sensors. Section Three illustrates the novel Peniel method and its associated hardware and software. Section Four describes the results. Finally, section five presents the conclusions and future work.

2. Material identification with ultrasonic sensors

The material identification using ultrasonic techniques is very important for a plethora of sectors of the industry, research, security, health, among others. In contrast with other methods for the identification of materials, those based on ultrasonic signals not only are cheaper but also some of them let to identify objects which are occluded by others.

Following is the general procedure to achieve this:

1. It is used a specific configuration for the receiver and transmitter ultrasonic transducers. Some of them are: Pulse-Echo, Through-Transmission and Pitch-Catch configuration [NASA, 2007]
2. It is sent an ultrasonic wave to the inspected sample
3. Once the wave has interacted with the object, it returns as an echo to its source and there it is captured by means of an ultrasonic transducer (receiver).
4. Then, the electrical signal goes to a signal conditioning system, which is formed by amplifiers, filters, digital to analog converters (DAC), and so on.
5. Finally, there is the processing module, which is in charge to perform the most important stage of the system, i.e., the signal processing. In this case, it can be used microcontrollers, microprocessors, FPGAs, computers, among other devices. In this processing module the algorithm is implemented, in order to process the respective signal patterns and indicates the material under inspection.

In the next lines is illustrated some methods proposed in the state of the art for the material identification and is mentioned their strengths and more important limitations.

In [Thomas et al, 1991] is developed an ultrasonic perception system for robots for the task of identification of materials. This system is based on the neural network Multi Layer Perceptron (MLP), which was trained with the characteristics of the frequency spectrum and the time domain of the ultrasonic echo. Its most relevant advantage is that the sample was inspected from different positions; additionally the ultrasonic transducer did not require making contact with the inspected sample. Its more remarkable limitations are that it identifies the material in a stochastic manner, i.e. sometimes the material is correctly identified and other not, and the issue that the sensor was never implemented in a robotic system.

In [Gunarathne et al, 2002] is proposed an implementation to identify the material of nonplanar cylindrical surface profiles (e.g. petroleum pipelines) with a prediction accuracy of the result. The method is mainly based on the feature extraction by curve fitting. The curve fitting consists in a computationally generated artificial signal which fits the shape of the experimental signal. The optimal parameters chosen for the artificial signal (AS) are those which make the AS to fit the experimental signal with a low error; those are used for the task of material identification.

Furthermore, for this case is used a database which contains the typical values of the parameters correspondent to specific materials. Its more relevant strengths are the identification of the materials of objects of non-planar cylindrical surface profiles and the accuracy of the results. On the other hand, its more significant limitations are the expensive computational cost of its processing algorithm, the absence of autonomy and the restrictions to the position of the transducer.

In [Ohtanil et al, 2006], is implemented an array of ultrasonic sensors and a MLP neural network for the materials identification. Some experiments were performed on copper, aluminium, pasteboard and acrylic, at different angles with respect to the x and z axes. The maximum distance between the sensor array and the target was 30cm. The main strength of the system is that it works in air over a wide range of distances between the target and the sensor system. Additionally, the sensor model is automatically adjusted based on this distance. The results are quite accurate. Its more significant limitation is the big size of the system (only appropriated for a huge robot), the computing requirements and the dependence on the position of the target to achieve the identification of the material.

In [Zhao et al, 2003] is used the ultrasonic identification of materials for the quality assurance in the production of canned products. The configuration of the ultrasonic system is pulse-echo. Water is the coupling medium between the transducer and the target. By means of this set up, it can be detected any shift from a specific value of the acoustic impedance, which is the result of foreign bodies within the bottle. Its main strengths are the accuracy of the system and the high sensibility. Its more relevant weaknesses are the absence of mobility and autonomy.

In [Pallav et al, 2009] is illustrated a system which has the same purpose of the previous mentioned system, i.e. also by means of the material identification performs quality control in canned food. The system uses an ultrasonic sensor to identify the acoustic impedance shift within the canned food which is related to foreign bodies within the can. The main difference in comparison to the previous mentioned system [Zhao et al, 2003] is the fact that the Through-Transmission configuration is used, and that the couplant is not water but air. Its more relevant strengths are the successful detection of foreign bodies and the mobility achieve in the X and Y axes. Its more important limitations are the high requirements of voltage, the dependence of the object's position and the narrow range of mobility.

In [Stepanić et al, 2003] is used the identification of materials to differentiate between common buried objects in the soil and antipersonnel landmines. For this case is used a tip with an ultrasonic transducer in its border and this latter is acoustically coupled with the target. The procedure to operate the system is: 1. Select a zone where is suspected the landmine is. 2. Introduced the tip in this zone in order to touch the suspecting object 3. Take some measurements. Based on this measurements the material can be identified and as a consequence it can be detected the presence or absence of landmines. Its more important limitations are the lack of autonomy, because it requires a human operator, and the dangerous fact which involves making contact with the landmine.

From the previous mentioned approaches the most significant limitation which forbids the automation in small robotic agents is the expensive computational cost of the processing algorithms. This problem was addressed in a master thesis and the results are exposed in this article.

3. Peniel method

In the following lines is exposed the Peniel Method, which is a method of low computational cost and therefore its associated algorithm and circuit can be implemented in a small robot. It is important to highlight that only one microcontroller is used to implement this method.

3.1 Mathematical model

In [Gunarathne et al, 1997, 1998, 2002; Gonzalez & Jiménez, 2010b] is exposed that the reverberations within a plate follow the next expression.

$$A(t) = Ae^{-B(t-C)} + D \tag{1}$$

where A is related to the first echo amplitude, B to the rate of decay of reverberations, C to the timing of the first echo from the start of the scan and D to the signal-to-noise ratio.

In [Gunarathne et al, 2002] is mentioned that knowing the decay coefficient (B) the material can be identified. Moreover, in [Allin, 2002] is illustrated that there are several methods that are based on the decay coefficient to identify the material. This variable is important because it only depends on the attenuation and acoustic impedance of the material, if the thickness is kept constant. In most of the solids, the attenuation is low and its effect is negligible in comparison with the acoustic impedance. Based on this the material can be identified if the acoustic impedance is known. In the following lines we present a novel method developed to estimate indirectly the decay coefficient B and therefore it can be used to identify the material of solid plates.

Without losing generality, in (1) we assume that D=0 and C=0. If it is sought the time interval during *A(t)* is greater than or equal to a value *w*, then the equation which express this is:

$$A(t) \geq w \tag{2}$$

Replacing (1) in (2) and taking into account the considerations that were made with respect C and D, we obtained:

$$\leftrightarrow Ae^{-B\cdot t} \geq w \tag{3}$$

After some algebra we obtained:

$$\leftrightarrow t \leq \frac{\ln\left(\frac{A}{w}\right)}{B} \tag{4}$$

It means that in the time interval $\left[0, \frac{\ln\left(\frac{A}{w}\right)}{B}\right]$ the exponential function is greater than or equal to *w*. The duration of this time interval is defined as t_{di}, and then for the previous case it can be said that:

$$t_{di} \leq \frac{\ln\left(\frac{A}{w}\right)}{B} \tag{5}$$

The left side term in the expression (3) is an artificial signal which fits the received ultrasonic echo (see page 3). If the received ultrasonic echo is amplified by a Gain G, as will the exponential function (3). Thus it is obtained from (3):

$$G \cdot Ae^{-B \cdot t} \geq w \tag{6}$$

It leads to expressions similar to those obtained in (4) and (5):

$$t \leq \frac{\ln\left(\frac{G \cdot A}{w}\right)}{B} \tag{7}$$

$$t_{di} = \frac{\ln\left(\frac{G \cdot A}{w}\right)}{B} \tag{8}$$

Therefore, for two signals with similar A values, the t_{di} duration is inversely proportional to the decay coefficient B. It can be seen in figure 1 for the case when A = 1, w = 0.5, and B = 1000, 2000, 3000 and 4000. Additionally, it is shown the behavior with G of the difference between t_{di3} and t_{di4}. This let us to conclude that t_{di} can be used to characterize a material if G, A and w are known.

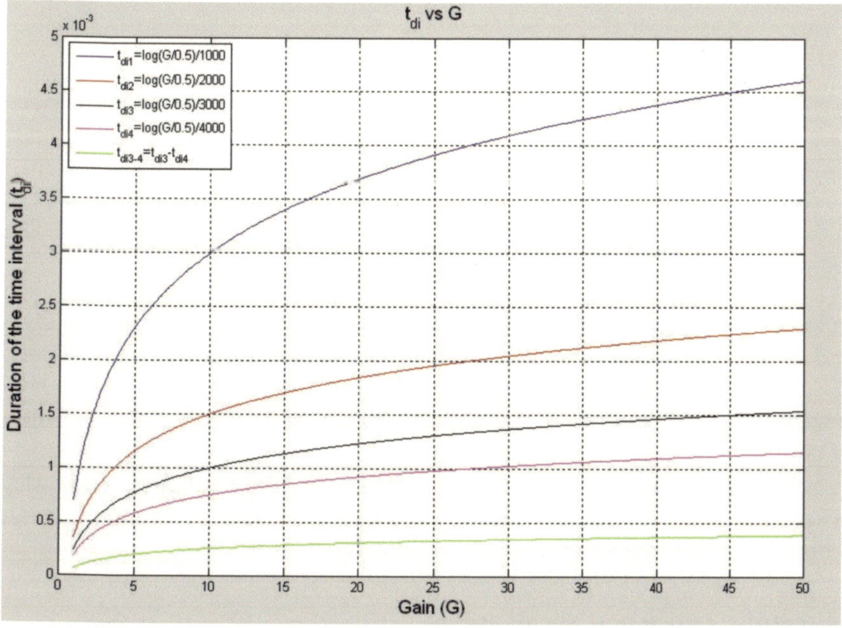

Fig. 1. The variation of the time interval duration depending on the gain. For this case, the time interval is the time during an exponential signal is above 0.5 (w)

Now, if it is used two different gain values (G1 and G2) for the same signal, the equation (8) yields:

$$t_{di1} = \frac{\ln\left(\frac{G_1 \cdot A}{w}\right)}{B} \tag{9}$$

$$t_{di2} = \frac{\ln\left(\frac{G_2 \cdot A}{w}\right)}{B} \tag{10}$$

If $G_1<G_2$ and (9) is subtracted from (10), it is obtained:

$$t_{di2} - t_{di1} = \frac{\ln\left(\frac{G_2 \cdot A}{w}\right)}{B} - \frac{\ln\left(\frac{G_1 \cdot A}{w}\right)}{B} \tag{11}$$

$$t_{di2} - t_{di1} = \frac{\ln\left(\frac{G_2 \cdot A}{w}\right) - \ln\left(\frac{G_1 \cdot A}{w}\right)}{B} \tag{12}$$

$$t_{di2} - t_{di1} = \frac{\ln\left(\frac{\frac{G_2 \cdot A}{w}}{\frac{G_1 \cdot A}{w}}\right)}{B} \tag{13}$$

$$t_{di2} - t_{di1} = \Delta t_1 = \frac{\ln\left(\frac{G_2}{G_1}\right)}{B} \tag{14}$$

$$\Delta t_1 = \frac{\ln(G_r)}{B} \tag{15}$$

where $G_r = \frac{G_2}{G_1}$

As it can be seen (14) does not depend on A and therefore is only necessary to know G_2 and G_1 to find B from Δt_1. Moreover, if there are two signals from different materials, even though G_2 and G_1 are not known, the material can be identified from the difference between the correspondent Δts. In figure 2 is shown the behavior of Δt as a function of B for different values of G_r.

As it can be seen in the figure 2 the duration increment of the time interval (Δt) depends on B if G_r is kept constant. Also, it can be recognized that if B is kept constant, high values of G_r result in high increases on the duration of the time interval (Δt).

3.2 The electronic circuit and the algorithm

One of the conditions to identify the material using the expression (8) of the Peniel Method, is that those signals compared in terms of the behavior of the time interval durations should have a similar value A (see equation 1). To meet this requirement it has been used the circuit in the figure 3. This circuit is called **clustering circuit**. Despite the fact that by means of (14) the material can be identified without knowing A, the clustering circuit is used to make more robust the method.

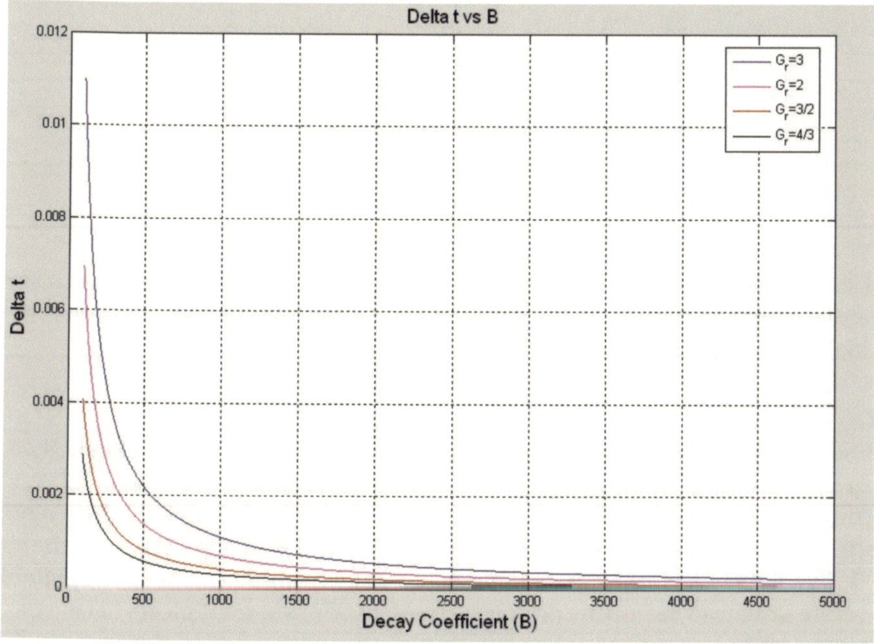

Fig. 2. Behavior of the duration increment of the time interval, (Δt), in function of the decay coefficient for values of G_1=10, 20, 30 and G_2= 20, 30, 40

3.2.1 Clustering algorithm and circuit

This circuit is formed by four amplifiers, four comparators and a microcontroller. Each amplifier has different gains. The gain grows in the direction of AMP4, i.e., the smaller and higher gains belong to the AMP1 and AMP4 amplifiers, respectively. Each amplifier represents a group, in this manner AMP3 represents the group 3 and group 2 is represented by AMP2

The output of the amplifiers goes to the respective comparators. The threshold (*w*) is the same for the last three comparators and a little bit higher for the first comparator (AMP1). The echo signal belongs to the group which has the amplifier with lower gain but with its comparator activated. In this manner a signal which activates the comparator three only belongs to the group three if the respective comparators of AMP1 and AMP2 are not activated.

The output of each comparator goes to the microcontroller and once this detects low levels on any comparator, it identifies what is the amplifier with lower gain which has its comparator activated and therefore it identifies the group the signal belongs to.

Consequently, those signals with similar peak amplitude, A (see equation 1), will be in the same group. In this manner is met the requirement that the A value should be similar between signals which are compared in terms of the duration of the time interval.

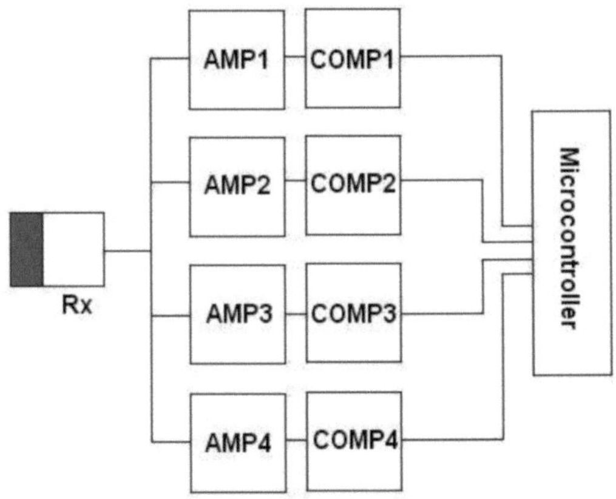

Fig. 3. Clustering circuit to assure that the A value is similar in signals which are compared in terms of duration of the time interval. The abbreviation Amp refers to the amplifier and Comp to the comparator

3.2.2 The Peak voltage detection circuit and the algorithm

In addition to the previous mentioned circuit also the system has a circuit that allows knowing the peak value of the received echo. This peak value is an estimation of A, which is also a parameter of the expression (1) and therefore can be used to identify the material. The associated circuit is shown in Figure 4. The AMP1, 2, 3 and 4 are the same as in the circuit of Figure 3. The ENV1, 2, 3 and 4 are envelope detectors, whose function is conditioning the signal before is sent to the microcontroller. The CD4052 is a multiplexer. The exit of each comparator goes to a specific input of the CD4052, and the output of this latter in the pin 3 goes to the microcontroller.

As mentioned above, the clustering circuit identifies which group the signal belongs to; this information is used by the microcontroller to select which of the four envelope detectors should be used to find the peak amplitude. This selection is done by means of the CD4052. Once the microcontroller has chosen the right envelope detector and captured the correspondent signal, it proceeds to search the peak value by successive analog to digital conversions and comparison.

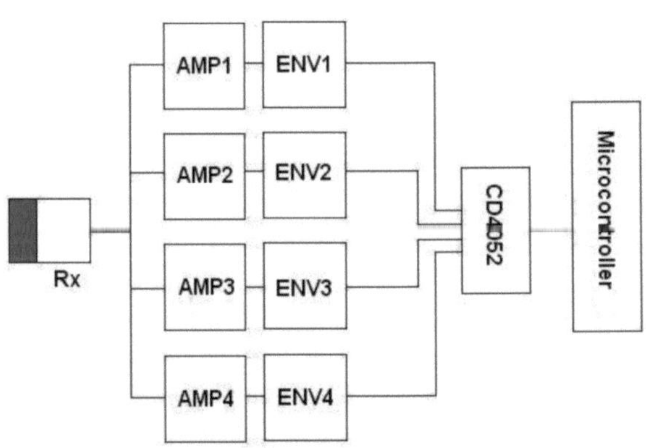

Fig. 4. Electronic circuit for calculating an estimated value of A. AMP1, AMP2, AMP3 and AMP4 are the same as in Figure 3. ENV means envelope detector

3.2.3 The Circuit and the algorithm for the identification of the changes in the durations

As it was mentioned in section 3.1 the material can be identified by finding the difference between the two durations which are produced for the same signal when it is amplified by two different gain values, i.e. finding Δt.

In order to do this, the circuit of Figure 5 was implemented. In this case AMP1-4 correspond to the same amplifiers mentioned in Figure 3. Amp1.1, Amp2.1, Amp3.1 and Amp4.1 are additional amplifiers. COMP5, COMP6, COMP7 and COMP8 are comparators which even with an amplitude modulated sinusoidal signal, such as the ultrasonic echo, remains activated as long as the envelope of the signal is above the threshold of 2.5V. While the output of this comparator is quite stable, still it has some noisy fluctuations which are necessary to be removed and this is the reason to use the comparators COMP5.1, COMP6.1, COMP7.1 and COMP8.1. The output of these comparators is connected to a specific input in the CD4052, and the output in the pin 13 of this multiplexer goes to the microcontroller.

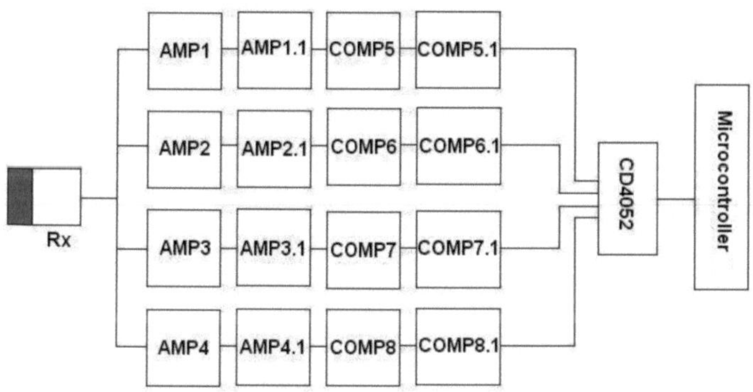

Fig. 5. The electronic circuit for calculating the durations associated to different gain values

4. Results and discussion

In order to prove the effectiveness of the Peniel Method, it was performed some experiments. The characteristics of the experiments are mentioned below:

The transducer used for the measurements was the AT120 (120KHz) of Airmar Technology, which is a low cost transducer commonly used for obstacle avoidance or level measurement but not for NDT. This transducer was used both for transmission and reception in the Through-Transmission configuration which was used to inspect the samples.

The transmitter circuit is the T1 *development board* of Airmar Technology (Airmar Technology, 2011). With this transmitter is sent a 130KHz toneburst which has a 40μs duration and a pulse rate of 10Hz.

The chosen samples for the inspection were Acrylic (16mm thickness), Aluminium (16mm thickness) and Glass (14mm thickness). Oil was the couplant used between the transducers and the inspected sample.

The receiver circuit is the result of integrating the three circuits mentioned in the section 3.2. The circuit is fed by four 9volts batteries. Two of them are for the positive source and the other two are for the negative. The data were processed by the main board of the mom-robot of the kit TEAC^{2}H-RI (González J.J. *et al*, 2010a).which was developed by the authors. In the main board, the microcontroller MC68HC908JK8 is the device in charge to assign tasks or

process the data of the sensors. The six outputs from the receiver circuit go to the main board and two outputs of the main board go to the receiver circuit. These two outputs control the CD4052.
The microcontroller sends the duration and the peak voltage of the signals to the PC. There, the signal is processed by Matlab and the results are plotted.
The whole set up of the experiments can be seen in figure 6.
In order to assure the repeatability of the results the following procedure was followed:
Procedure 1

1. Clean the inspected sample and the transducers face
2. Spread oil over the surface of both transducer faces
3. Push the transducers against the inspected sample
4. Take ten different measurements
5. Repeat this procedure for five times.

The ten different measurements refer to send in different moments a toneburst to the inspected sample and capture the received echo. This is done ten times.
For the different materials used in this experiment this procedure was followed and the results are in figure 7.

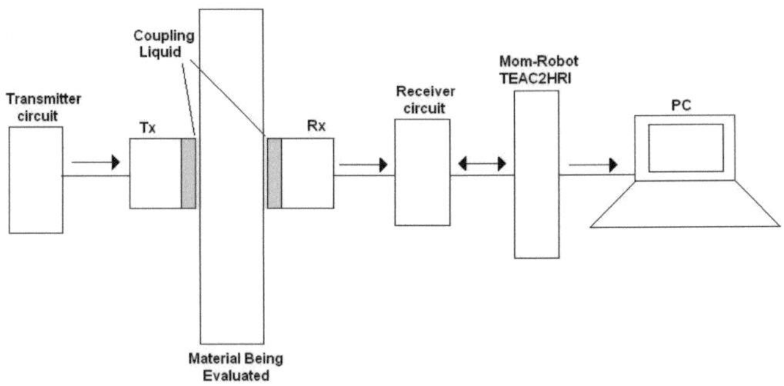

Fig. 6. Diagram of the Experimental set up used in procedure 1. The couplant is oil. Tx and Rx refer to the Transmitter and Receiver, respectively

In figure 7 Delta t_1 refers to t_{di2} - t_{di1}, Delta t_2 refers to t_{di3} - t_{di2}, and so on. t_{di1} refers to the duration of the signal at the output of the comparator which corresponds to the group signal and the other t_{di}, belong to the comparators of the groups with amplifiers with higher gain than the amplifier of the group of the signal, e.g. if the signal belongs to group 2 the t_{di1} refers to the duration of Comp6.1, and t_{di2} and t_{di3} belong to Comp7.1 and Comp8.1, respectively. As it can be seen, for the higher order groups the number of t_{di} will be fewer than for the groups of lower order.
As it can be seen from figures 7a-7h, the Δt is not constant for any material. However, it can be said that this value is always within a defined interval. In [Gunarathne et al, 2002] is mentioned that the decay coefficient B does not have a fixed value from measurement to measurement but it takes different values within an interval. Because of this is defined a Gaussian Probability Density Function for the B value of each material, then the value can be any of those values within the PDF. This fact, confirms that the results obtained here with Δt are correct because they are congruent with the theory. Though, for the current case it was not necessary to create a PDF but only intervals where the correspondent Δt belongs. This fact simplifies the material identification task. Following, the intervals are chosen.

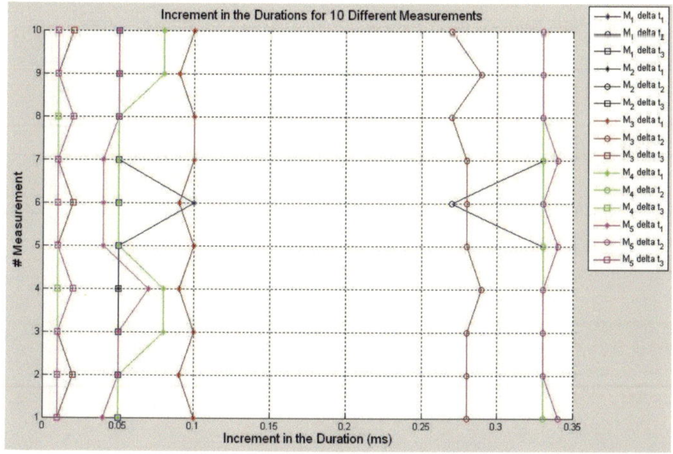

(a)

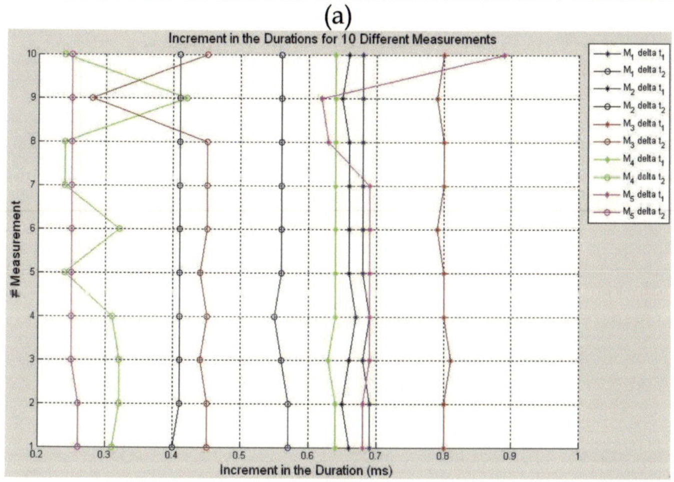

(b)

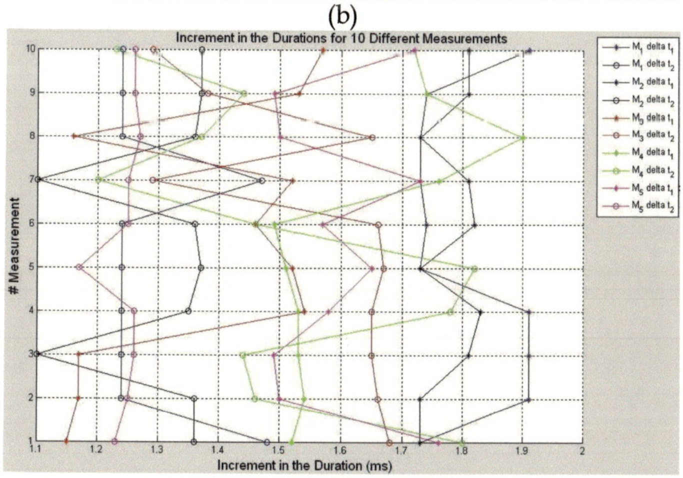

(c)

Fig. 7. Duration increment of the time interval for the results obtained using the procedure 1. The materials inspected were a) Acrylic b) Glass and c) Aluminum

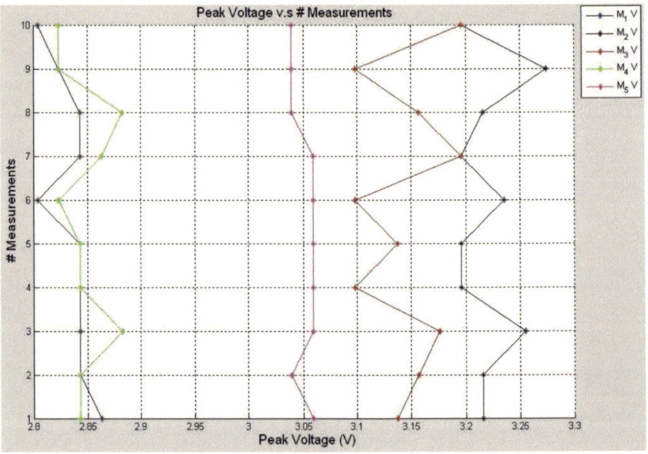

(a)

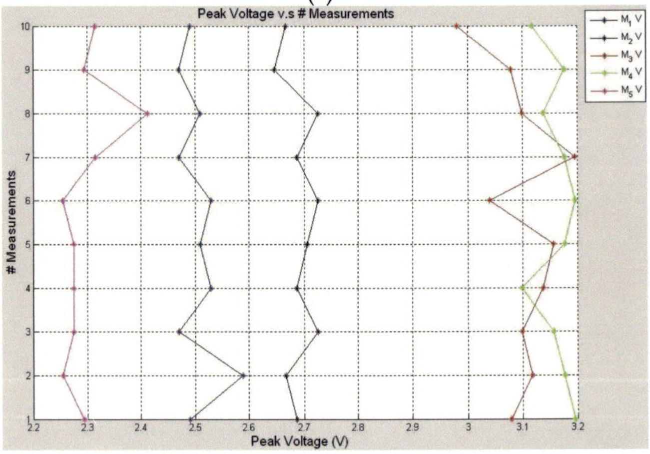

(b)

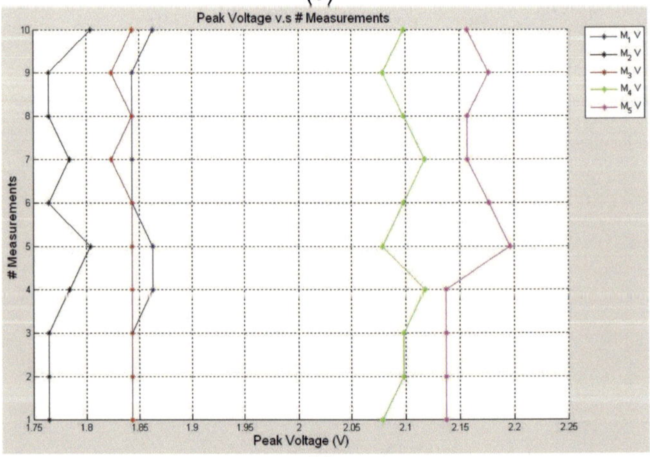

(c)

Fig. 8. Peak voltage for the results obtained using the procedure 1. The materials inspected were a) Acrylic b) Glass and c) Aluminum

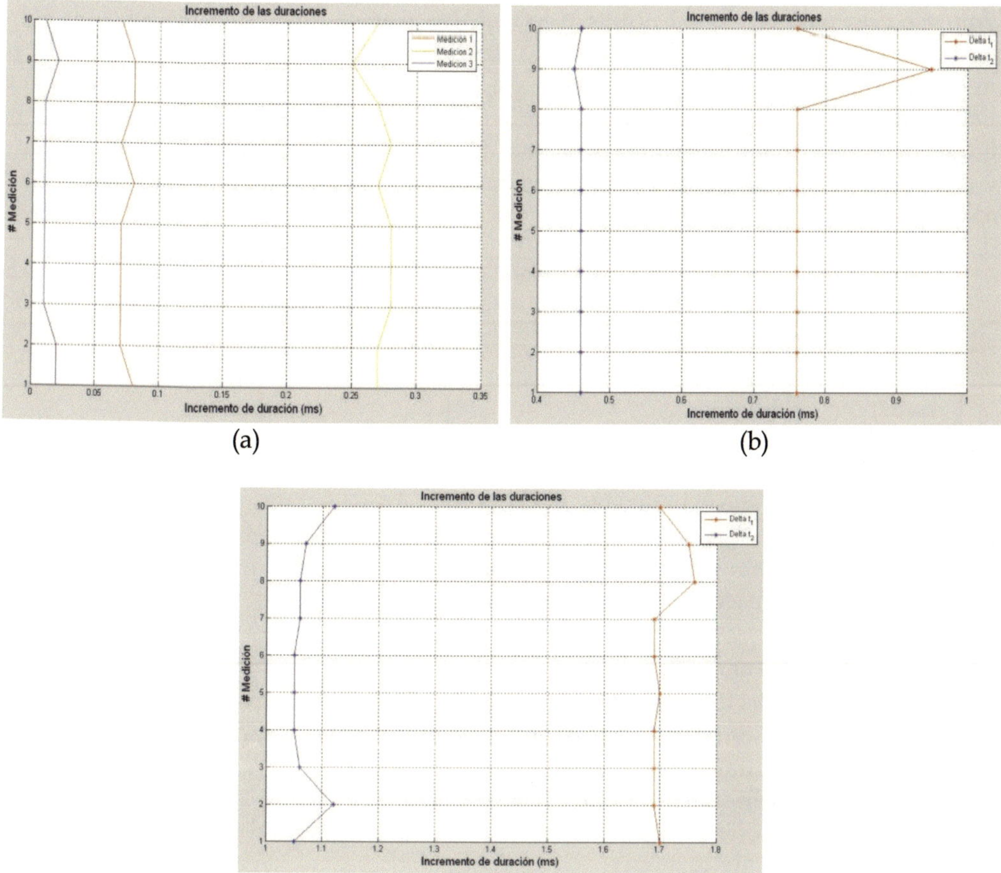

(a) (b)

(c)

Fig. 9. Duration increment of the time interval for the results obtained using the procedure 2. The materials inspected were a) Acrylic b) Glass and c) Aluminum

From figure 7a it can be seen that the Δt_1 is always within the range (0.025ms, 0.15ms), while Δt_2 and Δt_3 are in the ranges (0.16ms, 0.4ms) and (0, 0.1ms), respectively, then these intervals are chosen to identify the acrylic sample.

From figure 7b it can be seen that the Δt_1 and Δt_2 are always between (0.6ms, 1ms) and (0.2ms, 0.65ms), respectively. Then, those are the intervals for the glass sample.

From figure 7c it can be seen that the Δt_1 and Δt_2 are always between (1ms, 2ms) and (0.8ms, 1.9ms), respectively. Then, those are the intervals for the aluminium sample.

In figure 8 is shown the peak voltages for the correspondent measurements in figure 7. It can be seen from this figure that the peak voltage also varies in a wide range from measurement to measurement. Then, if this parameter is going to be implemented to identify the material also have to be defined a voltage interval for each material. This parameter can be used to make more robust the measurements but for this moment only the Δt variable will be used for the identification of the materials.

It is important to mention that both, the peak voltage and Δt change, are within an interval from measurement to measurement because those variables are very dependent on the characteristics of the couplant used between the transducer face and the inspected sample.

From the previous mentioned results the algorithm was modified and the intervals were used to identify the material. With this new algorithm a procedure (procedure 2) similar to procedure 1 was repeated for all the materials. The only difference in this new procedure is that the average of the ten measurements is taken and this result is compared with the different intervals. In figure 9 can be seen one of the five repetitions of the new procedure.
In table 1 can be seen the performance of the developed sensor for the material identification task.

Material	Trials	Correct Identification	Wrong Identification	% Accuracy
Acrylic	5	5	0	100%
Glass	5	5	0	100%
Aluminum	5	5	0	100%

Table 1. Accuracy of the developed sensor to identify the material

As it can be seen in these experiments the accuracy of the identification is 100% for all the materials. This fact is very important because with a quite easy method and a low cost computational algorithm was performed a task which requires of more complex methods and algorithms of more expensive computational cost.

4.1 The robotic kit TEAC^{2}H-RI

In the previous sections, the signal was processed using the main board of the mom-robot of the kit TEAC^{2}H-RI (González J.J. et al, 2010a) joint with the software Matlab. Then, the Peniel Method was partially implemented in a robotic system.
In future work the sensor will be fully implemented in a robotic system formed by two robots of the kit TEAC^{2}H-RI. These robots are the mom-robot and the son-robot. Both of them can be seen in figure 10

Fig. 10. Son-robot and mom-robot exploring the environment (González J.J. *et al*, 2010a)

5. Conclusion and future work

In this book chapter was proposed for the first time in the literature the shift in Δt as a variable which can be used to identify the material. The method based on this variable, Peniel Method, is much easier and much less computationally expensive than other methods in the literature. This fact is the most important approach of this research because it lets to implement the ultrasonic material identification sensor in two robots. This last result was not found in the literature reviewed.

The cost of the ultrasonic AT120 transducer is only $35US which is very low in comparison with other options (>$400US) in the market which are used for the identification of materials. Then, it can be concluded that low cost ultrasonic transducers can be used instead of more expensive transducers if the proper method is used. This is another important strength of the method because it can be used in industrial processes to automate the material identification task at a low cost.

Another important conclusion of this work is that if the appropriate methodology is followed to design a sensor, it could be save considerable economical investments in the transducer, the processing system and the other elements necessary to design the sensor. In this case the methodology followed to develop the sensor consists of seven steps: 1) to study in depth the model or physical-mathematical models proposed in the literature 2) to perform experiments to confirm the theory and identify new signal patterns which let to develop new models 3) to redefine or create a new physical-mathematical model which must be simpler than the others and fits to the specific characteristics of the application, which in this case is two small robots to identify materials 4) This new model is the basis of the new method, but again it is necessary to also consider the characteristics of the specific application and all the constraints in order to design the electronic circuit and the algorithm of the method 5) to build the sensor with the proper devices in terms of cost, size and functionality 6) to design the appropriate experiments for the verification of the method 7) to validate the performance of the sensor using the proposed experiments and debug both the algorithm and circuit for optimum performance.

As it can be seen not only is necessary to design or redefined the software or hardware but also the mathematical model in order to obtain the sensor which fulfils the requirements of the application at the lowest cost.

The next step of this work is to implement completely the sensor in the mom-robot and son-robot of the kit TEAC^{2}H-RI without depending on the processing of the PC to identify the material.

Also more materials will be evaluated with the sensor in order to provide to the robot a wider database of materials.

Also in future work, will be used the peak voltage as another factor to identify the materials or to differentiate between materials of very similar acoustic properties.

6. Acknowledgment

We would like to acknowledge to the Science, Technology and Innovation Management Department, COLCIENCIAS, for the economical support to the master student Juan José González España within the Youth Researchers and Innovators Program- 2009.

7. References

Allin M. J. (2002) Disbond *Detection in Adhesive Joints Using Low Frequency Ultrasound.* PhD Thesis. Imperial College of Science Technology and Medicine. University of London. 2002

Airmar Technology. (2011). http://www.airmartechnology.com/ Accesed 4th August 2011

González J.J.; Jiménez J.A.; Ovalle D.A. (2010) *TEAC2H-RI: Educational Robotic Platform for Improving Teaching-Learning Processes of Technology in Developing Countries.* Technological Developments in Networking, Education and Automation 2010, pp 71-76

González J., Jiménez J. *Algoritmo para la Identificación de Materiales en Agentes Robóticos (Spanish).* Quinto Congreso Colombiano de Computación (5CCC) Cartagena-Colombia (2010)

Gunarathne, G. P. P. (1997) *Measurement and monitoring techniques for scale deposits in petroleum pipelines.* In: Proc. IEEE Instrum. Meas. Technol. Conf., Ottawa, ON, Canada, May 19–21, 1997, pp. 841–847. DOI: 10.1109/IMTC.1997.610200

Gunarathne G.P.P ; Zhou Q. ; Christidis K. (1998) *Ultrasonic feature extraction techniques for characterization and quantification of scales in petroleum pipelines.* In: Proceedings IEEE Ultrasonic Symposium, vol. 1; 1998, p. 859–64 DOI: 10.1109/ULTSYM.1998.762279

Gunarathne G.P.P ; Christidis K. (2002) *Material Characterization in situ Using ultrasound measurements.* In: Proceedings of the IEEE Transactions on Instrumentation and Measurement. Vol. 51, pp. 368-373. 2002 DOI: 10.1109/19.997839

NASA (2007) *Ultrasonic Testing of Aerospace Materials.* URL: klabs.org/DEI/References/design_guidelines/test_series/1422msfc.pdf Accessed on: August 2011.

Ohtanil K., Baba M. (2006) *An Identification Approach for Object Shapes and Materials Using an Ultrasonic Sensor Array.* Proceedings of the SICE-ICASE International Joint Conference, pp. 1676-1681. 2006

Pallav P., Hutchins D., Gan T. (2009) *Air-coupled ultrasonic evaluation of food materials.* Ultrasonics 49 (2009) 244-253

Stepanić, H. Wüstenberg, V. Krstelj, H. Mrasek (2003) *Contribution to classification of buried objects based on acoustic impedance matching,* Ultrasonics, 41(2), pp. 115-123, 2003

Thomas, S.M, Bull D.R, (1991) *Neural Processing of Airborne Sonar for Mobile Robot Applications.* Proceedings of the Second International Conference on Artificial Neural Networks. pp. 267-270. 1991.

Zhao B., Bashir O.A., Mittal G.S., (2003) *Detection of metal, glass, plastic pieces in bottled beverages using ultrasound,* Food Research International 36 (2003) 513–521.

Part 4

Programming and Algorithms

23

Robotic Software Systems: From Code-Driven to Model-Driven Software Development

Christian Schlegel, Andreas Steck and Alex Lotz
Computer Science Department, University of Applied Sciences Ulm
Germany

1. Introduction

Advances in robotics and cognitive sciences have stimulated expectations for emergence of new generations of robotic devices that interact and cooperate with people in ordinary human environments (robot companion, elder care, home health care), that seamlessly integrate themselves into complex environments (domestic, outdoor, public spaces), that fit into different levels of system hierarchies (human-robot co-working, hyper-flexible production cells, cognitive factory), that can fulfill different tasks (multi-purpose systems) and that are able to adapt themselves to different situations and changing conditions (dynamic environments, varying availability and accessibility of internal and external resources, coordination and collaboration with other agents).

Unfortunately, so far, steady improvements in specific robot abilities and robot hardware have not been matched by corresponding robot performance in real-world environments. On the one hand, simple robotic devices for tasks such as cleaning floors and cutting the grass have met with growing commercial success. Robustness and single purpose design is the key quality factor of these simple systems. At the same time, more sophisticated robotic devices such as *Care-O-Bot 3* (Reiser et al., 2009) and *PR2* (Willow Garage, 2011) have not yet met commercial success. Hardware and software complexity is their distinguishing factor.

Advanced robotic systems are systems of systems and their complexity is tremendous. Complex means they are built by integrating an increasingly larger body of heterogeneous (robotics, cognitive, computational, algorithmic) resources. The need for these resources arises from the overwhelming number of different situations an advanced robot is faced with during execution of multitude tasks. Despite the expended effort, even sophisticated systems are still not able to perform at an expected and appropriate level of overall quality of service in complex scenarios in real-world environments. By quality of service we mean the set of system level non-functional properties that a robotic system should exhibit to appropriately operate in an open-ended environment, such as robustness to exceptional situations, performance despite of limited resources and aliveness for long periods of time.

Since vital functions of advanced robotic systems are provided by software and software dominance is still growing, the above challenges of system complexity are closely related to the need of mastering software complexity. Mastering software complexity becomes pivotal towards exploiting the capabilities of advanced robotic components and algorithms. Tailoring modern approaches of software engineering to the needs of robotics is seen as decisive towards significant progress in system integration for advanced robotic systems.

2. Software engineering in robotics

Complex systems are rarely built from scratch but their design is typically partitioned according to the variety of technological concerns. In robotics, these are among others mechanics, sensors and actuators, control and algorithms, computational infrastructure and software systems. In general, successful engineering of complex systems heavily relies on the *divide and conquer* principle in order to reduce complexity. Successful markets typically come up with precise role assignments for participants and stakeholders ranging from component developers over system integrators and experts of an application domain to business consultants and end-users.

Sensors, actuators, computers and mechanical parts are readily available as commercial off-the-shelf black-box components with precisely specified characteristics. They can be re-used in different systems and they are provided by various dedicated suppliers. In contrast, most robotics software systems are still based on proprietarily designed software architectures. Very often, robotics software is tightly bound to specific robot hardware, processing platforms, or communication infrastructures. In addition, assumptions and constraints about tasks, operational environments, and robotic hardware are hidden and hard-coded in the software implementation.

Software for robotics is typically embedded, concurrent, real-time, distributed, data-intensive and must meet specific requirements, such as safety, reliability and fault-tolerance. From this point of view, software requirements of advanced robots are similar to those of software systems in other domains, such as avionics, automotive, factory automation, telecommunication and even large scale information systems. In these domains, modern software engineering principles are rigorously applied to separate roles and responsibilities in order to cope with the overall system complexity.

In robotics, tremendous code-bases (libraries, middleware, etc.) coexist without being interoperable and each tool has attributes that favors its use. Although one would like to reuse existing and matured software building blocks in order to reduce development time and costs, increase robustness and take advantage from specialized and second source suppliers, up to now this is not possible. Typically, experts for application domains need to become experts for robotics software to make use of robotics technology in their domain. So far, robotics software systems even do not enforce separation of roles for component developers and system integrators.

The current situation in software for robotics is caused by the lack of separation of concerns. In consequence, role assignments for robotics software are not possible, there is nothing like a software component market for robotic systems, there is no separation between component developers and system integrators and even no separation between experts in robotics and experts in application domains. This is seen as a major and serious obstacle towards developing a market of advanced robotic systems (for example, all kinds of cognitive robots, companion systems, service robots)

The current situation in software for robotics can be compared with the early times of the *World Wide Web (WWW)* where one had to be a computer engineer to setup web pages. The *WWW* turned into a universal medium only since the availability of tools which have made it accessible and which support separation of concerns: domain experts like journalists can now easily provide content without bothering with technical details and there is a variety of specialized, competing and interoperable tools available provided by computer engineers, designers and others. These can be used to provide and access any kind of content and to support any kind of application domain.

Based on these observations, we assume that the next big step in advanced robotic systems towards mastering their complexity and their overall integration into any kind of environment and systems depends on separation of concerns. Since software plays a pivotal role in advanced robotic systems, we illustrate how to tailor a service-oriented component-based software approach to robotics, how to support it by a model-driven approach and according tools and how this allows separation of concerns which so far is not yet addressed appropriately in robotics software systems.
Experienced software engineers should get insights into the specifics of robotics and should better understand what is in the robotics community needed and expected from the software engineering community. *Experienced roboticists* should get detailed insights into how model-driven software development *(MDSD)* and its design abstraction is an approach towards system-level complexity handling and towards decoupling of robotics knowledge from implementational technologies. *Practitioners* should get insights into how separation of concerns in robotics is supported by a service-oriented component-based software approach and that according tools are already matured enough to make life easier for developers of robotics software and system integrators. *Experts in application domains* and *business consultants* should gain insights into maturity levels of robotic software systems and according approaches under a short-term, medium-term and long-term perspective. *Students* should understand how design abstraction as recurrent principle of computer science applied to software systems results in *MDSD*, how *MDSD* can be applied to robotics, how it provides a perspective to overcome the vicious circle of robotics software starting from scratch again and again and how software engineering and robotics can cross-fertilize each other.

2.1 Separation of concerns

Separation of concerns is one of the most fundamental principles in software engineering (Chris, 1989; Dijkstra, 1976; Parnas, 1972). It states that a given problem involves different kinds of concerns, promotes their identification and separation in order to solve them separately without requiring detailed knowledge of the other parts, and finally combining them into one result. It is a general problem solving strategy which breaks the problem complexity into loosely-coupled subproblems. The solutions to the subproblems can be composed relatively easily to yield a solution to the original problem (Mili et al., 2004). This allows to cope with complexity and thereby achieving the required engineering quality factors such as robustness, adaptability, maintainability, and reusability.
Despite a common agreement on the necessity of the application of the separation of concerns principle, there is not a well-established understanding of the notion of concern. Indeed, *concern* can be thought of as a unit of modularity (Blogspot, 2008). Progress towards separation of concerns is typically achieved through modularity of programming and encapsulation (or *transparency* of operation), with the help of information hiding. Advanced uses of this principle allow for simultaneous decomposition according to multiple kinds of (overlapping and interacting) concerns (Tarr et al., 2000).
In practice, the principle of separation of concerns should drive the identification of the right decomposition or modularization of a problem. Obviously, there are both: (i) generic and domain-independent patterns of how to decompose and modularize certain problems in a suitable way as well as (ii) patterns driven by domain-specific best practices and use-cases.
In most engineering approaches as well as in robotics, at least the following are dominant dimensions of concerns which should be kept apart (Björkelund et al., 2011; Radestock & Eisenbach, 1996):

Computation provides the functionality of an entity and can be implemented in different ways (software and/or hardware). Computation activities require communication to access required data and to provide computed results to other entities.

Communication exchanges data between entities (ranging from hardware devices to interfaces for real-world access over software entities to user interfaces etc.).

Configuration comprises the binding of configurable parameters of individual entities. It also comprises the binding of configurable parameters at a system level like, for example, connections between entities.

Coordination is about when is something being done. It determines how the activities of all entities in a system should work together. It relates to orchestration and resource management.

According to (Björkelund et al., 2011), this is in line with results published in (Delamer & Lastra, 2007; Gelernter & Carriero, 1992; Lastra & Delamer, 2006) although variations exist which split *configuration* (into *connection* and *configuration*) or treat *configuration* and *coordination* in the same way (Andrade et al., 2002; Bruyninckx, 2011).

It is important to recognize that there are cross-cutting concerns like *quality of service (QoS)* that have instantiations within the above dimensions of concerns. Facets of *QoS for computation* can manifest with respect to time (best effort computation, hard real-time computation) or anytime algorithms (explicated relationship between assigned computing resources and achieved quality of result). Facets of *QoS for communication* are, for example, response times, latencies and bandwidth.

It is also important to recognize that various concerns need to be addressed at different stages of the lifecycle of a system and by different stakeholders. For example, *configuration* is part of the *design phase* (a component developer provides dedicated configurable parameters, a system integrator binds some of them for deployment) *and* of the *runtime phase* (the task coordination mechanism of a robot modifies parameter settings and changes the connections between entities according to the current situation and task to fulfill).

It is perfectly safe to say that robotics should take advantage from insights and successful approaches for complexity handling readily available in other but similar domains like, for example, automotive and avionics industry or embedded systems in general. Instead, robotics often reinvents the wheel instead of exploiting cross-fertilization between robotics and communities like software engineering and middleware systems. The interesting question is whether there are differences in robotics compared to other domains which hinder roboticists from jumping onto already existing and approved solutions. One should also examine whether or not these solutions are tailorable to robotics needs.

2.2 Specifics in robotics

The difference of robotics compared to other domains like automotive and avionics is neither the huge variety of different sensors and actuators nor the number of different disciplines being involved nor the diversity of hardware-platforms and software-platforms. In many domains, developers need to deal with heterogeneous hardware devices and are obliged to deploy their software on computers which are often constrained in terms of memory and computational power.

We are convinced that differences of robotics compared to other domains originate from the need of a robot to cope with open-ended environments while having only limited resources at its disposal.

Limited resources require decisions: when to assign which resources to what activity taking into account perceived situation, current context and tasks to be fulfilled. Finding adequate solutions for this major challenge of engineering robotic systems is difficult for two reasons:

- the *problem space* is huge: as uncertainty of the environment and the number and type of resources available to a robot increase, the definition of the best matching between current situation and correct robot resource exploitation becomes an overwhelming endeavour even for the most skilled robot engineer,
- the *solution space* is huge: in order to enhance overall quality of service like robustness of complex robotic systems in real-world environments, robotic system engineers should master highly heterogeneous technologies, need to integrate them in a consistent and effective way and need to adequately exploit the huge variety of robotic-specific resources.

In consequence, it is impossible to statically assign resources in advance in such a way that all potential situations arising at runtime are properly covered. Due to open-ended real-world environments, there will always be a deviation between *design-time optimality* and *runtime optimality* with respect to resource assignments. Therefore, there is a need for dynamic resource assignments at runtime which arises from the enormous sizes of the problem space and the solution space.
For example, a robot designer cannot foresee how crowded an elevator will be. Thus, a robot will need to decide by its own and at runtime whether it is possible and convenient to exploit the elevator resource. The robot has to trade the risk of hitting an elevator's user with the risk of arriving late at the next destination. To match the level of safety committed at design-time, the runtime trade-off has to come up with parameters for speed and safety margins whose risk is within the design-time committed boundaries while still implementing the intent to enter the elevator.

2.2.1 Model-centric robotic systems

The above example illustrates why we have to think of engineering advanced robotic systems differently compared to other complex systems. A complex robotic system cannot be treated as design-time finalizable system. At runtime, system configurations need to be changeable according to current situation and context including prioritized assignments of resources to activities, (de)activations of components as well as changes to the wiring between components. At runtime, the robot has to analyze and to decide for the most appropriate configuration. For example, if the current processor load does not allow to run the navigation component at the highest level of quality, the component should be configured to a lower level of navigation quality. A reasonable option to prepare a component to cope with reduced resource assignments might be to reduce the maximum velocity of the robot in order to still guarantee the same level of navigation safety.
In consequence, we need to support design-time reasoning (at least by the system engineer) as well as runtime reasoning (by the robot itself) about both, the problem space and the solution space. This can be achieved by raising the level of abstraction at which relevant properties and characteristics of a robotics system are expressed. As for every engineering endeavour, this means to rely on the power of models and asks for an overall different design approach as illustrated in figure 1:

- The solution space can be managed by providing advanced design tools for robot software development to design reconfigurable and adaptive robotic systems. Different stakeholders involved in the development of a robotic system need the ability to formally

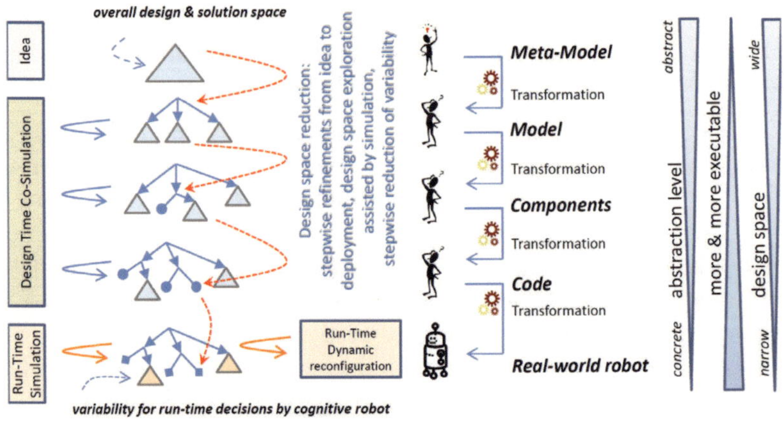

Fig. 1. Novel workflow bridging design-time and runtime model-usage: at design-time variation points are purposefully left open and allow for runtime decisions (Schlegel et al., 2010).

model and relate different views relevant to robotic system design. A major issue is the support of separation of concerns taking into account the specific needs of robotics.

- The problem space can be mastered by giving the robot the ability to reconfigure its internal structure and to adapt the way its resources are exploited according to its understanding of the current situation.

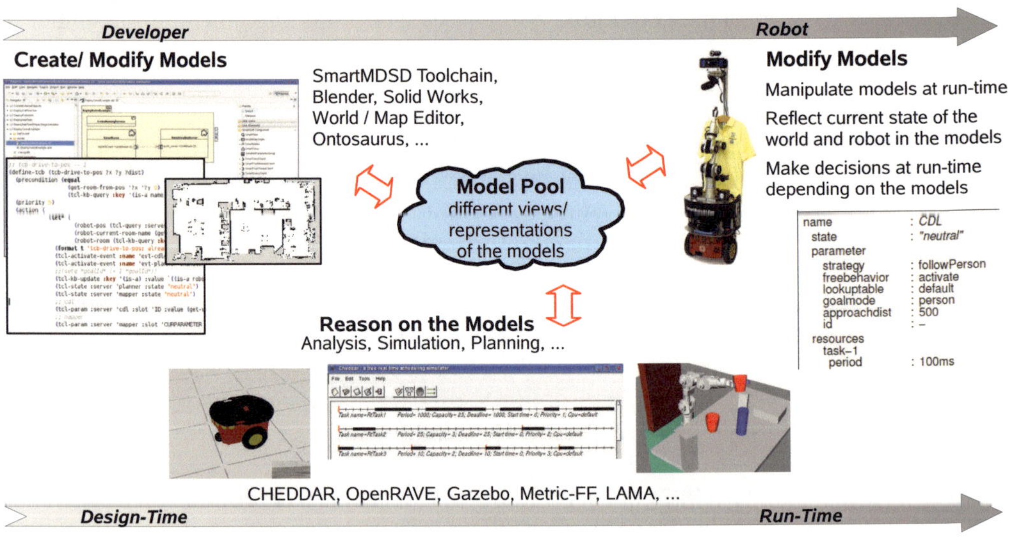

Fig. 2. Separation of concerns and design abstraction: models created at design-time are used and manipulated at runtime by the robot (Steck & Schlegel, 2011).

We coin the term *model-centric robotic systems* (Steck & Schlegel, 2011) for the new approach of using models to cover and support the whole life-cycle of robotic systems. Such a model-centric view puts models into focus and bridges design-time and runtime model-usage.

During the whole lifecycle, models are refined and enriched step-by-step until finally they become executable. Models comprise variation points which support alternative solutions. Some variation points are purposefully left open at design time and even can be bound earliest at runtime after a specific context and situation dependent information is available. In consequence, models need to be interpretable not only by a human designer but also by a computer program. At design-time, software tools should understand the models and support designers in their transformations. At runtime, adaptation algorithms should exploit the models to automatically reconfigure the control system according to the operational context (see figure 2).
The need to explicitly support the design for runtime adaptability adds robotic-specific requirements on software structures and software engineering processes, gives guidance on how to separate concerns in robotics and allows to understand where the robotics domain needs extended solutions compared to other and at first glance similar domains.

2.2.2 User roles and requirements

Another strong influence on robotic software systems besides technical challenges comes from the involved individuals and their needs. We can distinguish several user roles that all put a different focus on complexity management, on separation of concerns and on software engineering in robotics:

End users operate applications based on the provided user interface. They focus on the functionality of readily provided systems. They do not care on how the application has been built and mainly expect reliable operation, easy usage and reasonable value for money.

Application builders / system integrators assemble applications out of approved, standardized and reusable off-the-shelf components. Any non trivial robotic application requires the orchestration of several components such as computer vision, sensor fusion, human machine interaction, object recognition, manipulation, localization and mapping, control of multiple hardware devices, etc. Once these parts work together, we call it a *system*. This part of the development process is called, therefore, *system integration*. Components can be provided by different vendors. Application builders and system integrators consider components as black boxes and depend on precise specifications and explications of all relevant properties for smooth composition, resource assignments and mappings to target platforms. Components are customized during system level composition by adjusting parameters or filling in application dependent parts at so-called *hot spots* via plug-in interfaces. Application builders expect support for system-level engineering.

Component builders focus on the specification and implementation of a single component. They want to focus on algorithms and component functionality without being restricted too much with respect to component internals. They don't want to bother with integration issues and expect a framework to support their implementation efforts such that the resulting component is conformant to a system-level black box view.

Framework builders / tool providers prepare and provide tools that allow the different users to focus on their role. They implement the software frameworks and the domain-specific add-ons on top of state-of-the-art and standard software systems (like middleware systems), use latest software technology and make these available to the benefit of robotics.

The robotics community provides domain-specific concepts, best practices and design patterns of robotics. These are independent of any software technology and implementational technology. They form the *body of knowledge of robotics* and provide the domain-specific ground for the above roles.

The essence of the work of the *component builder* is to design reusable components which can seamlessly be integrated into multiple systems and different hardware platforms. A component is considered as a black box. The developer can achieve this abstraction only if he is strictly limited in his knowledge and assumptions about what happens outside his component and what happens inside other components.
On the other hand, the methodology and the purpose of the *system integrator* is opposite: he knows exactly the application of the software system, the platform where it will be deployed and its constraints. For this reason, he is able to take the right decision about the kind of components to be used, how to connect them together and how to configure their parameters and the quality of service of each of them to orchestrate their behavior. The work of the system integrator is rarely reusable by others, because it is intrinsically related to a specific hardware platform and a well-defined and sometimes unique use-case. We don't want the system integrator to modify a component or to understand the internal structure and implementation of the components he assembles.

2.2.3 Separation of roles from an industrial perspective

This distinction between the development of single components and system integration is important (figure 3). So far, reuse in robotics software is mainly possible at the level of libraries and/or complete frameworks which require system integrators to be component developers and vice versa. A formal separation between *component building* and *system integration* introduces another and intermediate level of abstraction for reuse which will make it possible to

- create commercial off-the-shelf (COTS) robotic software: when components become independent of any specific robot application, it becomes possible to integrate them quickly into different robotic systems. This abstraction allows the component developer to sell its robotic software component to a system integrator;
- overcome the need for the system integrator to be also an expert of robotic algorithms and software development. We want companies devoted to system integration (often SMEs) to take care of the Business-to-Client part of the value chain, but this will be possible only when their work will become less challenging;
- establish dedicated system integrators (specific to industrial branches and application domains) apart from experts for robotic components (like navigation, localization, object recognition, speech interaction, etc.);
- provide plug-and-play robotic hardware: so far the effort of the integration of the hardware into the platform was undertaken by the system integrator. If manufacturers start providing ready-to-use drivers which work seamlessly in a component-driven environment, robotic applications can be deployed faster and become cheaper.

This separation of roles will eventually have a positive impact in robotics: it will potentially allow the creation of a robotics industry, that is an ecosystem of small, medium and large enterprises which can profitably and symbiotically coexist to provide business-to-business

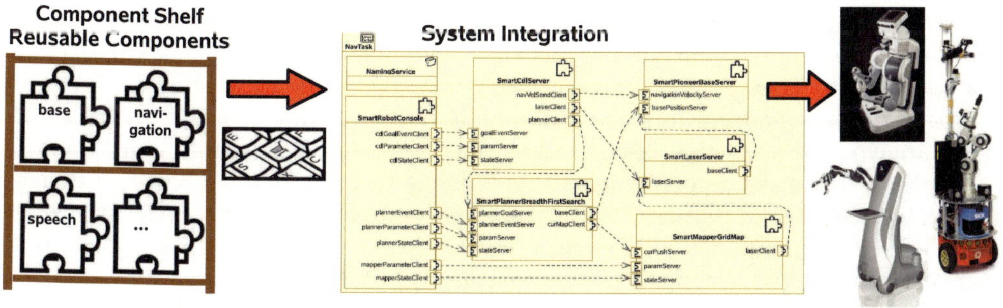

Fig. 3. Building robotic systems out of readily-available and reusable software components: separation of the roles of component development and system integration.

and business-to-client products such as hardware parts, software applications and complete robots with a specific application and deliver a valuable product to the customer.
To understand better what a *robotics industry* means, we draw an analogy to the personal computer industry. Apart from very few exceptions, we can identify several companies involved in the manufacturing of a single and very specific part of the final product: single hardware components (memories, hard drives, CPU, mother boards, screens, power supplies, graphic cards, etc.), operating systems (Windows, commercial Linux distributions), software applications (CAD, word processing, video games, etc.) and system integrators which provide ready-to-use platforms to the end user.

2.3 Service-oriented software components to master system complexity

Software engineering provides three major approaches that help to address the above challenges, that is *component-based software engineering (CBSE)*, *service-oriented architectures (SOA)* and *model-driven software development (MDSD)*.
CBSE separates the component development process from the system development process and aims at component reusability. *MDSD* separates domain knowledge (formally specified by domain experts) from how it is being implemented (defined by software experts using model transformations). *SOA* is about the right level of granularity for offering functionality and strictly separates service providers and consumers.

2.3.1 Component-based software engineering

CBSE (Heineman & Councill, 2001) is an approach that has arisen in the software engineering community in the last decade. It shifts the emphasis in system-building from traditional requirements analysis, system design and implementation to composing software systems from a mixture of reusable off-the-shelf and custom-built components. A compact and widely accepted definition of a software component is the following one:

> "A **software component** is a unit of composition with contractually specified interfaces and explicit context dependencies only. A software component can be developed independently and is subject to composition by third parties." (Szyperski, 2002).

Software components explicitly consider reusable pieces of software including notions of independence and late composition. *CBSE* promises the benefits of increased reuse, reduced production cost, and shorter time to market. In order to realize these benefits, it is vital to have components that are easy to reuse and composition mechanisms that can be applied systematically. Composition can take place during different stages of the lifecycle of components that is during the design phase (design and implementation), the deployment phase (system integration) and even the runtime phase (dynamic wiring of data flow according to situation and context). *CBSE* is based on the explication of all relevant information of a component to make it usable by other software elements whose authors are not known. The key properties of *encapsulation* and *composability* result in the following seven criteria that make a good component: "(i) may be used by other software elements (clients), (ii) may be used by clients without the intervention of the component's developers, (iii) includes a specification of all dependencies (hardware and software platform, versions, other components), (iv) includes a precise specification of the functionalities it offers, (v) is usable on the sole basis of that specification, (vi) is composable with other components, (vii) can be integrated into a system quickly and smoothly" (Meyer, 2000).

2.3.2 Service-oriented architectures

Another generally accepted view of a software component is that it is a software unit with *provided services* and *required services*. In component models, where components are architectural units, *services* are represented as *ports* (Lau & Wang, 2007). This view puts the focus on the question of a proper level of abstraction of offered functionalities. Services "combine information and behavior, hide the internal workings from outside intrusion and present a relatively simple interface to the rest of the program" (Sprott & Wilkes, 2004). The (CBDI Forum, 2011) recommends to define *service-oriented architectures (SOA)* as follows:

SOA are "the policies, practices, frameworks that enable application functionality to be provided and consumed as sets of services published at a granularity relevant to the service consumer. Services can be invoked, published and discovered, and are abstracted away from the implementation using a single, standards-based form of interface" (Sprott & Wilkes, 2004).

Service is the key to communication between providers and consumers and key properties of good service design are summarized as in table 1. *SOA* is all about style (policy, practice, frameworks) which makes process matters an essential consideration. A *SOA* has to ensure that services don't get reduced to the status of interfaces, rather they have an identity of their own. With *SOA*, it is critical to implement processes that ensure that there are at least two different and separate processes - for providers and consumers (Sprott & Wilkes, 2004).

reusable	use of service, not reuse by copying of code/implementation
abstracted	service is abstracted from the implementation
published	precise, published specification functionality of service interface, not implementation
formal	formal contract between endpoints places obligations on provider and consumer
relevant	functionality is presented at a granularity recognized by the user as a meaningful service

Table 1. Principles of good service design enabled by characteristics of *SOA* as formulated in (Sprott & Wilkes, 2004).

2.3.3 Model-driven software development

MDSD is a technology that introduces significant efficiencies and rigor to the theory and practice of software development. It provides a design abstraction as illustrated in figure

4. Abstractions are provided by models (Beydeda et al., 2005). Abstraction is a core principle of software engineering.

"A **model** is a simplified representation of a system intended to enhance our ability to understand, predict and possibly control the behavior of the system" (Neelamkavil, 1987).

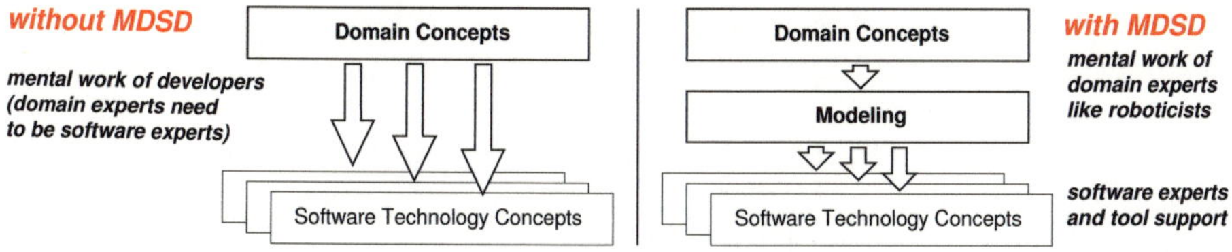

Fig. 4. Design abstraction of model-driven software development.

In *MDSD*, models are used for many purposes, including reasoning about problem and solution domains and documenting the stages of the software lifecycle; the result is improved software quality, improved time-to-value and reduced costs (IBM, 2006).

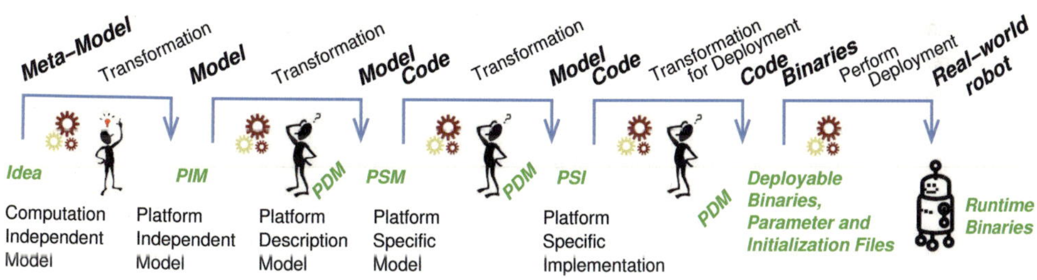

Fig. 5. Model-driven software development at a glance.

The standard workflow of a model-driven software development process is illustrated in figure 5. This workflow is supported by tools like the *Eclipse Modeling Project* (Eclipse Modeling Project, 2010) which provide means to express model-to-model and model-to-code transformations. They import standardized textual *XMI* representations of the models and can parse them according to the used meta-model. Thus, one can easily introduce domain-specific concepts to forward information from a model-level to the model transformations and code generators. Tools like *Papyrus* (PAPYRUS UML, 2011) allow for a graphical representation of the various models and can export them into the *XMI* format. Overall, there is a complete toolchain for graphical modelling and transformation steps available that can be tailored to the domain specific needs of robotics.

MDSD is much more than code generation for different platforms to address the technology change problem and to make development more efficient by automatically generating repetitive code. The benefits of *MDSD* are manifold (Stahl & Völter, 2006; Völter, 2006): (i) models are free of implementation artefacts and directly represent reusable domain knowledge including best practices, (ii) domain experts can play a direct role and are not requested to translate their knowledge into software representations, (iii) design patterns, sophisticated & optimized software structures and approved software solutions can be made available to domain experts and enforced by embedding them in templates for use by highly optimized code generators such that even novices can immediately take advantage from a coded immense experience, (iv) parameters and properties of components required for system level composition and the adaptation to different target systems are explicated and can be modified within a model-based toolchain.

2.4 Stable structures and freedom from choice

In robotics, we believe that the cornerstone is a *component model* based on *service-orientation* for its provided and required interactions represented in an abstract way in form of *models*.
A *robotics component model* needs to provide *component level* as well as *system level* concepts, structures and building blocks to support separation of concerns while at the same time ensuring composability based on a composition theory: (i) building blocks out of which one composes a component, (ii) patterns of how to design a well-formed component to achieve system level conformance, (iii) guidance towards providing a suitable granularity of services, (iv) specification of the behavior of interactions and (v) best practices and solutions of domain-specific problems. *MDSD* can then provide toolchains and thereby support separation of concerns and separation of roles.
The above approach asks for the identification of *stable structures* versus *variation points* (Webber & Gomaa, 2004). A robotics component model has to provide guidance via stable structures where these are required to support separation of concerns and to ensure system level conformance. At the same time, it has to allow for freedom wherever possible. The distinction between stable structures and variation points is of relevance at all levels (operating system interfaces, library interfaces, component internal structures, provided and required services etc.). In fact, identified and enforced stable structures come along with restrictions. However, one has to notice that well thought out limitations are not a universal negative and *freedom from choice* (Lee & Seshia, 2011) gives guidance and assurance of properties beyond one's responsibilities in order to ensure separation of concerns.
As detailed in (Schlegel et al., 2011), stable structures with respect to a service-oriented component-based approach can be identified. These are illustrated in figure 6.

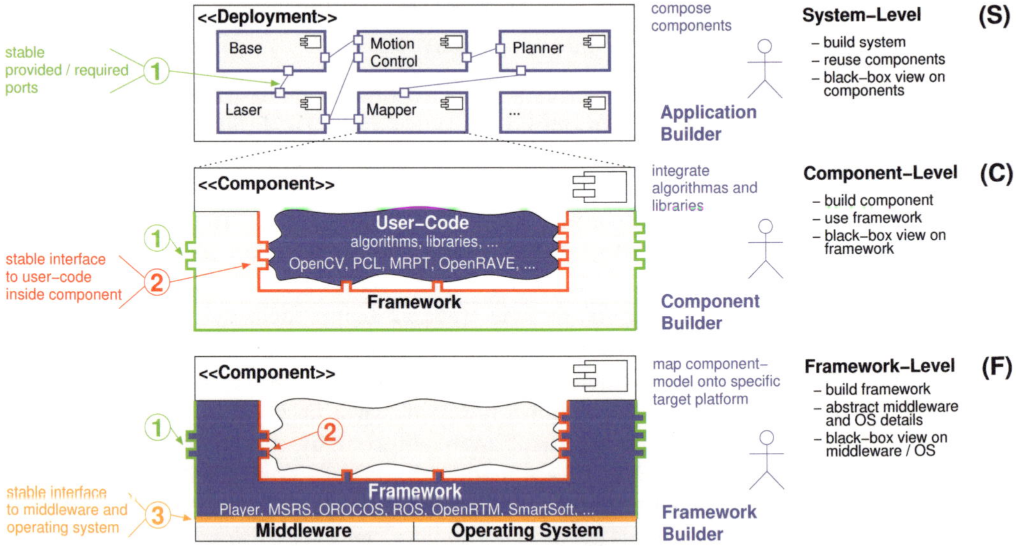

Fig. 6. Stable structures and different roles in a component-based software approach.

At the *system level (S)*, *provided* and *required* service ports ① of a component form a stable interface for the *application builder*. In an ideal situation, all relevant properties of a component are made explicit to support a black box view. Hence, system level properties like resource conformance of the component mapping to the computing platform can be checked during system composition and deployment.

At the *component level (C)*, the *component builder* wants to rely on a stable interface to the component framework ②. In an ideal situation, the component framework can be considered as black box hiding all operating system and middleware aspects from the user code. The component framework adds the execution container to the user code such that the resulting component is conformant to a black box component view.
At the *framework level (F)*, two stable interfaces exist: (i) between the framework and the user code of the component builder ② and (ii) between the framework and the underlying middleware & operating system ③. The stable interface ② ensures that no middleware and operating system specifics are unnecessarily passed on to the component builder. The stable interface ③ ensures that the framework can be mapped onto different implementational technologies (middleware, operating systems) without reimplementing the framework in its entirety. The *framework builder* maintains the framework which links the stable interfaces ② and ③ and maps the framework onto different implementational technologies via the interface ③.

3. The SMARTSOFT-approach

The basic idea behind SMARTSOFT (Schlegel, 2011) is to master the component hull and thereby achieve *separation of concerns* as well as *separation of roles*. Figure 7 illustrates the SMARTSOFT component model and how its component hull links the stable interfaces ①, ② and ③.

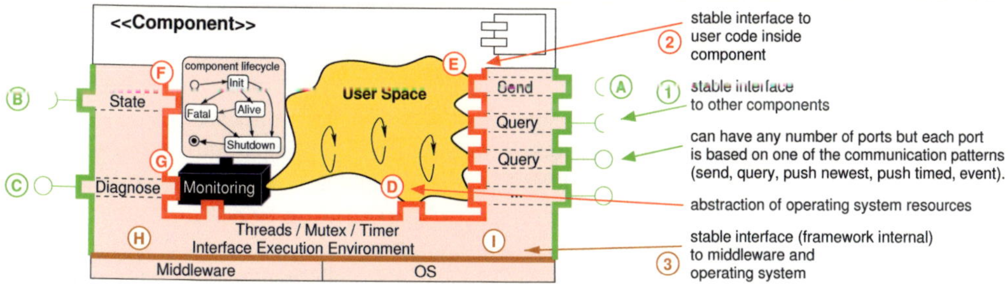

Fig. 7. The structure of a SMARTSOFT component and its stable interfaces.

Pattern	Description
send	one-way communication
query	two-way request
push newest	1-to-n distribution
push timed	1-to-n distribution
event	asynchronous notification

Service	Description
param	component configuration
state	activate/deactivate component services
wiring	dynamic component wiring
diagnose	introspection of components
(internally based on communication patterns)	

Table 2. The set of patterns and services of SMARTMARS.

The link between ① and ② is realized by *communication patterns*. Binding a communication pattern with the type of data to be transmitted results in an externally visible service represented as port. The small set of generic and predefined communication patterns listed in the left part of table 2 are the only ones to define externally visible services. Thus, the behavior and usage of a service is immediately evident as soon as one knows its underlying communication pattern.

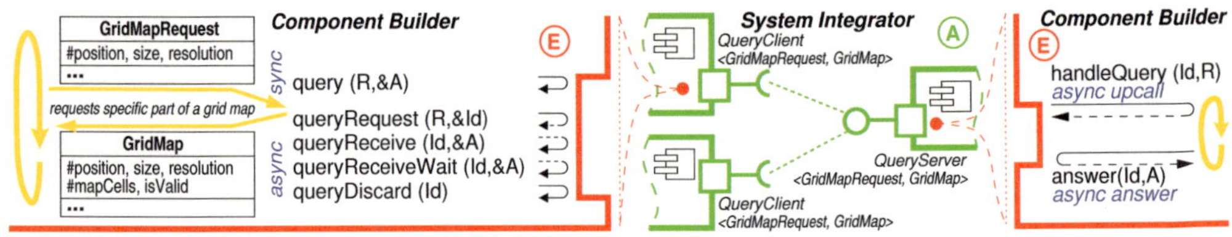

Fig. 8. The views of a component builder and a system integrator on services by the example of a grid map service based on a query communication pattern.

Figure 8 illustrates this concept by means of the *query* communication pattern which consists of a *query client* and a *query server*. The query pattern expects two *communication objects* to define a service: a request object and an answer object. Communication objects are transmitted *by-value* to ensure decoupling of the lifecycles of the client side and the server side of a service. They are arbitrary objects enriched by a unique identifier and get/set-methods. Hidden from the user and inside the communication patterns, the content of a communication object provided via *E* gets extracted and forwarded to the middleware interface *H*. Incoming content at *H* is put into a new instance of the according communication object before providing access to it via *E*.

In the example, the system integrator sees a provided port based on a *query server* with the communication objects *GridMapRequest* and *GridMap*. The map service might be provided by a map building component. Each component with a port consisting out of a *query client* with the same communication objects can use that service. For example, a path planning component might need a grid map and expose a required port for that service. The *GridMapRequest* object provides the parameters of the individual request (for example, the size of the requested map patch, its origin and resolution) and the *GridMap* returns the answer. The answer is self-contained comprising all the parameters describing the provided map. That allows to interpret the map independently of the current settings of the service providing component and gives the service provider the chance to return a map as close to but different from the requested parameters in case he cannot handle them exactly.

A component builder uses the stable interface *E*. In case of the client side of a service based on the query pattern, it always consists of the same synchronous as well as asynchronous access modes independent from the used communication objects and the underlying middleware. They can be used from any number of threads in any order. The server side in this example always consists of an asynchronous handler upcall for incoming requests and a separated answer method. This separation is important since it does not require the upcall to wait until the answer is available before returning. We can now implement any kind of processing model inside a component, even a processing pipeline where the last thread calls the answer method, without blocking or wasting system resources of the upcall or be obliged to live with the threading models behind the upcall.

In the example, the upcall at the service provider either directly processes the incoming *GridMapRequest* object or forwards it to a separate processing thread. The requested map patch is put into a *GridMap* object which then is provided as answer via the *answer* method.

It can be seen that the client side is not just a proxy for the server side. Both sides of a communication pattern are completely standalone entities providing stable interfaces *A* and *E* by completely hiding all the specifics of *H* and *I* (see figure 7). One can neither expose arbitrary member functions at the outside component hull nor can one dilute the semantics and behavior of ports. The different communication patterns and their internals are explained in detail in (Schlegel, 2007).

Besides the services defined by the component builder (*A*), several predefined services exist to support system level concerns (Lotz et al., 2011). Each component needs to provide a *state service* to support system level orchestration (outside view *B*: activation, deactivation, reconfiguration; inside view *F*: manage transitions between service activations, support housekeeping activities by entry/exit actions). An optional *diagnostic service* (*C*, *G*) supports runtime monitoring of the component. The optional *param service* manages parameters by name/value-pairs and allows to change them at runtime. The optional *wiring service* allows to wire required services of a component at runtime from outside the component. This is needed for task and context dependent composition of behaviors.

3.1 The SMARTMARS meta-model

All the stable interfaces, concepts and structures as well as knowledge about which ingredients and structures form a well-formed SMARTSOFT component and a well-formed system of SMARTSOFT components are explicated in the SMARTMARS meta-model (figure 9). The meta-model is abstract, universally valid and independent from implementation technologies (e.g. UML profile (Fuentes-Fernández & Vallecillo-Moreno, 2004), eCore (Gronback, 2009)). It provides the input for tool support for the different roles (like component developer, system integrator etc.), explicates separation of concerns and can be mapped onto different software technologies (e.g. different types of middleware like CORBA, ACE (Schmidt, 2011) and different types of operating systems).

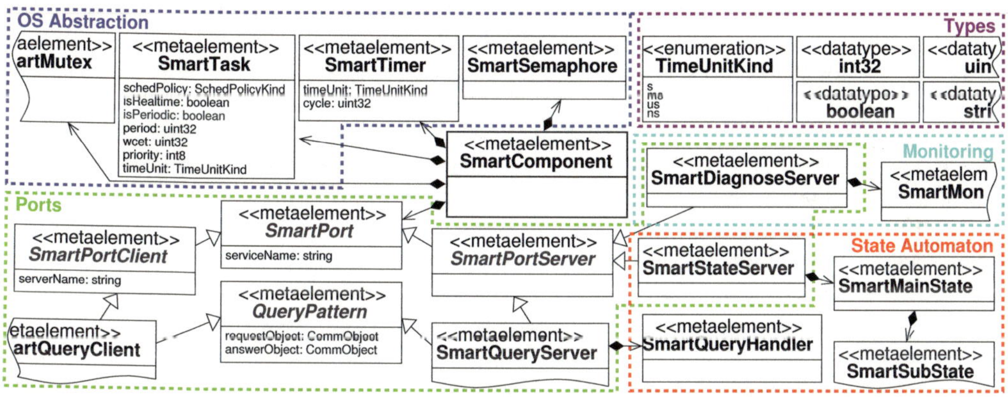

Fig. 9. Excerpt of the SMARTMARS meta-model.

3.2 Policies and strategies behind SMARTSOFT services

A major part of the SMARTSOFT approach are the policies and strategies that manifest themselves in the structure of the component model, explain its building blocks and guide their usage.

Separation of the roles of a component developer and a system integrator requires to control the interface between the inner part of a component and its outer part and to control this boundary. As soon as one gains control over the component hull, one can make sure that all relevant properties and parameters needed for the black box view of the system integrator become explicated at the component hull. One can also make sure that a component developer has no chance to expose component internals to the outside. SMARTSOFT achieves this via predefined communication patterns as the only building blocks to define externally visible services and by further guidelines on how to build good services.

A basic principle is that clients of services are not allowed to make any assumptions about offered services beyond the announced characteristics and that service providers are not allowed to make any assumptions about service requestors (like e.g. their maximum rate of requests).

This principle results in simple and precise guidelines of how to apply the communication patterns in order to come up with well-formed services. As long as a service is being offered, the service provider has to accept all incoming requests and has to respond to them according to its announced quality-of-service parameters.

We illustrate this principle by means of the *query pattern*. As long as there are no further quality-of-service attributes, the service provider accepts all incoming requests and guarantees to answer all accepted requests. However, only the service provider knows about its resources available to process incoming requests and clients are not allowed to impose constraints on the service provider (a request might provide further non-committal hints to the service provider like a request priority). Thus, the service provider is allowed to provide a nil answer (the flag *is valid* is set to false in the answer) in case he is running out of resources to answer a particular request. In consequence, all service requestors always must be prepared to get a nil answer. A service requestor is also not allowed to make any assumptions about the response time as long as according quality-of-service attributes are not set by the service provider. However, if a service provider announces to answer requests within a certain time limit, one can rely on getting at least a nil answer before the deadline. If a service requestor depends on a maximum response time although this quality-of-service attribute is not offered by the service provider, he needs to use client-side timeouts with his request. This overall principle ensures (i) loose coupling of services, (ii) prevents clients from imposing constraints on service providers and (iii) gives service providers the means to arbitrate requests in case of limited resources.

It now also becomes evident why SMARTSOFT offers more than just a request/response and a publish/subscribe pattern which would be sufficient to cover all communicational needs. The *send* pattern explicates a one-way communication although one can emulate it via a query pattern with a void answer object. However, practical experience proved that a much better clarity for services with this characteristic is achieved when offering a separate pattern. The same holds true for the *push newest* and the *push timed* pattern. In principle, the push timed pattern is a push newest pattern with a regular update. However, in case of a push newest pattern, service requestors rely on having the latest data available at any time. This is different from a push timed pattern where the focus is on the service provider guaranteeing a regular time interval (in some cases even providing the same data). Although one could cover some of these aspects by quality-of-service attributes, they also have an impact on the kind of perception of its usage by a component developer. Again, achieving clarity and making the characteristics easily recognizable is of particular importance for the strict separation of the roles of component developers and system integrators. This also becomes obvious with the *event* pattern. In contrast to the push patterns, service requestors get informed only in case a server side event predicate (service requestors individually parametrize each event activation) becomes true. This tremendously saves bandwidth compared to publishing latest changes to all clients since one then always would have to publish a snapshot of the overall context needed to evaluate the predicate at the client side instead of just the information when an event fired.

3.3 A robotics example illustrating the SMARTSOFT concepts

Figure 10 illustrates how the SMARTSOFT component model and its meta-elements provided by SMARTMARS structure and partition a typical robotics use-case, namely the navigation of a mobile platform. Besides access to sensor data and to the mobile base, algorithmic building blocks of a navigation system are map building, path planning, motion execution and self localization. Since these building blocks are generic for navigation systems independently of the used algorithms, it makes sense to come up with an according component structure and services (or expect readily available components and services).

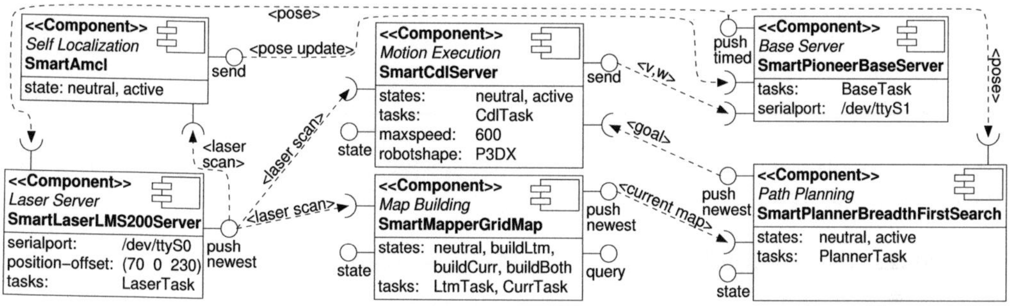

Fig. 10. Structure of a navigation task based on the SMARTSOFT component model.

The SmartLaserLMS200Server component provides the latest laser scan via a push newest port. Thus, all subscribed clients always get an update as soon as a new laserscan is available. It is subscribed to the pose service of the robot base to label laser scans with pose stamps. The component comprises a SmartTask to handle the internal communication with the laser hardware. This way, the aliveness of the overall component and its services is not affected by flaws on the laser hardware interface. Parameters like position-offset and serialport are used to customize the component to the target robotic system. These parameters have to be set by the application builder during the deployment step. The SmartMapperGridMap component requires a laser scan to build the longterm and the current map. The current map is provided by a push newest server port (as soon as a new map is available, it is provided to subscribed clients which makes sense since path planning depends on latest maps) and the longterm map by a query server port (since it is not needed regularly, it makes sense to provide it only on a per-request basis). The state port is used to set the component into different states depending on which services are needed in the current situation: build no map at all (neutral), build the current map only (buildCurr), build the longterm map only (buildLtm) or build both maps (buildBoth). The push newest server publishes the current map only in the states buildCurr and buildBoth. Requests for a longterm map are answered as long as the component and its services are alive but with an invalid map in case it is in the states neutral or buildCurr (valid flag of answer object set to false). Accordingly, the SmartPlannerBreadthFirstSearch component provides its intermediate waypoints by a push newest server (update the motion execution component as soon as new information is available). The motion execution component regularly commands new velocities to the robot base via a send service. The motion execution component is also subscribed to the laser scan service to be able to immediately react to obstacles in dynamic environments. This way, the different services interact to build various control loops to combine goal directed and reactive navigation while at the same time allowing for replacement of components.

3.4 State-of-the-art and related work

The historical need in robotics to be responsible for creation of the application logic and to be at the same time the system integrator generated a poor understanding in the robotics community that these two roles ought to be separated. In consequence, most robotics frameworks don't make this distinction and consequently they don't offer any clear guideline to the developer on how to achieve separation of roles.

For example, *ROS* (Quigley et al., 2009) is a currently widely-used framework in robotics providing a huge and valuable codebase. However, it lacks guidance for component developers to ensure system level conformance for composability. Instead, its focus is on side-by-side existence of all kinds of overlapping concepts without an abstract representation of its core features and properties in a way independent of any implementation.

The only approach in line with the presented concepts is the *RTC Specification* (OMG, 2008) which is considered the most advanced concept of *MDSD* in robotics. However, it is strongly influenced by use-cases requiring a data-flow architecture and they do not yet considerably take into account requirements imposed by runtime adaptability.

4. Reference implementation of the SMARTMDSD TOOLCHAIN

The reference implementation of the SMARTMDSD TOOLCHAIN implements the SMARTMARS meta-model within a particular MDSD-toolchain. It is used in real world operation to develop components and to compose complex systems out of them. The focus of this section is on technical details of the implementation of a meta-model. Another focus is on the role-specific view and the support a MDSD-toolchain provides. We illustrate the reference implementation of the toolchain along the different roles of the stakeholders and their views on the toolchain.

4.1 Decisions and tools behind the reference implementation - framework builder view

The reference implementation of our SMARTMDSD TOOLCHAIN is based on the *Eclipse Modeling Project (EMP)* (Eclipse Modeling Project, 2010) and *Papyrus UML* (PAPYRUS UML, 2011).

Papyrus UML is used as graphical modeling tool in our toolchain. Therefore, it is customized by the framework builder for the development of SMARTSOFT Components (component builder) and deployments of components (application builder). This includes for example a customized wizard to create communication objects, components as well as deployments. The modeling view of *Papyrus UML* is enriched with a customized set of meta-elements to create the models. The model transformation and code generation steps are developed with *Xpand* and *Xtend* (Efftinge et al., 2008) which are part of the *EMP*. These internals are not visible to the component builder and the application builder. They just see the graphical modeling tool to create their models and use the *CDT Eclipse Plugin* (Eclipse CDT, 2011) to extend the source code and to compile binaries. The SMARTMARS meta-model is implemented as a *UML Profile* (Fuentes-Fernández & Vallecillo-Moreno, 2004) using *Papyrus UML*.

The decision to use *UML Profiles* and *Papyrus UML* to implement our toolchain is motivated by the reduced effort to come up with a graphical modeling environment customized to the robotics domain and its requirements by reusing available tools from other communities. Although some shortcomings have to be accepted and taken into account we were not caught in the huge effort related to implementing a full-fledged *GMF*-based development environment. This allowed us to early come up with our toolchain and to gain deeper insights and more experience on the different levels of abstraction. However, the major drawbacks of *UML Profiles* are:

- *UML* is a general purpose modeling language covering aspects of several domains and is thus complex. Using profiles, it is only possible to enrich *UML*, but not to remove elements.
- Deployment and instantiations of components are not adequately supported.
- *UML Profiles* provide just a lightweight extension of *UML*. That means, the structure of *UML* itself cannot be modified. The elements can be customized only by stereotypes and tagged values.

To counter the drawbacks of *UML Profiles*, we only support the usage of the stereotyped elements provided by SMARTMARS to create the models of the components and deployments. Directly using pure *UML* elements in the diagrams is not supported. Thus, the models are created using just the meta-elements provided by SMARTMARS. Restricting the usage to SMARTMARS meta-elements, a mapping to another meta-model implementation technology like *eCore* (Gronback, 2009) is straightforward. The stereotyped elements can be mapped onto *eCore* without taking into account *UML* and its structure. In the current implementation of our toolchain, the restriction to only use SMARTMARS meta-elements is enforced with *check* (Efftinge et al., 2008), the *EMP* implementation of *OCL* (Object Management Group, 2010). In the model transformation and code generation steps of our toolchain pure *UML* elements are ignored. Another approach would be to customize the diagrams by removing the *UML* elements from the palette (see fig. 12) and thus restricting their usage. The latter approach is on the agenda of the *Papyrus UML* project and will be supported by future releases.

4.2 Development of components – component builder view

Figure 11 illustrates the roles of the framework builder and the component builder. The component builder creates a model of the component using the Eclipse based toolchain, focusing on the component hull. Pushing the button he receives the source files where to integrate the business logic (algorithms, libraries) of the component. During this process the component builder is supported and guided by the toolchain. The internals of the model transformation and code generation steps implemented by the framework builder are not visible to the component builder.

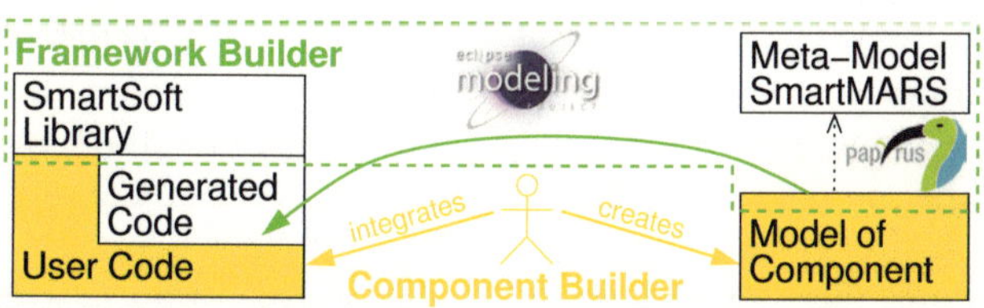

Fig. 11. The component builder models a component, gets the source code of its overall structure (component hull, tasks, etc.) generated by the toolchain and can then integrate user-code into these structures.

The view of the component builder on the toolchain is depicted in figure 12. It is illustrated by a face recognition component which is a building block in many service robotics scenarios as part of the human-robot interface (detection, identification and memorization of persons). In its active state, the component shall receive camera images, apply face recognition algorithms and report detected and recognized persons. Thus, besides the standard ports for setting states (active, neutral) and parameters, we need to specify a port to receive the latest camera images (based on a push newest client) and another one to report on the results (based on an event server). The component shall run the face recognition based on a commercially available library within one thread and optional visualization mechanisms within a second and separated thread. Thus, we need to specify two tasks within the component.

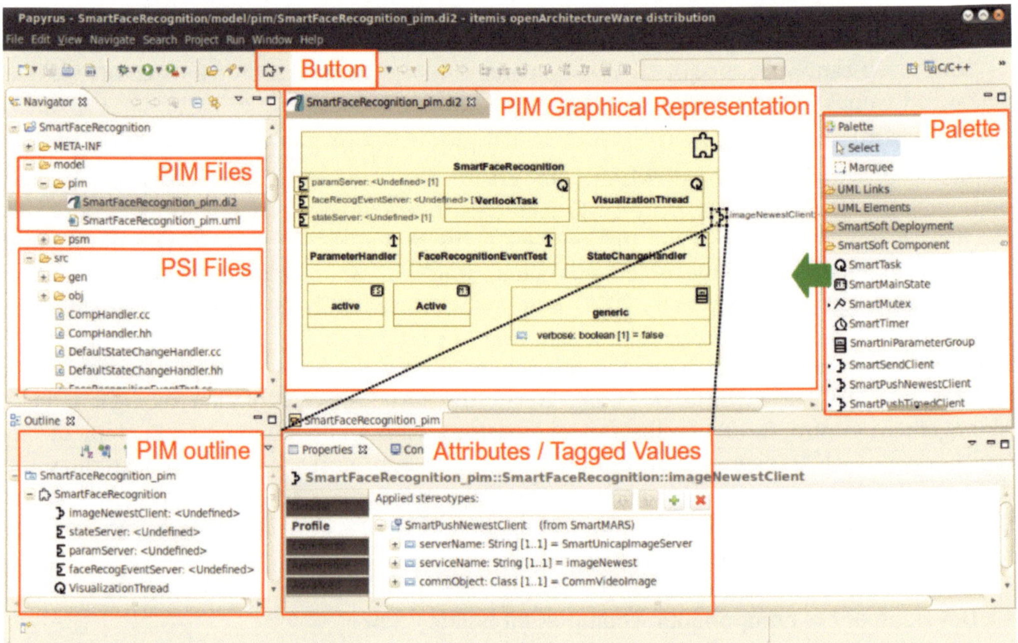

Fig. 12. Screenshot of our toolchain showing the view of the Component Builder.

To create the model the component builder uses the SMARTMARS meta-elements offered in the *palette*. The elements of the created model can be accessed either in the outline view or directly in the graphical representation. Several of the meta-element attributes (tagged values) can be customized and modified in the properties tab (e.g. customizing services to ports, specifying properties of tasks, etc.). The model is stored in files specific to *Papyrus UML*. Pushing the button, the workflow is started and the *PSI* (source) files are generated. The user code files are directly accessible in the *src* folder. The component builder integrates his business logic into these files (in our example, the interaction with the face recognition library). The generated files the component builder must not modify are stored in the *gen* folder. These files are generated and overwritten each time the workflow is executed. For the further processing of the source files, the *Eclipse CDT plugin* is used (*Makefile Project*). The makefile is also generated by the workflow specific to the model properties. User modifications in the makefile can be done inside of *protected regions* (Gronback, 2009).

4.3 Development of components – framework builder view

Taking a look behind the scenes of the toolchain, the workflow (fig. 13) appears as a two step transformation according to the *OMG MDA* (Object Management Group & Soley, 2000). The *Platform Independent Model (PIM)*, which is created by the component builder using the meta-elements provided by the *PIM UML Profile*, specifies the component independently of the implementation technology. The first step in the workflow is the model-to-model

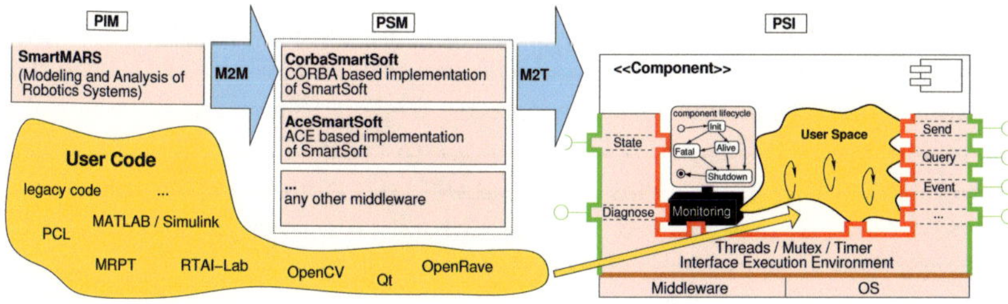

Fig. 13. Two step transformation workflow: Framework Builder view.

(*M2M*) transformation (encoded with *Xtend*) from the *PIM* into a *Platform Specific Model (PSM)*. In this step the elements of the *PIM* are transformed into corresponding elements of the *PSM* according to the selected target platform. The second step is the model-to-text (*M2T*) transformation (encoded with *Xpand* and *Xtend*) from the *PSM* into a *Platform Specific Implementation (PSI)*. This transformation is based on customizable code templates.

4.3.1 The SmartMARS UML profiles (PIM/PSM)

The abstract SMARTMARS meta-model is implemented by the framework builder as *UML Profile* using *Papyrus UML*. Therefore, standard *UML* elements (e.g. *Component*, *Class*, *Port*) are extended by stereotypes (e.g. SMARTCOMPONENT, SMARTTASK, SMARTQUERYSERVER) to give the meta-elements a new meaning according to the SMARTMARS concept. To distinguish and highlight the new element, it has its own icon attached. Tagged values are used to enrich the meta-element by new attributes which are not provided by the base *UML* element.
In fact there are two *UML Profiles*: one for the *PIM* and one for the *PSM*. The *PIM UML Profile* is visible to the component builder and is used by him to create the models of the components. For each SMARTSOFT implementation (e.g. *CORBA*, *ACE*), a *PSM UML Profile* has to be provided covering the specifics of the implementation. For example, the *CORBA*-based *PSM* supports *RTAI* linux to provide hard realtime tasks. This is represented by the meta-element *RTAITask*. The *PSM UML Profile* is not visible to the component builder and only used by the transformation steps inside the toolchain.

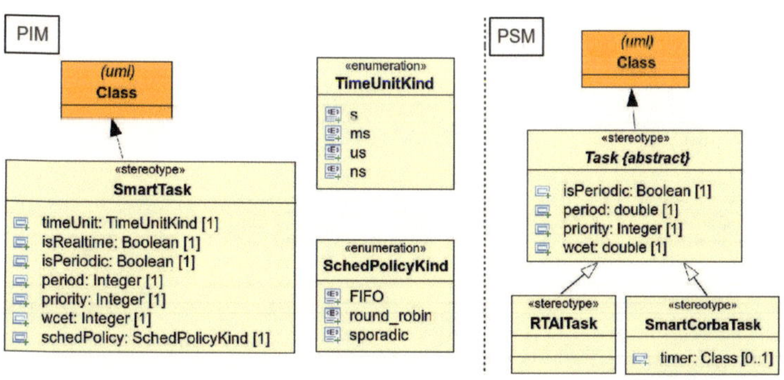

Fig. 14. Screenshots of excerpt of the UML Profiles created with *Papyrus UML* showing the metaelements dedicated to the SMARTTASK. *Left: PIM*; *Right: PSM* with the two variants (1) standard task and (2) *RTAI* task.

An excerpt of the *UML Profiles* is illustrated in figure 14. In the *UML Profile* for the *PIM*, the SMARTTASK extends the *UML class* and enriches it with attributes (tagged values) like *isPeriodic*, *isRealtime*, *period* and *timeUnit*. For the *timeUnit* an enumeration (*TimeUnitKind*) is used to specify the unit in which time values are annotated. In the *UML Profile* for the *CORBA-based PSM*, an abstract task is specified (cannot be instantiated) and the two variants (1) standard task and (2) realtime task are derived from it. They are both not abstract and can thus be instantiated by the component builder to create the model. The standard task adds an optional attribute referencing to a SMARTTIMER meta-element. This is used to emulate periodic non-realtime tasks which are not natively supported by standard tasks of the *CORBA*-based SMARTSOFT implementation.

4.3.2 Model transformation and code generation steps

The *M2M* transformation maps the platform independent elements of the *PIM* onto platform specific elements of the selected target platform. Such a mapping is illustrated by the example of the *SmartTask* (fig. 15 *left*) and the *CORBA*-based *PSM*. The SMARTTASK comprises several elements which are necessary to describe a task behavior and its characteristics.

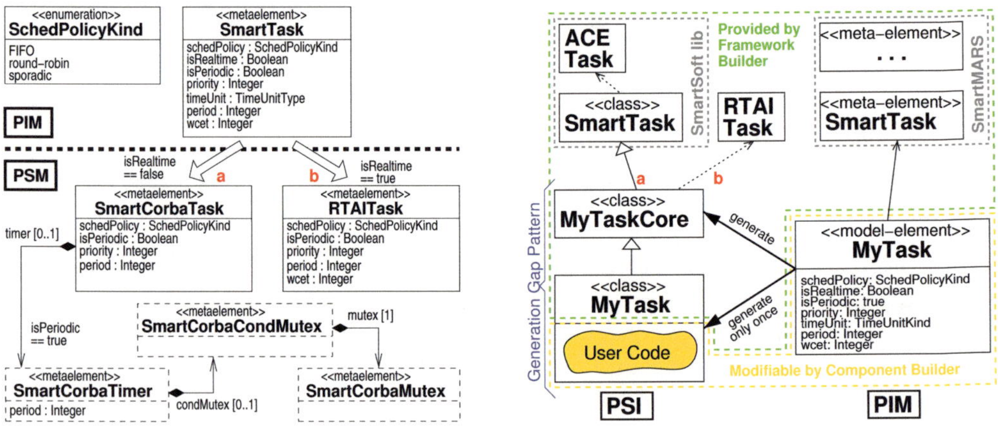

Fig. 15. Model transformation and code generation steps illustrated by the example of the SMARTTASK. *Left:* Transformation of the PIM into a PSM. *Right:* Code generation and Generation Gap Pattern.

```
create uml::Class this addSmartTask(SmartMARS::SmartTask tsk, uml::Component cmp) :
    cmp.packagedElement.add(this) ->
    this.setName(tsk.name) ->
    if( tsk.isRealtime == true) then
    {
        this.applyStereotype("CorbaSmartSoft::RTAITask") ->
        setTaggedValue(this, "CorbaSmartSoft::RTAITask", "isPeriodic", tsk.isPeriodic) ->
        setTaggedValue(this, "CorbaSmartSoft::RTAITask", "wcet", tsk.wcet.toSecond(tsk.timeUnit.name)) ->
        setTaggedValue(this, "CorbaSmartSoft::RTAITask", "period", tsk.period.toSecond(tsk.timeUnit.name)) ->
        setTaggedValue(this, "CorbaSmartSoft::RTAITask", "priority", tsk.priority)
    }
    else
    {
        this.applyStereotype("CorbaSmartSoft::SmartCorbaTask") ->
        setTaggedValue(this, "CorbaSmartSoft::SmartCorbaTask", "isPeriodic", tsk.isPeriodic) ->
        setTaggedValue(this, "CorbaSmartSoft::SmartCorbaTask", "wcet", tsk.wcet.toSecond(tsk.timeUnit.name)) ->
        setTaggedValue(this, "CorbaSmartSoft::SmartCorbaTask", "period", tsk.period.toSecond(tsk.timeUnit.name)) ->
        setTaggedValue(this, "CorbaSmartSoft::SmartCorbaTask", "priority", tsk.priority) ->
        if( tsk.isPeriodic == true ) then
        {
            setTaggedValue(this, "CorbaSmartSoft::SmartCorbaTask", "timer", cmp.addTimer(tsk.name, tsk.period, tsk.timeUnit.name))
        }
    };
```

Fig. 16. PIM to PSM model transformation of the SMARTTASK depending on the attribute *isRealtime*.

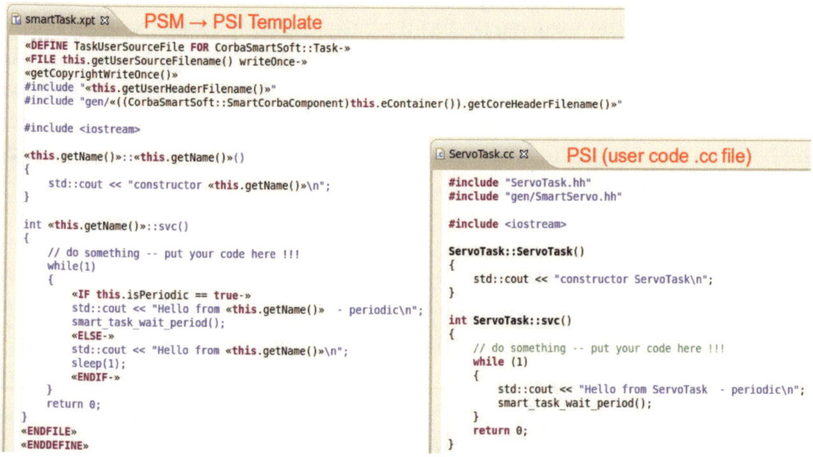

Fig. 17. *PSM* to *PSI* transformation of the SMARTTASK. *Left:* Excerpt of the transformation template (*xPand*) generating the PSI of a standard task. *Right:* The generated code where the user adds the business logic of the task.

Depending on the attribute *isRealtime* the SMARTTASK is either mapped onto a RTAITASK or a non-realtime SMARTCORBATASK[1]. The *Xtend* transformation rule to transform the *PIM* SMARTTASK into the appropriate *PSM* element is depicted in figure 16.
In case the attributes specify a non-realtime, periodic SMARTTASK, the toolchain extends the *PSM* by the elements needed to emulate periodic tasks (as this feature is not covered by standard tasks). In each case the user integrates his algorithms and libraries into the stable interface provided by the SMARTTASK (component builder view) independent of the hidden internal mapping of the SMARTTASK (generated code). Figure 17 depicts the *Xpand* template to generate the user code file for the task in the *PSI*. The figure shows the template on the left and the generated code on the right.
The *PSI* consists of the SMARTSOFT library, the generated code and the user code (fig. 15 *right*). To be able to re-generate parts of the component source code according to modified parameters in the model without affecting the source code parts added by the component builder, the generation gap pattern (Vlissides, 2009) is used. It is based on inheritance – the user code inherits from the generated code[2]. The source files called *generated code* are generated each time the transformation workflow in the toolchain is executed. These files contain the logic which is generated behind the scenes according to the model parameters and must not be modified by the component builder. The source files called *user code* are just generated if they do not already exist. They are intended for the component builder to add the algorithms and libraries. The generation of the user code files is more for the convenience of the component builder to have a code template as starting point. These files are in the full responsibility of the component builder and are never modified or overwritten by the transformation workflow of the toolchain. In this context *generate once* means that the file is only generated if it does not already exist. This is typically the case if the workflow is executed for the first time. The clear separation of generated code and user code by the generation gap pattern allows on the one hand to reflect modifications of the model in the generated source

[1] *Corba* in element names indicates that the element belongs to the *CORBA* specific *PSM*.
[2] The pattern could also be used in the opposite inheritance ordering so that the generated code inherits from the user code.

code without overwriting the user parts. On the other hand it gives the user the freedom to structure his source code according to his needs and does not restrict the structure as would be the case with, for example, *protected regions*. Consequently, the component builder can modify the *period*, *priority* or even the *isRealtime* attribute of the task in the model, re-generate and compile the code without requiring any modification in the user code files. The modification in the model just affects the generated code part of the *PSI*.

4.4 Deployment of components – application builder view

The deployment is used to compose a robotic system out of available components. The application builder imports the desired components and places them onto the target platform. Furthermore, he defines the initial wiring of the components by connecting the ports with the meta-element *Connection*. Figure 18 illustrates the composition of navigation

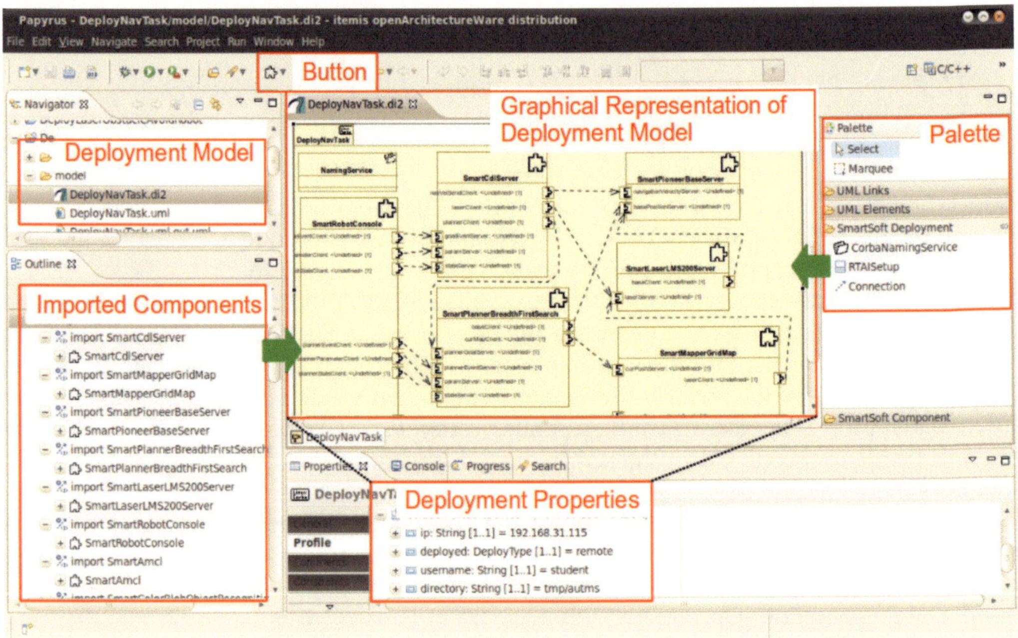

Fig. 18. Screenshot of our toolchain showing the deployment of components to build a robotic system.

components. In this example, the application builder (system integrator) imports components specific to a particular robot platform (SmartPioneerBaseServer) and specific to a particular sensor (SmartLaserLMS200Server). The navigation components (SmartMapperGridMap, SmartPlannerBreadthFirstSearch, SmartCDLServer) can be used across different mobile robots. The SmartRobotConsole provides a user interface to command the robot.

The components are presented to the application builder as black boxes with dedicated variation points. These have to be bound during the deployment step and can be specified according to system level requirements. For example, a laser ranger component might need the coordinates of its mounting point relative to the robot coordinate system. One might also reduce the maximum scanning frequency to save computing resources. Parameters also need to be bound for the target system. For example, in case *RTAI* is used inside of a component, the *RTAI* scheduler parameters (timer model underlying RTAI: periodic, oneshot) of the target

RTAI system have to be specified. If the application builder forgets to bind required settings, this absence is reported to him by the toolchain.

The application builder can identify the provided and required services of a component via its ports. He can inspect its characteristics by clicking on the port icon which opens a property view. That comprises the communication pattern type, the used communication objects and further characteristics like service name and also port specific information like update frequencies. The initial wiring is done within the graphical representation of the model. In case the application builder wants to connect incompatible ports, the toolchain refuses the connection and gives further hints on the reasons.

If the *CORBA*-based implementation of SMARTSOFT is used, the *CORBA* naming service properties *IP*-address and *port*-number have to be set. Furthermore, the deployment type (*local, remote*) has to be selected. For a remote deployment, the *IP*-address, *username* and *target folder* of the target computer have to be specified. The deployed system is copied to the target computer and can be executed there. In case of a local deployment, the system is customized to run on the local machine of the application builder. This is, for example, the case if no real robot is used and the deployed system uses simulation components (e.g. *Gazebo*). Depending on the initial wiring, parameter files are generated and also copied into the deployment folder. These parameter files contain application specific adjustments of the components. In addition, a shell script to start the system is generated out of the deployment model.

4.5 Deployment of components – framework builder view

To implement the deployment of components, some meta-elements are added by the framework builder to the *UML Profile* (fig. 19). This section focuses on the *CORBA*-based deployment.

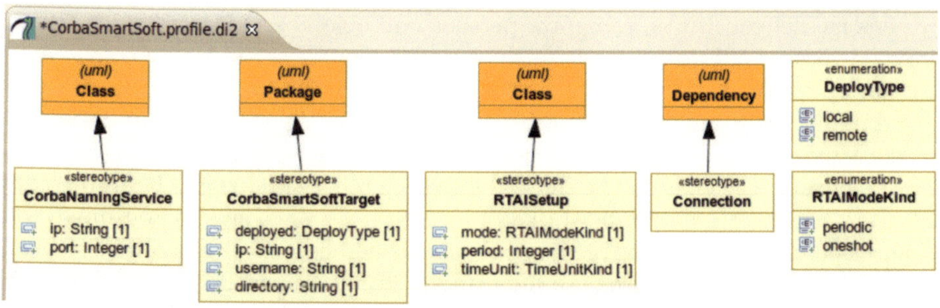

Fig. 19. Meta-elements to support the deployment of components.

The deployment model contains relevant information like the initial wiring between components (*Connection*), naming service properties (*CorbaNamingService*), scheduler properties (*RTAISetup*) and parameters about the deployment itself (*CorbaSmartSoftTarget*). The models of the components are made available to the deployment model using the *UML import* mechanism. This allows to access the internal structure of the components. Out of the deployment model the parameter files and a start script are generated (*M2T*) using *Xpand* and *Xtend* in a similar way as these transformation languages are used to generate code for the components. Based on the deployment model several analysis and simulation models can be generated to get feedback from 3rd-party tools. For example, one can extract parameters of all realtime tasks mapped onto a specific processor to perform hard realtime schedulability analysis (CHEDDAR (Cheddar, 2010)) (Schlegel et al., 2010).

Fig. 20. The clean-up scenario. (1) Kate approaches the table; (2/3) Kate stacks objects into each other; (4) Kate throws cups into the kitchen sink.

As deployments and especially instantiations of components are not sufficiently supported by *UML*, a few workarounds are necessary as long as the SMARTMARS meta-model is implemented as *UML Profile*. For example, a robot with two laser range finders (front, rear) requires two instances of the same component. Each laser instance requires its individual parameters (e.g. serial port, pose on robot). These parameters are assigned to the deployment model by the application builder specifically for each component. In the implementation based on the *UML Profile*, we hence work on copies of components. Individual instances with their own parameter sets are considered in the abstract SMARTMARS meta-model and are also covered in the SMARTSOFT implementation, thus switching to a different meta-model implementation technology would allow for instances. This has not yet been done due to the huge manpower needed compared to just reusing *UML* tools.

5. Example / scenario

The work presented has been used to build and run several real-world scenarios, including the participation at the RoboCup@Home challenge. Among other tasks our robot "Kate" can follow persons, deliver drinks, recognize persons and objects and interact with humans by gestures and speech.

In the clean-up scenario [3] (fig. 20) the robot approaches a table, recognizes the objects which are placed on the table and cleans the table either by throwing the objects into the trash bin or into the kitchen sink. There are different objects, like cups, beverage cans and different types of crisp cans. The cups can be stacked into each other and have to be thrown into the kitchen sink. Beverage cans can be stacked into crisp cans and have to be thrown into the trash bin. Depending on the type of crisp can, one or two beverage cans can be stacked into one crisp can. After throwing some of the objects into the correct disposal the robot has to decide whether to drive back to the table to clean up the remaining objects (if existing) or to drive to the operator and announce the result of the cleaning task. The robot reports whether all objects on the table could be cleaned up or, in case any problems occurred, reports how many objects are still left.

Such complex and different scenarios can neither be developed from scratch nor can their overall system complexity be handled without using appropriate software engineering methods. Due to their overall complexity and richness, they are considered as convincing stress test for the proposed approach. In the following the development of the cleanup example scenario is illustrated according to the different roles.

The **framework builder** provides the tools to develop SMARTSOFT components as well as to perform deployments of components to build a robotic system. In the described example this includes the *CORBA*-based implementation of the SMARTSOFT framework

[3] http://www.youtube.com/roboticsathsulm#p/u/0/xtLK-655v7k

and the SMARTMDSD toolchain which are both available on *Sourceforge* (`http://smart-robotics.sourceforge.net`).
The component builder view of the SMARTMDSD toolchain supports **component builders** to develop their components independently of each other, but based on agreed interfaces. These components are independent of the concrete implementation technology of SMARTSOFT. Component builders provide their components in a component shelf. The models of the components include all information to allow a black-box view of the components (e.g. services, properties, resources). The explication of such information about the components is required by the application builder to compose robotic systems in a systematic way. To orchestrate the components at run-time, the task coordination language SMARTTCL (Steck & Schlegel, 2010) is used. Therefore, SMARTTCL is wrapped by a SMARTSOFT component and is also provided in the component shelf. The SMARTTCL component provides reusable action plots which can be composed and extended to form the desired behavior of the robot.
The **application builder** uses the application builder view of the SMARTMDSD toolchain. He composes already existing components to build the complete robotic system. In the above described cleanup scenario, 17 components (e.g. mapping, path planning, collision avoidance, laser ranger, robot-base) are reused from the component shelf. It is worth noting that the components were not particularly developed for the cleanup scenario, but can be used in the cleanup scenario due to the generic services they provide. The SMARTTCL sequencer component is customized according to the desired behavior of the cleanup scenario. Therefore, several of the already existing action plots can be reused. Application specific extensions are added by the application builder.
At run-time the SMARTTCL sequencer component coordinates the software components of the **robot** by modifying the configuration and parametrization as well as the wiring between the components. As SMARTTCL can access the information (e.g. parameters, resources) explicated in the models of the components at run-time, this information can be taken into account by the decision making process. That allows the robot not only to take the current situation and context into account, but also the configuration and resource usage of the components. In the described scenario, the sequencer manages the resources of the overall system, for example, by switching off components which are not required in the current situation. While the robot is manipulating objects on the table and requires all available computational resources for the trajectory planning of the manipulator, the components for navigation are switched off.

6. Conclusion

The service-oriented component-based software approach allows separation of roles and is an important step towards the overall vision of a robotics software component shelf. The feasibility of the overall approach has been demonstrated by an Eclipse-based toolchain and its application within complex Robocup@Home scenarios. Next steps towards model-centric robotic systems that comprehensively bridge design-time and runtime model usage now become viable.

7. References

Andrade, L., Fiadeiro, J. L., Gouveia, J. & Koutsoukos, G. (2002). Separating computation, coordination and configuration, *Journal of Software Maintenance* 14(5): 353–369.
Beydeda, S., Book, M. & Gruhn, V. (eds) (2005). *Model-Driven Software Development*, Springer.

Björkelund, A., Edström, L., Haage, M., Malec, J., Nilsson, K., Nugues, P., Robertz, S. G., Störkle, D., Blomdell, A., Johansson, R., Linderoth, M., Nilsson, A., Robertsson, A., Stolt, A. & Bruyninckx, H. (2011). On the integration of skilled robot motions for productivity in manufacturing, *Proc. IEEE Int. Symposium on Assembly and Manufacturing*, Tampere, Finland.

Blogspot (2008). Discussion of Aspect oriented programming(AOP). URL: *http://programmingaspects.blogspot.com/*

Bruyninckx, H. (2011). Separation of Concerns: The 5Cs - Levels of Complexity, Lecture Notes, Embedded Control Systems. URL: *http://people.mech.kuleuven.be/ bruyninc/ecs/LevelsOfComplexity-5C-20110223.pdf*

CBDI Forum (2011). CBDI Service Oriented Architecture Practice Portal - Independent Guidance for Service Architecture and Engineering. URL: *http://everware-cbdi.com/cbdi-forum*

Cheddar (2010). A free real time scheduling analyzer. URL: *http://beru.univ-brest.fr/ singhoff/cheddar/*

Chris, R. (1989). *Elements of functional programming*, Addison-Wesley Longman Publishing Co, Boston, MA.

Delamer, I. & Lastra, J. (2007). Loosely-coupled automation systems using device-level SOA, *5th IEEE International Conference on Industrial Informatics*, Vol. 2, pp. 743–748.

Dijkstra, E. (1976). *A Discipline of Programming*, Prentice Hall, Englewood Cliffs, NJ.

Eclipse CDT (2011). C/C++ Development Tooling for Eclipse. URL: *http://www.eclipse.org/cdt/*

Eclipse Modeling Project (2010). Modeling framework and code generation facility. URL: *http://www.eclipse.org/modeling/*

Efftinge, S., Friese, P., Haase, A., Hübner, D., Kadura, C., Kolb, B., Köhnlein, J., Moroff, D., Thoms, K., Völter, M., Schönbach, P., Eysholdt, M. & Reinisch, S. (2008). openArchitectureWare User Guide 4.3.1.

Fuentes-Fernández, L. & Vallecillo-Moreno, A. (2004). An Introduction to UML Profiles, *UPGRADE* Volume V(2): 6–13.

Gelernter, D. & Carriero, N. (1992). Coordination languages and their significance, *Commun. ACM* 35(2): 97–107.

Gronback, R. C. (2009). *Eclipse Modeling Project: A Domain-Specific Language (DSL) Toolkit*, Addison-Wesley, Upper Saddle River, NJ.

Heineman, G. T. & Councill, W. T. (eds) (2001). *Component-Based Software Engineering: Putting the Pieces Together*, Addison-Wesley Professional.

IBM (2006). Model-Driven Software Development, *Systems Journal* 45(3).

Lastra, J. L. M. & Delamer, I. M. (2006). Semantic web services in factory automation: Fundamental insights and research roadmap, *IEEE Trans. Ind. Informatics* 2: 1–11.

Lau, K.-K. & Wang, Z. (2007). Software component models, *IEEE Transactions on Software Engineering* 33: 709–724.

Lee, E. A. & Seshia, S. A. (2011). *Introduction to Embedded Systems - A Cyber-Physical Systems Approach*, ISBN 978-0-557-70857-4. URL: *http://LeeSeshia.org*

Lotz, A., Steck, A. & Schlegel, C. (2011). Runtime monitoring of robotics software components: Increasing robustness of service robotic systems, *Proc. 15th Int. Conference on Advanced Robotics (ICAR)*, Tallinn, Estland.

Meyer, B. (2000). What to compose, *Software Development* 8(3): 59, 71, 74–75.

Mili, H., Elkharraz, A. & Mcheick, H. (2004). Understanding separation of concerns, *Proc. Workshop Early Aspects: Aspect-Oriented Requirements Engineering and Architecture Design*, Vancouver, Canada, pp. 75–84.

Neelamkavil, F. (1987). *Computer simulation and modeling*, John Wiley & Sons Inc.

Object Management Group (2010). Object Constraint Language (OCL). URL: *http://www.omg.org/spec/OCL/*

Object Management Group & Soley, R. (2000). Model-Driven Architecture (MDA). URL: *http://www.omg.org/mda*

OMG (2008). Robotic Technology Component (RTC). URL: *http://www.omg.org/spec/RTC/*

PAPYRUS UML (2011). Graphical editing tool for uml. URL: *http://www.eclipse.org/modeling/mdt/papyrus/*

Parnas, D. (1972). On the criteria to be used in decomposing systems into modules, *Communications of the ACM* 15(12).

Quigley, M., Gerkey, B., Conley, K., Faust, J., Foote, T., Leibs, J., Berger, E., Wheeler, R. & Ng, A. (2009). ROS: An open-source Robot Operating System, *ICRA Workshop on Open Source Software.*

Radestock, M. & Eisenbach, S. (1996). Coordination in evolving systems, *Trends in Distributed Systems – CORBA and Beyond*, Springer-Verlag, pp. 162–176.

Reiser, U., Connette, C., Fischer, J., Kubacki, J., Bubeck, A., Weisshardt, F., Jacobs, T., Parlitz, C., Hägele, M. & Verl, A. (2009). Care-O-bot 3 – Creating a product vision for service robot applications by integrating design and technology, *Proc. IEEE/RSJ International Conference on Intelligent Robots and Systems (ICRA)*, St. Louis, USA, pp. 1992–1997.

Schlegel, C. (2007). Communication patterns as key towards component interoperability, *in* D. Brugali (ed.), *Software Engineering for Experimental Robotics, STAR 30*, Springer, pp. 183–210.

Schlegel, C. (2011). SMARTSOFT – Components and Toolchain for Robotics. URL: *http://smart-robotics.sf.net/*

Schlegel, C., Steck, A., Brugali, D. & Knoll, A. (2010). Design abstraction and processes in robotics: From code-driven to model-driven engineering, *2nd Int. Conf. on Simulation, Modeling, and Programming for Autonomous Robots (SIMPAR)*, Springer LNAI 6472, pp. 324–335.

Schlegel, C., Steck, A. & Lotz, A. (2011). Model-driven software development in robotics: Communication patterns as key for a robotics component model, *Introduction to Modern Robotics*, iConcept Press.

Schmidt, D. (2011). The ADAPTIVE Communication Environment. URL: *http://www.cs.wustl.edu/ schmidt/ACE.html*

Sprott, D. & Wilkes, L. (2004). CBDI Forum. URL: *http://msdn.microsoft.com/en-us/library/aa480021.aspx*

Stahl, T. & Völter, M. (2006). *Model-Driven Software Development: Technology, Engineering, Management*, Wiley.

Steck, A. & Schlegel, C. (2010). SmartTCL: An Execution Language for Conditional Reactive Task Execution in a Three Layer Architecture for Service Robots, *Int. Workshop on DYnamic languages for RObotic and Sensors systems (DYROS/SIMPAR)*, Germany, pp. 274–277.

Steck, A. & Schlegel, C. (2011). Managing execution variants in task coordination by exploiting design-time models at run-time, *Proc. IEEE/RSJ International Conference on Intelligent Robots and Systems (IROS)*, San Francisco, CA.

Szyperski, C. (2002). *Component-Software: Beyond Object-Oriented Programming*, Addison-Wesley Professional, ISBN 0-201-74572-0, Boston.

Tarr, P., Harrison, W., Finkelstein, A., Nuseibeh, B. & Perry, D. (eds) (2000). *Proc. of the Workshop on Multi-Dimensional Separation of Concerns in Software Engineering (ICSE 2000)*, Limerick, Ireland.

Vlissides, J. (2009). Pattern Hatching – Generation Gap Pattern. URL: *http://researchweb.watson.ibm.com/designpatterns/pubs/gg.html*

Völter, M. (2006). MDSD Benefits - Technical and Economical. URL: *http://www.voelter.de/data/presentations/mdsd-tutorial/02_Benefits.pdf*

Webber, D. L. & Gomaa, H. (2004). Modeling variability in software product lines with the variation point model, *Science of Computer Programming - Software Variability Management* 53(3): 305–331.

Willow Garage (2011). PR2: Robot platform for experimentation and innovation. URL: *http://www.willowgarage.com/pages/pr2/overview*

24

Using Ontologies for Configuring Architectures of Industrial Robotics in Logistic Processes

Matthias Burwinkel and Bernd Scholz-Reiter
BIBA Bremen, University of Bremen
Germany

1. Introduction

The provision of goods and services accomplishes a transition to greater value-added-oriented logistics processes. The philosophy of logistics is changing to a cross-disciplinary function. Therefore it becomes a critical success factor for competitive companies (Göpfert, 2009). Thus logistics assumes the task of a modern management concept. It provides for the development, design, control and implementation of more effective and efficient flows of goods. Further, on aspects of information, money and financing flows are crucial for for the development of enterprise-wide and company-comprehensive success.
According to (Scheid, 2010a) this can be ensuring, by the automation of logistic processes. Based (Granlund, 2008), the necessity for automated logistics processes raises the focus on logistics by existing dominant factors of uncertainty and rapid changes in the business area environment. Therefore, the adoption of flexible automation systems is essential. Here robotics appears very promising due to its universal character as a handling machine. This is how (Suppa & Hofschulte, 2010) characterizes the development of industrial robotics: '[...] increasingly in the direction of flexible systems, which take over new fields with sensors and innovative fiscal and regulatory approaches.' Here, logistics represents a major application field. (Westkämper & Verl, 2009) describe the broad applications for logistics and demonstrate the capability for flexibility with examples from industry and research. Besides the technological feasibility, there is also the existing demand by logistics firms concerning the need for their application.
These representations demonstrate the interaction of robotics-logistics regarding the design of technical systems for the operator strongly driven by the manufacturer (technology push) and the technological standardization of the system. Robotic-logistics concentrates on the development and integration of products. Accordingly, standardization activities of the technical systems focus on components and sub-systems that represent the manufacturer-oriented perspective.
The main goal concentrates on the planning, implementation, and monitoring of enterprise-wide process chains of technological systems under the consideration of economic criteria. In this context, the interaction of the two domains 'process' and 'technology' are essential. Thus, the configuration design of technological layouts or machines is crucial. The harmonization of the two domains requires a systematic description framework concerning their exchange of information and knowledge. A high-level abstraction of knowledge representation in the description of the relationships and connections is essential. It also

allows the description of implicit relationships such as comparative relationship notations. This applies to both qualitative and quantitative types of relationships. The outcome is a framework that is available to represent an object dependency between process and technology and to serve the described requirements for flexibility regarding logistics cargo, throughput and machine- and process-layout.
Thus, there is the need for qualitative description of relationship between process and technology by means of specific parameters and properties on a high-level abstraction.

2. Robotics-Logistics: Challenges and potentials

Since the 1970s, there has been a multifaceted development of the basic understanding of logistics. The origin of 'logistics' refers to the Greek 'logos' (reason, arithmetic) and the Romanesque-French ('providing', 'supporting'). In the past logistics were understood in delimited functions. Nowadays logistics are global networks, which are necessary to optimize. The understanding of the task itself changed from a pure functional perspective through process chains to value-adding networks:
Fig. 1 shows the historical development starting in the 1970s. Today's logistics is characterized by its value and integration in the appropriate process chains. The Federal Logistics Association designates logistic processes to the areas of procurement, production, distribution, disposal and transport logistics. (Arnold, 2006) designates differentiated performance-oriented processes as transport and storage processes. Storage processes are the processes of handling, order picking, and packing. Logistics services are evaluated based on delivery time, delivery reliability, inventory availability, delivery quality, and delivery flexibility. These are the objectives for both intra-logistics and extra-logistics. Logistics institutions, such as logistics service providers, provide value-added benefits to this process. These services are dependent of the collection and the output of their product 'commodity'. Finishing or outer packaging operations are examples here.
The logistics of the future will be essentially determined by the automation of material and information flows. In this area, automation systems in logistics already exist for several years. Application areas for these systems, such as de-palletizing and palletizing, sorting, and picking, are 'technically feasible and tested for decades' (Scheid; 2010b). The complete automation of the so-called intra-logistics is technically feasible. However, this situation is not encountered in practice due to the singular character of isolated applications. In future material flow technologies will be more modularized as (Straube & Rösch, 2008) identified. Modular automation systems maximize flexibility in the logistics systems and enable the re-utilization of technical components of handling and storage technology. To summarize the research requirements concerning these technologies (Straube & Rösch, 2008) ask for new modular constructions, which can combine different techniques based on their standardized modular features. This simplifies the integration into new systems. They describe a weakening tendency in new features for the components of conveying and storage technology. The focus is set on the configuration of system architectures composed of existing commercial components. This approach leads to process-specific integrated systems.
From an industrial point of view, multiple logistics areas display a high potential for the automation of processes (Scheid, 2010b). Thus, a high potential exists for the processes 'transport," 'storage' and 'de-palletizing." Transport processes will be automated in 2015 by nearly 30 percent. The reasons for the limiting borders for straightening the degree of automation are lying in the characteristics of the material and information flows. The existing

process dynamics and process volatility are a handicap for standardized processes. The continuous automation of specific and individual processes appears to be difficult due to these reasons. Machine application requires great flexibility for adapting changing parameters.

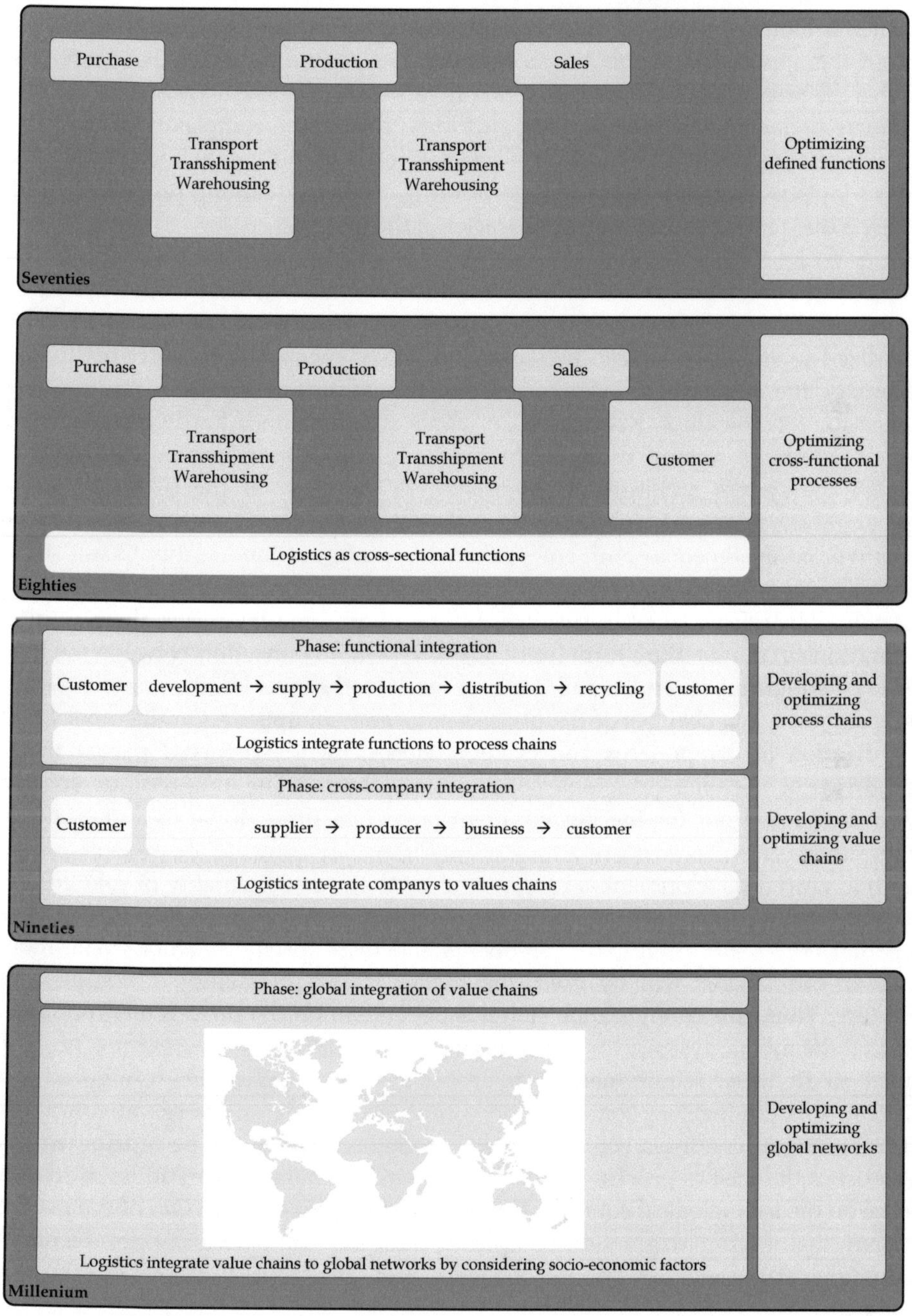

Fig. 1. Historical development of logistics philosophy [source: authors illustration following (Baumgarten, 2008)]

A fundamental role belongs to robotics. By definition, industrial robot systems are a central success factor in process automation due to their universality. The application of robotics systems in logistics factories should be designed flexibly. The automation of the processes under the customers' existing requirements can allow individually designed systems (Günthner & Hompel, 2009). In their recent study, the European Initiative (EUROP, 2009) identified the application of robotics in logistics as a central issue for the future. Thus, it highlights the broad range of application and diverse functions in this area. The operation of the systems under limited process standardization due to the complexity of the processes because of heterogeneous and manifold variables, leads to individual and special solutions in today's logistics factories. The adaption of the systems to changing process environment is hindered due to the process specific character of the systems.

(Fritsch & Wöltje, 2006) identifies the necessity for a paradigm shift from such individual system configuration and underline the relevance of standardized robot systems. This necessity establishes (Elger & Hausser, 2009) by describing the demand of more standardized solutions, which can also serve individual needs. The initiative (EUROP, 2009) characterizes the standardization of components and technical systems as an essential challenge for the so-called 'Robotics 2020." This concerns both hardware and software, and their interfaces among these components. In the authors' view, this requirement influences the system architecture essentially. In the view of the (EUROP, 2009), the system architecture accords robotics a central role. In the future architectures for robotic systems will be designed to both comprehensive configuration conditions and technical subsystems and components. They can be assigned from comparable and very different applications. Therefore, robotics systems will be more modularized in their architecture configurations in the medium term (until 2015). The interconnection between the modules is weakly configured in an overall perspective. On the one hand, this allows a rapid reconfiguration when changes of the process environment appear. On the other hand, the standardization of components and systems is besides the repeated partial usage the second driver of so-called 'adaptable configuration status." The long-term perspective for the year 2020 looks out for the development of architectures down to autonomous self-configurations.

The second crucial development is represented by the compositionality of robotic systems. A robotic system is compositional when the complexity of system architecture is based on compilation of the subsystems or components and their specific functions. The more sub systems or components will be used, the higher is the probability of complex system architecture. Thus, this configuration status is dependent on the process environment. This means that the robotic systems for self-changing or complicated processes must be explicitly designed to fit these requirements. The robotic system has to be configured process orientated. Robotics-logistics configuration conditions appear to diverge in comparison to the configuration condition of production robotics. This can be attributed to the characteristics of logistics processes. The process environment appears to have an essential influence on the technological configuration status of robotic systems. Out of the perspective of system theory, the degree of complexity can be influenced by its technological configuration status of the robotic systems and the characteristics of the process environment.

Thus, complicated processes often require robotic system architectures, which are composed of numerous components and are individually configured. This relationship can result in complicated or complex systems on the process- and on the technology-level. Additionally, procedural complexity influences technical complexity. The necessary reaction possibilities with technical components to procedurally dynamic events are the main driver here. The individual solutions counteract the intended economic standard solutions. Standardization serves to reduce complexity and have to integrate both the process environment and systems engineering. Robotic systems can be standardized by considering the two perspectives of the configuration.

The description of this relational structure is represented by an approach that works with a qualitative logical description on an abstract level. Current approaches to system modeling appear too formal. Ontological approaches with their level of abstraction are an interesting alternative. Despite the standardization of system architecture, a process-orientated configuration is to be ensured. The necessary flexibility intends to serve the dynamic and volatile processes. The construction and structure of the architecture has to be monitored and planned in its modular basic approach. To cover the historical, actual and future usage of technical systems, modular robotic systems are essential.

This book chapter describes a conceptually basic approach procedure for the representation of the relational structure between process and technology through an ontological vocabulary.

3. State of the art - modelling approaches for system representation

Examples of traditional modeling methods for representing systems where relationships between entities are described are: 'entity-relationship model," 'Petri nets,' and 'event-driven process chains' (Kastens et al., 2008, Seidlmeier, 2002, Siegert, 1996).

The Entity Relationship Model (ER-model) was developed in 1976 by Chen. It allows delimited systems to be represented in a way which is intelligible for all involved. The entities (objects) and the relationships between the objects form the basis of the modeling. Regarding the purpose of the modeling, only objects, relationships, and attributes are described (Chen, 1976). The method of Petri nets represents structural coherence between sets of events (Kiencke, 1997). In general, a Petri net is a graphic description where the transaction of the generation of sequences of event-driven networks is represented. It consists of types of nodes, which are representative of a so-called position or transition conditions and events. A directed edge connects a position with a transition. Petri nets are capable of describing a large class of possible processes (Tabeling, 2006). Event-Driven Process Chains (EPC) modeling is a process-oriented perspective on functions, events, organizational units, and information object systems. A process chain is defined by modeling rules and operators (Staud, 2006).

Systems can be also modeled by using ontologies. The concept of ontology originates from philosophy and describes the 'science of being." Many authors define ontology from different perspectives. (Gruber, 1993) describes ontologies as the explicit specification of a conceptualization. The abstract level has the advantage that many basic approaches of different research areas are defined. For example, linguistically and mathematically oriented ontologies are combined due to this definition. (Stuckenschmidt, 2009) establishes the common reference to this definition by many authors. (Studer et al., 1998) take it as a basis and defines ontologies from their formal logic: 'An ontology is a formal, explicit specification of a shared conceptualization.' They emphasize the machine-readable formality of ontology. (Neches et al., 1991) specifies this idea and describes ontologies as 'basic terms and relations comprising the vocabulary of a topic area, as well as the rules for combining terms and relations, to define extensions to the vocabulary.' According to this understanding, concepts are defined through basic distinctions of objects and their rule-based relationships to each other. (Bunge, 1977) describes ontology as the only area of science besides the fields of natural and social sciences, which focuses on concrete objects and concrete reality. Ontologies are to be assigned based on philosophy since they stress the basic principles of virtual science, which cannot be proven or refused by experiments.

Ontologies represent knowledge, which is structured and provided with information technologies. They can be a crucial part of knowledge management. According to (Staab, 2002) knowledge management has the goal to optimize the requirements for employees' performance. The following factors 'persons", 'culture", 'organization" and 'basic organization processes" are the major success criteria for knowledge management. According to (Gruber, 1993) ontologies can facilitate the sharing and exchange of knowledge.

There are many kinds and types of ontologies. Depending on their internal structure, ontologies vary in their complexity, as represented in Fig. 2:

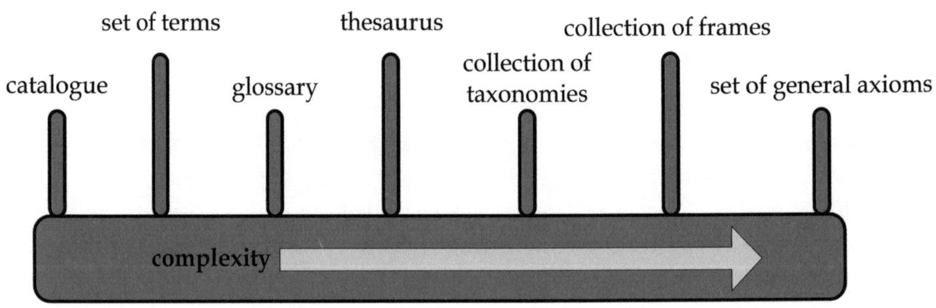

Fig. 2. Types of ontologies organized by increasing complexity [source: authors illustration following (Herb, 2006)]

Examples for trivial complex ontologies are simple catalogues or collections of concepts. Maximally complex ontologies contain an amount of general and weak-defined axioms. An interesting type is taxonomies, which can be defined as a hierarchical classification of concepts in categories.

Taxonomies are also considered as an attenuated definition of ontology. According to (Herb, 2006), they include a series of concepts that are interlinked by hereditary structures. Depending on their nature, ontologies can be applied and re-applied with different levels of intensity (Gómez-Pérez, Fernández-López, Corcho 2004). Ontologies can be classified in so-called 'lightweight ontologies' and in 'heavyweight ontologies.' 'Lightweight ontologies' describe notions (concepts), taxonomies, and relationships and properties between terms. Additionally to these properties, 'heavyweight ontologies' also consider axioms and constraints.

The ontological modeling of systems is possible through the application of existing relationships and rules. (Steinmann & Nejdl, 1999) describe the two tasks of ontologies. In the first sense, ontologies describe the nature of the constituents and the principles. He designates these as 'grammar of reality'. In the second sense, ontologies establish the objects and connections, which (Steinmann & Nejdl, 1999) designate as the 'encyclopedia of reality'. In the first sense, they function as meta-models. Abstract modeling concepts are described and provide the framework for the ontology. Specific meta-model-oriented ontologies will be designated as representation ontologies. For example, the frame-ontology in Ontolingua can be mentioned here, according to (Gruber, 1993). The ontology provides a grammar composed of concepts, relations and attributes. In the second sense, ontologies describe conceptual models are based on structures and correlation of the area of a specific application. Examples of existing conceptual models are legal texts, integration of application systems or open systems.'

Comparing classic and ontological methods, some differences can be identified. Ontologies describe the composition of reality. Traditional modeling approaches assume this information to be known. In this context (Herb, 2006) ascertains, that ontologies are applied for concept-based structuring of information. In his view, ontologies are essentially for more detailed structured information than conventional sources. (Stuckenschmidt, 2009) describes the existence of objects and items and the representation form. (Steinmann & Nejdl, 1999) detail this approach and describe it as a central factor for understanding of items. He concludes that ontologies always are based on a highly abstract level in comparison to model-based approaches.

The authors also indicate the borders of ontological modeling. The crucial difficulties are inconsistencies in classification of meta-data in ontologies, application of meta-data and the distinct classification and structuring of information. Therefore, these aspects are attributable to the highly abstract level of ontologies. Abstract notations lead to such assignment, classification, and structuring issues.

Ontologies can be differentiated in two aspects, conceptual and formal logical nature. According to (Swartout & Tate, 1999), the first aspect has the task of depicting and composing structures of reality. The second addresses the creation of the semantic framework with the definition of objects, classes and contexts. There are many basic approaches to different ontologies in the literature, oriented to the areas of application. (Bateman, 1993) describes the existence of basic types of interconnected entities and describes the so-called 'design patterns.' These are entities that can be differentiated according to the types 'endurant', 'perdurant/occurrence,' 'quality' and 'abstract.' While entities of the 'endurant' type have a continuous and predictable nature, entities of the type 'perdurant/occurrence' describe events that occur unexpectedly and unpredictably. Entities of the types 'quality' and 'abstract' unite properties, attributes, relations, and comparatives.

The ontology DOLCE is an example of the application of these basic types. DOLCE was developed by the Institute of Cognitive Science and Technology in Trento, Italy and stands for 'Descriptive Ontology for Linguistic and Cognitive Engineering.' DOLCE attempts to impart meanings to things and events. Here, entities deal with the meanings through use of agents in order to obtain consensus among all entities regarding to the meaning. (Gangemi et al., 2002) treated this principle in a plausible way. Further examples of conceptual ontologies are WordNet, the 'Unified Medical Language' ontology, 'Suggested Upper Merged Ontology," the ontology of 'ε-Connection' (Kutz et al., 2004), the ontology of 'Process Specification Language,' and the ontology 'OntoClean."

Another key component of conceptual ontologies is ontology engineering. Ontology engineering is concerned with the process design of ontology development, in order to create and to apply ontologies. There are multiple methods here. (Wiedemann, 2008) lists these as follows:

- Ontology Development
- Ontology Re-Engineering
- Ontology Learning
- Ontology Alignment/Merging
- Collaborative Ontology Construction

Ontology Development deals with the question of methodological development and the composition of ontologies. Ontology Re-Engineering focuses on existing approaches and adapts them to the current task. Ontology Learning focuses on approaches for semi- or fully-automatic knowledge acquisition. The Collaborative Ontology Construction issued guidelines for the generation of consensual knowledge. Ontology Merging combines two or more ontologies in order to depict various domains. This method allows handling knowledge that is brought together from different worlds.

(Gruninger, 2002) describes formal logical ontologies as communication, automatic conclusion and representation and re-utilization of knowledge. Formal logical ontologies aim to depict a semantic domain through syntax. The concept of semantics is to be classed in semiotics and describes the theory of signs. Semantics can be also defined as the 'theory of the relationships among the signs and the things in the world, which they denote' (Erdmann, 2001). Semantics are relevant for formal logical ontologies for modeling and generating calculations on a mathematical foundation. This basic approach with its syntax performs a key relevance by providing the mathematical grammar and the concretely denotable model. Exemplary syntaxes are algebraic terms, logical formulas or informational programs. Formal logic provides a language for formalizing the description of the real world and the tool for representing ontologies. It is differentiated according to propositional logic and predicate logic. In propositional logic, there exist exactly two possible truth-values: true or false. Predicate logic consists of terms and describes real world objects in an abstract manner by means of variables and functions. (Stuckenschmidt, 2009) presents methods and techniques of the notation.

Formal logic ontologies do not allow automatic proofs. Only computer-based evidence for sub-problems is possible. Examples of formal logical ontologies are OntoSpace, DiaSpace, OASIS-IP, CASL, OIL, and OWL.

In summary, it can be stated that both ontology types can be classified in different types according to (Guarino, 1998): 'top-level ontologies," 'domain ontologies' and 'application ontologies" which already represent known data and class models. 'Top-level ontologies' describe fundamental and generally applicable basic approaches which are independent of a specific real world. Their level of abstraction is high that allows a wide range of users

'Domain ontologies' focus on a specific application area and describe these fundamental events and activities by specifying the syntax of 'top-level' ontologies. 'Application ontologies' make use of known data or class models which apply to a specific application area.

The following table finally summarizes the described ontologies and compares them according to the presented properties and characteristics. Furthermore, the relevance of ontologies applicable for robotic logistics is specified and the ontologies of 'Process Specification Language' and the ontology 'OntoClean' are highlighted:

ontology	author	application area	characteristics	relevance
DOLCE	Institute of cognitive Science a Technology, Italien	semantic Web	cognitive basis	partially relevant; wide knowledge, uses cognitive aspects
Onto Clean	Laboratory for Applied Ontology, Trento	hierarchical strucutre of knowledge	checking inconsistencies automatically	relevant for structuring processes and robotics technologies
PSL	National Institute of Standards and Technology, Gaithersburg	neutral representation of process knowledge	modular d	relevant for describving relations between processes and robotics
WordNet	Princeton University	representation of natural languages in IT applications	lexical database	not relevant
UMLS	National Libary of Health	database for communicating medical terminology	terminology for medical applications	not relevant; especcially for biomedicine
SUMO	Teknowledge Corporation	providing information in databases and in the internet	combined out of multiple ontologies	not relevant; designed for automated verification
E-Connection	University of Liverpool	complex correlation of different domains	connecting different domains in a formal and logical way	relevant; complex
CASL	Common Framework Initiative	first-Order-logics for subsuming specific languages	modular concept	not relevant; formal approach
F-Logic	Stony Brook University	deductive database	conceptional and object oriented	not relevant, formal approach
OIL	Vrje Universität, Amsterdam	web based language	formal infrastructure for semantic web	not relevant, formal approach
Ontology Web Language	World wide Web Consortium	representation of correlation in the semantic web	base for integration of software	not relevant, formal approach

Table 1. Comparison of selected ontologies in the context of Robotic Logistics [source: author's illustration]

4. Logical ontologies for configuration of individual system architectures

4.1 Required ontological framework

'Robotic-Logistics' formulates the central expectations to the ontology for configuration robotic system architectures. The input and output variables of the environment due to the reference process have to be defined. On this basis, the relevant domains 'technology' and 'process' can be described as to contents. Classes and variables structure them. On the process side, the reference process is addressed. In this domain, the direct upstream and downstream processes of the reference process are also relevant. The output of the upstream process provides the input of the reference process. The output of the reference process provides the input of the downstream process. The relevant technical systems and components of robotic-logistics will be structured in the technology perspective. The regulatory framework has described the following entities. Fig. 3 gives an overview of the hierarchical structure of the domains 'process' and 'technology':

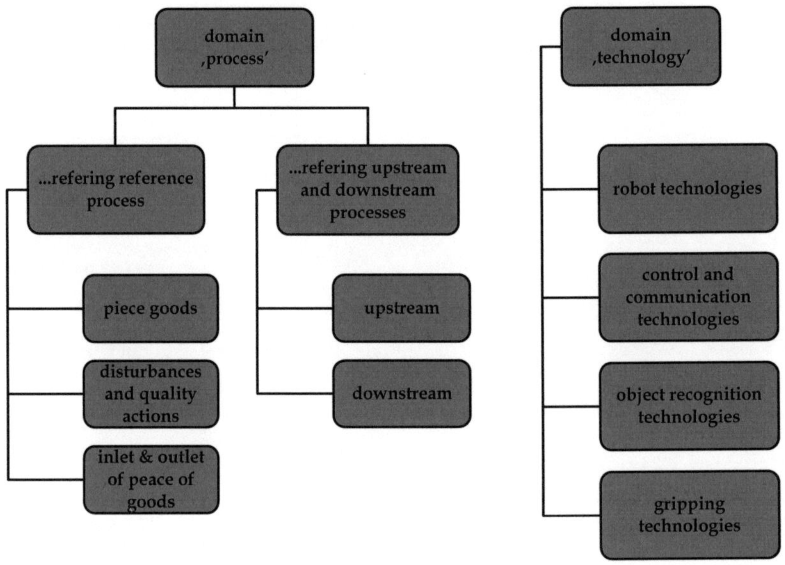

Fig. 3. Class structure of the 'process' and 'technology' domains [source: authors illustration]

The entities in these two class structures are the main processes of meta-model, which will be presented in this chapter. On this basis, process modules and process elements of the reference process are derived with application of the regulatory framework.

The reference process is situated in its systemic environment. It influences the reference process with input and output variables. Thereby the input describes the general framework and restrictions, which are valid for the robotic system. The output results directly from the target dimension of the automation task. The parameters cover technical, organizational, and economic aspects. (Kahraman et al., 2007) define a multi-criteria system for the evaluation of robotic systems, which provides a multiple key factors for the evaluation of robotic systems.

With the development of entity structures of the two domains of the reference process and the inputs and outputs of the environment, the fundamentals of ontology development are set. Based on these structures the hierarchical ontological taxonomies are created with the aid of the ontology OntoClean. This is necessary in order to be able to describe the relations between the two domains through ontology, the Process Specification Language.

4.2 Conceptional ontology for descriptive process technology relations

This section introduces a two-stage approach. In the first phase, the hierarchical structures of the respective domains are composed. The procedure model of (Stuckenschmidt, 2009) offers advantages for the creation of these taxonomies. This approach forms the taxonomies through the OntoClean ontology and analyzes potential sources of error. In the lowest level of taxonomy elements, properties and attributes of the process elements are denoted. They define the reference process. Thus, for example, the process module 'piece goods' with the process element 'bulk" displays the property 'five kilos". Due to this definition, the reference process is individualized and specified.

The second phase provides the combination of the two domains. The description of these relations is done through the ontology of 'Process Specification Language.' It is based on the descriptive notation of functions and processes through its manifold concepts and relations in different levels of detail. Each participant in a pair of relationship is standardized and the relationship is jointly depicted. Due to the functional and procedural point of view of the ontology, the relationship can be well illustrated. The representation is done by focusing on process elements of one domain that cause an impact on the process elements of the second domain.

For preparation, the conceptual framework is defined as the delimitation of the considered environment to be covered. It is defined according to the procedure model developed by (Figgener & Hompel, 2007). They describe a regulatory framework for reference processes:

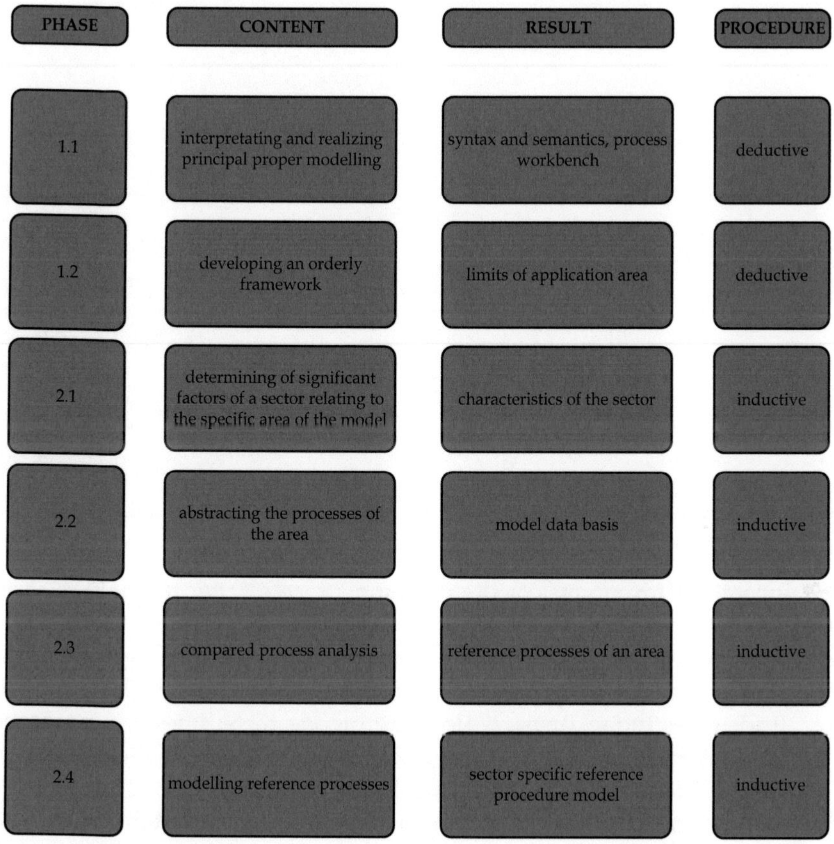

Fig. 4. Procedure model for the creation of reference process models [source: authors illustration following (Figgener & Hompel, 2007)]

The aspects of an application area are defined. Thus, displayed in fig. 4, six phases for the generation of a reference model are described. For the existing problem phase 1.2, phase 2.1, and phase 2.2 are especially relevant. Phase 1.2 describes the regulatory framework. The process modules and process elements are defined in phase 2.1 and 2.2. This distinction allows the reduction and control of the complexity and expenditure for model creation through the ontology.

With these results, both phases of the ontology model can be completed.

Fig. 5 shows the interdependence of both ontologies:

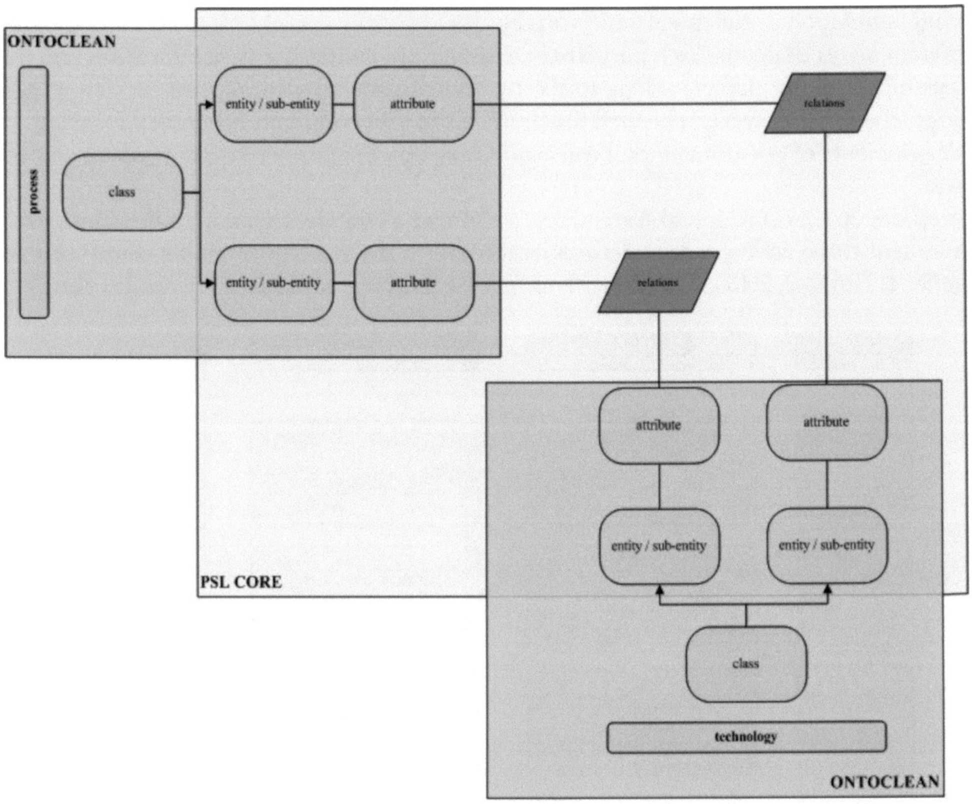

Fig. 5. Spheres of action of the ontologies 'OntoClean' and 'Process Specification Language'[source: author's illustration]

In the first sphere, the taxonomies of the domains 'process" and 'technology" are defined with 'OntoClean'. Depending on the reference process, the process taxonomies will be characterized by properties. They customize the taxonomies. The fig. also displays the sphere of action of the first part namely 'PSL Core' of the ontology 'Process Specification Language." It identifies the relations, which exist among the entities and properties of the process domain and the entities and attributes of the technology domains. Summarized the figure works out 2 phases of the ontology model.

The first phase develops the taxonomies using OntoClean. The taxonomies are denoted and structured. Based on the notation of the meta-properties, the accuracy of the taxonomies is analyzed. Inconsistencies regarding the clearness of the hierarchies arise when relations are utilized incorrectly. This leads to incorrect and misleading interpretations of the ontology (Herb, 2006). The OntoClean process examines existing subsuming structures existing between classes by using meta-features.

The second phase depicts relations between the domains by using the ontology of 'Process Specification Language'. Fixed taxonomies for the domain 'process" and the domain 'technology" for a specific reference process, are the basis for representation of the interaction between the two domains. The main goal of the ontology is to figure out parameters or components of a domain affecting the second domain. Besides the demonstration of the existing relations, the description of the quality of the relationship is an essential aspect. Thereby both the direction of the relationship and the qualitative description will be identified. With the ontology model, the defined requirements will be satisfied as follows:

requirement	ontology	contribution
structuring domains	OntoClean	- identification of relevant terms - structuring by using taxonomies
notation relations	PSL	- identification and notation of relations between process and technology
description relationship	PSL	- description of the relations

Table 2. Handling the requirements through the ontology model [source: author's illustration]

The ontology OntoClean structures the domains. It defines the concepts and composes the taxonomies of the domains 'process" and 'technology". The 'Process Specification Language' note the relations between the parameters. The worked out ontology model is the basis for the individual process modularization and configuration of technical robotic systems.

4.2.1 Definition of taxonomies using the ontology 'OntoClean'

The structuring of the domains is done by defining taxonomies. The usage of taxonomies joins and collects concepts and entities and forms a base frame for these ontologies by structuring them. Here, the relationships are developed associatively mutually. Descending rules work out the taxonomy structures. Due to the qualitative character, the taxonomies are often incorrectly distinguished. The process of 'OntoClean' creates taxonomies and checks the consistency and accuracy of the structures.

The procedure involves the definition of taxonomies and their meta-properties. It aims at overcoming the frequent deficit of false descent of entities in the taxonomic structure. These erroneous subsuming structures will be avoided by a philosophy-based distinction of the entities and classes with meta-properties. (Herb, 2006) describes comprehensively the meta-properties 'identity,' 'essence and rigidity,' 'dependency,' and 'unity.' Using these meta-properties, the taxonomies are distinctly defined through the concepts of class, entity, instance, and property. Entities describe the objects of taxonomy, which are collected in classes. Entities, which have a common property, instantiate a class and will be defined as instances. This is of great relevance. Especially the representation is challenging due to multiple components and parameters, which are displayed in both the domain 'technology' and the domain 'process'. Table 3 provides an overview of the conceptions.

The concepts are the basis of the taxonomies to be created. They are based on the entity structures. For a specific reference process, the entities are reviewed and adapted individually. The procedural taxonomies are developed based on the meta-models of process standardization developed by (Figgener & Hompel, 2007). Depending on the type of process, the main processes, process modules and process elements are applied. Based on the structured system techniques in the domain 'technology', the technical taxonomies are defined due to the commercial state of the art.

The claim of universality is not maintained. The conception is defined to each reference process specifically. This increases the risk of erroneous and inconsistent definition and description of the concepts. Due to this circumstance, the analysis and validation of the developed taxonomies is an essential part of ontology development.

term	definition of procedure model due to fig. 4	commentary
class	main process	structuring of classes
entity	process module	entities which are subsumed in one class
instance	process module with same attributes	all entities with same attributes in one process module
property	process element	characterization and individualization of the reference process

Table 3. Definition of conceptions in the framework of 'OntoClean' [source: authors illustration]

The OntoClean process provides a procedure of subsuming, shown in fig. 6:

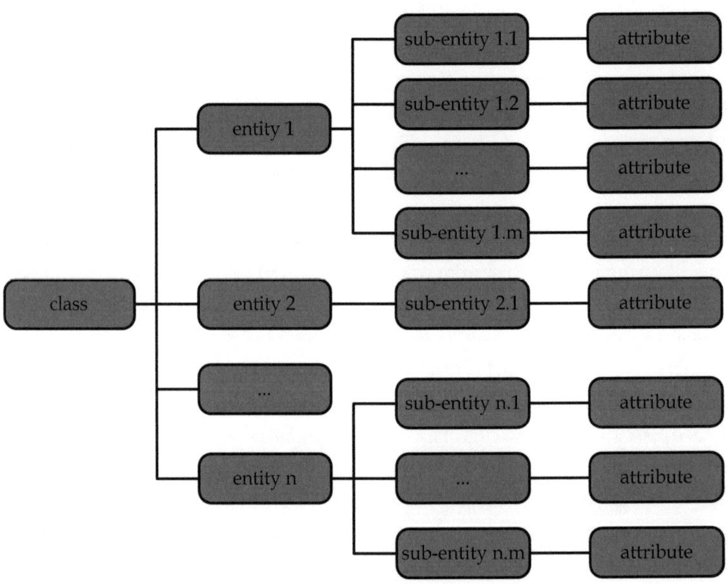

Fig. 6. General construction of taxonomy [source: author's illustration]

A class A subsumes a class B if all instances of class B are always also instances of class A. Fig. 6 presents a class with n entities. Each entity can display further lower-level hierarchical entities, known as sub-entities. Thus, the number of vertical levels is unlimited. On the lowest vertical level, the reference process is individualized by distinct properties. They provide the specific information about the reference process. These may be quantitative or qualitative. As an example here, for a procedural taxonomy, a sub-entity of type 'mass' can be specified with the quantitative property of '22 kg'.

A special case is presented due to the class structure 'environment'. Here, both are structured the environmental framework conditions and the target dimensions. This taxonomy is independent of the reference process, and provides an example, shown in its basic structure, as follows:

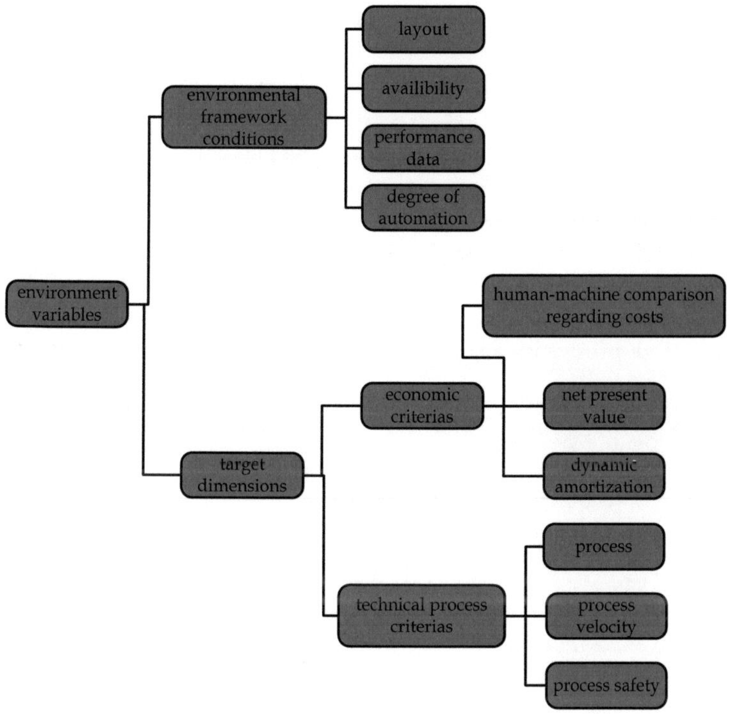

Fig. 7. Taxonomy of the environment variables [source: author's illustration]

The environmentally framework conditions define the technical requirements, such as availability and performance data. The listed properties are specifically defined for each reference process. They customize the process. By means of the target dimensions, technical and economic criteria are carried out. In this context, process safety or process velocity on the technical side are determined. On the economic side, capital value or amortization time is identified. The target dimensions represent the criteria for success of the realization of the robotic system in an ex-post manner.

For each taxonomy and its class K, a notation is defined with the property M. K is denoted as +M, when M applies to all instances of K. The notation -M is used, if not all instances of K have the property M. If M is not valid for any instance of the class K, this relationship is denoted by ~M. Each of the four meta-properties will be reviewed to that effect for each class and entity. This describes the meta-property, 'essence and rigidity' by fixing essence in general first. Second, the specific form of the essence, the rigidity, is described. A property is essential for an entity, if it occurs in every possible situation in the entity. In the next step, a property is rigid, if it is essential for all instances (+R). Non-rigid properties are referred with -R. They describe properties for those entities, which are needed not, but may be instances of the class. Anti-rigidity (~R) is available if there is no instance of associated class instance of the corresponding class.

The meta-property 'identity' describes criteria, which distinctly identify classes and differentiates instances from each other. Both classes and upper classes can provide these identity criteria. The upper classes inherit the criteria. In the first case, the classes are marked with +I. Thus, the identity criterion has been inherited by an upper class. In the second case, the criterion of identity is first defined in an upper class and is marked with +O. Classes that require a further identity criterion as restrictions for distinct definition are denoted with –I.

The third meta-property 'unity' is related to the property 'identity' and describes the affiliation of certain entities to a class. A unity criterion defines a unifying relation of all entities, which are interconnected. The corresponding classes are distinguished with +U. ~U denotes those entities of a class that cannot be distinctly described. If there is no unity criterion provided, the class is described with –U.

The fourth meta-property 'dependence' describes the dependence of a class to other. This fact is relevant if an instance of a class may not be an instance of a second class. Dependent classes will be listed with +D, while independent classes are notated with –D.

In summary, the meta-properties are defined as follows, according to (Herb, 2006):

Meta-property notification	definition
+R	a property is essential for all valid instances
-R	a property that has not inevitable an entity that is an instance of its class
~R	a property where an instance of an allowing class belongs to an instance of a regarded class
+I	classes which differentiate due to the criteria of the allowing instances
-I	class that does not have an identity criteria
+O	identity criteria that is defined for the first time and is not transmitted
+U	unity criteria that denotes connected entities
-U	none unity criteria is existing
~U	connection of entities which cannot described definitely
+D	dependent classes
-D	independent classes

Table 4. Summary definition of meta-properties, [source: authors illustration following (Herb, 2006)]

The review of meta-properties shows incorrect taxonomy structures and makes their correction possible. The next step involves reviewing the consistency of the meta-properties with each other. This will determine whether there are inadmissible combinations of meta-properties. , An example for such a combination is +O und –U. The next step focuses at the removal of all non-rigid classes from the taxonomy.

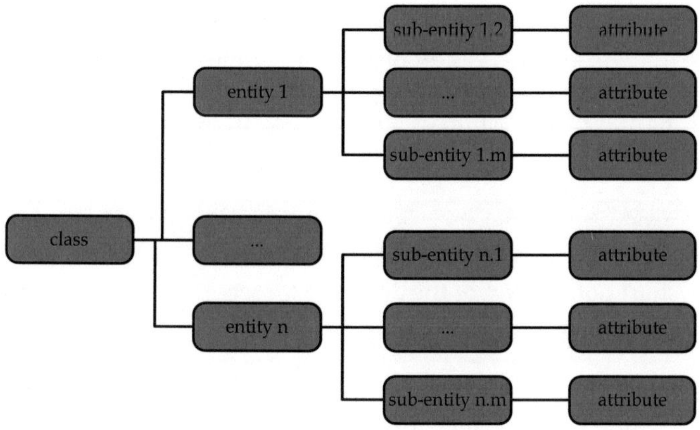

Fig. 8. Backbone taxonomy [source: author's illustration]

That figure points out for an exemplary illustration the removal of the non-rigid 'sub-entity 1.1' and the non-rigid 'entity 2'. This procedure results in the so-called backbone taxonomy. In the next step, subsuming structure has to be examined. It checks any violations of subsuming restrictions. Subsuming is described with the relation 'is-a' and is visualized by arrows. For instance, further relations are described with the notation type 'has.' To avoid false distinctions the arrows are inscribed with the relation name. The hierarchies can be described in the following ways:

- *Have*: The relationship type 'have' connects an attribute with a concept. Thereby, the Attribute is a type of the concept.
- *Att*: This type of relationship describes properties of elements. A concept can take on several properties simultaneously. They do not need to be met simultaneously at all.
- *Is a*: The relationship describes traditional subset relations (subsuming relations).

In the last step, the non-rigid classes and entities are added. Within this last step, the taxonomy is completed.

4.2.2 Description of the interaction by using the ontology of 'Process Specification Language'

In this section, the interacting entities and attributes have to be identified between the two domains by using the created taxonomies. In order to give these interactions a qualitative meaning, (Schlenoff et al., 1999) propose an approach to denote interactions of processes in independents worlds. Therefore, he develops the terminology 'Process Specification Language' (PSL). PSL is a neutral, standard language for process specification for the integration of multiple process applications within the product life cycle. The language is versatile in application and uses multiple concepts, classes, and functions to describe complex processes. Through its manifold applications and many years of further development, the language has diversified and expanded. PSL consists of several modules. The principle fundamental concepts are set in the first module which is called 'PSL-Core'. The module provides four concepts with corresponding functions. According to (Schlenoff et al., 1999), the aim of this module is to fix axioms to describe trivial process connections using a set of semantic relations.

The description of further and more complex processes is carried out with other modules, the so-called extensions. PSL offers in total three extensions: 'outer core', 'generic activities' and 'schedules'. The module 'PSL outer core' deals with generic and broadly based concepts regarding to their applicability. The module 'generic activities' defines a terminology to describe generic activities and their relations. The module 'schedules' describes the application and allocation of resources to activities under the premise of satisfying the temporary restrictions:

term	definition
PSL	short notification of the ontology 'Process Specification Language'
PSL module	group of concepts of the PSL
relation	interrelation between an entity couple of 'process domain' and 'technology domain'. None of the entities is a ad-hoc activity.
activity	process or technical entity or sub-activity that is continuous and relates to the second domain.
ad-hoc activity	process or technical (sub-) entity including its attribute that existence is not calculable
concept	first and highest level of a PSL
class	second level of PSL
function	third and lowest level of PSL
ad-hoc relation	relation of an entity couple of the process and technology domain. Minimum one activity is an ad-hoc activity.

Table 5. Definition of relevant terms of PSL [source: author's illustration]

The definitions are based on the adaptation of the ontology to the current requirements. With these concepts, the individual concepts of this approach will be presented and adapted to this task. With the creation of taxonomies, the relations of taxonomy properties of both domains are identified and described.

This section describes the identification of existing relations and their corresponding notation using the vocabulary of the first module 'PSL Core' of the ontology 'Process Specification Language'. The module contains three concepts. The first concept 'activity' describes general activities, which appear to be predictable and manageable. They do not have to be determined detailed. For instance, 'activities' may be standard processes of a recurring nature. The concept exhibits two different types of functions for further concepts. Function one focuses on further planned activities ('activity') ('is-occurring-at'). Function two describes the connection to unpredictable and unplanned activities ('occurrence-of').

The second concept, 'activity occurrence' describes a unique activity that proceeds unforeseen and unplanned. The concept can also exhibit two different functions for other concepts. The function 'occurrence of' is analogous to the second function of the first concept and describes the initiation of a second type of unpredictable activity of the type 'activity occurrence.' The second function describes the relationship to a concept of the type 'object'. This function expresses the impart of the concept 'activity occurrence' with a none further defined significance to the second concept named 'object.'

The third concept, 'object' describes all activities which do not correspond to any of the above concepts. The concept has two functions. The first relation of the type 'participate in' describes the concept 'object' which receives a non further defined relevance for a concept 'activity.' The second relation 'exists-at' describes an existing relevance to a particular point of time.

The entities are, inclusive of their properties, distinguished from the taxonomies of process and technology domains with these concepts. Here, procedural entities and properties can exist which are either calculable or definable. These activities relate to the concept 'activity'. Unpredictable, indefinable or changing conditions can be described as ad-hoc activities and assigned the concept of 'activity occurrence.' Other logistical or technical objects are called objects and assigned to the concept 'object.' An example describes the entity 'general cargo' as an activity (code 1.1) with its property 'cubic' and the entity 'stock situation' for a concept named ad-hoc activities (code 1.2) with the property 'chaotic'. An example of an object (code 1.3) is a technical process such as the process of recognizing the cargo. The following table summarizes the results of the relevant vocabulary:

PSL module	concept	definition	relation	definition	modification for robotics-logistics	code
PSL Core						1
	acitvity	a general non-defined acititvity			defined and calculable activity that notes an entity or an attribute of taxonomy.	1.1
			is-occurring-at	a primary activity generates a secondary activity at a defined time	the concept 1.1 generates a concept 1.1	1.1.1
			occurrence-of	the primary concept generates a secondary non-expected activity	the concept 1.1 generates a concept 1.2	1.1.2
	activity occurrence	a temporary activity and specific activity that occurs nonrecurring			a non-calculable and changing ad-hoc-activity that notes an entity or attribute of a taxonomy	1.2
			occurrence-of	the primary activity generates a secondary non-expected activity	the concept 1.2 initiates a new concept 1.2	1.2.1
			participates-in	a primary activity generates a non-definable relevance for an object	the concept 1.2 generates a non-defined relevance for an object at a specific time	1.2.2
	object	all entities that are not an activity or activity occurrence			entity or attribute of a taxonomy that are not concepts 1.1 or 1.2	1.3
			participates-in	a primary activity assigns a non-defined relevance to an object in a specific time	the concept 1.1 assigns a relevance to a concept 1.3	1.3.1
			exists-at	an object exists to a specific time	the concept is relevant in a specific time	1.3.2

Table 6. Vocabulary PSL module 'Core' [source: author's illustration]

In a first step, the implementation of ontologies for a specific reference process is associated with concepts and properties of the valid entities. The second step identifies and denotes the relations between the concepts. A matrix representation is provided which is shown in the general structure in tab. 7. The columns show the entities and properties of the technology domains. The lines depict the process domains. The individual hierarchy steps of taxonomies are presented. As described, they were indicated by the hierarchic structure. The coding of the lines and columns indicates the respective levels of the hierarchic structure. Additionally, the identified concepts of the respective sub-entities and properties are noted on the lowest structural level. In the cells, the interaction from tab. 6 are noted and distinguished by means of the coding. For example, the procedural sub-entity 1.1.1 affects the technical components 1.1.1 through the relationship 'object-participates-in' (code 1.3.1).

Process Specification Language				Code	T.1	T.1.1	T.1.1.1	...	...	T.n	T.n.m	T.n.m.o
				technical- taxonomy	system technique 1			...	...	system technique n		
						system technique 1.1		...	...		system technique n.m	
PSL Core							component 1.1.1	...	...			component n.m.o
Code	process taxonomy			PSL-concept			concept 1.z	...	...			concept 1.z
P.1	Class 1							...	...			
P.1.1		entity 1.1						...	...			
P.1.1.1			sub-entity 1.1.1	concept 1.x			1.3.1	...	...			1.x.y
...	...	...	...	...	...	...	...	...	...	...	...	...
...	...	...	...	...	...	...	...	...	...	...	...	...
...	...	...	...	...	...	...	...	...	...	...	...	...
P.n	class n							...	...			
Pn.m		entity n.m						...	...			
P.n.m.o			sub-entity n.m.o	concept 1.x			1.x.y	...	...			1.x.y

Table 7. General matrix representation of the process-technology relations in accordance with PSL module 'core' [source: author's illustration]

The vocabulary allows the description of the relational structure for a dedicated reference process, which describes the relations among the procedural entities and the technical components.

4.3 Industrial application: Depalletizing plastic boxes with a robotic system

This section presents the robot based automation of a simple industrial application by using the presented ontological framework. The presented example focuses on the interaction between 'piece good' of the 'process domain' and 'gripper' of the 'technology domain'. Here the automation of a logistics process by using the ontological framework will be presented. Online books shops package their goods in plastic boxes. Logistics Providers handle these boxes for delivering to the customer. Hence, the boxes are send on pallets in swap bodies by using trucks. The logistics provider unloads the trucks and imports them in their distribution center which operates with a high degree of automation. Therefore the boxes have to be depalletized and brought onto the conveyor technology system. In general this separation is done manually. A robotic systems was configured and integrated by using the ontological framework to automate this reference process. The following figure displays the process with the implemented robotic system:

Fig. 9. Industrial application of a robotic system for depalletizing plastic boxes: result of the ontological configuration [source: author's illustration]

The illustration points out that the configuration of the technical system depends on the parameters of the process. The upstream conveyor supply box pallets including a buffer function. The downstream conveyor conveyors the single boxes into the distribution cycle. The task of an robot-based automation system focuses on the handling of single or multiple boxes and the lay down onto the roller conveyor. Using the presented framework for configuring the robotics architecture, the first step of generating the procedural and technical taxonomies has to be executed. The class structure with its entities and attributes of the 'process domain' can be assigned with attributed as followed:

Entity	meta- property	sub-entity	attribute	meta-property
geometry	+R, +I, -U, +D	form	cubic	+R, -I, -U, +D
		dimension (min, max)	length = 55 [cm] width = 40[cm] height = 30 [cm]	+R, +I, -U, +D
		volume	V = 27 [l]	-R, -I, -U, -D
		surface	closed	+R, -I, -U, -D
material	+R, -I, -U, -D	art	plastic	+R, +I, +U, +D
		stability	high	-R, -I, -U, -D
packaging	+R, +I, -U, +D	strapping	1	-R, +I, -U, -D
		type of packaging	single	-R, +I, +U, +D
mass	+R, +I, -U, +D	weight	28 [kg]	+R, +I, -U, +D

Table 8. Entities and attributes of the class structure 'piece goods' of the 'process domain' [source: author's illustration]

The table displays the various entities with its attribute regarding the class structure 'piece goods'. For instance, the meta-properties define the taxonomy of this class. Also the hierarchical structure is defined. For instance, incorrect assignments of sub-entities will be avoided. Additionally the table assigns relevant attributes of the reference process to entities. The meta-properties figure out that the sub-entities 'geometry' and 'mass' are quite important due to their essential (+R). Furthermore they give identity to their class (+I). Finally, the corresponding taxonomy is presented in figure 10:

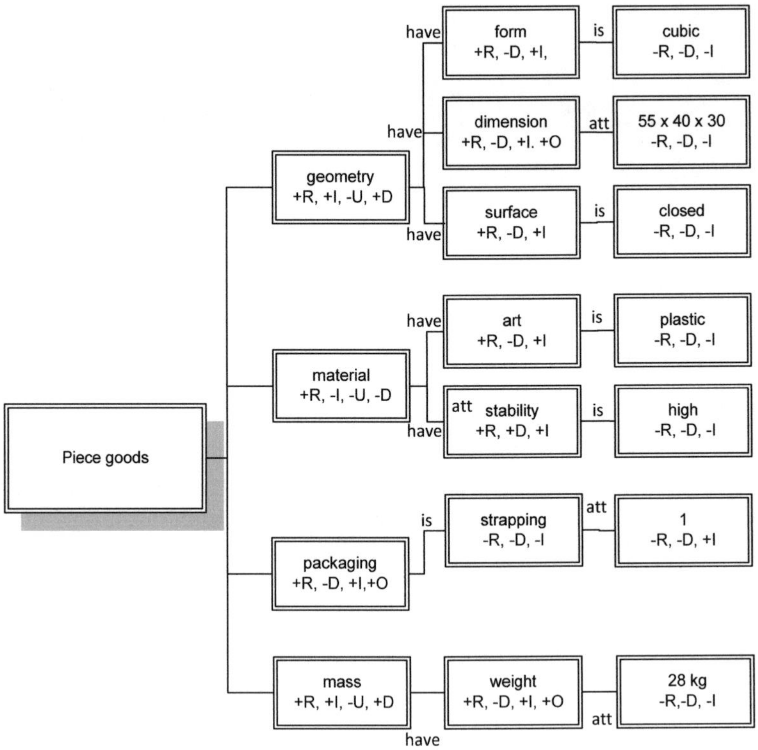

Fig. 10. Backbone taxonomy of the class structure "piece goods" of the "Process domain" [source: author's illustration]

It displays the corrected taxonomy of the exemplary class structure of the industrial application. The class "piece goods" consists of four entities (2nd level) and seven sub-entities. These are related with to attributes. The relations 'subsumption' (is) or 'attribute' (att) describes the connection to the entities. Subsumption are given if the attribute is part of the sub-entity.

The following step focuses on interactions between the "process domain" and "technology domain". Therefore, the relations are noted. Table 9 exhibits the relationship to the system technology "robotics" which unite the entities "kinematics", "geometry", "load", "accuracy" and "installation". Table 9 displays the relation codes between the two domains. For instance, there are some relations of the type "ad-hoc-activity" and "activity". For instance, the strapping has an influence to the accuracy of the robot. Also the mass defines the type of the robot. The table resumes these relations:

Process Specification Language	**PSL Core**	T.1 robotics	T.1.1 kinematics	T.1.2 geometry	T.1.3 load	T.1.4 accuracy	T.1.5 installation
P.1	piece goods						
P.1.1	geometry		1.1.2	1.1.1			
P.1.2	material		1.1.1		1.2.1	1.1.1	
P.1.3	packaging					1.1.1	
P.1.4	mass				1.1.1		

Table 9. Entities and attributes of the class structure 'piece goods' of the 'process domain' [source: author's illustration]

This example clarify the potential of the ontological framework. The framework offers a general and systemic knowledge to configure the best technical components and modules for the specific application due to the system technologies "robotics", "gripping technology", "pattern recognition" and "robot control and communication".

5. Conclusion

The paper presents an ontological approach to standardize robotic systems in logistic processes. Ontologies allow the systematic depiction of the technical systems in the procedural environment. Through their high level of abstraction, this chapter describes the conceptualization and elaboration of an ontological vocabulary for configuration process customized robotic architectures. The vocabulary allows the description of the relational structure for a dedicated reference process. It describes the relations among the procedural entities and the technical components. This ontology framework is the basis for the formation of modules and the configuration of the modules in robotics architectures.

The main goal provides a descriptive approach to the relationship between process and technology. Here, representations of conceptual ontologies were consulted. Due to the conceptual approach, the notation is on an abstract level, so that an automatic conclusion through formal ontologies is realistic. The representation of a dedicated solution space of possible technical configuration states of robotics system architectures is feasible, too.

In further research requirements, the development of formal ontologies in the context of this scope reduces the level of abstraction and enables the mechanical and automatic generation of ontologies.

In this way, interpretation and manipulation opportunities will be reduced and the interconnections of relationships between process and technology detailed. In this connection, formal ontologies allow the development of so-called architecture Configurator. They are based on the provided procedural and technical information and the possible ontological interrelationships. With this information, automatically development, including economic criteria, prioritizes configurations of robotic system architectures for dedicated reference processes. This approach can also serve as an appreciation of the nature of a 'Rapid Configuration Robotics' approach, which can digitally review prototyping activities such as technical feasibility and economic usefulness. The requirement for this type of IT-based configuration planning is shown by the RoboScan10 survey:

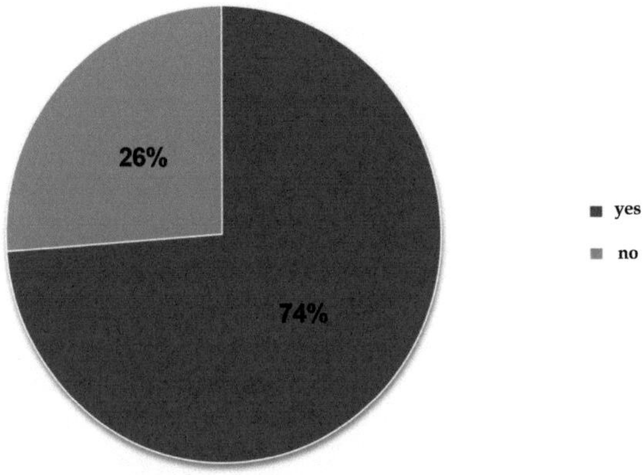

Fig. 11. Study RoboScan10: Answers about necessity for an IT-based system that plans the configuration of robotic systems [source: (Burwinkel, 2011)]

Fig. 11 shows the field of opinions in the context of RoboScan10 about the necessity for IT-based configuration planning of robotic systems. The question asked was: 'from a planning perspective: Could you envisage using an IT-based planning tool, which enables the configuration of both single robot systems and multi robot systems? 75% of all respondents could envisage the application of such tools.

6. References

Arnold, D. (2006). *Intralogistik : Potentiale, Perspektiven, Prognosen,* Springer, ISBN 978-3-540-29657-7, Berlin, Germany

Bateman, J. (1993): On the relationship between ontology construction and natural language: a socio-semiotic view. *International Journal of Human-Computer Studie,* Vol.43, No.5/6, pp.929-944, ISSN 1071-5819

Baumgarten, H. (2008). *Das Beste der Logistik : Innovationen, Strategien, Umsetzungen,* Springer, ISBN 978-3-540-78405-0, Berlin, Germany

Bunge, M. (1977). *Treatise on basic philosophy,* Reidel, ISBN 9027728399, Dordrecht, Netherlands

Burwinkel, M. & Pfeffermann, N. (2010). Die Zukunft der Robotik-Logistik liegt in der Modularisierung : Studienergebnisse "RoboScan'10". *Logistik für Unternehmen,* Vol.24, No.10, pp.21-23, ISSN 0930-7834

Chen, P. (1976). The entity-relationship model - toward a unified view of data. *ACM Trans. Database Syst,* Vol. 1, No. 4, pp. 9–36

Elger, J., & Haußener, C. (2009). Entwicklungen in der Automatisierungstechnik. In: *Internet der Dinge in der Intralogistik,* W. Günthner (Ed.), 23–27, Springer, ISBN 978-3-642-04895-1, Berlin, Germany

Erdmann, M. (2001). *Ontologien zur konzeptuellen Modellierung der Semantik von XML,* University Karlsruhe, ISBN 3-8311-2635-6, Karlsruhe, Germany

EUROP. (2009). Robotics Visions to 2020 And Beyond : The Strategic Research Agenda For Robotics in Europe, In: *European Robotics Technology Platform,* 20.06.2011, available from: www.robotics-platform.eu

Figgener, O. & Hompel, M. *Beitrag zur Prozessstandardisierung in der Intralogistik.* In: *Logistics Journal* (2007). ISSN 1860-5923 S. 1–12

Fritsch, D. & Wöltje, K. (2006). Roboter in der Intralogistik : Von der Speziallösung zum wirtschaftlichen Standardprodukt. *wt Werkstattsstechnik Online,* Vol.96, No.9, pp. 623–630, ISSN 1436-4980

Gangemi, A., Guarino, N., Masolo, C., Oltramari, A. & Schneider, L. (2002). Sweetening Ontologies with DOLCE, In: *Knowledge Engineering and Knowledge Management: Ontologies and the Semantic Web,* Gómez-Pérez, A. & Benjamins, V. (Ed.),. pp.223–233, Springer, ISBN 978-3-540-44268-4, Berlin, Germany

Gómez-Pérez, A., Fernández-López, M., Corcho, O. (2004). *Ontological engineering: With examples from the areas of knowledge management, e-commerce and the semantic web,* Springer, ISBN 978-1-852-33551-9, London, England Göpfert, I. (2009). *Logistik der Zukunft - Logistics for the future,* Gabler, ISBN 978-3-834-91082-0, Wiesbaden, Germany

Gruber, T. (1993). A translation approach to portable ontology specifications. *Knowledge Acquisition,* Vol. 5, No.2, pp.199-220, ISSN 1042-8143

Gruninger, M. (2002). Ontology - Applications and Design. *Communication of the ACM,* Vol.45, No.2, pp.39–41, ISSN 00010782

Guarino, N. (1998). *Formal ontology in information systems: Proceedings of the first international conference (FOIS '98), June 6 - 8, Trento, Italy,* FOIS, ISBN 905-1-993-994, Amsterdam, Netherlands

Günthner, W. & HompelM. (2009.). *Internet der Dinge in der Intralogistik,* Springer, ISBN 978-3-642-04895-1, Berlin, Germany

Herb, M. (2006). Ontology Engineering mit OntoClean. In: *IPD University Karlsruhe,* 10.06.2011, available from:
http://www.ipd.uni-karlsruhe.de/~oosem/S2D2/material/1-Herb.pdf

Kahraman, C., Cevik, S., Ates, N.Y. & Gulbay, M. (2007). Fuzzy multi-criteria evaluation of industrial robotic systems. *Computers & Industrial Engineering,* Vol.52, No.4, pp. 414-433, ISSN 0360-8352

Kastens, U. & Kleine Büning, H. (2008). *Modellierung : Grundlagen und formale Methoden,* Hanser, ISBN 978-3-446-41537-9, München, Germany

Kiencke, U. (1997). *Ereignisdiskrete Systeme: Modellierung und Steuerung verteilter Systeme,* Oldenbourg, ISBN 348-6-241-508, München, Germany

Kutz, O., Lutz, C., Wolter, F., Zakharyaschev, M. (2004). ε-connections of abstract description systems. *Artif. Intelligence,* Vol.156, No.1, pp.1-73, ISSN 0004-3702Neches, R., Fikes, R., Finin, T., Gruber, T., Patil, R., Senator, T. & Swartout, W. (1991). Enabling technology for knowledge sharing. *AI Mag,* Vol.12, No.3, pp.36–56

Scheid, W. (2010). Perspektiven zur Automatisierung in der Logistik : Teil 1 - Ansätze und Umfeld. *Hebezeuge Fördermittel - Fachzeitschrift für technische Logistik,* Vol.50, No.9, pp. 406–409, ISSN 0017-9442

Scheid, W. (2010): Perspektiven zur Automatisierung in der Logistik : Teil 2 - Praktische Umsetzung. *Hebezeuge Fördermittel - Fachzeitschrift für technische Logistik* Vol.50, No.10, pp. 482–483, ISSN 0017-9442

Schlenoff, C., Gruninger, M., Tissot, F., Valois, J. Lubell, J. & Lee, J. (1999). The Process Specification Language (PSL) Overview and Version 1.0 Specification, In: *www.mel.nist.gov/psl/*, 12.06.2011, available from: http://www.mel.nist.gov/msidlibrary/doc/nistir6459

Seidlmeier, H. (2002). *Prozessmodellierung mit ARIS® : Eine beispielorientierte Einführung für Studium und Praxis,* Vieweg, ISBN 352-8-058-048, Braunschweig, Germany

Siegert, H. (1996). *Robotik: Programmierung intelligenter Roboter,* Springer, ISBN 3540606653, Berlin, Germany

Staab, S. (2002). Wissensmanagement mit Ontologien und Metadaten. *Informatik-Spektrum,* Vol.25, No.3, pp.194–209, ISSN 0170-6012

Staud, J. (2006). *Geschäftsprozessanalyse : Ereignisgesteuerte Prozessketten und objektorientierte Geschäftsprozessmodellierung für betriebswirtschaftliche Standardsoftware,* Germany, ISBN 978-3-540-24510-0, Berlin, Germany

Steinmann, F. & Nejdl, W. (1999). Modellierung und Ontologie, In: Institut für Rechnergestützte Wissensverarbeitung, 25.05.2011, available from: www.kbs.uni-hannover.de/Arbeiten/Publikationen/1999/M%26O.pdf

Straube, F. & Rösch, F. (2008): *Logistik im produzierenden Gewerbe,* TU Berlin, ISBN 978-3-000-24165-9, Berlin, Germany

Stuckenschmidt, H. (2009). *Ontologien : Konzepte, Technologien und Anwendungen,* Springer, ISBN 978-3-540-79330-4, Berlin, Germany

Studer, R., Benjamins, V., Fensel, D. (1998). Knowledge engineering: Principles and methods. *Data & Knowledge Engineering,* Vol.25, , No.1-2, pp.161–197, ISSN 0169-023X

Suppa, M. & Hofschulte, J. (2010). Industrial Robotics. *at – Automatisierungstechnik,* Vol.58, No. 12., pp.663–664,

Swartout, W. , Tate, A. (1999). Ontologies. *IEEE Intelligent Systems and their Applications,* Vol.14, No.1, pp.18–19, ISSN 1094-7167

Tabeling, P. (2006): *Softwaresysteme und ihre Modellierung : Grundlagen, Methoden und Techniken,* Springer, ISBN 978-3-540-25828-5, Berlin, Germany

Westkämper, E., Verl,A. (2009). *Roboter in der Intralogistik : Aktuelle Trends - Neue Technologien - Moderne Anwendungen,* Verein zur Förderung produktionstechnischer Forschung, Stuttgart, Germany

Wiedemann, G. (2008). Ontologien und Ontology Engineering, In: Seminar 'Semantic Web Technologien', 12.06.2011, available from: http://www.informatik.uni-leipzig.de/~loebe/teaching/2008ss-seweb/08v-ontengineering-gwiedemann.pdf

Programming of Intelligent Service Robots with the Process Model "FRIEND::Process" and Configurable Task-Knowledge

Oliver Prenzel[1], Uwe Lange[2], Henning Kampe[2], Christian Martens[1] and Axel Gräser[2]
[1]*Rheinmetall Defence Electronics,*
[2]*University of Bremen*
Germany

1. Introduction

In Alex Proyas's science fiction movie "I, Robot" (2004) a detective suspects a robot as murderer. This robot is a representative of a new generation of personal assistants that help and entertain people during daily life activities. In opposition to the public opinion the detective proclaimed that the robot is able to follow his own will and is not forced to Isaac Asimov's three main rules of robotics (Asimov, 1991). In the end this assumption turned out to be the truth.

Even though the technological part of this story is still far beyond realization, the idea of a personal robotic assistant is still requested. Experts predicted robotic solutions to be ready to break through in domestic and other non-industrial domains (Engelberger, 1989) within the next years. But up to now, only rather simple robotic assistants like lawn mowers and vacuum cleaners are available on the market. As stated in (Gräfe & Bischoff, 2003), all these systems have in common that they only show traces of intelligence and are specialists, designed for mostly a particular task. Robots being able to solve more complex tasks have not yet left the prototypical status. This is due to the large number of scientific and technical challenges that have to be coped with in the domain of robots acting and interacting in human environments (Kemp et al., 2007).

The focus of this paper is to describe a tool based process model, called the "FRIEND::Process"[1], which supports the development of intelligent robots in the domain of personal assistants. The paper concentrates on the interaction and close relation between the FRIEND::Process and configurable task-knowledge, the so called process-structures. Process-structures are embedded in different layers of abstraction within the layered control architecture MASSiVE[2] (Martens et al., 2007). Even though the usage of layered control architectures for service robots is not a novel idea and has been proposed earlier (Schlegel &

---

[1] The name FRIEND::Process is related to the FRIEND projects (Martens et al., 2007). It has been developed within the scope of these projects, but is also applicable to other service robots.

[2] MASSiVE – Multilayer Control Architecture for Semi-Autonomous Service Robots with Verified Task Execution

Woerz, 1999; Schreckenghost et al., 1998; Simmons & Apfelbaum, 1998), MASSiVE is tailored for process-structures and thus is the vehicle for the realization of verified intelligent task execution for service robots, as it is shown in the following. The advantages of using process-structures shall be anticipated here:

- **Determinism**: Process-structures represent the complete finite sequence of actions that have to be carried out during the execution of a task. Due to the possibility of a bijective transformation from process-structures to Petri-Nets, a-priori verification with respect to deadlocks, reachability and liveness becomes possible. Thus, the task planner and executor, as part of the layered architecture, operate deterministically when using verified task-knowledge.
- **Real-time capability**: Additionally, the complexity of the task planning process satisfies real-time execution requirements, because this process is reduced to a graph search problem within the state-graph of the associated Petri-Net.
- **Fault-Tolerance**: Erroneous execution results are explicitly modeled within process-structures. Additionally, redundant behavior is programmatically foreseen. If an alternative robotic operation, which shall cope with the unexpected result, is not available, the user is included as part of a semi-autonomous task execution process.

To be able to provide a user-friendly configuration of process-structures and to guarantee consistency throughout all abstraction levels of task-knowledge, a tool-based process model - the FRIEND::Process - has been developing. The process model, on the one hand, guides the **development and programming** of intelligent behavior for service robots with process-structures. On the other hand, process-structures can be seen as a **process model for the service robot** itself, which guides the task execution of the robot during runtime. The unique feature of the FRIEND::Process in comparison to other frameworks (Gostai, 2011; Microsoft, 2011; Quigley et al., 2009) and the above mentioned control architectures is to completely rely on configurable process-structures and thus on determinism, real-time capability and fault tolerance.

The FRIEND::Process consists of the following development steps:

- **Analysis of Scenario and Task Sequence:** A scenario is split up into a sequence of tasks.
- **Configuration of Object Templates and Abstract Process-Structures:** The task participating objects are specified as Object Templates and pictographic process-structures on the symbolic (abstract) level are configured and verified.
- **Configuration of Elementary Process-Structures:** Process-structures on the level of system resources and sub-symbolic (geometric) information are configured and verified with the help of function block networks.
- **Configuration and Testing of Reactive Process-Structures:** Process-structures on the level of algorithms and closed loop control, operating sensors and actuators, are configured and tested, also with configurable function blocks.
- **Task Testing:** Task planning and execution is applied on all levels of process-structures and a complete and complex task execution is tested.

In the following Section 2, the motivation for the introduction of process-structures is explained in more detail by discussing the complexity of task planning for service robots with the help of an exemplary scenario. The description of the FRIEND::Process development steps is subject of Section 3. Throughout this description, exemplary process-structures of the sample scenario of Section 2 are introduced for each development step.

Finally, Section 4 summarizes and concludes the description of the FRIEND::Process for programming intelligent service robots.

2. Task planning on basis of process-structures

In this section, the complexity of classical task planning approaches is discussed first, before the introduction of process-structures is motivated. The discussion is carried out with the help of task execution examples from the field of rehabilitation robotics and the rehabilitation robot FRIEND III (IAT, 2009; Martens et al., 2007).

2.1 The complexity of classical task-planning approaches

With respect to one exemplary task - a service robot is supporting the preparation and the eating of a meal by a disabled person - the complexity of robotic task execution shall be illustrated. For this purpose the figures Fig. 1 to Fig. 3 are introduced. Fig. 1 shows the rehabilitation robot FRIEND III which is used as exemplary target system. In Fig. 2 snapshots of the task sequence "Meal preparation and eating assistance" are depicted. Finally, Fig. 3 shows the decomposition of this task sequence according to the principles to be presented in detail in this paper.

FRIEND III is a general purpose semi-autonomous rehabilitation robot suitable for the implementation of a wide range of support tasks. As depicted, FRIEND III consists of an electrical wheelchair which is equipped with several sensors and actuators: A stereo camera system mounted on a pan-tilt-head, force torque sensor, robotic arm and gripper with force control. FRIEND III has been developed by an interdisciplinary team of engineers, therapists and designers and has been tested with disabled users within the AMaRob project (IAT, 2009).

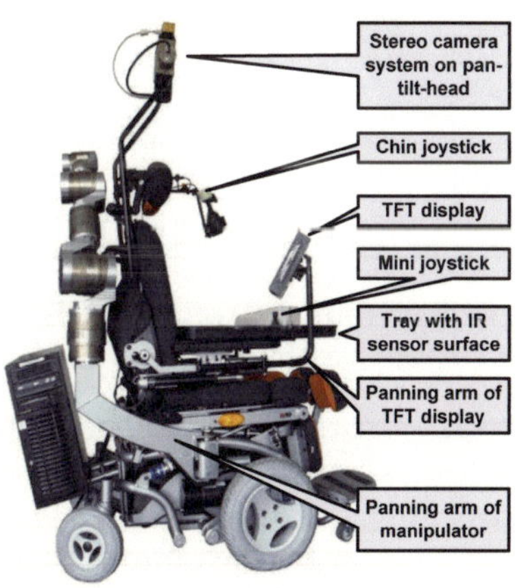

Fig. 1. FRIEND III rehabilitation robot

To perform "meal preparation and eating assistance", the robot system has to execute the following actions:

- Locate the refrigerator, open the refrigerator door, locate the meal inside the refrigerator, grasp and retrieve the meal from the refrigerator, close the refrigerator door
- Open the microwave-oven, insert the meal, close the oven, start the heating process

- Open the microwave-oven door again, grasp and retrieve the meal, close the microwave-oven door
- Place the meal in front of the user, take away the lid
- In a cycle, take food with the spoon and serve it near the user's mouth, finally put the spoon back to the meal-tray
- Clear the wheelchair tray

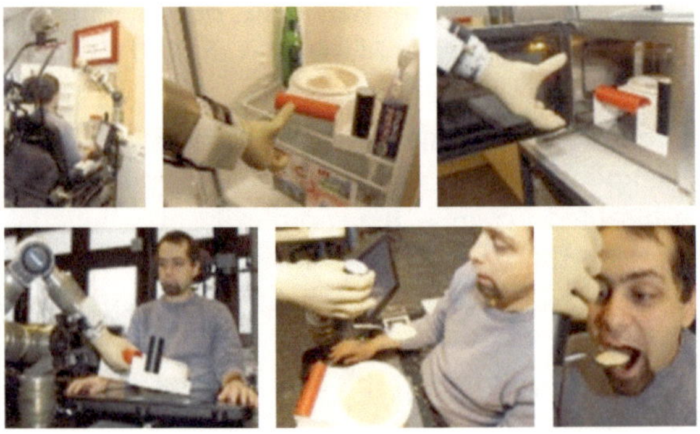

Fig. 2. Task sequence for meal preparation and eating assistance

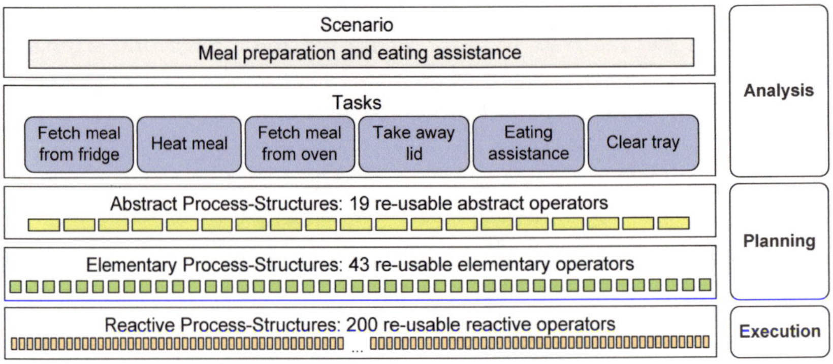

Fig. 3. Decomposition of a scenario on four abstraction levels, illustrated with the sample scenario "Meal preparation and eating assistance"

As shown in Fig. 3, the overall scenario is decomposed into tasks, abstract operators, elementary operators and reactive operators according to the layered control architecture MASSiVE. Abstract process-structures (PS_A[3]) model behavior on task planning level and elementary process-structures (PS_E) model behavior on system planning level. The reactive process-structures (PS_R) define reactive operations on the executable algorithmic level. From viewpoint of task planning, the "meal preparation and eating assistance" scenario is split up into 6 tasks, 19 abstract operators and 43 elementary task planning operators. Additionally, a large set of reactive operators is required within the execution layer.

In typical human environments, it is impossible to predefine a static sequence of operators beforehand. Many dynamic aspects resulting from dynamic environmental changes have to be

[3] Find all abbreviations in the glossary at the end of this paper.

considered, e. g. caused by changing lighting conditions, arbitrarily placed and filled objects, changing locations of objects and the robotic platform, various obstacles, and many more. Consequently, a strategy to plan a sequence of actions that fulfills a certain task is mandatory.
Many task planners are based upon deliberative approaches according to classical artificial intelligence. Typically, the robotic system models the world with the help of symbolic facts (e. g. first order predicate logic, (Russel & Norvig, 2003)), where each node of a graph represents a state (snapshot) of the world. The planner has to find a sequence of operations which transforms a given initial state into a desired target state. In the worst cases this leads to NP-complete problems, as there is an exponential complexity of classical search algorithms (Russel & Norvig, 2003). If we consider breadth-first search as a simple example, a calculation time of hours results at search depth 8; and with a depth of 14, hundreds of years are required for exhaustive search (branching factor 10 and calculation time of 10.000 nodes/s are assumed). The search depth is related to the number of required operators for a certain task and the branching factor results from the number of applicable operators in one node. Compared to the number of required and available operations shown in Fig. 3 it becomes obvious that only trivial problems can be solved on this basis. Certainly, the mean search time can be improved in comparison to breadth-first search, with e. g. heuristic approaches like A*, with hierarchical planning, search in the space of plans or successive reduction of abstraction (Russel & Norvig, 2003; Weld, 1999), but in worst cases a planning complexity as mentioned has to be faced. Even though the improvements of deliberative task planners are notable, it is still questionable whether they are efficient (real-time capable) and robust (deterministic and fault-tolerant) enough for the application in real world domains (Cao & Sanderson, 1998; Dario et al., 2004; Russel & Norvig, 2003).

2.2 Process-structures as alternative to classical planning approaches

An alternative to deliberative systems are assembly planning systems. Cao and Sanderson proposed such an approach for the application to service robotics (Cao & Sanderson, 1998). Based on this idea, Martens developed a software-technical framework (Martens, 2003) that operates on pre-structured task-knowledge, called *process-structures*. Table 1 summarizes the concept of process-structures and the distinction of task level, system level and algorithmic level.
Fig. 4 shows an example of an abstract process-structure that models the fetching of a cup from a container. The object constellations (OC) model the physical contact situation of the involved objects box (B), container (C), gripper (G) and table (T). The object constellations are connected via composed operators (COPs). These are in most cases (i. e. where this is physically meaningful) bi-directional operators. To be able to perform task planning based on an abstract process-structure, a set of OCs defines an initial situation and another set of OCs defines the target situation. Thus, task planning on abstract level means to find a sequence of COPs from initial to target situation. The initial situation is usually dynamically determined at runtime with the help of an initial monitoring procedure (Prenzel, 2005). The target situation is pre-determined for a certain PS_A.
A process-structure contains a context-related subset of task-knowledge. The finite size of a process-structure makes planning in real-time with short time intervals as well as a priori verification possible. The logical correctness of a structure is checked against a set of rules. A positive result of this check guarantees that no system resource conflicts exist. It also guarantees the correct control and data flow. Altogether, the concept of process-structures is the basis for a robust system runtime behavior. Despite pre-structuring, the process-

structures are still flexible to adapt to diverse objects, so that their re-usability in different scenarios is achieved. Technical details of process-structures beyond this summarized concept description can be found in (Martens et al., 2007).

PS_A	**Task Level**
Defines what happens	Models e. g. the fetching of an object
Is configured by:	Non-technical personnel or the user
PS_E	**System Level**
Defines how something happens from system perspective	Models the usage of system resources and the control and data flow
Is configured by:	System programmer
PS_R	**Algorithmic Level**
Defines how something happens from perspective of reactive algorithms	Models the combined usage of hardware sensors and actuators
Is configured by:	System programmer

Table 1. Summarized concept of process-structures

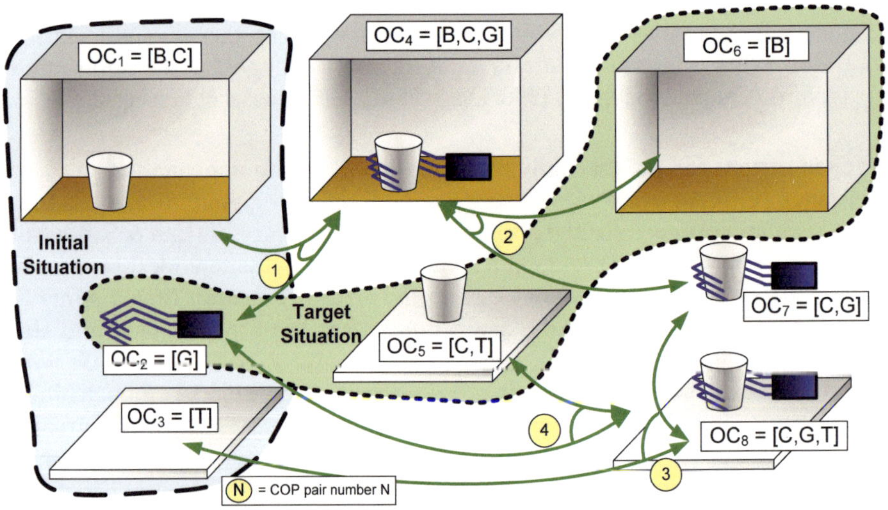

Fig. 4. Schematic illustration of an abstract process-structure (PS_A) which models the fetching of a cup from a container-like place as e. g. a fridge or a cupboard

The applicability of process-structures for the programming of service robots has been shown in (Martens, 2003) with the help of several representative rehabilitation robotic scenarios. As anticipated in the introduction this approach has been extended during the AMaRob project (2006 - 2009) and within (Prenzel, 2009) to embed the process-structure-based programming into a process model - the FRIEND::Process. From task analysis to final testing of implemented system capabilities, the FRIEND::Process guides through the complete development cycle of a service robot based on a closed chain of user-friendly configuration tools. Enhancements of the FRIEND::Process are matter of ongoing developments.

3. The FRIEND::Process

Process models structure complex processes in manifold application areas. With respect to system- and software-engineering, a process model shall organize the steps of development, the tools to be used and finally the artifacts to be produced throughout the different development stages. The overall scheme of the FRIEND::Process is depicted in Fig. 5. Central elements of the process and consequently the specialty in comparison to other process models are the process-structures. Within the development steps, the building blocks of process-structures are decomposed as shown in Table 2. In the following sections the five development steps of the FRIEND::Process are discussed in detail. Thus, the contents of Table 2, i. e. the composition of process-structures and the decomposition on the next level as well as the abbreviations will be explained. Also, the application of the FRIEND::Process for the development of the sample task of "meal preparation and eating assistance" is shown in each step.

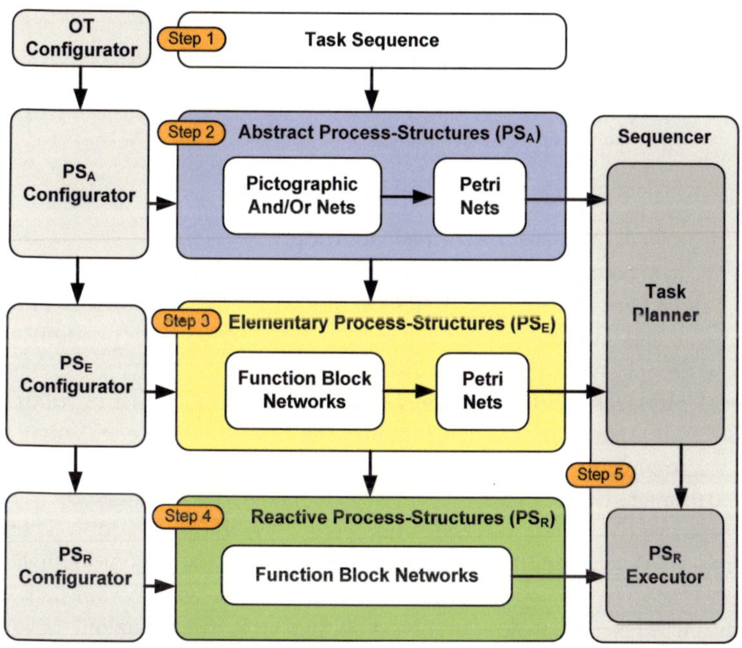

Fig. 5. Scheme of the FRIEND::Process with five development steps and the respective process-structure levels as well as the involved tools for configuration, planning and execution

Process-Structure Decomposition	Process-Structure Building Blocks
Scenario → Task Sequence	Tasks
Task → PS_A	System, Object Templates (OTs), Object Constellations (OCs), Facts, Composed Operators (COPs)
COP → PS_E	System, Object Templates (OTs), Facts, Skills
Skill → PS_R	System, Object Templates (OTs), Reactive Blocks

Table 2. Decomposition and building blocks of process-structures

3.1 FRIEND::Process step 1: Analysis of scenario and task sequence

Development according to the FRIEND::Process starts with the "Scenario Analysis" as step 1. Unlike the subsequent steps, this step is not (yet) tool-supported. The scenario analysis splits up a complex scenario like "meal preparation and eating assistance" into a sequence of re-usable tasks. Also, a structured collection of the objects takes place that are in the focus of a certain scenario.

3.1.1 Description of the process step

The development step 1 is dedicated to a first analysis of the desired task execution scenario. As shown in Fig. 6 a sequence of re-usable tasks is specified. Besides the strictly sequential concatenation of tasks, cyclic repetitions are also possible, as e. g. required for the eating assistance scenario introduced at the beginning of the paper.

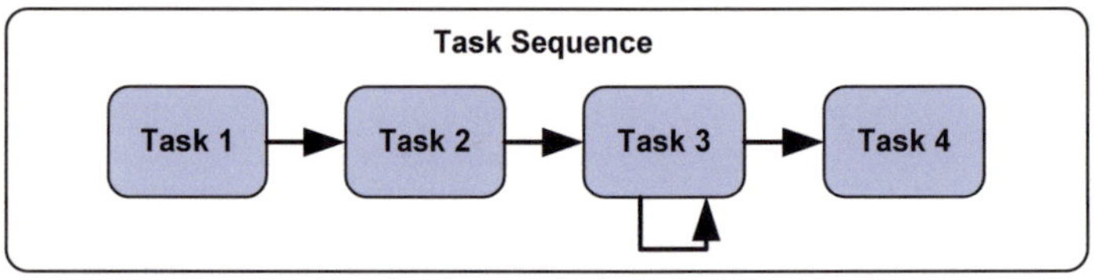

Fig. 6. A complex task sequence consists of several tasks

The FRIEND::Process defines criteria for task splitting:

- **Modularity, low complexity and re-usability:** One task is focusing on a set of objects. This set shall be kept as small as possible to limit the task's complexity and to ensure re-usability of a task. It shall be possible to use the tasks independently, but also to concatenate them to more complex action sequences.
- **The typical physical location of the objects:** If movement of the robotic platform is required, this is a clear indicator to switch the task context, e. g. when moving from fridge to microwave oven in the meal preparation scenario. After moving the platform, relative locations between platform and objects have to be re-assessed.

Currently, the process step 1 is not yet supported by a dedicated tool. Therefore, to still achieve a certain level of formality, the results of scenario analysis are collected in a UML use case diagram as seen in Fig. 7. For each task a use case with verbal task description is specified. This includes the objects involved in the task, the so-called task participating objects (TPO).

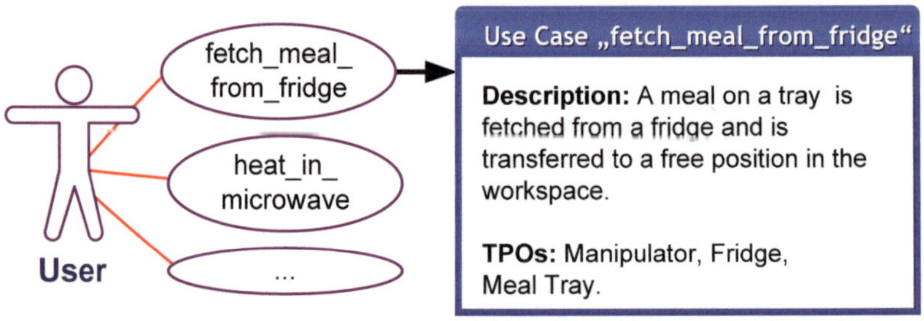

Fig. 7. Use case diagram with tasks (use cases) of the sample scenario. For each task, a detailed description as well as the set of task participating objects (TPO) is specified

The objects involved in task execution are the elements that are relevant in all subsequent development steps. To follow the principle of re-usable task-knowledge, the TPOs are specified as abstract object classes. For example, a task that describes the fetching of an object from a container-like place (see Fig. 4) can be re-used to fetch either a bottle or a meal from the refrigerator. In the FRIEND::Process the re-usable classes of objects are specified as hierarchical UML ontology. An exemplary ontology for the scenario "meal preparation and assistance to eat" is depicted in Fig. 8. It is depicted that the TPOs are constructed from basic geometric bodies (cuboid and cylinder) and more complex objects are created with inheritance and aggregation.

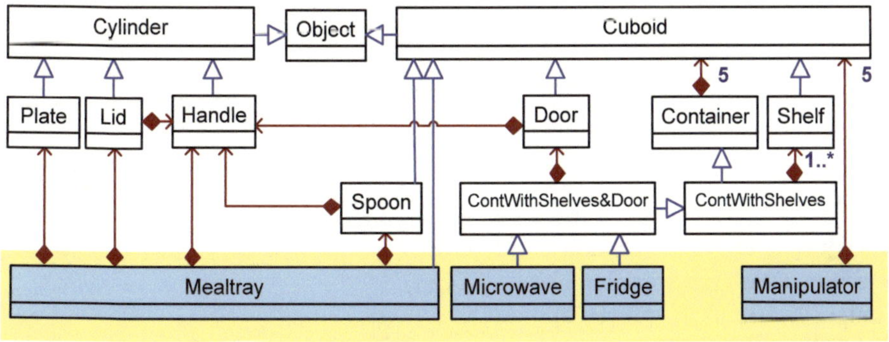

Fig. 8. Ontology of task participating objects (TPO) for the scenario "Meal preparation and assistance to eat"

To embed the TPOs in the tool-chain that covers all further development steps, the concept of "Object Templates" (OT) has been introduced (Kampe & Gräser, 2010). The configuration of Object Templates and their integration into the different levels of process-structure configuration will be discussed in more detail within the following process steps.

3.2 FRIEND::Process step 2: Configuration of object templates and abstract process-structures

In this development step the task participating objects are formally specified and configured with the help of Object Templates. Subsequently, an abstract process-structure (PS_A) is configured based on pictographic And/Or-Nets. This means that physical object constellations (OC) and physical transitions between the object constellations are specified. Besides configuration of PS_A, the logical correctness of the abstract process-structures is guaranteed by the configuration tool. Finally, the pictographic PS_A are converted to Petri-Nets according to (Cao & Sanderson, 1998) for the input into the task planner.
In the following, a description of the process step is introduced first. Afterwards, the configuration concept for Object Templates is shown. Finally, the configuration of an abstract process-structure is exemplified.

3.2.1 Description of the process step

As shown in Fig. 9 the FRIEND::Process decomposes each task into an abstract process-structure (PS_A). A schematic exemplary pictographic PS_A for the task "Fetch cup from container" has already been introduced and discussed in Fig. 4. Within the FRIEND::Process, the configuration of PS_A is carried out within a pictographic configuration environment, the so-called PS_A-Configurator. Fig. 10 shows the PS_A-Configurator with the PS_A "Fetch meal from fridge".

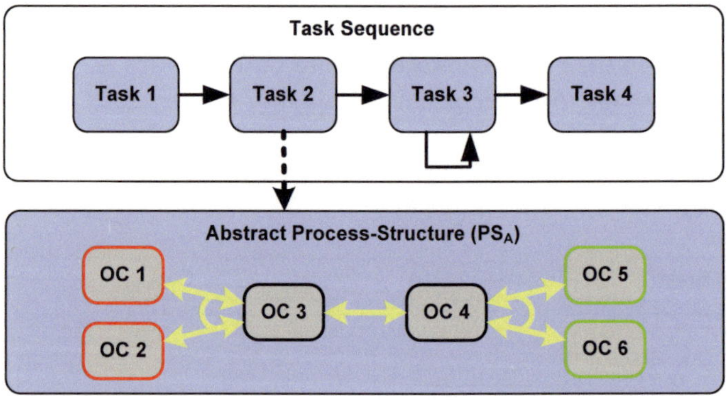

Fig. 9. Decomposition of a task as abstract process-structure with object constellations (OC) and composed operators (COP)

The procedure of PS_A configuration is as follows:

- Selection of task participating objects (TPOs)
- Composition of object constellations (OCs)
- Connection of OCs via composed operators (COPs)
- Selection of default initial and default target situation

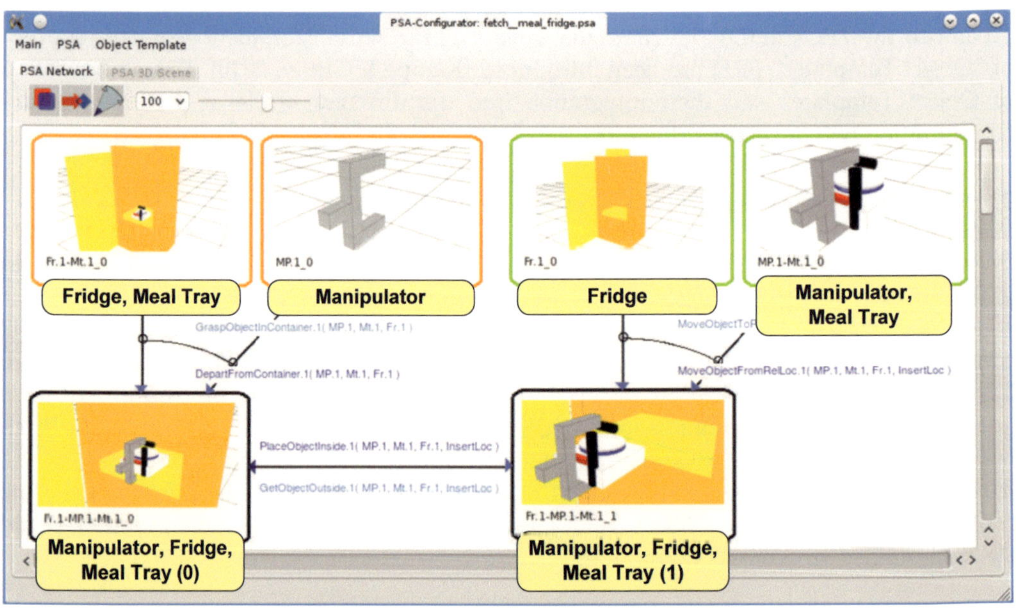

Fig. 10. PS_A-Configurator with the pictographic abstract process-structure modeling the Task "Fetch meal from fridge"[4]

---

[4] For better readability, overlays have been added in this illustration.

The pictographic representation of an OC is configured within a sub-dialog within the PS_A-Configurator. Within this configuration dialog, the predicate logic facts, which are assigned to an OC, can be inspected. These facts are the pre- and post-condition facts of the COPs that interconnect the OCs. Within the constraints given by the COP facts, the pictographic appearance of an OC can be adjusted within the PS_A-Configurator within a 3D scene. The rendering of object constellations is based on "Object Templates".

3.2.2 Object templates

Objects play a central role in process-structures. The different levels of process-structures model different aspects of objects. On abstract level, a symbol is associated with an object for the purpose of task planning (e. g. "Mt.1" for the meal tray in the sample scenario). On system level, i. e. on the level of elementary process-structures, so-called sub-symbolic (i. e. geometric) object information is processed. With respect to the meal tray this is, for instance, the location to grasp the tray. To model the different aspects of objects and to assure an information consistency throughout the different information layers, the concept of Object Templates has been introduced.
Object Templates comprise the following aspects:

- A 3D model of the object, used for pictographic rendering of object constellations on PS_A level as well as for motion planning and collision avoidance on PS_R level
- Associated sub-symbolic (geometric) data for planning and execution on PS_E and PS_R level, e. g. the grasping location of an object
- Complex objects can be composed of simpler objects; e. g. a meal tray consists of a tray, a plate, a lid and a spoon
- Object Templates are configured with natural parameters of the composed object, e. g. width, height, depth and wall thickness for a container, instead of separate specification of all geometric primitives
- The 3D appearance of Object Templates is associated with task-knowledge like symbolic facts and characteristics. For example the fact "IsAccessible(MicrowaveOven)" renders the opening status of the door of the oven's 3D model.

An exemplary Object Template is the meal tray depicted in Fig. 11. It consists of a base tray, a plate with a lid and a spoon. Both the lid and the spoon are detachable from the meal tray. The different stages of separation are depicted in Fig. 12.

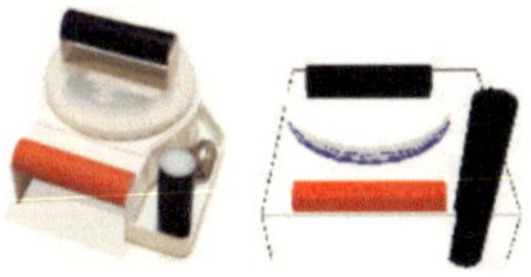

Fig. 11. The meal tray of the eating scenario as photo (left) and modeled by means of an Object Template (right)

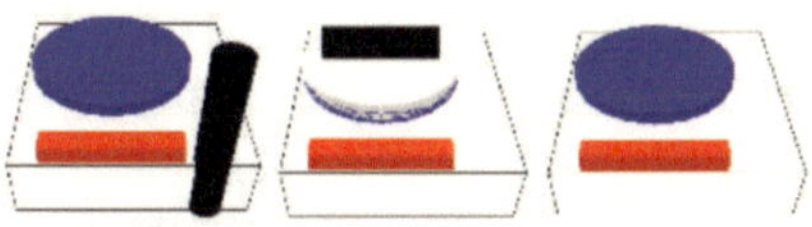

Fig. 12. The different separation stages (detached lid, detached spoon, both lid and spoon detached) of the meal tray

The configuration of Object Templates takes place within the Object-Template-Configurator (OT-Configurator) which is part of the PS_A-Configurator as shown in Fig. 13.

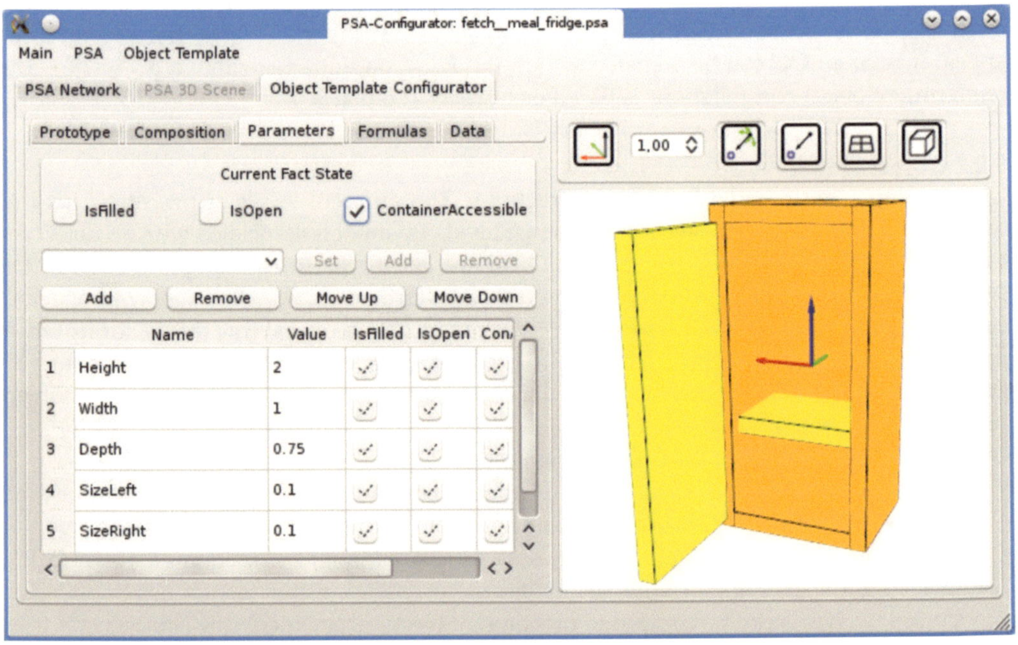

Fig. 13. Object-Template-Configurator (OT-Configurator) as part of the PS_A-Configurator

Within the screenshot in Fig. 13 the Object Template of the refrigerator is modeled. On the left side the parameters and their association to symbolic facts are specified. On the right side the 3D model of the object is rendered according to the current configuration. To render the 3D model of a composed object, the aggregated sub-objects are composed with formulas within the Object-Template-Configurator tool. Frequently required and complex formulas like alignment and rotation of Object Templates are provided with the help of assistive functions.

Certain aspects of the 3D geometry have a fix association with object characteristics as given in the following table:

Characteristic	Associated sub-symbolic element
IsGrippable	Coordinates to grasp the object
IsPlatform	Limits to place other objects onto this object
IsContainer	Limits to place other objects within this object

Table 3. Relations between characteristics and sub-symbolic elements

3.2.3 Exemplary abstract Process-Structure: Fetch meal tray from fridge

The exemplary PS_A that shall be discussed in detail has already been introduced within the PS_A-Configurator frontend in Fig. 10. In this PS_A the task participating object are a fridge (symbol "Fr" with instance number "1" → "Fr.1"), a meal tray ("Mt.1"), the manipulator ("MP.1") and the abstract symbol for a relative location ("InsertLoc"). In this PS_A the initial situation consists of two object constellations. The first one models the manipulator in a free position in the workspace (instance number "0" is assigned to this object constellation → "MP.1_0"). The second object constellation models the already opened fridge containing the meal tray ("Fr.1-Mt.1_0"). The two OCs are connected via the assembly COP "GraspObjectInContainer(MP.1, Mt.1, Fr.1)". If physically possible, a complementary disassembly operator is assigned to model the reverse operation for re-usage of the PS_A in another scenario context. In this case this is the COP "DepartFromContainer(MP.1, Mt.1, Fr.1)". The assembled object constellation is depicted on the bottom left side and the associated abstract planning symbol is "Fr.1-MP.1-Mt.1_0". Due to the associated symbolic facts, which are imposed within the object constellation by the post-condition facts of the COP, the pictographic representation is rendered so that the manipulator grasps the meal tray in the fridge.

Besides assembly and disassembly operators, the And/Or-Net syntax provides operators modeling the internal state transition (IST) of object constellations (IST COPs). IST COPs are applied when the physical contact state of the involved objects is not changed. From the viewpoint of planning on abstract level, objects being in close relative locations to each other are considered to be in a physical contact situation. Therefore, the IST COP "GetObjectOutside(MP.1, Mt.1, Fr.1, InsertLoc)" is applied to transform the OC "Fr.1-MP.1-Mt.1_0" on the left side into the OC "Fr.1-MP.1-Mt.1_1" on the right side. Finally, the COP "MoveObjectFromRelLoc(MP.1, Mt.1, Fr.1, InsertLoc)" models the disassembly operation and results in two object constellations which model the target situation of this abstract process-structure: "Fr.1_0" is the empty fridge and "MP.1-Mt.1_0" is the manipulator with the gripped meal tray in a free position in the work space.

To be able to develop and verify the three levels of process-structures independently, i. e. in a modular manner, the consistency of task-knowledge on all levels has to be assured. This is achieved with common building blocks of the different process-structures as shown in the decomposition chain in Table 2. The common elements are the interfaces to the next level of process-structures. The important interfacing elements between PS_A and PS_E are the pre- and post-condition facts of the COP to be decomposed as PS_E in the next process step. For the COP "GraspObjectInContainer" the facts are shown in Table 4.

Pre-Facts	Post-Facts
HoldsNothing(Manipulator) = True	HoldsNothing(Manipulator) = False
IsInFreePos(Manipulator) = True	IsInFreePos(Manipulator) = False
-	IsGripped(Manipulator, Object) = True
ContainerAccessible(Container) = True	-
IsInsideContainer(Object, Container) = True	-

Table 4. Pre- and Post facts of COP "GraspObjectInContainer(Manipulator, Object, Container)"

3.3 FRIEND::Process step 3: Configuration of elementary process-structures

In the third process step, each composed operator (COP) of an abstract process-structure (PS_A) is decomposed into an elementary process-structure (PS_E). To achieve user-friendly configuration of PS_E, configurable function blocks are assembled to function block networks (FBN). Each function block models a reactive robot system operation, also called skill. A priori verification of task-knowledge on this level takes place with the help of Petri-Nets, which result from automatic conversion of FBNs.

3.3.1 Description of the process step

Fig. 14 depicts the decomposition principle of COPs into elementary process-structures, consisting of skill blocks. An elementary process-structure, as first introduced by (Martens, 2003), is a Petri-Net with enhanced syntax and superordinated construction rules. The advantage of Petri-Nets is their ability to model parallel activities. This is useful for the behavioral modeling on robot system level, for instance, if a manipulator action is guided by input from a camera system or another sensor. Furthermore, Petri-Net-based PS_E offer mathematical methods for analysis of the reachability of a certain system state, for verification of the correctness of control and dataflow and for the exclusion of resource conflicts (Martens, 2003).

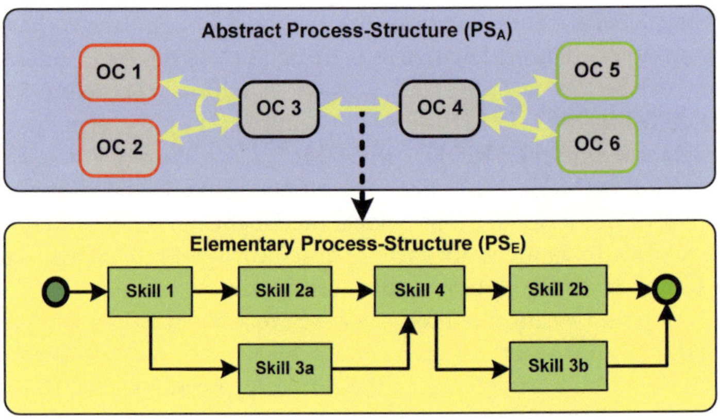

Fig. 14. Decomposition of a composed operator (COP) as elementary process-structure

Besides these conceptual advantages, from the viewpoint of implementation it turned out that the programming of elementary process-structures with Petri-Nets is a time consuming and error prone procedure. The setup of a correctly verified Petri-net-PS_E usually takes several hours. Even with strong modularization of the networks, the large number of places and transitions leads to hardly manageable Petri-Nets in real-life applications. This is the reason why the FRIEND::Process introduces the configuration of PS_E on the basis of function block networks (FBN). Similar to the PS_A-Configurator, a configuration frontend, called PS_E-Configurator, has been created. This tool subsumes all logical and syntactical rules that are required for PS_E-configuration. Furthermore, a conversion algorithm has been developed (Prenzel et al., 2008), which converts an FBN into a Petri-Net for automatic execution of verification routines, like a reachability analysis. A screenshot of the PS_E-Configurator with the PS_E "GraspObjectInContainer" is given in Fig. 15.

With respect to their representative function for Petri-Nets, the control flow within the FBN structures is token-oriented. The execution starts from the "Start" block and ends at the "Target Success" block. In-between, reactive skills are executed, including manipulative operations as well as sensor operations or user interactions. Each function block has one input port, and several output ports according to the possible execution results of the skill (see e. g. block "CoarseApproachToObjectInContainer" in Fig. 15 with the output ports "Success", "Failure", "Abort" and "UserTakeOver"). The output port "Abort" is not explicitly connected to an abort block to increase the readability of the network structure. The typical construction rule for a semi-autonomous system (like FRIEND) is to provide user interactions as redundant action for autonomous system operations. As shown in Fig. 15, the failure of an autonomous operation (e. g. "AcquireObjectBySCam") is linked to the user interaction "DetermineObjectBySCam", replacing the failed system action.

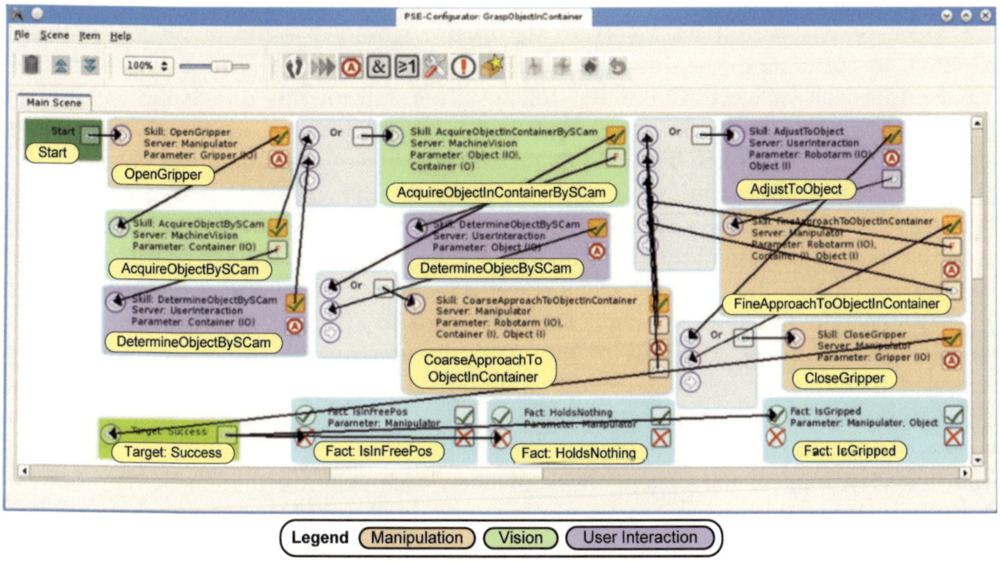

Fig. 15. PS_E-Configurator with elementary process-structure as function block network, modeling the COP "GraspObjectInContainer".[5]

The configuration of PS_E on the basis of function block networks does not only achieve a decisive increase of development comfort (configuration instead of programming), but it also decreases the required task-knowledge engineering time significantly. By building the PS_E directly in the correct manner, the time-consumption for the construction of one PS_E is reduced from hours to 10-15 minutes per network. On this basis, real world problems like the "Meal preparation and assistance" task, become manageable in their complexity.

3.3.2 Exemplary elementary process-structure: Manipulator grasps meal tray in fridge

The exemplary PS_E "GraspObjectInContainer", as shown in Fig. 15, models the grasping of an object in a container-like place in a general way. In the sample scenario "meal preparation" this PS_E is applied to fetch the meal tray from the refrigerator and also from the microwave oven after heating of the meal.

The objects (Object Templates) "Manipulator", "Object" and "Container", which are involved in this PS_E, are the input artifacts handed over as COP parameters from the previous step of the FRIEND::Process. The first skill block that follows the "Start" block is the manipulator skill "OpenGripper". Subsequently, the container (fridge) is located with the help of the vision skill "AcquireObjectBySCam(Object)". This skill calculates the location and size of the given object with the help of a stereo camera (SCam). In the sample scenario the COP parameter "Container" (i. e. the fridge) is inserted at the skill's placeholder

[5] For better readability, overlays have been added in this illustration.

"Object" according to the principle of type-conform parameter replacement (Martens, 2003). The Object Template of a fridge provides the according two sub-symbolic parameters location and size. A successful execution of the skill guarantees that the container's location and size are stored in the system's world model and can serve as input parameters for subsequent skills. After verification of the associated Petri-Net of this PS_E the correctness of the data flow between all skill blocks is assured. If the recognition of the fridge is successful and the user has not to be involved, the skill "AcquireObjectInContainerBySCam" is executed to determine the location of the meal tray in the fridge. Afterwards, a "CoarseApproachToObjectInContainer" follows. This skill roughly directs the manipulator in front of the meal tray in the fridge based on the location information calculated beforehand. Fig. 15 depicts that this manipulator skill is followed by an enforced user interaction, since all output ports are connected to the Or-block preceding the user interaction. The confirmation by the user is included at this place due to testing purposes to assure a correct execution of the first skill. For real task-execution a quick reconfiguration of the PS_E will change the system behavior and directly execute the next manipulator skill "FineApproachToObjectInContainer". This skill leads to a final grasping of the meal tray handle, while avoiding collisions of the manipulator with the fridge with the help of dedicated methods for collision avoidance and path planning (Ojdanic, 2009). The final action necessary to complete the grasping is to close the gripper. The PS_E ends with setting the post-facts of the COP as specified in Table 4.

From the viewpoint of the system's task planner, each skill-function-block represents an elementary (executable) operation. Within the execution level of the system, the operations are not seen as atomic units. The execution of one skill means to activate reactive system functionality, for instance the sensor-controlled approach of an object to be grasped in the skill "FineApproachToObjectInContainer". These basic system skills have to couple sensors and actuators on the algorithmic level. To pursue the paradigm of configurable process-structures also on this level, the FRIEND::Process introduces reactive process-structures.

3.4 FRIEND::Process step 4: Configuration and testing of reactive process-structures

Historically, during the elaboration of the FRIEND::Process, the elementary operators (skills) have been implemented directly in C++. Subsequently, when appropriate CASE tools became available, the elementary operators have been implemented with model driven development techniques (Schmidt, 2006) as executable UML models. Then, a configuration tool has been developed, which makes user-friendly configuration of process-structures possible also on this development level. With the help of this tool it is assured that the verified interfaces from the PS_E-layer are respected and the robustness assertion throughout the complete system architecture is maintained.

3.4.1 Description of the process step

Fig. 16 depicts the decomposition of a skill block from PS_E-layer into a reactive process-structure (PS_R) consisting of algorithmic blocks. Similar to the PS_E function blocks, PS_R are also based on configurable function block networks. The PS_R-Configurator tool results from the Open-Source Image Nets Framework[6], which originally has been developed for configurable image processing algorithms.

[6] http://imagenets.sourceforge.net/

The PS_R Configuration Framework consists of the following five parts (see Fig. 17):

- PS_R-Configurator,
- Embedding of PS_R into any C++ code via PS_R-Executor,
- Reactive process-structures (PS_R), which are executable function block networks,
- Extensible set of Plug-Ins and
- Configurable function blocks

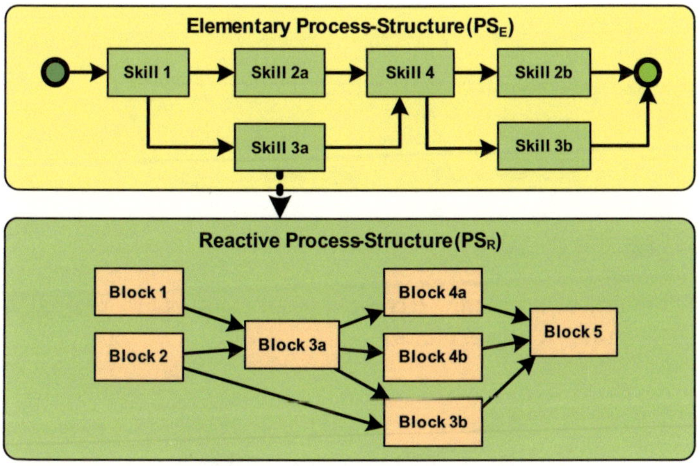

Fig. 16. Decomposition of a skill into a reactive process-structure

The "PSR-Configurator" is a graphical user interface, which can be used to rapidly create a "function block network", namely a reactive process-structure (PS_R). With the PS_R-Executor, it is possible to load and execute the previously configured PS_R. The PS_R itself is a directed graph, connecting configurable "function blocks". Each block can execute code to process its input (image data or other data) and save its outputs. One or more blocks are grouped in a "Plug-In" and an arbitrary number of Plug-Ins can be loaded dynamically by the PS_R. In this way, the PS_R-Framework can be easily extended by new independent Plug-Ins. This independency of the algorithmic modules results in completely independent development within a team of developers. In addition, the strong modularization leads to a technically manageable amount of code within a single block and reduces the time of inspecting an erroneous block. The PS_R execution library can save a PS_R in human readable XML format. Thus, on the one hand the PS_R-Configurator can configure, load and save a PS_R, but on the other hand also external C++ code can load a PS_R file.

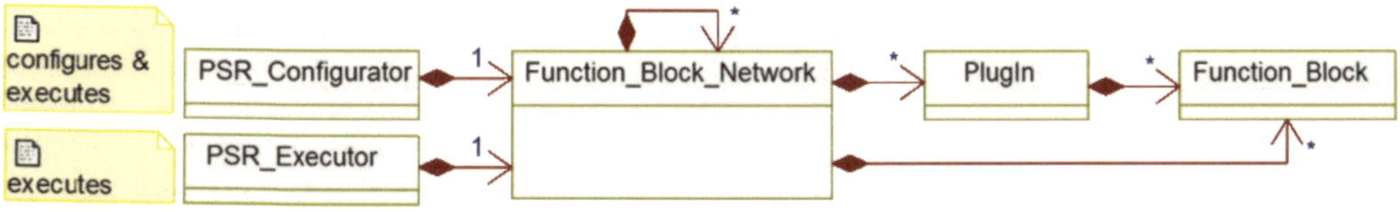

Fig. 17. The UML structure of the PS_R-Configuration Framework

Hierarchical modeling is a common method to subdivide algorithms into separate parts - it breaks down the complexity and facilitates reusability. In the PS_R-Framework, parts can be constructed as separate PS_R and can be combined afterwards to constitute a complete algorithm. The PS$_R$-Executor is in fact also a function block, which can load and process a PS$_R$. The connection between the PS$_R$ inside an Executor and the outer net is established by special input and output blocks. For example the PS$_R$ "Color2Color3D" shown in Fig. 18 calculates a colored point cloud out of a stereo image pair. On the left side there are two input blocks, which hand over the images from the block in orange. This block only exists in this PS$_R$ for testing the net and will be ignored on execution if this PS$_R$ is loaded by a PS$_R$-Executor (see Fig. 19, right side).

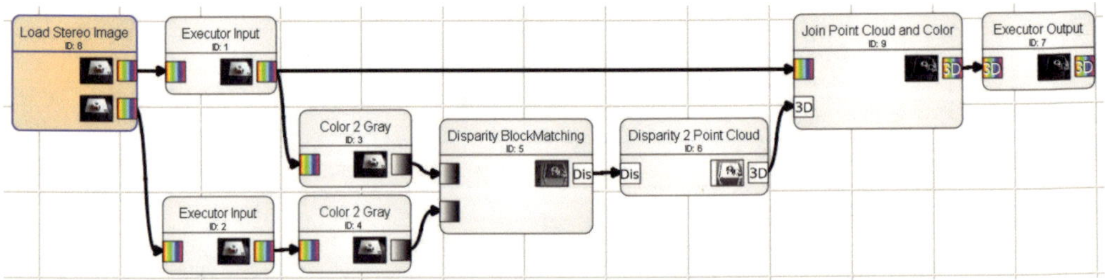

Fig. 18. The functionality of calculating a colored point cloud out of a stereo image pair is depicted in this PS$_R$, called "Color2Color3D"

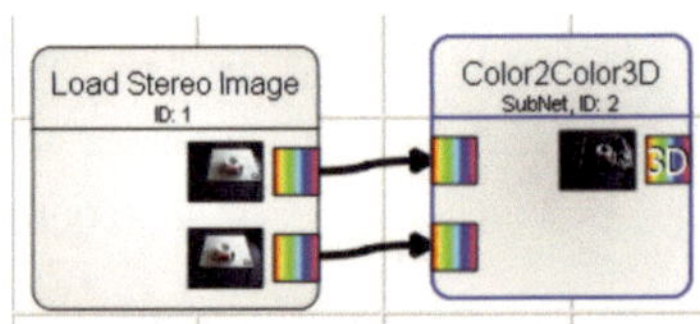

Fig. 19. The previously shown PS$_R$ can be loaded as one PS$_R$-Executor block

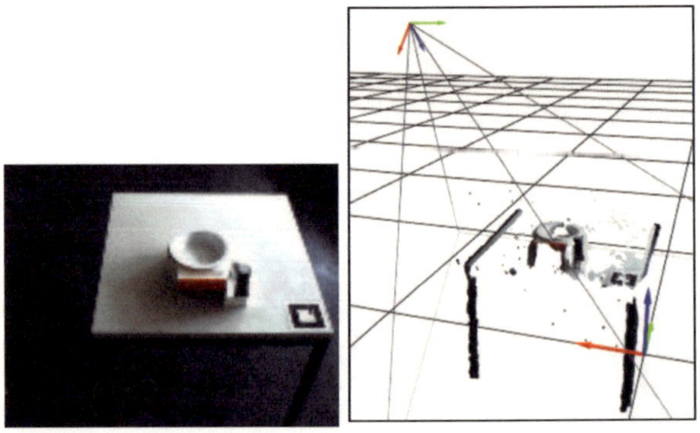

Fig. 20. Left: original image, right: resulting point cloud of the stereo camera images visualized in 3D by the PS$_R$-Configurator

The PS$_R$ in Fig. 19 shows the use of a subnet of an image acquisition together with the calculation of the extrinsic matrices of a stereo camera, which describe the relation of the cameras to the robot. These matrices depend on an invariant transformation frame inside the pan-tilt-head (see Fig. 1) and its rotation angles. By combining the two subnets, a live view of the stereo camera's point cloud can be calculated (depicted in Fig. 20, in the center of the images the meal tray can be seen). As a visually guided robot is a real world object, which moves in the three dimensional Cartesian space, it is useful to display the vision

results in the same space. While configuring a PS_R with the PS_R-Configurator, intermediate results can be visualized in two and three dimensions; depending on the data type, for example scalar values can only be visualized in 2D, camera matrices can be visualized in 2D and 3D (using OpenGL (Wright et al., 2010), see Fig. 20, right).
To be able to execute a PS_R as skill block within the context of the PS_E layer and to guarantee that the PS_E interfaces are respected, a special type of "Verified PS_R-Executor block" is created. During configuration of this kind of block, the PS_R-Configurator checks that the used resources as well as input and output parameters match the specification of a certain PS_E skill to be modeled as PS_R. For example in the case of the PS_R "AquireObjectBySCam(Object)" the allowed resource is the stereo camera system. The input parameter is the Object Template of the given object and the output parameter are the return values "Success" and "Failure".

3.4.2 Exemplary reactive process-structure: Acquire meal tray by stereo camera

To show the capabilities of the reactive process-structures, a simplified example is discussed in the following, namely the machine vision skill to acquire an object by the stereo camera with the configuration "Meal Tray". This example of a PS_R is non-reactive, as no actor is involved. Though, in a more complex PS_R, it is possible to combine the camera and the robot in a feedback loop to implement visual servoing to achieve reactive behaviour.
In Fig. 21 several general blocks are used to find the red meal tray handle in an image. The processing chain starts with the detection of highly saturated, red parts. It is followed by a 9x9 closing operation to eliminate noise. Afterwards, contours are detected and filtered according to a priori knowledge of the size of the handle. Then, the minimum rectangles around the contours are determined and the major axes and their end points are calculated. For testing the current PS_R, again the orange blocks have been added to visualize intermediate testing results and they are not executed during task execution.

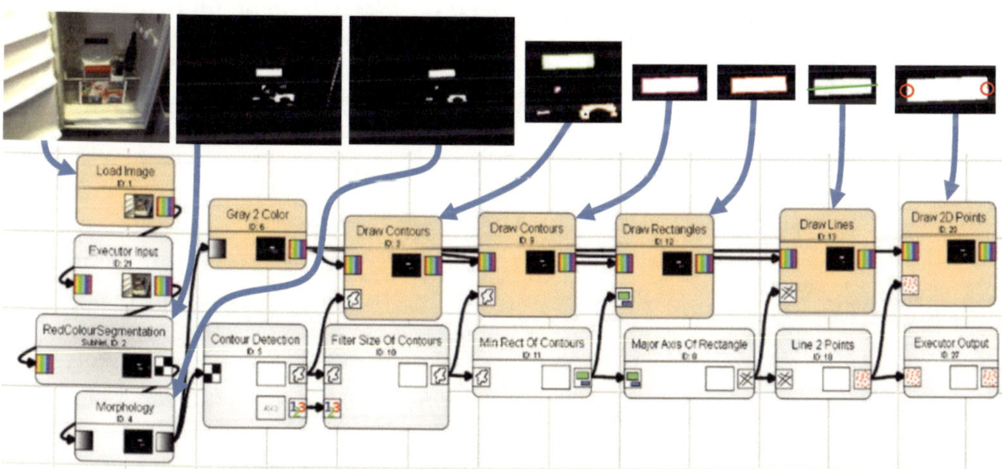

Fig. 21. PS_R "MajorAxisPoints" which detects red areas of a certain size and calculates the major axes of these areas. Orange blocks are omitted when this PS_R is used in a PS_R-Executor

To grasp the meal tray handle with the manipulator, the determination of its location in 3D is required. Thus, a 2D detection of the meal tray is not sufficient. However, the previously created and tested PS_R "MajorAxisPoints" can be used twice, one for each image of the stereo camera. Fig. 22 depicts the usage of the previous net to calculate the 3D line, describing the handle of the meal tray. The block *Optimal Stereo Triangulation* computes a 3D contour based on key feature points, extracted from a stereo image. With the known camera matrices and the 2D feature correspondences, the 3D points are found by the intersection of two projection lines in the 3D space using optimal stereo triangulation, as described in (Natarajan et al., 2011).

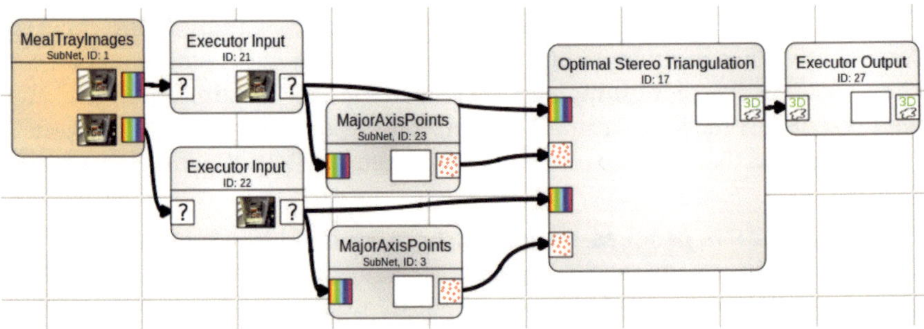

Fig. 22. PS_R which detects the meal tray handle in 3D

Next, the 3D line of the 3D handle detection is used to calculate a transformation frame, having the position of the right 3D point and the rotations to point the y-axis in line direction. Using the a priori knowledge that the meal tray should be parallel to the world coordinate system, only rotation around z-axis has to be calculated. Fig. 23 displays (top, from left to right) the 3D line, the calculated frame, the meal tray Object Template and the placed meal tray, based on the frame. For the fulfillment of the specification of the calling PS_E, the Object Template has to be written to the World Model (a service to read from and write data to) with the "Write to World Model" block. This ensures that the detected object is globally available for later processing steps and is the actual result of this PS_R.

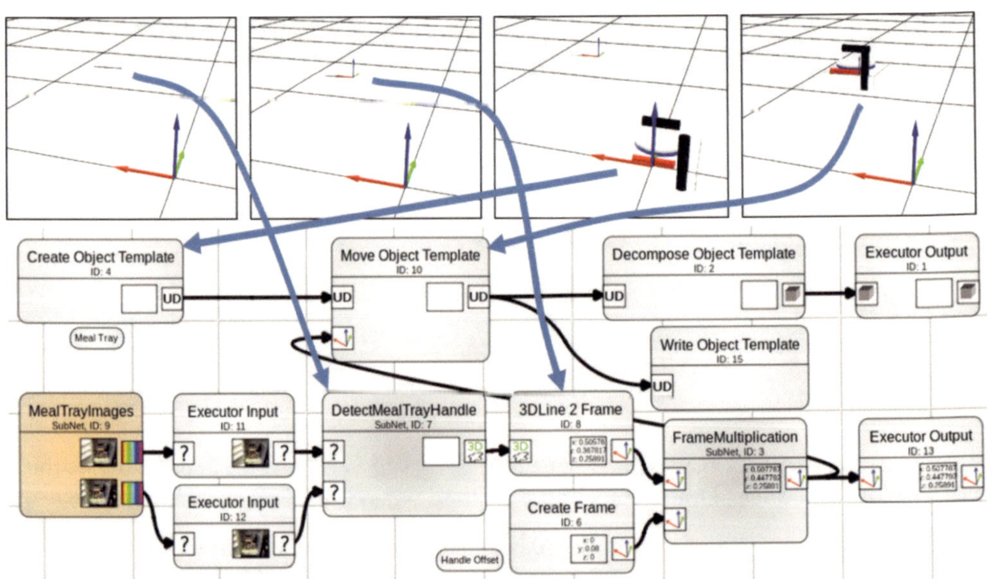

Fig. 23. Meal tray detection and Object Template placement based on 3D handle detection, frame calculation and Object Template movement (UD = user data)

For simulation and PS_R unit testing in the PS_R-Configurator, the fridge, the static environment (wheelchair, monitor and user) and the robot with its current configuration can be placed in the same 3D scene with the meal tray. Fig. 24 and Fig. 25 show the real scene and the 3D simulation result in comparison.

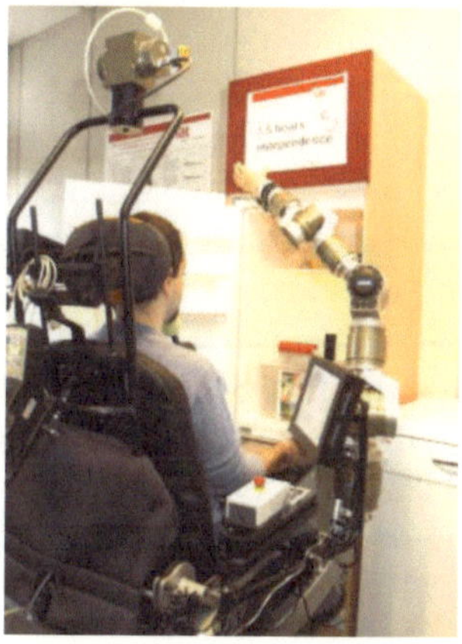

Fig. 24. Real scene of this PS_R

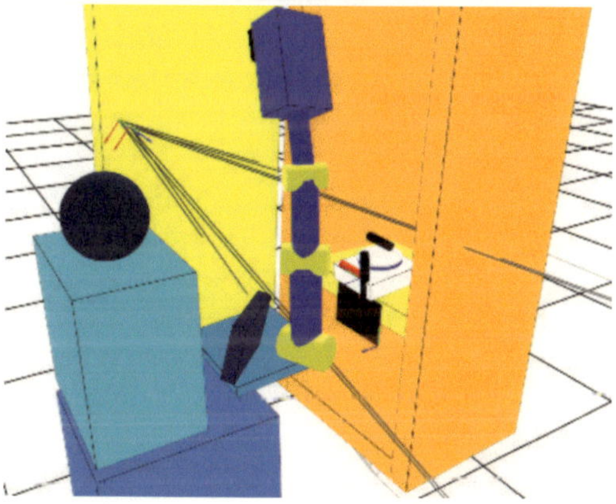

Fig. 25. Simulated scene of this PS_R

3.5 FRIEND::Process step 5: Task testing

After finishing the configuration of process-structures on all three levels, the planning and execution of a task (PS_A) has to be tested. The modularly configured, verified and tested process-structures of lower abstraction (PS_E and PS_R) are involved in this final process step.

3.5.1 Description of the process step

For the purpose of task testing the "Sequencer" is used, which embeds a task planner for process-structures and the PS_R-Executor (see Fig. 5). The Sequencer is part of the process-structure-based control architecture MASSiVE mentioned in Section 1. The Sequencer interacts with skill servers, which offer the functionality that has been configured and

verified as reactive process-structures beforehand. The layered system architecture organizes a hardware abstraction via skill layer, so that there is a unique access-point on the sensors and actuators from a certain responsible skill server.
Task tests can be performed in the following execution modes:

- *Probabilistic simulation:* the skill interfaces and the communication infrastructure are tested and skill return values are simulated,
- *Skill simulation:* the skill's functional core is simulated,
- *Motion simulation:* the motion governed by manipulative skills is simulated and visualized within a virtual 3D space as shown in Fig. 25,
- *Hardware simulation:* the sensors and actuators are simulated,
- *Real execution:* the skill is executed with access of sensors and actuators.

Based on the process-structures, a complete task is planned and executed in one of the listed skill execution modes. This means that the Sequencer first plans a sequence of COPs and subsequently decomposes each COP into an elementary process-structure. Planning on this level results in a sequence of skills to be executed then. Step by step and based on the execution result of each skill, the once planned skill sequence is pursued, or re-planning takes place if an unexpected result is obtained.

4. Conclusion

As shown in Section 2.1 it is a challenging task to establish intelligent behavior of service robots operating in human environments. Typical operation sequences of support tasks in daily life activities seem to be simple from human understanding. However, to realize them with a robotic system, a huge complexity arises due to the variability and unpredictability of human environments.
In this paper the FRIEND::Process - an engineering approach for programming robust intelligent robotic behavior - has been presented. This approach is an alternative solution in contrast to other existing approaches, since it builds on configurable process-structures as central development elements. Process-structures comprise a finite-sized and context-related set of task knowledge. This allows a priori verification of the programmed system behavior and leads to deterministic, fault-tolerant and real-time capable robotic systems.
The FRIEND::Process organizes the different stages of development and leads to consistent development artifacts. This is achieved with the help of a tool chain for user-friendly configuration of process-structures.
The applicability of the here proposed methods has been proven throughout the realization of the AMaRob project (IAT, 2009) where task execution in three complex scenarios for the support of disabled persons in daily life activities has been solved. One of theses scenarios is the "Meal preparation and eating assistance" scenario, used for exemplification throughout this paper. The most error prone and thus challenging action in this scenario is the correct recognition of smaller objects (e. g. the handle of the meal tray) under extreme lighting conditions. However, with the inclusion of redundant skills in the elementary process-structures, the system's robustness has been raised in an evolutionary manner. In cases where even redundant autonomous skills did not execute successfully, the accomplishment of the desired task was achieved via inclusion of the user within a user interaction skill.
Currently, the methods and tools discussed in this paper are continuously developed further and are applied in the project ReIntegraRob (IAT, 2011). The mid-term objective is to integrate the different configuration tools for process-structures into one integrated

configuration environment. The PS_R Configuration Framework, which is the most elaborated tool, will build the basis for this.

5. Glossary

COP	**C**omposed **Op**erator
FBN	**F**unction **B**lock **N**etwork
FRIEND	**F**unctional **R**obotarm with user-fr**IEN**dly interface for **D**isabled people
MASSiVE	**M**ultilayer Control **A**rchitecture for **S**emi-Autonomous **S**ervice Robots with **V**erified Task **E**xecution
OC	**O**bject **C**onstellation
OT	**O**bject **T**emplate
PS	**P**rocess-**S**tructure
PS_A	**A**bstract **P**rocess-**S**tructure
PS_E	**E**lementary **P**rocess-**S**tructure
PS_R	**R**eactive **P**rocess-**S**tructure
TPO	**T**ask **P**articipating **O**bject

6. References

Asimov, I. (1991). Robot Visions, Roc (Reissue 5th March 1991), ISBN-10: 0451450647

Cao, T. & Sanderson, A. C. (1998). AND/OR net representation for robotic task sequence planning, In: *IEEE Transactions on Systems, Man, and Cybernetics - part C: Applications and Reviews*, 28(2)

Dario, P., Dillman, R., and Christensen, H. I. (2004). EURON research roadmaps. Key area 1 on 'Research coordination', Available from http://www.euron.org

Engelberger, J. F. (1989), *Robotics in Service*, MIT Press, Cambridge, MA, USA, 1st ed, 1989

Gostai. (2010). Urbi 2.0. Available from http://www.gostai.com

Gräfe, V. & Bischoff, R. (2003). Past, present and future of intelligent robots, *Proceedings of the 2003 IEEE International Symposium on Computional Intelligence*, In: *Robotics and Automation* (CIRA 2003), volume 2, ISBN 0-7803-7866-0, Kobe, Japan

IAT (2009). *AMaRob Project*, Institute of Automation, University of Bremen, Germany. Available from http://www.amarob.de

IAT (2011). *ReIntegraRob Project*, Institute of Automation, University of Bremen, Germany. Available from http://www.iat.uni-bremen.de/sixcms/detail.php?id=1268

Kampe, H. & Gräser, A. (2010). Integral modelling of objects for service robotic systems, Proceedings for the joint conference of ISR 2010 (41st International Symposium on Robotics) und ROBOTIK 2010 (6th German Conference on Robotics), 978-3-8007-3273-9, Munich, Germany

Kemp, C. C., Edsinger, A. & Torres-Jara, E. (2007). Challenges for robot manipulation in human environments, In: *IEEE Robotics and Automation Magazine*, vol. 14, pp. 20-29

Martens, C. (2003). Teilautonome Aufgabenbearbeitung bei Rehabilitations-robotern mit Manipulator - Konzeption und Realisierung eines software-technischen und algorithmischen Rahmenwerks, PhD dissertation, University of Bremen, Faculty of Physics / Electrical Engineering, (in German)

Martens, C., Prenzel, O. & Gräser, A. (2007). The rehabilitation robots FRIEND-I & II: Daily life independency through semi-autonomous task-execution, In: *Rehabilitation Robotics* (Sashi S Kommu, Ed.), pp. 137-162., I-Tech Education and Publishing, Vienna, Austria, Available from http://www.intechopen.com/books/show/title/rehabilitation_robotics

Microsoft. (2011). Microsoft Robotic Studio, Available from http://www.microsoft.com/robotics

Natarajan, S.K., Ristic-Durrant, D., Leu, A., Gräser, A. (2011). Robust stereo-vision based 3D-modeling of real-world objects for assistive robotic applications, in *Proc. of IEEE/RSJ International Conference on Robots and Systems (IROS), San Francisco, USA*

Ojdanic, D. (2009). Using cartesian space for manipulator motion planning - application in service robotics, PhD dissertation, University of Bremen, Faculty of Physics and Electrical Engineering

Prenzel, O. (2005). Semi-autonomous object anchoring for service-robots, *in B. Lohmann (Ed.), A. Gräser, Methods and Applications in Automation,* pp. 57 - 68, Shaker-Verlag, Aachen, 2005, ISBN 3-8322-4502-2

Prenzel, O., Boit, A. and Kampe H. (2008) Ergonomic programming of service robot behavior with function block networks, in *Methods and Applications in Automation,* Shaker-Verlag, pp. 31-42

Prenzel, O. (2009). *Process model for the development of semi-autonomous service robots,* PhD dissertation, University of Bremen, Faculty of Physics and Electrical Engineering

Quigley, M., Conley, K., Gerkey, B. P., Faust, J., Foote, T., Leibs, J., Wheeler, R., Ng, A. Y. (2009) ROS: an open-source Robot Operating System, In: *Proc. Of ICRA Workshop on Open Source Software*

Russel, S., and Norvig, P. (2003). *Articial Intelligence - A Modern Approach,* Prentice Hall, Upper Saddle River, New Jersey, 2nd ed.

Schlegel, C., and Woerz, R. (1999) The software framework SmartSoft for implementing sensorimotor systems, In: *Proc. of the IEEE/RSJ International Conference on Intelligent Robots and Systems (IROS),* pp. 1610-1616

Schmidt, D. C. (2006). Model-driven-engineering, In: *guest editor's introduction,* pp. 25-31, IEEE Computer

Schreckenghost, D., Bonasso, R., Kortenkamp, D., Ryan D. (1998) Three tier architecture for controlling space life support systems, In: *Proc. of IEEE SIS'98,* Washington DC, USA

Simmons, R., Apfelbaum, D. (1998) A task description language for robot control, In: *Proc. of Conference on Intelligent Robotics and Systems*

Weld, D. S. (1999). Recent advances in AI planning, in *AI Magazine,* vol 20, pp. 93-123

Wright, R. S., Lipchak, B., Haemel, N. & Sellers, G. (2010). *OpenGL SuperBible: Comprehensive Tutorial and Reference* (5th Edition), Addison-Wesley, ISBN 978-0321712615

26

Performance Evaluation of Fault-Tolerant Controllers in Robotic Manipulators

Claudio Urrea[1], John Kern[1,2] and Holman Ortiz[2]
[1]*Departamento de Ingeniería Eléctrica, DIE, Universidad de Santiago de Chile, USACH, Santiago*
[2]*Escuela de Ingeniería Electrónica y Computación, Universidad Iberoamericana de Ciencias y Tecnología, UNICIT, Santiago Chile*

1. Introduction

Thanks to the incorporation of robotic systems, the development of industrial processes has generated a great increase in productivity, yield and product quality. Nevertheless, as far as technological advancement permits a greater automation level, system complexity also increases, with greater number of components, therefore rising the probability of failures or anomalous operation. This can result in operator's hazard, difficulties for users, economic losses, etc. Robotic automatic systems, even if helped in minimizing human operation in control and manual intervention tasks, haven't freed them from failure occurrences. Although such failures can´t be eliminated, they can be properly managed through an adequate control system, allowing to reduce degraded performance in industrial processes.

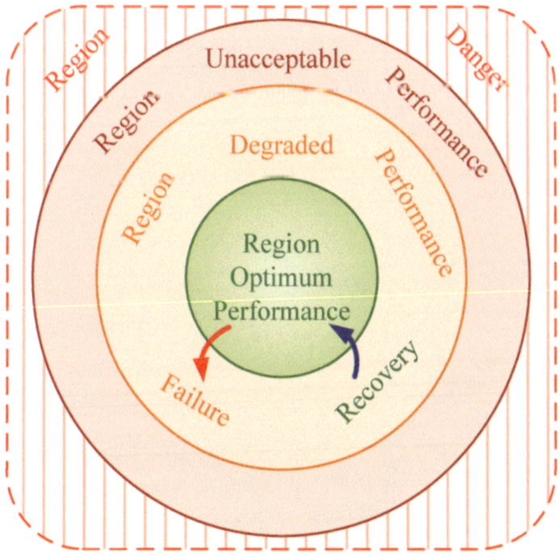

Fig. 1. Performance regions under failure occurrence

In figure 1 we see a scheme showing the different performance regions a given system can adopt when a failure occurs. If the system deviates to a degraded performance region in, presence of a failure, it can recover itself moving into an optimum performance region, or near to it. These systems are called fault tolerant systems and have become increasingly important for robot manipulators, especially those performing tasks in remote or hazardous environments, like outer space, underwater or nuclear environments.

In this chapter we will address the concept of fault tolerance applied to a robotic manipulator. We will consider the first three degrees of freedom of a redundant SCARA-type robot, which is intended to follow a Cartesian test trajectory composed by a combination of linear segments. We developed three fault-tolerant controllers by using classic control laws: *hyperbolic sine-cosine*, calculated torque and adaptive inertia. The essays for such controllers will be run in a simulation environment developed through MatLab/Simulink software. As a performance requirement for those controllers, we considered the application of a failure consisting in blocking one of the manipulator's actuators during trajectory execution. Finally, we present a performance evaluation for each one of the above mentioned fault-tolerant controllers, through joint and Cartesian errors, by means of graphics and rms rates.

2. Fault tolerant control

The concept of fault tolerant control (Zhang & Jiang, 2003) comes first from airplane fault tolerant control; although at scientific level it appears later, as a basic aim in the first congress of IFAC SAFEPROCESS 1991, with an especially stronger development since the beginning of 21th century. Fault tolerant control can be considered both under an active or passive approach, as seen in figure 2a. Passive tolerant control is based on the ability of feedback systems to compensate perturbations, changes in system dynamics and even system failures (Puig, Quevedo, Escobet, Morcego, & C., 2004). Passive tolerant control considers a robust design of the feedback control system in order to immunize it from some specific failures (Patton, 1997). Active tolerant control is centered in on-line failure, that is, the ability to identify the failing component, determine the kind of damage, its magnitude and moment of appearance and, from this information, to activate some mechanism for

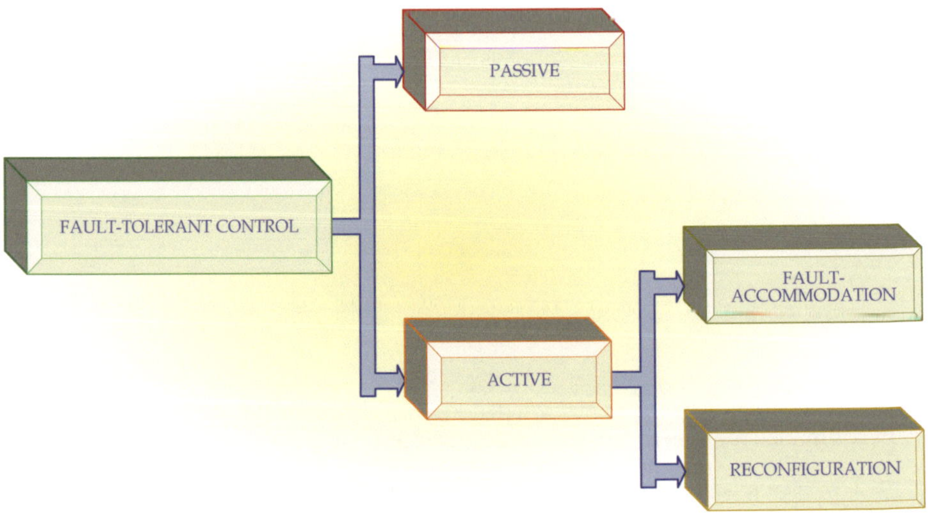

Fig. 2a. Types of fault tolerant control

rearrangement or control reconfiguration, even stopping the whole system, depending on the severity of the problem (Puig, Quevedo, Escobet, Morcego, & C., 2004).
Fault tolerant control systems (being of hybrid nature) consider the application of a series of techniques like: component and structure analysis; detection, isolation and quantification of failures; physical or virtual redundancy of sensors and/or actuators; integrated real-time supervision of all tasks performed by the fault tolerant control, as we can see in figure 2b (Blanke, Kinnaert, Lunze, & Staroswiecki, 2000).

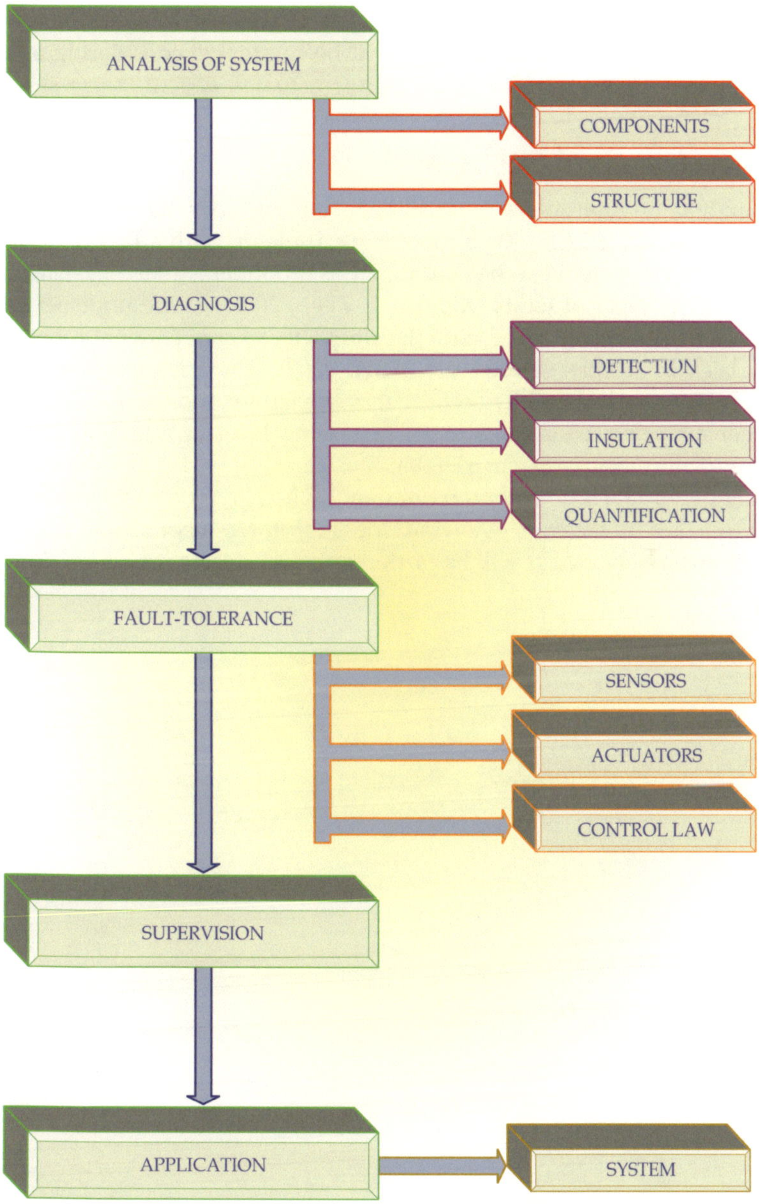

Fig. 2b. Stages included in the design of a fault tolerant control system

In the evaluation of fault tolerant controllers it is assumed that a robotic manipulator where a failure has arisen in one or more actuators, can be considered as an underactuated system, that is, a system with less actuators than the number of joints (El-Salam, El-Haweet, & and Pertew, 2005). Those underactuated systems present a greater degree of complexity compared with the simplicity of conventional robot control, being not so profoundly studied yet (Rubí, 2002). The advantages of underactuated systems have been recognized mainly because they are lighter and cheaper, with less energy consumption. Therefore, a great deal of concern is being focused on those underactuated robots (Xiujuan & Zhen, 2007). In figure 3 it is shown a diagram displaying the first three degrees of freedom of a SCARA type redundant manipulator, upon which essays will be conducted considering a failure in the second actuator, making the robot become an underactuated system.

3. SCARA-type redundant manipulator

For the evaluation of fault-tolerant controllers, we consider the first three degrees of freedom of a redundant SCARA-type robotic manipulator, with a failure occurring in one of its actuators; such a system can be considered as an underactuated system, *i.e.*, with less actuators than the number of joints (Xiujuan & Zhen, 2007). Those underactuated systems have a greater complexity compared with the simplicity of conventional robots control, and they haven't been so deeply studied yet (Rubí, 2002). The advantages of underactuated systems have been remarked mainly because they are lighter and less expensive; also having less energy consumption, consequently an increasing level of attention is being paid to underactuated robots (Xiujuan & Zhen, 2007).
In figure 3 it is shown the scheme of a redundant SCARA-type robotic manipulator, and in figure 4 we can see a diagram showing the first three degrees of freedom of such manipulator, on which the essays will be carried on.

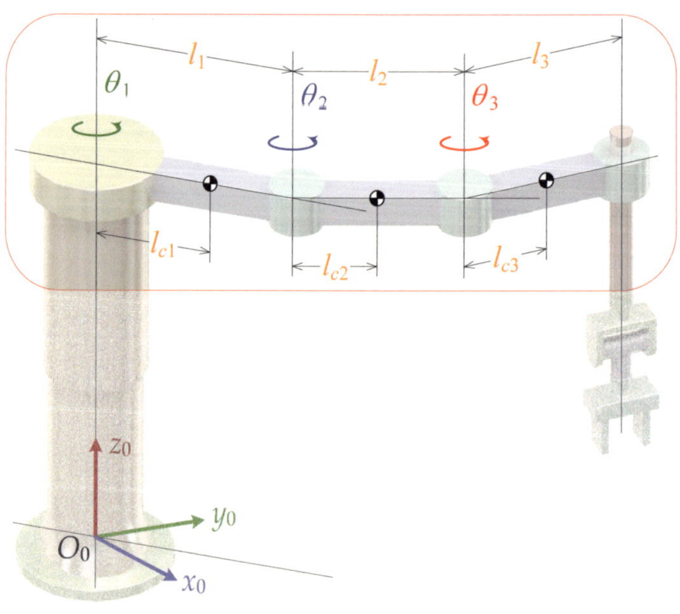

Fig. 3. Scheme of a SCARA-type redundant manipulator

The considered failure is the blocking of the second actuator, what makes this robot an underactuated system.
Having in mind the exposed manipulator, it is necessary to obtain its model; therefore we will consider that the dynamic model of a manipulator with n joints can be expressed through equation (1):

$$\boldsymbol{\tau} = \mathbf{M}(q)\ddot{\mathbf{q}} + \mathbf{C}(q,\dot{q}) + \mathbf{G}(q) + \mathbf{F}(\dot{q}) \tag{1}$$

where:

$\boldsymbol{\tau}$: Vector of generalized forces ($n\times1$ dimension).
$\mathbf{M}$: Inertia matrix ($n\times n$ dimension).
$\mathbf{C}$: Centrifugal and Coriolis forces vector ($n\times1$dimension).
q : Components of joint position vector.
$\dot{q}$: Components of joint speed vector.
$\mathbf{G}$: Gravity force vector ($n\times1$dimension).
$\ddot{\mathbf{q}}$: Joint acceleration vector ($n\times1$dimension).
$\mathbf{F}$: Friction forces vector ($n\times1$dimension).

Under failure conditions in actuator number 2, that is, it's blocking, the component 2 of equation (1) becomes a constant.

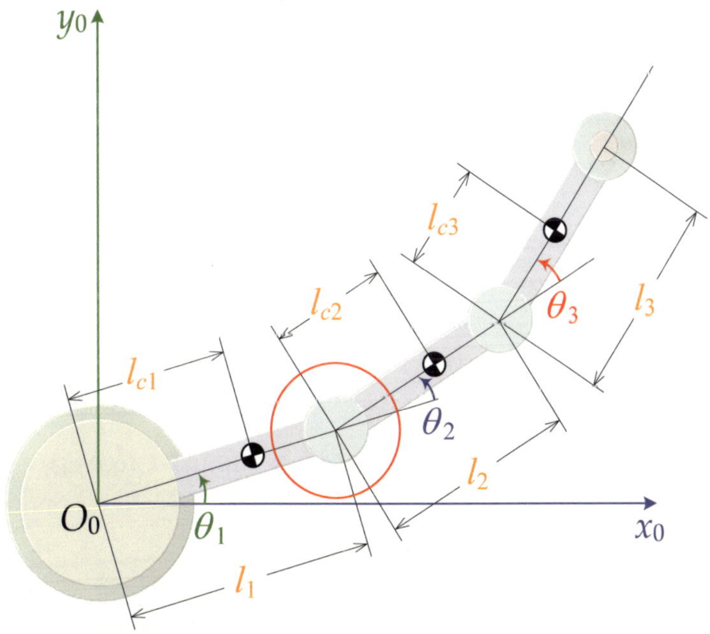

Fig. 4. Scheme of the three first DOF of a redundant SCARA-type robotic manipulator

4. Considered controllers

Considering the hybrid nature of fault tolerant control, it is proposed an active fault tolerant control having a different control law according to the status of the robotic manipulator, *i.e.*

normal or failing, with on-line sensing of possible failures and, in correspondence with this, reconfiguring the controller by selecting the most adequate control law (changing inputs and outputs).
Next, we will present a summary of the controllers considered for performance evaluation when a failure occurs in the second actuator of the previously described manipulator.

5. Fault tolerant controller: *hyperbolic sine* and *cosine*

This controller is based on the classic controller *hyperbolic sine-cosine* presented in (Barahona, Espinosa, & L., 2002), composed by a proportional part based on sine and *hyperbolic cosine* functions, a derivative part based on hyperbolic sine and gravity compensation, as shown in equation (2). The proposed fault tolerant control law includes two classic *hyperbolic sine-cosine* controllers that are "switched" to reconfigure the fault tolerant controller.

$$\tau = \mathbf{K_p}\mathbf{sinh}(\mathbf{q_e})\mathbf{cosh}(\mathbf{q_e}) - \mathbf{K_v}\mathbf{sinh}(\dot{\mathbf{q}}) + \mathbf{G}(\mathbf{q}) \tag{2}$$

$$\mathbf{q_e} = \mathbf{q_d} - \mathbf{q} \tag{3}$$

According to equations (2) and (3):
$\mathbf{K_p}$: Proportional gain, diagonal definite positivematrix (nxn dimension).
$\mathbf{K_v}$: Derivative gain, diagonal definite positive matrix ($n\times n$ dimension).
$\mathbf{q_e}$: Joint position error vector ($n\times 1$dimension).
$\mathbf{q_d}$: Desired joint position vector ($n\times 1$dimension).
In (Barahona, Espinosa, & L., 2002) it is established that robotic manipulator's joint position error will tend asymptotically to zero as long as time approaches to infinite:

$$\lim_{t\to\infty} \mathbf{q_e} \to 0 \tag{4}$$

This behavior is proved analyzing equation (5) and pointing that the only equilibrium point for the system is the origin (0,0).

$$\begin{aligned} &\frac{d}{dt}\begin{bmatrix}\mathbf{q_e}\\ \dot{\mathbf{q}}\end{bmatrix} = \begin{bmatrix}a_1\\ a_2\end{bmatrix} \\ &a_1 = -\dot{\mathbf{q}} \\ &a_2 = \mathbf{M}(\mathbf{q})^{-1}\left(\mathbf{K_p}\mathbf{sinh}(\mathbf{q_e})\mathbf{cosh}(\mathbf{q_e}) - \mathbf{K_v}\mathbf{sinh}(\dot{\mathbf{q}}) - \mathbf{C}(\mathbf{q},\dot{\mathbf{q}})\dot{\mathbf{q}}\right) \end{aligned} \tag{5}$$

6. Fault tolerant controller: Computed torque

Another active fault tolerant controller analyzed here uses a control law by computed torque, consisting in the application of a torque in order to compensate the centrifugal, Coriolis, gravity and friction effects, as shown in equation (6).

$$\tau = \hat{\mathbf{M}}(q)\left(\ddot{\mathbf{q}}_\mathbf{d} + \mathbf{K_v}\dot{\mathbf{q}}_\mathbf{e} + \mathbf{K_p}\mathbf{q_e}\right) + \hat{\mathbf{C}}(q,\dot{q}) + \hat{\mathbf{G}}(q) + \hat{\mathbf{F}}(\dot{q}) \tag{6}$$

where:
$\hat{\mathbf{M}}$: Estimation of inertia matrix ($n\times n$ dimension).

$\hat{\mathbf{C}}$: Estimation of centrifugal and Coriolis forces vector (n×1dimension).
$\hat{\mathbf{G}}$: Estimation of gravity force vector (n×1dimension).
$\hat{\mathbf{F}}$: Estimation of friction forces vector (n×1dimension).

$$\mathrm{K_v} = \begin{bmatrix} \mathrm{K}_{v1} & & & \\ & \mathrm{K}_{v2} & & \\ & & \ddots & \\ & & & \mathrm{K}_{vn} \end{bmatrix} \tag{7}$$

$\mathbf{K_v}$: Diagonal definite positive matrix (n×n dimension).

$$\mathrm{K_p} = \begin{bmatrix} \mathrm{K}_{p1} & & & \\ & \mathrm{K}_{p2} & & \\ & & \ddots & \\ & & & \mathrm{K}_{pn} \end{bmatrix} \tag{8}$$

$\mathbf{K_p}$: Diagonal definite positive matrix (n×n dimension).
$\ddot{\mathbf{q}}_\mathbf{d}$: Desired joint acceleration vector (n×1 dimension).

$$\dot{\mathbf{q}}_\mathbf{e} = \dot{\mathbf{q}}_\mathbf{d} - \dot{\mathbf{q}} \tag{9}$$

$\dot{\mathbf{q}}_\mathbf{e}$: Joint speed error vector (n×1dimension).
If estimation errors are little, joint errors near to a linear equation, as shown in equation (10).

$$\ddot{\mathbf{q}}_\mathbf{e} + \mathbf{K_v}\dot{\mathbf{q}}_\mathbf{e} + \mathbf{K_p}\mathbf{q}_\mathbf{e} \approx 0 \tag{10}$$

7. Fault tolerant controller: Adaptive inertia

The fault tolerant control under examination is based on an adaptive control law, namely: adaptive inertia (Lewis, Dawson, & Abdallah, 2004), (Siciliano & Khatib, 2008), for what it is necessary to consider the manipulator dynamic model in the form expressed in equation (11). The term corresponding to centrifugal and Coriolis forces is expressed through a matrix $\mathbf{V_m}$.

$$\tau = \mathbf{M}(q)\ddot{\mathbf{q}} + \mathbf{V_m}(q,\dot{q})\dot{\mathbf{q}} + \mathbf{G}(q) + \mathbf{F}(\dot{q}) \tag{11}$$

In this case, we define an auxiliary error signal $\mathbf{r}$ and its derivative$\dot{\mathbf{r}}$, as shown in equations (12) and (13), respectively:

$$\mathbf{r} = \Lambda\mathbf{q}_\mathbf{e} + \dot{\mathbf{q}}_\mathbf{e} \tag{12}$$

$$\dot{\mathbf{r}} = \Lambda\dot{\mathbf{q}}_\mathbf{e} + \ddot{\mathbf{q}}_\mathbf{e} \tag{13}$$

where:
Λ : Diagonal definite positive matrix (n×n dimension).

$$\mathbf{\Lambda} = \begin{bmatrix} \lambda_1 & & & \\ & \lambda_2 & & \\ & & \ddots & \\ & & & \lambda_n \end{bmatrix} \tag{14}$$

When replacing equations (3), (9), (12) and (13) into expression (11), we obtain:

$$\boldsymbol{\tau} = \mathbf{M}(q)(\ddot{\mathbf{q}}_\mathbf{d} + \mathbf{\Lambda}\dot{\mathbf{q}}_\mathbf{e}) + \mathbf{V}_\mathbf{m}(q,\dot{q})(\dot{\mathbf{q}}_\mathbf{d} + \mathbf{\Lambda}\mathbf{q}_\mathbf{e}) + \mathbf{G}(q) + \mathbf{F}(\dot{q}) - \mathbf{M}(q)\dot{\mathbf{r}} - \mathbf{V}_\mathbf{m}(q,\dot{q})\mathbf{r} \tag{15}$$

And making the following matching:

$$\mathbf{Y}(\bullet)\boldsymbol{\varphi} = \mathbf{M}(\mathbf{q})(\ddot{\mathbf{q}}_\mathbf{d} + \mathbf{\Lambda}\dot{\mathbf{q}}_\mathbf{e}) + \mathbf{V}_\mathbf{m}(q,\dot{q})(\dot{\mathbf{q}}_\mathbf{d} + \mathbf{\Lambda}\mathbf{q}_\mathbf{e}) + \mathbf{G}(q) + \mathbf{F}(\dot{q}) \tag{16}$$

where:

$$\mathbf{Y}(q,\dot{q},q_d,\dot{q}_d,\ddot{q}_d) = \begin{bmatrix} Y_{11} & Y_{12} & \cdots & Y_{1n} \\ Y_{21} & Y_{22} & \cdots & Y_{2n} \\ \vdots & \vdots & \ddots & \vdots \\ Y_{n1} & Y_{n2} & \cdots & Y_{nn} \end{bmatrix} \tag{17}$$

$\mathbf{Y}(\bullet)$: Regression matrix ($n\times n$ dimension).
$\boldsymbol{\varphi}$: Parameter vector ($n\times 1$ dimension).
With these relationships, expression (15) can be rewritten in the following way:

$$\boldsymbol{\tau} = \mathbf{Y}(\bullet)\boldsymbol{\varphi} - \mathbf{M}(q)\dot{\mathbf{r}} - \mathbf{V}_\mathbf{m}(q,\dot{q})\mathbf{r}\ \mathbf{Y}(\bullet) \tag{18}$$

And the control torque is expressed through equation (19):

$$\boldsymbol{\tau} = \mathbf{Y}(\bullet)\hat{\boldsymbol{\varphi}} + \mathbf{K}_\mathbf{v}\mathbf{r} \tag{19}$$

where:
$\hat{\boldsymbol{\varphi}}$: Parameter estimation vector ($n\times 1$dimension).
$\mathbf{K}_\mathbf{v}$: Diagonal definite positive matrix ($n\times n$ dimension).
The adaptive control updating rule can be expressed by:

$$\dot{\hat{\boldsymbol{\varphi}}} = -\dot{\tilde{\boldsymbol{\varphi}}} = \mathbf{\Gamma}\mathbf{Y}^T(\cdot)\mathbf{r} \tag{20}$$

where:
$\mathbf{\Gamma}$: Diagonal definite positive matrix ($n\times n$ dimension).

8. Fault tolerant control simulator

The three above mentioned control laws, along with the dynamic model of the redundant SCARA-type manipulator considering the first three degrees of freedom (Addendum A), are run under the simulation structure shown in figure 5, where we can see the hybrid nature of this kind of controller.

In Addendum B we show the set of parameter values employed in the manipulator dynamic model, and the gains considered for each kind of fault tolerant controller.

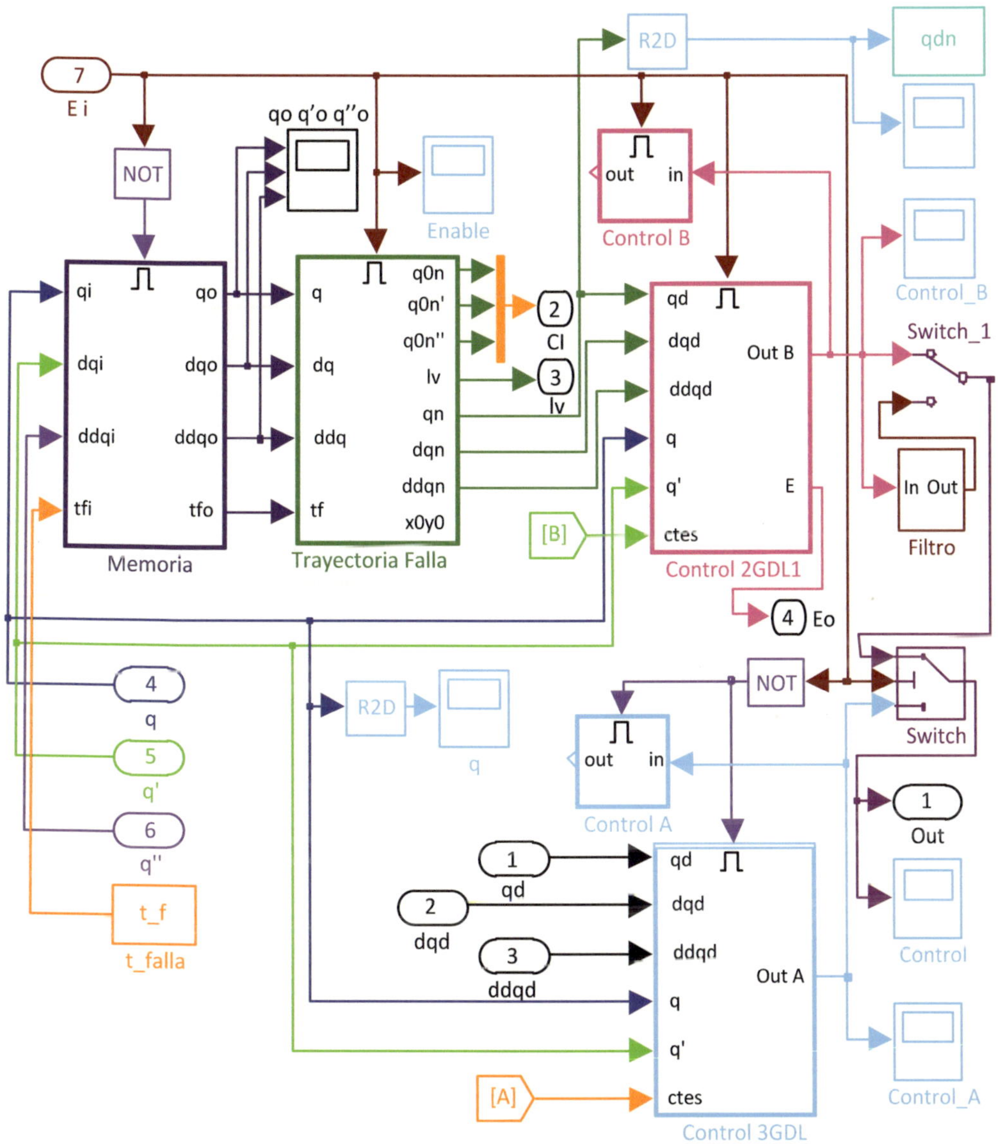

Fig. 5. Block diagram of the structure of the fault tolerant controller used to test the above mentioned control laws

9. Results

After establishing the control laws being utilized, we determine the trajectory to be entered in the control system to carry out the corresponding performance tests of fault tolerant control algorithms. This trajectory is displayed in figure 6.

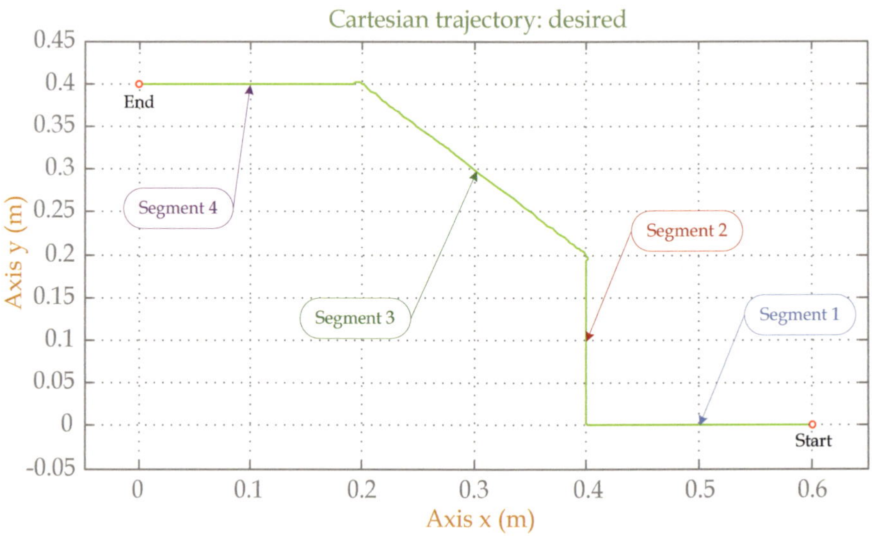

Fig. 6. Cartesian test trajectory

Figures 7 and 8a show the curves corresponding to the differences between desired and real joint trajectories, and between desired and real Cartesian trajectories, respectively, all this under *hyperbolic sine-cosine* fault tolerant control when there is a failure in actuator 2 at 0.5 sec from initiating movement.

Where:

e_{q1} : Joint trajectory error, joint 1.

e_{q2} : Joint trajectory error, joint 2.

e_{q3} : Joint trajectory error, joint 3.

e_x : Cartesian trajectory error, *x* axis.

e_y : Cartesian trajectory error, *y* axis.

Fig. 7. Joint trajectory error with fault control using *hyperbolic sine-cosine* controller

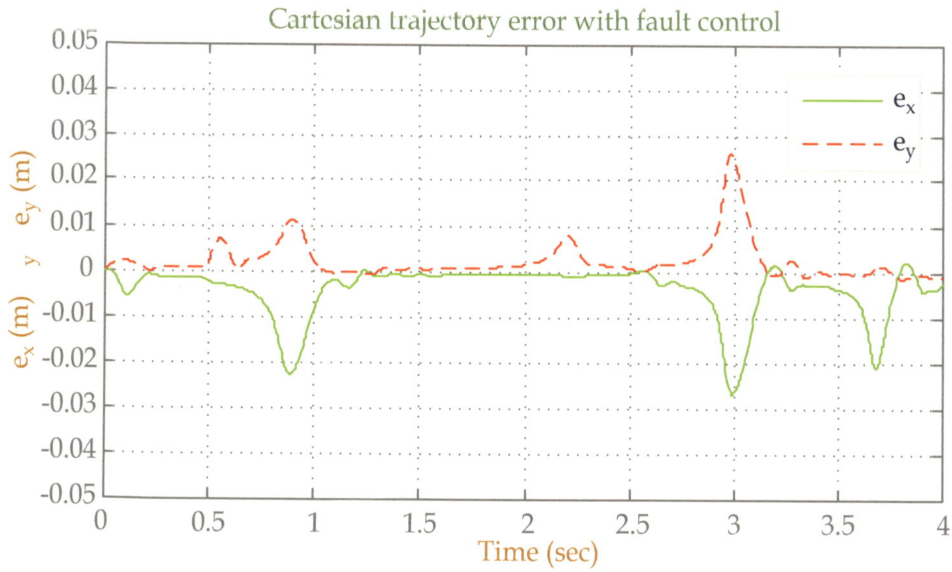

Fig. 8a. Cartesian trajectory error with fault control using *hyperbolic sine-cosine* controller

The performance of fault tolerant controller by computed torque is shown in figures 8b and 9, displaying the curves for joint and Cartesian errors under the same failure conditions than the previous case.

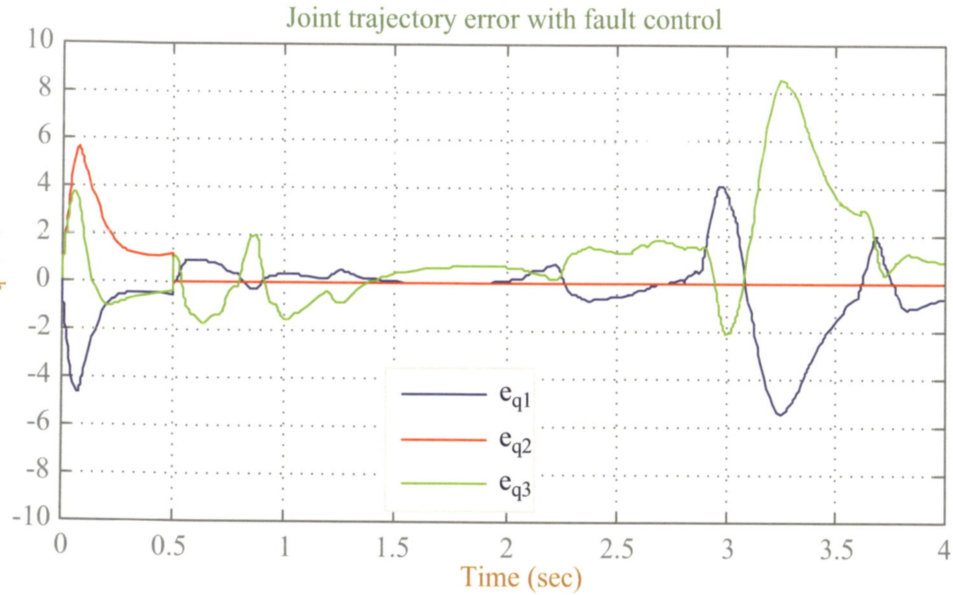

Fig. 8b. Joint trajectory error with fault control using computed torque controller

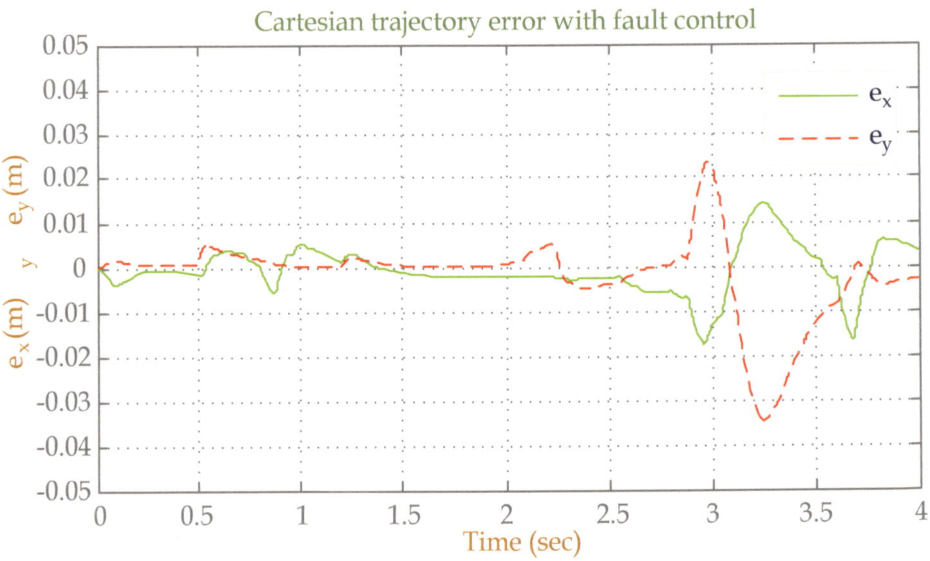

Fig. 9. Cartesian trajectory error with fault control using computed torque controller

In figures 10 and 11 we can see charts displaying respectively joint and Cartesian errors corresponding to the performance of fault tolerant controller by adaptive inertia, under the same failure conditions imposed to the previous controllers.

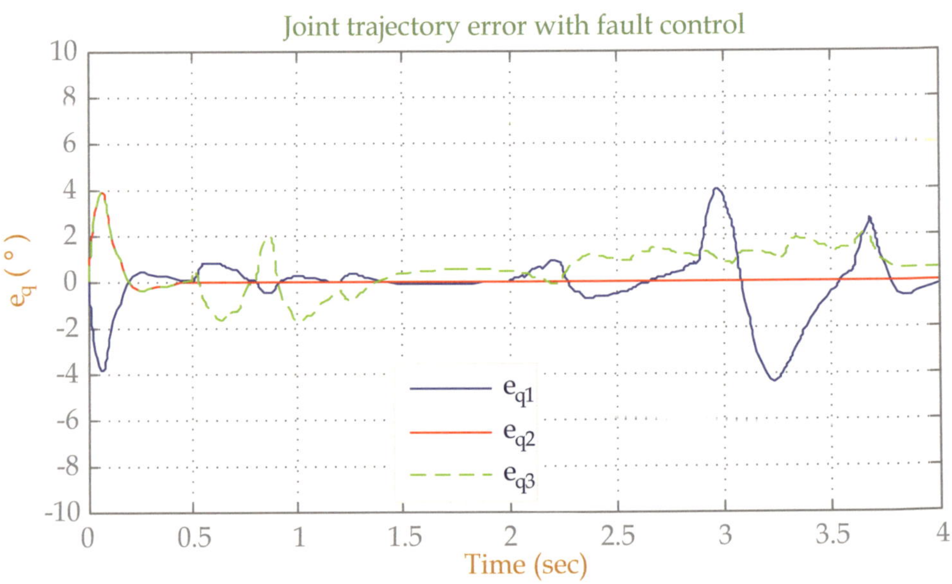

Fig. 10. Joint trajectory error with fault control using adaptive inertia controller

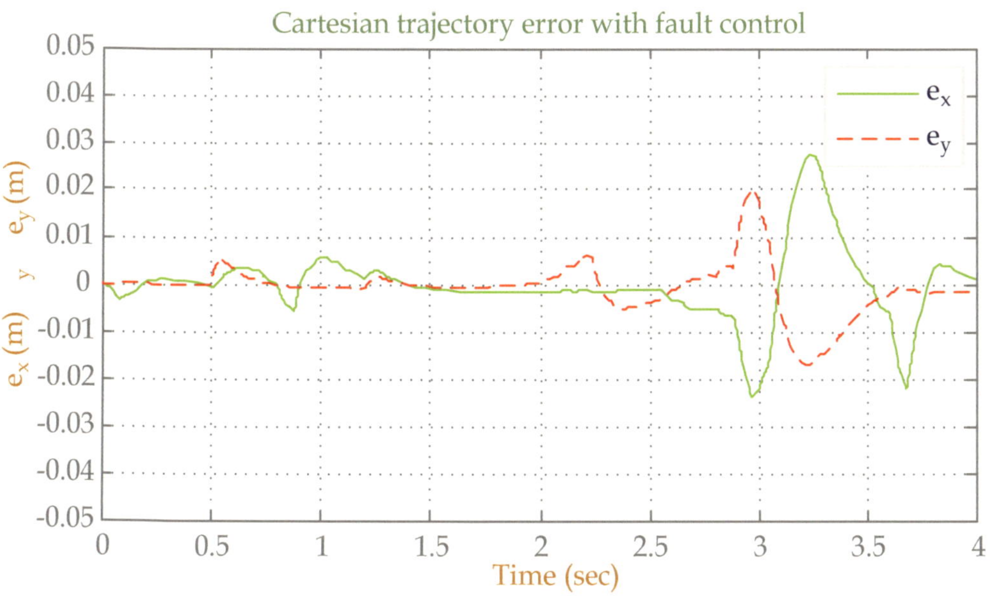

Fig. 11. Cartesian trajectory error with fault control using adaptive inertia controller

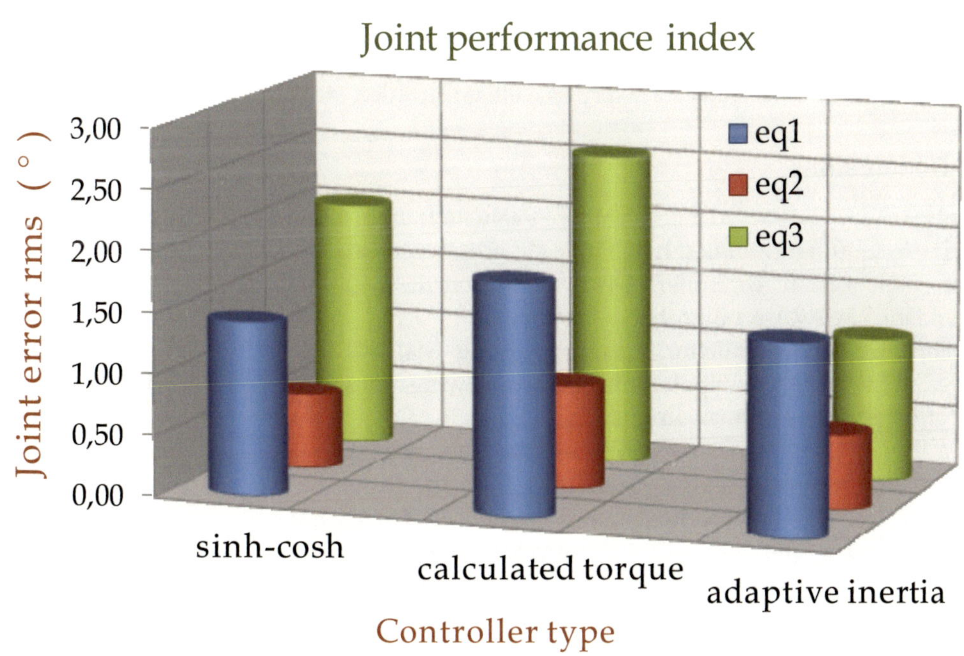

Fig. 12. Performance index corresponding to joint trajectory

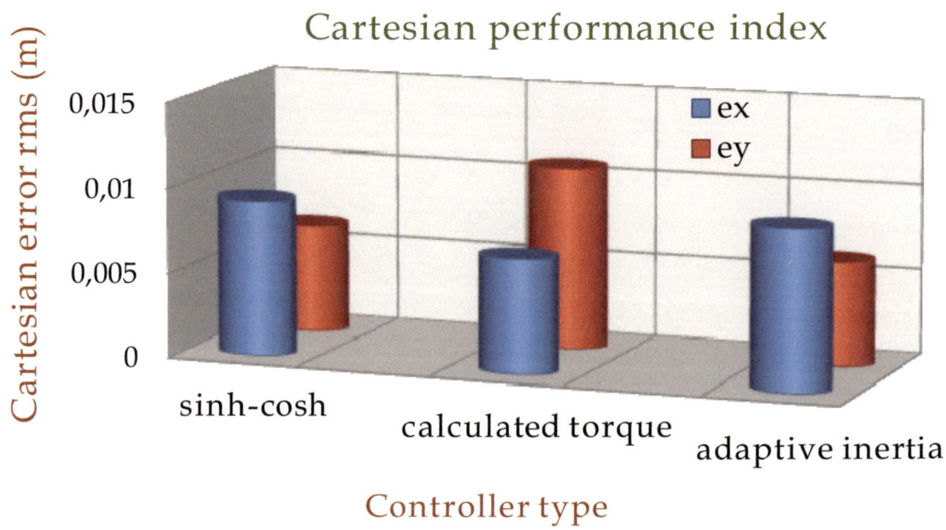

Fig. 13. Performance index corresponding to Cartesian trajectory

Finally, in figures 12 and 13 it is shown a performance summarization of the analyzed fault tolerant controllers in terms of joint and Cartesian *mean square root* errors, accordingly to equation (21)

$$rms = \sqrt{\frac{1}{n}\sum_{i=1}^{n} e_i^2} \tag{21}$$

Where e_i represents articular trajectory as well as Cartesian errors.

10. Conclusions

In this work we presented a performance evaluation of three fault tolerant controllers based on classic control techniques: hyperbolic sine-cosine, calculated torque and adaptive inertia. Those fault tolerant controllers were applied on the first three degrees of freedom of a redundant SCARA-type robotic manipulator. The different system stages were implemented in a simulator developed using MatLab/Simulink *software*, allowing to represent the robotic manipulator behavior following a desired trajectory, when blocking of one of its actuators occurs. In this way we obtained the corresponding simulation curves. From the obtained results, we observed that the adaptive inertia fault tolerant controller have errors with less severe maximums than the other controllers, resulting in more homogeneous manipulator movements. We noticed that greater errors were produced with the calculated torque fault tolerant controller, both for maximums and *rms*. Consequently, the best performance is obtained when using the adaptive inertia controller, as shown in figures 14 and 15. It is also remarkable that the hyperbolic sine-cosine fault tolerant controller have a lesser implementation complexity, since it does not require the second derivative of joint position. This can be a decisive factor in the case of not having high performance processors.

11. Further developments

Thanks to the development of this work, from the implemented simulation tools and the obtained results, fault tolerant control systems essays are being currently carried out, in order to apply them to actual robotic systems, with and without link redundancy, like the SCARA-type robots shown in figure 14 and figure 15, respectively.

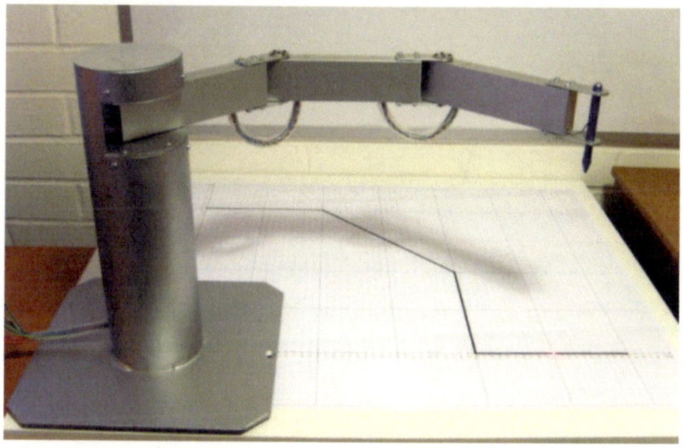

Fig. 14. SCARA-type redundant robot, DIE-USACH

Fig. 15. SCARA-type robot, DIE-USACH

12. Addendum A: Manipulator's dynamic model

The manipulator's dynamic model is given by equations a1 to a14.

$$\mathbf{M} = \begin{bmatrix} M_{11} & M_{12} & M_{13} \\ M_{21} & M_{22} & M_{23} \\ M_{31} & M_{32} & M_{33} \end{bmatrix} \tag{a1}$$

$$M_{11}=I_{1zz}+I_{2zz}+I_{3zz}+m_1 l_{c1}^2+m_2\left(l_1^2+l_{c2}^2\right)+m_2 2l_1 l_{c2}\cos\theta_2+\ldots$$
$$m_3\left(l_1^2+l_2^2+l_{c3}^2+2l_1 l_2\cos\theta_2+2l_2 l_{c3}\cos\theta_3+2l_1 l_{c3}\cos\left(\theta_2+\theta_3\right)\right) \tag{a2}$$

$$M_{21}=M_{12}=I_{2zz}+I_{3zz}+m_2\left(l_{c2}^2+l_1 l_{c2}\cos\theta_2\right)+\ldots$$
$$m_3\left(l_2^2+l_{c3}^2+l_1 l_2\cos\theta_2+2l_2 l_{c3}\cos\theta_3+l_1 l_{c3}\cos\left(\theta_2+\theta_3\right)\right) \tag{a3}$$

$$M_{31}=M_{13}=I_{3zz}+m_3\left(l_{c3}^2+l_2 l_{c3}\cos\theta_3\right)+m_3 l_1 l_{c3}\cos\left(\theta_2+\theta_3\right) \tag{a4}$$

$$M_{22}=I_{2zz}+I_{3zz}+m_2 l_{c2}^2+m_3\left(l_2^2+l_{c3}^2+2l_2 l_{c3}\cos\theta_3\right) \tag{a5}$$

$$M_{32}=M_{23}=I_{3zz}+m_3\left(l_{c3}^2+l_2 l_{c3}\cos\theta_3\right) \tag{a6}$$

$$M_{33}=I_{3zz}+m_3 l_{c3}^2 \tag{a7}$$

$$\mathbf{C}=\left[C_{11}\quad C_{21}\quad C_{31}\right]^T \tag{a8}$$

$$C_{11}=-2l_1\left(m_2 l_{c2}\sin\theta_2+m_3 l_2\sin\theta_2\right)\dot{\theta}_1\dot{\theta}_2-2l_1 m_3 l_{c3}\sin\left(\theta_2+\theta_3\right)\dot{\theta}_1\dot{\theta}_2+\ldots$$
$$-m_2 l_1 l_{c2}\sin\theta_2\cdot\dot{\theta}_2^2+m_3\left(l_1 l_2\sin\theta_2+l_1 l_{c3}\sin\left(\theta_2+\theta_3\right)\right)\dot{\theta}_2^2+\ldots$$
$$-2l_{c3}m_3\left(l_2\sin\theta_3+l_1\sin\left(\theta_2+\theta_3\right)\right)\dot{\theta}_1\dot{\theta}_3+\ldots$$
$$-2m_3 l_{c3}\left(l_2\sin\theta_3+l_1\sin\left(\theta_2+\theta_3\right)\right)\dot{\theta}_2\dot{\theta}_3+\ldots$$
$$m_3\left(-l_2 l_{c3}\sin\theta_3-l_1 l_{c3}\sin\left(\theta_2+\theta_3\right)\right)\dot{\theta}_3^2 \tag{a9}$$

$$C_{21}=m_3\left(l_1 l_2\sin\theta_2+l_1 l_{c3}\sin\left(\theta_2+\theta_3\right)\right)\dot{\theta}_1^2+m_2 l_1 l_{c2}\sin\theta_2\cdot\dot{\theta}_1^2+\ldots$$
$$-2m_3 l_2 l_{c3}\sin\theta_3\cdot\dot{\theta}_1\dot{\theta}_3-2m_3 l_2 l_{c3}\sin\theta_3\cdot\dot{\theta}_2\dot{\theta}_3-m_3 l_2 l_{c3}\sin\theta_3\cdot\dot{\theta}_3^2 \tag{a10}$$

$$C_{31}=m_3\left(l_2 l_{c3}\sin\theta_3+l_1 l_{c3}\sin\left(\theta_2+\theta_3\right)\right)\dot{\theta}_1^2+\ldots$$
$$2m_3 l_2 l_{c3}\sin\theta_3\cdot\dot{\theta}_1\dot{\theta}_2+m_3 l_2 l_{c3}\sin\theta_3\cdot\dot{\theta}_2^2 \tag{a11}$$

$$C_{31}=m_3\left(l_2 l_{c3}\sin\theta_3+l_1 l_{c3}\sin\left(\theta_2+\theta_3\right)\right)\dot{\theta}_1^2+2m_3 l_2 l_{c3}\sin\theta_3\cdot\dot{\theta}_1\dot{\theta}_2+\ldots$$
$$m_3 l_2 l_{c3}\sin\theta_3\cdot\dot{\theta}_2^2 \tag{a12}$$

$$\mathbf{G}=\left[0\quad 0\quad 0\right]^T \tag{a13}$$

$$\mathbf{F}=\left[F_{11}\quad F_{21}\quad F_{31}\right]^T \tag{a14}$$

where:

m_1 : First link mass.

m_2 : Second link mass.

m_3 : Third link mass.
l_1 : First link length.
l_2 : Second link length.
l_3 : Third link length.
l_{c1} : Length from 1st link origin to its centroid.
l_{c2} : Length from 2nd link origin to its centroid.
l_{c3} : Length from 3rd link origin to its centroid.
I_{1zz} : 1st link inertial momentum with respect to the first z axis of its joint.
I_{2zz} : 2nd link inertial momentum with respect to the first z axis of its joint.
I_{3zz} : 3rd link inertial momentum with respect to the first z axis of its joint.

13. Addendum B: Considered parameter values

Parameter values considered for the manipulator as well as controller gains values are shown in tables B1 and B2, respectively.

Link 1	Link 2	Link 3	Units
l_1 = 0.2	l_2 = 0.2	l_3 = 0.2	$[\text{m}]$
l_{c1} = 0.0229	l_{c2} = 0.0229	l_{c3} = 0.0983	$[\text{m}]$
m_1 = 2.0458	m_2 = 2.0458	m_3 = 6.5225	$[\text{kg}]$
I_{1zz} = 0.0116	I_{2zz} = 0.0116	I_{3zz} = 0.1213	$[\text{kg}\cdot\text{m}^2]$
F_{v1} = 0.025	F_{v2} = 0.025	F_{v3} = 0.025	$\left[\frac{\text{N}\cdot\text{m}\cdot\text{s}}{\text{rad}}\right]$
F_{eca1} = 0.05	F_{eca2} = 0.05	F_{eca3} = 0.05	$[\text{N}\cdot\text{m}]$
F_{ecb1} = -0.05	F_{ecb2} = -0.05	F_{ecb3} = -0.05	$[\text{N}\cdot\text{m}]$

Table B1. Considered parameters for the manipulator

	Controller Type		
Constants	Hyperbolic Sine-Cosine	Computed Torque	Adaptive Inertia
K_{p1}, K_{p2}, K_{p3}	400, 300, 200	800, 800, 800	—
K_{v1}, K_{v2}, K_{v3}	3, 2, 1	140, 140, 140	20, 20, 20
$\lambda_1, \lambda_2, \lambda_3$	—	—	8, 8, 8
$\gamma_1, \gamma_2, \gamma_3$	—	—	0.1, 0.1, 0.1

Table B2. Controller gains

14. References

Barahona, J., Espinosa & L., C.F., 2002. Evaluación Experimental de Controladores de Posición tipo Saturados para Robot Manipuladores. In *Congreso Nacional de Electrónica, Centro de Convenciones William o Jenkins*. Puebla. México, 2002.

Blanke, M., Kinnaert, M., Lunze, J. & Staroswiecki, M., 2000. What is Fault-Tolerant Control. In *IFAC Symposium on Fault Detection, Supervision and Safety for Technical Process - SAFEPROCESS 2000*. Budapest, 2000. Springer-Verlag Berlin Heidelberg.

El-Salam, A., El-Haweet, W. & and Pertew, A., 2005. Fault Tolerant Kinematic Controller Design for Underactuated Robot Manipulators. In *The Automatic Control and Systems Engineering Conference*. Cairo, 2005.

Lewis, F., Dawson, D. & Abdallah, C., 2004. *Robot Manipulator Control Theory and Practice.* New York: Marcel Dekker, Inc.

Patton, R.J., 1997. Fault-Tolerant Control: The 1997 Situation. In *Proc. IFAC Symposium Safeprocess*. Kingston Upon Hull, 1997.

Puig, V. et al., 2004. Control Tolerante a Fallos (Parte I): Fundamentos y Diagnóstico de Fallos. *Revista Iberoamericana de Automática e Informática Industrial*, 1(1), pp.15-31.

Rubí, J., 2002. *Cinemática, Dinámica y Control de Robots Redundantes y Robots Subactuados*. Tesis Doctoral. San Sebastián, España: Universidad de Navarra.

Siciliano, B. & Khatib, O., 2008. *Handbook of Robotics*. Berlin, Heidelberg: Springer-Verlag.

Xiujuan, D. & Zhen, L., 2007. *Underactuated Robot Dynamic Modeling and Control Based on Embedding Model*. In *12th IFToMM World Congress*. Besançon. France, 2007.

Zhang, Y. & Jiang, J., 2003. Bibliographical Review on Reconfigurable Fault-Tolerant Control Systems. In *Proceedings IFAC SAFEPROCESS*. Washington, D.C., USA, 2003.

An Approach to Distributed Component-Based Software for Robotics*

A. C. Domínguez-Brito, J. Cabrera-Gámez,
J. D. Hernández-Sosa, J. Isern-González and E. Fernández-Perdomo
Instituto Universitario SIANI & the Departamento de Informática y Sistemas,
Universidad de Las Palmas de Gran Canaria
Spain

1. Introduction

Programming robotic systems is not an easy task, even developing software for simple systems may be difficult, or at least cumbersome and error prone. Those systems are usually multi-threaded and multi process, so synchronization problems associated with processes and threads must be faced. In addition distributed systems in network environments are also very common, and coordination between processes and threads in different machines increases programming complexity, specially if network environments are not completely reliable like a wireless network. Hardware abstraction is another question to take into account, uncommon hardware for an ordinary computer user is found in robotics systems, sensors and actuators with APIs (Applications Programming Interfaces) in occasions not very stable from version to version, and many times not well supported on the most common operating systems. Besides, it is not rare that even sensors or actuators with the same functionality (i.e. range sensors, cameras, etc.) are endowed with APIs with very different semantics. Moreover, many robotic systems must operate in hard real time conditions in order to warrant system and operation integrity, so it is necessary that the software behaves obeying strictly specific response times, deadlines and high frequencies of operation. Software integration and portability are also important problems in those systems, since many times only in one of them we may find a variety of machines, operating systems, drivers and libraries which we have to cope to. Last but not least, we want robotic systems to behave "autonomous" and "intelligently", and to carry out complex tasks like mapping a building, navigating safely in a cluttered, dynamic and crowded environment or driving a car safely in a motorway, to name a few.

Despite there is no established standard methodology or solution to the situation described in the previous paragraph, in the last ten years many approaches have blossomed in the robotics community in order to tackle with it. In fact, many software engineering techniques and experiences coming from other areas of computer science are being applied to the specific area of robotic control software. A review of the state of the art of software engineering applied

*This work has been partially supported by the Research Project *TIN2008-06068* funded by the Ministerio de Ciencia y Educación, Gobierno de España, Spain, and by the Research Project *ProID20100062* funded by the Agencia Canaria de Investigación, Innovación y Sociedad de la Información (ACIISI), Gobierno de Canarias, Spain, and by the European Union's FEDER funds.

specifically to robotics can be found in [Brugali (2007)]. Many of the approaches that have come up in these last years, albeit different, are either based completely, or follow or share to a certain extend some of the fundamentals ideas of the CBSE (Component Based Software Engineering) [George T. Heineman & William T. Councill (2001)] paradigm as a principle of design for robotic software.
Some of the significant approaches freely available within the robotics community based on the CBSE paradigm are G^{en}oM/BIP [Mallet et al. (2010)][Basu et al. (2006)][Bensalem et al. (2009)], Smartsoft [Schlegel et al. (2009)], OROCOS [*The Orocos Project* (2011)], project ORCA [Brooks et al. (2005)], OpenRTM-aist [Ando et al. (2008)] and Willow Garage's ROS project [*ROS: Robot Operating System* (2011)]. All these approaches, in general incompatible among them, use many similar abstractions in order to build robotic software out of a set of software components. Each of these approaches usually solve or deal with many of the mentioned problems faced when programming robotic systems (hard real-time operation, distributed, multithread and multiprocess programming, hardware abstractions, portability, etc.), and using any of them implies to get used to its own methodology, abstractions and software tools to develop robotic software. Our group have also developed an approach to tackle with the problem of programming robotic systems. It is some years already that we have been using a CBSE C++ distributed framework designed and developed in our laboratory, termed CoolBOT [Antonio C. Domínguez-Brito et al. (2007)], which is also aimed at easing software development for robotic systems. Along several years of use acquiring experience programming mobile robotic systems, we have ended up integrating in CoolBOT some new developments in order to improve their use and operation. Those new improvements have been focused mainly in two main questions, namely: transparent distributed computation, and "deeper" interface decoupling; in next sections they will be presented more deeply. In order to do so this paper is organized as follows. In section 2 we will introduce briefly an overview about CoolBOT. Next, we will pass to focus on each one of the mentioned topics respectively, in sections 4 and 5. The last section is devoted to presenting the conclusions of this work.

2. CoolBOT. Overview

CoolBOT [Antonio C. Domínguez-Brito et al. (2007)] is a C++ component oriented programming framework aimed to robotics, developed at our laboratory some years ago [Domínguez-Brito et al. (2004)], which is normally used to develop the control software for the mobile robots we have available at our institution. It is a programming framework that follows the CBSE paradigm for software development. The key concept in the CBSE paradigm is the concept of *software component* which is a unit of integration, composition and software reuse [Brugali & Scandurra (2009)][Brugali & Shakhimardanov (2010)]. Complex systems might be composed of several ready-to-use components. Ideally, interconnecting available components out of a repository of components previously developed, we can program a complete system. Thus, it should be only necessary a graphical interface or alike, to set up a system. Hence, being CoolBOT CBSE oriented, it also makes use of this central concept to build software systems.
Fig. 1 gives a global view of a typical system developed using CoolBOT. As we can observe, there are five CoolBOT *components* and two CoolBOT *views*, forming all them four CoolBOT *integrations* involving three different machines sharing a computer network. In addition, hosted by one of the machines, there is a non CoolBOT application which uses a CoolBOT *probe* to interact with one of the components of the system. Thus, in CoolBOT we can find three types of software components: *components*, *views* and *probes*. All these three types

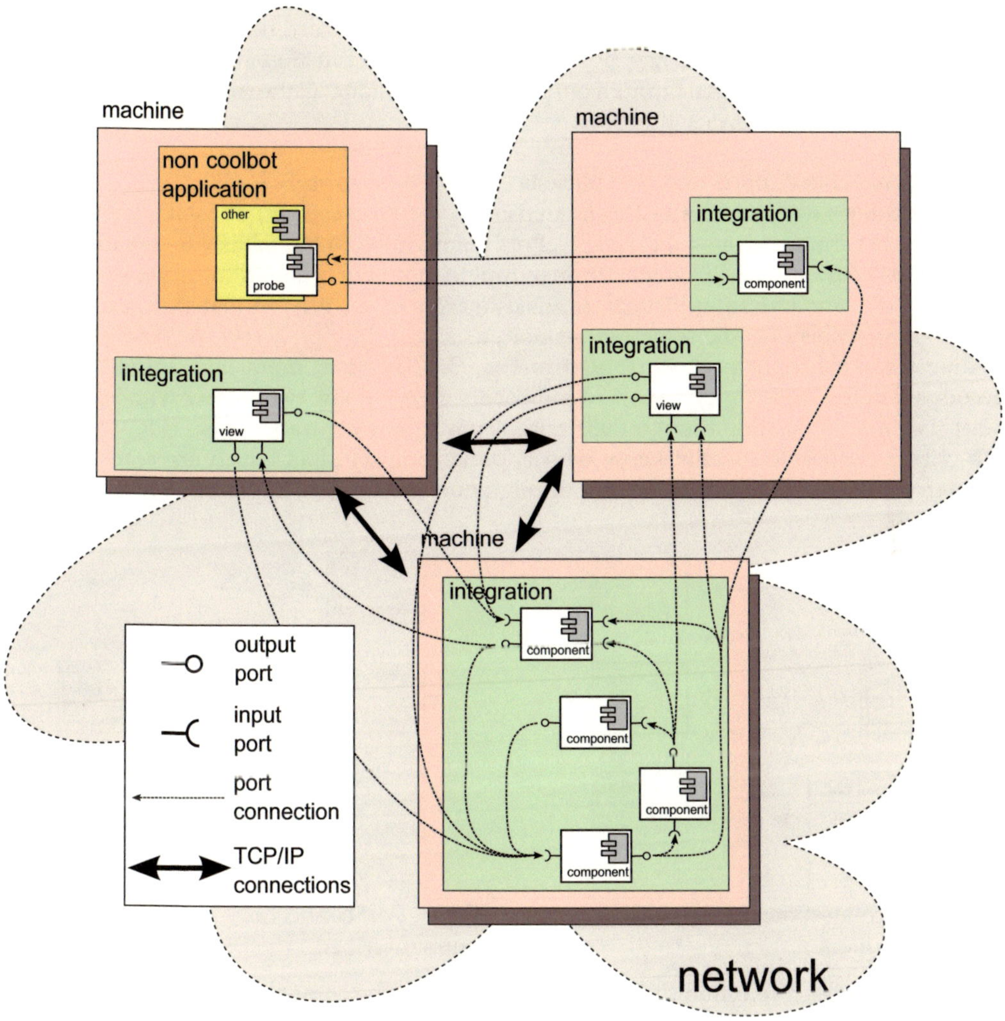

Fig. 1. Diagram of the elements and interconnections of a system designed with CoolBOT.

are software components in the whole sense, since we can compose them indistinctly and arbitrarily without changing their implementation to built up a given system. The main difference among them is that *views* and *probes* are "light-weight" software components in relation to CoolBOT *components*. *Views* are software components which implement graphical control and monitoring interfaces for CoolBOT systems, which are completely decoupled from them. On the other side, *probes* mainly allow to implement decoupled interfaces for interoperation of CoolBOT systems with non CoolBOT applications, as depicted in the figure. Both will be explained in more detail in section 5.

In CoolBOT, systems are made of CoolBOT *components* (components for short). A component is an *active entity*, in the sense that it has its own unit of execution. It also presents a clear separation between its external interface and its internals. Components intercommunicate among them only through their external interfaces which are formed by *input and output ports*. When connected, they form *port connections*, as depicted on Fig. 1. Through them, components interchange discrete units of information termed *port packets*. *Views* and *probes* have a similar external interface of input and output ports, hence, they can also be interconnected among them and with components using port connections. The functionality of a whole system comes up from the interaction through port connections among all the components integrating the system, including views and probes.

2.1 Port connections, ports and port packets

CoolBOT components interact among them using *port connections*. A *port connection* is defined by an output port and an input port. Port connections are established dynamically in runtime, and they are unidirectional (from output to input port), and follow a *publish/subscribe* paradigm of communication [Brugali & Shakhimardanov (2010)]. In that way, we can have multiple subscribers for the same output port, as shown in Fig. 2, and multiple publishers feeding the same input port, illustrated in Fig. 3. Note that input and output ports are decoupled in the sense that component publishers do not know necessarily who is receiving what they are publishing through their output ports. The contrary is also true, component subscribers do not necessarily know who is publishing the data which are reaching them through their input ports. Data are sent through port connections in discrete units called *port packets*.

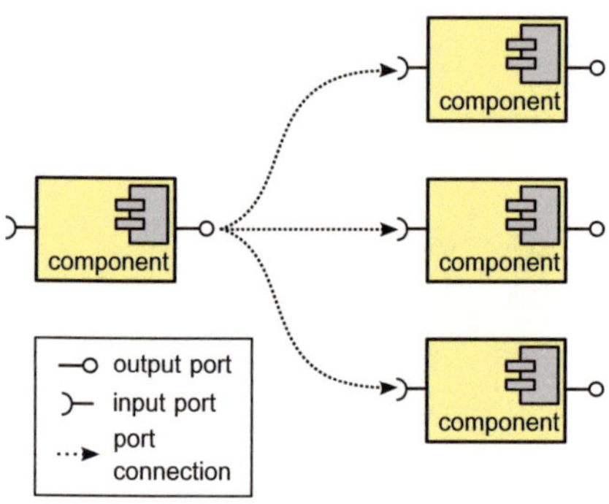

Fig. 2. Port connections: one publisher, many subscribers.

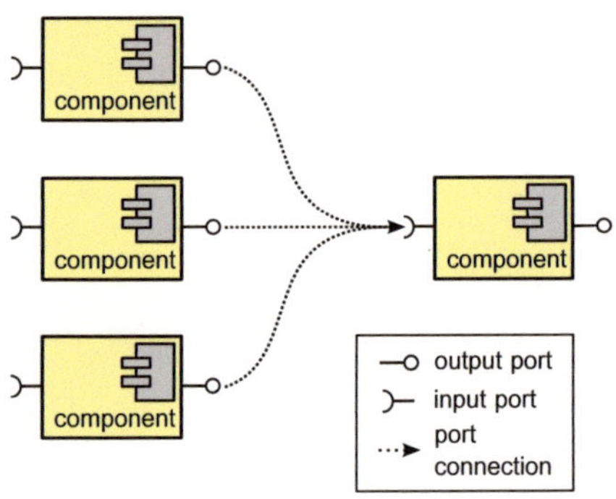

Fig. 3. Port connections: many publishers, one subscriber.

To establish a port connection the ports involved should be compatible, i.e., they must have compatible types, and should transport the same types of port packets. In particular, when defining an output or input port we have to specify three aspects for them, namely:

1. **An identifier**: This is an identifier or name for the port. It has to be unique at component (or *view* or *probe*) scope. The port identifier is what we use to refer to a specific port when establishing/de-establishing port connections.
2. **A port type**: This is the type of the port. There are several typologies available for input and output ports, and depending on how we combine them, we can establish a different model of communication for each connection. The typologies of the input and output ports involved in a connection determine the pattern and semantics of the communication through it, following the same philosophy that the communications patterns of Smartsoft [Schlegel et al. (2009)], and the interfaces for component communication available in OROCOS [*The Orocos Project* (2011)]. On Tables 1 and 2 we can see all the types of connections we have available in CoolBOT right now, we will elaborate on this later.
3. **Port packet types**: Those are the types of port packets accepted by the output or input port. Most input and output port types only accept one type of port packet through them, although we have also some of them that accept a set of different port packet types.

Bear in mind that in CoolBOT port connections are established dynamically, but the definition of each input and output port for each software component, whether a component, a view or a probe, is static. Thus, for a given component we define statically its external interface of input and output ports, each one with its identifier, port type and accepted port packet types. In opposition, port connections are established at runtime. Only a compatible pair of output and input ports can form a port connection, and we say that they are compatible when two conditions fulfill: first, they have compatible port types (the compatible combinations are shown in Tables 1 and 2); and second, the port packets the pair of port accepts also match.
As commented, Tables 1 and 2 show all the possible types and combinations of output and input ports available in CoolBOT. As we can observe we have two groups of types of port connections depending on the types of the output and input ports we connect, namely:

- ***Active Publisher/Passive Subscriber (AP/PS)* connections**. In this kind of connections we say the publisher is the *active* part of the communication through the connection, since it is the publisher (the sender) who invest more computational resources in the communication. More specifically, in these connections, there is a buffer (a cache, a memory) in the input port where incoming port packets get stored, when they get to the subscriber (the receiver) end. We say the publisher is active, because the copy of port packets in the input port buffers is made by the publisher's threads of execution. Those memories get signaled for the subscribers to access them at their own pace. Evidently, if the output port has several subscribers, the publisher has to make a copy for all of them, so the computational cost for copies increases and this cost is afforded by the publisher. Table 1 enumerates all the available types of port connections following this model of communication.

- ***Passive Publisher/Active Subscriber (PP/AS)* connections**. Those connections follow a model of communication where the subscriber plays the *active* role in the communication, in the sense that in opposition to the previous ones, the subscriber is the part of the communication which invests more computational resources. In this type of connections, we have buffers at both ends of the port connection. When the publisher sends (publishes) a port packet through the connection, it gets stores in a buffer in the output port, and the input port gets signaled, in order to notify the subscriber that there are new data on the connection. Note that the publisher does not copy the port packet on the subscriber's input port buffer. It is the subscriber the one who copies the port packet on its input port buffer when it access its input port in order to get fresh port packets, stored at the other end of the port connection. In this way, when we have several subscribers and one publisher the computational cost of copies is afforded by each one of the subscribers separately.

Apart from the computational cost of using AP/PS connections versus PP/AS connections, there is another important aspect to take into account when using any of them. PP/AS connections are *persistent* in the sense that, as explained in Table 2 the last data (port packet) sent by the publisher through the output port is stored there, in the output port buffer, so subscribers which are connected to this output port once the last port packet was sent, can access those data posteriorly. On the contrary AP/PS connections are not persistent, because port packets are stored on the subscriber end, so packets, once they have been sent, are not available for new subscribers, because port packets only gets to those subscribers who are connected to the output port right at the moment of sending them.
Other important aspect to take into account with respect to port connections is that they follow an asynchronous model of communications. Note that they are unidirectional, port packets go from the publisher's output port to the subscriber's input port, the publisher sends port

Active Publisher/Passive Subscriber (AP/PS)		
Output Port	**Input Port**	**Port Connection Type**
tick	*tick*	*tick* connections: those connections do not transport any port packet, they only communicate the occurrence of an event.
generic	*last*	*last* connections: the input port stores always the *last* port packet sent through the connection by the publisher (publishers). Only one type of port packet is accepted through the port connection.
	fifo	*fifo* connections: at the input port there is a circular fifo with a specific length. Port packets sent through the port connection by publishers, get stored there. Only one type of port packet is accepted through the port connection.
	ufifo	*unbounded fifo* connections: at the input port there is a fifo with a specific length. Port packets sent through the port connection by publishers, get stored there. When the fifo is full and port packets keep coming the fifo grows in order to store them. Only one type of port packet is accepted through the port connection.
multipacket	*multipacket*	*multipacket* connections: those connections accept a set of port packet types. There is a different buffer for each accepted port packet type, the last port packet of each type which is sent through the connection by publishers gets stored on them.
lazymultipacket		*lazymultipacket* connections: those connections accept a set of port packet types. At the input port there is a different buffer for each accepted port packet type, the last port packet of each type which is sent through the connection by publishers gets stored on them. On the output port, port packets get stored in a queue of port packets to send, they are really sent to the other end when a *flush* operation is applied by the publisher on the output port.

Table 1. Available port connections types: *active publisher/passive subscriber (AP/PS)* connections.

Passive Publisher/Active Subscriber (PP/AS)		
Output Port	**Input Port**	**Port Connection Type**
poster	*poster*	*poster* connections: there is a buffer at the output port where the last packet send by the publisher gets stored. There is another buffer at the input port which is "synchronized" with the output port buffer when the subscriber accesses its input port in order to get the last port packet sent through the connection. Therefore, the port packet gets copied to the input port only when a new port packet has been stored at the output port end. Only one type of port packet is accepted through the port connection.

Table 2. Available port connections types: *passive publisher/ active subscriber (PP/AS)* connections.

packets and keeps doing something different, the subscriber gets packets at its own pace and not necessarily at the right moment they get to its input ports.

As to port packets, when defining an output or input port, we have to specify which port packet type or types (depending on the port type being defined), the port will accept. In general, port packet types are defined by the user, as we will see in section 3, we may also use port packets types provided by CoolBOT itself (the available ones are shown in Fig. 3), port packet types previously developed, or third party port packet types.

Port packet type	Description
`PacketUChar`	Transports a C++ `unsigned char`.
`PacketInt`	Transports a C++ `int`.
`PacketLong`	Transports a C++ `long`.
`PacketDouble`	Transports a C++ `double`.
`PacketTime`	Transports a CoolBOT `Time` value (a time-stamp).
`PacketCoordinates2D`	Transports a CoolBOT `Coordinates2D` value (stores a 2D point).
`PacketFrame2D`	Transports a CoolBOT `Frame2D` value (stores a 2D frame).
`PacketCoordinates3D`	Transports a CoolBOT `Coordinates3D` value (stores a 3D point).
`PacketFrame3D`	Transports a CoolBOT `Frame3D` value (stores a 3D frame).

Table 3. Available port packet types provided by CoolBOT itself.

2.2 CoolBOT components

CoolBOT components are *active objects* [Ellis & Gibbs (1989)], as [Brugali & Shakhimardanov (2010)] states: "a component is a computation unit that encapsulates data structures and operations to manipulate them". Moreover, in CoolBOT, components can be seen as "data-flow machines", since they process data when they dispose of it at their input ports. Otherwise, they stay idle, waiting for incoming port packets. On the other side, components send processed data in form of port packets through their output ports. All in all, the model of computation of CoolBOT systems follows the *Flow Based Programming* (FBP) paradigm according to [J. Paul Morrison (2010)], so systems can be built as networks of components interconnected by means of port connections. More formally, CoolBOT components are modeled as *port automata* [Steenstrup et al. (1983)][Stewart et al. (1997)]. Fig. 4 provides a view of the structure of a component in CoolBOT. There is a clear separation between its external interface and its internal structure. Externally, a component can only communicate with other components (and views and probes) through its input and output ports. Thence, a component's external interface comprises all its input and output ports, its types, and the port packets it accepts through them. As we can see on the figure there are two special ports in any component: the *control port* and the *monitoring port*, the rest of the ports are user defined. The *control port* allows to modify component's *controllable variables*. Through the *monitoring port* components publish their *observable variables*. Both ports allows an external supervisor (i.e. another component, a view or a probe) to observe and modify the execution and configuration of a given component.

Internally a component is organized as an automaton, as illustrated in Fig. 4. All components follow the same automaton structure. The part of this automaton which is common to all components is called the *default automaton*, and comprises several states, namely: *starting*, *ready*, *suspended*, *end*, and four states for component exception handling. This structure allows for an external supervisor to control the execution of any component in a system using an standard protocol, likewise in an operating system where threads and processes transit among different states during their lifetime. To complete the automaton of a component the user defines the *user automaton* which is specific for each component and implements its functionality. This is represented in Fig. 4 with a dotted circle as the meta-state *running*. Transitions among component's automaton states are triggered by incoming port packets through any of its input ports, and also by internal events (timers, empty transition, a control variable that has been modified by an external supervisor, entering or exiting a state, etc.). The user can associate C++ callbacks to transitions, much like the *codels* for G^{en}oM [Mallet et al. (2010)] modules.

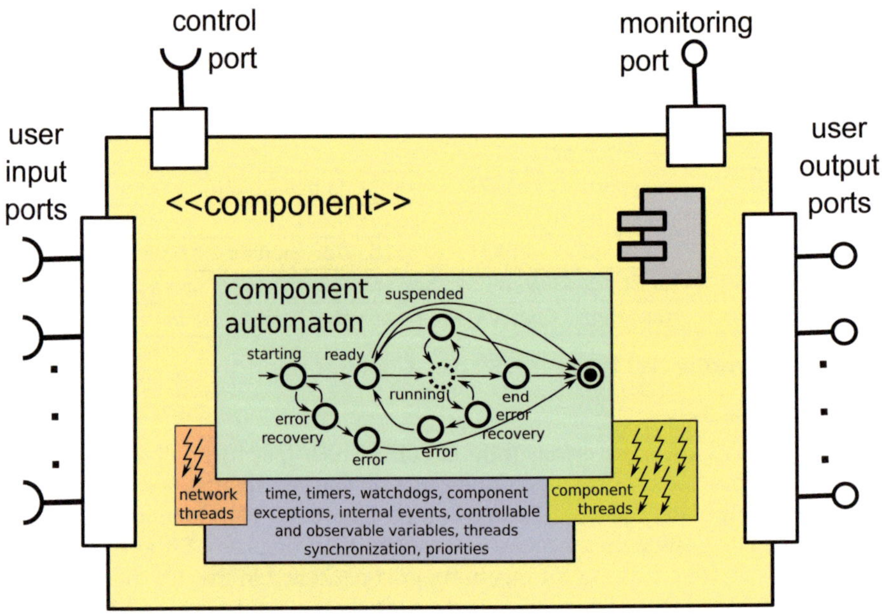

Fig. 4. CoolBOT. Component structure.

A key design principle for CoolBOT components is to take advantage of multithreaded and multicore capacities of mainstream operating systems, and the infrastructure they provide for multithreaded programming. Another key principle of design for components was to separate functional logic from thread synchronization logic. The user should be only worried about the functional logic, synchronization issues should be completely transparent to the developer. CoolBOT should be responsible for them behind the scenes. As active objects, CoolBOT components can organize its execution using multiple threads of execution as depicted on Fig. 4. Those threads are mapped on the underlying operating system (see Fig. 5). Thus, when developing a component the user assigns disjointly threads to automaton states, and to input ports and internal events provoking transitions Those transitions, i.e. their associated callbacks, will be executed by the specific threads being assigned. The synchronization among them is guaranteed by the underlying framework infrastructure. All components are endowed at least with one thread of execution; the rest, if any, are user defined.

As depicted in Fig. 1, CoolBOT provides means for distributed computation. A given system can be mapped on a set formed by different machines sharing a computer network. Port connections among components, views and probes are transparently multiplexed using TCP/IP connections (see section 4). Furthermore, each machine can host one or several CoolBOT *integrations*. A CoolBOT *integration* is an application (a process) which integrates instances of CoolBOT components, views and probes. Integrations can be instantiated in any machine, and are user defined using a description language as we will see in next section.

3. CoolBOT development tools

CoolBOT provides several tools for helping developers. Fig. 6 shows the main ones, namely: `coolbot-ske` and `coolbot-c`. The former one, `coolbot-ske`, is used to create a directory structure for development of CoolBOT components, probes, port packets, views and integrations. It also generates CMake [Kitware, Inc. (2010)] template files for compiling them, description language template files for `coolbot-c`, and test programs for components. The

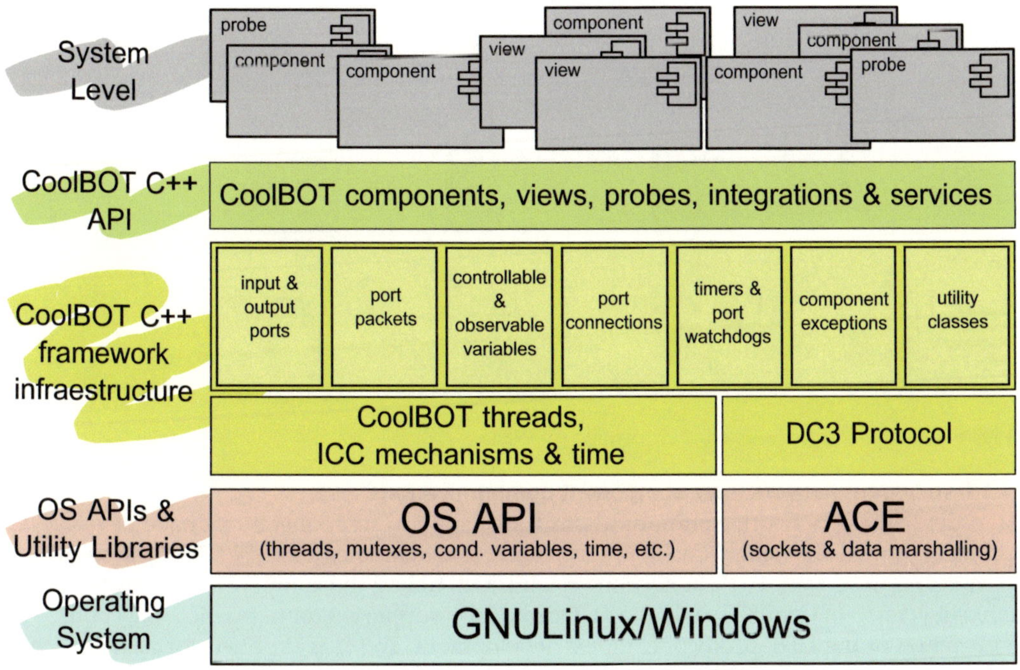

Fig. 5. Abstraction layers in CoolBOT.

latter tool, the CoolBOT compiler `coolbot-c`, generates C++ skeletons for components, port packets, views and integrations, and for each component it also generates its corresponding probe. All, the probe and the skeletons are C++ classes, and CoolBOT uses a description language as source code to generate those C++ classes. Except for probes, which are complete functional C++ classes, `coolbot-c` generates incomplete C++ classes which constitute the mentioned C++ skeletons. They are incomplete in the sense that they lack functionality, the user is responsible for completing them. Once completed, and using the CMake templates provided by `coolbot-ske`, they can be compiled. Components, probes, port packets, and views compile yielding dynamic libraries, integrations compile yielding executable programs. Moreover, the `coolbot-c` compiler preserves information when recompiling description files which have been modified, in such a way that, all C++ code introduced by the user into the skeletons is preserved.

4. Transparent distributed computation

Transparent distributed computation is the first development we have integrated on CoolBOT in order to improve its use and operation. The main idea was to make network communications as transparent as possible to developers (and components). We wanted CoolBOT to be responsible for them on behalf of components. Thus, at system level, to connect two component instances instantiated in different CoolBOT integrations should be as easy as connecting them when instantiated in the same integration. In particular, we follow three main principles related to transparent distributed computation facilities: transparent network inter component communications, network decoupling of component's functional logic, and incremental design.

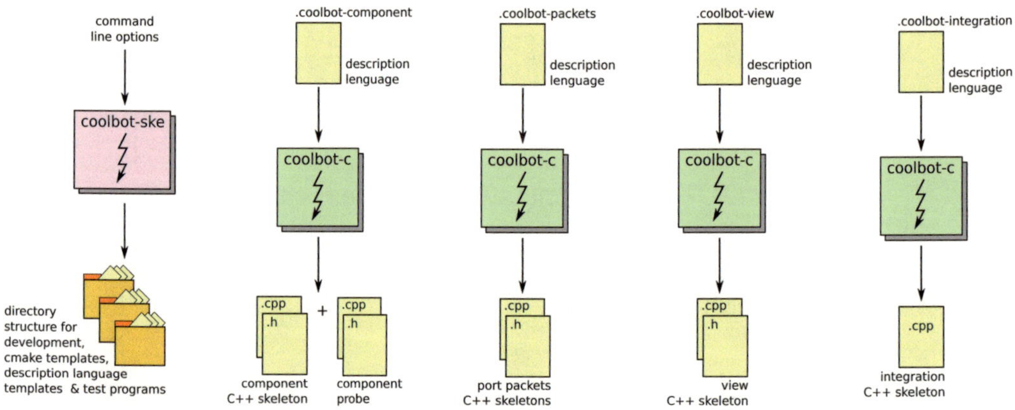

Fig. 6. CoolBOT's software development process.

4.1 Transparent network inter component communications

In order to make network communications transparent to components, we have developed a protocol termed *Distributed CoolBOT Component Communication Protocol (DC3P)* to multiplex port connections over TCP connections established among the components involved. In the current version of CoolBOT, only the TCP protocol is supported for network connections. The integration of the UDP protocol is under development, and it is expected for next CoolBOT version. DC3P has been implemented using the TCP/IP socket wrappers and the marshalling facilities provided by the ACE library [Douglas C. Schmidt (2010)], illustrated in Fig. 5. The protocol consists of the following packets:

- *Port Type Info* (request & response): For asking type information about input and output ports through network connections. This allows port connection compatibility verification when establishing a port connection through the network.
- *Connect* (request & response): For establishing a port connection over TCP/IP.
- *Disconnect* (request & response): To disconnect a previous established port connection.
- *Data Sending*: Once a port connection is established over TCP/IP, port packets are sent through it using this DC3P packet.
- *Remote Connect* (request & response): For establishing port connections between two remote component instances. Permits to connect component instances remotely.
- *Remote Disconnect* (request & response): To disconnect port connections previously established between two remote component instances.
- *Echo Request & Response*: Those DC3P packets are useful to verify that the other end in a network communication is active and responding.

All DC3P packets and port packets sent through port connections are marshalled and unmarshalled in order to be correctly sent through the network. We have used the facilities ACE provides for marshalling/demarshalling based on the OMG Common Data Representation (CDR) [Object Management Group (2002b)]. In general, port packets are user defined. In order to make their marshalling/demarshalling as easy and transparent as possible for developers, the description language accepted by the `coolbot-c` compiler includes sentences for describing port packets (as we can observe in Fig. 6), much like CORBA IDL [Object Management Group (2002a)]. The compiler generates a C++ skeleton class for each port packet where the code for marshalling/demarshalling is part of the skeleton's code generated automatically. In addition, we have endowed also CoolBOT with a rich set of C++ templates and classes to support marshalling and demarshalling of port packets (or any other arbitrary C++ class).

4.2 Network decoupling of component's functional logic

Another important aspect for network communication transparency is the decoupling of network communication logic from the functional logic of the component. Fig. 4 illustrates how this decoupling has been put into practice. Each component is endowed with a pair of network threads, and *output network thread*, and an *input network thread*, which are responsible for network communications using DC3P. CoolBOT guarantees transparently thread synchronization between them and the functional threads of the component. The network threads are mainly idle, waiting to have port packets to send through open network port connections, or to receive incoming port packets that should be redirected to the corresponding component's input ports. At instantiation time, it is possible to deactivate the network support for a component instance (and also for views and probes instances). In this manner, the component is not reachable from outside the integration where it has been instantiated, and evidently network threads and the resources they have associated are not allocated.

4.3 Incremental design

In future versions of CoolBOT, it is very possible that the set of DC3P protocol packets grow with new ones. In order to allow an easy integration of new DC3P packets in CoolBOT, we have applied the *composite* and *prototype* patterns [Gamma et al. (1995)] to their design. Those design patterns, jointly with the C++ templates and classes to support marshalling and demarshalling provide a systematic and easy manner of integrating new DC3P packets in future versions of the framework.

5. Deeper interface decoupling: Views and probes

Inspired by one of the "good practices" proposed by the authors of Carnegie Mellon's Navigation Toolkit CARMEN [Montemerlo et al. (2003)]: "one important design principle of CARMEN is the separation of robot control from graphical displays", we have introduced in CoolBOT the concept of *view* as an integrable, composite and reusable graphical interface available for CoolBOT system integrators and developers. Thus, CoolBOT *views* are graphical interfaces which, as software components, may be interconnected with any other component, view or probe in a CoolBOT system. In Fig. 7 is depicted the structure of a view in CoolBOT. As shown, CoolBOT views are also endowed with an external interface of input and output ports. Through this interface the view can communicate with other components, views and probes present in a given system. Identically to components, views are provided with the same network thread support which allows transparent and decoupled network communications through port connections. Internally, a view is a graphical interface, in fact, the current views already developed and operational which are available in CoolBOT have been implemented using the GTK graphical library [*The GTK+ Project* (2010)]. As shown in Fig. 6, C++ skeletons for views are generated using the `coolbot-c` compiler. The part which should be completed by the user is precisely the graphical implementation, which can be done using directly the GTK library or a GUI graphical programming software for designing window-based interfaces like Glade [*Glade - A User Interface Designer* (2010)].
As depicted in Fig. 7 a CoolBOT *probe* is provided with an external interface of input and output ports, and likewise component and views, as software components, this allows them to

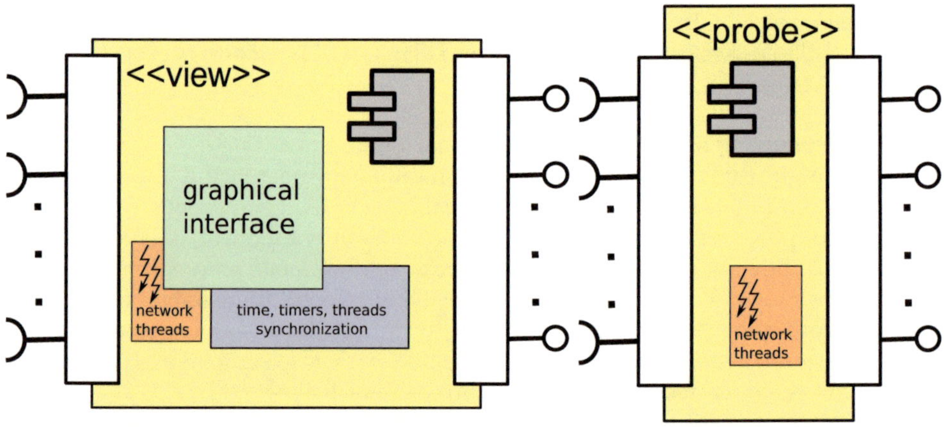

Fig. 7. CoolBOT. View and probe structures.

be interconnected with other components, views or probes. Equally they implement the same network decoupled support of threads for transparent network communications. In CoolBOT, *probes* are devised as interfaces for interoperability with non CoolBOT software, as illustrated graphically in Fig. 1. A complete functional C++ class implementing a probe is generated when a component is compiled by `coolbot-c`. The probe implements the complementary external interface of their corresponding component. Those probes generated automatically can be seen as automatic refactorings of external component interfaces in order to support interoperability of CoolBOT components with non CoolBOT software. As mentioned in [Makarenko et al. (2007)] this is an important factor in order to facilitate integration of different development robotic software approaches.

6. A real integration

In its current operating version, CoolBOT has been mainly used to control mobile robotic systems with the platforms we have available at our laboratory: Pioneer mobile robots models 3-DX, and P3-AT from Adept Mobile Robots [*Adept Mobile Robots* (2011)] (former Activ Media Robotics).

In this section we will show next a real robotic system using one of our Pioneer 3-DX mobile robots, in order to give a glimpse of a real system controlled using CoolBOT. The system is illustrated in Fig. 8. The example shows a secure navigation system for an indoor mobile robot. This is a real application we have usually in operation on the robots we have at the laboratory. The system implements a secure navigation system based on the ND+ algorithm for obstacle avoidance [Minguez et al. (2004)]. It has been implemented attending to [Montesano et al. (2006)]. In the figure, input ports, output ports, and port connections, have been reduced for the sake of clarification. Some of them represent several connections and ports in the real system.

The system is organized using two CoolBOT integrations, one only formed by CoolBOT component instances, and the other one containing CoolBOT view instances. The former one is really the integration which implements the navigation system. As we can observe, it consists of five component instances, namely: `PlayerRobot` (this is a wrapper component for hardware abstraction using the Player/Stage project framework [Vaughan et al. (2003)]), `MbICP` (this is a component which implements the MbICP scan matching algorithm based on laser range sensor data [Minguez et al. (2006)]), `GridMap` (this component maintains a grid map of the surroundings of the robot built using robot range laser scans, it also generates

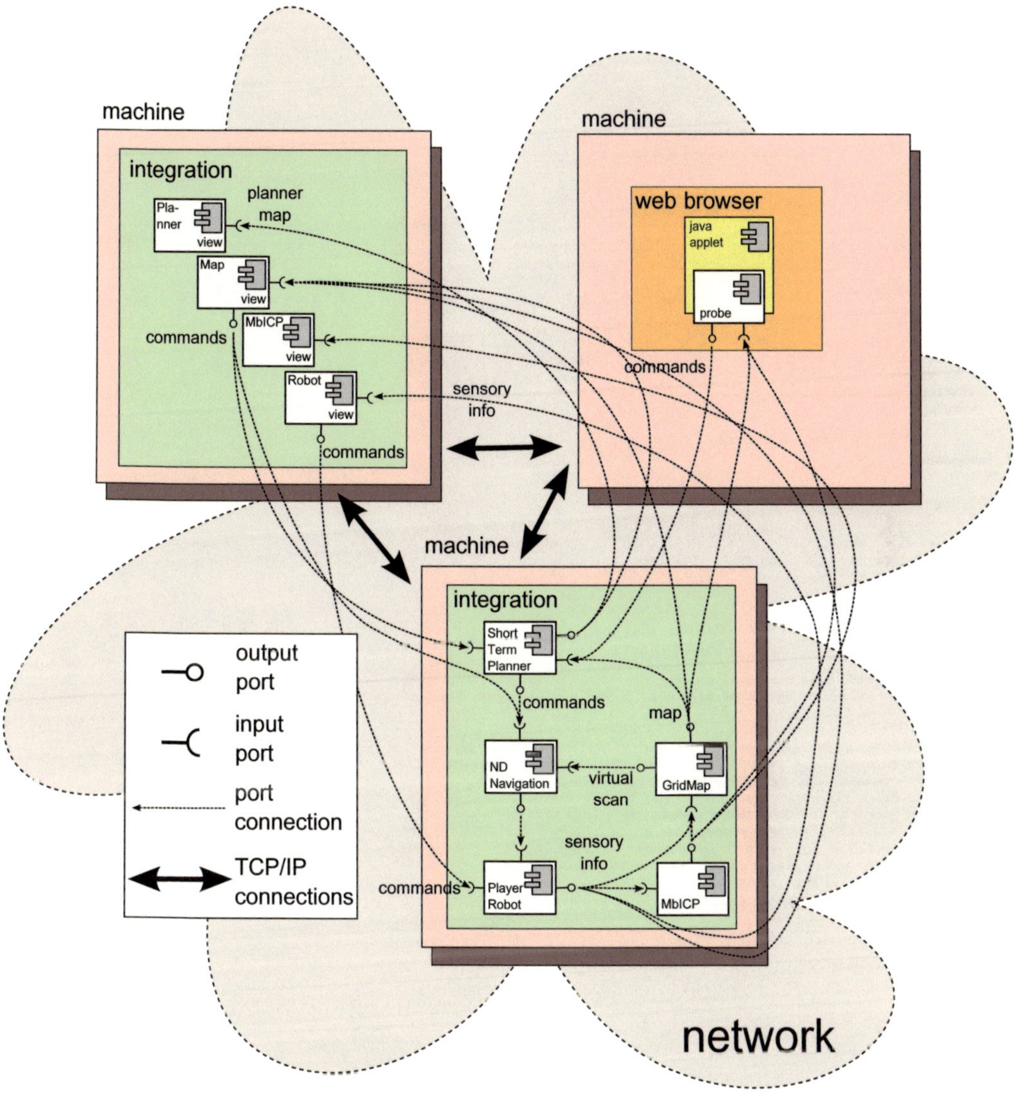

Fig. 8. CoolBOT. Secure Navigation System.

periodically a 360° virtual scan for the ND+ algorithm), `NDNavigation` (implements the ND+ algorithm) and `ShortTermPlanner` (a planner which uses the grid map for planning paths in the robot surroundings using a modification of the numerical navigation function NF2 found in [Jean-Claude Latombe (1991)]). On other machine another integration is shown hosting four view instances through which we can control and monitor the system remotely. In addition, in another machine, there is a web browser hosting a Java applet using a CoolBOT *probe* to connect to some of the components of the system.

In order to clarify how the integration of Fig. 8 has been built, and also to clarify the process of development of each of its components, in next paragraphs we will have a look at the description files used to generate some of them, including the whole integration shown in the figure. Thus, in Fig. 9 we can see the description file accepted by `coolbot-c` of one of the

```
/*
 * File: player-robot.coolbot
 * Description: description file for PlayerRobot component
 * Date: 02 June 2010
 * Generated by coolbot-ske
 */

component PlayerRobot
{
  header
  {
    author "Antonio Carlos Domínguez Brito <adominguez@iusiani.ulpgc.es>";
    description "PlayerRobot component";
    institution "IUSIANI - Universidad de Las Palmas de Gran Canaria";
    version "0.1"
  };

  constants
  {
    LASER_MAX_RANGE="LaserPacket::LASER_MAX_RANGE";
    SONAR_MAX_RANGE=5000; // millimeters

    private FIFO_LENGTH=5;
    private ROBOT_DATA_INCOMING_FREQUENCY= 10; // Hz
    private LASER_MIN_ANGLE= -90; // degrees

    ...
  };

  // input ports
  input port Commands type fifo port packet CommandPacket length FIFO_LENGTH;
  input port NavigationCommands type fifo port packet NDNavigation::CommandPacket length FIFO_LENGTH;

  //output ports
  output port RobotConfig type poster port packet ConfigPacket;
  output port Odometry type generic port packet OdometryPacket network buffer FIFO_LENGTH;
  output port OdometryReset type generic port packet PacketTime;
  output port BumpersGeometry type poster port packet "BumperGeometryPacket";
  output port Bumpers type generic port packet BumperPacket;
  output port SonarGeometry type poster port packet "SonarGeometryPacket";
  output port SonarScan type generic port packet "SonarPacket";
  output port LaserGeometry type poster port packet PacketFrame3D;
  output port LaserScan type generic port packet LaserPacket;
  output port Power type generic port packet PacketDouble;
  output port CameraImage type poster port packet CameraImagePacket;
  output port PTZJoints type generic port packet "PacketPTZJoints";

  exception RobotConnection
  {
    description "Robot connection failed.";
  };

  exception NoPositionProxy
  {
    description "Position proxy not available in this robot.";
  };

  exception InternalPlayerException
  {
    description "A Player library exception has been thrown.";
  };

  entry state Main
  {
    transition on Commands,NavigationCommands,Timer;
  };
};
```

Fig. 9. `player-robot.coolbot-component`: `PlayerRobot`'s description file.

components of Fig. 8, file `player-robot.coolbot-component`, corresponding to component `PlayerRobot`. As to views, in Fig. 10, we can see the description file for one of the view instances of Fig. 8, concretely for the `Map` view in the figure, which is an instance of view `GridGtk`. As we can observe, the description file specifies mainly the view's external interface formed by input and output ports.

```
/*
 * File: grid-gtk.coolbot-view
 * Description: description file for GridGtk view
 * Date: 29 April 2011
 * Generated by coolbot-ske
 */

view GridGtk
{
  header
  {
    author "Antonio Carlos Domínguez-Brito <adominguez@iusiani.ulpgc.es>";
    description "GridGtk View";
    institution "IUSIANI - ULPGC (Spain)";
    version "0.1"
  };

  constants
  {
    private DEFAULT_REFRESHING_PERIOD=500; // milliseconds
    ...
  };

  // input ports
  input port ROBOT_CONFIG type poster port packet PlayerRobot::ConfigPacket;
  input port GRID_MAP type poster port packet GridMap::GridMapPacket;
  input port PLANNER_PATH type last port packet ShortTermPlanner::PlannerPathPacket;

  // output ports
  output port PLANNER_COMMANDS type generic port packet ShortTermPlanner::CommandPacket;
  output port ND_COMMANDS type generic port packet NDNavigation::CommandPacket;
};
```

Fig. 10. `grid-gtk.coolbot-view`: `GridGtk`'s description file.

In Fig. 11 it is shown a snapshot of the view in runtime. Once developed CoolBOT views are graphical components we can integrate in a window-based application like the one shown in the figure. In particular, in the application we can see, views are plugged in as "pages" of the GTK widget `notepad` (a container of "pages" with "tabs") we can observe in the figure. In fact, the GUI application shown integrates the four view instances of Fig. 8, whose C++ skeleton has been generated using also `coolbot-c` from a description file (the `.coolbot-integration` file in Fig. 6). In Fig. 12 we can see part of this file. Notice that `coolbot-c` generates C++ skeletons for integrations where the static instantiation and interconnection among components and views are generated automatically. If we want to build a dynamic integration, in terms of dynamic instantiation of components and views, and also in terms of dynamic interconnections among them establishing port connections, we must complete the dynamic part of the skeleton generated by `coolbot-c` using the C++ runtime services provided by the framework (Fig. 5).
With respect to CoolBOT probes, as of now we have used them to interoperate with Java applets inserted in a web browser, as shown in Fig. 8. More specifically we have used SWIG [*SWIG* (2011)] in order to have access to the probe C++ classes in Java with the aim of implementing several Java GUI interfaces in Java equivalent to some of the CoolBOT views we have already developed. In Fig. 13 we can see a snapshot of the Java equivalent of a view to represent the range sensor information of a mobile robot.

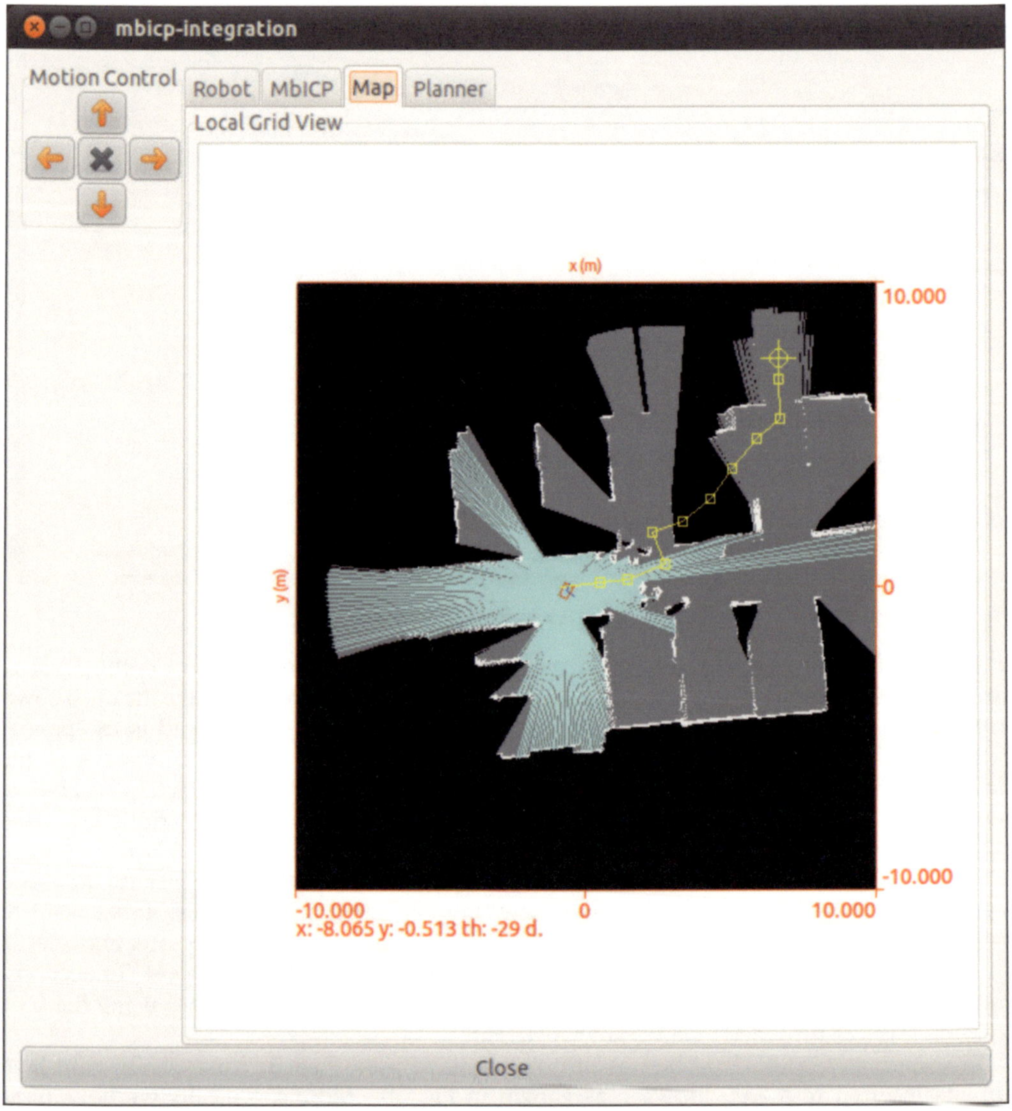

Fig. 11. View `GridGtk`'s snapshot.

```
/*
 * File: mbicp-integration.coolbot-integration
 * Description: description file for mbicp-integration integration.
 * Date: 29 April 2011
 * Generated by coolbot-ske
 */

integration mbicp_integration
{
  header
  {
    author "Antonio Carlos Domínguez-Brito <adominguez@iusiani.ulpgc.es>";
    description "MbICP's views integration";
    institution "IUSIANI - ULPGC (Spain)";
    version "0.1"
  };

  machine addresses
  {
    local my_machine: "127.0.0.1";
    the_other_machine: "...";
  };

  listening ports // TCP/IP ports
  {
    ROBOT_PORT: 1950;
    MBICP_PORT: 1965;
    NAVIGATION_MAP_PORT: 1970;
    ND_PORT: 1980;
    NAVIGATION_PLANNER_PORT: 1990;

    ROBOT_VIEW_PORT: 1955;
    MBICP_VIEW_PORT: 1985;
    NAVIGATION_MAP_VIEW_PORT: 1975;
    NAVIGATION_PLANNER_VIEW_PORT: 1995;
  };

  local instances
  {
    view robotView:PlayerRobotGtk listening on ROBOT_VIEW_PORT with description "Robot";
    view mbicpView:MbICPGtk listening on MBICP_VIEW_PORT with description "MbICP";
    view mapView:GridGtk listening on NAVIGATION_MAP_VIEW_PORT with description "Map";
    view navigationPlannerView:PlannerGtk listening on NAVIGATION_PLANNER_VIEW_PORT with description "Planner";
  };

  remote instances on the_other_machine;
  {
    component robot:PlayerRobot listening on ROBOT_VIEW_PORT;
    component mbicpInstance:MbICPCorrector listening on MBICP_PORT;
    component navigationMap:GridMap listening on NAVIGATION_MAP_PORT;
    component nd:NDNavigation listening on ND_PORT;
    component navigationPlanner:ShortTermPlanner listening on NAVIGATION_PLANNER_VIEW_PORT;
  };

  port connections // static connections
  {
    connect robot:ODOMETRY to robotView:ODOMETRY;
    connect robot:LASERGEOMETRY to robotView:LASER_GEOMETRY;
    connect robot:LASERSCAN to robotView:LASER_SCAN;
    connect robot:POWER to robotView:POWER;
    connect robot:SONARGEOMETRY to robotView:SONAR_GEOMETRY;
    connect robot:SONARSCAN to robotView:SONAR_SCAN;

    connect robot:ODOMETRY to mbicpInstance:ODOMETRY;
    connect robot:LASERGEOMETRY to mbicpInstance:LASER_GEOMETRY;
    connect robot:LASERSCAN to mbicpInstance:LASER_SCAN;

    connect robotView:COMMANDS to robot:COMMANDS;

    ...
  };
};
```

Fig. 12. The integration containing `Robot`. `MbICP`, `Map` and `Planner` views of Fig. 8

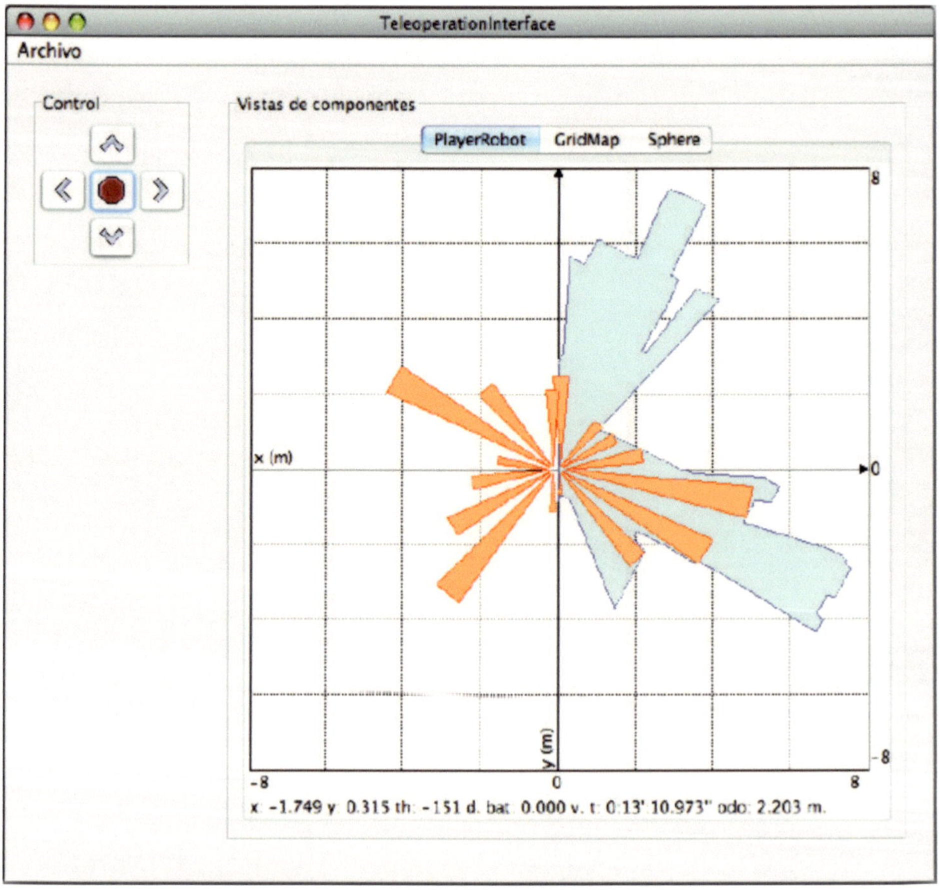

Fig. 13. Java view implemented using a probe to access range sensor information for a mobile robot. A snapshot.

7. Conclusions

In this document we have presented the last developments which have been integrated in the last operating version of CoolBOT. The developments have been aimed mainly to two questions: transparent distributed computation, and "deeper" interface decoupling. It is our opinion that the use and operation of CoolBOT has improved considerably. CoolBOT is an open source initiative supported by our laboratory which is freely available via **www.coolbotproject.org**, including the secure navigation system depicted in Fig. 8.

8. References

Adept Mobile Robots (2011). http://www.mobilerobots.com/.

Ando, N., Suehiro, T. & Kotoku, T. (2008). A Software Platform for Component Based RT-System Development: OpenRTM-Aist, *in* S. Carpin, I. Noda, E. Pagello, M. Reggiani & O. von Stryk (eds), *Simulation, Modeling, and Programming for Autonomous Robots*, Vol. 5325 of *Lecture Notes in Computer Science*, Springer Berlin / Heidelberg, pp. 87–98.

Antonio C. Domínguez-Brito, Daniel Hernández-Sosa, José Isern-González & Jorge Cabrera-Gámez (2007). *Software Engineering for Experimental Robotics*, Vol. 30 of *Springer Tracts in Advanced Robotics Series*, Springer, chapter CoolBOT: a Component Model and Software Infrastructure for Robotics, pp. 143–168.

Basu, A., Bozga, M. & Sifakis, J. (2006). Modeling heterogeneous real-time components in BIP, In Fourth IEEE International Conference on Software Engineering and Formal Methods, pages 3-12, Pune (India).

Bensalem, S., Gallien, M., Ingrand, F., Kahloul, I. & Thanh-Hung, N. (2009). Designing autonomous robots, *IEEE Robotics and Automation Magazine* 16(1): 67–77.

Brooks, A., Kaupp, T., Makarenko, A., S.Williams & Oreback, A. (2005). Towards component-based robotics, *In IEEE International Conference on Intelligent Robots and Systems*, Tsukuba, Japan, pp. 163–168.

Brugali, D. (ed.) (2007). *Software Engineering for Experimental Robotics*, Springer Tracts in Advanced Robotics, Springer.

Brugali, D. & Scandurra, P. (2009). Component-based robotic engineering (part i) [tutorial], *Robotics Automation Magazine, IEEE* 16(4): 84 –96.

Brugali, D. & Shakhimardanov, A. (2010). Component-based robotic engineering (part ii), *Robotics Automation Magazine, IEEE* 17(1): 100 –112.

Domínguez-Brito, A. C., Hernández-Sosa, D., Isern-González, J. & Cabrera-Gámez, J. (2004). Integrating Robotics Software, IEEE International Conference on Robotics and Automation, New Orleans, USA.

Douglas C. Schmidt (2010). The Adaptive Communication Environment (ACE), www.cs.wustl.edu/~schmidt/ACE.html.

Ellis, C. & Gibbs, S. (1989). *Object-Oriented Concepts, Databases, and Applications*, ACM Press, Addison-Wesley, chapter Active Objects: Realities and Possibilities.

Gamma, E., Helm, R., Johnson, R. & Vlissides, J. (1995). *Design Patterns: Elements of Reusable Object-Oriented Software*, Addison-Wesley Professional Computing Series, Addison-Wesley.

George T. Heineman & William T. Councill (2001). *Component-Based Software Engineering*, Addison-Wesley.

Glade - A User Interface Designer (2010). glade.gnome.org.

J. Paul Morrison (2010). *Flow-Based Programming, 2nd Edition: A New Approach to Application Development*, CreateSpace.

Jean-Claude Latombe (1991). *Robot-motion planning*, The Kluwer International Series in Engineering and Computer Science, Kluwer Academic.

Kitware, Inc. (2010). The CMake Open Source Build System, www.cmake.org.

Makarenko, A., Brooks, A. & Kaupp, T. (2007). On the benefits of making robotic software frameworks thin, IEEE/RSJ Int. Conf. on Intelligent Robots and Systems (IROS'07), San Diego CA, USA.

Mallet, A., Pasteur, C., Herrb, M., Lemaignan, S. & Ingrand, F. (2010). GenoM3: Building middleware-independent robotic components, IEEE International Conference on Robotics and Automation.

Minguez, J., Montesano, L. & Lamiraux, F. (2006). Metric-based iterative closest point scan matching for sensor displacement estimation, *Robotics, IEEE Transactions on* 22(5): 1047 –1054.

Minguez, J., Osuna, J. & Montano, L. (2004). A "Divide and Conquer" Strategy based on Situations to achieve Reactive Collision Avoidance in Troublesome Scenarios, IEEE International Conference on Robotics and Automation, New Orleans, USA.

Montemerlo, M., Roy, N. & Thrun, S. (2003). Perspectives on standardization in mobile robot programming: the carnegie mellon navigation (carmen) toolkit, Vol. 3, pp. 2436 – 2441 vol.3.

Montesano, L., Minguez, J. & Montano, L. (2006). Lessons Learned in Integration for Sensor-Based Robot Navigation Systems, *International Journal of Advanced Robotic Systems* 3(1): 85–91.

Object Management Group (2002a). OMG IDL: Details, (http://www.omg.org/gettingstarted/omg_ idl.htm).

Object Management Group (2002b). The Common Object Request Broker: Architecture and Specification, Ch. 15, Sec. 1-3. (http://www.omg.org/cgi-bin/doc?formal/02-06-01).

ROS: Robot Operating System (2011). http://www.ros.org.

Schlegel, C., Haßler, T., Lotz, A. & Steck, A. (2009). Robotic Software Systems: From Code-Driven to Model-Driven Designs, In Proc. 14th Int. Conf. on Advanced Robotics (ICAR), Munich.

Steenstrup, M., Arbib, M. A. & Manes, E. G. (1983). Port automata and the algebra of concurrent processes, *Journal of Computer and System Sciences* 27: 29–50.

Stewart, D. B., Volpe, R. A. & Khosla, P. (1997). Design of Dynamically Reconfigurable Real-Time Software Using Port-Based Objects, *IEEE Transactions on Software Engineering* 23(12): 759–776.

SWIG (2011). http://www.swig.org/.

The GTK+ Project (2010). www.gtk.org.

The Orocos Project (2011). http://www.orocos.org.

Vaughan, R. T., Gerkey, B. & Howard, A. (2003). On Device Abstractions For Portable, Reusable Robot Code, *IEEE/RSJ International Conference on Intelligent Robot Systems (IROS 2003), Las Vegas, USA, October 2003*, pp. 2121–2427.

Sequential and Simultaneous Algorithms to Solve the Collision-Free Trajectory Planning Problem for Industrial Robots – Impact of Interpolation Functions and the Characteristics of the Actuators on Robot Performance

Francisco J. Rubio, Francisco J. Valero, Antonio J. Besa and Ana M. Pedrosa
Centro de Investigación de Tecnología de Vehículos, Universitat Politècnica de València
Spain

1. Introduction

Trajectory planning for robots is a very important issue in those industrial activities which have been automated. The introduction of robots into industry seeks to upgrade not only the standards of quality but also productivity as the working time is increased and the useless or wasted time is reduced. Therefore, trajectory planning has an important role to play in achieving these objectives (the motion of robot arms will have an influence on the work done).

Formally, the trajectory planning problem aims to find the force inputs (control $u(t)$) to move the actuators so that the robot follows a trajectory $q(t)$ that enables it to go from the initial configuration to the final one while avoiding obstacles. This is also known as the complete motion planning problem compared with the path planning problem in which the temporal evolution of motion is neglected.

An important part of obtaining an efficient trajectory plan lies with both the interpolation function used to help obtain the trajectory and the robot actuators. Ultimately actuators will generate the robot motion, and it is very important for robot behavior to be smooth. Therefore, the trajectory planning algorithms should take into account the characteristics of the actuators without forgetting the interpolation functions which also have an impact on the resulting motion. As well as smooth robot motion, it is also necessary to monitor some working parameters to verify the efficiency of the process, because most of the time the user seeks to optimize certain objective functions. Among the most important working parameters and variables are the time required to get the trajectory done, the input torques, the energy consumed and the power transmitted. The kinematic properties of the robot´s links, such as the velocities, accelerations and jerks are also important.

The trajectory algorithm should also not overlook the presence of possible obstacles in the workspace. Therefore it is very important to model both the workspace and the obstacles efficiently. The quality of the collision avoidance procedure will depend on this modelization.

2. A brief look at previous work

Trajectory planning for industrial robots is a very important topic in the field of robotics and has attracted a great number of researchers so that there are at the moment a variety of methodologies for its resolution.
By studying the work done by other researchers on this topic it is easy to deduce that the problem has mainly been tackled with two different approaches: direct and indirect methods. Some authors who have analyzed this topic using indirect methods are Saramago, 2001; Valero et al., 2006; Gasparetto and Zanotto, 2007 ; du Plessis et al., 2003.
Other authors, on the other hand, have implemented the direct method such as Chettibi et al, 2002; Macfarlane, 2003; Abdel-Malek et al. 2006. However, in these examples the obstacles have been neglected which is a drawback.
Over the years, the algorithms have been improved and the study of the robotic system has become more and more realistic. One way of achieving that is to analyze the complete behavior of the robotic system, which in turn leads us to optimize some of the working parameters mentioned earlier by means of the appropriate objective functions. The most widely used optimization criteria can be classified as follows:

1. Minimum time required, which is bounded to productivity.
2. Minimum jerk, which is bounded to the quality of work, accuracy and equipment maintenance.
3. Minimum energy consumed or minimum actuator effort, both linked to savings.
4. Hybrid criteria, e.g. minimum time and energy.

The early algorithms that solved the trajectory planning problem tried to minimize the time needed for performing the task (see Bobrow et al., 1985; Shin et al., 1985; Chen et al., 1989). In those studies, the authors impose smooth trajectories to be followed, such as spline functions.
Another way of tackling the trajectory planning problem was based on searching for jerk-optimal trajectories. Jerks are essential for working with precision and without vibrations. They also affects the control system and the wearing of joints and bars. Jerk constraints were introduced by Kyriakopoulos (see Kyriakopoulos et al.,1988). Later, Constantinescou introduces (Constantinescu et all, 2000) a method for determining smooth and time optimal path-constrained trajectories for robotic manipulators imposing limits on the actuator jerks.
Another different approach to solving the trajectory planning problem is based on minimizing the torque and the energy consumed instead of the execution time or the jerk. An early example is seen in Garg et al., 1992. Similarly, Hirakawa and Kawamura searched for the minimum energy consumed (Hirakawa et al., 1996). In Field and Stepanenko, 1996, the authors plan minimum energy consumption trajectories for robotic manipulators. In Saramago and Steffen, 2000, the authors considered not only the minimum time but also the minimum mechanical energy of the actuators. They built a multi-objective function and the results obtained depended on the associated weighting factor. The subject of energy minimization continues to be of interest in the field of robotics and automated manufacturing processes (Cho et al., 2006).
Later, new approaches appear for solving the trajectory planning problem. The idea of using a weighted objective function to optimize the operating parameters of the robot arises (Chettibi et al., 2004). Gasparetto and Zanotto also use a weighted objective function (see Gasparetto and Zanotto, 2010). In this chapter we will introduce an indirect method which has been called the "sequential algorithm".

In this chapter we will describe two algorithms for solving the collision-free trajectory planning for industrial robots that we have developed. We have called them "sequential" and "simultaneous" algorithms. The first is an indirect method while the second is a direct one. The "sequential" algorithm considers the main properties of the actuators (torque, power, jerk and energy consumed). The "simultaneous" algorithm analyzes what is the best interpolation function to be used to generate the trajectory considering a simple actuator (only the torque required). The chapter content is based on the previous work done by the authors (see Valero et al., 2006, and Rubio et al., 2007). Specifically, the two approaches to solving the trajectory planning problem are explained.

3. Robot modelling

The robot model used henceforth is the wire model corresponding to the PUMA 560 robot shown in Fig. 1. The robot involves rigid links that are joined by the corresponding kinematic joints (revolution).The robot has F degrees of freedom and each robot´s configuration C^j can be unequivocally set using the Cartesian coordinates of N points, which are called significant points. These points, defined as α^j_i ($x^j_{3(i-1)+1}$,$x^j_{3(i-1)+2}$, $x^j_{3(i-1)+3}$, i=1..F, j=number of configuration) are chosen systematically. Therefore, ultimately, every configuration will be expressed in Cartesian coordinates by means of the significant points, i.e. $C^j = C^j(\alpha^j_i)$, which represent the specifics of the robot under study. It is important to point out that they do not constitute an independent set of coordinates. Besides the significant points, some other points p^j_k, called interesting points, will be used to improve the efficiency of the algorithms, the coordinates of which are obtained from the significant points and the geometric characteristics of the robot.

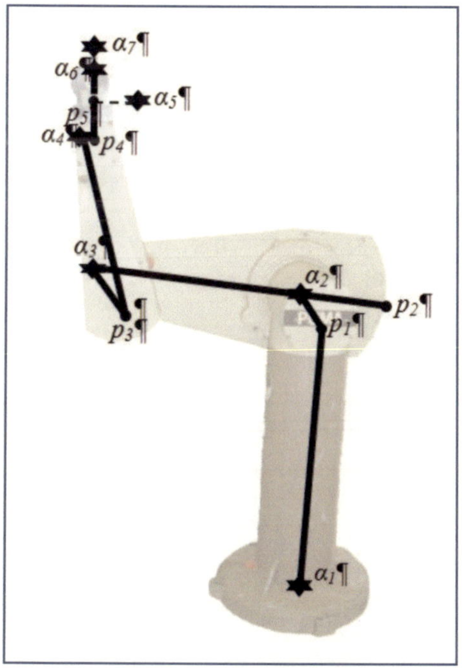

Fig. 1. Model of robot PUMA 560. Significant and Interesting Points (mobile base)

The PUMA 560 robot can be modelled with a movable base or a fixed base. The mobile-based robot is shown in Fig. 1 with the seven significant points used (α_1, α_2, α_3, α_4, α_5, α_6 and α_7), together with another five interesting points $p^j{}_k$. As a result of this, the configuration C^j is determined by twenty-one variables corresponding to the coordinates of the significant points. These variables are connected through fourteen constraint equations relative to the geometric characteristics of the robot (length of links, geometric constraints and range of motion). See Rubio et al, 2009 for more details. It must be noted that any other industrial robot can be modelled in this way by just selecting and choosing appropriately those significant points that best describe it.
This property is very important as far as the effectiveness of the algorithm is concerned.

4. Workspace modelling

The workspace is modelled as a rectangular prism with its edges parallel to the axes of the Cartesian reference system. The work environment is defined by the obstacles bound to produce collisions when the robot moves within the workspace. The obstacles are considered static, i.e. their positions do not vary over time and they are represented by means of unions of patterned obstacles.
The fact of working with patterned obstacles introduces two fundamental advantages:

1. It allows the modelling of any generic obstacle so that collisions with the robot´s links can be avoided.
2. It permits working with a reduced number of patterned obstacles in order to model a complex geometric environment so that its use is efficient. It means that a small number of constraints are introduced into the optimization problem when obtaining collision-free adjacent configurations.

The patterned obstacles have a geometry based on simple three-dimensional figures, particularly spheres, rectangular prisms and cylinders. Any obstacle present in the workspace could be represented as a combination of these geometric figures.
The definition of a patterned obstacle is made in the following way:

- Spherical obstacle SO_i, is defined when the position of its centre and its radius are known. It is characterized by means of
- Centre of Sphere *i*: $c_i^{SO} = \left(c_{xi}^{SO}, c_{yi}^{SO}, c_{zi}^{SO} \right)$
- Radius of Sphere *i*: r_i^{SO}

Therefore $SO_i = \left(c_i^{SO}, r_i^{SO} \right)$. See Fig. 2

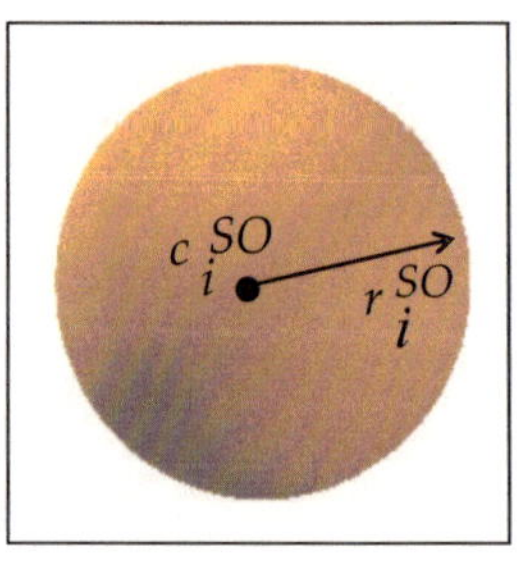

Fig. 2. Generic Spherical obstacle SO_i

- Cylindrical obstacle CO_k, is defined when the coordinates of the centres of its bases and its radius are known. It is characterized by means of
- Centre of base 1 for cylinder k: $c_{1k}^{CO} = \left(c_{1xk}^{CO}, c_{1yk}^{CO}, c_{1zk}^{CO}\right)$
- Centre of base 2 for cylinder k: $c_{2k}^{CO} = \left(c_{2xk}^{CO}, c_{2yk}^{CO}, c_{2zk}^{CO}\right)$
- Radius of cylinder k: r_k^{CO}

Therefore $CO_k = \left(c_{1k}^{CO}, c_{2k}^{CO}, r_k^{CO}\right)$. See Fig. 3

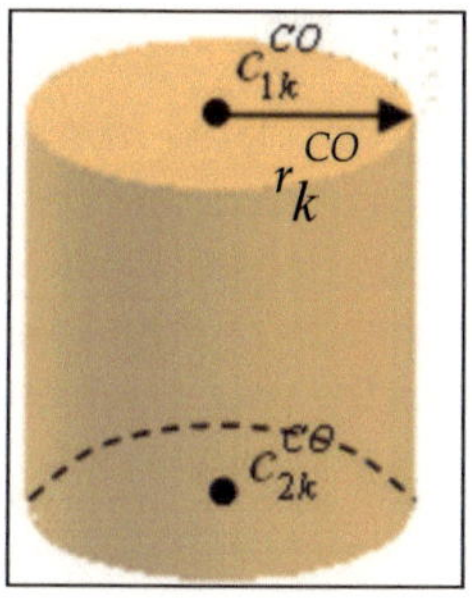

Fig. 3. Generic Cylindrical obstacle CO_k

- Prismatic obstacle PO_l, is defined when four points located in the vertices of the rectangular prism are known so that vectors that are perpendicular to each other can be drawn up. It is characterized by means of
- Point a of prism l: $a_l^{PO} = \left(a_{xl}^{PO}, a_{yl}^{PO}, a_{zl}^{PO}\right)$
- Point q_1 of prism l: $q_{1l}^{PO} = \left(q_{1xl}^{PO}, q_{1yl}^{PO}, q_{1zl}^{PO}\right)$
- Point q_2 of prism l: $q_{2l}^{PO} = \left(q_{2xl}^{PO}, q_{2yl}^{PO}, q_{2zl}^{PO}\right)$
- Point q_3 of prism l: $q_{3l}^{PO} = \left(q_{3xl}^{PO}, q_{3yl}^{PO}, q_{3zl}^{PO}\right)$

Therefore $PO_l = \left(a_l^{PO}, q_{1l}^{PO}, q_{2l}^{PO}, q_{3l}^{PO}\right)$. See Fig. 4

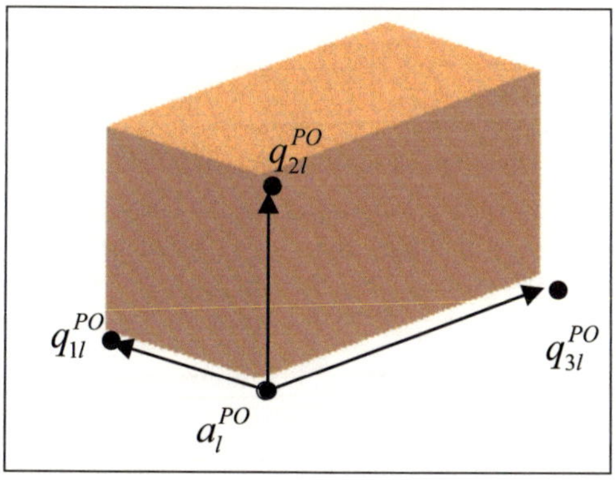

Fig. 4. Generic Prismatic obstacle PO_l

5. Discretizing the workspace

With the purpose of working with a limited number of configurations, the generation of a discrete workspace that represents the possible positions of the end-effector of the robot is considered. To do this, a rectangular prism with its edges parallel to the axes of the

Cartesian reference system is created and whose opposite vertices correspond to the positions of the end-effector of the robot in the initial and final configurations from which the connecting path is calculated. The set of positions that the end-effector of the robot can adopt within the prism is restricted to a finite number of points resulting from the discretization of the prism according to the following increases:

$$\Delta x = \frac{\left|\alpha_{nx}^{f} - \alpha_{nx}^{i}\right|}{N_x} \quad \Delta y = \frac{\left|\alpha_{ny}^{f} - \alpha_{ny}^{i}\right|}{N_y} \quad \Delta z = \frac{\left|\alpha_{nz}^{f} - \alpha_{nz}^{i}\right|}{N_z} \tag{1}$$

Where the values of Δx , Δy and Δz are calculated from the values of the number of intervals N_x , N_y and N_z in which the prism is discretized, and those increments should be smaller than the smallest dimension of the obstacle modelled in the workspace. Points ($\alpha_{nx}{}^{f}$, $\alpha_{ny}{}^{f}$, $\alpha_{nx}{}^{f}$) and ($\alpha_{nx}{}^{i}$, $\alpha_{ny}{}^{i}$, $\alpha_{nz}{}^{i}$) correspond to the coordinates of the end-effector of the robot for the initial and final configurations. Fig. 6 demonstrates the way in which the prism that gives rise to the set of nodes that the end-effector of the PUMA 560 robot with a mobile base can adopt is discretized.

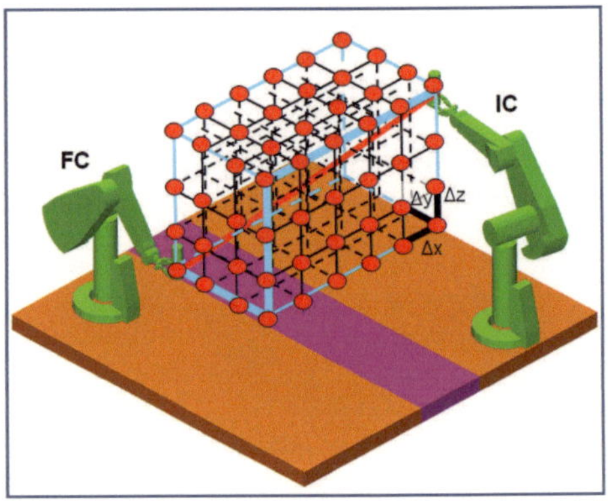

Fig. 5. Rectangular prism with edges parallel to the axes of the Cartesian reference system

6. Obstacle avoidance

By controlling the distance from the different patterned obstacles to the cylinders that cover the robot links, collision avoidance between the robot and the obstacles is possible. Distances are constraints in the optimization problem. They serve to calculate collision-free adjacent configurations (for adjacent configuration see Section 7).

6.1 Calculation of distances

Each robot´s link is modelled as a cylinder and it is characterized as $RC_i = \left(c_{1i}^{RC}, c_{2i}^{RC}, r_i^{RC}\right)$ (see section 4). The application of the procedure to calculate distances between link i of the robot and the patterned obstacle j (which may be a cylinder, sphere or a prism), can give rise to three different cases to prevent collisions:

A) Cylinder-Sphere

See Fig. 6. Here we compute the distance between a line segment (cylinder) to a point (centre of the sphere). Let AB be a line segment specified by the endpoints A and B. Given an arbitrary point C, the problem is to determine the point P on AB closest to C. Then we calculate the distance between these two points.

Projecting *C* onto the extended line through *AB* provides the solution. If the projection point *P* lies within the segment, then *P* itself is the correct answer.
If *P* lies outside the segment, then the segment endpoint closest to C is instead the closest point (A or B).

B) Cylinder-Cylinder

See Fig. 6. Here we compute the distance between two line segments. The problem of determining the closest points between two line segments S_1 (P_1Q_1) and S_2 (P_2Q_2) (and therefore the distance) is more complicated than computing the closest points of the lines L_1 and L_2 of which the segments are a part. Only when the closest points of L_1 and L_2 happen to lie on the segments does the method for closest points between lines apply. For the case in which the closest points between L_1 and L_2 lie outside one or both segments, a common misconception is that it is sufficient to clamp the outside points to the nearest segment endpoint. It can be shown that if just one of the closest points between the lines is outside its corresponding segment, that point can be clamped to the appropriate endpoint of the segment and the point on the other segment closest to the endpoint is computed.
If both points are outside their respective segments, the same clamping procedure must be repeated twice.

C) Cylinder-Prism

See Fig. 6. The prismatic surfaces are divided into triangles. In this case we compute the distance between a line segment and a triangle. The closest pair of points between a line segment *PQ* and a triangle is not necessarily unique. When the line segment is parallel to the plane of the triangle, there may be an infinite number of equally close pairs. However, regardless of whether the segment is parallel to the plane or not, it is always possible to locate a point such that the minimum distance falls either between the end point of the segment and the interior of the triangle or between the segment and an edge of the triangle. Thus, the closest pair of points (and therefore the distance) can be found by computing the closest pairs of points between the following entities:

- Segment *PQ* and triangle edge *AB*.
- Segment *PQ* and triangle edge *BC*.
- Segment *PQ* and triangle edge *CA*.
- Segment endpoint *P* and plane of triangle (when *P* projects inside *ABC*)
- Segment endpoint *Q* and plane of triangle (when *Q* projects inside *ABC*).

The number of tests required to calculate the distance can be reduced in some cases.

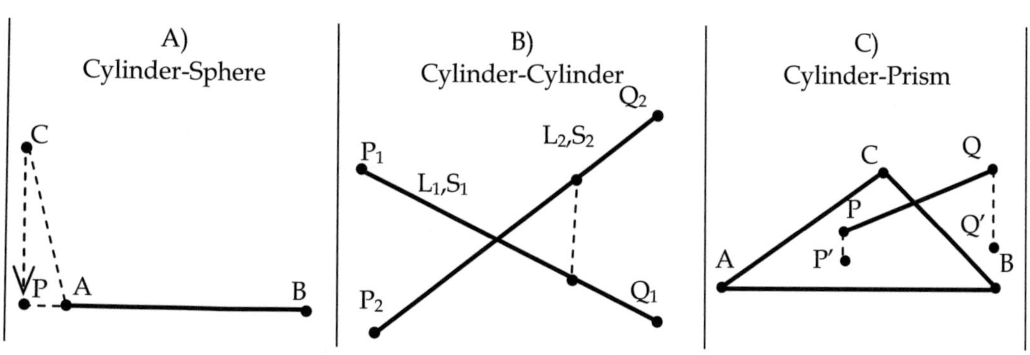

Fig. 6. Three different cases to calculate distances (and prevent collisions)

7. Obtaining adjacent configurations

The discrete configuration space is obtained by means of generating adjacent configurations. Given a feasible configuration C^k, it is said that a new configuration C^p is adjacent to the first if it is also feasible (i.e. it fulfils the characteristics associated to the robot modelling and avoids collisions with the obstacles), and in addition the following three properties are fulfilled:

1. The position of the end-effector that corresponds to a node of the discrete workspace is at a distance of one unit with respect to the position of the end-effector of configuration C^k. That means that at least one of the following conditions has to be fulfilled:

$$\left|\alpha_{nx}^{k}-\alpha_{nx}^{p}\right|=\Delta x \qquad \left|\alpha_{ny}^{k}-\alpha_{ny}^{p}\right|=\Delta y \qquad \left|\alpha_{nz}^{k}-\alpha_{nz}^{p}\right|=\Delta z \tag{2}$$

   n being the subscript corresponding to the significant point associated to the end-effector of the robot. For PUMA 560 with mobility in the base, n=7, as can be seen in Fig. 1.
   What we obtain is a sequence of configurations that is contained in the path, so that by using interpolation we can obtain a collision-free and continuous path.
2. Verification of the absence of obstacles between adjacent configurations C^k and C^p. Since the algorithm works in a discrete space it is necessary to verify that there are no obstacles between adjacent configurations, for which the following condition is set out:

$$\left|\overrightarrow{\alpha_i^k \alpha_i^p}\right| \le 2 \cdot min\left(r_j\right) \tag{3}$$

   where r_j is the characteristic dimension of the smallest patterned obstacle. This condition is necessary to guarantee that the distance for each link between two adjacent configurations is less than the characteristic dimension of the smallest patterned obstacle.
3. Configuration C^p must be such that it minimizes the following objective function:

$$\left\|C^p - C^f\right\| = \sum_{i=1}^{n}\left(\left(\alpha_{ri}^{p}-\alpha_{xi}^{f}\right)^2+\left(\alpha_{yi}^{p}-\alpha_{yi}^{f}\right)^2+\left(\alpha_{zi}^{p}-\alpha_{zi}^{f}\right)^2\right) \tag{4}$$

   n being the number of significant points of the robot. This third property facilitates the final configuration to be reached even for redundant robots, i.e. the robot´s end-effector should not be part, at the final node, of a configuration different from the desired one. On the other hand, this property has an influence on the configurations generated, facilitating the configurations in the neighbourhoods at the end so that they are compatible with the end.

An optimization procedure is set by using a sequential quadratic programming method (SQP). This method serves to minimize a quadratic objective function subject to a set of constraints which might include those from a simple limit to the values of the variables, linear restrictions and nonlinear continuous constraints. This is an iterative method.

Applying this procedure to the path planning problem, the objective function used is given by Eq. (4). The constraints are associated to the geometry of the robot, the limits of the actuators and the avoidance of collision. And the configuration C^k is used as an initial estimation for its resolution. The solution of this optimization problem gives the adjacent configuration C^p looked for. By repeating the obtaining of adjacent configuration, the discrete configuration space of the robot is obtained. These configurations are recorded in a graph.

8. "Sequential" algorithm applied to solving the trajectory planning problem. Problem statement

8.1 Introduction

The "sequential" algorithm is based on an indirect approach to solving the trajectory planning problem. The algorithm takes into account the characteristics of the actuators (torque, power, jerk and consumed energy), the interpolation functions and the obstacles in the workspace. It generates the configuration space. Then, a graph is associated to the previously obtained configuration space, which allows a collision-free path to be obtained between the initial and final configurations. Once the path is available, the dynamic characteristics of the robot are included, setting an optimal trajectory planning problem which aims to obtain the minimum time trajectory that is compatible with the robot features and the actuator capabilities (torque, jerk and consumed energy constraints).

8.2 Obtaining a path

First, the algorithm solves the path planning problem, obtaining the discrete configuration space of the robot (the discrete configuration space is generated by means of adjacent configurations, see Section 7) and then the minimum distance path is calculated. This path (a sequence of m configurations) is obtained by associating a weighted graph to the discrete configuration space and looking for the minimum weighted path. In the graph, the nodes correspond to the robot configurations and the arcs are related to joint displacements between adjacent configurations.

The weight corresponding to the arc that goes from node k (C^k robot configuration) to node p (C^p robot configuration), can be given as:

$$a(k,p) = \sum_{i=1}^{3(F-1)} \left(x_i^p - x_i^k\right)^2 \tag{5}$$

when that C^k and C^p are adjacent. In addition C^k and C^p must satisfy both type (3) constraints that avoid the obstacles between configurations and the angle increased from C^k to C^p must be smaller than the magnitude of the forbidden zone for that joint, so that large displacements are avoided for movement between adjacent configurations.

In case the points above mentioned are not satisfied, then we consider that $a(k,p)=\infty$. Finally, the searching is started in the weighed graph with the path that joins the node corresponding to the initial configuration to the node corresponding to the final configuration. Since the arcs satisfy that $a(k,p)\geq 0$, the Dijkstra´s algorithm is used to obtain the path that minimises the distance between the initial and final configurations. If this path exists, it is easy to obtain a sequence of m robot configurations S.

8.3 Interpolation function

Once the path has been obtained (at this point, the algorithm uses Cartesian coordinates), we have a sequence of m robot configurations, $S = \{S_1(q_{i1}), S_2(q_{i2})... S_m(q_{im})\}$. These configurations are expressed now in joint coordinates. And the objective now is to look for a minimum time trajectory (t_{min}), that contains them. The path is decomposed into m-1 intervals, so the time needed to reach the S_{j+1} configuration from the initial S_1 is t_j, and the time spent in the segment j (between S_j and S_{j+1} configurations) will be t_j-t_{j-1}. Cubic interpolation functions have been used for joint trajectories. They are defined by means of joint variables between successive configurations, so that for the segment j it is:

$\forall t \in \left[t_{j-1}, t_j\right[\Rightarrow q_{ij} = a_{ij} + b_{ij}t + c_{ij}t^2 + d_{ij}t^3$ for i=1,…,dof (dof being the degrees of freedoom of the robot) and j=1,…,m-1. (m is the number of the robot configuration)

To ensure motion continuity between configurations, the following conditions associated to the given configurations are considered.

- Position: it gives a total of (*2dof (m-1)*) equations:

$$q_{ij}\left(t_{j-1}\right) = a_{ij} + b_{ij}t_{j-1} + c_{ij}t_{j-1}^2 + d_{ij}t_{j-1}^3 \tag{6}$$

$$q_{ij}\left(t_j\right) = a_{ij} + b_{ij}t_j + c_{ij}t_j^2 + d_{ij}t_j^3 \tag{7}$$

- Velocity: for each interval, the initial and final velocity is zero, the velocity condition gives place to (*2dof*) equations:

$$\dot{q}_{i1}\left(t_0\right) = 0 \tag{8}$$

$$\dot{q}_{im}\left(t_m\right) = 0 \tag{9}$$

When passing through each configuration, the final velocity of the previous configuration should be equal to the initial velocity of the next configuration, leading to (*dof (m-2)*) equations

$$\dot{q}_{ij}\left(t_j\right) = \dot{q}_{ij+1}\left(t_j\right) \tag{10}$$

- Acceleration: For each intermediate configuration, the final acceleration of the previous configuration should be equal to the initial acceleration of the next configuration, giving rise to (*dof*(*m*-2)) equations:

$$\ddot{q}_{ij}\left(t_j\right) = \ddot{q}_{ij+1}\left(t_j\right) \tag{11}$$

In addition, the minimum time trajectory must meet the following constraints:

- Maximum torque on the actuators,

$$\tau_i^{\min} \le \tau_i\left(t\right) \le \tau_i^{\max} \quad \forall t \in \left[0, t_{\min}\right], i = 1 \ldots dof \tag{12}$$

- Maximum power on the actuators,

$$P_i^{\min} \le P_i\left(t\right) \le P_i^{\max} \quad \forall t \in \left[0, t_{\min}\right], \mathrm{i} = 1 \ldots dof \tag{13}$$

- Maximum jerk on the actuators,

$$\dddot{q}_i^{\min} \le \dddot{q}_i\left(t\right) \le \dddot{q}_i^{\max} \quad \forall t \in \left[0, t_{\min}\right], \ \mathrm{i} = 1 \ldots dof \tag{14}$$

- Consumed Energy,

$$\sum_{j=1}^{m-1}\left(\sum_{i=1}^{dof} \varepsilon_{ij}\right) \le \mathrm{E} \ , \tag{15}$$

ε_{ij} being the energy consumed by the *i* actuator between configurations *j* and *j*+1

Given the large number of iterations required by the process, the technique used for obtaining the coefficients is crucial. The first task is to normalize the polynomials that define the stages (see Suñer et al., 2007).In short, the optimization problem is set by using incremental time variables in each interval, so that in the interval between S_j and S_{j+1}, the time variable should be $\Delta t_j = t_j - t_{j-1}$, and the objective function,

$$\sum_{j=1}^{m-1} \Delta t_j = t_{\min} \tag{16}$$

The solution is obtained by means of SQP procedures, so that at each iterative step it is necessary to obtain the above mentioned polynomials coefficients from the estimation of the variables of the problem.

8.4 Obtaining a trajectory

The trajectory is obtained when the optimization problem posed has been solved. The solution (and therefore the trajectory) is achieved by solving an optimization problem whose objective function is the trajectory total time and the constraints are the maximum torques in the robot actuators, maximum power, maximum jerk and the consumed energy. The solution of the optimization problem is approached by means of a SQP algorithm of Fortran mathematical library NAG. In each iterative step is necessary to obtain the coefficients of the previously mentioned polynomials from an estimation of the variables (t_j). Notice that the previous conditions above mentioned define a system of ($4N_{dof}(m-1)$) independent linear equations. Since the complete trajectory has ($4N_{dof}(m-1)$) unknowns corresponding to the coefficients of the polynomials, the linear system can be solved, obtaining the complete trajectory. And this linear system is solved in each iteration whithin the optimization problem. These coefficients are necessary to calculate the maximum torque, power, jerk and consumed energy for each one of the actuators by means of solving the inverse dynamic problem in each interval.

Finally, when the optimization problem has been solved we obtain the minimum time trayectory (subject to the mentioned constraints) and also all the kinematic properties of the robotic system.

8.5 Impact of interpolation function

The impact of the interpolation function is very important from the point of view of the robot´s performance. Polynomial interpolation functions have been used in the "sequential" algorithm. It has been noticed during the resolution of the examples that they extract the maximum dynamic capabilities of the robot´s actuators, so that the robot moves faster than if any other interpolation function is used (harmonic functions, etc). Therefore when the polynomial interpolation functions are used the algorithm gives the best results from the point of view of the time requiered to do the tasks.

8.6 Application and examples solved

Different examples have been solved for a PUMA 560 robot. The examples have been solved with sequences of different initial and final configurations. The trajectories calculated meet constraints on torque, power, jerk and energy consumed and the goal is to analyze impact of these constraints on the generation of minimum time collision-free trajectories for industrial robots. The results obtained show that constraints on the energy consumed must enable the manipulator to exceed the requirements associated with potential energy, as the algorithm works on the assumption that the energy can be dissipated but not recovered. Also, an increase in the severity of energy constraints results in longer time trajectories with more soft power requirements. When constraints are not very severe, efficient trajectories can be obtained without high penalties on the working time cycle. An increase in the severity of the jerk constraints involves longer time trajectories with more soft power requirements and lower energy consumed. When constraints are very severe, times are also severely penalized even the jerk might appear. To obtain competitive results in the balance between time cycle and energy consumed, the actuators should work with the maximum admissible value of the jerk so that the robot can work with the desired accuracy.

9. "Simultaneous" algorithm applied to solving the trajectory planning problem. Problem statement

9.1 Introduction

The "simultaneous" algorithm is based on a direct approach to solving the trajectory planning problem in which the path planning problem and the problem of determining the time history of motion are treated as one instead of treating them separately as the indirect methods do. The algorithm is called "simultaneous" because of the simultaneous generation of discrete configuration space and the minimum distance path, making use of the information that the objective function is generating when new configurations are obtained. The algorithm works on a discretized configuration space which is generated gradually as the direct procedure solution evolves. It uses Cartesian coordinates (to specify the motion of the end-effector) and joint coordinates (to solve the inverse dynamic problem). An important role is played by the generation of adjacent configurations using techniques described by Valero et al. ,2006 . The resolution of the inverse dynamic problem has been done using Gibbs-Appel´s equations, as proposed by Sebastian Provenzano (see Provenzano, 2001). Any obstacle can be modelled using simple obstacle patterns: sphere, cylinder and prism. This helps calculate distances and avoid collisions.
The algorithm takes into account the torque required by the actuators, analyses the best interpolation function and consider the obstacles in the workspace. To obtain a new adjacent configuration C^k , a first optimization problem has to be solved which can be stated as follows:

$$\text{Find } C^k\text{, minimizing } Min\left(\left\|C^k - C^f\right\|\right) = Min\left(\sqrt{\sum_{i=1}^{n}\left(x_i^k - x_i^f\right)}\right) \tag{17}$$

and subject to:

a. Geometrical constraints of the robot structure;
b. Constraints on the mobility of robot joints;
c. Collision avoidance within the robot workspace;

(where x_i^k and x_i^f are the Cartesian coordinates of intermediate and final configurations C^k and C^k respectively).
The process for calculating the whole trajectory between initial and final configurations (C^1 and C^f) is based on a second and different optimization problem which can be stated as:

$$\text{Find} \quad q(t), \tau(t), t_f \text{ between each two configurations} \tag{18}$$

$$\text{Minimizing } \min_{\tau\in\Omega} J = \int_0^{t_f} 1 \cdot dt \tag{19}$$

Subject to the robot dynamics

$$M(q(t))\ddot{q}(t) + C(q(t), \dot{q}(t))\dot{q}(t) + g(q(t)) = \tau(t) \tag{20}$$

Unknown boundary conditions for intermediate configurations a priori

$$\begin{aligned} q(t_{int-1}) &= q_{int-1} \ ; \ q(t_{int}) = q_{int} \\ \dot{q}(t_{int-1}) &= \dot{q}_{int-1} \ ; \ \dot{q}(t_{int}) = \dot{q}_{int-1} \\ \ddot{q}(t_{int-1}) &= \ddot{q}_{int-1} \ ; \ \ddot{q}(t_{int}) = \ddot{q}_{int-1} \end{aligned} \tag{21}$$

Boundary conditions for initial and final configuration (used to solve the first and final step)

$$\begin{aligned} q(0) &= q_o \ ; q(t_f) = q_f \\ \dot{q}(0) &= 0 \ ; \dot{q}(t_f) = 0 \end{aligned} \tag{22}$$

Actuator torque rate limits

$$\tau_{\min} \le \tau(t) \le \tau_{\max} \tag{23}$$

Collision avoidance within the robot workspace

$$d_{ij} \ge r_j + w_j \tag{24}$$

where d_{ij} is the distance from any obstacle pattern *j* (sphere, cylinder or prism) to link *i*; r_j is the characteristic radius of the obstacle pattern and w_j is the radius of the smallest cylinder that contains the link *i*.

As well, $q(t)\in R^n$ is the vector of joint positions (*n* being the number of degrees of freedom of the robot), $\tau(t)\in R^n$ is the vector of actuator torques, $M(q(t))\in R^{nxn}$ is the inertia matrix of the robot, $C(q(t),\dot{q}(t))\in R^{nxn}$ is a third-order tensor representing the coefficients of centrifugal and Coriolis forces and $G(q(t))\in R^n$ is the vector of gravity terms, and Ω is the space state in which the vector of actuator torques is feasible. Each time a new adjacent configuration C^k is generated, an uncertainty to be overcome lies in the fact that at this stage we do not know its kinematic characteristics (particularly velocity and acceleration), although we know they should be compatible with the dynamic characteristics of the robot.

It should also be noted when calculating the minimum time between two adjacent configurations that each step starts from a configuration with its kinematic properties known, obtaining the time and the kinematic properties at the end configuration, so that if the dynamic capabilities of the actuators had been exhausted ($\tau_i(t_{\text{int}}) \cong \tau_{\min}$ or $\tau_i(t_{\text{int}}) \cong \tau_{\max}$) due to the kinematic properties generated at the end configuration ($q(t_{\text{int}})$, $\dot{q}(t_{\text{int}})$ and $\ddot{q}(t_{\text{int}})$), it would have been impossible to observe constraints on the next generation step of

the trajectory $\tau_{min} \le \tau(t_{int+1}) \le \tau_{max}$. Finally, by connecting adjacent configurations, the whole trajectory is generated.
The process explained is applied repeatedly to generate adjacent configurations until reaching the final configuration. Finally, by connecting adjacent configurations, the whole trajectory is generated.

9.2 Interpolation function

It must be noticed that three types of trajectory spans should be distinguished because of their different boundary conditions: the initial (which contains the initial configuration C^1), the final (which contains the final configuration C^f), and the intermediate (which does not contain either the initial or the final configuration).
Each pair of adjacent configurations is interpolated using harmonic functions in order to limit the kinematic characteristics of goal configuration so that progression to the following step should be admissible without breaking the dynamic properties. In that way, it is not necessary to previously impose kinematic constraints onto the process. Now, starting from the initial configuration, the harmonic function leads to the knowledge of the kinematic characteristics of the configurations adjacent to it, and so on. And therefore the process of obtaining adjacent configurations can continue until reaching the end. It is true that the results are influenced by the use of different interpolation functions between adjacent configurations. We use harmonic functions because they are capable of limiting the maximum values of velocity and acceleration required for the actuators. So, values for velocities and accelerations are limited. This important trait is deduced from the properties of Fourier series because of the harmonic functions used as interpolation functions and can be expressed by means of their Fourier series, which can ultimately be expressed as

$$f(t) = C_0 + C_1 \cos(t + \theta_1) \tag{25}$$

whose C_1 coefficient is the value of the amplitude for the fundamental component and θ_1 is the phase angle. It can be demonstrated that the values of the function are limited to the interval $[C_1, -C_1]$ (the coefficients of the cosine terms). Analyzing the harmonic function on the basis of the type of trajectory span we distinguish three types. We use different interpolation functions to determine their impact on the characteristics of the solution generated. The cases analyzed and interpolation functions for each case are as follows.

a. Initial span: Cases A, B and C

In all three cases we have used the same interpolation function for the first span, therefore the procedure to calculate the constants is identical

$$q_{i1} = a_{i1} \cdot sin(t) - b_{i1} \cdot cos(t) + c_{i1} \tag{26}$$

with $i = 1..N_{dof}$ and 1 is for the first span. N_{dof} is the number of the robot´s degrees of freedom. For this type of interpolation function, velocity and acceleration values are limited by the coefficients a_{ij}, b_{ij}. The known boundary conditions are three: the initial and final configuration of the interval and the initial velocity. They allow the set of coefficients a_{ij}, b_{ij} and c_{ij} to be obtained, which are dependent on time.

b. Intermediate span.

Three different interpolation functions corresponding to cases A, B and C have been used. To calculate the constants in each case we have proceeded as follows:

b1) Case A

The interpolation function is

$$q_{ij} = a_{ij} \cdot sin(t) - b_{ij} \cdot cos(2 \cdot t) + c_{ij} \cdot sin(3 \cdot t) - d_{ij} \cdot cos(4 \cdot t) \quad (27)$$

with i=1..N_{dof} and j =1..N_{span}. . N_{span} is the number of the span that is being analyzed.
From experience in the resolution of a great number of cases, a polynomial term has been added to ensure the boundary conditions of velocity and acceleration along the trajectory in this span. Velocity and acceleration equations are

$$\dot{q}_{ij} = a_{ij} \cdot cos(t) + 2 \cdot b_{ij} \cdot sin(2 \cdot t) + 3 \cdot c_{ij} \cdot cos(3 \cdot t) + 4 \cdot d_{ij} \cdot sin(4 \cdot t) \quad (28)$$

$$\ddot{q}_{ij} = -a_{ij} \cdot sin(t) + 4 \cdot b_{ij} \cdot cos(2 \cdot t) - 9 \cdot c_{ij} \cdot sin(3 \cdot t) + 16 \cdot d_{ij} \cdot cos(4 \cdot t) \quad (29)$$

Their values are limited by the coefficients a_{ij}, b_{ij}, c_{ij} and d_{ij}. The known boundary conditions are four: the initial and final configurations of the span and the velocities and accelerations at the beginning, and they allow the expressions for the constants a_{ij}, b_{ij}, c_{ij} and d_{ij} to be determined which, as in the previous case, are dependent on time.

b2) Case B

The interpolation function is

$$q_{ij} = cos(t) \cdot (sin(t) \cdot (a_{ij} \cdot sin(t) + b_{ij}) + c_{ij}) + d_{ij} \quad (30)$$

Velocity and acceleration equations are

$$\begin{aligned} \dot{q}_{ij} = & cos(t) \cdot (cos(t) \cdot (a_{ij} \cdot sin(t) + b_{ij}) + a_{ij} \cdot cos(t) \cdot sin(t)) - sin(t) \cdot \\ & (sin(t) \cdot (a_{ij} \cdot sin(t) + b_{ij}) + c_{ij}) \end{aligned} \quad (31)$$

$$\begin{aligned} \ddot{q}_{ij} = & cos(t) \cdot (- a_{ij} \cdot sin^2(t) - sin(t) \cdot (a_{ij} \cdot sin(t) + b_{ij}) + 2 \cdot a_{ij} \cdot cos^2(t)) - \\ & cos(t) \cdot (sin(t) \cdot (a_{ij} \cdot sin(t) + b_{ij}) + c_{ij}) - 2 \cdot sin(t) \cdot (cos(t) \cdot (a_{ij} \cdot sin(t) \\ & + b_{ij}) + a_{ij} \cdot cos(t) \cdot sin(t)) \end{aligned} \quad (32)$$

Their values are limited by the new coefficients a_{ij}, b_{ij}, c_{ij} and d_{ij}. The known boundary conditions are also four: the initial and final configurations of the span and the velocities and accelerations at the beginning. Therefore the constants a_{ij}, b_{ij}, c_{ij} and d_{ij} can be determined which, as in the previous case, are dependent on time.

b3. Case C

The interpolation function is

$$q_{ij} = sin(t) \cdot (cos(t) \cdot (a_{ij} \cdot sin(t) + b_{ij}) + c_{ij}) + d_{ij} \quad (33)$$

Velocity and acceleration equations are

$$\dot{q}_{ij} = sin(t) \, (a_{ij} \cdot cos^2(t) - sin(t) \cdot (a_{ij} \cdot sin(t) + b_{ij})) + cos(t) \cdot (cos(t) \cdot (a_{ij} \cdot sin(t) + b_{ij}) + c_{ij}) \quad (34)$$

$$\begin{aligned} \ddot{q}_{ij} = & 2 \cdot cos(t) \cdot (a_{ij} \cdot cos^2(t) - sin(t) \cdot (a_{ij} \cdot sin(t) + b_{ij})) - sin(t) \cdot (cos(t) \cdot (a_{ij} \cdot sin(t) + b_{ij}) + \\ & c_{ij}) + sin(t) \cdot (- cos(t) \cdot (a_{ij} \cdot sin(t) + b_{ij}) - 3 \cdot a_{ij} \cdot cos(t) \cdot sin(t)) \end{aligned} \quad (35)$$

Their values are limited by the new coefficients a_{ij}, b_{ij}, c_{ij} and d_{ij}. The known boundary conditions are also four: the initial and final configurations of the span and the velocities and accelerations at the beginning. Therefore the constants a_{ij}, b_{ij}, c_{ij} and d_{ij} can be determined which, as in the previous case, are dependent on time.

c. Final span: Cases A, B and C

In all three cases we used the same interpolation function for the last span and therefore the procedure to calculate the constants is identical

$$q_{iF} = a_{iF} \cdot sin(t) + b_{iF} \cdot cos(t) + c_{iF} \cdot sin(t)^2 + d_{jF} \cdot t + e_{iF} \cdot t^2 \tag{36}$$

with i=1..N_{dof} and F is for the final trajectory span.

In this type of span a polynomial term has been introduced, in this case of grade 2, which would ensure the continuity of velocity and acceleration. The velocity and acceleration equations are

$$\dot{q}_{iF} = a_{iF} \cdot cos(t) - b_{iF} \cdot sin(t) + 2 \cdot c_{iF} \cdot sin(t) \cdot cos(t) + d_{iF} + 2 \cdot e_{iF} \cdot t \tag{37}$$

$$\ddot{q}_{iF} = -a_{iF} \cdot sin(t) - b_{iF} \cdot cos(t) + 2 \cdot c_{iF} \cdot (cos(t)^2 - sin(t)^2) + 2 \cdot e_{iF} \tag{38}$$

Their values are limited by the coefficients a_{ij}, b_{ij}, c_{ij}, d_{ij} and e_{ij}. The known boundary conditions are five: the initial and final configuration in the last span or interval, the velocity and acceleration at the beginning of the interval and the velocity at the end. These boundary conditions enable the coefficients a_{ij}, b_{ij}, c_{ij}, d_{ij} and e_{ij} to be obtained.

Whenever a new adjacent configuration is generated by solving Eq. (4), a new trajectory span will also be created (by solving the second optimization problem Eq. (17)), and the necessary time t_j to perform the span is then obtained. The joint positions are adjusted using the corresponding harmonic interpolation function again. The solution of equations is obtained by iteration using quadratic sequential programming techniques (SQP) through the mathematical commercial software NAG (Numerical Algorithms Group). An each step of the iterative process it is necessary to recalculate the coefficients of the harmonic interpolation functions used, since they are time functions. To facilitate calculations, each span has been discretized using ten subintervals, so that the kinematic and dynamic characteristics are to be calculated at this discrete set of points. The solution of the optimization process provides the minimum time t_j to go from one configuration to its adjacent one and consequently the joint positions $q(t)$ that must be followed between these two configurations, as well as the necessary torques in the actuators $\tau(t)$ and the corresponding kinematic characteristics $\dot{q}(t)$ and $\ddot{q}(t)$.

9.3 Impact of interpolation function

As it was said earlier, the impact of the interpolation function is very important from the point of view of the robot´s performance. Three types of interpolation functions have been used (A, B and C) for the computation of intermediate configurations (harmonic functions) when using the "simultaneous algorithm. Pure polynomial interpolation functions have been excluded because they exceeded the dynamic capabilities of the actuators and therefore the algorithm failed to reach any solution. Therefore, after having analysed all kinds of interpolation functions, we state that the best of all them C (notice that each actuator has been characterized by the maximum and minimum torque it can provide, see Eq. (23). Nonetheless, both the computational and execution time are very high compared with the results obtained using the "sequential algorithm".

9.4 Cost function

An important point of the algorithm is to understand the process by which the algorithm is gradually creating the trajectory. The algorithm works in a discretised workspace (see Rubio et al.,2009) , looking for a trajectory that joins the initial and final configurations by starting from the initial configuration and, on the basis of generating adjacent configurations and branching out from the most promising one, obtaining new configurations until reaching the final one. Therefore, the trajectory contains a discrete set of intermediate configurations. To ensure that the process moves from one configuration to another, that is, that the algorithm branches out from a general intermediate configuration to generate more new adjacent configurations, the uniform cost function is used. The discrete configuration space is analysed as a graph, where the configurations generated are the nodes and the arc between nodes (arc $(i, j) = time(i,j)$) is calculated as the time necessary to perform the motion between adjacent configurations. It is desirable that the number of configurations generated is not high and, in addition, that these configurations enable efficient trajectories to be obtained. The process followed to achieve the growth of the configuration space in the search for the final configuration is as follows:

Let $CC = \{C^1, C^2, \ldots, C^k\}$ be the set of existing configurations at a given instant, and *CR* the subgroup of *CC* that contains *r* ($r < k$) configurations that have still not been used to branch out. Now, it is necessary to follow what is called a branching strategy or searching strategy to select a C^p configuration pertaining to *CR*, from which the algorithm tries to generate another six new adjacent configurations C^{p+1}, C^{p+2}, C^{p+3}, C^{p+4}, C^{p+5} and C^{p+6} (according to the technique explained in Valero (2006)), which are new configurations belonging to *CR*, while C^p is taken out of this subgroup. The process finishes when the final configuration is reached.

The cost function *c (p)* used to select a new configuration to branch out from is defined as follows

- Uniform Cost: the time function *c (j)* associated with the configuration C^j is defined as the minimum sum of arcs that permit the node *j* to be reached from the initial node

$$c(j) = time(1, j) \tag{39}$$

And the new branching is started from configuration C^p, which meets

$$c(p) = min\left[c(j)\right], \forall j \in CR \tag{40}$$

When a set of adjacent configurations has been created, Eq. (40) is used to select that one from which the process is expected to branch out again. Given two adjacent configurations the minimum time between them is calculated as explained in Section 4. Time is used to select the new configuration as just explained in Section 5, which is used to repeat the branching process and this in turn is repeated until the final configuration is reached.

9.5 Obtaining the trajectory

When the final configuration is reached we know not only the robot configuration through the joint positions $q(t)$ but also the necessary torques $\tau(t)$ and the kinematic characteristics of the motion $\dot{q}(t)$ and $\ddot{q}(t)$. The trajectory obtained is of minimum time on the graph generated. To obtain the global minimum time, the process should be repeated with different discretization sizes. The global minimum time is the smallest of all times calculated.

9.6 Application and examples solved

This algorithm has been applied to the PUMA 560 robot type, and a great number of examples have been analysed. Four important operational parameters have been monitored: the computational time used in generating a solution, the execution time, the distance travelled (which corresponds to the sum of the whole distance travelled by each significant point throughout the path to go from the initial to the final configuration, measured in meters) and the number of configurations generated. Though the examples, the behaviour of those four operational parameters mentioned earlier when the simultaneous algorithm and the different interpolation functions have been use can be analized. The results obtained show that the worst computational time is achieved when using the interpolation function of case A. Case B and C yield similar results. Also, the results show that the smallest execution time is achieved when using the interpolation function of case C. The smallest distance travelled is achieved when using the interpolation function of case C as well as the smallest number of configurations generated are achieved when using the interpolation function of case C.

10. Conclusion

In this paper, two algorithms that solve the trajectory planning problem for industrial robots in an environment with obstacles have been introduced and summarized. They have been called "sequential" and "simultaneous" algorithm respectively. Both are off-line algorithms. The first one is based on an indirect methodology because it solves the trajectory planning in two sequential steps (first a path is generated and once the path is known, a trajectory is adjusted to it). Polynomial interpolation functions have been in this algorithm because they yield the best results. Besides, the trajectories calculated meet constraints on torque, power, jerk and energy consumed. The second algorithm is a direct method, which solves the equations in the state space of the robot. Unlike other direct methods, it does not use previously defined paths, which enables working with mobile obstacles although the obstacles used in this chapter are statics. Three types of interpolation functions have been used for the computation of intermediate configurations (harmonic functions). Polynomial interpolation functions have been excluded from this algorithm because during the resolution phase of the examples, because converge problems in the optimization problem have come up.

The main conclusions are summarized as follows:

a. The algorithms solve the trajectory planning problem for industrial robots in environments with obstacles therefore avoiding collisions.
b. It can be applied to any industrial robot.
c. "Sequential" algorithm:
    c.1. Constraints on the energy consumed must be compatible with the robot´s demanded potential energy, as energy recovery is not considered, as the algorithm works on the assumption that the energy can be dissipated but not recovered.

c.2. To obtain competitive results in the balance between time cycle and energy consumed, the actuators should work with the maximum admissible value of the jerk so that the robot can work with the desired accuracy.

c.3. The cubic interpolation function gives the best computational and execution time.

d. "Simultaneous" algorithm: as for the peculiarities of the interpolation functions in relation to the four monitored operating parameters (computational time, execution time, distance travelled and number of configurations generated), the main point is that the best results are obtained when using the interpolation function of case C (taking into account that each actuator has been characterized by the maximum and minimum torque it can provide). With this algorithm the cubic interpolation function does not work because during the resolution phase of the examples, they exceeded the dynamic capabilities of the actuators and therefore the algorithm failed to reach any solution.

11. Acknowledgment

This paper has been possible thanks to the funding of Science and Innovation Ministry of the Spain Government by means of the Researching and Technologic Development Project DPI2010-20814-C02-01 (IDEMOV).

12. References

Abdel-Malek, K., Mi, Z., Yang, J.Z. & Nebel, K. (2006), Optimization-based trajectory planning of the human upper body, *Robotica*, Vol. 24, nº 6, pp. (683-696).

Bobrow, J.E., Dubowsky, S. & Gibson, J.S. (1985), Time-Optimal Control of Robotic Manipulators Along Specied Paths, *International Journal of Robotics Research*, Vol. 4, nº 3, pp. (3-17).

Chen, Y. & Desrochers, A.A. (1989), Structure of minimum time control law for robotic manipulators with constrained paths, *IEEE Int Conf Robot Automat*, ISBN: 0-8186-1938-4, pp. (971-976), Scottsdale, USA, 1989.

Chettibi, T., Lehtihet, H.E., Haddad, M. & Hanchi, S. (2002), Optimal pose trajectory planning for robot manipulators. *Mechanism and Machine Theory*, vol 37, nº 10, pp. (1063-1086).

Chettibi, T., Lehtihet, H.E., Haddad, M. & Hanchi, S. (2004), Minimum cost trajectory planning for industrial robots. European Journal of Mechanics a-solids 23 (4): 703-715.

Cho, B. H., Choi, B. S. & Lee, J. M. (2006), Time-optimal trajectory planning for a robot system under torque and impulse constraints, *International Journal of Control, Automation, and Systems*, Vol. 4, nº 1, pp. (10-16).

Constantinescu, D. & Croft, E.A. (2000), Smooth and time-optimal trajectory planning for industrial manipulators along specified paths, *Journal of Robotic Systems*, Vol. 17, no 5, pp. (233-249).

du Plessis, L. J. & Snyman, J. A. (2003), Trajectory-planning through interpolation by overlapping cubic arcs and cubic splines. *International Journal for Numerical Methods in Engineering*, Vol. 57, nº 11, pp. (1615-1641).

Field, G. & Stepanenko, Y. (1996), Iterative dynamic programming: an approach to minimum energy trajectory planning for robotic manipulators, *Proc. of the IEEE*

International Conference on Robotics and Automation, ISBN: 0-7803-2988-0, pp. (2755-2760), Minneapolis, USA, 1996.

Garg, D. & Ruengcharungpong, C. (1992), Force balance and energy optimization in cooperating manipulators, *Proceedings of the 23rd Annual Pittsburgh Modeling and Simulation Conference,* pp. (2017-2024), Pittsburgh, USA, 1992.

Gasparetto, A. & Zanotto, V. (2007),A new method for smooth trajectory planning of robot manipulators, *Mechanism and Machine Theory,* Vol. 42, nº 4, pp. (455-471).

Gasparetto, A. & Zanotto, V. (2010), Optimal trajectory planning for industrial robots, *Advances in Engineering Software,* vol. 41, No 4, pp. 548-556.

Hirakawa, A. & Kawamura, A. (1996), Proposal of trajectory generation for redundant manipulators using variational approach applied to minimization of consumed electrical energy, *Proceedings of the Fourth International Workshop on Advanced Motion Control,* ISBN: 0-7803-3219-9, pp. (687-692), Mie, Japan, 1996.

Kyriakopoulos, K.J. & Saridis, G.N. (1988), Minimum jerk path generation, *in IEEE international conference on robotics and automation,* ISBN: 0-8186-0852-8, pp. (364-369), Philadelphia, USA, 1988.

Macfarlane, S. & Croft, E. A. (2003), Jerk-bounded manipulator trajectory planning: Design for real-time applications. *IEEE Transactions on Robotics and Automation,* Vol. 19, nº 1, pp. (42-52).

Provenzano, S. E. (2001), Aplicación de las ecuaciones de Gibbs-Appell a la dinámica de robots. Doctoral thesis, Universidad Politécnica de Valencia, Spain, 2001.

Rubio, F.J., Valero, F.J., Suñer, J.L. & Mata, V. (2009), Direct step-by-step method for industrial robot path planning, *Industrial Robot: An International Journal,* Vol. 36, nº 6, pp. (594-607).

Rubio, F.J., Valero, F.J., Suñer, J.L. and & Mata, V. (2007), Técnicas globales para la planificación de caminos de robots industriales, *VIII Congreso Iberoamericano de Ingeniería Mecánica in Cusco,* ISBN: 978-9972-2885-31, Cuzco (Peru), Octubre, 2007

Saramago, S.F.P. & Steffen, V. Jr. (2000), Optimal trajectory planning of robot manipulators in the presence of moving obstacles, *Mechanism and Machine Theory,* Vol. 35, pp. (1079-1094).

Saramago, S. F. & Steffen Jr,V. (2001), Trajectory modelling of robot manipulators in the presence of obstacles. Journal of optimization theory and applications. 110(1), 17-34.

Shin, K.G. & McKay, N.D. (1985), Minimum-time control of robotic manipulators with geometric path constraints, *IEEE Transactions on Automatic Control,* ISSN: 0018-9286, pp. (531-541).

Suñer, J.L., Valero, F.J., Ródenas, J.J., & Besa, A. (2007), Comparación entre procedimientos de solución de la interpolación por funciones splines para la planificación de trayectorias de robots industriales, *VIII Congreso Iberoamericano de Ingeniería Mecánica in Cusco,* ISBN: 978-9972-2885-31, Cuzco (Peru), Octubre, 2007

Valero,F. J., Mata V. & Besa A. (2006), Trajectory planning in workspaces with obstacles taking into account the dynamic robot behaviour. *Mechanism and Machine Theory.* Vol. 41, pp. (525-536).

Methodology for System Adaptation Based on Characteristic Patterns

Eva Volná[1], Michal Janošek[1], Václav Kocian[1],
Martin Kotyrba[1] and Zuzana Oplatková[2]
[1]*University of Ostrava*
[2]*Tomas Bata University in Zlín*
Czech Republic

1. Introduction

This paper describes the methodology for system description and application so that the system can be managed using real time system adaptation. The term system here can represent any structure regardless its size or complexity (industrial robots, mobile robot navigation, stock market, systems of production, control systems, etc.). The methodology describes the whole development process from system requirements to software tool that will be able to execute a specific system adaptation.

In this work, we propose approaches relying on machine learning methods (Bishop, 2006), which would enable to characterize key patterns and detect them in real time and in their altered form as well. Then, based on the pattern recognized, it is possible to apply a suitable intervention to system inputs so that the system responds in the desired way. Our aim is to develop and apply a hybrid approach based on machine learning methods, particularly based on soft-computing methods to identify patterns successfully and for the subsequent adaptation of the system. The main goal of the paper is to recognize important pattern and adapt the system's behaviour based on the pattern desired way.

The paper is arranged as follows: Section 1 introduces the critical topic of the article. Section 2 details the feature extraction process in order to optimize the patterns used as inputs into experiments. The pattern recognition algorithms using machine learning methods are discussed in section 3. Section 4 describes the used data-sets and covers the experimental results and a conclusion is given in section 5. We focus on reliability of recognition made by the described algorithms with optimized patterns based on the reduction of the calculation costs. All results are compared mutually.

1.1 The methodology for system description

Gershenson (Gershenson, 2007) proposed a methodology called *The General Methodology* for system description necessary to manage a system. It presents a conceptual framework for describing systems as self-organizing and consists of five steps: representation, modelling, simulation, application and evaluation. Our goal is to use and adapt this methodology for our specific needs. Basically we would like to describe a methodology that the designer should be able to use to describe his system, find key patterns in its behaviour based on the

observation and prepare suitable response to these patterns that emerge from time to time and adapt to any deviation in the system's behaviour.
As we are using Gershenson's methodology we are not going to describe it in detail because detailed info can be found in his book (Gershenson, 2007). Let's mention crucial parts of his methodology that is important to our work. The methodology is useful for designing and controlling complex systems. Basically a complex system consists of two or more interconnected components and these components react together and it is very complicated to separate them. So the system's behaviour is impossible to deduce from the behaviour of its individual components. This deduction becomes more complicated how more components $\#\bar{E}$ and more interactions $\#\bar{I}$ the system has (C_{sys} corresponds with system complexity; C_e corresponds with element complexity; C_i corresponds with interaction complexity).

$$C_{sys} \sim \begin{cases} \bar{E} \\ \bar{I} \\ \sum_{j=0}^{\#\bar{E}} C_{e_j} \\ \sum_{k=0}^{\#\bar{I}} C_{i_k} \end{cases} \quad (1)$$

Imagine a manufacturing factory. We can describe the manufacturing factory as a complex system. Now it is important to realize that we can have several levels of abstraction starting from the manufacturing line to the whole factory complex. The manufacturing line can consist of many components. There can be robots, which perform the main job. Conveyor belts, roller beds, jigs, hangers and other equipment responsible for the product or material transport and other equipments. All the interactions are some way related to the material or product. Although it is our best interest to run all the processes smoothly there will be always some incidents we cannot predict exactly. The supply of the material can be interrupted or delayed, any equipment can have a multifunction and it is hard to predict when and how long will it takes. Because there are interactions among many of these components we can call manufacturing factory a complex system.
If we want to characterize a system we should create its model. Gershenson (Gershenson, 2002) proposes two types of models, absolute and relative. The absolute model (abs-model) refers to what the thing actually is, independently of the observer. The relative model (rel-model) refers to the properties of the thing as distinguished by an observer within a context. We can say that the rel-model is a model, while the abs-model is modelled. Since we are all limited observers, it becomes clear that we can speak about reality only with rel-beings/models (Gershenson, 2007).
So how we can model a complex system? Any complex system can be modelled using multi-agent system (MAS) where each system's component is represented by an agent and any interactions among system's components are represented as interactions among agents. If we take into consideration *The General Methodology* thus any system can be modelled as group of agents trying to satisfy their goals. There is a question. Can we describe a systems modelling as a group of agents as self-organizing? We think that we can say *Yes*. As the agents in the MAS try to satisfy their goals, same as components in self-organizing systems interact with

each other to achieve desired state or behaviour. If we determine the state as a self-organizing state, we can call that system self-organizing and define our complex self-organizing system. In our example with manufacturing line our self-organizing state will be a state where the production runs smoothly without any production delays. But how can we achieve that?
Still using Gerhenson's General Methodology we can label fulfilling agent's goal as its satisfaction $\sigma \in [0,1]$. Then the system's satisfaction σ_{sys} (2) can be represented as function $f:\mathbf{R} \rightarrow [0,1]$ and it is a satisfaction of its individual components.

$$\sigma_{\mathrm{sys}} = f(\sigma_1, \sigma_2, \ldots, \sigma_n, w_0, w_1, w_2, \ldots, w_n) \quad (2)$$

w_0 represents bias and other weights w_i represents an importance given to each σ_i.
Components, which decrease σ_{sys} and increase their σ_i shouldn't be considered as a part of the system. Of course it is hard to say if for higher system's satisfaction it is sufficient to increase satisfaction of each individual component because some components can use others fulfilling their goals. For maximization of σ_{sys} we should minimize the friction among components and increase their synergy. A mediator arbitrates among elements of a system, to minimize conflict, interferences and frictions; and to maximize cooperation and synergy. So we have two types of agents in the MAS. Regular agents fulfil their goals and mediator agents streamline their behaviour. Using that simple agent's division we can build quite adaptive system.

1.2 Patterns as a system's behaviour description

Every system has its unique characteristics that can be described as patterns. Using patterns we would like to characterize particular system and its key characteristics. Generally a system can sense a lot of data using its sensors. If we put the sensor's data into some form, a set or a graph then a lot of patterns can be recognized and further processed. When every system's component has some sensor then the system can produce some patterns in its behaviour. Some sensor reads data about its environment so we can find some patterns of the environment, where the system is located. If we combine several sensors data, we would be able to recognise some patterns in the whole system's behaviour. It is important to realize that everything, which we observe is relative from our point of view. When we search for the pattern, we want to choose such pattern, which represents the system reliably and define its important properties. Every pattern, which we find, is always misrepresented with our point of view.
We can imagine a pattern as some object with same or similar properties. There are many ways how to recognize and sort them. When we perform pattern recognition, we assign a pre-defined output value to an input value. For some purpose, we can use a particular pattern recognition algorithm, which is introduced in (Ciskowski & Zaton, 2010). In this case we try to assign each input value to the one of the output sets of values. Some input value can be any data regardless its origin as a text, audio, image or any other data. When patterns repeat in the same or altered forms then can be classified into predefined classes of patterns. Since we are working on computers, the input data and all patterns can be represented in a binary form without the loss of generality. Such approach can work nearly with any system, which we would like to describe. But that is a very wide frame content.
Although theory of regulation and control (Armstrong & Porter, 2006) is mainly focused on methods of automatic control, it also includes methods for adaptive and fuzzy controls. In general, through the control or regulation we guide the system's behaviour in the desired direction. For our purposes, it suffices to regulate the system behaviour based on the predefined target and compensate any deviation in desired direction. So we search for key

patterns in system's behaviour a try to adapt to any changes. However, in order to react quickly and appropriately, it is good to have at least an expectation of what may happen and which reaction would be appropriate, i.e. what to anticipate. Expectations are subjective probabilities that we learn from experience: the more often pattern B appears after pattern A, or the more successful action B is in solving problem A, the stronger the association A → B becomes. The next time we encounter A (or a pattern similar to A), we will be prepared, and more likely to react adequately. The simple ordering of options according to the probability that they would be relevant immensely decreases the complexity of decision-making (Heylighen, 1994).

Agents are appropriate for defining, creating, maintaining, and operating the software of distributed systems in a flexible manner, independent of service location and technology. Systems of agents are complex in part because both the structural form and the behaviour patterns of the system change over time, with changing circumstances. By structural form, we mean the set of active agents and inter-agent relationships at a particular time. This form changes over time as a result of inter-agent negotiations that determine how to deal with new circumstances or events. We call such changing structural form morphing, by analogy with morphing in computer animation. By behaviour patterns, we mean the collaborative behaviour of a set of active agents in achieving some overall purpose. In this sense, behaviour patterns are properties of the whole system, above the level of the internal agent detail or of pair wise, inter-agent interactions. Descriptions of whole system behaviour patterns need to be above this level of detail to avoid becoming lost in the detail, because agents are, in general, large grained system components with lots of internal detail, and because agents may engage in detailed sequences of interactions that easily obscure the big picture. In agent systems, behaviour patterns and morphing are inseparable, because they both occur on the same time scale, as part of normal operation. Use case maps (UCMs) (Burth & Hubbard, 1997) are descriptions of large grained behaviour patterns in systems of collaborating large grained components.

1.3 System adaptation vs. prediction

Let's say we have built pattern recognition system and it is working properly to meet our requirements. We are able to recognize certain patterns reliably. What can we do next? Basically, we can predict systems behaviour or we can adapt to any change that emerge.

It is possible to try to predict what will happen, but more or less it is a lottery. We will never be able to predict such systems' behaviour completely. This doesn't mean it is not possible to build a system based on prediction (Gershenson, 2007). But there is another approach that tries to adapt to any change by reflecting current situation. To adapt on any change (expected or unexpected) it should be sufficient to compensate any deviation from desired course. In case that response to a deviation comes quickly enough that way of regulation can be very effective. It does not matter how complicated system is (how many factors and interactions has) in case we have efficient means of control (Armstrong & Porter, 2006). To respond quickly and flexible it is desirable to have some expectation what can happen and what kind of response will be appropriate. We can learn such expectation through experiences.

2. Feature extraction process in order to optimize the patterns

Identification problems involving time-series data (or waveforms) constitute a subset of pattern recognition applications that is of particular interest because of the large number of

domains that involve such data. The recognition of structural shapes plays a central role in distinguishing particular system behaviour. Sometimes just one structural form (a bump, an abrupt peak or a sinusoidal component), is enough to identify a specific phenomenon. There is not a general rule to describe the structure - or structure combinations - of various phenomena, so specific knowledge about their characteristics has to be taken into account. In other words, signal structural shape may be not enough for a complete description of system properties. Therefore, domain knowledge has to be added to the structural information.
However, the goal of our approach is not knowledge extraction but to provide users with an easy tool to perform a first data screening. In this sense, the interest is focused on searching for specific patterns within waveforms (Dormido-Canto et al., 2006). The algorithms used in pattern recognition systems are commonly divided into two tasks, as shown in Fig. 1. The description task transforms data collected from the environment into features (primitives).

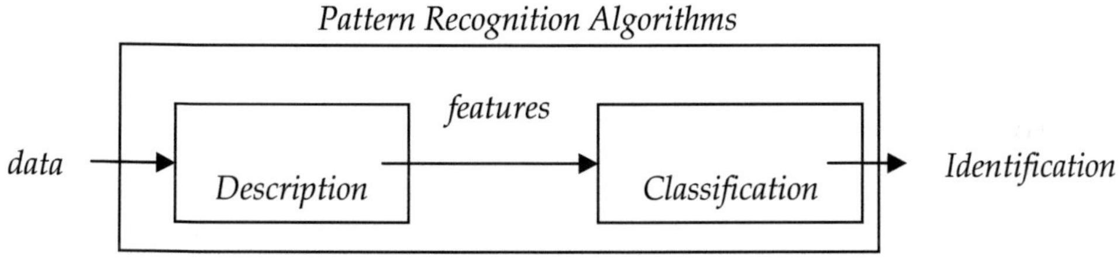

Fig. 1. Tasks in the pattern recognition systems

The classification task arrives at an identification of patterns based on the features provided by the description task. There is no general solution for extracting structural features from data. The selection of primitives by which the patterns of interest are going to be described depends upon the type of data and the associated application. The features are generally designed making use of the experience and intuition of the designer.
The input data can be presented to the system in various forms. In principle we can distinguish two basic possibilities:

- The numeric representation of monitored parameters
- Image data - using the methods of machine vision

Figures 2 and 3 show an image and a numerical expression of one particular section of OHLC data. The image expression contains only information from the third to the sixth column of the table (Fig.3). In spite of the fact, the pattern size (number of pixels) equals to 7440. In contrast to it, a table expression with 15 rows and 7 columns of 16-bit numbers takes only.

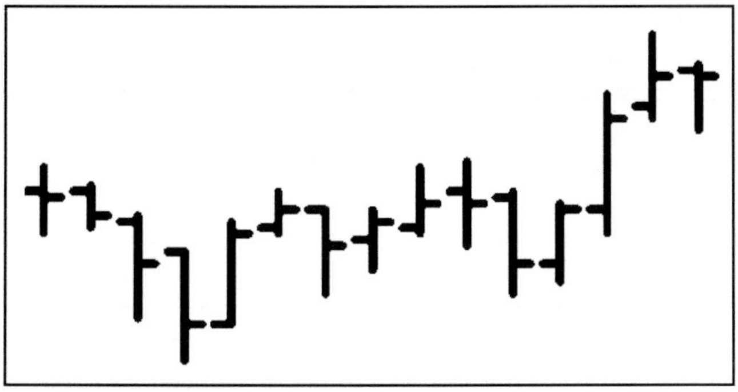

Fig. 2. Visual representations of pattern

2010.06.02	08:15	1.22220	1.22260	1.22140	1.22210	107
2010.06.02	08:20	1.22220	1.22230	1.22150	1.22170	76
2010.06.02	08:25	1.22160	1.22170	1.21990	1.22090	71
2010.06.02	08:30	1.22110	1.22110	1.21910	1.21970	85
2010.06.02	08:35	1.21980	1.22160	1.21970	1.22140	78
2010.06.02	08:40	1.22150	1.22220	1.22140	1.22190	61
2010.06.02	08:45	1.22180	1.22190	1.22030	1.22120	84
2010.06.02	08:50	1.22130	1.22180	1.22070	1.22160	71
2010.06.02	08:55	1.22150	1.22260	1.22140	1.22200	61
2010.06.02	09:00	1.22220	1.22270	1.22120	1.22200	91
2010.06.02	09:05	1.22210	1.22220	1.22030	1.22090	53
2010.06.02	09:10	1.22080	1.22200	1.22050	1.22190	87
2010.06.02	09:15	1.22180	1.22390	1.22140	1.22350	90
2010.06.02	09:20	1.22370	1.22500	1.22350	1.22430	87
2010.06.02	09:25	1.22440	1.22450	1.22330	1.22430	77

Fig. 3. Tabular expression of pattern

The image data better correspond to an intuitive human idea of patterns recognitions, which is their main advantage. We also have to remember that even table data must be transferred into binary (image) form before their processing.

Image data are always two-dimensional. Generally, tabular patterns can have more dimensions. Graphical representation of OHLC data (Lai, 2005) in Fig.2 is a good example of the expression of multidimensional data projection to two-dimensional space. Fig.2 shows a visual representation of 4-dimensional vector in time, which corresponds to 5 - dimensions. In this article, we consider experiments only over two-dimensional data (time series). Extending the principles of multidimensional vectors (random processes) will be the subject of our future projects.

The intuitive concept of "pattern" corresponds to the two-dimensional shapes. This way allows showing a progress of a scalar variable. In the case that a system has more than one parameter, the graphic representation is not trivial anymore.

3. Pattern recognition algorithms

Classification is one of the most frequently encountered decision making tasks of human activity. A classification problem occurs when an object needs to be assigned into a predefined group or class based on a number of observed attributes related to that object. Pattern recognition is concerned with making decisions from complex patterns of information. The goal has always been to tackle those tasks presently undertaken by humans, for instance to recognize faces, buy or sell stocks or to decide on the next move in a chess game. Rather simpler tasks have been considered by us. We have defined a set of classes, which we plan to assign patterns to, and the task is to classify a future pattern as one of these classes. Such tasks are called classification or supervised pattern recognition. Clearly someone had to determine the classes in the first phase. Seeking the groupings of patterns is called cluster analysis or unsupervised pattern recognition. Patterns are made up of features, which are measurements used as inputs to the classification system. In case that patterns are images, the major part of the design of a pattern recognition system is to select suitable features; choosing the right features can be even more important than what is done with them subsequently.

3.1 Artificial neural networks

Neural networks that allow so-called supervised learning process (i.e. approach, in which the neural network is familiar with prototype of patterns) use to be regarded as the best choice for pattern recognition tasks. After adaptation, it is expected that the network is able to recognise learned (known) or similar patterns in input vectors. Generally, it is true - the more training patterns (prototypes), the better network ability to solve the problem. On the other hand, too many training patterns could lead to exceeding a memory capacity of the network. We used typical representative of neural networks, namely:

- Hebb network
- Backpropagation network

Our aim was to test two networks with extreme qualities. In other words, we chose such neural networks, which promised the greatest possible differences among achieved results.

3.1.1 Hebb network

Hebb network is the simplest and also the "cheapest" neural network, which adaptation runs in one cycle. Both adaptive and inactive modes work with integer numbers. These properties allow very easy training set modification namely in applications that work with very large input vectors (e.g. image data).
Hebbian learning in its simplest form (Fausett, 1994) is given by the weights update rule (3)

$$\Delta w_{ij} = \eta\, a_i\, a_j \qquad (3)$$

where w_{ij} is the change in the strength of the connection from unit j to unit i, a_i and a_j are the activations of units i and j respectively, and η is a learning rate. When training a network to classify patterns with this rule, it is necessary to have some method of forcing a unit to respond strongly to a particular pattern. Consider a set of data divided into classes $C_1, C_2,\ldots,C_m$.
Each data point x is represented by the vector of inputs $(x_1, x_2, \ldots, x_n)$. A possible network for learning is given in Figure 4. All units are linear. During training the class inputs $c_1, c_2, \ldots, c_m$ for a point x are set as follows (4):

$$\begin{aligned} c_i &= 1 \quad \boldsymbol{x} \in C_i \\ c_i &= 0 \quad \boldsymbol{x} \notin C_i \end{aligned} \qquad (4)$$

Each of the class inputs is connected to just one corresponding output unit, i.e. c_i connects to o_i only for $i = 1, 2, \ldots, m$. There is full interconnection from the data inputs $x_1, x_2, \ldots, x_n$ to each of these outputs.

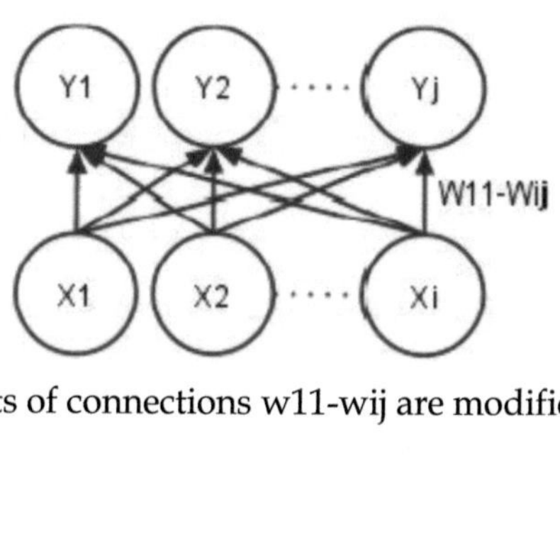

Fig. 4. Hebb network. Weights of connections w11-wij are modified in accordance with the Hebbian learning rule

3.1.2 Backpropagation network

Back propagation network is one of the most complex neural networks for supervised learning. Its ability to learning and recognition are much higher than Hebb network, but its disadvantage is relatively lengthy processes of adaptation, which may in some cases (complex input vectors) significantly prolong the network adaptation to new training sets.

Backpropagaton network is a multilayer feedforward neural network. See Fig. 5, usually a fully connected variant is used, so that each neuron from the *n-th* layer is connected to all neurons in the *(n+1)-th* layer, but it is not necessary and in general some connections may be missing - see dashed lines, however, there are no connections between neurons of the same layer. A subset of input units has no input connections from other units; their states are fixed by the problem. Another subset of units is designated as output units; their states are considered the result of the computation. Units that are neither input nor output are known as hidden units.

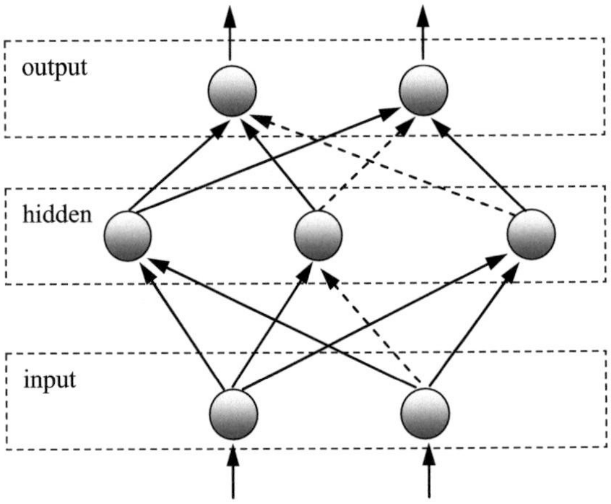

Fig. 5. A general three-layer neural network

Backpropagation algorithm belongs to a group called "gradient descent methods". An intuitive definition is that such an algorithm searches for the global minimum of the weight landscape by descending downhill in the most precipitous direction. The initial position is set at random selecting the weights of the network from some range (typically from -1 to 1 or from 0 to 1). Considering the different points, it is clear, that backpropagation using a fully connected neural network is not a deterministic algorithm. The basic backpropagation algorithm can be summed up in the following equation (the *delta rule*) for the change to the weight w_{ji} from node i to node j (5):

$$\underset{\text{weight change}}{\Delta w_{ji}} = \underset{\text{learning rate}}{\eta} \times \underset{\text{local gradient}}{\delta_j} \times \underset{\text{input signal to node } j}{y_i} \tag{5}$$

where the local gradient δ_j is defined as follows: (Seung, 2002):

1. If node j is an output node, then δ_j is the product of $\varphi'(v_j)$ and the error signal e_j, where $\varphi(\_)$ is the logistic function and v_j is the total input to node j (i.e. $\Sigma_i w_{ji}y_i$), and e_j is the error signal for node j (i.e. the difference between the desired output and the actual output);

2. If node j is a hidden node, then δ_j is the product of $\varphi'(v_j)$ and the weighted sum of the δ's computed for the nodes in the next hidden or output layer that are connected to node j.

[The actual formula is $\delta_j = \varphi'(v_j)$ Σ$_k$ $\delta_k w_{kj}$ where k ranges over those nodes for which w_{kj} is non-zero (i.e. nodes k that actually have connections from node j. The δ_k values have already been computed as they are in the output layer (or a layer closer to the output layer than node j).]

3.2 Analytic programming

Basic principles of the analytic programming (AP) were developed in 2001 (Zelinka, 2002). Until that time only genetic programming (GP) and grammatical evolution (GE) had existed. GP uses genetic algorithms while AP can be used with any evolutionary algorithm, independently on individual representation. To avoid any confusion, based on use of names according to the used algorithm, the name - Analytic Programming was chosen, since AP represents synthesis of analytical solution by means of evolutionary algorithms.

The core of AP is based on a special set of mathematical objects and operations. The set of mathematical objects is set of functions, operators and so-called terminals (as well as in GP), which are usually constants or independent variables. This set of variables is usually mixed together and consists of functions with different number of arguments. Because of a variability of the content of this set, it is called here "general functional set" - GFS. The structure of GFS is created by subsets of functions according to the number of their arguments. For example *GFSall* is a set of all functions, operators and terminals, *GFS3arg* is a subset containing functions with only three arguments, *GFS0arg* represents only terminals, etc. The subset structure presence in GFS is vitally important for AP. It is used to avoid synthesis of pathological programs, i.e. programs containing functions without arguments, etc. The content of GFS is dependent only on the user. Various functions and terminals can be mixed together (Zelinka, 2002; Oplatková, 2009).

The second part of the AP core is a sequence of mathematical operations, which are used for the program synthesis. These operations are used to transform an individual of a population into a suitable program. Mathematically stated, it is a mapping from an individual domain into a program domain. This mapping consists of two main parts. The first part is called discrete set handling (DSH), see Fig. 6 (Zelinka, 2002) and the second one stands for security procedures which do not allow synthesizing pathological programs. The method of DSH, when used, allows handling arbitrary objects including nonnumeric objects like linguistic terms {hot, cold, dark…}, logic terms (True, False) or other user defined functions. In the AP DSH is used to map an individual into GFS and together with security procedures creates the above mentioned mapping which transforms arbitrary individual into a program.

AP needs some evolutionary algorithm (Zelinka, 2004) that consists of population of individuals for its run. Individuals in the population consist of integer parameters, i.e. an individual is an integer index pointing into GFS. The creation of the program can be schematically observed in Fig. 7. The individual contains numbers which are indices into GFS. The detailed description is represented in (Zelinka, 2002; Oplatková, 2009).

AP exists in 3 versions - basic without constant estimation, APnf - estimation by means of nonlinear fitting package in *Mathematica* environment and APmeta - constant estimation by means of another evolutionary algorithms; meta means metaevolution.

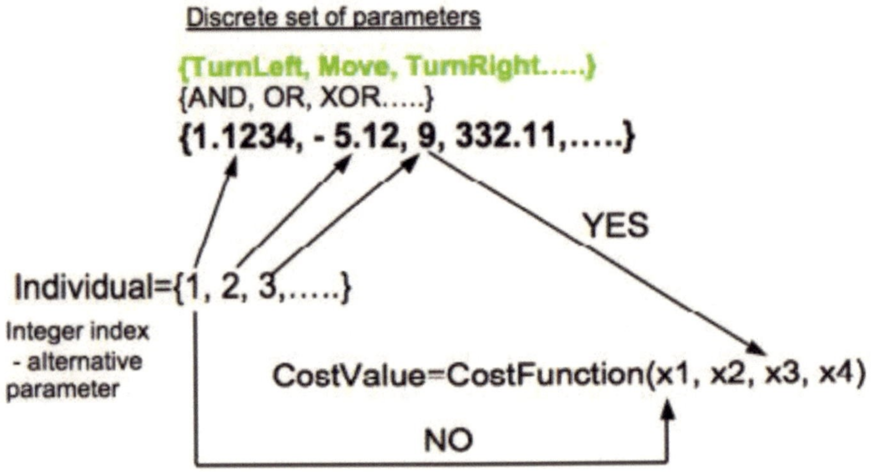

Fig. 6. Discrete set handling

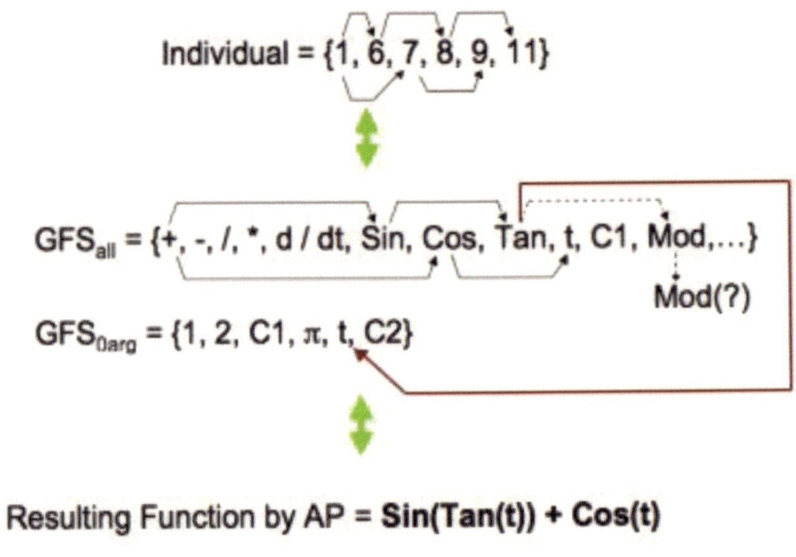

Fig. 7. Main principles of AP

4. Experimental results

4.1 Used datasets

This approach allows a search of structural shapes (patterns) inside time-series. Patterns are composed of simpler sub-patterns. The most elementary ones are known as primitives. Feature extraction is carried out by dividing the initial waveform into segments, which are encoded. Search for patterns is accomplished process, which is performed manually by the

user. In order to test the efficiency of pattern recognition, we applied a database downloaded from (Google finance, 2010). We used time series, which shows development of the market value of U.S. company Google and represents the minute time series from 29 October 2010, see Fig. 8.
Used algorithms need for their adaptation training sets. In all experimental works, the training set consists of 100 samples (e.g. training pars of input and corresponding output vectors) and it is made from the time series and contains three peaks, which are indicated by vertical lines and they are shown in Figure 8. Samples obtained in this way are always adjusted for the needs of the specific algorithm. Data, which were tested in our experimental works, contains only one peak, which is indicated by vertical lines and it is shown in Fig. 9.

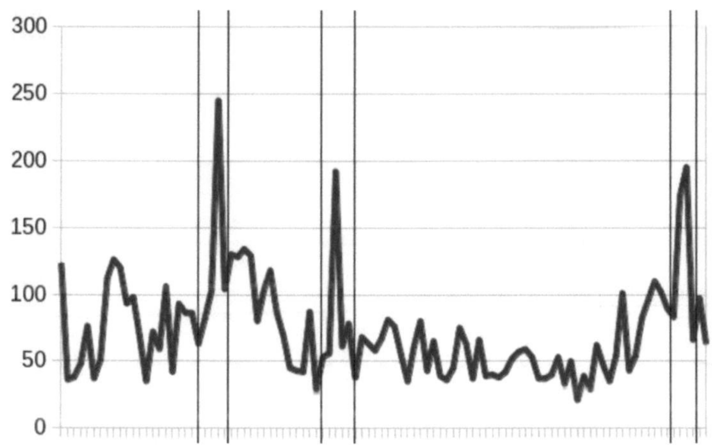

Fig. 8. The training set with three marked peaks

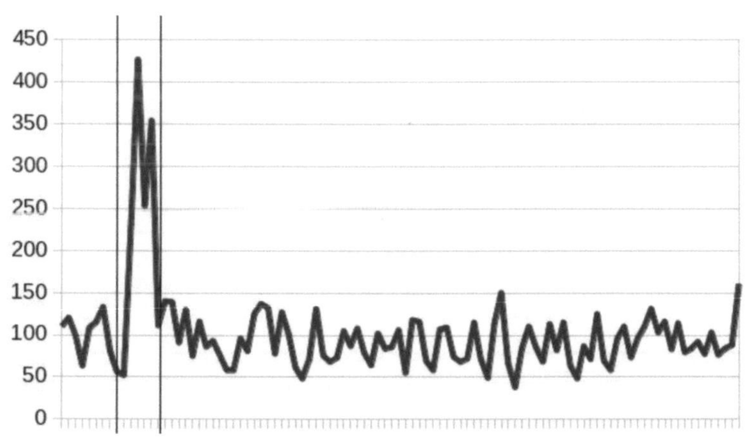

Fig. 9. The test set with one marked peak, which is searched

4.2 Pattern recognition via artificial neural networks

The aim of this experiment was to adapt neural network so that it could find one kind of pattern (peak) in the test data. We have used two sets of values, which are graphically depicted in Figure 10 (training patterns) and Figure 11 (test patterns) in our experiments. Training set always contained all define peaks, which were completed by four randomly selected parts out of peaks. These randomly selected parts were used to network can learn to recognize what is or what is not a search pattern (peak). All patterns were normalized to the square of a bitmap of the edge of size $a = 10$. The effort is always to choose the size of training set as small as possible, because especially backpropagation networks increases their computational complexity with the size of a training set.

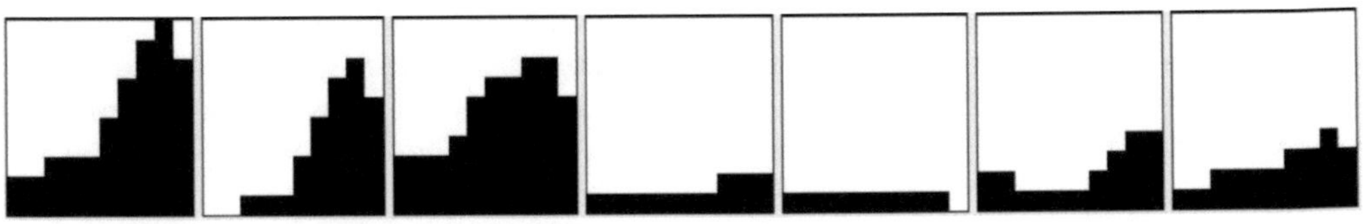

Fig. 10. Graphic representation of learning patterns (**S** vectors) that have been made by selection from training data set. The first three patterns represent peaks. Next four patterns are representatives of non-peak "not-interested" segments of values

<table>
<tr><th>No.</th><th>S</th><th>T</th></tr>
<tr><td>0.</td><td>--------+-|-------++-|-------+++|------++++|------++++|
-----+++++|-----+++++|--++++++++|++++++++++|++++++++++</td><td>-+</td></tr>
<tr><td>1.</td><td>----------|----------|--------+-|-------++-|-------+++|
------++++|------++++|-----+++++|-----+++++|--++++++++</td><td>-+</td></tr>
<tr><td>2.</td><td>----------|----------|-------++-|-----++++-|----++++++|
----++++++|---+++++++|++++++++++|++++++++++|++++++++++</td><td>-+</td></tr>
<tr><td>3.</td><td>----------|----------|----------|----------|----------|
----------|----------|----------|-------+++|++++++++++</td><td>+-</td></tr>
<tr><td>4.</td><td>----------|----------|----------|----------|----------|
----------|----------|----------|----------|+++++++++-</td><td>+-</td></tr>
<tr><td>5.</td><td>----------|----------|[illegible]|----------|----------|
----------|--------++|-------+++|++----++++|++++++++++</td><td>+-</td></tr>
<tr><td>6.</td><td>----------|----------|----------|----------|----------|
----------|--------+-|------++++|--++++++++|++++++++++</td><td>+-</td></tr>
</table>

Table 1. Vectors **T** and **S** from the learning pattern set. Values of '-1' are written using the character '-' and values of '+1' are written using the character '+' because of better clarity

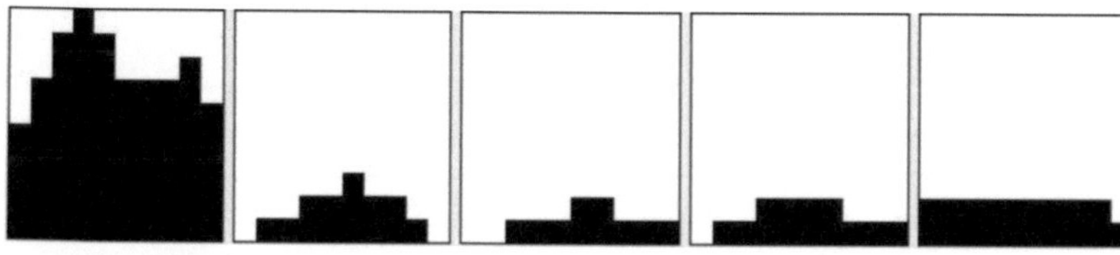

Fig. 11. Graphic representation of test patterns (**S** vectors) that have been made by selection from the test data set. The first pattern represents the peak. Next four patterns are representatives of non-peak "not-interested" segments of values

<table>
<tr><th>No.</th><th>S</th><th>T</th></tr>
<tr><td>0.</td><td>---+------|--+++-----|--+++---+-|-++++++++-|-+++++++++|
++++++++++|++++++++++|++++++++++|++++++++++|++++++++++</td><td>-+</td></tr>
<tr><td>1.</td><td>----------|----------|----------|----------|----------|
----------|----------|-----+----|---+++++--|-++++++++-</td><td>+-</td></tr>
<tr><td>2.</td><td>----------|----------|----------|----------|----------|
----------|----------|----------|-----++---|--++++++++</td><td>+-</td></tr>
<tr><td>3.</td><td>----------|----------|----------|----------|----------|
----------|----------|----------|---++++---|-+++++++++</td><td>+-</td></tr>
<tr><td>4.</td><td>----------|----------|----------|----------|----------|
----------|----------|----------|+++++++++-|++++++++++</td><td>+-</td></tr>
</table>

Table 2. Vectors **T** and **S** from the test pattern set. Values of '-1' are written using the character '-' and values of '+1' are written using the character '+' because of better clarity

Two types of classifiers: Backpropagation and classifier based on Hebb learning were used in our experimental part. Both used networks classified input patterns into two classes. Backpropagation network was adapted according the training set (Fig.10, Tab. 1) in 7 cycles. After its adaptation, the network was able to also correctly classify all five patterns from the test set (Fig. 11, Tab. 2), e.g. the network was able to correctly identify the peak and "uninteresting" data segments too. Other experiments gave similar results too.

Backpropagation network configuration:

Number of input neurons:	100
Number of output neurons:	2
Number of hidden layers:	1

Number of hidden neurons:	3
α - learning parameter:	0.4
Weight initialization algorithm:	Nguyen-Widrow
Weight initialization range:	(-0.5; +0.5)
Type of I/O values:	bipolar

Hebb network in its basic configuration was not able to adapt given training set (Fig.10, Tab. 1), therefore we used modified version of the network removing useless components from input vectors (Kocian & Volná & Janošek & Kotyrba, 2011). Then, the modified Hebb network was able to adapt all training patters (Fig. 12) and in addition to that the network correctly classified all the patterns from the test set (Fig. 11, Tab. 2), e.g. the network was able to correctly identify the peak and "uninteresting" data segments too. Other experiments gave similar results too.

Hebbian-learning-based-classifier configuration:

Number of input neurons:	100
Number of output neurons:	2
Type of I/O values:	bipolar

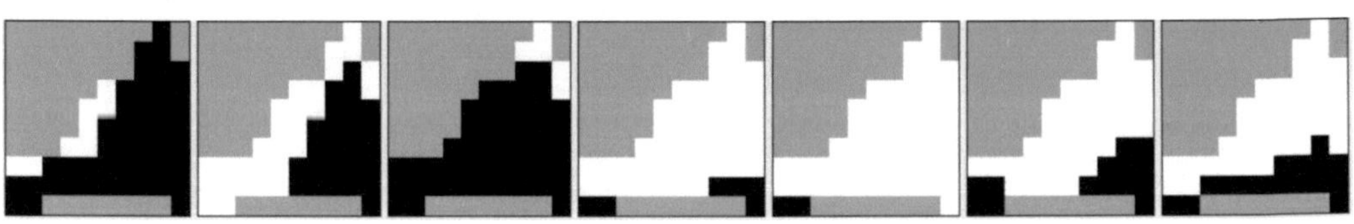

Fig. 12. Learning patterns from Fig. 10 with uncovered redundant components (gray colour). The redundant components prevented the Hebbian-learning-based-classifier in its default variant to learn patterns properly. So the modified variant had to be used

4.2 Pattern recognition via analytic programming

As an evolutionary algorithm used in our experimental work was differential evolution (DE). DE is a population-based optimization method that works on real-number-coded individuals (Price, 1999). For each individual $\vec{x}_{i,G}$ in the current generation (G), DE generates a new trial individual $\vec{x}'_{i,G}$ by adding the weighted difference between two randomly selected individuals $\vec{x}_{r1,G}$ and $\vec{x}_{r2,G}$ to a randomly selected third individual $\vec{x}_{r3,G}$. The resulting individual $\vec{x}'_{i,G}$ is crossed-over with the original individual $\vec{x}_{i,G}$. The fitness of the resulting individual, referred to as a perturbed vector $\vec{u}_{i,G+1}$, is then compared with the fitness of $\vec{x}_{i,G}$. If the fitness of $\vec{u}_{i,G+1}$ is greater than the fitness of $\vec{x}_{i,G}$, then $\vec{x}_{i,G}$ is replaced with $\vec{u}_{i,G+1}$; otherwise, $\vec{x}_{i,G}$ remains in the population as $\vec{x}_{i,G+1}$. DE is quite robust, fast, and

effective, with global optimization ability. It does not require the objective function to be differentiable, and it works well even with noisy and time-dependent objective functions.
The technique for the solving of this problem by means of analytic programming was inspired in neural networks. The method in this case study used input values and future output values - similarly as training set for the neural network and the whole structure which transfer input to output was synthesized by analytic programming. The final solution of the analytic programming is based on evolutionary process which selects only the required components from the basic sets of operators (Fig. 6 and Fig 7). Fig. 13 shows analytic programming experimental result for exact modelling during training phase.

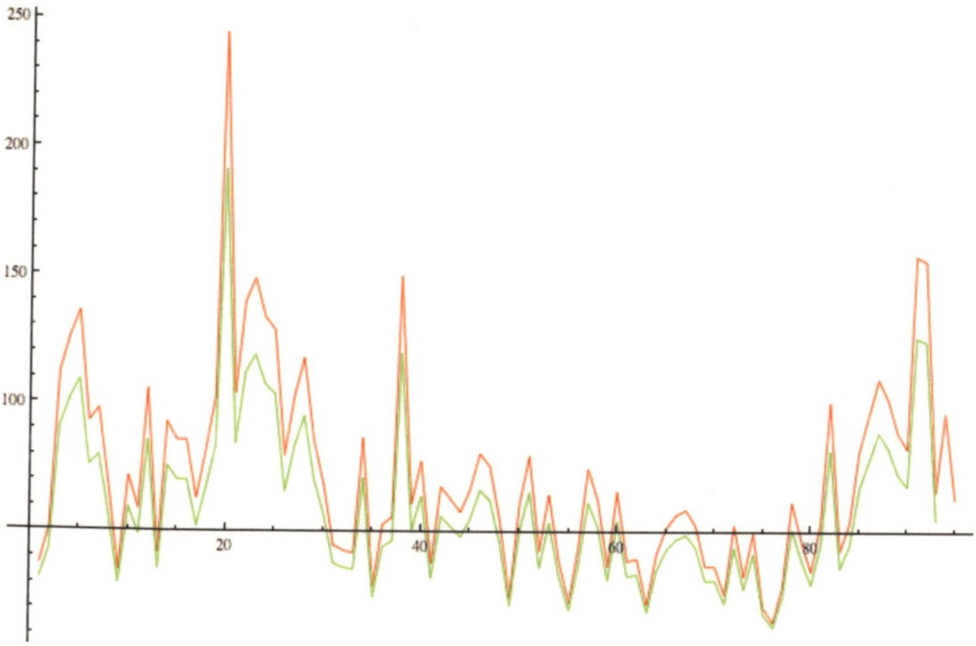

Fig. 13. Analytic programming experimental result for exact modelling during training phase. Red colour represents original data from training set (Fig. 8), while green colour represents modelling data using formula (6)

The resulting formula, which calculates the output value x_n was developed using AP (6):

$$x_n = 85.999 \cdot e^{(17.1502 - x_{n-3})^{0.010009 \cdot x_{n-2}}} \tag{6}$$

Analytic programming experimental results are shown in Fig. 14. Equation (6) also represents the behaviour of training set so that the given pattern was also successfully identified in the test set (Fig. 9). Other experiments gave similar results too.
The operators used in GFS were (see Fig. 7): +, -, /, *, Sin, Cos, K, x_{n-1} to x_{n-4}, exp, power. As the main algorithm for AP and also for constants estimation in meta-evolutionary process differential evolution was used. The final solution of the analytic programming is based on evolutionary process which selects only the required components from the basic sets of operators. In this case, not all components have to be selected as can be seen in one of solutions presented in (6).

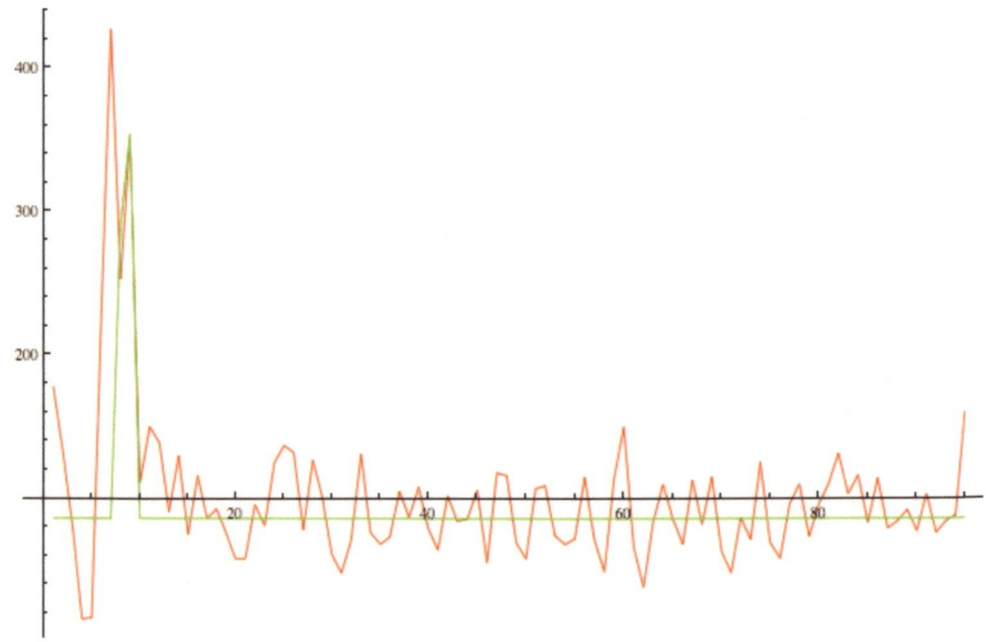

Fig. 14. Analytic programming experimental result. Red colour represents original data from test set (Fig. 9), while green colour represents modelling data using formula (6)

5. Conclusion

In this chapter, a short introduction into the field of pattern recognition using system adaptation, which is represented via time series, has been given. Two possible approaches were used from the framework of softcomputing methods. The first approach was based on analytic programming and the second one was based on artificial neural networks. Both types of used neural networks (e.g. Hebb and backpropagation networks) as well as analytic programming demonstrated ability to manage to learn and recognize given patterns in time series, which represents our system behaviour. Our experimental results suggest that for the given class of tasks can be acceptable simple classifiers (we tested the simplest type of Hebb learning). The advantage of simple neural networks is very easy implementation and quick adaptation. Easy implementation allows to realize them at low-performance computers (PLC) and their fast adaptation facilitates the process of testing and finding the appropriate type of network for the given application.

The method of analytic programming described here is universal (from point of view of used evolutionary algorithm), relatively simple, easy to implement and easy to use. Analytic programming can be regarded as an equivalent of genetic programming in program synthesis and new universal method, which can be used by arbitrary evolutionary algorithm. AP is also independent of computer platform (PC, Apple, ...) and operation system (Windows, Linux, Mac OS,...) because analytic programming can be realized for example in the Mathematica® environment or in other computer languages. It allows manipulation with symbolic terms and final programs are synthesised by AP of mapping, therefore main benefit of analytic programming is the fact that symbolic regression can be done by arbitrary evolutionary algorithm, as was proofed by comparative study.

According to the results of experimental studies, it can be stated that pattern recognition in our system behaviour using all presented methods was successful. It is not possible to say with certainty, which of them reaches the better results, whether neural networks or analytic programming. Both approaches have an important role in the tasks of pattern recognition.

In the future, we would like to apply pattern recognition tasks with the followed system adaptation methods in SIMATIC environment. SIMATIC (SIMATIC, 2010) is an appropriate application environment for industrial control and automation. SIMATIC platform can be applied at the operational, management and the lowest, physical level. At an operational level, it particularly works as a control of the running processes and monitoring of the production. On the management and physical level it can be used to receive any production instructions from the MES system (Manufacturing Execution System - the corporate ERP system set between customers' orders and manufacturing systems, lines and robots). At the physical level it is mainly used as links among various sensors and actuators, which are physically involved in the production process (Janošek, 2010). The core consists of the SIMATIC programmable logic computers with sensors and actuators. This system collects information about its surroundings through sensors. Data from the sensors can be provided (e.g. via Ethernet) to proposed and created software tools for pattern recognition in real time, which runs on a powerful computer.

6. Acknowledgment

The research described here has been financially supported by University of Ostrava grant SGS23/PRF/2011. It was also supported by the grant NO. MSM 7088352101 of the Ministry of Education of the Czech Republic, by grant of Grant Agency of Czech Republic GACR 102/09/1680 and by the European Regional Development Fund under the Project CEBIA-Tech No. CZ.1.05/2.1.00/03.0089. Any opinions, findings and conclusions or recommendations expressed in this material are those of the authors and do not necessarily reflect the views of the sponsors.

7. References

Armstrong, M. and Porter, R. ed. (2006): *Handbook of Industrial Organization, vol. III.* New York and Amsterdam: North-Holland.

Bishop, C. (2006) *Pattern Recognition and Machine Learning*. Springer, 2006.

Buhr, R.J.A. and Hubbard, A. (1997) Use Case Maps for Engineering Real Time and Distributed Computer Systems: A Case Study of an ACE-Framework Application. In *Hawaii International Conference on System Sciences*, Jan 7-10, 1997, Wailea, Hawaii, Available from http://www.sce.carletonca/ftp/pub/UseCaseMaps/hicss-final-public.ps

Ciskowski, P. and Zaton, M. (2010) Neural Pattern Recognition with Self-organizing Maps for Efficient Processing of Forex Market Data Streams. In *Artificial Intelligence and Soft Computing*, Volume 6113/2010, pp. 307-314, DOI: 10.1007/978-3-642-13208-7_39

Dormido-Canto, S., Farias, G., Vega, J., Dormido, R., Sánchez, J. and N. Duro *et al.* (2006) *Rev. Sci. Instrum.* 77 (10), p. F514.

Fausett, L.V. (1994) *Fundamentals of neural networks: architectures, algorithms and applications, first edition*. Prentice Hall. ISBN 978-953-7619-24-4

Gershenson, C. (2007): *Design and Control of Self-organizing Systems*. Mexico: CopIt ArXives, ISBN: 978-0-9831172-3-0.

Gershenson, C. (2002) Complex philosophy. In: *Proceedings of the 1st Biennial Seminar on Philosophical, Methodological & Epistemological Implications of Complexity Theory*. La Habana, Cuba. 14.02.2011, Available from http://uk.arXiv.org/abs/nlin.AO/0108001

Gogle finance [online], http://www.google.com/finance?q=NASDAQ:GOOG, 10.8. 2010

Heylighen, F. (1994) Fitness as default: the evolutionary basis for cognitive complexity reduction. In Trappl (Ed.) *Proceedings of Cybernetics and Systems '94*, R. Singapore: World Science, pp. 1595–1602, 1994.

Janošek, M. (2010) Systémy Simatic a jejich využití ve výzkumu. In: *Studentská vědecká konference 2010*. Ostrava: Ostravská univerzita, pp. 177-180. ISBN 978-80-7368-719-9

Kocian, V., Volná, E., Janošek, M. and Kotyrba, M. (2011) Optimizatinon of training sets for Hebbian-learningbased classifiers. In R. Matoušek (ed.): *Proceedings of the 17th International Conference on Soft Computing, Mendel 2011*, Brno, Czech Republic, pp. 185-190. ISBN 978-80-214-4302-0, ISSN 1803-3814.

Lai, K.K., Yu, L. and Wang, S: A (2005) Neural Network and Web-Based Decision Support System for Forex Forecasting and Trading. In *Data Mining and Knowledge Management*, Volume 3327/2005, pp. 243-253, DOI: 10.1007/978-3-540-30537-8_27.

Oplatkova, Z. (2009) Metaevolution - Synthesis of Optimization Algorithms by means of Symbolic. In *Regression and Evolutionary Algorithms, Lambert-Publishing*, ISBN 978-8383-1808-0.

Price, K. (1999) An Introduction to Differential Evolution, In: (D. Corne, M. Dorigo and F. Glover, eds.) *New Ideas in Optimization*, pp. 79–108, London: McGraw-Hill.

Seung, S. (2002). Multilayer perceptrons and backpropagation learning. 9.641 Lecture4. 1-6. Available from: http://hebb.mit.edu/courses/9.641/2002/lectures/lecture04.pdf

SIMATIC (2010) [online]. SIMATIC Controller, Available from http://www.automation.siemens.com/salesmaterial-as/brochure/en/brochure_simatic-controller_en.pdf

Zelinka, I. (2002) Analytic programming by Means of Soma Algorithm. Mendel '02, In: *Proc. Mendel'02*, Brno, Czech Republic, 2002, 93-101., ISBN 80-214-2135-5

Zelinka, I. (2004) SOMA - Self Organizing Migrating Algorithm", In: B.V. Babu, G. Onwubolu (eds), *New Optimization Techniques in Engineering Springer-Verlag*, 2004, ISBN 3-540-20167X